Handbook of Lifespan Cognitive Behavioral Therapy

Handbook of Lifespan Cognitive Behavioral Therapy

Childhood, Adolescence, Pregnancy, Adulthood, and Aging

Edited by

Colin R. Martin
Professor of Clinical Psychobiology and Applied Psychoneuroimmunology & Clinical Director, Institute for Health and Wellbeing, University of Suffolk, Ipswich, United Kingdom

Vinood B. Patel
Reader in Clinical Biochemistry, University of Westminster, London, United Kingdom

Victor R. Preedy
Emeritus Professor of Nutritional Biochemistry, King's College London, London, United Kingdom
Professor of Clinical Biochemistry and Pathology (Hon), King's College Hospital, London, United Kingdom

ELSEVIER

ACADEMIC PRESS
An imprint of Elsevier

Academic Press is an imprint of Elsevier
125 London Wall, London EC2Y 5AS, United Kingdom
525 B Street, Suite 1650, San Diego, CA 92101, United States
50 Hampshire Street, 5th Floor, Cambridge, MA 02139, United States
The Boulevard, Langford Lane, Kidlington, Oxford OX5 1GB, United Kingdom

ISBN 978-0-323-85757-4

SET ISBN 978-0-443-18556-4

For information on all Academic Press publications
visit our website at https://www.elsevier.com/books-and-journals

Publisher: Nikki P. Levy
Acquisitions Editor: Natalie Farra
Editorial Project Manager: Timothy Bennett
Production Project Manager: Swapna Srinivasan
Cover Designer: Miles Hitchen

Typeset by STRAIVE, India

Dedication

I dedicate this book to my wonderful daughter,
Dr. Caragh Brien, of whom I am so incredibly proud.
Colin R. Martin

Contents

Contributors

Numbers in parentheses indicate the pages on which the authors' contributions begin.

Yakoub Aw Aden Abdi (395), Yacquub Clinic, Hargeisa, Somalia

Samridhi Ahuja (377), Department of Psychology, Bharathiar University, Coimbatore, India

Devin R. Barlaan (441), Department of Psychology, University of Tulsa, Tulsa, OK, United States

Mariana Branquinho (473), Faculty of Psychology and Educational Sciences, Center for Research in Neuropsychology and Cognitive-Behavioral Intervention, University of Coimbra, Coimbra, Portugal

Tara R. Buck (441), Department of Psychiatry, University of Oklahoma School of Community Medicine, Tulsa, OK, United States

Michele Burn (23), Parent-Infant Research Institute, Heidelberg Heights, VIC, Australia

E. Thomaseo Burton (193), Department of Pediatrics, College of Medicine, University of Tennessee Health Science Center; Children's Foundation Research Institute, Le Bonheur Children's Hospital, Memphis, TN, United States

Abigail E. Candelari (249), Department of Psychiatry and Behavioral Sciences, Baylor College of Medicine, Houston, TX, United States

Sampurna Chakraborty (11), Project Stree Manoraksha, Department of Psychiatry, NIMHANS, Bangalore; Department of Clinical Psychology, Central Institute of Psychiatry, Ranchi, India

Lisa D. Cromer (441), Department of Psychology, University of Tulsa, Tulsa, OK, United States

Abdifatah Haji Daud (395), Yacquub Clinic, Hargeisa, Somalia

Joanne L. Davis (441), Department of Psychology, University of Tulsa, Tulsa, OK, United States

Joseph M. Dzierzewski (313), Department of Psychology, Virginia Commonwealth University, Richmond, VA, United States

Kristen H. Erps (171), Department of Psychiatry and Behavioral Sciences, Baylor College of Medicine, Houston, TX, United States

J.C. Escher (237), Department of Pediatric Gastroenterology, Erasmus Medical Centre Sophia's Children Hospital, Rotterdam, The Netherlands

Mahdi Razmara Ferezghi (81), Department of Dermatology, Imamreza Hospital of Mashhad, Mashhad University of Medical Sciences, Mashhad, Iran

Sergio Fernández-Artamendi (145), Department of Psychology, Universidad Loyola Andalucía, Sevilla, Andalucía, Spain

Madison Fitzpatrick (287), Bio Behavioral Institute, Great Neck, NY, United States

Ana Fonseca (53), Faculty of Psychology and Educational Sciences, Center for Research in Neuropsychology and Cognitive Behavioral Intervention, University of Coimbra, Coimbra, Portugal

Yvette Fruchter (421,431), Bio Behavioral Institute, Great Neck, NY, United States

Danielle H. Gardini (421,431), Bio Behavioral Institute, Great Neck, NY, United States

Daniel A. Geller (249), Massachusetts General Hospital Department of Psychiatry, Harvard Medical School, Boston, MA, United States

Alan W. Gemmill (23), Parent-Infant Research Institute, Heidelberg Heights, VIC, Australia

Debasruti Ghosh (299), Department of Psychology, MDDM College (Babasaheb Bhimrao Ambedkar Bihar University), Muzaffarpur, Bihar, India

Alba González-Roz (145), Addictive Behaviors Research Group, Department of Psychology, Faculty of Psychology, University of Oviedo, Oviedo, Principality of Asturias, Spain

Sasha Gorrell (219), Department of Psychiatry and Behavioral Sciences, University of California, San Francisco, CA, United States

Andrew G. Guzick (171), Department of Psychiatry and Behavioral Sciences, Baylor College of Medicine, Houston, TX, United States

Susmita Halder (11,465), Department of Psychology, St. Xavier's University, Kolkata; Amity Institute of Behavioural Health and Allied Sciences, Amity University Kolkata, Kolkata, West Bengal, India

Daniel Hamiel (205,485), Baruch Ivcher School of Psychology, IDC Herzliya, Herzliya; Cohen-Harris Resilience Center; Tel-Aviv-Brüll Community Mental Health Center, Clalit Health Services, Tel-Aviv, Israel

Bente Storm Mowatt Haugland (123), Department of Clinical Psychology, Faculty of Psychology, University of Bergen, Bergen, Norway

Yafit Hirshler (23), Parent-Infant Research Institute, Heidelberg Heights, VIC, Australia

Davíð R.M.A. Højgaard (109,275,413), Department of Child and Adolescent Psychiatry, Aarhus University Hospital, Psychiatry, Aarhus, Denmark

Kathryn M. Huryk (219), Department of Psychiatry and Behavioral Sciences, University of California, San Francisco, CA, United States

Elisabeth Bakke Husabø (123), Regional Centre for Child and Youth Mental Health and Child Welfare, Norwegian Research Centre, Bergen, Norway

Katja Anna Hybel (275,413), Department of Child and Adolescent Psychiatry, Aarhus University Hospital, Psychiatry, Aarhus, Denmark

Shin-ichi Ishikawa (359), Faculty of Psychology, Doshisha University, Kyotanabe City, Kyoto, Japan

Tord Ivarsson (275), Institute of Neurosciences and Physiology, University of Gothenburg, Gothenburg, Sweden

Matthew Jacofsky (287), Bio Behavioral Institute, Great Neck, NY, United States

Dilshad Jaff (459), Gillings School of Global Public Health, University of North Carolina at Chapel Hill, Chapel Hill, NC, United States

Emily R. Jellinek (171), Department of Psychiatry and Behavioral Sciences, Baylor College of Medicine, Houston, TX, United States

Sanne Jensen (413), Department of Child and Adolescent Psychiatry, Aarhus University Hospital, Psychiatry, Aarhus, Denmark

Amanda R. Johnson (181), Department of Education, University of California, Los Angeles, Los Angeles, CA, United States

Madelyn Johnson (219), Department of Psychiatry and Behavioral Sciences, University of California, San Francisco, CA, United States

Julie Jomeen (41), School of Health and Human Sciences, Southern Cross University, Bilinga, QLD, Australia

Catriona Jones (41), Institute for Clinical and Applied Health Research, University of Hull, Hull, United Kingdom

Rebecca C. Kamody (193), Child Study Center; Department of Pediatrics; Department of Psychiatry, Yale School of Medicine, New Haven, CT, United States

Anowra Khan (403), Department of Behavioral Sciences, School of Social Sciences, National University of Science and Technology, Islamabad, Pakistan

Suncica Lah (449), School of Psychology, The University of Sydney, Brain and Mind Centre, NSW, Australia

Lindsey N. Landry (171), Department of Psychiatry and Behavioral Sciences, Baylor College of Medicine, Houston, TX, United States

Daniel Le Grange (219), Department of Psychiatry and Behavioral Sciences, University of California, San Francisco, CA; Department of Psychiatry, The University of Chicago (Emeritus), Chicago, IL, United States

Tai Wa Liu (345), School of Nursing and Health Studies, Hong Kong Metropolitan University, Ho Man Tin, Hong Kong (SAR)

Ananya Mahapatra (377), Department of Psychiatry, Employees' State Insurance Hospital and Dental College, New Delhi, India

Akash Kumar Mahato (465), Amity Institute of Behavioural Health and Allied Sciences, Amity University Kolkata, Kolkata, West Bengal, India

Zahid Mahmood (395), School of Professional Psychology, University of Management and Technology, Lahore, Pakistan

Tamkeen Ashraf Malik (403), Department of Behavioral Sciences, School of Social Sciences, National University of Science and Technology, Islamabad, Pakistan

Claire Marshall (41), Department of Psychological Medicine, Humber Teaching NHS Foundation Trust, Princes Court, Hull, United Kingdom

Colin R. Martin (41), Institute for Clinical and Applied Health Research, University of Hull, Hull, United Kingdom

Verónica Martínez-Borba (473), Department of Basic and Clinical Psychology and Psychobiology, Faculty of Psychology, Jaume I University, Castelló de la Plana, Spain

Víctor Martínez-Loredo (145), Department of Psychology and Sociology, University of Zaragoza, Teruel, Spain

Anne C. Miers (89), Unit Developmental and Educational Psychology, Institute of Psychology, Leiden University, Leiden, The Netherlands

Jeannette Milgrom (23), Parent-Infant Research Institute and Melbourne School of Psychological Sciences, University of Melbourne, Melbourne, VIC, Australia

Shalini Mittal (299), Northcap University, Gurgaon, Haryana, India

Virginia Sklar Muscatello (181), Department of Education, University of California, Los Angeles, Los Angeles, CA, United States

Fugen Neziroglu (287,421,431), Bio Behavioral Institute, Great Neck, NY, United States

Shamay S.M. Ng (345), Department of Rehabilitation Sciences, The Hong Kong Polytechnic University, Hung Hom, Hong Kong (SAR)

Thomas Enias Pasipanodya (299), Experienced Professional, Psychological Recovery Clinic, Toronto, ON, Canada

Vinood B. Patel (1,491), School of Life Sciences, University of Westminster, London, United Kingdom

Elliottnell Perez (313), Department of Psychology, Virginia Commonwealth University, Richmond, VA, United States

Andrea Pozza (101,481), Department of Medical Sciences, Surgery and Neurosciences, University of Siena, Siena, Italy

Victor R. Preedy (1,491), Faculty of Life Science and Medicine, King's College London, London, United Kingdom

Kristi Pruiksma (441), Department of Psychiatry and Behavioral Science, University of Texas Health Science Center at San Antonio, San Antonio, TX, United States

Saurabh Raj (299), Department of Psychology, Ramdayalu Singh College (Babasaheb Bhimrao Ambedkar Bihar University), Muzaffarpur, Bihar, India

Rajkumar Rajendram (1,491), College of Medicine, King Saud bin Abdulaziz University for Health Sciences; Department of Medicine, King Abdulaziz Medical City, King Abdullah International Medical Research Center, Ministry of National Guard Health Affairs, Riyadh, Saudi Arabia

Scott G. Ravyts (313), Department of Psychology, Virginia Commonwealth University, Richmond, VA, United States

Andre L. Rodrigues (23), Parent-Infant Research Institute, Heidelberg Heights, VIC, Australia

Maryam Saeidi Sardabi (81), Faculty of Educational Science and Psychology, Shahid Chamran University of Ahvaz, Ahvaz, Iran

Sophie C. Schneider (171), Department of Psychiatry and Behavioral Sciences, Baylor College of Medicine, Houston, TX, United States

Roberto Secades-Villa (145), Addictive Behaviors Research Group, Department of Psychology, Faculty of Psychology, University of Oviedo, Oviedo, Principality of Asturias, Spain

Pegah AM Seidi (459), Research Center, University of Garmian, Kalar, Iraq

Jaiganesh Selvapandiyan (137), Department of Psychiatry, All India Institute of Medical Sciences, Vijayawada, Andhra Pradesh, India

Omer Shaked (205,485), Pediatric Endocrine and Diabetes Unit, Sheba Medical Center, Edmond and Lily Safra Children's Hospital; The Louis and Gabi Weisfeld School of Social Work, Bar-Ilan University, Ramat-Gan, Israel

Pragya Sharma (377), Department of Clinical Psychology, Dr Ram Manohar Lohia Hospital, New Delhi, India

Gemma Sicouri (157), Black Dog Institute, University of New South Wales, Randwick, NSW, Australia

Tushar Singh (299), Department of Psychology, Faculty of Social Sciences, Banaras Hindu University, Varananasi, UP, India

Gudmundur Skarphedinsson (109,275), Faculty of Psychology, School of Health Sciences, University of Iceland, Reykjavík, Iceland

Pablo Soto (313), Department of Psychology, Virginia Commonwealth University, Richmond, VA, United States

Paakhi Srivastava (377), Center for Weight, Eating, and Lifestyle Science, Drexel University, Philadelphia, PA, United States

L. Stapersma (237), Department of Child and Adolescent Psychiatry/Psychology, Erasmus Medical Centre Sophia's Children Hospital, Rotterdam; Levvel, Center for Child Psychiatry/Department of Child and Adolescent Psychiatry, Academic Medical Center, Amsterdam, The Netherlands

Elizabeth Stewart (449), School of Psychology, The University of Sydney, Brain and Mind Centre, NSW, Australia

Eric A. Storch (171,249), Department of Psychiatry and Behavioral Sciences, Baylor College of Medicine, Houston, TX, United States

Ponnusamy Subramaniam (337), Faculty of Health Sciences, Universiti Kebangsaan Malaysia, Federal Territory of Kuala Lumpur, Malaysia

Katsutoshi Tanaka (323), Department of Occupational Mental Health, Graduate School of Medical Sciences, Kitasato University, Sagamihara, Kanagawa, Japan

Mika Tanaka (323), Department of Nursing, School of Nursing, Kitasato University, Sagamihara, Kanagawa, Japan

Arezoo Moradi Tavallaei (81), Faculty of Educational Science and Psychology, Department of Psychology, Ferdowsi University of Mashhad, Mashhad, Iran

Kok-Wai Tay (337), Department of Psychology and Counseling, Faculty of Arts and Social Science, Tunku Abdul Rahman University, Kampar, Perak, Malaysia

Per Hove Thomsen (275), Department of Child and Adolescent Psychiatry, Aarhus University Hospital, Psychiatry, Aarhus, Denmark

Ingrid S. Tien (181), Department of Education, University of California, Los Angeles, Los Angeles, CA, United States

Guido G. Urizar, Jr. (65), Department of Psychology, California State University, Long Beach, CA, United States

E.M.W.J. Utens (237), Levvel, Center for Child Psychiatry/Department of Child and Adolescent Psychiatry, Academic Medical Center; Research Institute of Child Development and Education, University of Amsterdam, Amsterdam, The Netherlands

Daniëlle Van der Giessen (259), Department of Child Development and Education, University of Amsterdam, Amsterdam, Netherlands

Francisca J.A. van Steensel (259), Department of Child Development and Education, University of Amsterdam, Amsterdam, Netherlands

Carrie Masia Warner (89), Psychology, Montclair State University, Montclair, NJ, United States

Gro Janne Wergeland (123), Department of Child and Adolescent Psychiatry, Division of Psychiatry, Haukeland, University Hospital, Bergen, Norway

Samara Wolpe (181), Department of Education, University of California, Los Angeles, Los Angeles, CA, United States

Jeffrey J. Wood (181), Department of Education, University of California, Los Angeles, Los Angeles, CA, United States

Foreword

Cognitive behavioral therapy (CBT) has a long history and, importantly, within the era of evidence-based practice, has been systematically subject to research scrutiny and efficacy across a range of mental health-related clinical presentations. It is quite extraordinary, in many respects, that CBT has been found to be so effective, particularly for the treatment of anxiety-based clinical presentations, given that the principles behind the approach were originally straightforward, easily learned, and based on models that can be generalized to all behavior, whether maladaptive or not. However, over more recent years—and particularly within contemporary times—the reach of CBT has extended to more sophisticated clinical presentations and also embraces the notion of comorbidities. For example, application to a treatment anxiety in relation to a chronic medical condition requiring ongoing intervention such as kidney disease and the burden of hemodialysis. Thus, there are many examples of the application of CBT in ever more complex structures in which we may deliver clinical care. Furthermore, the focus of the primary presenting problem has expanded over recent years to include mental health presentations that may previously have been inaccessible or believed to be inappropriate for CBT.

The success of CBT as an intervention in these contexts has been realized by the fundamental aspect of the approach that assessment and outcome evaluation are embedded within the whole treatment plan, and that these will invariably include robust measures that are both clinically relevant and can be used in research. Therefore, compared to other types of psychotherapy, for example, psychodynamic psychotherapy, CBT is fundamentally evaluated by many of the same measures used in assessment, over a defined time period of the treatment plan and with a predictable treatment outcome. Interestingly, though a focus of CBT over recent years has been to extend the approach into new areas for appraisal of its efficacy in novel presentations, there is little in the way of a comprehensive and evidence-based resource that brings these innovations together across the life course. Professor Martin, Dr. Patel, and Professor Preedy have in my view addressed this need with this exciting new book, the *Handbook of Lifespan Cognitive Behavioral Therapy*, which is complemented with the second book in this series *Handbook of Cognitive Behavioral Therapy by Disorders.*

Bringing together leading experts and specialists from around the world, the editors have contextualized both traditional areas where CBT has been established to be effective, as well as exciting contemporary areas where CBT is proving to be highly efficacious within the course of the lifespan. Chapters within the book thus examine focused CBT interventions applied to specific conditions at various stages of the life course, ranging from children to the elderly. The editors are thus to be congratulated on addressing this gap with a broad, accessible, and applied compendium on the application of CBT across diverse clinical, social, and lifespan contexts. I am sure this timely and innovative new book will be of interest to practitioners and researchers alike as CBT becomes evermore enmeshed and pervasive in routine clinical practice for conditions where the approach is indicated and evidenced.

Helen Langton
Professor Helen Langton is Vice Chancellor at the University of Suffolk and a professor in Children's Cancer Nursing Education. Helen's early career was as a nurse specializing in children and young people with cancer. After several years in clinical practice, Helen moved into higher education, initially to teach nursing and, subsequently, Helen has held a number of senior posts in universities in the United Kingdom and has a range of experience across health and social care, science, social science, and education. Helen has been a member of a number of national boards and bodies including the Nursing and Midwifery Council, Department of Health, Health Education England, and the Quality Assurance Agency and is a trustee of East Anglia Children's Hospice. Internationally, Helen has held board membership for the Academy of Health Care Improvement and the De Souza Nursing Institute in Toronto, Canada, and is an editor of the *International Journal of Child Health Care*.

Preface

Overall, the disease burden of mental illness is considerable. Recent estimates suggest that the global burden of mental illness is just over 10% of disability adjusted life years (DALYs) and about 30% of years lived with a disability (YLDs). There are a variety of conditions, disorders, and phobias associated with these burdens and include those within early life stages (e.g., pregnancy), childhood, adolescence, adulthood, and aging. Mental illness impacts not only the individual but also the family unit, the community, and society at large.

Hitherto, some conditions were treated by pharmacological agents or conventional psychotherapy. More recently, however, cognitive behavioral therapy has seen extensive usage. The material on cognitive behavioral therapy disseminated in the last 5 years outstrips the material produced in the previous 30 years.

Cognitive behavioral therapy has been shown to be more cost- and outcome-effective than conventional or drug-based therapies. However, there are many ways in which cognitive behavioral therapy can be carried out. Particular practitioners have their own techniques. The response rates and outcome indices can also be variable depending on the condition being treated. Furthermore, there are unexplored translational aspects from different studies. For example, evidence-based information, limitations, and practitioner-based experiences in one particular situation may be applicable to other conditions of mental ill-health. However, the material on cognitive behavioral therapy is published across vastly different scientific domains. Furthermore, some texts are devoted exclusively for the expert or those from a particular discipline. As a consequence, important information from one particular scenario may not be readily available to other practitioners. Thus, different cognitive behavioral therapy approaches provide the foundation for future studies.

To address the aforementioned issues above, the editors have compiled the **Handbook of Cognitive Behavioral Therapy**.

The **Handbook of Cognitive Behavioral Therapy** is divided into two books

Handbook of Cognitive Behavioral Therapy by Disorder: Case Studies and Application for Adults
Handbook of Lifespan Cognitive Behavioral Therapy: Childhood, Adolescence, Pregnancy, Adulthood, and Aging

This book, *Handbook of Lifespan Cognitive Behavioral Therapy*, comprises five sections. Section A: Pregnancy and childbirth covers topics on cognitive behavior therapy in female infertility, perinatal depression, fear of childbirth, and postnatal depression. Section B: Children and adolescents covers topics on cognitive behavioral therapy in adolescent social anxiety, skin picking and hair pulling disorders, Asperger syndrome, depression, anxiety and autism spectrum disorder, inflammatory bowel disease, obsessive-compulsive disorder, and body dysmorphic disorder. Section C: Aging, the elderly, and related conditions covers topics on cognitive behavioral therapy in late-life insomnia, dementia and anxiety, and stroke. Section D: International aspects covers topics on cognitive behavioral therapy in children and anxiety, adolescents with depression, and adolescents with anxiety. Section E: Case studies and resources covers topics on cognitive behavioral therapy in body dysmorphic disorder, children with academic difficulties, death anxiety in children, obsessive-compulsive disorder, and postnatal depression.

This book not only bridges the intellectual and disciplinary divides but also describes translational aspects. It has a far thinking and evidence-based ethos. Each chapter has the following six sections:

- Abstract (published online)
- Practice and procedures
- Applications to other areas
- Key facts
- Mini-dictionary of terms
- Summary points

The section Key facts covers focused areas of knowledge written for the novice. The Mini-dictionary of terms section explains terms that are frequently used in the chapter. The section Applications to other areas describes how cognitive behavioral therapy is applied to other fields that the chapter may be relevant to. The Practice and procedures section covers a cognitive behavioral therapy methodology or technique that is applied in the chapter. The section Summary points encapsulates each chapter in a succinct way.

Contributors to the *Handbook of Lifespan Cognitive Behavioral Therapy* are innovators, authors of international and national standing, and leaders in the field. Emerging fields of science and important discoveries relating to cognitive behavioral therapy are incorporated in this book. This book represents indispensable reading for practitioners of cognitive behavioral therapy, physicians, psychologists, psychiatrists, behavioral scientists, councilors, and health scientists. It is also suitable for novice students and experts, including lecturers, professors, and institutional leaders. The book will be a valuable resource for academic libraries, medical departments, and colleges.

Editors
Colin R. Martin
Vinood B. Patel
Victor R. Preedy

Chapter 1

Life span-related mental health disorders and cognitive behavioral therapy

Rajkumar Rajendram[a,b], Vinood B. Patel[c], and Victor R. Preedy[d]
[a]*College of Medicine, King Saud bin Abdulaziz University for Health Sciences, Riyadh, Saudi Arabia,* [b]*Department of Medicine, King Abdulaziz Medical City, King Abdullah International Medical Research Center, Ministry of National Guard Health Affairs, Riyadh, Saudi Arabia,* [c]*School of Life Sciences, University of Westminster, London, United Kingdom,* [d]*Faculty of Life Science and Medicine, King's College London, London, United Kingdom*

Introduction

Mental disorders can occur at any time from childhood to seniority. However, select individuals may be highest risk of developing certain mental disorders at specific points in their life span. These points include adolescence (Ghafari, Nadi, Bahadivand-Chegini, & Doosti-Irani, 2022), pregnancy (Jha, Salve, Goswami, Sagar, & Kant, 2018), and ageing (Ribeiro et al., 2020). There are significant amounts of information are available to guide the management of disorders of mental health across the life span.

The diagnostic and statistical manual of mental disorders (DSM) (American Psychiatric Association, 2013), and indeed other instruments, can be used to define and diagnose these diseases across the life span. Furthermore, various institutions have investigated the incidence of mental disorders at different phases of life. The United Nations Children's Fund (UNICEF), for example, has produced data which demonstrate the scale of mental disorders in young people (UNICEF, 2021).

Incidence of disorders of mental health in young people

Fig. 1 illustrates the percentage of young people (i.e., aged 10–19 years old) living in Europe with a mental disorder in 2019. The country-level data presented were derived from the State of the World's Children report (UNICEF, 2021). These data suggest that between 1 in 5 and 9 young people in Europe are affected by a disorder of mental health. Yet, there is considerable variation between individual countries. For example, more than 1 in 5 young people in Spain are afflicted by a mental disorder while around 1 in 10 young people in Poland are similarly affected.

Fig. 2 shows the various mental disorders that affect young people. Anxiety and depression account for more than 40% of those mental disorders that affect young people. However, the incidence of mental disorders such as depression in the young varies significantly between countries (Table 1). The incidence varies from 10% to 30% but is highest in Cameroon and lowest in Japan. The age of onset of disorders of mental health also varies significantly.

Age of onset of disorders of mental health

Fig. 3 illustrates the age of onset of various disorders of mental health. While there are many peaks and troughs, the onset of many mental disorders peak between 15 and 21 years of age. These illnesses (e.g., anxiety, depression, psychoses, and developmental disorders) can cause significant morbidity and mortality during childhood, adolescence, pregnancy, and in the elderly population. Several studies (e.g., The Global Burden of Disease study) have investigated this.

Burden of disease attributable to disorders of across the life span

Fig. 4 illustrates the changes in the total burden of disease for different age groups worldwide between 1990 and 2019. The burden of disease is presented as disability adjusted life years (DALY). One DALY equals one lost year of healthy

Handbook of Lifespan Cognitive Behavioral Therapy. https://doi.org/10.1016/B978-0-323-85757-4.00003-1

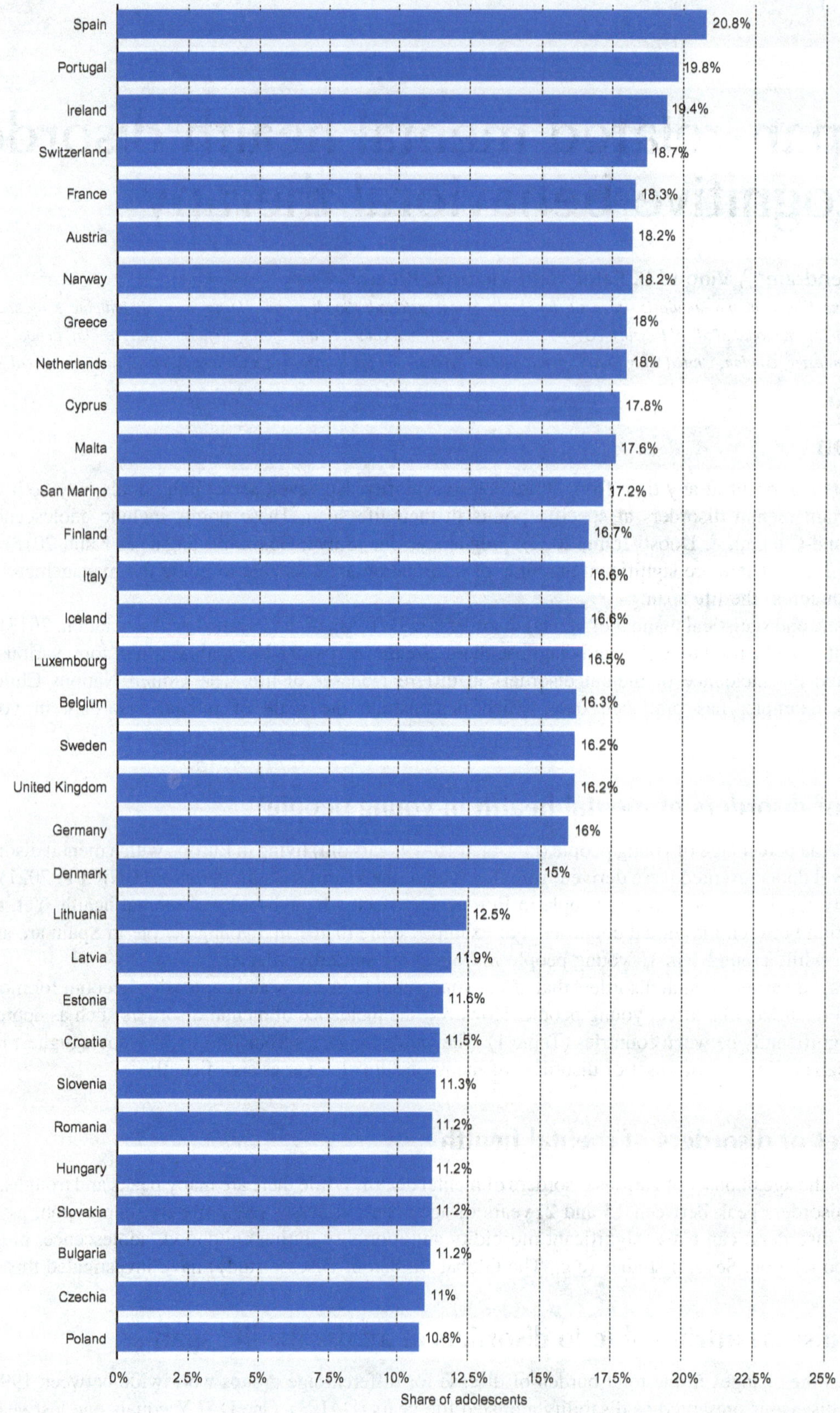

FIG. 1 Percentage of young people living with mental disorders in Europe. Data pertain to 2019 data for young people (i.e., those who are 10–19 years old). *(Figure from Statista. Source UNICEF. (2021). United Nations Children's Fund. The State of the World's Children 2021: On my mind—Promoting, protecting and caring for children's mental health. New York: UNICEF. October 2021. The State of the World's Children 2021 www.Unicef.org.)*

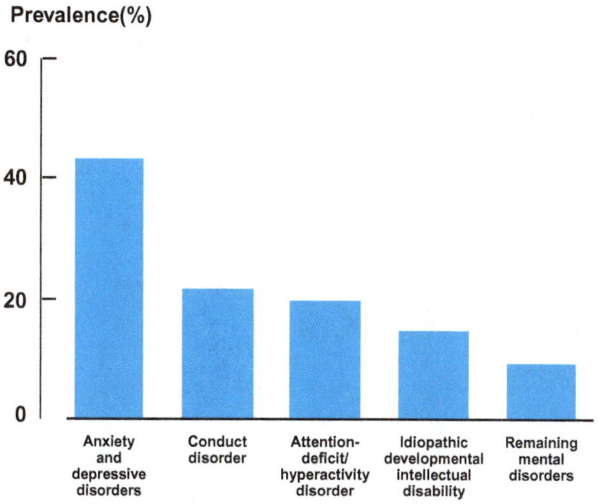

FIG. 2 Mental disorders among adolescents (10–19 years old boys and girls) globally. *(From UNICEF. (2021). United Nations Children's Fund. The State of the World's Children 2021: On my mind—Promoting, protecting and caring for children's mental health. New York: UNICEF. October 2021. As some adolescents may have more than one mental disorder, the total exceeds 100%.)*

life. The data presented in Fig. 4 contextualize those shown in Figs. 5–7. Fig. 5 presents the global age-stratified DALYs rate per 100,000 people caused by disorders of mental health and substance misuse. Fig. 6 presents the age-stratified DALYs rate per 100,000 people worldwide caused by anxiety. Fig. 7 presents the age-stratified DALYs rate per 100,000 people worldwide caused by eating disorders. These data highlight that people of all ages are significantly affected by disorders of mental health.

Thus, disorders of mental health clearly cause significant morbidity and mortality across the life span. Yet, many clinicians are pessimistic when faced with an adolescent or a pregnant patient. Indeed, the term "Sisyphean task" has been used by some authors to describe the treatment of severe mental disorders (Daugherty, Warburton, & Stahl, 2020) and is reiterated by us (Rajendram, Patel, & Preedy, 2022).

Patients with traumatic brain injury or spinal cord injury may induce a similar melancholy in inexperienced physicians (Rajendram, Patel, & Preedy, 2021a, 2021b). This is particularly troubling because those who sustain injuries to the spinal cord or the brain are often young and have a high incidence of mental disorders as a consequence. Fortunately, the interest of clinicians and scientists in life span-related mental disorders has been piqued.

Research relevant to cognitive behavioral therapy for life span-related mental disease

Rajendram et al. (2022) describe the trends in publications pertaining to cognitive behavioral therapy over approximately 40 years. Essentially, there has been a huge increase in CBT-related publications in recent years (Rajendram et al., 2022). While the analyses presented by Rajendram et al. (2022) could be considered crude, the data provide information on changing trends in recent years.

The application of cognitive behavioral therapy and other psychotherapies for young people, for example, has clearly advanced and there is considerable discourse into its effectiveness (for example, see Bahji et al., 2021; Benjamin et al., 2011; Braito et al., 2021; Duffy, Sharpe, & Schwannauer, 2019; Mestre et al., 2021; Zhou et al., 2020). Similarly, cognitive behavioral therapy and other psychotherapies have advanced for conditions or scenarios in pregnancy (Golshani, Hasanpour, Mirghafourvand, & Esmaeilpour, 2021; Green et al., 2021; Lau, Yen, Wong, Cheng, & Cheng, 2022; Li et al., 2022; Markin, 2017) and old age (Ayda, Anna, Franziska, & Frank, 2022; Chan et al., 2021; Tay, Subramaniam, & Oei, 2019) and even caregivers (Hopkinson, Reavell, Lane, & Mallikarjun, 2019; Verreault, Turcotte, Ouellet, Robichaud, & Hudon, 2021).

Therapeutic nihilism toward disorders of mental health that develop during childhood, adolescence, pregnancy, or old age is clearly inappropriate. Yet, more data are needed to increase the efficacy of psychotherapies such as CBT across the life span. The use of CBT across the life span are summarized and reviewed in many of the chapters in this book (for pregnancy, for example, see Fonseca, 2022; Halder and Chakraborty, 2022; Martin, Jones, and Marshall, 2022; Milgrom, Michele Burn, Rodrigues, Gemmill, and Hirshler, 2022; Urizar, 2022).

TABLE 1 Young people feeling depressed or having little interest in doing things (percent).

Cameroon	32
Mali	31
Indonesia	29
Zimbabwe	27
France	24
Germany	24
United States	24
Brazil	22
Lebanon	21
United Kingdom	20
Argentina	19
Kenya	19
Peru	16
Bangladesh	14
India	14
Morocco	14
Nigeria	14
Ukraine	12
Spain	11
Ethiopia	10
Japan	10
21-country median	19

From UNICEF. (2021). United Nations Children's Fund. The State of the World's Children 2021: On my mind—Promoting, protecting and caring for children's mental health. New York: UNICEF. October 2021. The data are from a survey of approximately 10,000 people aged 15–24 years by telephone (landline or mobile). As the study was carried out via telephone, the data can be considered selective.

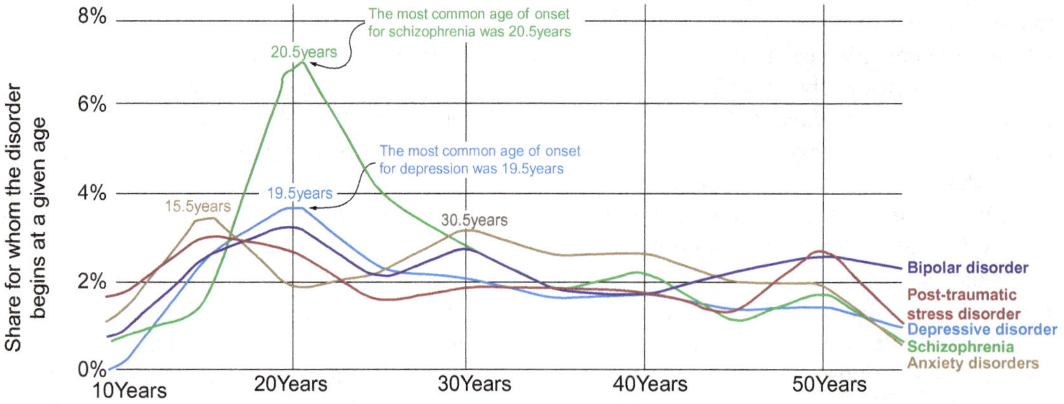

FIG. 3 Age of onset for mental health conditions. *(Data from Solmi, M., et al. (2021). Age of onset of mental disorders worldwide: large scale meta analysis of 192 epidemiological studies.* Nature Molecular. *https://doi.org.10.1038/s41380-021-01161-7. Figure from Dattani. (2022). OurWorldinData.org.)*

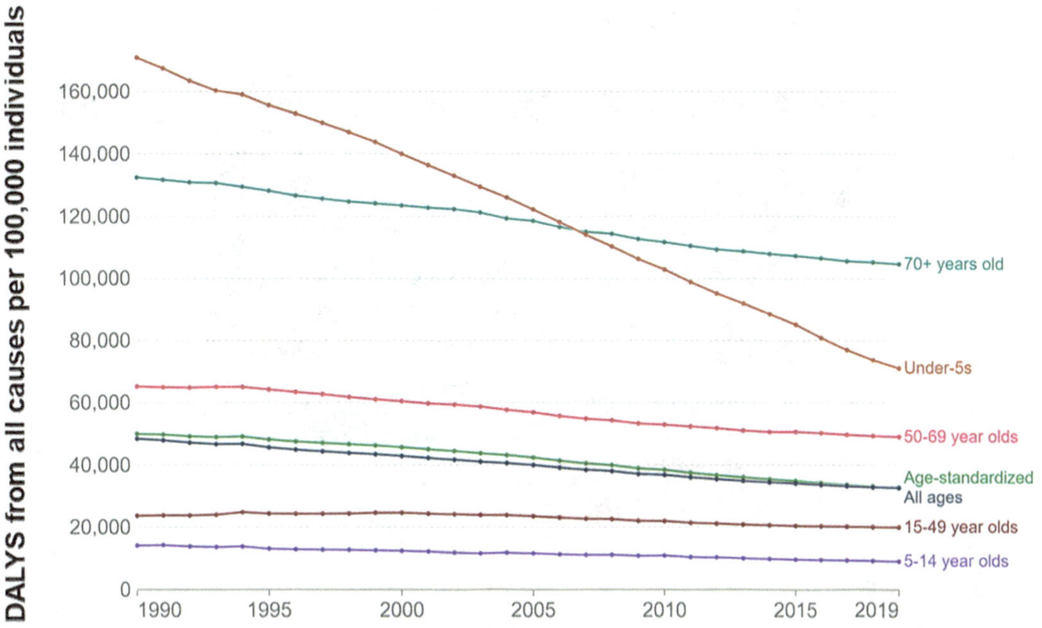

FIG. 4 Changes in the burden of disease for different age groups: global data for 1990–2019. *(Data from Global Burden of Disease Collaborative Network. Global Burden of Disease Study 2019 (Global Burden of Disease (GBD). (2019). Global burden of disease collaborative network. Global Burden of Disease Study 2019 (GBD). Seattle, United States: Institute for Health Metrics and Evaluation (IHME), http://ghdx.healthdata.org/gbd-results-tool). Seattle, United States: Institute for Health Metrics and Evaluation (IHME), 2021. DALYS: Disability Adjusted Life Years.)*

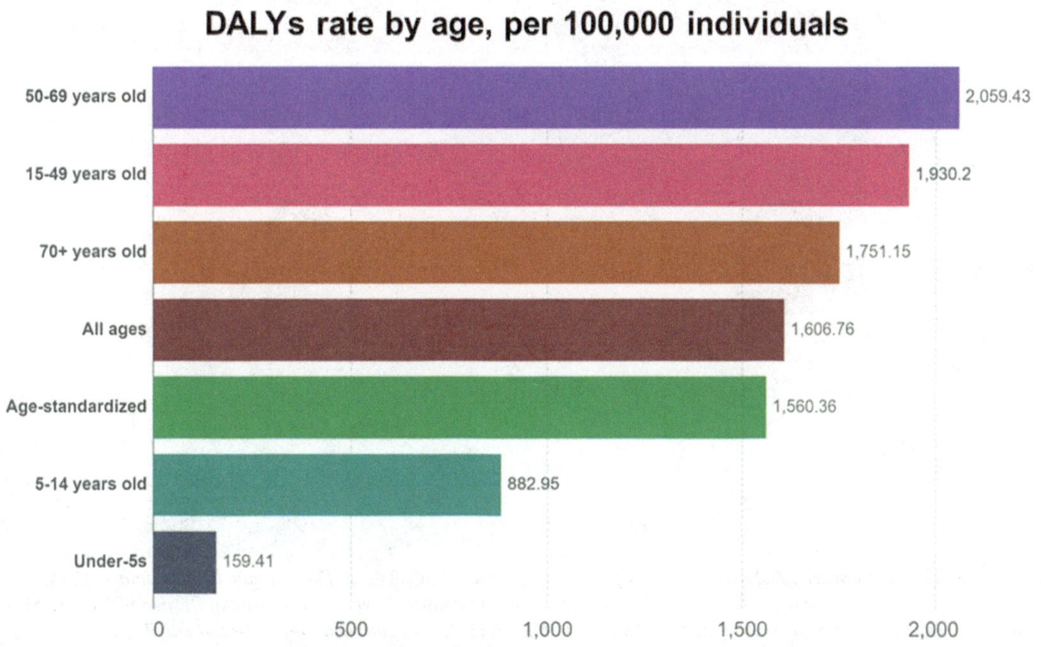

FIG. 5 Age-stratified DALYs rate per 100,000 people caused by disorders of mental health and substance misuse: Global data. *(Data pertains to 2016. DALYS: Disability Adjusted Life Years (Global Burden of Disease Collaborative Network. (2017). Global Burden of Disease Study 2016 (GBD 2016) Results. Seattle, United States: Institute for Health Metrics and Evaluation (IHME). http://ghdx.healthdata.org/gbd-results-tool).)*

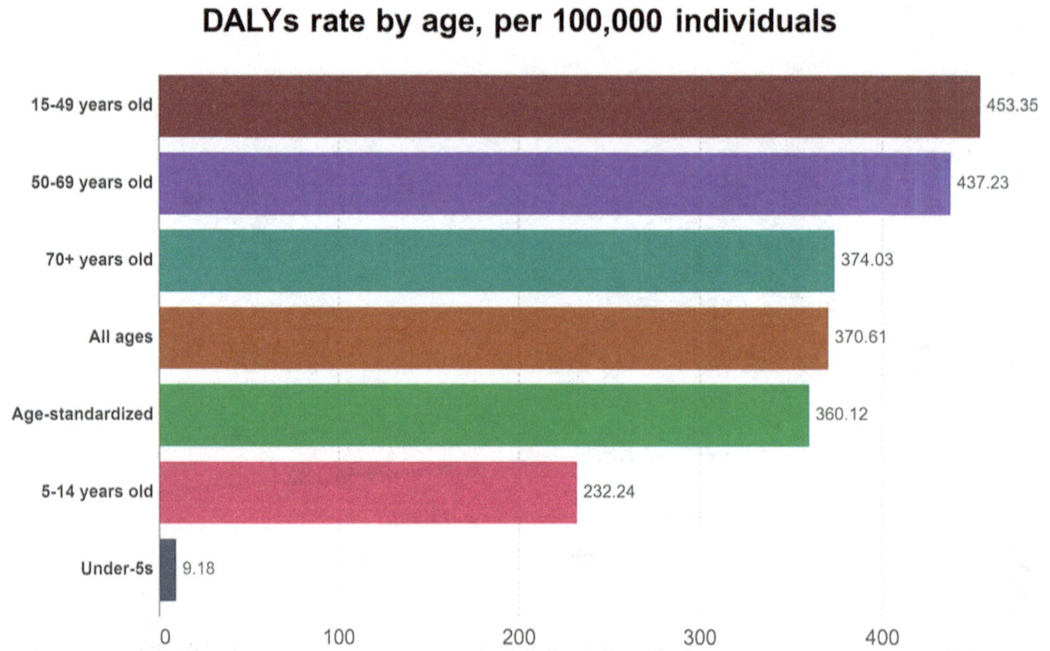

FIG. 6 Age-stratified DALYs rates per 100,000 people caused by anxiety disorders: Global data. *(Data pertain to 2019 data. DALYS: Disability Adjusted Life Years (Global Burden of Disease (GBD). (2019). Global burden of disease collaborative network. Global Burden of Disease Study 2019 (GBD). Seattle, United States: Institute for Health Metrics and Evaluation (IHME), http://ghdx.healthdata.org/gbd-results-tool).)*

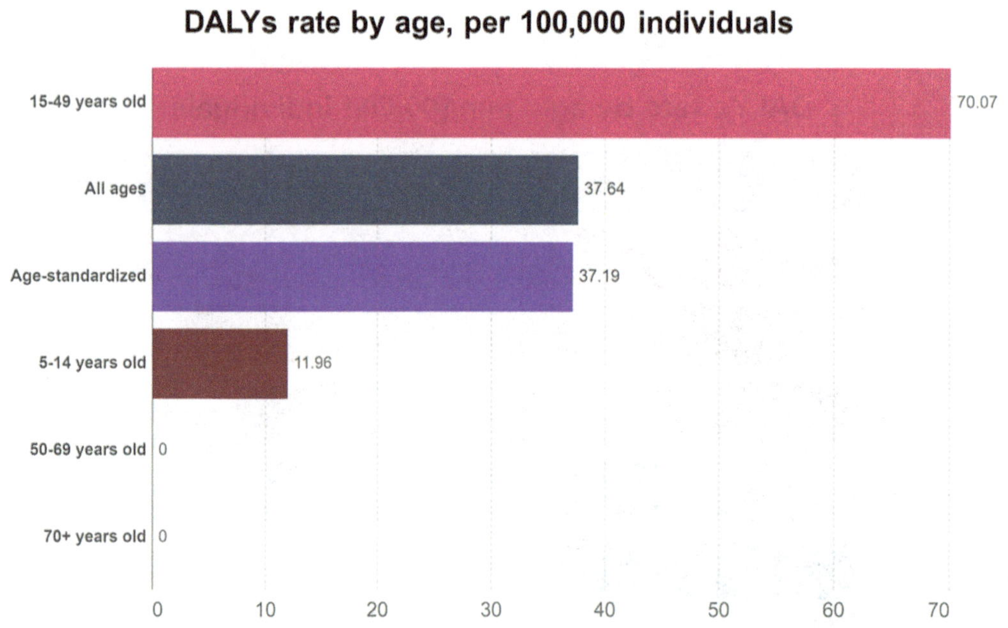

FIG. 7 Age-stratified DALYs rates per 100,000 people caused by eating disorders: Global data. *(Data pertain to 2019 data. DALYS: Disability Adjusted Life Years (Global Burden of Disease (GBD). (2019). Global burden of disease collaborative network. Global Burden of Disease Study 2019 (GBD). Seattle, United States: Institute for Health Metrics and Evaluation (IHME), http://ghdx.healthdata.org/gbd-results-tool).)*

Applications to other areas

Understanding the nuances of mental disease in childhood, adolescence, pregnancy, and seniority may provide insight into the principles that underpin the disorders of mental health that can occur at any age.

Key facts of disability-adjusted life years (DALYs)

- Disability-adjusted life years (DALYs) are a way to measure the impact or burden of disease in a quantitative measure.
- The World Health Organization has defined a *DALY* as: *"a loss equivalent to a whole year of full health. To calculate the DALYs for a health condition the years of life lost to due to premature mortality (YLLs) and years lived with a disability (YLDs) due to that condition are summed."*
- DALYs are a measure of societal impact.
- DALYs may be applied to common communicable and noncommunicable disease as well as accidents or violence.
- DALYs may be calculated for rare or infrequent conditions or disease.
- More DALYs occur if a disorder of mental health develops during adolescence compared to more senior and older age groups.

Summary points

- Disorders of mental health may strike during any of the life stages: childhood, adolescence, pregnancy, and seniority.
- Many mental disorders are most likely to develop from 15 to 21 years of age.
- More DALYs occur if a disorder of mental health develops during adolescence compared to more senior and older age groups.
- Therapeutic nihilism is unwarranted and psychotherapies such as cognitive behavioral therapy can be used to treat of mental disease across the life span.
- Much remains unknown about disorders of mental health, cognitive behavioral therapy, and other psychotherapies.

References

American Psychiatric Association. (2013). *Diagnostic and statistical manual of mental disorders* (5th ed.). Arlington, VA: American Psychiatric Association.

Ayda, R., Anna, K., Franziska, K., & Frank, J. (2022). Psychotherapeutic interventions in individuals at risk for Alzheimer's dementia: A systematic review. *Alzheimer's Research & Therapy*, *14*(1).

Bahji, A., Pierce, M., Wong, J., Roberge, J. N., Ortega, I., & Patten, S. (2021). Comparative efficacy and acceptability of psychotherapies for self-harm and suicidal behavior among children and adolescents: A systematic review and network meta-analysis. *JAMA Network Open*, *4*(4).

Benjamin, C. L., Puleo, C. M., Settipani, C. A., Brodman, D. M., Edmunds, J. M., Cummings, C. M., & Kendall, P. C. (2011). History of cognitive-behavioral therapy in youth. *Child and Adolescent Psychiatric Clinics of North America*, *20*(2), 179–189.

Braito, I., Rudd, T., Buyuktaskin, D., Ahmed, M., Glancy, C., & Mulligan, A. (2021). Review: Systematic review of effectiveness of art psychotherapy in children with mental health disorders. *Irish Journal of Medical Science*. https://doi.org/10.1007/s11845-021-02688-y.

Chan, P., Bhar, S., Davison, T. E., Doyle, C., Knight, B. G., Koder, D., et al. (2021). Characteristics and effectiveness of cognitive behavioral therapy for older adults living in residential care: A systematic review. *Aging & Mental Health*, *25*(2), 187–205.

Daugherty, B., Warburton, K., & Stahl, S. M. (2020). A social history of serious mental illness. *CNS Spectrums*, *25*(5), 584–592.

Duffy, F., Sharpe, H., & Schwannauer, M. (2019). Review: The effectiveness of interpersonal psychotherapy for adolescents with depression—A systematic review and meta-analysis. *Child and Adolescent Mental Health*, *24*(4), 307–317.

Fonseca, A. (2022). Postnatal depression: Cognitive-behavioral conceptualization and prevention and treatment approaches. In C. R. Martin, V. B. Patel, & V. R. Preedy (Eds.), *Handbook of lifespan cognitive behavioral therapy* (pp. 53–64). Cambridge, MA USA: Elsevier.

Ghafari, M., Nadi, T., Bahadivand-Chegini, S., & Doosti-Irani, A. (2022). Global prevalence of unmet need for mental health care among adolescents: A systematic review and meta-analysis. *Archives of Psychiatric Nursing*, *36*, 1–6.

Golshani, F., Hasanpour, S., Mirghafourvand, M., & Esmaeilpour, K. (2021). Effect of cognitive behavioral therapy-based counseling on perceived stress in pregnant women with history of primary infertility: A controlled randomized clinical trial. *BMC Psychiatry*, *21*(1).

Green, S. M., Donegan, E., McCabe, R. E., Streiner, D. L., Furtado, M., Noble, L., et al. (2021). Cognitive behavior therapy for women with generalized anxiety disorder in the perinatal period: Impact on problematic behaviors. *Behavior Therapy*, *52*(4), 907–916.

Halder, S., & Chakraborty, S. (2022). Cognitive-behavioral therapy in female infertility. In C. R. Martin, V. B. Patel, & V. R. Preedy (Eds.), *Handbook of lifespan cognitive behavioral therapy* (pp. 11–22). Cambridge, MA, USA: Elsevier.

Hopkinson, M. D., Reavell, J., Lane, D. A., & Mallikarjun, P. (2019). Cognitive behavioral therapy for depression, anxiety, and stress in caregivers of dementia patients: A systematic review and meta-analysis. *The Gerontologist*, *59*(4), e343–e362.

Jha, S., Salve, H. R., Goswami, K., Sagar, R., & Kant, S. (2018). Burden of common mental disorders among pregnant women: A systematic review. *Asian Journal of Psychiatry*, *36*, 46–53.

Lau, Y., Yen, K. Y., Wong, S. H., Cheng, J. Y., & Cheng, L. J. (2022). Effect of digital cognitive behavioral therapy on psychological symptoms among perinatal women in high income-countries: A systematic review and meta-regression. *Journal of Psychiatric Research*, *146*, 234–248.

Li, X., Laplante, D. P., Paquin, V., Lafortune, S., Elgbeili, G., & King, S. (2022). Effectiveness of cognitive behavioral therapy for perinatal maternal depression, anxiety and stress: A systematic review and meta-analysis of randomized controlled trials. *Clinical Psychology Review*, *92*.

Markin, R. D. (2017). An introduction to the special section on psychotherapy for pregnancy loss: Review of issues, clinical applications, and future research direction. *Psychotherapy*, *54*(4), 367–372.

Martin, C. R., Jones, C., Marshall, C., & Jomeen, J. (2022). Childbirth-related fear, tokophobia, and cognitive behavioral therapy. In C. R. Martin, V. B. Patel, & V. R. Preedy (Eds.), *Handbook of lifespan cognitive behavioral therapy* (pp. 41–51). Cambridge, MA, USA: Elsevier.

Mestre, J. M., Taubner, S., Mota, C. P., Rangel, H. M., Saliba, A., Heinonen, E., et al. (2021). Theories of change and mediators of psychotherapy effectiveness in adolescents with externalising behaviours: A systematic review. *Frontiers in Psychiatry*, *12*.

Milgrom, J., Burn, M., Rodrigues, A. L., Gemmill, A. W., & Hirshler, Y. (2022). Cognitive behavioral therapy for perinatal depression—Tailored face-to-face treatment and translation for online delivery of MumMoodBooster. In C. R. Martin, V. B. Patel, & V. R. Preedy (Eds.), *Handbook of lifespan cognitive behavioral therapy* (pp. 23–39). Cambridge, MA, USA: Elsevier.

Rajendram, R., Patel, V. B., & Preedy, V. R. (2021a). Recommended resources and sites for the neuroscience of traumatic brain injury. In R. Rajendram, V. B. Patel, & V. R. Preedy (Eds.), *Neuroscience of traumatic brain injury*. USA: Elsevier.

Rajendram, R., Patel, V. B., & Preedy, V. R. (2021b). Recommended resources and sites for the neuroscience of spina cord injury. In R. Rajendram, V. B. Patel, & V. R. Preedy (Eds.), *Neuroscience of spinal cord injury*. USA: Elsevier.

Rajendram, R., Patel, V. B., & Preedy, V. R. (2022). Recommended resources for cognitive-behavioral therapy across the lifespan. In C. R. Martin, V. B. Patel, & V. R. Preedy (Eds.), *Handbook of lifespan cognitive behavioral therapy* (pp. 491–500). USA: Elsevier.

Ribeiro, G. C. A., Vieira, W. A., Herval, A. M., Rodrigues, R. P. C. B., Agostini, B. A., Flores-Mir, C., et al. (2020). Prevalence of mental disorders among elderly men: A systematic review and meta-analysis. *São Paulo Medical Journal*, *138*(3), 190–200.

Tay, K.-W., Subramaniam, P., & Oei, T. P. (2019). Cognitive behavioural therapy can be effective in treating anxiety and depression in persons with dementia: A systematic review. *Psychogeriatrics*, *19*(3), 264–275.

UNICEF. (2021). *United Nations Children's Fund. The State of the World's Children 2021: On my mind—Promoting, protecting and caring for children's mental health*. New York: UNICEF. October 2021.

Urizar, G. G., Jr. (2022). Stress and cortisol regulation during pregnancy: Implications for cognitive behavioral stress management among low-income women. In C. R. Martin, V. B. Patel, & V. R. Preedy (Eds.), *Handbook of lifespan cognitive behavioral therapy* (pp. 65–77). Cambridge, MA, USA: Elsevier.

Verreault, P., Turcotte, V., Ouellet, M.-C., Robichaud, L.-A., & Hudon, C. (2021). Efficacy of cognitive-behavioural therapy interventions on reducing burden for caregivers of older adults with a neurocognitive disorder: A systematic review and meta-analysis. *Cognitive Behaviour Therapy*, *50*(1), 19–46.

Zhou, X., Teng, T., Zhang, Y., Del Giovane, C., Furukawa, T. A., Weisz, J. R., et al. (2020). Comparative efficacy and acceptability of antidepressants, psychotherapies, and their combination for acute treatment of children and adolescents with depressive disorder: A systematic review and network meta-analysis. *The Lancet Psychiatry*, *7*(7), 581–601.

Further reading

Hall, J., Kellett, S., Berrios, R., Bains, M. K., & Scott, S. (2016). Efficacy of cognitive behavioral therapy for generalized anxiety disorder in older adults: Systematic review, meta-analysis, and meta-regression. *American Journal of Geriatric Psychiatry*, *24*(11), 1063–1073.

Roser, M., & Ritchie, H. (2022). *Burden of disease*. Published online at OurWorldInData.org. Retrieved from: https://ourworldindata.org/burden-of-disease (Online Resource).

Thoma, N., Pilecki, B., & McKay, D. (2015). Contemporary cognitive behavior therapy: A review of theory, history, and evidence. *Psychodynamic Psychiatry*, *43*(3), 423–461.

Section A

Pregnancy and childbirth

Chapter 2

Cognitive-behavioral therapy in female infertility

Susmita Halder[a] and Sampurna Chakraborty[b,c]

[a]Department of Psychology, St. Xavier's University, Kolkata, India [b]Project Stree Manoraksha, Department of Psychiatry, NIMHANS, Bangalore, India
[c]Department of Clinical Psychology, Central Institute of Psychiatry, Ranchi, India

Abbreviations

CBT cognitive-behavioral therapy
GAD generalized anxiety disorder
HPA hypothalamic pituitary adrenal
IVF in vitro fertilization
SUD Subjective Units of Distress

Introduction

Infertility is a global challenge for many married couples across the world and is recognized as a major health problem across cultures. It has been estimated that in 2010 there were 48.5 million infertile couples worldwide (Verkuijlen, Verhaak, Nelen, Wilkinson, & Farquhar, 2016). According to World Health Organization, infertility is known as the inability to achieve a pregnancy after 12 months or more of without contraception, regular sexual intercourse (World Health Organization, 2016). Considering the prevalence rate of infertility, it can be said that it is most common in the South East Asia and Sub-Saharan Africa countries (Mascarenhas, Flaxman, Boerma, Vanderpoel, & Stevens, 2012).

Infertility can be classified into two types namely, primary and secondary. The World Health Organization (2013) defines primary infertility as inability to become pregnant at any point of life, while secondary infertility is the inability to conceive a child following a previous successful pregnancy (Verkuijlen et al., 2016). The treatment and intervention of infertility is a growing field in medicine, and some of the effective treatment options for infertility include fertility-enhancing medications, intrauterine insemination (IUI), in vitro fertilization (IVF), donor eggs and embryos, intracytoplasmic sperm injection (ICSI), gamete intrafallopian transfer (GIFT), and zygote intrafallopian transfer (ZIFT) (World Health Organization, 2016). The diagnosis of infertility is often experienced as a life crisis for individuals and can lead to feelings of anxiety, depression, and relational strain (Verkuijlen et al., 2016). There are a large number of studies indicating the psychological ramifications of infertility and various studies have demonstrated the importance of the mind-body connection and fertility, thereby hinting at how stress and psychological factors can also contribute to infertility (Cwikel, Gidron, & Sheiner, 2004). However, despite these findings the psychosocial aspects of infertility have not been adequately addressed in fundamental research and clinical trials.

The experience of infertility

Childbearing and child rearing are perceived as a gift and fulfilling experience by most of the people; as a result, the experience of infertility comes with major implications. The diagnosis of infertility, acceptance and living with it entails many aftereffects. The fertility treatment has also been perceived as a chronic stress and the individuals undergoing it has reported serious psychological strain. It was found that approximately half of women undergoing fertility treatment rated infertility as the most stressful experience of their life (Freeman, Boxer, Rickels, Tureck, & Mastroianni, 1985). The psychological impact of experiencing infertility can be summed as the following:

Handbook of Lifespan Cognitive Behavioral Therapy. https://doi.org/10.1016/B978-0-323-85757-4.00014-6

Failure to fulfill a major life event

Childbearing is a normal transition implying it to be one of the most obvious outcomes of marriage and a major life event of adult life. To find out that one is infertile thwarts the natural growth and progression to adulthood development (Gibson & Myers, 2002). It is often perceived as a failure to attain a milestone and may trigger beliefs that their life goals have been disrupted and that they will not be able to achieve a major milestone (Butler & Koraleski, 1990). Family members like grandparents (Cooper-Hilbert, 2001; Forrest & Gilbert, 1992), may feel disappointed by inability to have a child which adds up to the burden and feeling that they have failed in their role as adults (Atwood & Dobkin, 1992; Loftus & Andriot, 2012; Ulbrich, Coyle, & Llabre, 1990).

Emotional roller coaster

Infertile females experience extreme shifts in their emotions. They can sometimes feel intense jealousy of a woman who is pregnant or has children (Callan & Hennessey, 1989; Daniluk, 2001) and betrayed by her body or become angry with her partner (Cooper-Hilbert, 2001) while sometimes they can experience intense guilt and hold oneself responsible for the failure to conceive. The feeling of guilt may increase if the woman had a planned abortion earlier or has taken birth control pills in the past, they may hold themselves responsible for the inability and thus increasing the chances of self-punishment and suicide (Atwood & Dobkin, 1992). Apart from these the female may experience shame relating to the inadequacy in self and may fear rejection from spouse and being blamed. All this emotional turmoil essentially impacts the relationship with the spouse and may result in a permanent strenuous marriage. There may also be physical changes taking place throughout the body and a number of somatic complaints such as nausea, heart palpitations, and headache and stomach aches may also be experienced by the woman (Klempner, 1992) that also impacts the emotional wellbeing of the individual. Overall, it can be understood that infertility results in an endless number of emotions ranging from feeling of helplessness and a sense of loss of control, to deep experience of failure and its impact in the form of disappointment, irritability and loneliness (Daniluk, 2001; Forrest & Gilbert, 1992; Williams, Bischoff, & Ludes, 1992). These emotions may eventually result in development of clinical conditions like anxiety, body image issues, depression, suicide and at times psychosis (Altshuler et al., 1996).

Stigmatization

Infertility in our society is mostly perceived as incomplete womanhood (McQuillan, Greil, White, & Jacob, 2003). Usually the stigma related to failed pregnancy is either felt stigma or enacted stigma. Felt stigma are common perceptions of the stigmatized person that they are being labeled, stereotyped, or isolated from/by others (Link & Phelan, 2001). As a result, this person may self-stigmatize and have negative beliefs and automatic thoughts about themselves, and eventually experience loss of self-esteem and self-efficacy, and thus tend to isolate themselves (Corrigan & Watson, 2002). Enacted stigma is actual status loss and specific acts of discrimination toward an individual. Women feel marginalized and are viewed as deviant or abnormal when they cannot or do not become pregnant and give birth to children (Greil, 1991; Greil, Slauson-Blevins, & McQuillan, 2010; McQuillan et al., 2003).

Mental health problems related to infertility

According to a recent literature review conducted by Lykeridou, Gourounti, Deltsidou, Loutradis, and Vaslamatzis (2009), infertility was rated as the fourth most distressing life experience for women. It was associated with feelings of grief, depression, guilt, and anxiety. McQuillan et al. (2003) found that women who experienced infertility endorsed significantly high levels of distress when compared to others, such as those with resolved infertility problems. When compared to fertile women, emotional distress in infertile women was found to be significantly higher and the experience of subjective well-being was evidently poorer (Domar et al., 2012). A study conducted in a primary care sample found that, women who endorsed conception difficulties reported higher levels of depressive and anxiety symptoms on the Patient Health Questionnaire (Jordan & Ferguson, 2006). In a qualitative study by Bell (2013), many women endorsed a sense of guilt that they were not able to become pregnant, regardless of the infertility etiology. The importance of parenthood and intent to have children tends to have a strong connection to feelings of disempowerment to bring about a life wish (Jacob, McQuillan, & Greil, 2007; Johnson & Fledderjohann, 2012). Brothers and Maddux (2003) studied women who were more likely to ruminate about attaining pregnancy and parenthood and they strongly linked that to their happiness and psychological well-being, therefore the rumination was predictive of higher distress levels, anxiety and depressive symptoms (Thompson,

Woodward, & Stanton, 2011). Engagement with other life goals was associated with fewer depressive symptoms. Research also suggests that women's appraisal of the infertility and its associated distress depended largely on the social pressure, and their gender role identity (Miles, Keitel, Jackson, Harris, & Licciardi, 2009).

Symptoms of depression may include loss of interest in usual activities including sexual activities, anxiety, thoughts of suicide resulting from excessive guilt and self-punishment, isolation and persistent feelings of worthlessness and hopelessness (Domar, 2004). In another study conducted by Loftus and Andriot (2012) the participants endorsed depression over the loss of their identities as women due to childlessness, which led them to feel inauthentic and inadequate compared to women who had biological children.

Anxiety disorders include panic disorder, obsessive-compulsive disorder, posttraumatic stress disorder, generalized anxiety disorder, specific phobias usually blood and injection phobia, social phobia, and/or somatoform disorders. Approximately 40 million or about 18.1% of people in any given year, approximately have an anxiety disorder related to infertility (Kessler, Chiu, Demler, & Walters, 2005). Anxiety disorders frequently cooccur with depressive disorders or sometimes even with substance abuse. Most people who start with one anxiety disorder also gradually develop comorbid anxiety and/or depressive disorder (Kessler et al., 2005). The levels of anxiety and depression observed among infertile women is often comparable to women with cancer, or with Chronic Heart Disease, but was found to be less than anxiety and depression levels observed among women with chronic pain syndromes or HIV (Domar, Zuttermeister, & Friedman, 1993). One study, presented with an interesting finding by relating self-blame and a proneness to avoidance coping as the best predictors of psychological distress like anxiety and depression in infertile women (Bell, 2013; Loftus & Andriot, 2012). Mental symptoms in infertile women and showed that the rates of depression and anxiety were higher in infertile women than the fertile women; in addition, they also found that the rates of depression, anxiety, and low self-confidence increased with aging and prolonged infertility (Guz et al., 2003). Another study reported high prevalence of mental problems, specifically generalized anxiety disorder (GAD), obsessive compulsive disorder and somatoform disorder among infertile women, which considerably affected their quality of life; that warrants care and intervention (Hasanpour, Bani, Mirghafourvand, & Kochaksarayie, 2014; Upkong & Orji, 2006).

Trying to understand the neuro-immuno-endocrinological pathways linking anxiety and depression or more loosely speaking of stress with infertility there are two main neuro-endocrinological pathways that mediate and is responsible for the effects of psychological factors (e.g., stress, anxiety, depression, etc.) on the reproductive system. The hypothalamic-pituitary adrenal (HPA) axis is regulated by the secretion of corticotropic-releasing hormone (CRH) from the hypothalamus, which triggers the release of adrenocorticotrophic hormone (ACTH) from the pituitary gland and finally the stress hormone, cortisol is secreted from the adrenal cortex. The sympathetic nervous system-adrenal cortex-medulla (SAM) axis is involved in stress-induced secretions of norepinephrine from the brain stem and epinephrine from the adrenal medulla. CRH-induced pro-opiomelanocortin (POMC) that are derivatives from the pituitary gland inhibits the control of the hypothalamus over the gonadal axis (Rivest & Rivier, 1995). In addition to this, it was also found that cortisol levels were higher throughout the cycle among infertile women compared to fertile women (Csemiczky, Landgren, & Collins, 2000).

Cognitive-behavioral therapy in infertility

Despite the numerous areas to address through psychotherapeutic interventions for individuals and couples experiencing infertility, there is limited research examining the efficacy of specific treatments. Boivin, Bunting, Collins, and Nygren (2007) conducted an examination of literature on psychosocial interventions for infertility, including psycho-analytic, psychodynamic, cognitive-behavioral therapy, information/education-giving, autogenic relaxation training, coping skills focused, and stress reduction training programs. The results showed that all the interventions were effective. However, the findings also indicated that the intervention were successful in reducing the negative emotions temporarily associated with infertility but failed or were less effective in changing fixed beliefs about self and the future therefore the treatment outcome in psychological wellbeing and interpersonal relationship was not adequately promising (Esselstrom, 2014).

Cognitive-behavioral therapy (CBT) as a psychological intervention aims at changing negative cognitions and beliefs (Verkuijlen et al., 2016). CBT module typically includes primarily helping individuals identify a challenging situation in their life, in this case it is infertility or coming to terms with the fact that the individual will not be able to bear her own biological child; then acknowledging the thoughts and emotions associated with the situation and finally to identify negative or inaccurate or more precisely maladaptive behavioral responses to this situation, and thereafter lastly reshape the negative or inaccurate thinking and beliefs that eventually reduces the negative and distressing emotions and helps in modifying the maladaptive behavior (Mayo Clinic, 2021). Research has shown that the type of faulty coping strategies and certain cognitive errors, including self-blame and labeling, rumination, discounting the positives and catastrophizing, that individuals use to handle infertility-related distress is directly associated with depressive and anxiety symptoms (Kraaij, Garnefski, Schroevers, Weijmer, & Helmerhorst, 2010; Kraaij, Garnefski, & Vlietstra, 2008). Focusing on cognitive copings skills to implement more positive coping may greatly improve infertility-related distress for individuals experiencing infertility (Choudhary & Halder, 2019).

Formulation and conceptualization

Cognitive-behavioral therapy is based on the behavioral and cognitive theories that emphasize on interactionist perspective regarding the determinants of human behavior and psychological well-being (Lehman & Salovey, 1990). The core belief of this perspective is that the interaction between the individual and the environment continuously determines the behavior, cognitions, and effect of the individual. Inner thoughts and beliefs and intrapersonal factors such as temperament and personality act on the environment, just as the environmental factors influence the individual's thought process, feeling state and behaviors and actions. Thus, central to the cognitive-behavioral therapy is a detailed functional assessment and relief of the immediate and current problems and use several empirically based techniques to achieve mutually determined goals (Peake, Limbert, & Whitehead, 2005). Cognitive-behavioral therapy therefore strives for symptom relief, improving coping skills and developing self-efficacy, and finally prevents any relapse. This is done through challenging faulty cognitions and their behavioral correlates which corrects the negative and distressing emotions (Cooper, 1992). Thus, it can be understood that CBT is an appropriate treatment for infertility patients who present as anxious, depressed, or fearful of medical interventions. Furthermore, cognitive-behavioral therapy can often be as effective or even a better alternative for the treatment of these conditions than psychopharmacological medications, especially for women who are attempting to achieve pregnancy through various treatment modalities (Applegarth, 2006) (refer Figs. 1 and 2).

Components of the therapy

After formulation and conceptualization of the therapeutic goals and structure, the therapy sessions may be started formally. Before the sessions it is important to do baseline assessment depending on the symptoms (e.g., Beck Depression Inventory (BDI), Beck Anxiety Inventory (BAI), Illness Anxiety Questionnaire, etc.). The first session starts with detailed psycho education explaining the problem behavior/illness; the course and prognosis; the role of the therapy and the process in which it will be conducted; the outcome to be expected from the therapy and by when to expect the outcome. The role of the patient or individual seeking treatment during the therapy needs to be explained elaborately because the therapeutic process is essentially about the collaborative empiricism and not a therapist guided protocol. Special emphasis is to be given to the short-term and long-term goals and certain ground rules of session timings, homework assignments, and confidentiality are to be discussed.

Thereafter it is important to understand how CBT as a therapy works. The patient needs to be explained about the connection between cognition (thought process), Affect (feeling state or pervasive mood)—for example, anxiety or depressed mood or irritability, etc. and behavior that can be overt or observable or covert **(A-B-C model)**. The presence and link with **somatic symptoms** need to be established here with help of mind-body principle. This can be done **using metaphors and real-life examples**. After the Affect-Behavior-Cognition loop is explained and how it's correlated then the foundation to start the identification of faulty/maladaptive thoughts can be started. The aim needs to be made clear to the patient that since thought, emotion and behavior is interrelated therefore our target in the therapy is to identify and modify the faulty thoughts which is triggering the distressing emotions and maladaptive behavior and by doing so we will be able to achieve symptom reduction. The mind-body connection needs to be established so that the person identifies the physical symptoms because of faulty thoughts or distressing and negative emotions (refer Fig. 3).

Following this the somatic complaints which usually is very distressing for most of the patients may be addressed by giving diaphragmatic breathing exercises followed by progressive muscle relaxation techniques. It is imperative to explain the HPA axis and how our sympathetic and parasympathetic nervous system works for the patient to fully follow and practice any kind of relaxation techniques. During this time the Subjective Unit of Distress (SUDS) must be noted down and regular feedback and SUDS to be noted for keeping the patient adhered to the therapy by regulating their motivation and at the same time keeping track of the improvement.

At the same time the patient needs to start identifying the faulty cognition or thoughts and note down the negative or associated emotions to the thoughts and the behavior as a result of the thought and emotion which can be an observable behavior or some mental act which is distressing to the patient or is maladaptive. This can be done by introducing thought record diary. Although, there is a fixed structure of a thought record diary is followed in most CBT sessions, some amount of modifications can be done depending on the needs and/or limitations of the patient (refer Table 1).

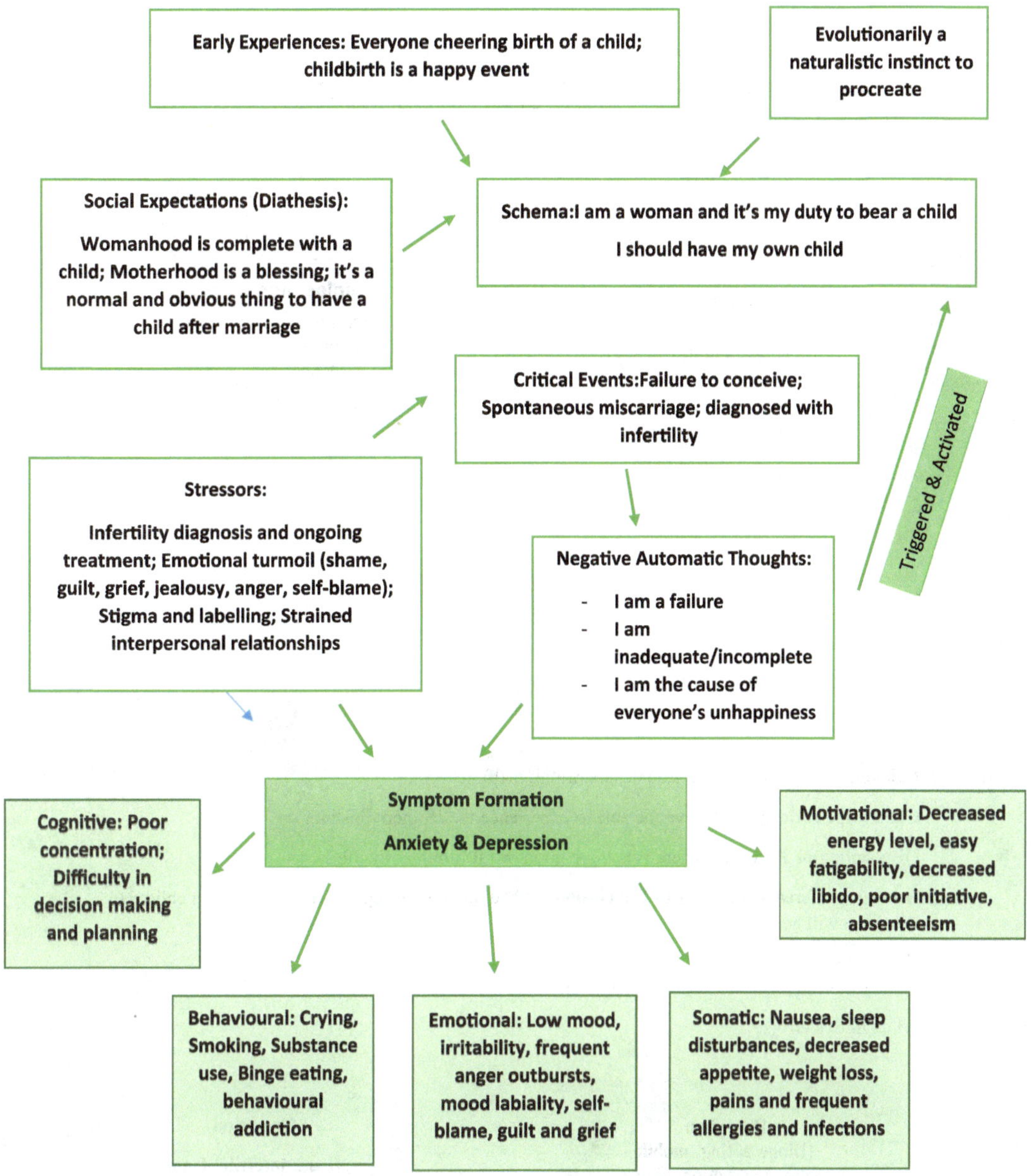

FIG. 1 CBT formulation of infertility and associated disorders.

It may take some time and multiple attempts for the patient to be able to correctly identify the thoughts and note down the associated emotions and behaviors. The therapist needs to be patient and explain it further with more use of examples. Once the patient can identify the thoughts, emotions, and behaviors then the next step would be to identify the cognitive errors. The patient is to be given the list of cognitive errors and show with example in identification and correct labeling of the cognitive bias that led to the faulty thought or experiencing that emotion. This process is to be done till the patient can identify the errors correctly. With this step, a new column is added where the patient identifies the various cognitive errors and discusses with the therapist in the session. Thought record diary and identification of cognitive errors are also given as homework assignment and is reviewed regularly in every session along with an account of the SUDS.

After the regular identification of thoughts, emotions, behaviors, and the cognitive errors it is time to move to the next phase of the therapy where the schema or belief system has to be identified. This can be down by various techniques like downward arrow technique, thought challenging, seeking evidence, providing contrary evidence, self-disclosure, giving examples and metaphors, strength weakness analysis, cost benefit analysis, etc. An example of some of the technique is given in Table 2.

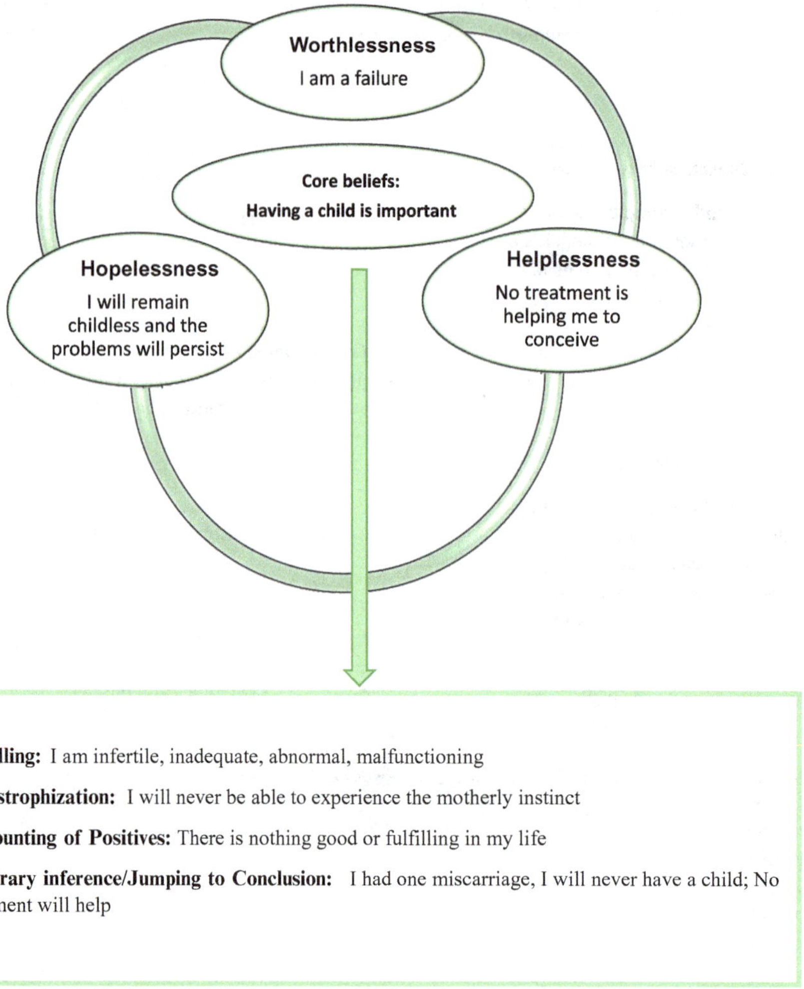

FIG. 2 Cognitive triad and cognitive errors.

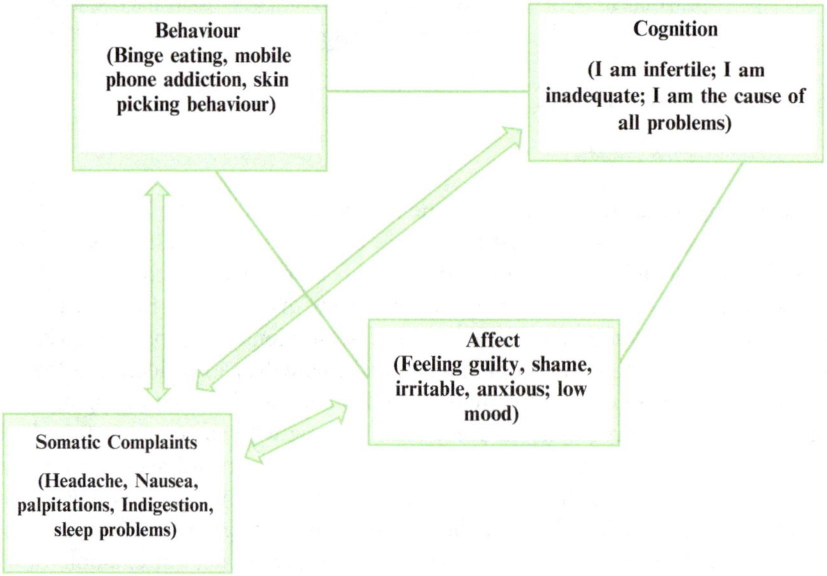

FIG. 3 Affect-behavior-cognition (A-B-C) model.

TABLE 1 Example of a thought record diary and identification of cognitive errors.

Date and time	Situation	Thought	Emotion	Behavior	Cognitive error
26.04.2019	My sister in law came home with her two children	I am infertile and defective I can never be a mother	Anger toward self	Misbehaved with sister in law Did not eat lunch	Labeling, arbitrary inference

TABLE 2 Showing thought challenging techniques and modifying core beliefs.

Patient-therapist illustrations	CBT techniques
Therapist (T): So, you know you are infertile, I understand the medical reason being endometriosis. Do you understand the process entirely? *Patient (P):* Yes, I do. And I feel terrible about it *Therapist:* When you say you feel terrible about it, what do you mean?	Downward arrow technique
P: My entire life changed; It will never be the same again **(Hopelessness)**; I am incapable of bearing a child; everyone around can have children, it's only me who is a failure **(Worthlessness)** *T:* I understand how you are feeling, and it must be very important to you to have a child; but when you say you are a failure what do you mean?	Downward arrow technique
P: I am defective and incompetent **(Old Core Belief)** *T:* How sure are you that you are defective? *P:* Almost 100% *T:* Is there any areas where you excel or manage things well and not be incompetent? Patient: Yes, I manage the job and the household; but I think I don't do that also very well	Evidence seeking
T: How sure are you about that? *P:* Somewhat *T:* How many times have you not met work deadlines; failed to pay bills; complete your household chores?	Thought challenging
P: No, I haven't mostly. But I am sure there are people who does it better than me *T:* There will always be someone better than you at everything; but how much out of 100 do you think you do these things well? *P:* 70%–80% *T:* How sure are you about it? Can you give me some examples	Evidence seeking
P: You can ask my family members if I have ever failed to give them food on time or not been there for them when they needed me. My colleagues also can depend on me for work *T:* That's brilliant. So how do you succeed at these? Considering that you are incompetent and inadequate and not good at anything?	Thought challenging
P: May be I am not a failure in everything *T:* Okay, true I think that is the case for most of us, for example I am good at doing finances but according to me I am a very bad cook **(Self-Disclosure)**. For example, if you are aware of Sheldon Cooper's character in the famous TV show, he is a brilliant physicist, but he can't drive a car or understand sarcasm or jokes **(Stories/Examples/ Metaphors)**. On that note then can we make a list of things that you are good at and things you don't do that well	Giving contrary evidence
P: (after making the list and **examining strengths and weaknesses**) May be I am not defective; I have failed to bear a child but succeeded in many things; I am like any other person with strengths and limitations **(New Core Belief)** *T:* Now with this development I would like you to note down all things you succeed at and fail at for the next 3 days before the next session	Homework assignment: Credit sheet

It can be seen from the tentative example in Table 2 how the thought challenging is done. After the schema is identified it is time to modify the faulty beliefs by generating alternate and adaptive beliefs. This can be done by brainstorming, role playing techniques, behavioral experiments, and acting "as if," restructuring early memories, etc. Also important is to develop coping skills. Problem solving skill needs to be enhanced and that is successful when it can be applied in various hypothetical problem situations initially and then in real life situation. Positive self-talk, reaffirmation, coping cards, etc.

techniques can be very useful in achieving this goal. The modification, application of new skills and improvement and success rate needs to be regularly monitored and validated. For this purpose, booster session is taken.

Finally, when the short-term and long-term goals of the planned therapy are met, the therapist may start the process of termination of therapy. Therapeutic blueprint and coping skill enhancement and various problem-solving skills needs to be reaffirmed to prevent relapse. Relapse prevention is highly important. Around the end of the therapy sessions post assessment needs to be done and the findings need to be discussed so that the patient gets a positive feedback on how much she has been able to work through her problems. Assessment in between sessions at regular intervals and noting down SUDS is also important for the course and success of the therapy. Alternatively, if the psychological issues are not as serious as anxiety or depression or other major distress and is only restricted to some adjustment problem in accepting the infertility diagnosis and in order to deal with it, a shorter and brief cognitive-behavioral therapy may be planned with the necessary cognitive and behavioral techniques.

Important consideration of the therapy

Allowing catharsis and validating the experience

Being infertile takes away the sense of control from the individual, and she may be trying to recreate a sense of meaning and purpose in her life or is not ready for the process. They may withdraw or continue to anticipate lives as parents (Atwood & Dobkin, 1992). Research has shown that the grief experienced with infertility or failed pregnancy is similar to that of a terminal illness or death (Cooper-Hilbert, 2001; McDaniel, Hepworth, & Doherty, 1992). The individual goes through feelings of shock, denial, guilt, anger, helplessness, isolation, depression, and loss. Atwood and Dobkin (1992) developed a breakdown of the grief cycle related to infertility. They identified four general phases that most infertile individuals/couples move through: disbelief and denial, anxiety and loss of control, isolation and guilt, and eventually resolution. It is extremely important to deeply explore the pain and grief associated with the inability to conceive biological children. In the therapeutic sessions, the therapist needs to provide comfortable accepting environment, with active listening to allow and facilitate the process of catharsis. When the grief and other distressing emotions with the disturbing thoughts of self and environment gets a venting channel and is adequately validated there is hope that resolution will come as the woman accepts the loss of control over fertility gradually and is able to take new productive directions in life (Cooper-Hilbert, 2001; Forrest & Gilbert, 1992; Klempner, 1992).

Normalize the experience with coping and ego strengthening skills

Primarily, the therapist needs to bear witness to the pain, acknowledge the stress and normalize the experiences of the client (Daniluk, 2001; Forrest & Gilbert, 1992; McDaniel et al., 1992). In the session they may need time to examine and analyses the meaning of their infertility or what it implies and how it impacts. For normalizing the experiences, educating the patient regarding typical responses and emotions relating to infertility may be helpful by providing them past records or examples of other patients (Cooper-Hilbert, 2001). Atwood and Dobkin (1992) suggested acknowledging the crisis through effective communication and empathic counseling skills is highly effective in treating the psychological aspects of infertility. They suggested exploring the woman's or couple's feelings of being at a loss of control and helping them express feelings of anger, sadness, and helplessness, therapists should provide the opportunity and time to express frustrations regarding the responses of their friends and family. If they are taking medication for the infertility, possible side effects such as headaches, nausea, moodiness, emotional outbursts, and distractibility (Cooper-Hilbert, 2001) needs to be discussed.Forrest and Gilbert (1992) proposed that, during the process of normalizing the experiences, it would be helpful to make a connection to another time when they have had similar emotions or been in an extremely challenged situation. This could provide insight regarding the coping skills that helped, and those strengths can then be drawn upon. Atwood and Dobkin (1992) also proposed that the therapist should address, challenge, provide contrary evidence and disconfirm myths such as "Failure to have a child means my womanhood is incomplete," "Women are made by nature to give birth and procreate," "It is my fault because I took birth control." Besides busting such myths an alternate identity can be developed, providing them with a sense of a role in society with or without children.

Self-acceptance

Often the woman feels that infertility is her personal failure and it is her inadequacy, or she is responsible for it. Many a times, the society as well reinforces this belief. The role of the therapist is to explore feelings related to self-blame.

Self-acceptance is found to be a prerequisite for a satisfying future and the infertility needs to be understood as an unwanted and unfortunate turn of events that has a medical origin (Cooper-Hilbert, 2001; McDaniel et al., 1992). Very important to note here is that infertility is a couple's challenge, not an individual's problem (Cooper-Hilbert, 2001; Daniluk, 1991). The goals of the therapy is to assist the patient to deal with anxiety or depression or other issues resulting from the infertility but at the same time to affirm it as a couple's hurdle and it's the job of a therapist to adequately reaffirm the patient's self-esteem (Butler & Koraleski, 1990) which will eventually allow them to forgive themselves and be more self-accepting (Williams et al., 1992).

Working on coping style

Therapists need to educate couples about the different coping styles used by men and women. As an individual's coping strategies are increased and improved, so is their ability to handle the crisis (Cooper-Hilbert, 2001). In most cases the woman needs to know that she is not grieving alone, and that infertility has impacted her partner equally. Coping strategies may include talking to others for support and understanding, this will help them understand that they are not responsible, and others are not blaming them. Assistance with communication skills can improve the couple's understanding of each other's experience. Effective communication can be accomplished through "I statement," listening, reflecting, and role playing of family needs (e.g., saying no to going to a baby shower (Atwood & Dobkin, 1992). The couple may also need to address if distancing has occurred between them regarding triangulation related to the hoped for a child situation (Cooper-Hilbert, 2001). Communication about the couple's sexuality concerns may be dealt with also. The couple may have been avoiding sexual intercourse because of the link to infertility and feelings of inadequacy. Sexuality can be addressed in therapy, with a focus on sexual intercourse for fun and closeness instead of for "procreation." Daniluk (2001) in his research work encouraged moving the focus of sexuality off failure and more toward sexual pleasure. When communication has been restored, the mourning process and acceptance process between the couple can be explored (Cooper-Hilbert, 2001; Forrest & Gilbert, 1992; McDaniel et al., 1992).

Gaining greater control over life

The most significant way the couple can regain control over the infertility situation is in the "decision-making" process. Daniluk (2001) pointed out that the woman must relinquish control over her fertility but gain control over her life choices. This includes deciding whether to try assisted birth procedures or opt for adoption procedures; whom to tell about their decision and when to tell them (Butler & Koraleski, 1990; Cooper-Hilbert, 2001; McDaniel et al., 1992). Role-playing and assertiveness training may be done in therapy sessions that could include having an impression who will be supportive and understanding. Gaining control will also help to reestablish the lost connections with family and friends (Cooper-Hilbert, 2001; McDaniel et al., 1992; Williams et al., 1992). In a study conducted by Gibson and Myers (2002) with infertile women a direct correlation was found between higher growth-fostering relationships and lower levels of stress. Therefore, having a strong social support is of great help in dealing with infertility. To elucidate the process of positive feedback, social support, and empathy, it was also proposed that the woman with the help of her partner may create a life story that did not include the problem of infertility. This new life story can include creating a new worldview with different meaning and focus for the couple while also remembering who they were before the infertility problem (Cooper-Hilbert, 2001; McDaniel et al., 1992). There are many resources that have been helpful for couples coping with infertility such as networking with other infertile couples (Atwood & Dobkin, 1992; McDaniel et al., 1992). Additional sources for relief and control include adopting pets, spending time with other children in the family, books, meditation, prayer, family, and friends (Atwood & Dobkin, 1992; Callan & Hennessey, 1989).

Future orientation

After the couple have achieved a sense of closure regarding the inability to have a biological child, establishing a new definition of family can be addressed. A reconstruction of the goals and motivation for the marriage will help the woman determine a new purpose in her life, and the desire for parenting can be readdressed (Cooper-Hilbert, 2001; Daniluk, 2001). Perhaps the woman could move from a focus on pregnancy and parenthood to a focus on couple relationships or parenting by adoption, forming a new construct of what family means for them and clarifying their own values related to the concept of family (Atwood & Dobkin, 1992; Matthews & Matthews, 1986; Williams et al., 1992). Forrest and Gilbert (1992) pointed out that adoption and foster care needs to be considered as a viable and alternate choice. There are costs and benefits involved with the decision that is made regarding whether or not to have children such as financial burdens, career choices, and life goals and all of that need to be prioritized and explored (Daniluk, 2001; Williams et al., 1992). It was recommended having the individual writing down two 5-year plans, one with a child and the other without. Once the woman has analyzed and incorporated all possibilities, she may be fit to make a decision about her future, and a strong sense of control may be gained.

Mini-dictionary of terms

A-B-C model: Cognitive-behavioral therapy includes the exploration of negative beliefs and alteration through more adaptive way. A-B-C techniques in CBT model refer to an approach to interpret the connection between the antecedent (A) or the triggering event/situation which leaded to irrational beliefs (B) in the person and indicated maladaptive or unexpected consequence (C) or behavior. A-B-C model explores the pattern of manifested distress.

Collaborative empiricism: In cognitive-behavioral therapeutic intervention, the term collaborative empiricism defined as a systematic process to the progression of therapy where the patient and therapist establish therapeutic goals collaboratively and facilitates the CBT outcome expectancy.

Infertility related distress: Infertility condition has an adverse impact over the psychological health and quality of life of the individual. It is associated with emotional, social, sexual, and interpersonal concerns with the individual with female sexual dysfunction. Infertile couple experience significant emotional turmoil and malfunctioning social behavior.

Self-acceptance: Self-acceptance is related to the awareness of own strength and weakness and being realistic about the expectations from one's own capability and self-deficiencies without blaming or doubting.

Coping skills: The capacity to deal, tolerate or accept a stressful situation and behaving in adaptive manner is known as coping. Skills which are constructed and performed by an individual to minimize the distress in different situation can be described as coping skills.

Key facts

- Infertility is a global challenge and crisis in married couple all over the world.
- Infertility could be precursor of major psychological problems like distress, anxiety, and depression.
- CBT model is found effective to deal with the psychological problems associated with infertility.
- Catharsis, enhancing ego strength, and self-acceptance are key features to overcome problems.
- Relapse prevention is highly important in therapeutic process.

Applications to other areas

Cognitive-behavioral therapy (CBT), has been found highly effective intervention technique with short-term and long-term effects, of reducing anxiety and depression in infertile couples and enables them to achieve general mental health through identifying and challenging their faulty thoughts and emotions of self-blame, guilt and disrupted interpersonal relationship (Williams, 2011). CBT helps individuals mourn the grief they experience and to develop acceptance of the events. Therefore, it also helps to bring about behavior change, development of relationships and support system, adaptive problem-solving strategies, and cognitive reconstruction required for acceptance of infertility or plan for adoption (Beck, 2008). The effectiveness of CBT has shown positive outcome in improving anxiety and depression in women dealing with infertility and undergoing various treatments like intrauterine insemination (Heidari, Latifnejad, Sahebi, Jahaniyan, & Mazloum, 2002). Another study showed that the continuous application of CBT techniques on infertile women, who did not have a medical or physical problem, reduced anxiety, depression, and psychological symptoms and even led to successful pregnancy in some cases (House, 2011).

Summary points

- Infertility is a condition of an individual and characterized by difficulties or failure to conceive in natural way after 12 months or more of regular unprotected sexual intercourse.
- The experience of the phase after the diagnosis of infertility includes chronic stress and serious psychological strain and the failure to fulfillment of a major life event which leads to extreme shifts of negative emotions. The infertility has also been perceived as a social stigma related to failed pregnancy.
- The infertility condition is related to most distressing life experience for women and often associated with feelings of grief, depression, guilt, and anxiety.

- Cognitive-behavioral therapy module mostly includes identification of a challenging situation in their life, in this case it is infertility or coming to terms with the fact that the individual will not be able to bear her own biological child and acknowledging the thoughts.
- Central to the cognitive-behavioral therapy is a detailed functional assessment and relief of the immediate and current problems and use a number of empirically based techniques to achieve mutually determined goals to replace negative thoughts and improve coping strategies.
- CBT includes the catharsis and validation of experience and feelings related to the experience and normalize the experience through improvement of coping skills, self-acceptance, and control.
- Cognitive-behavioral therapy (CBT), has been found highly effective intervention technique with short-term and long-term effects, of reducing anxiety and depression in infertile couples and enables them to achieve general mental health through identifying and challenging their faulty thoughts and emotions of self-blame, guilt and disrupted interpersonal relationship.

References

Altshuler, L. L., Cohen, L., Szuba, M. P., Burt, V. K., Gitlin, M., & Mintz, J. (1996). Pharmacologic management of psychiatric illness during pregnancy: Dilemmas and guidelines. *American Journal of Psychiatry.*, *153*(5), 592–606.

Applegarth, L. D. (2006). 20 Embryo donation: Counseling donors and recipients. In *Infertility counseling* (p. 356). Cambridge University Press.

Atwood, J. D., & Dobkin, S. (1992). Storm clouds are coming: Ways to help couples reconstruct the crisis of infertility. *Contemporary Family Therapy*, *14* (5), 385–403.

Beck, J. S. (2008). *Questions and answers about cognitive therapy* (p. 134). Beck Institute for Cognitive Therapy and Research.

Bell, K. (2013). Constructions of "infertility" and some lived experiences of involuntary childlessness. *Affilia*, *28*(3), 284–295.

Boivin, J., Bunting, L., Collins, J. A., & Nygren, K. G. (2007). International estimates of infertility prevalence and treatment-seeking: Potential need and demand for infertility medical care. *Human Reproduction*, *22*(6), 1506–1512.

Brothers, S. C., & Maddux, J. E. (2003). The goal of biological parenthood and emotional distress from infertility: Linking parenthood to happiness. *Journal of Applied Social Psychology*, *33*(2), 248–262.

Butler, R. R., & Koraleski, S. (1990). Infertility: A crisis with no resolution. *Journal of Mental Health Counseling.*, *12*(2), 151–163.

Callan, V. J., & Hennessey, J. F. (1989). Strategies for coping with infertility. *British Journal of Medical Psychology*, *62*(4), 343–354.

Choudhary, M., & Halder, S. (2019). Cognitive behavior therapy in management of psychosocial factors in female infertility. *Journal of Psychosocial Research*, *14*(1), 53–62.

Cooper, T. G. (1992). The epididymis as a site of contraceptive attack. In *Spermatogenesis—Fertilization—Contraception* (pp. 419–460). Berlin, Heidelberg: Springer.

Cooper-Hilbert, B. (2001). Helping couples through the crisis of infertility. *Clinical Update: The American Association for Marriage and Family Therapy*, *3*, 1–6.

Corrigan, P. W., & Watson, A. C. (2002). The paradox of self-stigma and mental illness. *Clinical Psychology: Science and Practice*, *9*(1), 35–53.

Csemiczky, G., Landgren, B. M., & Collins, A. (2000). The influence of stress and state anxiety on the outcome of IVF-treatment: Psychological and endocrinological assessment of Swedish women entering IVF-treatment. *Acta Obstetricia et Gynecologica Scandinavica: Original Article*, *79*(2), 113–118.

Cwikel, J., Gidron, Y., & Sheiner, E. (2004). Psychological interactions with infertility among women. *European Journal of Obstetrics & Gynecology and Reproductive Biology*, *117*(2), 126–131.

Daniluk, J. C. (1991). Strategies for counseling infertile couples. *Journal of Counseling & Development*, *69*(4), 317–320.

Daniluk, J. C. (2001). Reconstructing their lives: A longitudinal, qualitative analysis of the transition to biological childlessness for infertile couples. *Journal of Counseling and Development*, *79*(4), 439–449.

Domar, A. D. (2004). Impact of psychological factors on dropout rates in insured infertility patients. *Fertility and Sterility*, *81*(2), 271–273.

Domar, A., Gordon, K., Garcia-Velasco, J., La Marca, A., Barriere, P., & Beligotti, F. (2012). Understanding the perceptions of and emotional barriers to infertility treatment: A survey in four European countries. *Human Reproduction*, *27*(4), 1073–1079.

Domar, A. D., Zuttermeister, P. C., & Friedman, R. (1993). The psychological impact of infertility: A comparison with patients with other medical conditions. *Journal of Psychosomatic Obstetrics and Gynaecology*, *14*, 45–52.

Esselstrom, L. (2014). *A guide for the psychosocial treatment of infertility*. scholarsrepository.llu.edu.

Forrest, L., & Gilbert, M. S. (1992). Infertility: An unanticipated and prolonged life crisis. *Journal of Mental Health Counseling*, *14*(1), 42–58.

Freeman, E. W., Boxer, A. S., Rickels, K., Tureck, R., & Mastroianni, L., Jr. (1985). Psychological evaluation and support in a program of in vitro fertilization and embryo transfer. *Fertility and Sterility*, *43*(1), 48–53.

Gibson, D. M., & Myers, J. E. (2002). *The effect of social coping resources and growth-fostering relationships on infertility stress in women*. Doctoral dissertation University of North Carolina at Greensboro.

Greil, A. L. (1991). *Not yet pregnant: Infertile couples in contemporary America*. Rutgers University Press.

Greil, A. L., Slauson-Blevins, K., & McQuillan, J. (2010). The experience of infertility: A review of recent literature. *Sociology of Health & Illness*, *32*(1), 140–162.

Guz, H., Ozkan, A., Sarisoy, G., Yanik, F., Yanik, A., & Schuiling, G. A. (2003). Psychiatric symptoms in Turkish infertile women. *Journal of Psychosomatic Obstetrics and Gynecology*, *24*(4), 267–271.

Hasanpour, S., Bani, S., Mirghafourvand, M., & Kochaksarayie, F. Y. (2014). Mental health and its personal and social predictors in infertile women. *Journal of Caring Sciences*, *3*(1), 37.

Heidari, P., Latifnejad, R., Sahebi, A., Jahaniyan, M., & Mazloum, S. R. (2002). Impact of cognitive behaviour therapy on anxiety level of primary infertile women undergoing IUI. *Journal of Reproduction & Infertility*, *3*(3), 40–51.

House, S. H. (2011). Epigenetics in adaptive evolution and development: The interplay between evolving species and epigenetic mechanisms. In *Handbook of epigenetics* (pp. 425–445). Academic Press.

Jacob, M. C., McQuillan, J., & Greil, A. L. (2007). Psychological distress by type of fertility barrier. *Human Reproduction*, *22*(3), 885–894.

Johnson, K. M., & Fledderjohann, J. (2012). Revisiting "her" infertility: Medicalized embodiment, self-identification and distress. *Social Science & Medicine*, *75*(5), 883–891.

Jordan, C. B., & Ferguson, R. J. (2006). Infertility-related concerns in two family practice sites. *Families, Systems & Health*, *24*(1), 28.

Kessler, R. C., Chiu, W. T., Demler, O., & Walters, E. E. (2005). Prevalence, severity, and comorbidity of 12-month DSM-IV disorders in the National Comorbidity Survey Replication. *Archives of General Psychiatry*, *62*(6), 617–627.

Klempner, L. G. (1992). Infertility: Identifications and disruptions with the maternal object. *Clinical Social Work Journal*, *20*(2), 193–203.

Kraaij, V., Garnefski, N., Schroevers, M. J., Weijmer, J., & Helmerhorst, F. (2010). Cognitive coping, goal adjustment, and depressive and anxiety symptoms in people undergoing infertility treatment: A prospective study. *Journal of Health Psychology*, *15*(6), 876–886.

Kraaij, V., Garnefski, N., & Vlietstra, A. (2008). Cognitive coping and depressive symptoms in definitive infertility: A prospective study. *Journal of Psychosomatic Obstetrics and Gynecology*, *29*(1), 9–16.

Lehman, A. K., & Salovey, P. (1990). Psychotherapist orientation and expectations for liked and disliked patients. *Professional Psychology: Research and Practice*, *21*(5), 385.

Link, B. G., & Phelan, J. C. (2001). Conceptualizing stigma. *Annual Review of Sociology*, *27*(1), 363–385.

Loftus, J., & Andriot, A. L. (2012). "That's what makes a woman": Infertility and coping with a failed life course transition. *Sociological Spectrum*, *32*(3), 226–243.

Lykeridou, K., Gourounti, K., Deltsidou, A., Loutradis, D., & Vaslamatzis, G. (2009). The impact of infertility diagnosis on psychological status of women undergoing fertility treatment. *Journal of Reproductive and Infant Psychology*, *27*(3), 223–237.

Mascarenhas, M. N., Flaxman, S. R., Boerma, T., Vanderpoel, S., & Stevens, G. A. (2012). National, regional, and global trends in infertility prevalence since 1990: A systematic analysis of 277 health surveys. *PLoS Medicine*, *9*(12), e1001356.

Matthews, R., & Matthews, A. M. (1986). Infertility and involuntary childlessness: The transition to non-parenthood. *Journal of Marriage and the Family*, *48*, 641–649.

Mayo Clinic. (2021). *Male infertility*. https://www.mayoclinic.Org/diseases- conditions/male infertility/symptoms causes/syc-20374773. (Accessed April 2021).

McDaniel, S. H., Hepworth, J., & Doherty, W. (1992). Medical family therapy with couples facing infertility. *American Journal of Family Therapy*, *20*(2), 101–122.

McQuillan, J., Greil, A. L., White, L., & Jacob, M. C. (2003). Frustrated fertility: Infertility and psychological distress among women. *Journal of Marriage and Family*, *65*(4), 1007–1018.

Miles, L. M., Keitel, M., Jackson, M., Harris, A., & Licciardi, F. (2009). Predictors of distress in women being treated for infertility. *Journal of Reproductive and Infant Psychology*, *27*(3), 238–257.

Peake, K. J., Limbert, C., & Whitehead, L. (2005). Gone, but not forgotten: An examination of the factors associated with dropping out from treatment of eating disorders. *European Eating Disorders Review: The Professional Journal of the Eating Disorders Association*, *13*(5), 330–337.

Rivest, S., & Rivier, C. (1995). The role of corticotropin-releasing factor and interleukin-1 in the regulation of neurons controlling reproductive functions. *Endocrine Reviews*, *16*(2), 177–199.

Thompson, E. H., Woodward, J. T., & Stanton, A. L. (2011). Moving forward during major goal blockage: Situational goal adjustment in women facing infertility. *Journal of Behavioural Medicine*, *34*(4), 275–287.

Ulbrich, P. M., Coyle, A. T., & Llabre, M. M. (1990). Involuntary childlessness and marital adjustment: His and hers. *Journal of Sex & Marital Therapy*, *16* (3), 147–158.

Upkong, D., & Orji, E. (2006). Nijerya'dakiinfertilkadınlardaruhsağlığı. *Turk PsikiyatriDerg*, *17*, 259–265.

Verkuijlen, J., Verhaak, C., Nelen, W. L., Wilkinson, J., & Farquhar, C. (2016). Psychological and educational interventions for subfertile men and women. *Cochrane Database of Systematic Reviews*, *3*.

Williams, H. (2011). Is there a role for Psychological Wellbeing Practitioners and Primary Care Mental Health Workers in the delivery of low intensity cognitive behavioural therapy for individuals who self-harm? *Journal of Mental Health Training, Education and Practice*, *6*(4), 165.

Williams, L., Bischoff, R., & Ludes, J. (1992). A biopsychosocial model for treating infertility. *Contemporary Family Therapy*, *14*(4), 309–322.

World Health Organization (2013). Infertility definitions and terminology. http://www.who.int/reproductivehealth/topics/infertility/definitions/en/index.html (accessed 9 October 2013).

World Health Organization. (2016). *Infertility definitions and terminology*. World Health Organization.

Chapter 3

Cognitive behavioral therapy for perinatal depression—Tailored face-to-face treatment and translation for online delivery of MumMoodBooster

Jeannette Milgrom[a,b], Michele Burn[a], Andre L. Rodrigues[a], Alan W. Gemmill[a], and Yafit Hirshler[a]
[a]*Parent-Infant Research Institute, Heidelberg Heights, VIC, Australia,* [b]*Melbourne School of Psychological Sciences, University of Melbourne, Melbourne, VIC, Australia*

Abbreviations

BAI	Beck Anxiety Inventory
BBB	Beating the Blues before Birth
BDI	Beck Depression Inventory
CBT	Cognitive Behavioral Therapy
DASS-21	Depression Anxiety Stress Scale, 21-item
DSM	Diagnostic and Statistical Manual of mental disorders
EPDS	Edinburgh Postnatal Depression Scale
FTF	Face-to-Face
GAPND	Getting Ahead of Postnatal Depression
GP	General Practitioner
HDRS	Hamilton Depression Rating Scale
MMB	MumMoodBooster
PHQ-9	Patient Health Questionnaire-9
PIRI	Parent-Infant Research Institute
RCT	Randomized Controlled Trial

Introduction

Depression in new and expectant mothers is highly prevalent, and one of the most careful estimates has been provided by Gavin et al. (2005). This metaanalysis estimated a period prevalence of 18.4% in pregnancy and 19.2% by 3 months postpartum. Depression occurring in the period from conception to the first-year postbirth has commonly been called perinatal depression.

Left untreated, perinatal depression can have a serious, negative impact not only on the woman but also on her infant and partner (Letourneau et al., 2012). Early intervention is therefore indicated.

A considerable body of research supports the effectiveness of cognitive behavioral therapy (CBT) interventions for depression in the general adult population (Cuijpers, Cristea, Karyotaki, Reijnders, & Huibers, 2016), and CBT is often considered the gold standard for psychotherapy of depression (David, Cristea, & Hofmann, 2018).

CBT treatments delivered in the perinatal period have also been effective (Cuijpers & Karyotaki, 2021; Sockol, 2015), and anecdotal feedback received from mothers in our Getting Ahead of Postnatal Depression (GAPND) program has reflected this:

"I felt like thoughts and feelings swallowed me up, I now realise I can actually change them".

"I loved the group: sharing experiences and seeing myself in others...."

The challenges for women in the perinatal period may present differently in pregnancy and postnatally and treatment needs may differ. Support for the effectiveness of CBT for maternal depression in specific periods is detailed below.

Handbook of Lifespan Cognitive Behavioral Therapy. https://doi.org/10.1016/B978-0-323-85757-4.00025-0

Depression treatments delivered antenatally

A recent metaanalysis evaluated the effect of antenatally delivered CBT on depressive symptoms both immediately after treatment while still pregnant, and also when following up women postnatally (Li et al., 2020). CBT was superior to usual care in reducing depressive symptoms posttreatment and up to 4 months postnatally. Of the 13 RCTs included, 6 involved individual CBT (including 1 self-help, 1 iCBT, and 1 multimodal intervention with CBT elements) and 7 involved group CBT (including 1 iCBT and 2 multimodal interventions). The number of sessions varied from 3 to 12 delivered in a variety of settings, including obstetric clinics, women's home, or mixed (clinic and home) settings. Another strategy in which CBT has been utilized to mitigate perinatal depression is prevention, and research is ongoing. Dennis and Dowswell (2013) concluded in their metaanalysis that antenatal psychological interventions were most effective in preventing postnatal depression when targeted to women at risk.

Depression treatments delivered postnatally

Results of several metaanalyses have demonstrated the beneficial impact of postnatally initiated CBT on maternal depressive symptoms. A metaanalysis by Dennis, Ross, and Grigoriadis (2007) reported a beneficial effect of CBT on depressive symptoms (variously defined) when compared to usual care immediately posttreatment and within the first year postpartum. Of these five trials, four were RCTs and one was a quasirandomized trial. The number of CBT sessions delivered in these trials ranged from 6 to 18 and the sessions were delivered by one or two health visitors, psychologists, or trained early childhood nurses at the mothers' homes or elsewhere (not specified).

The metaanalysis of RCTs by Cuijpers, Brannmark, and van Straten (2008) demonstrated similar results.

Likewise, the metaanalysis by Sockol, Epperson, and Barber (2011) reported a beneficial effect of CBT on postnatal depressive symptoms compared to control conditions. This metaanalysis included eight RCTs and one open trial. CBT in these trials was delivered individually (six trials) or in a group setting (three trials) and the number of sessions varied from 6 to 12.

Other considerations when treating perinatal depression

Perinatal depression affects not only the woman but also the fetus, infant, partner, and her broader social relationships (Letourneau et al., 2012). This can be understood from a family systems perspective, which posits that interactions between parent, infant (child/ren), spouse, extended family, and the broader social context outside the family (e.g., work, school, and support services) are complex and reciprocal (Bronfenbrenner, 1994). Thus, the mental health of one family member influences the mental health and wellbeing of others in the family system. In addition, antenatal depression and anxiety may have a direct effect on the infant. These considerations reinforce the importance of treating perinatal depression not just to improve maternal wellbeing but to mitigate the effects on infants and partners.

The mother-infant relationship and the impact of depression on the child

Impact of antenatal depression on child development

A large body of evidence suggests that children whose mothers experienced depression while pregnant might be at risk of negative developmental outcomes, including behavioral and cognitive problems (Stein, Pearson, & Goodman, 2014). This deleterious impact of antenatal depression is thought to be exerted at least in part via fetal programming mechanisms (Lautarescu, Craig, & Glover, 2020).

Evidence on the efficacy of antenatal depression treatment in preventing adverse child development has been limited (Glover, 2014). Our pilot RCT has shown promising results indicating several medium to large effects in the behavior of 9-month-old children whose mothers received our Beating the Blues before Birth (BBB) antenatal depression CBT treatment (Milgrom et al., 2015). A follow-up study demonstrated medium effects in child behavior at 24 months favoring the intervention group (Milgrom et al., 2019). This approach is currently being replicated in an adequately powered RCT.

Impact of postnatal depression on child development and mother-infant relationship

It is increasingly recognized that the first 1000 days of life provide a critical foundation for child development, and postnatal depression has been shown to have a negative impact on cognitive, social, and emotional development of children (Murray, Halligan, & Cooper, 2018; Stein et al., 2014).

Depressed mothers experience difficulties providing infants with the contingent responsiveness necessary for a "good enough" relationship (Murray et al., 2018). Dysfunctional relationships between postnatally depressed mothers and their infants were found in 73% of dyads (Milgrom, Ericksen, McCarthy, & Gemmill, 2006). Furthermore, the quality of mother-infant relationship has been found to play a central role in moderating between adverse infant developmental outcomes and mothers' postnatal depression (Ahun & Côté, 2019; Milgrom, Westley, & Gemmill, 2004)

Treatment of maternal depression alone is not sufficient for the mother-infant relationship to recover in many instances (Forman et al., 2007; Milgrom et al., 2006). There is therefore increasing interest in the application of mother-infant interaction interventions. Previous RCTs, including our own studies, have produced promising results in improving mother-infant interactions (e.g., Beeber et al., 2013; Holt, Gentilleau, Gemmill, & Milgrom, 2021; Murray et al., 2003).

Maternal perinatal depression, the partner and the couple relationship

Poor couple relationship and lack of partner support are both known risk factors of maternal perinatal depression. Further, interventions involving the partner show improvement in couple relationships and lower depressive symptoms (Pilkington, Whelan, & Milne, 2015). Therefore, integrating couple sessions in the maternal depression treatment is essential. Our CBT interventions include between 1 (antenatal program) and 3 sessions (postnatal program) which are dedicated to the partner and the couple relationship (see section below for more information).

Perinatal depression can also be experienced by 10% of fathers (Rao et al., 2020). The prevalence of depression in men whose partners experience perinatal depression increases (Thiel, Pittelkow, Wittchen, & Garthus-Niegel, 2020), ranging from 24% to 50%, in the first year postpartum (Goodman, 2004). Thus there is an emerging consensus that attention should be given to treating men.

Despite this, there are almost no specialized treatments available (Rodrigues, Ericksen, Watson, Gemmill, & Milgrom, 2021), and we have recently developed an online CBT program for men experiencing postnatal depression, which is currently being evaluated in an RCT.

Anxiety

While perinatal depression has been a focus of treatments and many metaanalyses, it has long been recognized that anxiety is often comorbid (Fairbrother, Janssen, Antony, Tucker, & Young, 2016). At times, anxiety is present with minimal symptoms of depression and yet can be debilitating to maternal functioning. There is also interest in the concept of pregnancy-specific anxiety (Ayers, Coates, & Matthey, 2015). The prevalence of perinatal anxiety disorders has been estimated as: specific phobia (7.7%), social anxiety disorder (5.2%), obsessive-compulsive disorder (3.9%), generalized anxiety disorder (3.6%), panic disorder (1.6%), posttraumatic stress disorder (0.7%), and agoraphobia (0.3%) (Fairbrother et al., 2016). However, Wisner and colleagues (Wisner et al., 2013) reported much higher rates of generalized anxiety disorders in a large sample of postnatal women who had screened positive on the EPDS. In that sample, anxiety disorders were the most common comorbidity of depressed women (82.9%), and among those the most common anxiety disorder by far was generalized anxiety disorder (52.3%). The CBT programs described in this chapter incorporate anxiety management strategies due to this high comorbidity and transdiagnostic treatments are ideally provided in an integrated manner (Schaeuffele, Schulz, Knaevelsrud, Renneberg, & Boettcher, 2021).

Access, stigma and COVID-19 pandemic

Having a new baby makes clinic attendance particularly challenging. The lack of energy and difficulty making decisions that are a common feature of depression make it difficult to organize an infant and attend a clinic appointment. In addition, childcare may not be available, and not all therapists are open to the infant being present. Accessing treatment, therefore, remains a challenge and is often complicated by feelings of shame and stigma (Bilszta, Ericksen, Buist, & Milgrom, 2010). These feelings are not aided by societal myths that motherhood is a time of great joy.

The situation has become complicated during the COVID-19 pandemic which has had a significant impact on new and expectant parents. The social restrictions imposed and increased isolation have resulted in an increased need for mental health services and information (Fisher et al., 2020; Van Agteren et al., 2020).

Reduced healthcare access to face-to-face (FTF) services has also meant the availability of online prevention and treatment programs supporting the emotional health of new and expectant mothers is more important than ever.

In recent years there has been growing use of internet-based interventions to address perinatal depression (Lau, Htun, Wong, Tam, & Klainin-Yobas, 2017), and we report below on our successful roll-out in Australia. Being anonymous, accessible, affordable, and convenient, internet-based depression programs can help overcome some major barriers to treatment uptake.

Practice and procedures for the GAPND face-to-face program

Adjustments of traditional CBT programs for perinatal populations

We have adapted traditional CBT programs for women in the perinatal period. The challenges for women in pregnancy and after the birth of a child differ. As such, we have developed both a postnatal (GAPND) and an adapted antenatal (BBB) CBT program.

In the antenatal period, there are challenges in the transition to parenthood. After childbirth, the increased demands of an infant make it difficult for mothers to find time to practice strategies that are time consuming, or to follow through with homework. In addition, the constant tasks associated with infant care accentuate symptoms that accompany depression. For example, women may have more difficulty concentrating, so there is a need to introduce reinforcing activities very early on in the treatment to maintain engagement.

Our approach was informed both by best practice CBT models as well as our biopsychosocial model of postnatal depression, shown below in Fig. 1. The model takes into account risk factors, triggering factors and the societal factors that may contribute to depression and then considers the exacerbating and maintaining factors that may be important to suggest treatment targets. The model depicts what we know about factors that contribute to a woman's likelihood of developing postnatal depression. The major risk factors that consistently appear include previous history of depression, family history of depression, stressful life events (e.g., domestic violence), and social support (including partner support) (Dennis et al., 2007). Sociocultural factors can also result in additional stress due to the cultural myths that continue to be propagated, painting a simplistic picture of a positive motherhood experience. For some women, biological factors play a role by increasing vulnerability (e.g., due to genetic factors).

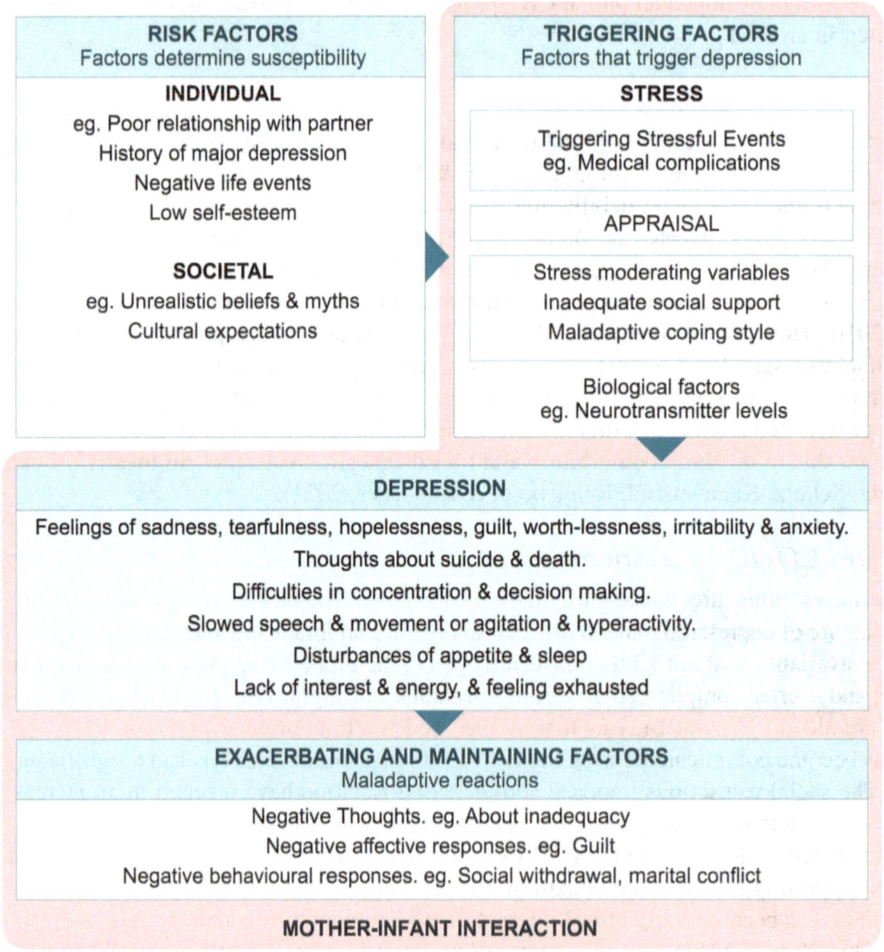

FIG. 1 A Biopsychosocial Model of Perinatal Depression. Our intervention was informed by best practice CBT models as well as our biopsychosocial model of postnatal depression. A review of the literature has identified a range of factors which play a key role in both predicting and maintaining depression. *(Adapted from Milgrom, J., Martin, P. R., & Negri, L. M. (1999).* Treating postnatal depression. A psychological approach for health care practitioners. *Chichester: Wiley.)*

Different factors in the model may be relevant for a woman based on her personal history. For example, a poor relationship with a partner may be a precipitating factor, exacerbated by her general lack of support and tendency to depressogenic thoughts. Feelings of hopelessness, self-blame, and low mood make it less likely that difficulties become resolved.

A central feature of the model is the relationship between triggering factors and maladaptive responses. The model highlights the importance of the appraisal of stressful events and how maladaptive coping styles may moderate this stress, whereas social support may protect and mitigate the effects. Depression is maintained by negative thoughts, feelings and behaviors so that the model provides a focus for our CBT treatment.

Finally, the mother-infant interaction and partner relationship are affected both directly by depressive symptoms and by triggering factors and ongoing negative thoughts and feelings. We therefore included partner sessions as well as a family-of-origin session, not commonly found in traditional CBT. Advances in interpersonal psychotherapy (Sockol, 2018) and the context of earlier family experiences make this an important addition. However, complex problems need to be dealt with outside the group program.

Description of the postnatal program

The GAPND program is a specialized treatment program for postnatal depression that we developed with the aim of providing a manualized treatment that is accessible and effective. The manual provides detailed step by step descriptions, including scripts, for each session of the program (Milgrom et al., 1999) suitable for a range of professionals with CBT expertise.

The program is suitable for women experiencing moderate to severe symptoms of depression, including those with a clinical diagnosis.

We built on the Coping with Depression Course (Lewinsohn, Antonuccio, Steinmetz, & Teri, 1984), which has been evaluated in numerous outcome studies, with demonstrated effectiveness (Cuijpers, Muñoz, Clarke, & Lewinsohn, 2009) and is generally considered a gold standard CBT treatment. Lewinsohn focuses on both behavioral (pleasant activities and relaxation) and cognitive skills, also stressing the importance of assertiveness, problem solving, communication and time management within a psychoeducational framework. Self-monitoring is seen as an important tool to create change. We retained these foci in our program and followed his example of the early introduction of response contingent positive reinforcement, introducing activities relevant to postnatal women. We also increased the number of cognitive sessions, but only after some success has been achieved through behavioral interventions. Our goal was to bring about positive changes in women's lives very early on. Many new mothers feel too overwhelmed to implement cognitive strategies, so these need to be introduced slowly. Tiredness and competing daily infant care tasks meant that we also reduced homework activities.

In addition, as in Lewinsohn's course, we delivered treatment as a group program, given the importance of social support for postnatal women (Milgrom, Hirshler, Reece, Holt, & Gemmill, 2019). Thus, each session includes a time segment for dealing with the group process. We offer women a 9-session program, and also include three additional partner sessions due to the critical role of partner support and the partner relationship (see Table 1). We drew not only on social learning theory but also on principles of interpersonal therapy (IPT), which has also proven effective for perinatal depression (Sockol, 2018). The partner sessions are generally offered in the evenings. Women are encouraged to invite another family member should they not be in a partner relationship, as the emphasis is on psychoeducation and building support.

Partner sessions details (see Table 1)

Session 1: Partners and postnatal depression

The focus of this session is psychoeducation to support a shared understanding of postnatal depression for women and partners. This includes depression symptomatology, etiology, course, evidence-based treatment options and prognosis. Postnatal depression is introduced using a biopsychosocial framework, through a systemic lens. Couples share their observations and reflections about their partner, and their own expectations and feelings about parenthood. Support strategies for partners are provided in the session, including tips for behavioral strategies and engagement with their support network.

Session 2: The couple relationship

This session provides psychoeducation on the changes that occur for couples after they have a baby including common difficulties in communication, role changes, intimacy and their sexual relationship. Strategies to support positive communication are provided, and partners are provided information on how to support each other and discuss their expectations about their intimacy and sexual relationship.

TABLE 1 Structure, content and procedures of the Getting Ahead of Postnatal Depression (GAPND) program.

Procedures for delivering the Getting Ahead of Postnatal Depression (GAPND) program in practice		
BEHAVIORAL Session 1: Understanding and managing postnatal depression	BEHAVIORAL Session 2: Pleasant activities—How can I find the time?	BEHAVIORAL Session 3: Relaxation on the run
Welcome and housekeeping Activity: Rapport and connections Group discussion: Sharing pregnancy and birth experience Psychoeducation: • What is postnatal depression? • Introduction to CBT, the relationship between thoughts, feelings and behavior, and the depressive mood spiral *Homework*	Psychoeducation: • Review of the cognitive triad (negative thoughts about the self, the world and the future) and effect on behavior Group Activity: • Increasing pleasant activities and monitoring effect on mood Group Discussion. *Homework* **FIRST PARTNER SESSION OCCURS THIS WEEK**	Review: Consolidating pleasant activities Psychoeducation: Relaxation training and relaxing on the run Group activity: deep muscle relaxation Group discussion: Portable stress busters Group activity: Relaxation plan for high tension times *Homework*
BEHAVIORAL Session 4: Assertiveness and self-esteem—telling others what I think and how I feel	COGNITIVE Session 5: Unrealistic expectations of parenting—influences from the past	COGNITIVE Session 6: My internal dialogue—the missing link
Psychoeducation: Assertiveness skills Group discussion: Styles of communication (passive, aggressive and assertive communication) Group activity: Assertiveness training Psychoeducation: Self-esteem *Homework:* Love letter	Psychoeducation: Parenting styles and connection with family of origin, for self and partner Activity: My family of origin Psychoeducation: Challenging unrealistic expectations of parenting *Homework* **SECOND PARTNER SESSION OCCURS THIS WEEK**	Psychoeducation: Self-instruction techniques Group discussion: Link to cognitive model and recognizing unhelpful thoughts and unrealistic expectations Group activity: Self Instruction (small group exercise) Preparation for termination *Homework* Preview Session 7 *Homework*
COGNITIVE Session 7: Developing a more helpful thinking style	COGNITIVE Session 8: Challenging my internal critic	COGNITIVE Session 9: Putting it all together—travelling on
Psychoeducation: Negative thoughts; feeling worthless and depressed Individual activity: Tuning into my thoughts and related feelings Group discussion Psychoeducation: Ways to reduce negative thoughts and increase positive thoughts Group discussion *Preview Session 8* *Homework*	Psychoeducation: Shifting cognitive distortions Activity: Identifying positive, negative or neutral thoughts and distortions concerning self and parenthood Group activity: My expectations of motherhood and parenting *Homework* **THIRD PARTNER SESSION OCCURS THIS WEEK**	Group discussion: What was helpful? Group discussion: What about another baby? Psychoeducation: Maintaining gains. Group discussion: Social support Individual exercise: My goals Feelings about treatment termination. *An optional session can be offered as a booster to build on relapse prevention, consolidate gains and plans for the future*

This table shows the structure and the content of the nine group GAPND sessions as delivered in practice.
Adapted from Milgrom, J., Martin, P. R., & Negri, L. M. (1999). *Treating postnatal depression. A psychological approach for health care practitioners.* Chichester: Wiley.

Session 3: Doing it ourselves

This session reviews depressive symptoms and changes that may have been observed by the woman and her partner, relating to woman's mood and their couple relationship. It is an opportunity to reflect on coping strategies that were learned during the course of the program and treatment gains for the women and the couple. The session concludes with discussion and planning for treatment termination, including relapse prevention and ongoing avenues of support.

Optional mother-infant intervention

As discussed earlier, improvement in depression does not always result in improvements in the mother-infant relationship. We developed an optional module, which could be delivered after the GAPND program. The idea was to only introduce targeted interventions, focusing on the infant, after the mother had shown some improvement in her depression to increase her availability to the infant and not engender further feelings of failure. We built upon the principles of a "good enough" parent interaction and the program was designed to target two key elements of mother-infant relationships that are amenable to change: (i) maternal communication, observation and responsiveness skills and (ii) distorted maternal internal representations.

A 3-session mother-infant intervention, happiness, understanding, giving and sharing (HUGS), was developed and published with the GAPND Program (Milgrom et al., 1999). A newer 4-session version has recently been evaluated in an RCT.

The 4-session program includes psychoeducation and interaction coaching: Session 1: Play and physical contact; Session 2: Observing and understanding baby's signals; Session 3: Parental responses to infant cues; Session 4: Consolidating gains. We have evaluated the impact of HUGS in an RCT involving women who initially received depression treatment (i.e., GAPND program). HUGS, which capitalized on the therapeutic momentum and skills acquired in the preceding 12-session GAPND program, was found to improve mother-infant relationships at 6-month postintervention in comparison to a control playgroup (Holt et al., 2021).

Case study

The following is a combined case example drawn from a number of mothers who participated in the GAPND program, and is designed to illustrate experiences of program participants.

Client description

Lily is a 30-year-old, first-time mother of 6-month-old Alicia. Lily is married to partner, Dean. She presented with a major depressive episode with postpartum onset, beginning 3 weeks postbirth but continuing until she presented for help when her baby was 6 months of age. Prior to this, she had visited various medical practitioners focusing on Alicia's sleeping difficulties but not sharing her feelings. Finally, as her mood deteriorated, she revealed her ongoing depression and was referred to our GAPND program.

Baseline depression

During an initial triage call, Lily completed the Patient Health Questionnaire-9 (PHQ-9) and her score of "10" was indicative of a moderate level of depressive symptoms. During assessment, it is important to ensure the safety of both mother and baby. While Lily reported very occasional thoughts of wishing there was a way this could all end, these thoughts were very infrequent and represented more "wishful thinking" to find a magical solution to her problems. On questioning, it was clear there was no intent or plan to act on these thoughts. She did not have any thoughts of harm towards her baby and the presence of her baby helped Lily disregard any fleeting thoughts, due to her wish to protect Alicia.

Existing stressors

Lily's current depressive episode was predisposed by antenatal anxiety, previous experiences of subclinical depressive symptoms and a prior diagnosis of Major Depressive Disorder in her twenties. She reported a strained relationship with her partner, mother and father, and limited social support.

Treatment summary

Lily attended all 12 sessions of the GAPND program, including nine mothers' group sessions and three couple sessions. Lily found the group program to be very supportive and her ability to share her experiences with other mothers was an important factor in her recovery. She enjoyed the structure of CBT therapy and engaged well. Lily developed the confidence to share her difficulties and stressors in group sessions, and made major gains in reducing her level of depression and improving her partner relationship.

Table 2 below describes the gains made by Lily as she progressed through the intervention.

TABLE 2 The case study of Lily.

Details of sessions		
Session 1	**Sessions 2 and 3**	**Session 4**
When psychoeducation was provided on postnatal depression, Lily expressed a sense of relief that what she had experienced was well recognized, and quite common. She was able to identify her mood symptoms, particularly anhedonia, feelings of guilt, and difficulty sleeping. She found the homework activities of mood monitoring particularly helpful	Lily initially found it difficult to schedule pleasurable activities and to fit it into her day. Despite recognizing the benefits of incorporating behavioral changes discussed in the group, e.g., self-care and relaxation, her day was too full to consider these. The barriers of time and limited practical support were explored and possible solutions were discussed, which Lily later reported as helpful	Lily identified that her communication style as predominately "passive." She felt it was particularly challenging to express her needs for managing infant care her way, to her parents and in-laws. Lily used a family scenario during the role play to practice asserting her needs using assertive communication skills, while still maintaining a relationship with her parents and in-laws. The importance of assertiveness training in a perinatal program has been reinforced by the many anecdotes of women sharing stories of unhelpful advice during the postnatal period, which they find difficult to manage
Session 5	**Session 6**	**Session 7**
Lily developed awareness of her unhelpful thinking patterns, particularly the relationship between her early childhood experiences of inconsistent caregiving, and her unhelpful expectations and evaluations of herself as a mother	The example of "High Expectations Helen," introduced in session, resonated with Lily. She identified with Helen who set goals for herself that were unreachable, was surprised that other mothers in thegroup shared similar challenges and was able to change her self-statements	During a sessional activity, Lily found that when asked to list the thoughts she had had that day, she had a 1:10 ratio of positive to negative thoughts. During the group discussion she became aware of a sense of hopelessness and was keen to practice strategies to reduce negative thoughts
Session 8	**Session 9**	**Couple sessions**
Lily identified negative automatic thoughts, including, should statements (e.g., "I should be enjoying Alicia"), mind reading (e.g., "they think I am not doing a goodjob") and polarized/black and white thinking, (e.g., "I've failed"). She engaged in disputing cognitive distortions and developed a more balanced, realistic self-talk, although at times she continued to find this challenging	Lily shared that her fluctuating periods of lowered mood were significantly less frequent and severe. She indicated that she had increased awareness of triggers of mood spirals and was more actively engaging with personal supports (e.g., mother's group friends and sister). She described some concern about planning to have another baby and these concerns and support options were discussed during discharge planning	Lily attended the couples' sessions with her husband, Dean. They shared some insights into the problems they experienced after transitioning from romantic partners to parents. At times, Dean appeared guarded and reserved during sessions. Despite this, Lily described improvements in their communication as a couple, and increased awareness of each other's needs. She felt they could continue to build upon these positive changes together
Outcome At the completion of the GAPND program, Lily scores on the PHQ-9 indicated minimal to mild depression. She had not experienced any fleeting thoughts of harming herself since beginning the program. Lily reported that she had found the group helpful and that the group format was very supportive and validating		

The table above is a combined case example drawn from a number of mothers who participated in the GAPND program, and is designed to illustrate experiences of program participants.

Description of the antenatal program

The antenatal CBT intervention, BBB, was adapted from the GAPND intervention (Milgrom et al., 1999; Milgrom, Negri, Gemmill, McNeil, & Martin, 2005) to accommodate the specific issues pregnant women were likely to struggle with, including time or physical health constraints preparing for the transition to parenthood as well as focusing on self-care and lifestyle. The basic structure of the postnatal program was retained, as well as addressing relationship issues with the partner and the elevated anxiety that is often associated with antenatal depression. The number of maternal sessions were reduced to seven 1-h weekly sessions, with reordering of some sessions and reducing the amount of content. The program was delivered individually to facilitate completion prior to giving birth.

We only included one additional partner session but communication tips were added and the session was designed to provide psychoeducation and support to partners and allow them to become more involved in the therapeutic process (Milgrom, Holt, et al., 2015).

Efficacy of our face-to-face programs

Trials of postnatal programs

A number of trials have evaluated the GAPND suite of interventions, which have proven effective against a range of comparator conditions in treating clinically diagnosed depressive disorders (Table 3).

The first study was a pilot RCT (Meager & Milgrom, 1996) in which GAPND produced significant reductions in depressive symptoms compared to a waitlist control. In the next, larger 4-arm RCT, both group GAPND and group or

TABLE 3 Evaluated postnatal and antenatal programs.

Study	CBT intervention evaluated	Participants	Study design	Main outcome measures	Results
Postnatal programs (GAPND and related programs)					
Meager & Milgrom, 1996	"Getting Ahead of Postnatal Depression" group therapy program	$N=20$ Depressed women <1 year postpartum	Parallel 2-group RCT. CBT Intervention versus Waitlist Control	Depressive symptoms: EPDS and BDI	CBT Intervention statistically superior to Waitlist Control in reducing depressive symptoms on EPDS and BDI
Milgrom et al., 2005	"Getting Ahead of Postnatal Depression" group therapy program	$N=192$ Depressed women <1 year postpartum	Parallel 4-group RCT CBT Intervention; individual counseling Intervention; group counseling Intervention; Routine Care Control	Depressive symptoms: BDI Anxiety symptoms: BAI	All psychological interventions statistically superior to Routine Care in reducing depressive and anxious symptoms on BDI and BAI
Milgrom et al., 2011	"Overcoming Depression," individual CBT-counseling program	$N=68$ Depressed women <1 year postpartum	Parallel 3-group RCT. CBT-counseling Intervention delivered by nurse; CBT-counseling Intervention delivered by Psychologist; Routine Care Control	Depressive symptoms: BDI Anxiety symptoms: DASS-21	Depressive and anxiety symptoms showed statistically significant reductions across time in all three groups The proportion of women who continued to have above-threshold symptoms on the BDI after treatment was significantly lower in the two CBT-counseling groups compared to the Routine Care Control

Continued

TABLE 3 Evaluated postnatal and antenatal programs—cont'd

Study	CBT intervention evaluated	Participants	Study design	Main outcome measures	Results
Milgrom, Holt, et al., 2015	"Getting Ahead of Postnatal Depression" group therapy program	*N*=45 Depressed women <1 year postpartum	Parallel 3-group RCT CBT Intervention versus two active treatment conditions: (i) sertraline monotherapy; (ii) combination therapy (CBT+sertraline)	Depressive symptoms: BDI Anxiety symptoms: BAI	Depressive and anxiety symptoms reductions across time in all three groups. For depression, the CBT Intervention was superior to combination therapy and sertraline in reducing the proportion of women who continued to have above-threshold symptoms on the BDI after treatment For anxiety, the single therapies (CBT Intervention and sertraline) demonstrated a significantly larger reduction than combination therapy. Also, a significant difference was found between CBT and sertraline, with the CBT Intervention demonstrating the higher decrease in anxiety symptoms
Antenatal program					
Milgrom, Gemmill, et al., 2015	Beating the Blues before Birth, individual CBT program	*N*=25 Depressed women <30 weeks pregnant	CBT Intervention, single group uncontrolled before-and-after design	Depressive symptoms: BDI Anxiety symptoms: BAI	Across time, for both depression and anxiety, large statistically significant reductions occurred from baseline to posttreatment
Milgrom, Gemmill, et al., 2015	Beating the Blues before Birth, individual CBT program	*N*=54 depressed women <30 weeks pregnant, and their infants at 9-months of age	Parallel 2-group RCT. CBT Intervention versus Routine Care Control	Depressive symptoms: BDI Anxiety symptoms: BAI Infant development: The Infant Behavior Questionnaire-Revised; Ages & Stages Questionnaires	During pregnancy, for both depression and anxiety, large effects favoring the CBT Intervention were observed from baseline to posttreatment. These effects were maintained at 9-month postnatal follow-up Nine-month infant outcomes showed several significant, medium-to-large effects favoring the CBT Intervention in domains including problem solving, self-regulation and stress reactivity

A number of trials have evaluated the effectiveness of the GAPND suite of interventions, which have proven effective against a range of comparator conditions.

individual counseling were clinically and statistically more effective in reducing depressive and anxiety symptoms than routine primary care (Milgrom et al., 2005).

In a further RCT, which compared adjunctive CBT-counseling from a nurse ("Overcoming Depression" based on GAPND), the same adjunctive CBT-counseling from a psychologist and routine GP management, depressive symptoms dropped significantly across all groups on a similar trajectory between baseline and postintervention. A significant overall drop in anxiety symptoms was also observed at postintervention. Women in the GP management group were significantly more likely to exhibit above-threshold depressive scores at postintervention compared to women receiving CBT-counseling from a nurse or a psychologist (Milgrom et al., 2011).

Finally, an RCT involving women with DSM-IV diagnoses of depression compared the combination of GAPND with sertraline against GAPND and sertraline monotherapies (Milgrom et al., 2015). Significant decreases in depressive symptoms from baseline to 12 weeks were evident in the three conditions (see table). CBT monotherapy also produced a significantly more rapid weekly drop in depressive symptoms than the other treatment conditions. Significant reductions in anxiety symptoms were found for both the monotherapies compared to combination therapy. CBT monotherapy was statistically superior to sertraline monotherapy (Milgrom, Gemmill, et al., 2015).

Taken together, the findings of these trials support the efficacy of the GAPND suite of programs in treating postnatal depression and the anxiety that often accompanies it.

Trials of the antenatal program

In a feasibility study, 25 depressed pregnant women were recruited with an average depression score within the severely depressed range. Large, significant effects were observed following BBB treatment for both depression and anxiety and were sustained over the following postnatal months. Participant feedback suggested that individual delivery was an acceptable format based on women's ratings of treatment effectiveness (mean = 7.8/10, SD 1.8), improvements in their condition, and satisfaction with the treatment. Average session attendance was 6.7 sessions (SD 1.9) of the possible 8.

Following this successful uncontrolled trial, a pilot RCT ($N = 54$) of the BBB program, found that depressive symptoms dropped from the "severe" range to the "minimal" range in the intervention group following treatment (moderately large treatment effect size). Anxiety symptoms decreased from the "moderate" range to the "mild" range in the intervention group, but remained relatively high in the usual care group. Treatment adherence was good and satisfaction ratings were high (Milgrom, Holt, et al., 2015).

Importantly, improvements in depression, anxiety and infant developmental outcomes favoring the intervention group were also evident at 9 months postpartum (moderately large effects) (Milgrom, Holt, et al., 2015) (Table 3).

Internet-based CBT interventions for perinatal depression

Internet-based CBT interventions have been shown to be effective for general depression (Andersson, Carlbring, Titov, & Lindefors, 2019) and offer several advantages to users, including anonymity, home access and time flexibility, all especially critical in the perinatal population. Several internet-based psychological interventions specifically for perinatal depression have emerged in recent years and their efficacy has been evaluated in RCTs (e.g., (Loughnan et al., 2019; Milgrom et al., 2016; O'Mahen, Himle, Fedock, Henshaw, & Flynn, 2013; O'Mahen et al., 2013; Pugh, Hadjistavropoulos, & Dirkse, 2016; Sheeber et al., 2012). These internet interventions, built upon CBT or behavioral activation fundamentals, have varied considerably in the number of sessions and in the extent and nature of human guidance offered (ranging from extended support by professional therapists to brief, low-intensity "technician" assistance). Metaanalyses of this evidence have reported significant improvement in depression (Loughnan, Joubert, Grierson, Andrews, & Newby, 2019; Roman, Constantin, & Bostan, 2020). We are the first to publish an RCT examining the efficacy of an internet-based intervention designed specifically for postnatal women with diagnosed depressive disorders compared directly to FTF psychological treatment (Milgrom et al., 2021).

Development of our online treatments

Content and structure of the MumMoodBooster program

The content of MumMoodBooster (MMB) was adapted from the GAPND program and further adjusted based on consumer feedback (Danaher et al., 2012), resulting in a unique interactive online program tailored to individual needs. MMB aims to deliver a CBT treatment that is as close to face-to-face therapy as possible. The resultant program includes the content

TABLE 4 Studies that have evaluated the MumMoodBooster treatment program.

Study	Online CBT Intervention Evaluated	Participants	Study Design	Main Outcome Measures	Results
Danaher et al., 2013	MumMoodBooster program	N=53 Depressed women <9 months postpartum	Feasibility trial	Depressive symptoms: PHQ-9	90% of users did not meet PHQ-9 threshold criteria for minor/major depression posttreatment compared to 55% of the users at baseline
Milgrom et al., 2016	MumMoodBooster program	N=43 Depressed women <1 year postpartum	Parallel 2-group RCT. CBT Intervention versus Treatment as Usual (TAU)	Depression: DSM diagnosis, PHQ-9, BDI Anxiety: DASS-21	80% of users in the Intervention group were not clinically depressed compared to 20% of users in the Treatment as Usual group Significant reductions in depression and anxiety symptoms were seen on the BDI, PHQ-9 and DASS-21
Milgrom et al., 2021	MumMoodBooster program	N=116 Clinically depressed women <1 year postpartum	Parallel 3-group RCT. Online CBT Intervention versus FTF-CBT and TAU	Depression diagnosis	Relative to TAU, participants in the MMB group were 47.7% less likely to have depressive episode diagnosis (not significantly different from FTF-CBT)
				Depressive symptoms: PHQ-9	MMB significantly reduced depressive symptoms compared to both FTF-CBT and TAU
				Anxiety symptoms: DASS-21	MMB significantly reduced anxiety symptoms compared to TAU
Solness, Kroska, Holdefer, & O'Hara, 2020	MomMoodBooster program (U.S. version)	N=326 Depressed veteran women <1.5 years postpartum	Comparison between rural women and their urban counterparts	Depressive symptoms: Inventory of depression and anxiety symptoms II (IDAS-II)	CBT Intervention significantly improved depressive symptoms over time No effect of rural/urban residence was found

MumMoodBooster has been extensively evaluated in the research setting. Central to the growing evidence base are two gold-standard RCTs by Milgrom et al. (2016, 2021).

described in Table 4, structured as 6 sequential CBT sessions, supplemented by a partner's website, browsing library, and low-intensity support from a telephone coach (Fig. 2).

Efficacy of the MumMoodBooster programs

MMB was developed with systematic consumer involvement and direct usability testing of a prototype ($N=22$) by the target demographic (Danaher et al., 2012).

A feasibility trial (Danaher et al., 2013) of the final program then tested MMB in a sample of postpartum women ($N=53$) with elevated scores on the EPDS or PHQ-9. Program adherence was high with all six sessions completed by 87% of participants. Participants visited MMB 20 separate times and on 15 distinct visit dates on average. At baseline, 55% of participants met PHQ-9 threshold criteria for minor or major depression. At 3 months, 90% did not.

MMB was next compared to a treatment as usual group in a RCT (Milgrom et al., 2016). This demonstrated a fourfold increase in remission from diagnosed depressive disorders at 3-month follow-up. While 80% of MMB users were no longer depressed, this compared to just 18% of those in treatment as usual.

To obtain a direct comparison of MMB with specialized, FTF treatment in a clinically diagnosed population, we conducted a second, larger RCT (Milgrom et al., 2021) which also compared both interventions with a treatment as usual

MumMoodBooster Online Program: Six Core Sequential Sessions

(1) Getting Started Symptoms in PND, Reasons, Myths & facts, CBT approach, Partner support, Practice change	**(2) Managing Mood** Mood rating form, Stress & anxiety, Relaxation, Personal goals	**(3) Increasing Pleasant Activities** Balancing your life, Choosing & tracking, Setting goals, Scheduling time, Viewing charts, Practice change
(4) Managing Negative Thoughts Negative thoughts, Healthy concerns, Extreme thoughts, Controlling, Practice change	**(5) Increasing Positive Thoughts** Why increase? Savoring thoughts, Anticipating thoughts, Other types, Practice change	**(6) Planning for the Future** Program concepts, Your strategies, New routine, Watch & respond, Commitment

FIG. 2 MumMoodBooster online program: Six Core Sequential Sessions. The MumMoodBooster program is structured around 6 core CBT sessions, supplemented by a partner's website, a browsing library, and individual, low-intensity support from a personal coach.

control group. The sample comprised depressed postpartum women ($N = 116$), 92% of whom had a major depressive disorder diagnosis. In this 3-arm RCT, MMB performed at least as well as specialized FTF-CBT in producing remission at 21 weeks follow-up. MMB was also significantly more rapid and effective than FTF-CBT in reducing depression, anxiety and perceived stress symptom severity over time, probably because its online nature meant that treatment could be commenced at once without delay. At follow-up, depression and anxiety scores for MMB participants were 50% lower than in the FTF-CBT and treatment as usual conditions.

The feasibility trial and the RCTs produced very similar effect sizes for depression and anxiety and similar completion rates, confirming that MMB is acceptable, well-adhered to by users, and clinically efficacious for diagnosed perinatal depressive disorders. The effect of a U.S. "MomMoodBooster" version of program on depressive symptoms was also recently evaluated successfully, in a veteran population (Solness et al., 2020).

We have developed and successfully tested an antenatal version, "Mum2BMoodBooster" (M2B), in pregnant women with elevated symptoms of depression, with results of similar magnitude to those found in our RCTs of the postnatal version (manuscript submitted for publication).

Real-world implementation and effectiveness

In late 2017, MMB and its sister program M2B were rolled-out into the real world as part of Australia's MumSpace initiative, delivering a digital stepped-care approach to prevention and early, self-help intervention to ameliorate the impact of perinatal depression and reduce the burden on existing services. A Clinician Portal allows General Practitioners, Nurses, Psychologists and others to refer patients directly to MMB and to monitor progress.

Since this translation into real-world implementation, the performance of the MMB and M2B programs in terms of symptom reduction has proven to be very similar to the benchmark effects and patterns of symptom change established in our RCTs. Users are asked to complete the PHQ-9 at weeks 1, 4 and 8. Our results show that while using the programs, the average user's depression symptoms show a steady decrease over time, moving from "moderately depressed" into the "mildly depressed" range by week 4 and week 8.

Concluding statement and future directions

In this chapter on perinatal depression we reviewed antenatal and postnatal depression CBT interventions and detailed our tailored CBT approach suitable for use by women in the perinatal period. Our GAPND program for postnatal depression was adapted from traditional CBT approaches and laid the foundation on which the BBB program was built to suit pregnant women. Both the GAPND and BBB programs have proven effective at reducing depressive and anxiety symptoms. Capitalizing on the beneficial impact of GAPND, we developed the online program MMB with the intention of achieving higher reach to women in the postnatal period. MMB has been found to be effective in reducing depression at least as well as FTF-CBT. Given the efficacy of CBT for perinatal depression, a future direction is not only implementing these widely but also broadening the areas of application (see sections below). Of note is the need to develop evidence-based treatments for perinatal depression in men as few studies exist (Rodrigues, Ericksen, Watson, Gemmill, & Milgrom, 2021). In addition, given the impact of perinatal depression on the mother-infant relationship there is a need for more evidence-based interventions that could be implemented into routine practice. These considerations are not firm recommendations in existing national practice guidelines, e.g., in Australia and the UK (Austin et al., 2017; NICE, 2014).

Finally, the value of e-solutions in delivering CBT for perinatal mental health issues is now especially relevant for increasing access, due the advent of the COVID-19 pandemic.

Mini-dictionary of terms

- ***Perinatal depression*** refers to depression from conception to the first year postpartum.

Key facts

- Prevalence of maternal antenatal and postnatal depression (defined as "perinatal") is approximately 20%.
- Depression diagnostic criteria are similar to other phases of the life cycle.
- Major risk factors for perinatal depression are a past history of a mental health condition (e.g., major depression or anxiety disorder), major stressful life events, lack of practical/emotional support and poor partner support.
- CBT treatment delivered antenatally or postnatally is effective at reducing maternal depressive symptoms.
- Current clinical guidelines recommend CBT as an effective treatment for antenatal and postnatal depression.
- Treatment uptake is poor due to the presence of an infant which impacts the mother's ability to focus, or attend traditional services, as well as other barriers such as stigma.
- Screening may not be sufficient to ensure subsequent treatment uptake.
- Maternal perinatal depression may impact the whole family, including the relationship with the partner and with the infant, and can also have a negative long-term effect on child development.
- Tailored treatments are useful for addressing the specific needs of perinatal women.
- Digital technology can improve access. For example, in Australia 99% of adults have internet access and telehealth/online consultations are increasing rapidly.

Applications to other areas

In this chapter on perinatal depression, we reviewed antenatal and postnatal depression treatments and detailed our tailored CBT approach (GAPND and BBB) suitable for use by women in the perinatal period. We also described the online adaptation, MMB which has a robust evidence base.

Core elements of these programs could apply, with some further modifications, to other areas including:

- Other populations that commonly experience depression in the perinatal period such as fathers/partners, young/teenage mothers, couples undergoing assisted reproduction, other special needs groups.
- Diverse communities (LGBTQI, migrant populations, Indigenous, CALD, and socio-economically disadvantaged communities, etc.)
- Consideration of depression occurring preconception, and ongoing for mothers of older children.
- Expanding depression and anxiety treatment to include modules dealing with other mental health problems presenting in the perinatal period (e.g., PTSD, childbirth trauma), common issues (e.g., sleep, pregnancy loss/still birth) and other parts of the family unit (e.g., fathers/partners or the parent-child relationship).
- While our programs have been largely designed for individuals in an outpatient setting, the programs can be delivered in other settings where perinatal women present, e.g., inpatient mother–baby units, maternal and child health centers.

Summary points

- Previous evidence supports the effectiveness of CBT interventions in treating and, to some extent, preventing perinatal depression.
- CBT interventions need to be tailored, taking into account the special needs of perinatal women (e.g., time demands of the infant) and the impact on partner and infant relationships.
- Our FTF CBT interventions (GAPND and BBB) are tailored to the needs of antenatal and postnatal women and include couple sessions; they have been proved to be effective in improving depression.
- Our online MMB program has been shown to be more effective in increasing remission of postnatal depression compared to treatment as usual and to be equally effective as specialized FTF-CBT. With respect to symptom reduction, MMB performed better than FTF-CBT.
- MMB postnatal and antenatal programs are publicly available through the MumSpace internet platform and continue to show similar effects to those demonstrated in our RCTs.

References

Ahun, M. N., & Côté, S. M. (2019). Maternal depressive symptoms and early childhood cognitive development: A review of putative environmental mediators. *Archives of Women's Mental Health*, *22*(1), 15–24.

Andersson, G., Carlbring, P., Titov, N., & Lindefors, N. (2019). Internet interventions for adults with anxiety and mood disorders: A narrative umbrella review of recent meta-analyses. *Canadian Journal of Psychiatry*, *64*(7), 465–470. https://doi.org/10.1177/0706743719839381.

Austin, M. P., Highet, N., & Expert Working Group. (2017). *Mental health care in the perinatal period: Australian clinical practice guideline*. Melbourne: Centre of Perinatal Excellence.

Ayers, S., Coates, R., & Matthey, S. (2015). Identifying perinatal anxiety. In J. Milgrom, & A. Gemmill (Eds.), *Identifying perinatal depression and anxiety* (pp. 93–107). John Wiley & Sons, Ltd.

Beeber, L. S., Schwartz, T. A., Holditch-Davis, D., Canuso, R., Lewis, V., & Hall, H. W. (2013). Parenting enhancement, interpersonal psychotherapy to reduce depression in low-income mothers of infants and toddlers: A randomized trial. *Nursing Research*, *62*(2), 82.

Bilszta, J., Ericksen, J., Buist, A., & Milgrom, J. (2010). Women's experiences of postnatal depression—Beleifs and attitudes as barriers to care. *Australian Journal of Advanced Nursing*, *27*(3), 44–54.

Bronfenbrenner, U. (1994). Ecological models of human development. *Readings on the Development of Children*, *2*(1), 37–43.

Cuijpers, P., Brannmark, J. G., & van Straten, A. (2008). Psychological treatment of postpartum depression: A meta-analysis. *Journal of Clinical Psychology*, *64*(1), 103–118.

Cuijpers, P., Cristea, I. A., Karyotaki, E., Reijnders, M., & Huibers, M. J. H. (2016). How effective are cognitive behavior therapies for major depression and anxiety disorders? A meta-analytic update of the evidence. *World Psychiatry*, *15*(3), 245–258. https://doi.org/10.1002/wps.20346.

Cuijpers, P., & Karyotaki, E. (2021). The effects of psychological treatment of perinatal depression: An overview. *Archives of Women's Mental Health*, *24* (5), 801–806. https://doi.org/10.1007/s00737-021-01159-8.

Cuijpers, P., Muñoz, R., Clarke, G., & Lewinsohn, P. (2009). Psychoeducational treatment and prevention of depression: The "coping with depression" course thirty years later. *Clinical Psychology Review*, *29*(5), 449–458.

Danaher, B. G., Milgrom, J., Seeley, J. R., Stuart, S., Schembri, C., Tyler, M. S., et al., & Lewinsohn, P. (2012). Web-based intervention for postpartum depression: Formative research and design of the MomMoodBooster program. *JMIR Research Protocols*, *1*(2).

Danaher, B. G., Milgrom, J., Seeley, J. R., Stuart, S., Schembri, C., Tyler, M. S., et al., & Lewinsohn, P. (2013). MomMoodBooster web-based intervention for postpartum depression: Feasibility trial results. *Journal of Medical Internet Research*, *15*(11).

David, D., Cristea, I., & Hofmann, S. G. (2018). Why cognitive behavioral therapy is the current gold standard of psychotherapy. *Frontiers in Psychiatry*, *9*, 4.

Dennis, C. L., & Dowswell, T. (2013). Psychosocial and psychological interventions for preventing postpartum depression. *Cochrane Database of Systematic Reviews*, *2*.

Dennis, C. L., Ross, L. E., & Grigoriadis, S. (2007). Psychosocial and psychological interventions for treating antenatal depression. *Cochrane Database of Systematic Reviews*, (3). https://doi.org/10.1002/14615818.CD006309.pub2.

Fairbrother, N., Janssen, P., Antony, M. M., Tucker, E., & Young, A. H. (2016). Perinatal anxiety disorder prevalence and incidence. *Journal of Affective Disorders*, *200*, 148–155.

Fisher, J. R., Tran, T. D., Hammarberg, K., Sastry, J., Nguyen, H., Rowe, H., … Kirkman, M. (2020). Mental health of people in Australia in the first month of COVID-19 restrictions: A national survey [Research Support, Non-U S Gov't]. *Medical Journal of Australia*, *213*(10), 458–464.

Forman, D. R., O'Hara, M. W., Stuart, S., Gorman, L. L., Larsen, K. E., & Coy, K. C. (2007). Effective treatment for postpartum depression is not sufficient to improve the developing mother-child relationship. *Development and Psychopathology*, *19*(2), 585–602.

Gavin, N. I., Gaynes, B. N., Lohr, K. N., Meltzer-Brody, S., Gartlehner, G., & Swinson, T. (2005). Perinatal depression: A systematic review of prevalence and incidence. *Obstetrics & Gynecology*, *106*(5 Pt 1), 1071–1083.

Glover, V. (2014). Maternal depression, anxiety and stress during pregnancy and child outcome; what needs to be done. *Best Practice & Research. Clinical Obstetrics & Gynaecology*, *28*(1), 25–35. https://doi.org/10.1016/j.bpobgyn.2013.08.017.

Goodman, J. H. (2004). Paternal postpartum depression, its relationship to maternal postpartum depression, and implications for family health. *Journal of Advanced Nursing*, *45*(1), 26–35.

Holt, C., Gentilleau, C., Gemmill, A. W., & Milgrom, J. (2021). Improving the mother-infant relationship following postnatal depression: A randomised controlled trial of a brief intervention (HUGS). *Archives of Women's Mental Health*, 1–11.

Lau, Y., Htun, T. P., Wong, S. N., Tam, W. S. W., & Klainin-Yobas, P. (2017). Therapist-supported internet-based cognitive behavior therapy for stress, anxiety, and depressive symptoms among postpartum women: A systematic review and Meta-analysis. *Journal of Medical Internet Research*, *19*(4).

Lautarescu, A., Craig, M. C., & Glover, V. (2020). Prenatal stress: Effects on fetal and child brain development. *International Review of Neurobiology*, *150*, 17–40.

Letourneau, N. L., Dennis, C.-L., Benzies, K., Duffett-Leger, L., Stewart, M., Tryphonopoulos, P. D., … Watson, W. (2012). Postpartum depression is a family affair: Addressing the impact on mothers, fathers, and children. *Issues in Mental Health Nursing*, *33*(7), 445–457.

Lewinsohn, P. M., Antonuccio, D. O., Steinmetz, J. L., & Teri, L. (1984). *The coping with depression course: A psycho-educational intervention for unipolar depression*. Eugene: Castalsa Publishing Company.

Li, Z., Liu, Y., Wang, J., Liu, J., Zhang, C., & Liu, Y. (2020). Effectiveness of cognitive behavioural therapy for perinatal depression: A systematic review and meta-analysis. *Journal of Clinical Nursing*, *29*(17–18), 3170–3182.

Loughnan, S. A., Butler, C., Sie, A. A., Grierson, A. B., Chen, A. Z., Hobbs, M. J., … Newby, J. M. (2019). A randomised controlled trial of 'MUMentum postnatal': Internet-delivered cognitive behavioural therapy for anxiety and depression in postpartum women. *Behaviour Research and Therapy*, *116*, 94–103.

Loughnan, S. A., Joubert, A. E., Grierson, A., Andrews, G., & Newby, J. M. (2019). Internet-delivered psychological interventions for clinical anxiety and depression in perinatal women: A systematic review and meta-analysis. *Archives of Women's Mental Health*, *17*(10), 019–00961.

Meager, I., & Milgrom, J. (1996). Group treatment for postpartum depression: A pilot study. *Australian and New Zealand Journal of Psychiatry*, *30*(6), 852–860.

Milgrom, J., Danaher, B., Holt, C., Holt, C. J., Seeley, J., Tyler, M. S., … Gemmill, A. W. (2016). Internet cognitive behavioural therapy for women with postnatal depression: A randomised controlled trial of MumMoodBooster. *Journal of Medical Internet Research*, *18*(3). https://doi.org/10.2196/jmir.4993, e54.

Milgrom, J., Danaher, B., Seeley, J., Holt, C., Holt, C., Ericksen, J., … Gemmill, A. (2021). Internet and face-to-face cognitive behavioral therapy for postnatal depression compared with treatment as usual: Randomized controlled trial of MumMoodBooster. *Journal of Medical Internet Research*, *23* (12), e17185.

Milgrom, J., Ericksen, J., McCarthy, R. M., & Gemmill, A. W. (2006). Stressful impact of depression on early mother-infant relations. *Stress and Health*, *22*(4), 229–238.

Milgrom, J., Gemmill, A. W., Ericksen, J., Burrows, G., Buist, A., & Reece, J. (2015). Treatment of postnatal depression with cognitive behavioural therapy, sertraline and combination therapy: A randomised controlled trial. *Australian and New Zealand Journal of Psychiatry*, *49*(3), 236–245. https://doi.org/10.1177/0004867414565474.

Milgrom, J., Hirshler, Y., Reece, J., Holt, C., & Gemmill, A. W. (2019). Social support—A protective factor for depressed perinatal women? *International Journal of Environmental Research and Public Health*, *16*(8), 1426.

Milgrom, J., Holt, C. J., Bleker, L. S., Holt, C., Ross, J., Ericksen, J., … Gemmill, A. W. (2019). Maternal antenatal mood and child development: An exploratory study of treatment effects on child outcomes up to 5 years. *Journal of Developmental Origins of Health and Disease*, 1–11. https://doi.org/10.1017/s2040174418000739.

Milgrom, J., Holt, C., Gemmill, A., Ericksen, J., Leigh, B., Buist, A., & Schembri, C. (2011). Treating postnatal depressive symptoms in primary care: A randomised controlled trial of GP management, with and without adjunctive counselling. *BMC Psychiatry*, *11*(1), 95.

Milgrom, J., Holt, C., Holt, C. J., Ross, J., Ericksen, J., & Gemmill, A. W. (2015). Feasibility study and pilot randomised trial of an antenatal depression treatment with infant follow-up. *Archives of Women's Mental Health*, *18*(5), 717–730. https://doi.org/10.1007/s00737-015-0512-5.

Milgrom, J., Martin, P. R., & Negri, L. M. (1999). *Treating postnatal depression. A psychological approach for health care practitioners*. Chichester: Wiley.

Milgrom, J., Negri, L. M., Gemmill, A. W., McNeil, M., & Martin, P. R. (2005). A randomized controlled trial of psychological interventions for postnatal depression. *British Journal of Clinical Psychology*, *44*(4), 529–542.

Milgrom, J., Westley, D., & Gemmill, A. W. (2004). The mediating role of maternal responsiveness in some longer-term effects of postnatal depression on infant development. *Infant Behavior & Development*, *27*, 443–454.

Murray, L., Cooper, P. J., Wilson, A., Romaniuk, H., Wilson, A., & Romaniuk, H. (2003). Controlled trial of the short- and long-term effect of psychological treatment of post-partum depression. 2. Impact on the mother-child relationship and child outcome. *British Journal of Psychiatry*, *182*, 420–427.

Murray, L., Halligan, S., & Cooper, P. (2018). Postnatal depression and young children's development. In *Handbook of infant mental health* (pp. 172–186). New York, NY: Guilford Press.

NICE. (2014). *National Institute for health and care excellence (NICE). Antenatal and postnatal mental health: Clinical management and service guidance: Guideline CG192*. National Institute for Health and Care Excellence.

O'Mahen, H., Himle, J. A., Fedock, G., Henshaw, E., & Flynn, H. (2013). A pilot randomised controlled trial of cognitive behavioural therapy for perinatal depression adapted for women with low incomes. *Depression and Anxiety*, *30*(7), 679–687.

O'Mahen, H. A., Woodford, J., McGinley, J., Warren, F. C., Richards, D. A., Lynch, T. R., & Taylor, R. S. (2013). Internet-based behavioral activation—Treatment for postnatal depression (Netmums): A randomized controlled trial. *Journal of Affective Disorders*, *150*(3), 814–822.

Pilkington, P. D., Whelan, T. A., & Milne, L. C. (2015). A review of partner-inclusive interventions for preventing postnatal depression and anxiety. *Clinical Psychologist*, *19*(2), 63–75. https://doi.org/10.1111/cp.12054.

Pugh, N. E., Hadjistavropoulos, H. D., & Dirkse, D. (2016). A randomised controlled trial of therapist-assisted, internet-delivered cognitive behavior therapy for women with maternal depression. *PLoS One*, *11*(3), e0149186. https://doi.org/10.1371/journal.pone.0149186.

Rao, W. W., Zhu, X. M., Zong, Q. Q., Zhang, Q., Hall, B. J., Ungvari, G., Xiang, Y. T., et al. (2020). Prevalence of prenatal and postpartum depression in fathers: A comprehensive meta-analysis of observational surveys. *Journal of Affective Disorders*, *263*, 491–499.

Rodrigues, A. L., Ericksen, J., Watson, B., Gemmill, A. W., & Milgrom, J. (2021). Interventions for perinatal depression and anxiety in fathers: A mini-review. *Frontiers in Psychology*, *12*.

Roman, M., Constantin, T., & Bostan, C. M. (2020). The efficiency of online cognitive-behavioral therapy for postpartum depressive symptomatology: A systematic review and meta-analysis. *Women & Health*, *60*(1), 99–112.

Schaeuffele, C., Schulz, A., Knaevelsrud, C., Renneberg, B., & Boettcher, J. (2021). CBT at the crossroads: The rise of transdiagnostic treatments. *International Journal of Cognitive Therapy*, *14*(1), 86–113. https://doi.org/10.1007/s41811-020-00095-2.

Sheeber, L. B., Seeley, J. R., Feil, E. G., Davis, B., Sorensen, E., Kosty, D. B., & Lewinsohn, P. M. (2012). Development and pilot evaluation of an internet-facilitated cognitive-behavioral intervention for maternal depression. *Journal of Consulting and Clinical Psychology*, *80*(5), 739–749.

Sockol, L. E. (2015). A systematic review of the efficacy of cognitive behavioral therapy for treating and preventing perinatal depression. *Journal of Affective Disorders*, *177*, 7–21.

Sockol, L. E. (2018). A systematic review and meta-analysis of interpersonal psychotherapy for perinatal women. *Journal of Affective Disorders*, *232*, 316–328.

Sockol, L. E., Epperson, C., & Barber, J. (2011). A meta-analysis of treatments for perinatal depression. *Clinical Psychology Review*, *31*, 839–849.

Solness, C. L., Kroska, E. B., Holdefer, P. J., & O'Hara, M. W. (2020). Treating postpartum depression in rural veterans using internet delivered CBT: Program evaluation of MomMoodBooster. *Journal of Behavioral Medicine*, 1–13.

Stein, A., Pearson, R. M., & Goodman, S. H. (2014). Effects of perinatal mental disorders on the fetus and child. *The Lancet*, *384*(9956), 1800–1819.

Thiel, F., Pittelkow, M.-M., Wittchen, H.-U., & Garthus-Niegel, S. (2020). The relationship between paternal and maternal depression during the perinatal period: A systematic review and meta-analysis. *Frontiers in Psychiatry*, *11*, 1072.

Van Agteren, J., Bartholomaeus, J., Fassnacht, D. B., Iasiello, M., Ali, K., Lo, L., & Kyrios, M. (2020). Using internet-based psychological measurement to capture the deteriorating community mental health profile during COVID-19: Observational study. *JMIR Mental Health*, *7*(6), e20696.

Wisner, K. L., Sit, D. K., McShea, M. C., Rizzo, D. M., Zoretich, R. A., Hughes, C. L., Eng, H. F., Luther, J. F., Wisniewski, S. R., Costantino, M. L., Confer, A. L., Moses-Kolko, E. L., Famy, C. S., & Hanusa, B. H. (2013). Onset timing, thoughts of self-harm, and diagnoses in postpartum women with screen-positive depression findings. *JAMA psychiatry*, *70*(5), 490–498.

Chapter 4

Childbirth-related fear, tokophobia, and cognitive behavioral therapy

Colin R. Martin[a], Catriona Jones[a], Claire Marshall[b], and Julie Jomeen[c]
[a]*Institute for Clinical and Applied Health Research, University of Hull, Hull, United Kingdom,* [b]*Department of Psychological Medicine, Humber Teaching NHS Foundation Trust, Princes Court, Hull, United Kingdom,* [c]*School of Health and Human Sciences, Southern Cross University, Bilinga, QLD, Australia*

Abbreviations

FOBS fear of birth scale
OWLS Oxford Worries about Labour Scale
SPECS Slade-Pais Expectations of Childbirth Scale
TSS Tokophobia Severity Scale
WDEQ Wijma Delivery Expectancy/Experience Questionnaire

Introduction and background

Emotional and psychological well-being is currently a prevailing theme in perinatal care (Martin, 2012), and its importance is recognized in national UK policies and guidelines (National Institute for Health and Care Excellence, 2014). Impaired well-being has serious and long-lasting implications for women, their babies, partners and the wider family as well as being important in its own right (Martin, 2012). Furthermore, perinatal mental health problems have a considerable economic impact (Bauer, Knapp, & Parsonage, 2016). The myriad of potential perinatal mental health issues is summarized in Fig. 1.

Fear of childbirth (FoC) is an under-recognized and an under-prioritized perinatal mental health issue. It is not unusual for expectant parents to have some level of fear in relation to labor and childbirth. To an extent, fear is a common feeling among pregnant women (Molgora et al., 2018). FoC exists on a scale, there are varying levels from normal to pathological (Jones, Marshall, Martin, & Jomeen, 2020). A pregnant woman with a high level of FoC believes that she will not be capable of successfully coping with birth and is unable to identify courses of action (Striebich, Mattern, & Ayerle, 2018). Tokophobia is a term used to describe a pathological fear of labor and childbirth and intense fear of pregnancy, leading to the avoidance of childbirth (Bhatia & Jhanjee, 2012). When fear becomes so intense that it affects daily function and overshadows a pregnancy, it requires additional clinical support beyond that of routine maternity care. Tokophobia may be an extreme form of FoC in terms of the continuum of anxiety associated with the presentation or potentially so profoundly psychologically incapacitating as to represent a distinct disorder (Jones et al., 2020; Martin et al., 2020).

Prevalence estimates for FoC and tokophobia vary. Previous studies have indicated a prevalence of FoC in up to 20% of pregnancies (Henriksen, Borgen, Rislokken, & Lukasse, 2020; Poggi, Goutaudier, Sejourne, & Chabrol, 2018); however, across the literature, there is a lack of consistency in the way FoC and tokophobia are defined and measured, and this creates problems when accurately evaluating prevalence rates. While rates in Western countries have been identified tokophobia as over 20% (Demsar et al., 2018), an Australian study has quoted a rate of 48% for moderate tokophobia and 26% for strong (Fenwick, Gamble, Nathan, Bayes, & Hauck, 2009). Other studies referring to a "pathological and disabling fear of birth" have used a range of 6%–10% to highlight the prevalence (Kjærgaard, Wijma, Dykes, & Alehagen, 2008; Searle, 1996). There is also the additional lack of clarity over whether these figures relate solely to pregnant women, or if they include women who have never entered maternity services due to the avoidance of pregnancy. Not all women with tokophobia are pregnant women. There are women for whom fear has impacted upon their lives to the extent that it prevents them from embarking upon a desired pregnancy (Onley, 2008). The most recent metaanalysis estimates a pooled prevalence of tokophobia in pregnant women at around 14% (O'Connell, Leahy-Warren, Khashan, Kenny, & O'Neill, 2017). However, this

Handbook of Lifespan Cognitive Behavioral Therapy. https://doi.org/10.1016/B978-0-323-85757-4.00043-2

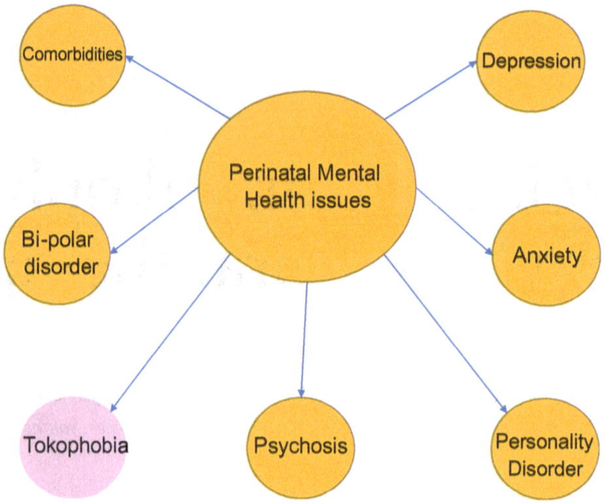

FIG. 1 Distinct psychological/psychiatric conditions associated with perinatal mental health. Perinatal mental health issues are broad with many distinct diagnoses related to psychological/psychiatric disturbance. Some conditions are specifically related to the perinatal period for example, postnatal depression whereas others are enduring and may be exacerbated by the perinatal period, for example personality disorder.

figure represents a group who feel able to discuss their levels of fear, but women presenting with tokophobia may not always articulate their fears openly.

Across the perinatal period, FoC can cause prolonged and enduring distress and anxiety (Hofberg & Ward, 2004). Heightened anxiety and stress during pregnancy may alter the uterine environment, with implications for child cognitive and emotional development (O'Connor, Heron, Glover, & Alspac Study, 2002). Extreme FoC has been associated with protracted and complicated labor (Adams, Eberhard-Gran, & Eskild, 2012), may contribute to postnatal depression (Mortazavi & Mehrabadi, 2021), posttraumatic stress disorder (Ayers, Bond, Bertullies, & Wijma, 2016), and more recently, perceived bonding difficulties at 3 months postpartum, even after adjusting for concurrent postpartum depressive symptoms and the presence of antenatal anxiety disorder (Challacombe, Nath, Trevillion, Pawlby, & Howard, 2020). It is an under recognized and under prioritized health issue. A meta synthesis of qualitative studies exploring women's experiences of FoC concludes that women who fear childbirth are in a severe existential situation of suffering, loneliness, lack of guarantees and understanding, and experience a loss of control when having to face their greatest fears (Wigert et al., 2020). Seminal work by Hofberg and Brockington (2000) based on case studies of 26 women with an "unreasoning dread of childbirth" (Hofberg & Brockington, 2000, p. 83), identified three types of tokophobia.

1. Primary tokophobia, affecting women with no history of previous childbirth related trauma.
2. Secondary tokophobia, occurring after a traumatic experience of childbirth.
3. Tokophobia as a symptom of depression.

Across the more contemporary body of literature, tokophobia is classified into primary (affecting nulliparous women) and secondary (affecting parous women, usually after a previous traumatic birth experience) tokophobia.

The etiology of tokophobia is said to be multifactorial (Demsar et al., 2018). Risk factors for tokophobia include anxiety and depression (Demsar et al., 2018; Hall et al., 2009). Fear of vaginal delivery however has been found to be associated with psychosocial factors such as dissatisfaction with partnership, lack of social support and unemployment (Saisto, Salmela-Aro, Nurmi, Kononen, & Halmesmaki, 2001).

Women's pregnancy and childbirth related fears vary; some are influenced by aspects associated with a previous childbearing experiences, while others may date from adolescence (Hofberg & Brockington, 2000), or relate to vicarious exposure to negative birth information (Sheen & Slade, 2018). Sheen and Slade (2018) use metasynthesis to identify the key elements of FoC as reported by women. Six key elements of women's fears are identified; the unknown nature of childbirth, the potential for complications, injury or harm, pain, their capacity to give birth, losing control during labor and interactions with care providers. Three factors were found to moderate fear; awareness of negative birth experiences, information about birth and support, and support from care providers. Key to women receiving appropriate support and treatment options is the need to provide an opportunity for women to discuss their feelings about the pregnancy and pending birth as part of routine antenatal care. Assessment for childbirth related fear is an important aspect of routine maternity care, not only because it can have bearings on antenatal anxiety symptoms, the experience of pregnancy, labor and childbirth but also because it could affect perceptions of mother-infant bonding and interactions.

Practice and procedures

Screening

A critical component in any care pathway in order for women to be availed of a clinical service is screening. However, before a consideration of the relative merits of screening tools for FoC, it is useful to consider a more philosophical aspect, which may impact on the screening veracity of the measures themselves. Importantly, there is little consensus on whether primary and secondary tokophobia are intrinsically the same presentation, or represent distinct and separable entities in terms of content, form and manifestation of symptoms. This represents a highly problematic situation within the context of screening where the same instrument is used for each of these presentations, there being at least one tool that is used in both presentation with minor modification (the tense the items are framed within). Since the threshold cut-off score will be the same, the assumption from a measurement perspective would be that primary and secondary tokophobia are the same. However, since there is no agreement that this is indeed the case, a single screen threshold in the context of these conditions being fundamentally different would imbue a screening decision within implicit error (Fig. 2).

Illuminating this point further, if primary and secondary tokophobia are distinct, it is reasonable to assume that the item content of a single screening measure will differentially map onto discrete aspects of the woman's experience of FoC and the assumption that a threshold score would be identical between these conditions would be difficult to reconcile or justify. This issue is highlighted in Fig. 3, where a screening threshold score is optimized for the desirable characteristics of sensitivity and specificity by adjusting the score between the two presentations.

A range of screening tools have been proposed for the use in this population (Martin et al., 2020). One of the most influential is the Wijma Delivery Expectancy/Experience Questionnaire (WDEQ:(Wijma, Wijma, & Zar, 1998), which

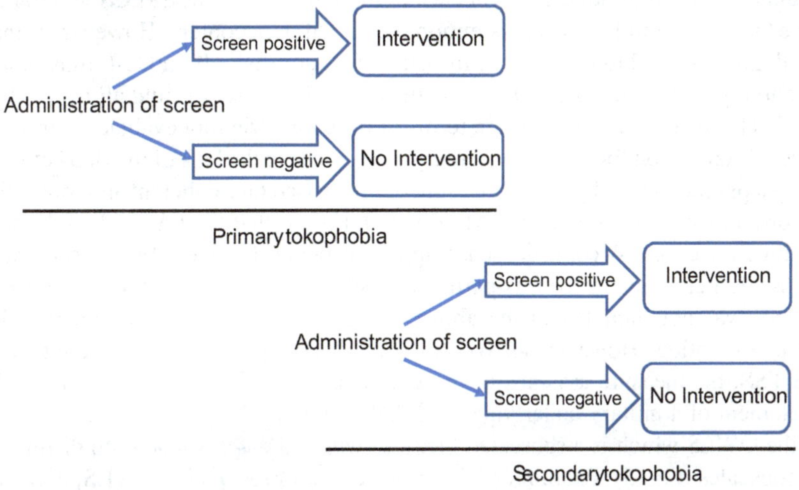

FIG. 2 Screening pathways for fear of childbirth differentiated by primary and secondary tokophobia. The screening process describes the outcome of a positive compared to a negative screen result for primary and secondary tokophobia. The process is the same in both cases and the distinction lies in the assumption that primary and secondary tokophobia are different conditions.

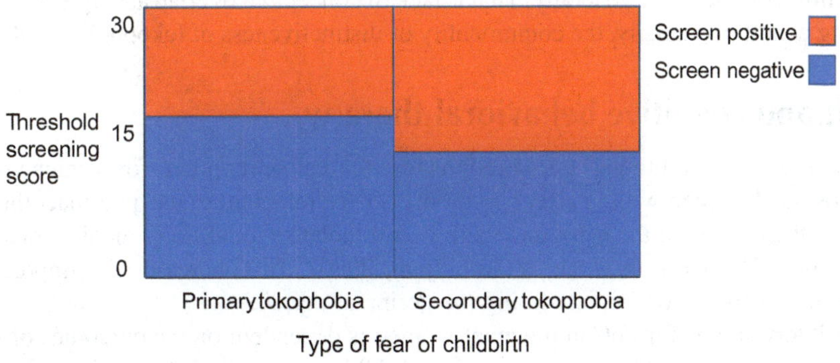

FIG. 3 Impact of cut-score threshold on primary and secondary tokophobia case detection rates. Impact of different cut-score threshold screening scores for primary and secondary tokophobia on detection rates, the difference between cut-scores would implicitly assume some differences between primary and secondary tokophobia.

is available in two main versions, one antenatal and one postnatal. The WDEQ has been used extensively, predominantly within the context of clinical research. One reason that the WDEQ hasn't penetrated mainstream clinical application is its length at 33 items. Clearly, within the context of screening, where a quick and unburdensome (to the woman) measure is desirable, the length of the WDEQ places this particular scale at a disadvantage for clinical application. One alternative is the fear of birth scale (FOBS;(Haines, Pallant, Karlstrom, & Hildingsson, 2011) which comprises just two questions and is easily administered and scored. Contrasting with the WDEQ, an issue for the FOBS is its absolute brevity, which will inevitably impact on the sensitivity and specificity of the tool. An alternative is the Slade–Pais Expectancy of Childbirth Scale (SPECS;(Slade, Pais, Fairlie, Simpson, & Sheen, 2016) which is a multidimensional subscaled tool, however, the SPECS, as far as the authors are aware, has not been widely used in clinical and research practice and therefore data on its measurement performance remains limited to the original paper describing its development. Another measure that has been suggested as potentially suitable to assess FoC is the Oxford Worries about Labour Scale (OWLS;(Redshaw, Martin, Rowe, & Hockley, 2009); however, although the OWLS has very good psychometric properties, it was never designed as a measure of FoC, therefore using it in this context is outside its instrument development validation. Therefore, to determine the suitability of the FoC in this population a further validation study is necessary. Finally, a recently developed measure is the Tokophobia Severity Scale (TSS;(Wootton, Davis, Moses, Moody, & Maguire, 2020); a 13-item measure asserted by its author as suitable for use in clinical practice. However, as far as the authors are aware, the TSS has not to date been used routinely in clinical practice and there has yet to be any research published regarding its applied use. Moreover, the TSS was developed in a nonpregnant, nonclinical sample; therefore, its performance in a clinical sample is yet to be evaluated.

A particularly innovative and elegant study was undertaken by Martin et al. (2020) and examined the overlap of item content between the WDEQ, FOBS, SPECS, and OWLS. Using a complex statistical approach to collate the appraisals of a number of clinical researchers of the individual items of each of these measures, the SPECS was found to be potentially the best instrument out of the battery of four in terms of common overlap of item content. However, as the authors of this study highlighted, the overall degree of overlap between all measures was surprisingly modest, thus though each measure has been suggested as a potentially useful measure of FoC, the measures themselves would all appear to be measuring something unique beyond FoC. This is important because in terms of psychometric tool evaluation, this "uniqueness" could be considered error variance. A conclusion for this study, largely in response to the lack of overlap between tools is that there is a pressing need to develop a psychometrically robust measure of FoC based on a coherent and robust theoretical framework. This notion is also supported by observations within Martin et al. (2020) that the OWLS had the weakest degree of item overlap with the other three measures. A common-sense appraisal being that the OWLS itself was never designed as a measure of FoC and thus the item-content for this particular tool were never developed with a presentation of phobic anxiety in mind. The TSS was not included in the above study since it was not available at the time Martin et al. (2020) undertook their investigation. However, as Wootton et al. (2020) advocate from their factor analysis work on the development of the TSS, the measure assumes a single dimension of FoC (Fig. 4) where all items are assumed to measure an individual element of a unitary underlying tokophobia domain.

This contrasts with the OWLS which is a subscaled measure where the items load onto distinct and separable factors which are then used as subscales (Fig. 5). Examining the subscale descriptors of the OWLS, none of the constituent subscales (pain/distress, uncertainty and interventions) could be considered conceptually parallel to FoC and again this highlights a further potential rationale for the lack of overlap of this measure in the instrument comparison study of Martin et al. (2020).

Summing up the inconsistencies and issues associated with screening measures, echoes, and resonates with the issues associated with the definition of FoC more generally. That is lack of conceptual overlap and agreement on the characteristic features of the presenting condition, namely, the commonality or distinctiveness of tokophobia, FoC, and anxiety (Fig. 6).

Fear of childbirth and cognitive behavioral therapy

Pregnancy and breastfeeding brings additional considerations for the treatment options for women with mental health conditions. Pregnant and postnatal women with anxiety symptoms can be reluctant to use pharmacotherapy due to concerns about negative effects for themselves or the baby, and there be a reluctance to take any medications during pregnancy, or while breastfeeding (Lemon, Vanderkruik, Arch, & Dimidjian, 2020). This increases the importance of good quality effective psychological interventions for women during the perinatal period.

The effectiveness of interventions for FoC in pregnant women is dependent on the outcomes of interest. Reduction in fear, improvements in a woman's self-efficacy in relation to childbirth, overall birth satisfaction, and mode of delivery, such as a reduction in rates of maternal request caesarean section, could all be considered important outcomes. However

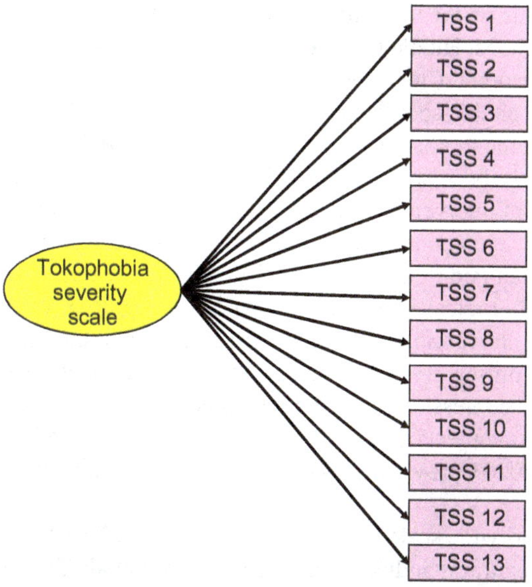

FIG. 4 Measurement model of the Tokophobia Severity Scale. The assumed measurement model of the Tokophobia Severity Scale derived from factor analysis assumes a single dimension of tokophobia. On this basis a single sum score is justified and used in practice. Should the instrument be developed for use in clinical practice for screening a threshold cut-off score would be required which would be based on the sum score of all items.

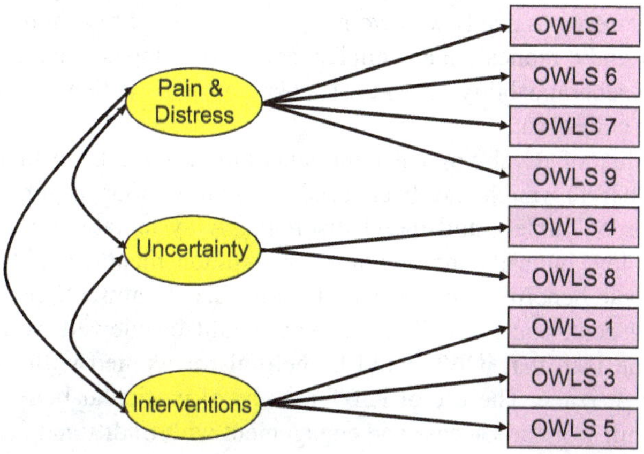

FIG. 5 Multidimensional model of the Oxford Worries about Labour Scale specified as a measurement model. There has been the suggestion that the Oxford Worries about Labour Scale might be a suitable measure for use in tokophobia despite the measure having been developed for a different purpose distinct from fear of childbirth. A potential issue in the application of the Oxford Worries about Labour Scale as a screening measure is its multidimensionality as revealed in its underlying measurement model of three related but distinct dimensions (and subscales) of pain/distress, uncertainty and interventions. This presents challenges for the use of this measure as a screening tool using a singular cut-score based on a single unitary sum score.

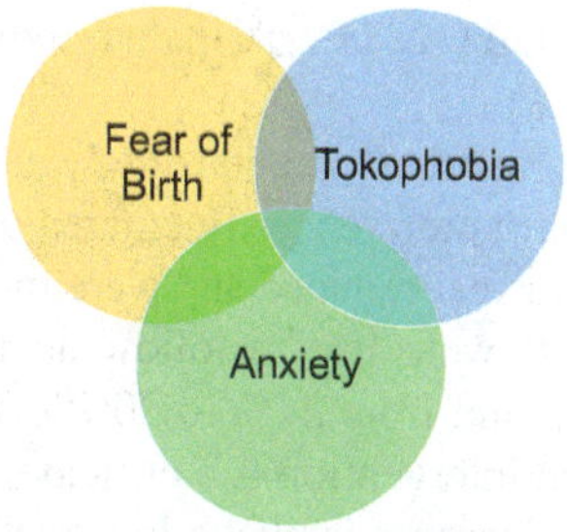

FIG. 6 Overlap between fear of childbirth, tokophobia and anxiety. A critical issue in understanding fear of childbirth and tokophobia in the development of evidence-based interventions, particularly CBT, is a fundamental understanding of the underlying psychological phenomena. The current evidence base is equivocal highlighting both overlap and distinctiveness between fear of childbirth, tokophobia, and anxiety. It is imperative for future research to clarify these relationships in order to have confidence in a theoretically robust model of the phenomena as a basis for the development of evidence-based interventions.

there appears to be a lack of consensus on the most appropriate interventions, and approaches to care vary widely and are not based on empirical evidence (Richens, Hindley, & Lavender, 2015), despite recent developments in care pathways and designated clinics for this group of women (Jones et al., 2020).

Tokophobia may be considered a phobia specific to pregnancy (Patel & Hollins, 2015), and pregnancy-related specific phobias can have adverse impacts on the mother and her baby if not treated before or during early pregnancy, however the effectiveness of treatments have rarely been studied (Nath, Busuulwa, Ryan, Challacombe, & Howard, 2020).

When assessing the effectiveness of interventions for pregnant women with tokophobia, it is critical to examine not only the fundamental and established clinical outcomes and indicators, such as vaginal birth rate but also the physical well-being and mental health status of the woman and in addition her birth experience an satisfaction (Striebich et al., 2018).

The most widely discussed psychological intervention for FoC is cognitive behavioral therapy (CBT). CBT is considered an effective and acceptable intervention within the perinatal population, and is the treatment of choice for many women with mental health problems during pregnancy and postpartum (National Institute for Health and Care Excellence, 2014). A recent systematic review on depression in pregnant women found a significant decrease in depression scores following CBT interventions, with improvements that were maintained long term and with high levels of patient satisfaction (Shortis, Warrington, & Whittaker, 2020), providing hope for the efficacy of this treatment in other perinatal mental health conditions.

There have been great interest and wide scale recommendations for CBT as an intervention for women with FoC, yet there is a lack of published studies regarding individualized CBT. Saisto et al. (2001) conducted a randomized controlled trial of intensive cognitive therapy delivered by an obstetrician compared to routine care in a group of women referred due to FoC. The main goal of this study was to evaluate the possibility of reducing requests for caesareans section by helping women choose vaginal delivery without fear. An obstetrician trained in cognitive therapy delivered the intensive intervention, which offered high levels of contact with the patients. Women randomized to the intensive treatment group over routine care experienced reduced fear of labor, shorter labor's for those delivering vaginally and a reduction in CS request. The patients included in the study were physically well women, at low obstetric risk, middle/upper social class. Despite the encouraging outcomes from this study, replication and clinical application appears unlikely due to the high cost nature of the intervention, and problems in generalizability across the standard obstetric patient group who tend to be high risk, often as a result of previous traumatic childbirth.

In recent years, the availability of psychological treatments online have been increasing, such as internet-based cognitive behavioral therapy (iCBT) which has been used to treat various psychological conditions and proven effective in lowering symptoms of anxiety and mood disorders. A systematic review and metaanalysis evaluated the effectiveness of iCBT as a low intensity intervention for common mental health disorders (Grist & Cavanagh, 2013), with iCBT having several benefits compared to face-to-face treatment, including high treatment fidelity, increased convenience, reduced cost, and scalability. As FoC could be viewed as a domain of anxiety (Rondung et al., 2018) it has been hypothesized that iCBT could be helpful for women with FoC and in turn reduce anxiety and promote a positive birth experience. The use of iCBT may be of particular benefit to this population of patients due to a range of barriers that limit women's access and engagement with traditional face to face interventions, such as extended waiting lists, geographical distance to services, childcare issues and perceived stigma associated with seeking help (Woolhouse, Brown, Krastev, Perlen, & Gunn, 2009).

A feasibility study of iCBT showed promising results with a significant reduction in FoC for 28 women who engaged in 8 weekly sessions of iCBT (Nieminen, Andersson, Wijma, Ryding, & Wijma, 2016). However, a randomized controlled trial that aimed to compare the efficacy of a guided iCBT with standard care on the levels of fear of birth in a sample of pregnant women reporting FoC found no significant differences between the groups, very low treatment adherence in the iCBT group. The authors concluded that guided iCBT, as offered in this study, did not seem to be a feasible or well-accepted approach for treating FoC (Rondung et al., 2018).

Larsson et al. (2017) conducted a multisite randomized controlled study of iCBT versus counseling, with childbirth preference as the primary outcome. Women's birth preferences fluctuated over the course of pregnancy and after the birth, regardless of treatment method. Women felt their fear reduced and were more satisfied with face-to-face counseling compared to iCBT. A higher percentage of patients were lost to follow-up in ICBT group. Caesarean section preference decreased from 34% to 12% in the ICBT group and from 24% to 20% in the counseling group.

Understanding the patient's experience of an intervention is a critical component in helping to determine whether the treatment is acceptable, and whether treatment adherence is likely. In a qualitative study of 15 participants, Nieminen et al. (2015) found that after iCBT women described a more realistic attitude towards childbirth, more self-confidence and active coping strategies. They perceived their partners, and the staff as more supportive, and they were more aware of their baby when giving birth.

Hildingsson and Rubertsson (2020) conducted qualitative interviews with a group of 19 women who had completed a program of iCBT for FoC and women reported a number of positive outcomes, such as helping them to think in other ways about the birth, and assisting in an understanding of their fears and ability to work with their emotions. However, there were some complaints that iCBT did not enable the women to clearly identify and pinpoint their fears. The model of iCBT did not meet all the women's needs and while some appreciated the privacy and flexibility, that iCBT afforded, others would have preferred face to face sessions with the therapist. Motivation and commitment to completing the therapy and associated tasks proved challenging for some women. The critical differences between these studies appear to be that Nieminen et al. (2015) used a group of self-selected women whereas the Hildingsson and Rubertsson (2020) study were randomized as part of a different study. Concluding that women who approach the intervention with a positive optimistic outlook, and choose the mode of intervention may be more likely to find benefit from it.

Practical issues related to the use of technology, access to the internet have also been reported across a number of studies of iCBT; however, within the population of childbearing women, unlike other patient groups such as older people, this is likely to be less of an issue overall, due to the age range of the population and acceptance of technology.

While there have been a number of studies investigating a group approach to deliver an intervention to reduce FoC, the main intervention researched in a group format is psychoeducation, most frequently delivered by midwives in the maternity setting. CBT interventions studied within a group format for women with FoC have had significant less attention, although a recent study showed promising results. Ucar and Golbasi (2019) conducted an experimental study comparing a group CBT educational program, with routine care to examine the effect of the group CBT intervention on FoC. The intervention group attended the educational program on coping with childbirth fears based on CBT, which took 3 weeks over six sessions. The intervention group had lower levels of pain during childbirth, and were more highly satisfied with their childbirth experience.

A number of other alternative approaches have been considered, including eye movement desensitization and reprocessing (EMDR). EMDR is a psychotherapeutic approach that focuses on the brains information processing system to help cope with the troubling physical and mental symptoms associated with traumatic events (Greenwald & Shapiro, 2010). EMDR helps to process traumatic memories using eye movements or other touch points on the body, and although tends to be associated with posttraumatic stress disorder, a recent metaanalysis found a good effect in reducing anxiety symptoms, including phobia and panic symptoms, which are closely associated with FoC (Yunitri et al., 2020). The OptiMUM study (Baas, Stramrood, Dijksman, de Jongh, & van Pampus, 2017) although not reported on to date, uses a two armed randomized controlled study design to determine whether EMDR therapy is an effective and safe treatment for pregnant women with childbirth-related PTSD or FoC.

Recent studies have focused on art therapy, yet, despite the intervention being well accepted by women (Wahlbeck, Kvist, & Landgren, 2018), no positive benefit from the intervention was found, in part due to slow and poor recruitment to the study (Wahlbeck, Kvist, & Landgren, 2020).

There has been considerable interest and research into the use of mindfulness to manage a number of mental health (Hofmann & Gomez, 2017) and physical health problems, including pain management (Creswell, Lindsay, Villalba, & Chin, 2019). The concept of Mindfulness is commonly defined as (1) the ongoing monitoring of present-moment experience (2) with an orientation of acceptance (Creswell et al., 2019).

Recent studies have shown promising results in the use of mindfulness techniques in reducing FoC. In a small pilot study of 12 participants (Byrne, Hauck, Fisher, Bayes, & Schutze, 2014) found that weekly blended childbirth education and mindfulness training improved self-confidence and efficacy in relation to childbirth and an increase in a sense of control relating to the birth. In the Prenatal Education about Reducing Labour Stress (PEARLS), a randomized controlled trial of 30 participants, randomized women to ether mindfulness training or standard birth preparation classes was delivered over a weekend, and found improvements in childbirth self-efficacy, reduction in pain catastrophizing, and a reduction in postnatal depressive symptoms in the mindfulness group. There was no difference in the use of pain relief during labor between the intervention groups (Duncan et al., 2017).

Application to other areas

The contemporary evidence base in relation to FoC and tokophobia highlights that excess anxiety is fundamental to the occurrence of the presentation and thus both prevention and treatment interventions are focused at least to some degree on an understanding of this component of the condition. Effective intervention innovation in this area therefore "borrows" from those interventions found to be effective in other anxiety-related conditions, including the use of CBT. Conversely, as the FoC/tokophobia evidence base matures, insights into this condition in terms of both etiological understanding and treatment innovation may likely be readily applied to other conditions where anxiety is a driver of the underlying clinical presentation.

Conclusion

A recent international consensus statement (Jomeen et al., 2021), highlighted that significant gaps remain within the FoC/tokophobia knowledge base. That the incidence and prevalence figures vary, is underpinned by a number of key issues, including the nature of the relationship between FoC and tokophobia, clarity of etiology, not just of tokophobia itself but of primary and secondary tokophobia, screening practices and the veracity of the screening tools for both FoC and tokophobia. Beyond that is an ethical challenge related to screening. The ability to identify any deviation from the norm in the perinatal period must be accompanied by the availability of effective interventions that can be situated into clear pathways of care. The evidence on interventions to date is limited and open to interpretation. The greatest number of studies have utilized CBT, results are not consistent and there are issues with adherence specifically with iCBT. While mindfulness appears to offer promise, studies to date have focused on FoC and as discussed, it may not be appropriate to assume that such interventions will demonstrate efficacy in women with tokophobia. Interest and related research into tokophobia has increased in recent years, yet it is clear that further research is necessary to explore the issues above to ensure that women receive optimal care, resulting in improved outcomes for them and their babies.

Mini-dictionary of terms

Screening by questionnaire: The process of accurate detection of a case of FoC/tokophobia by a valid, reliable, accurate and self-report questionnaire.

Psychometrics: The measurement characteristics of a measure including the main indices of validity, reliability and factor structure.

Factor analysis: The mathematical process of identifying the underlying measurement model of a questionnaire and associated factors.

Factor structure: The underlying measurement model or inherent domains embedded within a questionnaire measure.

Sensitivity: The proportion of true positives of all cases diagnosed with a disorder (antenatal depression in this context) in the population.

Specificity: The proportion of true negatives of all cases that are not diagnosed with a disorder (antenatal depression in this context) in the population.

Overlap: The degree of commonality of items across self-report questionnaire measures (in this context self-report FoC items).

Key facts

- FoC and tokophobia represent a distinct anxiety-related phenomena impacting deleteriously on the well-being and life choices of child-bearing age women.
- There remains disagreement as to whether FoC and tokophobia are distinct from one another or whether tokophobia represents severe FoC.
- Tokophobia is generally categorized as either primary (affecting women with no history of previous childbirth related trauma) or secondary (occurring after a traumatic experience of childbirth).
- There remains disagreement as to whether primary tokophobia and secondary tokophobia are distinct from one another or represent the same underlying clinical entity.
- Screening measure have been developed for FoC/tokophobia; however, there use is problematic due to inconsistent measurement characteristics between alternative screening instruments.
- CBT for FoC/tokophobia findings is equivocal in terms of therapeutic efficacy. Approaches based on mindfulness have showed some therapeutic promise.

Summary points

- Screening for FoC/tokophobia is a clinically important component of an effective care pathway.
- The evidence base for the effective of CBT for FoC/tokophobia is currently relatively weak.
- Screening measures for FoC/tokophobia vary greatly in item content and lack foundation in a cogent theoretically robust model of the condition.
- Identification and intervention for FoC/tokophobia is an important and contemporary component of perinatal mental health care.

- The area of FoC/tokophobia is fundamentally under-researched, therefore significant research enterprise is required to ameliorate the shortcomings of the current evidence base in order to improve and optimize clinical care.

References

Adams, S. S., Eberhard-Gran, M., & Eskild, A. (2012). Fear of childbirth and duration of labour: A study of 2206 women with intended vaginal delivery. *BJOG: An International Journal of Obstetrics and Gynaecology*, *119*(10), 1238–1246. https://doi.org/10.1111/j.1471-0528.2012.03433.x.

Ayers, S., Bond, R., Bertullies, S., & Wijma, K. (2016). The aetiology of post-traumatic stress following childbirth: A meta-analysis and theoretical framework. *Psychological Medicine*, *46*(6), 1121–1134. https://doi.org/10.1017/S0033291715002706.

Baas, M. A., Stramrood, C. A., Dijksman, L. M., de Jongh, A., & van Pampus, M. G. (2017). The OptiMUM-study: EMDR therapy in pregnant women with posttraumatic stress disorder after previous childbirth and pregnant women with fear of childbirth: Design of a multicentre randomized controlled trial. *European Journal of Psychotraumatology*, *8*(1), 1293315. https://doi.org/10.1080/20008198.2017.1293315.

Bauer, A., Knapp, M., & Parsonage, M. (2016). Lifetime costs of perinatal anxiety and depression. *Journal of Affective Disorders*, *192*, 83–90. https://doi.org/10.1016/j.jad.2015.12.005.

Bhatia, M. S., & Jhanjee, A. (2012). Tokophobia: A dread of pregnancy. *Industrial Psychiatry Journal*, *21*(2), 158–159. https://doi.org/10.4103/0972-6748.119649.

Byrne, J., Hauck, Y., Fisher, C., Bayes, S., & Schutze, R. (2014). Effectiveness of a mindfulness-based childbirth education pilot study on maternal self-efficacy and fear of childbirth. *Journal of Midwifery & Women's Health*, *59*(2), 192–197. https://doi.org/10.1111/jmwh.12075.

Challacombe, F. L., Nath, S., Trevillion, K., Pawlby, S., & Howard, L. M. (2020). Fear of childbirth during pregnancy: associations with observed mother-infant interactions and perceived bonding. *Archives of Women's Mental Health.* https://doi.org/10.1007/s00737-020-01098-w.

Creswell, J. D., Lindsay, E. K., Villalba, D. K., & Chin, B. (2019). Mindfulness training and physical health: Mechanisms and outcomes. *Psychosomatic Medicine*, *81*(3), 224–232. https://doi.org/10.1097/PSY.0000000000000675.

Demsar, K., Svetina, M., Verdenik, I., Tul, N., Blickstein, I., & Globevnik Velikonja, V. (2018). Tokophobia (fear of childbirth): Prevalence and risk factors. *Journal of Perinatal Medicine*, *46*(2), 151–154. https://doi.org/10.1515/jpm-2016-0282.

Duncan, L. G., Cohn, M. A., Chao, M. T., Cook, J. G., Riccobono, J., & Bardacke, N. (2017). Benefits of preparing for childbirth with mindfulness training: A randomized controlled trial with active comparison. *BMC Pregnancy and Childbirth*, *17*(1), 140. https://doi.org/10.1186/s12884-017-1319-3.

Fenwick, J., Gamble, J., Nathan, E., Bayes, S., & Hauck, Y. (2009). Pre- and postpartum levels of childbirth fear and the relationship to birth outcomes in a cohort of Australian women. *Journal of Clinical Nursing*, *18*(5), 667–677. https://doi.org/10.1111/j.1365-2702.2008.02568.x.

Greenwald, R., & Shapiro, F. (2010). What is EMDR?: Commentary by Greenwald and invited response by Shapiro. *Journal of EMDR Practice and Research*, *4*(4), 170–179. https://doi.org/10.1891/1933-3196.4.4.170.

Grist, R., & Cavanagh, K. (2013). Computerised cognitive behavioural therapy for common mental health disorders, what works, for whom under what circumstances? A systematic review and Meta-analysis. *Journal of Contemporary Psychotherapy*, *43*(4), 243–251. 2013/12/01) https://doi.org/10.1007/s10879-013-9243-y.

Haines, H., Pallant, J. F., Karlstrom, A., & Hildingsson, I. (2011). Cross-cultural comparison of levels of childbirth-related fear in an Australian and Swedish sample. *Midwifery*, *27*(4), 560–567. https://doi.org/10.1016/j.midw.2010.05.004.

Hall, W. A., Hauck, Y. L., Carty, E. M., Hutton, E. K., Fenwick, J., & Stoll, K. (2009). Childbirth fear, anxiety, fatigue, and sleep deprivation in pregnant women. *Journal of Obstetric, Gynecologic, and Neonatal Nursing*, *38*(5), 567–576. https://doi.org/10.1111/j.1552-6909.2009.01054.x.

Henriksen, L., Borgen, A., Rislokken, J., & Lukasse, M. (2020). Fear of birth: Prevalence, counselling and method of birth at five obstetrical units in Norway. *Women and Birth*, *33*(1), 97–104. https://doi.org/10.1016/j.wombi.2018.11.008.

Hildingsson, I., & Rubertsson, C. (2020). Childbirth experiences among women with fear of birth randomized to internet-based cognitive therapy or midwife counseling. *Journal of Psychosomatic Obstetrics and Gynecology*, *41*(3), 205–214. 2020/07/02) https://doi.org/10.1080/0167482X.2019.1634047.

Hofberg, K., & Brockington, I. (2000). Tokophobia: An unreasoning dread of childbirth. A series of 26 cases. *British Journal of Psychiatry*, *176*, 83–85. https://doi.org/10.1192/bjp.176.1.83.

Hofberg, K., & Ward, M. R. (2004). Fear of childbirth, tocophobia, and mental health in mothers: The obstetric-psychiatric interface. *Clinical Obstetrics and Gynecology*, *47*(3), 527–534. https://doi.org/10.1097/01.grf.0000132527.62504.ca.

Hofmann, S. G., & Gomez, A. F. (2017). Mindfulness-based interventions for anxiety and depression. *Psychiatric Clinics of North America*, *40*(4), 739–749. https://doi.org/10.1016/j.psc.2017.08.008.

Jomeen, J., Martin, C. R., Jones, C., Marshall, C., Ayers, S., Burt, K., … Thomson, G. (2021). Tokophobia and fear of birth: A workshop consensus statement on current issues and recommendations for future research. *Journal of Reproductive and Infant Psychology*, *39*(1), 2–15. 2021/01/01) https://doi.org/10.1080/02646838.2020.1843908.

Jones, C., Marshall, C., Martin, C. R., & Jomeen, J. (2020). Pregnant with fear. *Community Practitioner*, *2*. https://www.communitypractitioner.co.uk/features/2020/02/pregnant-fear.

Kjærgaard, H., Wijma, K., Dykes, A. K., & Alehagen, S. (2008). Fear of childbirth in obstetrically low-risk nulliparous women in Sweden and Denmark. *Journal of Reproductive and Infant Psychology*, *26*(4), 340–350. 2008/11/01) https://doi.org/10.1080/02646830802408498.

Larsson, B., Karlstrom, A., Rubertsson, C., Ternstrom, E., Ekdahl, J., Segebladh, B., & Hildingsson, I. (2017). Birth preference in women undergoing treatment for childbirth fear: A randomised controlled trial. *Women and Birth*, *30*(6), 460–467. https://doi.org/10.1016/j.wombi.2017.04.004.

Lemon, E., Vanderkruik, R., Arch, J. J., & Dimidjian, S. A. (2020). Treating anxiety during pregnancy: Patient concerns about pharmaceutical treatment. *Maternal and Child Health Journal*, *24*(4), 439–446. https://doi.org/10.1007/s10995-019-02873-7.

Martin, C. R. (2012). *Perinatal mental health: A clinical guide*. M&K Update.

Martin, C., Jones, C., Marshall, C. A., Huang, C., Reeve, J., Fleming, M. P., ... Jomeen, J. (2020). Fear of childbirth measurement: Appraisal of the content overlap of four instruments. *Journal of Reproductive and Infant Psychology*, *1-13*. https://doi.org/10.1080/02646838.2020.1861226.

Molgora, S., Fenaroli, V., Prino, L. E., Rolle, L., Sechi, C., Trovato, A., ... Saita, E. (2018). Fear of childbirth in primiparous Italian pregnant women: The role of anxiety, depression, and couple adjustment. *Women and Birth*, *31*(2), 117–123. https://doi.org/10.1016/j.wombi.2017.06.022.

Mortazavi, F., & Mehrabadi, M. (2021). Predictors of fear of childbirth and normal vaginal birth among Iranian postpartum women: A cross-sectional study. *BMC Pregnancy and Childbirth*, *21*(1), 316. https://doi.org/10.1186/s12884-021-03790-w.

Nath, S., Busuulwa, P., Ryan, E. G., Challacombe, F. L., & Howard, L. M. (2020). The characteristics and prevalence of phobias in pregnancy. *Midwifery*, *82*, 102590. https://doi.org/10.1016/j.midw.2019.102590.

National Institute for Health and Care Excellence. (2014). *Antenatal and postnatal mental health: clinical management and service guidance: Clinical guideline [CG192]*. www.nice.org.uk/guidance/cg192.

Nieminen, K., Andersson, G., Wijma, B., Ryding, E. L., & Wijma, K. (2016). Treatment of nulliparous women with severe fear of childbirth via the internet: A feasibility study. *Journal of Psychosomatic Obstetrics and Gynaecology*, *37*(2), 37–43. https://doi.org/10.3109/0167482X.2016.1140143.

Nieminen, K., Malmquist, A., Wijma, B., Ryding, E. L., Andersson, G., & Wijma, K. (2015). Nulliparous pregnant women's narratives of imminent childbirth before and after internet-based cognitive behavioural therapy for severe fear of childbirth: A qualitative study. *BJOG : An International Journal of Obstetrics and Gynaecology*, *122*(9), 1259–1265. https://doi.org/10.1111/1471-0528.13358.

O'Connell, M. A., Leahy-Warren, P., Khashan, A. S., Kenny, L. C., & O'Neill, S. M. (2017). Worldwide prevalence of tocophobia in pregnant women: Systematic review and meta-analysis. *Acta Obstetricia et Gynecologica Scandinavica*, *96*(8), 907–920. https://doi.org/10.1111/aogs.13138.

O'Connor, T. G., Heron, J., Glover, V., & Alspac Study, T. (2002). Antenatal anxiety predicts child behavioral/emotional problems independently of postnatal depression. *Journal of the American Academy of Child and Adolescent Psychiatry*, *41*(12), 1470–1477. https://doi.org/10.1097/00004583-200212000-00019.

Onley, D. (2008). *Investigating the psychological factors underlying Tokophobia in women following birth trauma, and the need for psychological counselling of women who fear and avoid childbirth Wolverhampton*. https://core.ac.uk/download/pdf/1933216.pdf.

Patel, R. R., & Hollins, K. (2015). Clinical report: The joint obstetric and psychiatric management of phobic anxiety disorders in pregnancy. *Journal of Psychosomatic Obstetrics and Gynaecology*, *36*(1), 10–14. https://doi.org/10.3109/0167482X.2014.993312.

Poggi, L., Goutaudier, N., Sejourne, N., & Chabrol, H. (2018). When fear of childbirth is pathological: The fear continuum. *Maternal and Child Health Journal*, *22*(5), 772–778. https://doi.org/10.1007/s10995-018-2447-8.

Redshaw, M., Martin, C., Rowe, R., & Hockley, C. (2009). The Oxford worries about Labour Scale: women's experience and measurement characteristics of a measure of maternal concern about labour and birth. *Psychology, Health & Medicine*, *14*(3), 354–366. https://doi.org/10.1080/13548500802707159.

Richens, Y., Hindley, C., & Lavender, T. (2015). A national online survey of UK maternity unit service provision for women with fear of birth. *British Journal of Midwifery*, *23*(8), 574–579. 2015/08/02) 10.12968/bjom.2015.23.8.574.

Rondung, E., Ternstrom, E., Hildingsson, I., Haines, H. M., Sundin, O., Ekdahl, J., ... Rubertsson, C. (2018). Comparing internet-based cognitive behavioral therapy with standard care for women with fear of birth: Randomized controlled trial. *JMIR Mental Health*, *5*(3), e10420. https://doi.org/10.2196/10420.

Saisto, T., Salmela-Aro, K., Nurmi, J. E., Kononen, T., & Halmesmaki, E. (2001). A randomized controlled trial of intervention in fear of childbirth. *Obstetrics and Gynecology*, *98*(5 Pt 1), 820–826. https://doi.org/10.1016/s0029-7844(01)01552-6.

Searle, J. (1996). Fearing the worst—Why do pregnant women feel 'at risk'? *Australian and New Zealand Journal of Obstetrics and Gynaecology*, *36*(3), 279–286. https://doi.org/10.1111/j.1479-828x.1996.tb02711.x.

Sheen, K., & Slade, P. (2018). Examining the content and moderators of women's fears for giving birth: A meta-synthesis. *Journal of Clinical Nursing*, *27*(13–14), 2523–2535. https://doi.org/10.1111/jocn.14219.

Shortis, E., Warrington, D., & Whittaker, P. (2020). The efficacy of cognitive behavioral therapy for the treatment of antenatal depression: A systematic review. *Journal of Affective Disorders*, *272*, 485–495. https://doi.org/10.1016/j.jad.2020.03.067.

Slade, P., Pais, T., Fairlie, F., Simpson, A., & Sheen, K. (2016). The development of the Slade-Pais Expectations of Childbirth Scale (SPECS). *Journal of Reproductive and Infant Psychology*, *34*(5), 495–510. 2016/10/19) https://doi.org/10.1080/02646838.2016.1209300.

Striebich, S., Mattern, E., & Ayerle, G. M. (2018). Support for pregnant women identified with fear of childbirth (FoC)/tokophobia—A systematic review of approaches and interventions. *Midwifery*, *61*, 97–115. https://doi.org/10.1016/j.midw.2018.02.013.

Ucar, T., & Golbasi, Z. (2019). Effect of an educational program based on cognitive behavioral techniques on fear of childbirth and the birth process. *Journal of Psychosomatic Obstetrics and Gynaecology*, *40*(2), 146–155. https://doi.org/10.1080/0167482X.2018.1453800.

Wahlbeck, H., Kvist, L. J., & Landgren, K. (2018). Gaining hope and self-confidence-an interview study of women's experience of treatment by art therapy for severe fear of childbirth. *Women and Birth*, *31*(4), 299–306. https://doi.org/10.1016/j.wombi.2017.10.008.

Wahlbeck, H., Kvist, L. J., & Landgren, K. (2020). Art therapy and counseling for fear of childbirth: A randomized controlled trial. *Art Therapy*, *37*(3), 123–130. 2020/07/02) https://doi.org/10.1080/07421656.2020.1721399.

Wigert, H., Nilsson, C., Dencker, A., Begley, C., Jangsten, E., Sparud-Lundin, C., ... Patel, H. (2020). Women's experiences of fear of childbirth: A metasynthesis of qualitative studies. *International Journal of Qualitative Studies on Health and Well-Being*, *15*(1), 1704484. https://doi.org/10.1080/17482631.2019.1704484.

Wijma, K., Wijma, B., & Zar, M. (1998). Psychometric aspects of the W-DEQ; a new questionnaire for the measurement of fear of childbirth. *Journal of Psychosomatic Obstetrics and Gynaecology*, *19*(2), 84–97. https://doi.org/10.3109/01674829809048501.

Woolhouse, H., Brown, S., Krastev, A., Perlen, S., & Gunn, J. (2009). Seeking help for anxiety and depression after childbirth: Results of the maternal health study. *Archives of Women's Mental Health*, *12*(2), 75–83. https://doi.org/10.1007/s00737-009-0049-6.

Wootton, B. M., Davis, E., Moses, K., Moody, A., & Maguire, P. (2020). The development and initial validation of the Tokophobia Severity Scale. *Clinical Psychologist*, *24*(3), 267–275. 2020/11/01) https://doi.org/10.1111/cp.12220.

Yunitri, N., Kao, C. C., Chu, H., Voss, J., Chiu, H. L., Liu, D., … Chou, K. R. (2020). The effectiveness of eye movement desensitization and reprocessing toward anxiety disorder: A meta-analysis of randomized controlled trials. *Journal of Psychiatric Research*, *123*, 102–113. https://doi.org/10.1016/j.jpsychires.2020.01.005.

Chapter 5

Postnatal depression: Cognitive-behavioral conceptualization and prevention and treatment approaches

Ana Fonseca
Faculty of Psychology and Educational Sciences, Center for Research in Neuropsychology and Cognitive Behavioral Intervention, University of Coimbra, Coimbra, Portugal

Abbreviations

CBT cognitive-behavioral therapy
PND postnatal depression

Introduction

Maternal mental health has increasingly been recognized to play an important role in the wellbeing of mothers, their babies, and their families, as well to be a critical component of quality perinatal care and the assurance of maternal safety (Kendig et al., 2017). Among perinatal mood and anxiety disorders, postnatal depression (PND) has assumed clinical significance. PND is a highly prevalent condition, particularly in low- and middle-income countries (18.7% vs 9.5% in high-income countries; Woody, Ferrari, Siskind, Whiteford, & Harris, 2017).

Although the Diagnostic and Statistical Manual of Mental Disorders (DSM-5; American Psychiatric Association (APA), 2013) includes a *peripartum onset specifier* if the onset of mood symptoms occurs during pregnancy or within 4 weeks postpartum, there is a clinical consensus that PND may be defined as a nonpsychotic depressive episode of mild to major severity that may occur during the first year after childbirth (Gelaye, Rondon, Araya, & Williams, 2016). In addition to the common symptoms of depressive episodes, women may experience other symptoms, such as anxiety symptoms, excessive concerns with the baby's wellbeing and with their own parenting skills, obsessive thoughts (e.g., thoughts about harming the baby) or being afraid of being alone with the baby (Riecher-Rossler & Hofecker, 2003). Recently, research has identified five different phenotypes of perinatal depression: (1) severe anxious depression, (2) moderate anxious depression, (3) anxious anhedonia, (4) pure anhedonia, and (5) resolved depression (Putnam et al., 2017). Moreover, a recent review reported that longitudinal studies have also indicated heterogeneity in patterns of PND symptom trajectories (Santos Junior, Tan, & Salomon, 2017).

The effects of PND on the family system seem to be universal across different cultures and countries (Field, 2010). A recent review of the consequences of PND provided evidence of the negative effects on mothers' physical health (e.g., body pain, role limitations) and psychological health, including poor quality of life, higher comorbidity with other psychiatric symptoms, interpersonal difficulties, and higher propensity to engage in risky behaviors (e.g., addictive behaviors, suicidal ideation; Slomian, Honvo, Emonts, Reginster, & Bruyère, 2019). Additionally, PND can negatively affect mother-child interactions (Field, 2010; Slomian et al., 2019) and it can result in short-term and long-term effects on the child's development, particularly at the socioemotional and cognitive levels (Slomian et al., 2019). These pervasive effects deprive many mothers of their basic right to health and wellbeing and prevent their children from reaching their full developmental potential (Rahman, Surkan, Cayetano, Rwagatare, & Dickson, 2013), resulting in high societal and economic costs (Epperson et al., 2020).

Handbook of Lifespan Cognitive Behavioral Therapy. https://doi.org/10.1016/B978-0-323-85757-4.00006-7

Conceptualizing PND within a developmental and cognitive-emotional-behavioral framework

A developmental psychopathology approach to PND

Developmental psychopathology has been mainly applied to the study of children and adolescents, but its main concepts may also offer important insights into adult psychopathology (Benjet, 2017), namely, antenatal depression (Goodman & Dimidjian, 2012) and PND. First, developmental psychopathology examines the interaction of biological, psychological, and sociocontextual factors that determine the trajectories of adaptative and maladaptive functioning over the life course. There is extensive research on the risk factors for PND. In addition to biological vulnerabilities (e.g., hormonal changes), PND has been associated with demographic (e.g., single mother, low socioeconomic status), clinical (e.g., history of psychopathology, depression/anxiety during pregnancy), psychological (e.g., neuroticism, perfectionism, negative attributional style) and sociocontextual (e.g., lack of social support, poor marital relationship, stressful life events) factors that increase the likelihood of women developing PND (O'Hara & McCabe, 2013; Xiao-hu & Zhi-hua, 2020). These risk factors rarely occur in isolation, and their effects tend to be synergistic (Goodman & Dimidjian, 2012).

Second, the idea that psychopathology occurs on a continuum is critical to the conceptualization of PND (Goodman & Dimidjian, 2012), namely in distinguishing PND from normative experiences in the postnatal period, as the somatic symptoms of depression may overlap with the experiences of many nondepressed postnatal women (e.g., fatigue, sleep disruptions and mood fluctuations). Third, consideration of the relationship between the women and their contexts is crucial to understand the ways in which the risk factors for PND operate under different conditions. On the one hand, women and their contexts mutually influence each other, from a transactional perspective rather than a unidirectional perspective (Goodman & Dimidjian, 2012). The interpersonal context (social support and relationships with one's partner and other family members) is relevant to the understanding of PND (O'Hara, 1994; Pilkington, Milne, Cairns, Lewis, & Whelan, 2015). Similarly, the cultural context and the "myths of perfect motherhood" seem to play a major role in the development of PND (Milgrom, Martin, & Negri, 1999). Fourth, the principles of equifinality and multifinality respectively serve as a reminder of the heterogeneity of potential pathways and the combinations of risk factors from multiple levels (e.g., biological, interpersonal) and their transactions over time, and as a reminder that the same set of risk factors and symptoms in different people may lead to different trajectories and outcomes (Goodman & Dimidjian, 2012).

A cognitive-behavioral model of PND

Based on empirical research and Beck's cognitive model of depression (Beck, 1987; Beck, Rush, Shaw, & Emery, 1979), Milgrom et al. (1999) presented the biopsychosocial model of PND, which was more recently adapted by Wenzel and Kleiman (2015). Based on these models, we can derive several principles to explain the etiology and maintenance of PND:

1. **Some women have vulnerability factors that make them more susceptible to the development of PND**: Vulnerability factors may include biological (genetic and neurochemical characteristics), psychological (e.g., family experiences in childhood, cognitive and personality styles, negative life events, psychiatric history) and environmental (e.g., low socioeconomic status) factors that predispose women to experience emotional distress (Milgrom et al., 1999; Wenzel & Kleiman, 2015).
2. **These vulnerability factors are often activated during periods of stress, such as the transition to motherhood**: Stress-inducing conditions during and after childbirth (e.g., obstetric complications, breastfeeding difficulties, high caregiving requirements and low support) can act as triggers (precipitating factors) for PND in the most vulnerable women (Milgrom et al., 1999; Wenzel & Kleiman, 2015).
3. **When women's vulnerability factors interact with triggering/precipitating factors (stress-inducing conditions), dysfunctional beliefs (both general and motherhood-specific) are activated, influencing the way women interpret the situations**: Women's dysfunctional beliefs may reflect global negative judgments of the self (e.g., "I am a failure"), others (e.g., "Others will reject me") or the future (e.g., "I have no hope in the future"), or motherhood-specific beliefs about parenting skills (e.g., "I am not a good mother"), maternal responsibility (e.g., "I should always know what my baby needs"), maternal role idealization ("I should be always happy") or others' judgment (e.g., "People will think I am a bad mother if I'm not able to calm down my baby"; Sockol, Epperson, & Barber, 2014). Although the general beliefs are typically in the background for many parental women, the motherhood-specific beliefs typically assume primary importance during the postnatal period (Wenzel & Kleiman, 2015).
4. **Women's dysfunctional beliefs may lead to negative interpretations of situations (negative automatic thoughts) and associated emotional and behavioral responses**: In times of stress, women's maladaptive/dysfunctional beliefs are activated, and they operate mainly as a function of such beliefs, with negative interpretations of situations (negative

automatic thoughts) and associated emotional, physiological, and behavioral responses, which may also influence each other (Milgrom et al., 1999; Wenzel & Kleiman, 2015).

5. **The way women and significant others respond to PND symptoms may have a role in its maintenance or exacerbation**: PND is experienced as a cumulative struggle in which the expression of negative feelings is not congruent with social constructions of motherhood (Knudson-Martin & Silverstein, 2009). For example, PND symptoms may generate negative cognitive (e.g., thoughts of parental inadequacy), emotional (e.g., feelings of guilt and shame) and behavioral (e.g., isolation) responses to these symptoms among women which can maintain or exacerbate them (Milgrom et al., 1999). As a result, women may also become overwhelmed with feelings of incompetence, and connection with others (their children, significant others and even themselves) may become difficult (Knudson-Martin & Silverstein, 2009). Moreover, the reaction of significant others to women's PND symptoms (e.g., marital conflicts, poor support) may also contribute to its maintenance or exacerbation (Milgrom et al., 1999). A support network in which negative aspects of mothering can be expressed and validated is an important part of positively addressing PND symptoms (Knudson-Martin & Silverstein, 2009).
6. **PND occurs in a broader social context**: Sociocultural factors may also play a role in women's PND, explaining either its etiology or its maintenance. Specifically, unrealistic myths about motherhood (e.g., the myth of perfect motherhood or the myth of the joy of motherhood; Milgrom et al., 1999) should be considered, as they play an important role in the development of motherhood-specific dysfunctional beliefs about maternal responsibility and maternal role idealization and may also exacerbate women's negative responses to PND symptoms (e.g., "I shouldn't feel like this, this means I am a bad mother").

Cognitive processes and PND

Research has shown the important role of cognitive processes in the occurrence and maintenance of PND. For example, increased levels of worries about the baby's health and caregiving (Austin, Tully, & Parker, 2007; Peñacoba-Puente, Marín-Morales, Carmona-Monge, & Furlong, 2016) and brooding rumination (DeJong, Fox, & Stein, 2016; Denis & Luminet, 2018) are significantly associated with PND symptoms. Moreover, while an external locus of control has been associated with PND and has been found to strengthen the relationship between perfectionism and PND (Jackman, Thorsteinsson, & McNeil, 2017; Peñacoba-Puente et al., 2016), an increased perception of control over situations seems to play a protective role in mothers' mental health, including PND (Keeton, Perry-Jankins, & Sayer, 2008).

Moreover, dysfunctional beliefs about motherhood were found to play a key role in the etiology and maintenance of PND symptoms (Fonseca & Canavarro, 2018; Fonseca, Monteiro, & Canavarro, 2018). A more dysfunctional cognitive style has been associated with higher PND rates (Madar, 2013; Sockol et al., 2014), with depressed postnatal women presenting less rational and more irrational beliefs than nondepressed women, including higher self-devaluation and higher needs for achievement and approval (Milgrom & Beatrice, 2003). Finally, expectations about motherhood also seem to be a relevant process associated with PND symptoms. Women with negative self-efficacy expectations show a higher likelihood of having depressive symptoms, which is increased even further if a woman has negative expectations about motherhood (Peñacoba-Puente et al., 2016). In a study conducted with first-time mothers, after early depressive symptoms and risk factors were accounted four, PND was significantly and positively associated with expectations or beliefs that parenting involves great change and sacrifice, was significantly and positively associated with beliefs that the infant's temperament reflects maternal skills, and was negatively associated with the expectation that motherhood is naturally fulfilling (Henshaw, Fried, Teeters, & Siskind, 2014).

Emotional regulation and PND

Women's emotion regulation processes are crucial for addressing the high demands (e.g., caregiving-related demands) of the postnatal period; women need to regulate their emotions to be able to regulate their children's emotional states (Rutherford, Wallace, Laurent, & Mayes, 2015) and women's emotion regulation is associated with their wellbeing. Specifically, there is evidence that emotion regulation difficulties, defined as the use of maladaptive emotion regulation strategies such as suppression of emotions (i.e., avoidance) or self-blame to deal with negative emotions are more related to the occurrence of PND (Edwards et al., 2017; Haga et al., 2012). Congruently, women presenting PND symptoms were found to report higher levels of nonacceptance of emotional responses and lack of emotional awareness, as well as more difficulty engaging in goal-directed behaviors when experiencing negative emotions than nondepressed postnatal women (Marques, Monteiro, Canavarro, & Fonseca, 2018).

Interpersonal processes and PND

A lack of social support and marital dissatisfaction/lack of marital support have been consistently established as risk factors for PND (Smorti, Ponti, & Pancetti, 2019) and they can also contribute to the maintenance of symptoms over time. For example, a recent study found that two components of social support are particularly important to women's mental health in the postnatal period, namely reassurance of worth (i.e., supportive reassurance, relationships in which the woman's skills and abilities are acknowledged) and reliable alliance (i.e., relationships in which the woman can count on another for assistance under any circumstances); the study also found that the absence of such support was linked to higher postnatal anxiety and depressive symptoms (Milgrom, Hirshler, Reece, Holt, & Gemmill, 2019). On the other hand, it is also important to distinguish between women's actual received support and women's perceived received support. While it may be true that the level of support is not sufficient to meet the needs of new mothers, it may also be that the problems are related to mothers' perception of social support (Heh, 2003). In fact, Logsdon and McBride (1994) found that the discrepancy between the actual received support and the expected support predicted the level of PND more accurately than the actual level of received support.

Regarding couple relationship, a recent systematic review found that poor communication, lower satisfaction with the division of household labor, lack of emotional closeness, high conflict and poor relationship and sexual satisfaction are associated with higher levels of PND symptoms (Pilkington et al., 2015). Moreover, a significant proportion of partners of women with PND experience fear, confusion, concern for their wives, and they often express difficulties in understanding their partners' experience of PND, feeling unable to help. This experience is exacerbated by the fact that women with PND generally have more difficulties in adequately expressing their needs. Consequently, this may contribute to increased stress in the couple's relationship, which may further exacerbate women's depressive symptoms (Davey, Dziurawiec, & O'Brien-Malone, 2006; Westall & Liamputtong, 2011).

Cognitive-behavioral therapy for PND prevention and treatment

Recent systematic reviews and meta-analyses focused on the efficacy of cognitive-behavioral therapy (CBT) for preventing and treating PND. Globally, this research has found strong evidence that CBT is effective in both the prevention (Sockol, 2015; Sockol, Epperson, & Barber, 2013; Yasuma et al., 2020) and treatment (Huang, Zhao, Qiang, & Fan, 2018; Nillni, Mehralizade, Mayer, & Milanovic, 2018; Sockol, 2015) of PND, with the magnitude of intervention effects being comparable to that found in CBT for depression in general populations (Sockol, 2015).

Concerning preventive CBT interventions, there is evidence that interventions initiated during the postpartum period were more effective than those initiated during pregnancy, because women may find the information more immediately relevant and helpful in the postpartum period (Sockol, 2015). A recent scoping review of preventive interventions for PND also suggested that interventions grounded in a theoretical framework such as CBT were more effective (Fonseca, Pereira, et al., 2018). In this scoping review, the authors examined the main content of the existing preventive interventions for PND, identifying five major topics: (a) education and skills regarding the transition to parenthood (e.g., women's education about the changes and reorganizations of motherhood and coping strategies to address them); (b) education about PND and cognitive-behavioral skills to address depressive and anxiety symptoms; (c) promotion of social support and couple relationship quality; and (d) stigma reduction, help-seeking and helping the family (Fonseca, Pereira, et al., 2018).

CBT to treat PND usually includes modules related to psychoeducation about PND, cognitive restructuring techniques, behavioral activation or assertiveness; modules related to promotion of interpersonal support and/or improvements in communication are often combined together with the more traditional CBT modules (Nillni et al., 2018). Another important point concerns the delivery formats of CBT prevention and treatment approaches for PND. There is evidence that both individual group and online delivery formats have been effective in preventing or treating PND, which suggests that CBT can be successfully implemented in a wider range of contexts outside of traditional clinic settings (Sockol, 2015).

Practice and procedures

In this section, we explore in depth some of the important topics related to the cognitive-behavioral assessment and conceptualization of PND, as well as the application of cognitive-behavioral techniques for PND.

Clinical assessment

Clinical assessment is very relevant to case conceptualization for women's PND symptoms. Some important topics and examples of questions to address in clinical assessment are as follows:

1. Current PND manifestations at the cognitive, emotional, behavioral, and physiological levels:
   - *Tell me about how you have been feeling in these last days. What are the most frequent emotions that you feel? How intense are these emotions? Do you feel these emotions in your body in some way? In which situations these strong emotions are most frequent? What do you usually do to deal with these emotions and thoughts?*
2. Beginning of PND symptoms: *In your opinion when did you start to feel these symptoms? Can you identify some relevant situations or circumstances that you think might have contributed to the beginning of your symptoms?*
3. Experience of maternity: *Can you describe what were the largest changes after your baby was born? Were these changes like what you expected or imagined? Do you think that the birth of your baby impacted other areas of your life (e.g., your couple relationship, your work)? In what ways?*
4. Cognitive processes:
   - *Can you describe a recent situation in which you felt very sad? What thoughts came up to your mind in that situation? Do you have those kinds of thoughts frequently?*
   - *Do you have frequent health-related or other concerns about your baby? Do you think you worry now more than you used to? When you feel sad, do you find yourself going through your thoughts repeatedly?*
   - *How would you define yourself as a mother? What means being a "good mother" to you? Do you sometimes feel pressure to be a "perfect mother"? If yes, how do you think it makes you feel?*
5. Emotion regulation processes: *It is easy for you to clearly identify how you feel, or do you sometimes find it difficult to label the emotions? How do you usually deal with unpleasant or negative emotions?*
6. Interpersonal processes:
   - *How do you characterize your current relationship with your partner? Do you feel that there were some changes in this relationship after the birth of your baby? Can you describe how your partner has been dealing with your current emotional symptoms? Do you think he understands how you are feeling right now? Do you feel your partner's support? In what ways?*
   - *Can you describe your relationship with your family (parents, brothers/sisters)? Do you feel supported by your family? Do you have other people who can help you during this period (e.g., in-laws, friends)?*

In addition to these topics, **preexisting vulnerabilities** (e.g., psychiatric history, family experiences, stressful events, personality characteristics) should be assessed, as they are important for case conceptualization. Moreover, in addition to clinical interviews, which are usually the main source of clinical information, the assessment should be complemented with self-report questionnaires (e.g., to assess rumination, dysfunctional beliefs about motherhood, and emotion regulation strategies) and self-monitoring records (e.g., thought records).

Case conceptualization

Case conceptualization is the process by which therapists apply cognitive-behavioral theory to understand the etiology and maintenance of a clinical disorder (Beck et al., 1979). Based on the factors mentioned above, it is important that case conceptualization for PND identify (a) preexisting vulnerability factors; (b) the precipitating event (or context) of the symptoms; (c) the general and motherhood-related dysfunctional beliefs activated by the stress-inducing context; (d) the links between thoughts, emotions, and behaviors in a specific situation; and (e) maintenance factors. An example of the clinical conceptualization of the Kate's case (Clinical vignette 1) is presented in Fig. 1.

Clinical vignette 1

Kate is 29 years old, is married and is the mother of one child (Sophie, 2 months). She was referred for psychological intervention by her doctor, who was worried about her depressed mood, lack of energy and fatigue. These symptoms began right after Sophie's birth.

Kate feels overwhelmed and exhausted with everyday activities, and notes "Sometimes, when I am not alone with the baby, I just lay in bed, I just don't have motivation to get up" She sleeps very few hours at night, feels constantly tired and cries frequently during the day, mentioning "I was supposed to feel happy" Despite feeling overwhelmed with the caregiving and household tasks, Kate has difficulty seeking help from others "I feel so overwhelmed that I don't know who can help me; I can't think."

Kate says breastfeeding is not going well: "I think my baby is rejecting me. He refuses to breastfeed. I don't know what I am doing wrong. If I do not breastfeed, I will not be a good mother." Kate says that every time she tries to breastfeed, she feels very sad and frustrated, which results in great difficulty in smoothing her baby. After breastfeeding, when she

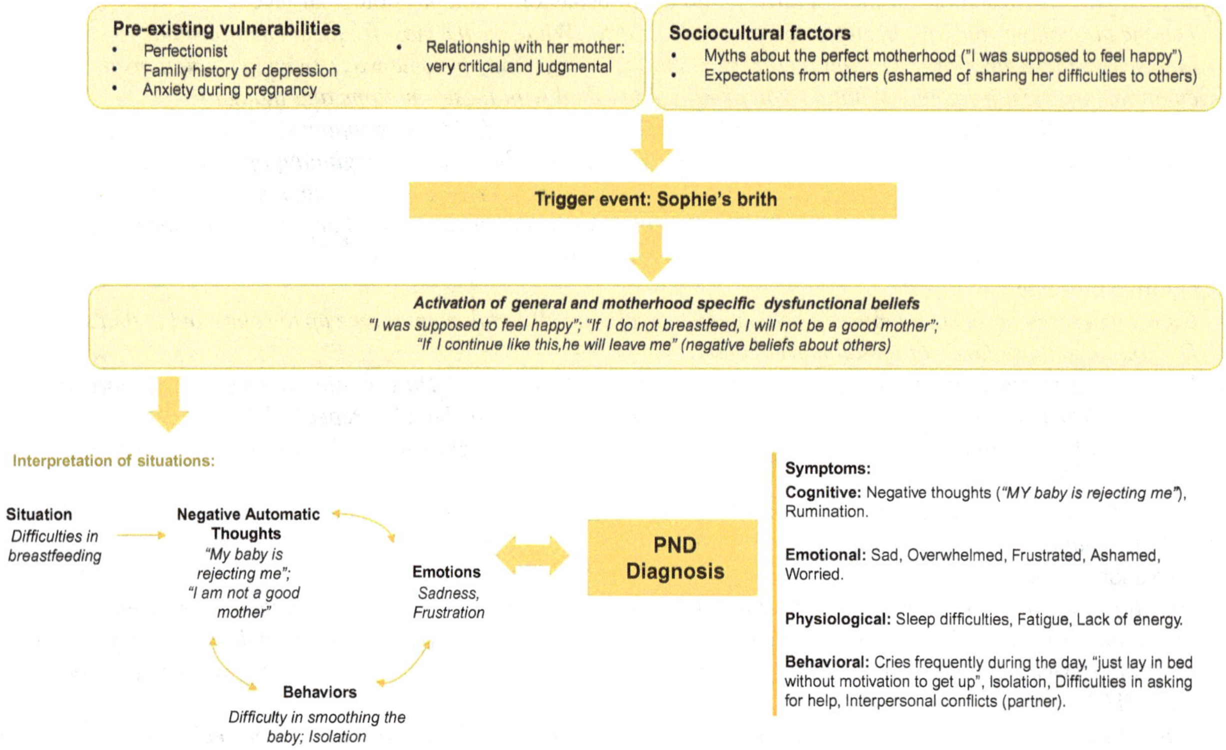

FIG. 1 Cognitive-behavioral case conceptualization of Kate's PND symptoms. Cognitive-behavioral conceptualization of a PND clinical case. Identification of preexisting vulnerabilities and sociocultural factors, trigger event, general and motherhood-specific beliefs, interpretation of situations and cognitive, emotional, physiological, and behavioral responses.

is not alone, she isolates herself for a while. Kate notes that these thoughts ("Why this happening to me?") constantly pop into her mind.

Kate stopped seeing and calling her friends, explaining that she felt ashamed to share her difficulties and thought that others would think that her baby deserved a better mother.

Kate's husband is often upset because, according to her, "He says I am always tired, that I do not want to go out, and that we do not have sexual relations anymore. He probably thinks I don't like him anymore; if I continue like this, he will leave me." Kate says her husband is supportive but that he does not understand her difficulties and that they cannot communicate anymore. Despite helping when she asks, her mother is "very critical and judgmental, as she was during my entire life," and Kate refers to difficulties in dealing with her mother's comments and opinions about the baby's care.

When asked about her previous history, Kate mentioned being fearful and worried since she learned she was pregnant, as the pregnancy was not planed. She notes that she did not have previous history of psychological problems, but some family members have had depression. Kates describes herself as perfectionist and talks about being worried about returning to work (Kate is an architect) and not being able to be as good of a worker as she used to be.

Identifying and challenging negative automatic thoughts

Identifying and challenging negative automatic thoughts constitute a key dimension of CBT. Therapists should start by educating women about the automatic nature of thoughts and explaining that they thoughts are only one way of interpreting a situation and that they are grounded in our prior experiences and beliefs, so two different individuals in the same situation may interpret the situation differently. Moreover, it is important to help women identify their automatic thoughts when they feel sad, anxious, or irritated, by using the thought records. Another important technique is to help women to distinguish between helpful and unhelpful thoughts. While helpful thoughts may aid us in finding solutions for concrete problems and motivate us to take action, unhelpful thoughts create "stories" about what is happening that are not based on real facts and that are associated with more destructive and devaluing feelings. An example of how to identify helpful and unhelpful thoughts is presented in the next clinical vignette.

Clinical vignette 2

Therapist: You were telling me that you felt really sad yesterday, when you were trying to breastfeed. What was running through your mind at that moment?

Kate: I think my baby is rejecting me because she refuses to breastfeed.

Therapist: It appears that this thought is causing you suffering, is that correct?

Kate: Yes ...

Therapist: I understand it feels that way. I would like to understand if this thought is helpful for you to be more effective or solve your current problems.

Kate: Well, no, I still cannot breastfeed. It is not helpful at all, maybe it evens worsens the situation.

Therapist: That is possible, yes. These thoughts, we call them negative or unhelpful thoughts, usually get in way of doing things. Now I would like you and try to look at the "big picture." What "real facts" show you that Sophie is rejecting you?

Kate: She does not want to breastfeed ... She spits the milk and starts crying. I must be doing something wrong.

Therapist: And what facts show that Sophie is not rejecting you?

Kate: For example, she plays with me and she smiles at me ...

Therapist: So, do you think these thoughts that your baby is rejecting you are based on real facts?

Kate: Well, I have never put it that way. Now it does not make sense, no.

Therapist: Great. I have been posing these questions for us to understand when a negative or unhelpful thought arises. So, when you notice your mood changing, you can stop and ask yourself, "What is going through my mind right now?" Whenever you are unsure whether a particular thought is negative or unhelpful, ask yourself these questions: "Is this thought based on real facts? Is this thought useful to me? Does this thought help me to be more effective or achieve my goals? Or is it causing me suffering?"

Moreover, challenging negative thoughts is another important cognitive technique that should be used in PND treatment. Challenging women's negative thoughts may help women increase their cognitive flexibility, i.e., perceive that their thoughts are only one possible interpretation of a situation, and they are not the "absolute truth." By challenging negative automatic thoughts, a therapist can help a mother reduce the strength with which she believes in such thoughts (see the clinical vignette for an example).

Clinical vignette 3

Therapist: You're telling me that you are worried about the relationship with your partner.

Kate: Yes, if I continue like this, he will leave me.

Therapist: Can you specify what you mean with like "this"?

Kate: If I stay depressed, if I do not get better.

Therapist: Well, you have been feeling bad, as you told me, and now you are just starting the treatment to get better.

Kate: Yes, that is true ... Actually, my husband encouraged me to get help.

Therapist: What do you think it shows you, the fact that he encouraged you to come here to get better?

Kate: Hmm, that he worries about me. I know he wants me to get better; I just don't know if he can handle this much longer.

Therapist: This thought that he will leave you, what evidence is there that this though is true?

Kate: He says he misses me, including being more intimate and having sexual relations, but I just can't right now ... He also keeps saying we should go out more. He is probably tired of being at home all the time, and caring for the baby when I just can't ...

Therapist: I understand, it looks like this is a very powerful thought. Are there any other possible ways to interpret these things he says?

Kate: I don't know ...

Therapist: For instance, when someone tells you she or he misses you, what could that mean?

Kate: Well, that he likes being with me ...

Therapist: Hmm hmm ... And when someone says you should go out, what possible interpretation could there be for this?

Kate: He says it would be good for me, to go to back to the yoga classes for example, that I used to like so much.

Therapist: Ok, so there is some evidence that goes against that idea that your husband will leave you, is that correct?

Kate: Yes.

Therapist: What kinds of things does he do or say that might show he will not leave you?
Kate: He actually never said that he would leave me …
Therapist: So, we see that there is little evidence that this thought is actually true, do you agree?
Kate: Well, yes.
Therapist: What would you say to a friend who, in the same situation, thought the same thing as you?
Kate: I would tell my friend that it did not make much sense, her husband loves her and is always present. I would also say that it is normal to have these negative thoughts sometimes, but that we shouldn't always listen to them.

Increasing social support

One of the struggles experienced by Kate has been related to her difficulty obtaining the necessary support during this period. One of the most frequent reasons for not receiving support may be difficulty in identifying the type of support needed and the people within one's social network who can provide the different types of support. One interesting technique to address this struggle is to propose that the woman create an S.O.S. List. See the following clinical vignette in which the therapist presents the exercise to Kate and Fig. 2.

Clinical vignette 4

Therapist: We have been talking about how to effectively ask for help, and we agreed that it is very important to clearly identify the help that we need when asking for help. Is that right?
Kate: Yes, but that's not always easy … sometimes, it's difficult for me to know exactly what I need. I feel overwhelmed, and I don't know how to start …
Therapist: Many women report these same difficulties. That's why I think it would be important for us to try to identify your support needs more clearly before you really need to ask for help. We may need different types of support. For example, mothers may need practical/instrumental support (someone who can help with groceries or take care of the baby for a while so the mother can rest a little bit), emotional support (someone with whom mothers can share their

S.O.S. List

Practical/Instrumental support (e.g., help to cook meals for the family, to take care of the baby, assistance with groceries shopping)	
Practical support that I need?	**Who can provide me practical support?**
Emotional support (e.g., someone who listens to me, someone who wants to talk with me about the experience of being a mother, someone to give me a hug, someone to praise me)	
Emotional support that I need?	**Who can provide me emotional support?**
Informative support (e.g., information on how to deal with my baby's colic, strategies to help with breastfeeding or baby's sleep)	
Informative support that I need?	**Who can provide me informative support?**

FIG. 2 Increasing social support: The S.O.S. List. The S.O.S. List is an exercise aimed to help women to identify their support needs (column one—practical, emotional, and informational support needs) and the persons in one's network who can meet the different types of needs (column two). This exercise was adapted with permission from Be a Mom©, a self-guided CBT program for PND developed by Ana Fonseca (author of this chapter) and by M. Cristina Canavarro.

emotional difficulties), and informative support (someone who can provide information about the baby's health concerns, such as the doctor or a nurse). I would like to ask you to look at this form (Fig. 2) and try to identify your support needs in each of these domains (first column). After that, please fill in the second column, including all the persons who you can think of who can offer you some help in each of these domains. Can you try to do that?

Kate: Ok, let's try.

Communicating with partner

Learning an assertive style of communication and putting it in practice in interpersonal relationships represent an important component of PND treatment, mostly when interpersonal issues are part of women's difficulties. The therapist should provide education about the different communication styles (passive, aggressive, and assertive) and start by helping women to identify what is the communication style they use most frequently, and how this has been impacting their mood and relations. After, the therapist should discuss with the woman the characteristics of an assertive communication style and help her to put these characteristics in practice when communicating with others. Frequently, it is useful to propose some role-playing during the session so the woman can experiment with this new communication style in a safe environment. See an example in the next clinical vignette.

Clinical vignette 5

Therapist: Whenever you need to communicate, ask for help or show that other people's opinion is not useful for you, it is important to speak clearly and openly, demonstrating how you feel and at the same time, respecting other people's opinions.

Kate: This is really hard for me. I just don't know how to do it …

Therapist: I understand you may feel this way. Assertiveness is a skill that we are not born with and that we can learn. Do you recognize any characteristics of being assertive in communication?

Kate: Hmm, I guess people who are assertive can say what they think in a calm and clear way.

Therapist: That's right! To be assertive and communicate clearly and openly, we should speak with a calm and relaxed posture, maintain eye contact, and be clear and specific. For instance, if you do not want to talk about breastfeeding with your mother, instead of saying, "I don't feel like talking!," you could say "I am not comfortable talking about breastfeeding, it is a very delicate subject for me."

Kate: Hmm, I think I could use these techniques to talk to my husband and explain to him that I do not feel comfortable with having sexual relations.

Therapist: Yes, that it is in fact a good situation to practice these skills, Kate. It is really important to say how you feel and use phrases beginning with "I." Instead of saying, "Whenever you talk about this, you're hurting me," say for example: "I feel sad when you tell me … ." Also, you should try to understand the other person's point of view, for example, by restating what he said, "I understand that you are feeling …," and not interrupting, even if you do not agree. Do you think it is clear?

Kate: Yes, it makes sense … Sometimes we start talking, and we just keep interrupting each other, we start arguing, and then I feel we can't talk.

Therapist: If you start fighting, sometimes it is better to take a break and talk again when both of you are calmer. To be assertive, it is also really important to show respect and consideration for others' opinions. Say, for example, "I appreciate your attention and concern, but right now I don't need any help."

Kate: It looks so simple, but in practice …

Therapist: I understand, and these skills need to be practiced to begin to be a part of your communication naturally. You said you would like to use the assertiveness skills when talking to your husband, is that right?

Kate: Yes.

Therapist: Ok, so let's start by identifying the topic you would like to discuss with your partner.

Kate: I think it would be about how I feel at this moment about sexual relations.

Therapist: Now, think about where and when you can have this conversation.

Kate: It could be later at night, when Sophie is sleeping.

Therapist: Great. Considering the strategies we have talked about, I would like you to practice what you could say to your partner, in an assertive and proper way. Can we practice this in a role play?

Kate: Ok.

Therapist: I'll play your husband, and you should communicate with me as if you were in that situation, applying the strategies we just reviewed, is that ok? You can start by communicating a positive aspect so that the other person will not be "defensive." Then, express your opinion and how you feel and end with another positive note, such as expressing that the relationship is really important to you or showing appreciation for the other person. What do you think?
Kate: I agree.
Therapist: Also, I would like you to think about how you could respond to your partner in a useful way for both of you, if he has a different opinion than yours. Let's try?

Mini-dictionary of terms

- ***Postnatal depression***. A nonpsychotic depressive episode of mild to major severity that may occur during the first year after childbirth.
- ***Risk factors***. The characteristics that, if present in a given individual, make it more likely that this individual, rather than someone selected at random from the general population, will develop a specific disorder.
- ***Case conceptualization***. Application of a psychological theoretical framework (e.g., CBT) to understand and explain the etiology and maintenance of a clinical disorder.
- ***Dysfunctional beliefs***. Negative beliefs about the self, others or the future that are grounded on the individual's early experiences and that, when activated by a specific trigger, influence the way in which the individual interprets the situations.
- ***Cognitive flexibility***. The practice of deliberately considering multiple interpretations or predictions about a situation rather than assuming that the first thought is accurate and useful.
- ***Social support***. Provision of assistance or comfort to others. Support may arise from any member of the individual's social network, involving family members, friends, neighbors, colleagues, caregivers, or other members of the community.
- ***Assertive communication style***. A way of communicating feelings, thoughts, and beliefs in an open and honest manner, without disrespecting the rights of others.

Key facts about PND treatment

- Few women seek professional assistance for PND.
- The main barriers to help-seeking are a lack of proper knowledge of symptoms (e.g., poor perinatal depression literacy), stigma about mental illness, and structural barriers (e.g., lack of specialized mental health professionals in the services, financial restrictions).
- Clinical guidelines (e.g., from the National Institute for Health and Clinical Excellence (NICE)) recommend CBT as the first-line treatment for mild to moderate PND, and as combined treatment (with medication) for severe PND.
- Women's decision-making process concerning treatment options may be complex, as many postnatal women show strong reluctance to pursue medication (refusing treatment or not complying with prescribed treatment) due to its potential teratogenic effects, and psychotherapy often has limited availability.
- Beyond being effective, web-based CBT interventions may improve women's accessibility to psychological treatment, addressing the mental health needs of women in the postnatal period.

Applications to other areas

In this chapter, we reviewed the cognitive-behavioral case conceptualization and prevention and treatment approaches for PND. Although specifically targeting PND, the topics addressed here can be applied and extended to antenatal depression. For example, the developmental psychopathology approach has also been applied to antenatal depression (Goodman & Dimidjian, 2012), and case conceptualization is grounded in the same major concepts, such as preexisting vulnerabilities (e.g., personality characteristics such as perfectionism) that interact with triggering/precipitating events (stress-inducing situations during pregnancy, e.g., medical complications, problems with the partner) to activate dysfunctional beliefs about the self and others, or even motherhood-specific dysfunctional beliefs, that influence the way women interpret the situations (Wenzel & Kleiman, 2015). Moreover, the CBT techniques discussed in this chapter (e.g., identifying and challenging unhelpful thoughts, increasing social support and promoting assertive communication) are also applicable in CBT treatment protocols for treating antenatal depression, which were also found to be effective in reducing depressive symptoms at postintervention and follow-up (Shortis, Warrington, & Whittaker, 2020). Notably, some of the same

CBT principles used in PND comprehension and treatment may even be extended to understanding and treating maternal depression in early childhood, although in the latter case, a deeper focus on the mother-child interaction and emotional regulation within parenting interaction is needed.

Summary points

- PND has been recognized as a significant public health problem, and is the most prevalent mental health condition during the postnatal period.
- PND has a negative impact on mothers' physical and psychological wellbeing, as well as on mother-child interactions and infant's development.
- Risk/vulnerability factors for PND (e.g., history of depression, personality traits such as neuroticism or perfectionism) interact with precipitating/triggering events to activate both general and motherhood-specific dysfunctional beliefs.
- Dysfunctional beliefs influence the way women give meaning to events, and consequently how they feel and behave in specific situations.
- The broader social context ("myths about perfect motherhood") and women's interpersonal contexts (relationships with partners and significant others) also play a crucial role in the occurrence and maintenance/exacerbation of PND.
- CBT has been found to be effective in both preventing and treating PND, regardless of the delivery format (individual face-to-face, group, and online).

References

American Psychiatric Association (APA). (2013). *Diagnostic and statistical manual of mental disorders* (5th ed.). Washington, DC: American Psychiatric Association.

Austin, M. P., Tully, L. M., & Parker, G. (2007). Examining the relationship between antenatal anxiety and postnatal depression. *Journal of Affective Disorders, 101*, 169–174.

Beck, A. T. (1987). Cognitive models of depression. *Journal of Cognitive Psychotherapy: An International Quarterly, 1*, 5–37.

Beck, A. T., Rush, A. J., Shaw, B. F., & Emery, G. (1979). *Cognitive therapy for depression.* New York: The Guilford Press.

Benjet, C. (2017). Developmental psychopathology and the epidemiology and expression of psychopathology from infancy through early adulthood. In J. Matson (Ed.), *Handbook of childhood psychopathology and development disabilities treatment* (pp. 11–23). Cham, Switzerland: Springer.

Davey, S., Dziurawiec, S., & O'Brien-Malone, A. (2006). Men's voices: Postnatal depression from the perspective of male partners. *Qualitative Health Research, 16*, 206–220.

DeJong, H., Fox, E., & Stein, A. (2016). Rumination and postnatal depression: A systematic review and a cognitive model. *Behavior Research and Therapy, 82*, 38–49.

Denis, A., & Luminet, O. (2018). Cognitive factors and post-partum depression: What is the influence of general personality traits, rumination, maternal self-esteem, and alexithymia? *Clinical Psychology & Psychotherapy, 25*, 359–367.

Edwards, E., Holzman, J., Burt, N., Rutherford, H., Mayes, L., & Bridgett, D. (2017). Maternal emotion regulation strategies, internalizing problems and infant negative affect. *Journal of Applied Developmental Psychology, 48*, 59–68.

Epperson, C., Huang, M., Cook, K., Gupta, D., Chawla, A., Greenberg, P., & Eldar-Lissai, A. (2020). Healthcare resource utilization and costs associated with postpartum depression among commercially insured households. *Current Medical Research and Opinion, 36*, 1707–1716.

Field, T. (2010). Postpartum depression effects on early interactions, parenting, and safety practices: A review. *Infant Behavior & Development, 33*(1), 1–6.

Fonseca, A., & Canavarro, M. C. (2018). Exploring the paths between dysfunctional attitudes towards motherhood and postpartum depressive symptoms: The moderating role of self-compassion. *Clinical Psychology & Psychotherapy, 25*, e96–e106.

Fonseca, A., Monteiro, F., & Canavarro, M. C. (2018). Dysfunctional beliefs towards motherhood and postpartum depressive and anxiety symptoms: Uncovering the role of experiential avoidance. *Journal of Clinical Psychology, 74*, 2134–2144.

Fonseca, A., Pereira, M., Araújo Pedrosa, A., Gorayeb, R., Moura-Ramos, M., & Canavarro, M. C. (2018). Be a Mom: Formative evaluation of a web-based psychological intervention to prevent postpartum depression. *Cognitive and Behavioral Practice, 25*, 473–495.

Gelaye, B., Rondon, M., Araya, R., & Williams, M. (2016). Epidemiology of maternal depression, risk factors, and child outcomes in low-income and middle-income countries. *Lancet Psychiatry, 3*, 973–982.

Goodman, S. H., & Dimidjian, S. (2012). The developmental psychopathology of perinatal depression: Implications for psychosocial treatment development and delivery in pregnancy. *Canadian Journal of Psychiatry. Revue Canadienne de Psychiatrie, 57*, 530–536.

Haga, S., Ulleberg, P., Slinning, K., Kraft, P., Steen, T., & Staff, A. (2012). A longitudinal study of postpartum depressive symptoms: Multilevel growth curve analyses of emotion regulation strategies, breastfeeding self-efficacy, and social support. *Archives of Women's Mental Health, 15*, 175–184.

Heh, S. (2003). Relationship between social support and postnatal depression. *Kaohsiung Journal of Medical Sciences, 19*, 491–496.

Henshaw, E., Fried, R., Teeters, J. B., & Siskind, E. (2014). Maternal expectations and postpartum emotional adjustment in first-time mothers: Results of a questionnaire survey. *Psychosomatic Obstetrics & Gynecology, 35*, 69–75.

Huang, L., Zhao, Y., Qiang, C., & Fan, B. (2018). Is cognitive behavioral therapy a better choice for women with postnatal depression? A systematic review and meta-analysis. *PLoS One, 13*. https://doi.org/10.1371/journal.pone.0205243, e0205243.

Jackman, L., Thorsteinsson, E., & McNeil, D. (2017). Perfect imperfections: Locus of control, perfectionism, and postpartum depression. *SAGE Open*. https://doi.org/10.1177/2158244017710689.

Keeton, C., Perry-Jankins, M., & Sayer, A. (2008). Sense of control predicts depressive and anxious symptoms across the transition to parenthood. *Journal of Family Psychology*, *22*, 212–221.

Kendig, S., Keats, J., Hoffman, C., Kay, L., Miller, E., Simas, T., … Lemieux, L. (2017). Consensus bundle on maternal mental health: Perinatal depression and anxiety. *Obstetrics and Gynecology*, *129*, 420–430.

Knudson-Martin, C., & Silverstein, R. (2009). Suffering in silence: A qualitative meta-data-analysis of postpartum depression. *Journal of Marital and Family Therapy*, *35*, 145–158.

Logsdon, M. C., & McBride, A. (1994). Social support and postpartum depression. *Research in Nursing & Health*, *17*(6), 449–457.

Madar, A. (2013). Maternal attitudes and beliefs scale: Development and piloting. *Procedia - Social and Behavioral Sciences*, *78*, 415–419.

Marques, R., Monteiro, F., Canavarro, M. C., & Fonseca, A. (2018). The role of emotion regulation difficulties in the relationship between attachment representations and depressive and anxiety symptoms in the postpartum period. *Journal of Affective Disorders*, *238*, 39–46.

Milgrom, J., & Beatrice, G. (2003). Coping with the stress of motherhood: Cognitive and defence style of women with postnatal depression. *Stress and Health*, *19*, 281–287.

Milgrom, J., Hirshler, Y., Reece, J., Holt, C., & Gemmill, A. W. (2019). Social support: A protective factor for depressed perinatal women? *International Journal of Research and Public Health*, *16*, 1426. https://doi.org/10.3390/ijerph16081426.

Milgrom, J., Martin, P. R., & Negri, L. M. (1999). *Treating postnatal depression: A psychological approach for health care practitioners*. West Sussex: Wiley.

Nillni, Y., Mehralizade, A., Mayer, L., & Milanovic, S. (2018). Treatment of depression, anxiety, and trauma-related disorders during the perinatal period: A systematic review. *Clinical Psychological Review*, *66*, 136–148.

O'Hara, M. W. (1994). *Postpartum depression: Causes and consequences*. New York: Springer-Verlag.

O'Hara, M. W., & McCabe, J. E. (2013). Postpartum depression: Current status and future directions. *Annual Review of Clinical Psychology*, *9*, 379–407.

Peñacoba-Puente, C., Marín-Morales, D., Carmona-Monge, F., & Furlong, L. (2016). Post-partum depression, personality, and cognitive-emotional factors: A longitudinal study on Spanish pregnant women. *Health Care for Women International*, *37*, 97–117.

Pilkington, P., Milne, L., Cairns, K., Lewis, J., & Whelan, T. (2015). Modifiable partner factors associated with perinatal depression and anxiety: A systematic review and meta-analysis. *Journal of Affective Disorders*, *178*, 165–180.

Putnam, K., Wilcox, M., Robertson-Blackmore, E., Sharkey, K., Bergink, V., Munk-Olsen, T., … Postpartum Depression: Action Towards Causes and Treatment (PACT) Consortium. (2017). Clinical phenotypes of perinatal depression and time of symptom onset: Analysis of data from an international consortium. *Lancet Psychiatry*, *4*, 477–485. https://doi.org/10.1016/S2215-0366(17)30136-0.

Rahman, A., Surkan, P. J., Cayetano, C., Rwagatare, P., & Dickson, K. (2013). Grand challenges: Integrating maternal mental health into maternal and child health programmes. *PLoS Medicine*, *10*. https://doi.org/10.1371/journal.pmed.1001442, e1001442.

Riecher-Rossler, A., & Hofecker, M. (2003). Postpartum depression: Do we still need this diagnostic term? *Acta Psychiatrica Scandinavica*, *108*, 51–56.

Rutherford, H., Wallace, N. S., Laurent, H., & Mayes, L. (2015). Emotion regulation in parenthood. *Developmental Review*, *36*, 1–14.

Santos Junior, H., Tan, X., & Salomon, R. (2017). Heterogeneity in perinatal depression: How far have we come? A systematic review. *Archives of Women's Mental Health*, *20*, 11–23.

Shortis, E., Warrington, D., & Whittaker, P. (2020). The efficacy of cognitive behavioral therapy for the treatment of antenatal depression: A systematic review. *Journal of Affective Disorders*, *272*, 485–495.

Slomian, J., Honvo, G., Emonts, P., Reginster, J.-Y., & Bruyère, O. (2019). Consequences of maternal postpartum depression: A systematic review of maternal and infant outcomes. *Women's Health*, *15*, 1–55.

Smorti, M., Ponti, L., & Pancetti, F. (2019). A comprehensive analysis of postpartum depression risk factors: The role of socio-demographic, individual, relational, and delivery characteristics. *Frontiers in Public Health*, *24*. https://doi.org/10.3389/fpubh.2019.00295.

Sockol, L. E. (2015). A systematic review of the efficacy of cognitive behavioral therapy for treating and preventing perinatal depression. *Journal of Affective Disorders*, *177*, 7–21.

Sockol, L., Epperson, C., & Barber, J. (2013). Preventing postpartum depression: A meta-analytic review. *Clinical Psychology Review*, *33*, 1205–1217.

Sockol, L., Epperson, C., & Barber, J. (2014). The relationship between maternal attitudes and symptoms of depression and anxiety among pregnant and postpartum first-time mothers. *Archives of Women's Mental Health*, *17*, 199–212.

Wenzel, A., & Kleiman, K. (2015). *Cognitive behavioral therapy for perinatal distress*. New York: Routledge.

Westall, C., & Liamputtong, P. (2011). Living with uncertainty: The partners' experiences of postnatal depression. In C. Westall, & P. Liamputtong (Eds.), *Motherhood and postnatal depression: Narratives of women and their partners* (pp. 123–141). Dordrecht: Springer Netherlands.

Woody, C., Ferrari, A., Siskind, D., Whiteford, H., & Harris, M. (2017). A systematic review and meta-regression of the prevalence and incidence of perinatal depression. *Journal of Affective Disorders*, *219*, 86–92.

Xiao-hu, Z., & Zhi-hua, Z. (2020). Risk factors for postpartum depression: An evidence-based systematic review of systematic reviews and meta-analyses. *Asian Journal of Psychiatry*, *53*, 102353. https://doi.org/10.1016/j.ajp.2020.102353.

Yasuma, N., Narita, Z., Sasaki, N., Obikani, E., Sekiya, J., Inagawa, T., … Nishi, D. (2020). Antenatal psychological interventions for universal prevention of antenatal and postnatal depression: A systematic review and meta-analysis. *Journal of Affective Disorders*, *273*, 231–239.

Chapter 6

Stress and cortisol regulation during pregnancy: Implications for cognitive behavioral stress management among low-income women

Guido G. Urizar, Jr.
Department of Psychology, California State University, Long Beach, CA, United States

Abbreviations

ACOG	American College of Obstetricians and Gynecologists
APGAR	appearance, pulse, grimace, activity, and respiration
CAR	cortisol awakening response
CBSM	cognitive behavioral stress management
CES-D	Center for Epidemiological Studies-Depression Scale
HIV	human immunodeficiency virus
HPA	hypothalamic-pituitary-adrenal axis
LRC	low risk comparison
SMART Moms	stress management and relaxation training for moms
UC	usual care
USPSTF	U.S. Preventive Services Task Force

Introduction

Prenatal stress effects on maternal and infant health

Although stress is generally considered to be an adaptive response to challenges in the environment, stressors occurring at critical junctures in one's life, such as when one becomes a mother, could affect not only the woman's health, but also that of her infant. Stress experienced during pregnancy has been implicated as a trigger for several physical and mental health problems for women, including poor nutrition, increased smoking and alcohol consumption, higher cardiovascular disease risk, increased anxiety, and postpartum depression (Dunkel Schetter, Niles, Guardino, Khaled, & Kramer, 2016; Lobel et al., 2008; Yim, Tanner Stapleton, Guardino, Hahn-Holbrook, & Dunkel Schetter, 2015). In fact, the greatest period of risk for a first major depressive episode among women occurs during their childbearing years (Marcus & Heringhausen, 2009). Depression during this susceptible period increases their chances of developing a subsequent depressive episode by 60%–80% (Accortt, Freeman, & Allen, 2008). Increased stress during pregnancy has also been linked to adverse birth outcomes, such as preterm births and low birthweight, which are presently the leading causes of mortality and morbidity among newborns and are associated with significantly higher rates of long-term neurodevelopmental impairments in children (Van den Bergh et al., 2020). Prenatal stress could therefore have long-term consequences for both the mother and her child.

The American College of Obstetricians and Gynecologists (ACOG) Committee on Health Care for Underserved Women prominently states the need to identify and assess for psychosocial stressors associated with adverse maternal and infant health outcomes, as well as help women manage their stressors through empirically based prenatal programs aimed at preventing stress-related disorders (ACOG, 2006). These guidelines are consistent with the objectives of *Healthy*

Handbook of Lifespan Cognitive Behavioral Therapy. https://doi.org/10.1016/B978-0-323-85757-4.00034-1

People 2030 to improve maternal and child health outcomes and have been emphasized amid recent data showing that rates of stress-related health complications, such as preterm birth and low birthweight, have continued to increase over the past 5 years despite increasing rates of prenatal care utilization (Martin, Hamilton, & Osterman, 2020). Economic estimates indicate that pre- and postnatal health complications cost the U.S. approximately $26 billion dollars per year in direct medical costs, including increased hospitalizations, physician visits, and use of medications (Beam et al., 2020; Institute of Medicine, 2006). Of particular concern are low-income women, who are more prone to experiencing a number of psychosocial stressors during pregnancy (e.g., poverty, low prenatal health education, single parenthood, poor health behaviors) and who are at greatest risk for several birth complications (Lefmann, Combs-Orme, & Orme, 2017; Reynolds, Labad, Buss, Ghaemmaghami, & Räikkönen, 2013). Prospective studies have shown that low-income women who experience chronic stressors during pregnancy are at increased risk for experiencing prenatal health complications and adverse birth outcomes, such as hypertension, preterm labor, low birthweight, low APGAR scores, premature birth, and increased incidence of infant hospitalizations (Blumenshine, Egerter, Barclay, Cubbin, & Braveman, 2010; Fox, Entringer, Buss, DeHaene, & Wadhwa, 2015; Larrabee Sonderlund, Schoenthaler, & Thilsing, 2021). Given the heightened health concerns among low-income pregnant women, studies examining potential biological mechanisms underlying the impact of prenatal stress on maternal and infant health outcomes are needed.

Cortisol during pregnancy

The stress hormone cortisol has been identified as a possible biological mechanism associated with the heightened risk for developing health complications during pregnancy and the postpartum period. Cortisol is a hormone that is modulated by the hypothalamic-pituitary-adrenal (HPA) axis, is released by the adrenal glands in response to stressors in our environment, and is considered one of the main markers associated with the biological stress response (Fries, Dettenborn, & Kirschbaum, 2009). An abrupt and dramatic change in maternal cortisol occurs during pregnancy and the early postpartum period, with cortisol levels rising throughout pregnancy (playing an important part in fetal organ development) and returning to prepregnancy levels after childbirth (Mastorakos & Ilias, 2003). Although the rise in cortisol during pregnancy is a normal occurrence, studies show that higher elevations in prenatal cortisol are related to several adverse birth outcomes, including increased rates of pre- and postmature births, a higher need for infant resuscitation assistance at birth, and infant brain cell damage, which can lead to long-term cognitive and emotional deficits later in childhood (Hodyl et al., 2017; Zijlmans, Riksen-Walraven, & de Weerth, 2015).

In addition to overall cortisol levels, researchers have examined women's cortisol awakening response (CAR) to better understand stress effects during pregnancy. The CAR refers to an approximate 50%–75% increase in cortisol levels that occur from awakening to about 30–45 min after awakening and is a measure of HPA axis reactivity (Wilhelm, Born, Kudielka, Schlotz, & Wüst, 2007). Because the hippocampus is involved in regulating HPA axis activity, it has been hypothesized that the CAR may be a form of biological stress response that reflects individuals' anticipated demands or stressors for the upcoming day (Fries et al., 2009). A higher CAR during pregnancy has been associated with shorter gestational periods (Buss et al., 2009) and a blunted or flat CAR with increased risk for postpartum depression (Scheyer & Urizar, 2016). These mixed results may be explained by variations in the duration of exposure to stressful experiences. Specifically, a greater CAR has been hypothesized to be more common among those that experience situational, acute stressors that lead to cortisol hyperactivity, whereas a lower CAR may be more common among those that experience longer lasting, chronic stress (e.g., poverty, discrimination; Duan et al., 2013; Fries et al., 2009).

Researchers have also examined diurnal cortisol patterns in relation to pre- and postnatal health outcomes. Diurnal cortisol patterns show the changes of cortisol throughout 1 day, with a normal diurnal cortisol pattern characterized by a peak in cortisol in the morning hours and a steady decline of cortisol throughout the day. This diurnal cortisol pattern is still observed throughout pregnancy (Mastorakos & Ilias, 2003). In contrast, an altered diurnal cortisol pattern may not show this steady decline and look flatter (i.e., smaller decrease in cortisol levels across the day). These flatter diurnal cortisol patterns during pregnancy have been associated with several health problems during the postpartum period, including increased postpartum depression in mothers and low birthweight and impaired sleep among infants (Kivlighan, DiPietro, Costigan, & Laudenslager, 2008; Scheyer & Urizar, 2016; Van den Bergh et al., 2020).

Collectively, these prenatal changes in maternal biochemistry (i.e., total cortisol output, CAR, diurnal cortisol) are highly significant in fetal development, as maternal cortisol readily crosses the placenta (Talge, Neal, & Glover, 2007). Consequently, infants of mothers who exhibit elevated cortisol levels during pregnancy have been shown to have high cortisol levels after birth, thus mimicking their mother's prenatal cortisol levels (Field, Diego, & Hernandez-Reif, 2006; Van den Bergh et al., 2020). Infants of low-income mothers who experience chronic stressors during pregnancy have also been shown to have higher cortisol levels shortly after birth (Keenan, Gunthorpe, & Grace, 2007; Luecken et al., 2013).

Further, higher cortisol levels have been observed among infants of mothers who experience chronic stress and depression during the early postpartum period, leading to the later development of behavioral, mental health, and cognitive deficits among preschool and elementary-school children (Gunnar & Quevedo, 2007; Sohr-Preston & Scaramella, 2006). These findings suggest that elevated cortisol levels during pregnancy and the early postpartum period negatively affect mothers and their infants. Furthermore, these findings highlight the urgent need to develop and test prenatal interventions that target optimal management of cortisol and stress levels during this critical period of human development, which could yield multiple benefits for the mother, her child, and subsequent generations.

Prenatal stress management approaches

Health providers of pre- and postnatal care are becoming more aware of the need to create clear guidelines for the management of stress during pregnancy given that up to 20% of women experience prenatal anxiety and depression (ACOG, 2006; Goodman & Tyer-Viola, 2010). The prevalence rates for these stress-related outcomes are even higher for low-income and ethnic minority women (30%–50%; Gennaro et al., 2020; Lara-Cinisomo, Clark, & Wood, 2018). For nonpregnant populations, common stress management strategies include, but are not limited to, relaxation training, imagery, meditation, yoga, cognitive behavioral therapies, and group support, all of which have been found to be effective in enhancing well-being and reducing distress (Clarke, Black, Stussman, Barnes, & Nahin, 2015; Esch & Stefano, 2010). Despite the availability of these stress management approaches, only 14% of pregnant women with mental health symptoms are referred for treatment by their health care providers due to having inadequate resources and/or lack of knowledge about available treatments (Cox, Sowa, Meltzer-Brody, & Gaynes, 2016; Price, Corder-Mabe, & Austin, 2012). To address this gap, the U.S. Preventive Services Task Force (USPSTF) recommends a collaborative care model that would help primary health and obstetric care providers to link their patients to available stress management services (Siu et al., 2016).

As part of these recommendations, the USPSTF prominently endorses group cognitive behavioral stress management (CBSM) interventions as one of the best nonpharmacological methods and empirically supported modalities for helping to prevent stress-related disorders during pregnancy so as to not harm the fetus, given their increasing acceptance and use among nonpregnant populations (Hofmann, Asnaani, Vonk, Sawyer, & Fang, 2012; Siu et al., 2016). CBSM interventions have largely consisted of standard cognitive behavioral strategies and relaxation techniques that teach stressed individuals various active coping strategies to enhance psychological functioning by: (a) increasing one's awareness of different stress responses and their effect on one's health; (b) teaching relaxation skills to reduce anxiety (e.g., diaphragmatic breathing, progressive muscle relaxation); (c) modifying one's maladaptive cognitive appraisals for different stressors using cognitive restructuring; (d) building effective coping skills and increasing emotional expressiveness through assertiveness training; and (e) reducing social isolation by building one's social support network (Antoni, 2013; Beck & Dozois, 2011).

CBSM interventions and stress outcomes

Over the past two decades, progress has been made toward examining the effectiveness of CBSM interventions on regulating biological markers of stress, such as cortisol. Previous randomized control studies have shown group-based CBSM interventions to be effective in producing short-term reductions (i.e., 2–10-week follow-up period) in stress and cortisol levels in patients with stressful medical conditions, such as HIV and breast cancer (Antoni, 2013; Antoni et al., 2005). One of the few studies to examine the effects of a prenatal CBSM intervention on cortisol demonstrated a lower CAR by the second trimester of pregnancy among women at high risk for anxiety and depression (Richter et al., 2012). Despite these promising results, limited resources and training have precluded the translation of the USPSTF's recommendations for systematically evaluating the effectiveness of group-based CBSM interventions in regulating stress and cortisol levels among low-income pregnant women that are in most need of these programs (Freeman, 2019). In fact, despite the disproportionate number of premature births and low birthweight babies observed among low-income mothers, less than 25% report actively engaging in stress management during pregnancy (Birdee, Kemper, Rothman, & Gardiner, 2014). This is partially due to a lack of awareness of how important managing one's stress levels can be for the health of the mother and her baby (Esperat, Feng, Zhang, & Owen, 2007). Given these disparities in birth outcomes and dissemination of preventive interventions, intervention studies are needed to reach underserved women during pregnancy and provide them with the necessary skills and knowledge to facilitate stress management. In the section below, we summarize our work supporting the use of prenatal CBSM interventions to regulate cortisol and optimize maternal and infant health outcomes among low-income families.

Prenatal CBSM interventions and cortisol regulation

Relaxation day pilot project

In one of our earliest studies examining the effects of cognitive behavioral strategies on regulating prenatal cortisol levels, we examined whether instructing women to engage in a day of relaxation would significantly reduce salivary cortisol and stress levels among 41 low-income pregnant women (81% Spanish-speaking Latina; 68% with an annual family income <$20,000; Urizar et al., 2004). More specifically, women were instructed to eliminate things that were stressful to them and/or participate in things that would increase their level of relaxation. In this experiment, women provided morning and evening salivary cortisol samples and self-reported their stress levels on two collection days: at baseline (prior to receiving these instructions) and on the day they followed the instructions to have a relaxing day (95% adhered to the study protocol). Compared to their baseline levels, women surprisingly demonstrated significant decreases in both prenatal cortisol and perceived stress after attempting to have a relaxing day (see Fig. 1). Interestingly, the majority of women in this study reported engaging in pleasant activities (e.g., walking outside), seeking out positive interactions with others, and modifying their thoughts (key cognitive behavioral strategies for managing stress) in order to reduce their stress levels during pregnancy. Similar cortisol and stress findings have been reported for pregnant women receiving massage therapy (Field et al., 2006) and for those taking part in community-based stress management programs (Wesley, 2006). These findings provide some insight into possible "active ingredients" for prenatal CBSM interventions that can be effective in regulating cortisol levels during pregnancy. However, since these are preexperimental studies, it is necessary to conduct randomized control trials to test the effects of CBSM interventions on cortisol regulation compared to a usual care or control group.

Mothers & Babies Project

Building upon this work, we conducted a pilot randomized trial to examine whether a CBSM intervention would be effective in regulating salivary cortisol levels among low-income mothers and their infants at 6 and 18 months postpartum, relative to two control groups (Urizar & Muñoz, 2011). Participants for this trial were recruited as part of the Mothers and Babies: Mood and Health Project, which was a pilot depression intervention study designed to prevent the onset of major depressive episodes among low-income, pregnant women who were identified as being at high risk for developing depression (Muñoz et al., 2007). The focus of this study was of particular importance given that prenatal depression has been shown to lead to serious, long-term health and developmental consequences for low-income, ethnic minority mothers and their children (Alhasanat & Giurgescu, 2017; Urizar & Muñoz, 2021). During their second trimester of pregnancy (mean gestational age $= 16 \pm 5$ weeks), 86 low-income mothers (80% Spanish-speaking Latinas; 87% with an annual

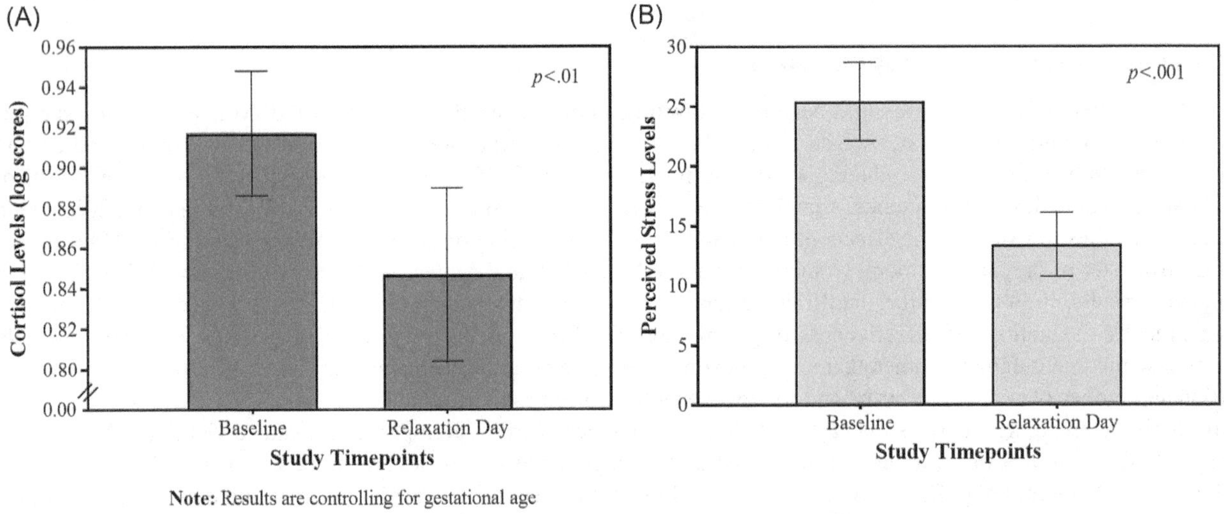

FIG. 1 Intervention effects (baseline vs relaxation day) on salivary cortisol and perceived stress levels among low-income pregnant women. (A) Baseline vs relaxation day cortisol levels. (B) Baseline vs relaxation day perceived stress levels. *(Adapted from Urizar, G. G., Jr., Milazzo, M., Le, H. N., Delucchi, K., Sotelo, R., & Muñoz, R. F. (2004). Impact of stress reduction instructions on stress and cortisol levels during pregnancy.* Biological Psychology, *67(3), 275–282. https://doi.org/10.1016/j.biopsycho.2003.11.001, with permission.)*

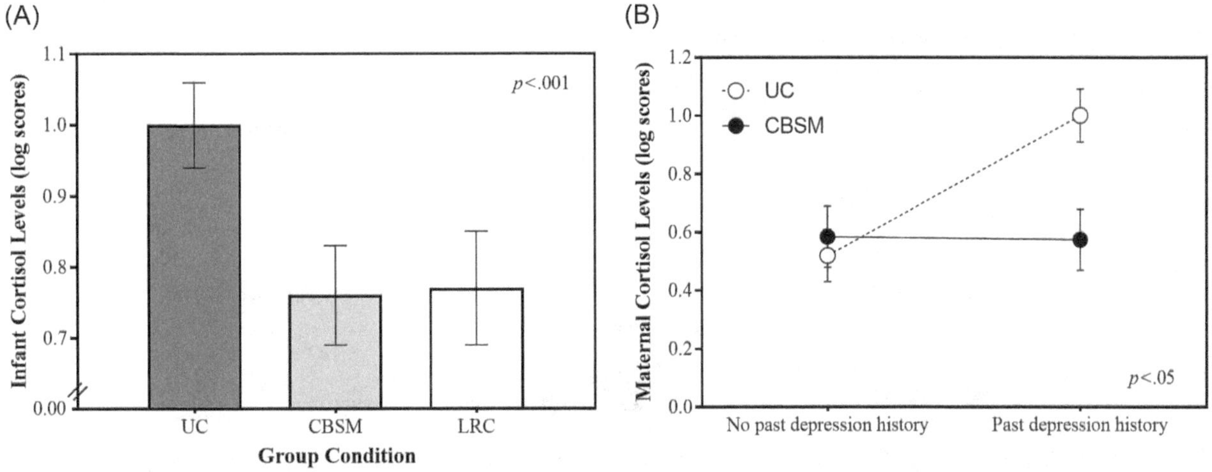

Note: Results are controlling for maternal cortisol and maternal age; UC = Usual Care, CBSM = Cognitive Behavioral Stress Management, LRC = Low Risk Comparison

Note: UC = Usual Care, CBSM = Cognitive Behavioral Stress Management

FIG. 2 Prenatal group condition effects (UC vs CBSM vs LRC) on maternal and infant postpartum cortisol levels. (A) Infant cortisol levels at 6 months postpartum by their mother's prenatal group condition (UC vs CBSM vs LRC). (B) Maternal cortisol levels at 18 months postpartum by randomization group (UC vs CBSM) and past history of depression. *(Adapted from Urizar, G. G., Jr., & Muñoz, R. F. (2011). Impact of a prenatal cognitive-behavioral stress management intervention on salivary cortisol levels in low-income mothers and their infants.* Psychoneuroendocrinology, *36(10), 1480–1494. https://doi.org/10.1016/j.psyneuen.2011.04.002, with permission.)*

family income <$30,000) were screened for depression. Mothers identified to be at high risk for depression (i.e., having either a past history of major depression or current elevated symptoms of depression (≥16 on CES-D)) were randomized to either a 12-week CBSM group (*n* = 24; focused on mood and stress management) or a usual care (UC) group (*n* = 33), while a low risk comparison (LRC) group (*n* = 29) was comprised of mothers not meeting either depression criteria. Although no significant between group differences on cortisol levels were observed at 6 months postpartum, mothers who had participated in CBSM had lower cortisol levels than women in the UC group at 18 months postpartum. The CBSM intervention seemed to be particularly effective in reducing cortisol levels among mothers with a past history of depression (see Fig. 2B). Interestingly, infants of mothers in the CBSM and LRC groups also had lower cortisol levels than infants of mothers in the UC group at 6 months postpartum (see Fig. 2A). Qualitative data indicated that women found the prenatal CBSM intervention helpful in managing their stress, improving their confidence as a mother, and improving the quality of the relationship with their baby (Muñoz et al., 2007). These promising findings provided preliminary support for the effectiveness of CBSM interventions in regulating cortisol levels among low-income mothers and their infants up to 18 months postpartum and highlighted the need to reach a wider range of low-income, ethnic minority women who experience stress and anxiety during pregnancy.

SMART Moms Project

As an extension of our previous work, we launched the Stress Management and Relaxation Training for Moms (SMART Moms) Project, which was a randomized controlled trial that examined whether a prenatal CBSM intervention, in comparison to a control group, was effective in reducing perceived stress levels and regulating salivary cortisol patterns among low-income pregnant women. The SMART Moms Project specifically focused on examining CBSM intervention effects on women's overall cortisol output (area under the curve), CAR, and their diurnal slope given that previous findings have found altered cortisol patterns for all three of these cortisol indices (i.e., higher cortisol concentrations, higher/flat CAR, flatter diurnal cortisol slope) to be associated with distinct adverse pre- and postnatal health outcomes for both mothers and their infants (Hodyl et al., 2017; Kivlighan et al., 2008; Scheyer & Urizar, 2016). During their first trimester of pregnancy (mean gestational age = 10 ± 4 weeks), 100 low-income women (71% Latina, 18% African-American; 76% with an annual family income <$20,000) were screened for pregnancy anxiety given that it has been associated with adverse stress-related health outcomes in infants more so than other psychological states during pregnancy (Dunkel Schetter & Tanner, 2012). Women were then randomized (stratified by low vs high pregnancy anxiety; a total score

TABLE 1 Cognitive behavioral stress management (CBSM) intervention topics and aims.[a]

Intervention topic	Aims
Class 1: Stress Awareness	• Increase awareness of different stress responses • Learn diaphragmatic breathing
Class 2: Thought Awareness	• Recognize how thoughts affect our emotions • Learn mindfulness
Class 3: Thought Replacement	• Learn how to balance overly negative/positive self-talk • Learn guided imagery
Class 4: Coping Awareness	• Define different types of coping • Learn progressive muscle relaxation
Class 5: Matching Coping	• Identify steps to matching coping strategies to stressors • Learn "letting go of tension"
Class 6: Social Support	• Recognize how our interactions with others affect our emotions • Learn "supportive imagery"
Class 7: Communication	• Define different types of communication • Learn "therapeutic touch"
Class 8: Review of Coping & Relaxation Skills	• Identify coping and relaxation skills learned • Use of skills learned in postpartum

[a]The SMART Moms (Stress Management and Relaxation Training for Moms) Course: A Prenatal Stress Management Program *(Urizar & Kofman, 2012). Based on the B-SMART Program for women with breast cancer (Antoni, 2003) and the Mothers and Babies Depression Prevention Course for pregnant women (Muñoz et al., 2007).*
From Urizar, G. G., Jr., Yim, I. S., Rodriguez, A., & Schetter, C. D. (2019). The SMART Moms Program: A randomized trial of the impact of stress management on perceived stress and cortisol in low-income pregnant women. *Psychoneuroendocrinology, 104,* 174–184. https://doi.org/10.1016/j.psyneuen.2019.02.022, with permission.

<17 or ≥ 17 on the Prenatal Anxiety Scale) to either an 8-week CBSM group ($n = 55$) or a control group ($n = 45$). Women randomized to CBSM were taught coping and relaxation skills for stress management in a group-based setting with other pregnant women in the clinic where they received prenatal care services (see Table 1). In contrast, women randomized to the control group received print-based prenatal health information that was mailed to them once per week for 8 weeks. Women in both groups self-reported their stress (using the Perceived Stress Scale) and provided salivary cortisol samples at baseline (first trimester of pregnancy), at postintervention (second trimester of pregnancy), during their third trimester of pregnancy (30–32 weeks gestation), and at 3 months postpartum.

Results showed that women receiving CBSM had lower perceived stress levels during pregnancy and early postpartum compared to women in the control group (Urizar, Yim, Rodriguez, & Schetter, 2019). Findings also revealed that the CBSM intervention was particularly effective in regulating cortisol patterns among African-American women and those with high pregnancy anxiety (Urizar, Yim, et al., 2019). Specifically, African-American women receiving CBSM had a lower CAR at 3 months postpartum compared to African-American women in the control group (see Fig. 3A). These results are particularly salient given that African-American women have been shown to be at risk for altered CAR patterns due to chronic stress related to racial discrimination, which in turn, have been associated with greater risk for preterm birth in this population (Adam et al., 2015; Glynn, Schetter, Chicz-DeMet, Hobel, & Sandman, 2007). Furthermore, results demonstrated that women with high pregnancy anxiety who received CBSM had a steeper decrease in diurnal cortisol levels at 3 months postpartum compared to those in the control group (see Fig. 3B). Women with high pregnancy anxiety have been shown to be at greater risk for having flatter diurnal cortisol patterns at 3 months postpartum, compared to women with low pregnancy anxiety, potentially due to greater concerns or fears about the health of their baby, childbirth, and the responsibilities associated with being a mother (Urizar, Yim, et al., 2019). Therefore, the coping and relaxation strategies taught in the CBSM intervention may have been significant in balancing these fears and regulating cortisol levels in this at-risk group. Collectively, these results have important implications for future prenatal CBSM intervention studies.

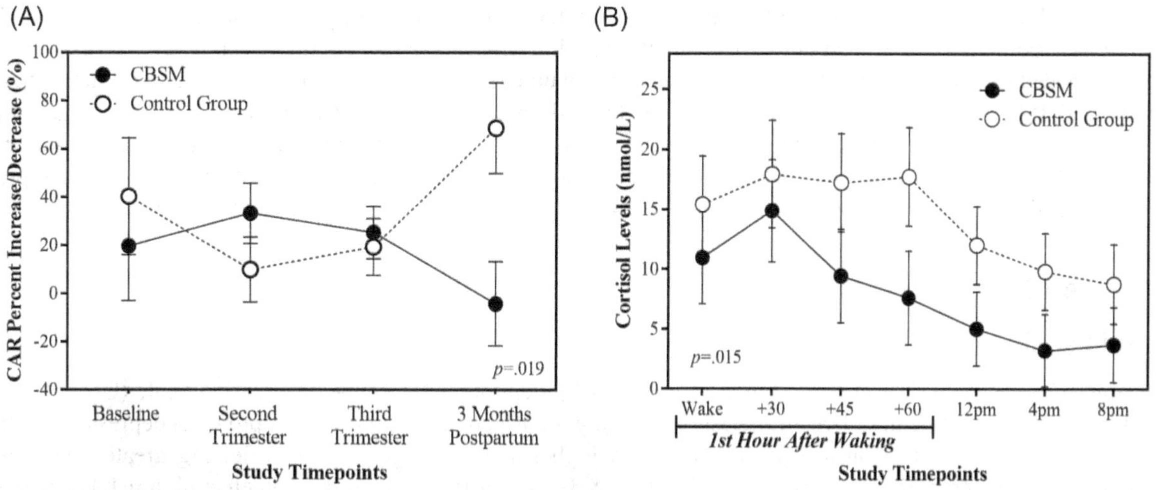

FIG. 3 Intervention effects (CBSM vs control group) on cortisol patterns (CAR and diurnal cortisol) among African-American women and those with high pregnancy anxiety. (A) African-American women's cortisol awakening response (CAR) (CBSM vs control group). (B) High pregnancy anxiety women's diurnal cortisol patterns at 3 months postpartum (CBSM vs control group). *(Adapted from Urizar, G. G., Jr., Yim, I. S., Rodriguez, A., & Schetter, C. D. (2019). The SMART Moms Program: A randomized trial of the impact of stress management on perceived stress and cortisol in low-income pregnant women.* Psychoneuroendocrinology, 104, *174–184. https://doi.org/10.1016/j.psyneuen.2019.02.022, with permission.)*

Implications and future directions

Our results to date support the efficacy of prenatal CBSM interventions in reducing perceived stress and regulating cortisol patterns among low-income women. Although CBSM did not appear to affect prenatal cortisol levels, it was effective in influencing cortisol after childbirth up to 18 months postpartum, particularly among ethnic minority women (African-American and Latina) and those at high risk for depression and pregnancy anxiety (Urizar & Muñoz, 2011; Urizar, Yim, et al., 2019). These delayed intervention effects may be due to normative increases in cortisol during pregnancy overshadowing any CBSM effects during this time period. This may also be due to women's continued practice of the coping and relaxation skills they learned during pregnancy leading to CBSM effects by the early postpartum period. However, additional studies are needed to understand the mechanisms by which these interventions work. In particular, process evaluation studies can help to identify CBSM mechanisms by assessing implementation fidelity, or the degree to which an intervention was delivered as intended, to better understand why an intervention was effective (Breitenstein et al., 2010). As an example, we recently conducted a process evaluation of the prenatal CBSM intervention for the SMART Moms Project to assess whether certain implementation fidelity factors (intervention delivery, receipt, and enactment) were associated with observed changes in perceived stress and cortisol. Results showed that women receiving CBSM reported feeling more confident in using the relaxation and coping skills they had learned at postintervention, with instructor adherence to the CBSM intervention content (delivery) and women's use of the intervention skills they learned in real-world scenarios (enactment) being the factors most associated with improvements in intervention outcomes (Urizar, Caliboso, Gearhart, Yim, & Dunkel Schetter, 2019). Such results help identify the mechanisms by which CBSM interventions help to improve pre- and postnatal health outcomes and enhance the reliability and validity of such interventions.

The CBSM intervention developed as part of the SMART Moms Project could help to address existing barriers (limited resources and training) to providing stress management interventions for English- and Spanish-speaking low-income women by providing a manualized prenatal program that can be delivered by prenatal health care professionals (e.g., social workers, OBGYNs). The SMART Moms Project was successfully implemented in a centrally located, community prenatal clinic where women were already receiving their prenatal care in collaboration with the clinic's director. These types of community partnerships are consistent with the USPSTF's recommendations to implement a collaborative care model that helps prenatal care providers more readily link their patients to available stress management services and builds trust for low-income ethnic minority women to participate in studies requiring collection of biological samples to help evaluate CBSM interventions (Kuhlman, Urizar, Robles, Yim, & Dunkel Schetter, 2019; Siu et al., 2016). Such collaborative studies are critical as future studies are needed to evaluate CBSM intervention effects on other stress and immune system biomarkers (e.g., alpha amylase, cytokines, telomere length) associated with short- (e.g., birth complications) and long-term (e.g., psychiatric illness, cognitive development) health outcomes among low-income mothers and their infants (Urizar &

Muñoz, 2021). Preliminary evidence points to the first trimester of pregnancy being the ideal time to implement CBSM interventions to potentially have a greater impact on fetal development (Sandman, Glynn, & Davis, 2013). However, more studies are needed to test different delivery modalities to reach underserved pregnant women who may be unable to attend group-based CBSM classes during this time period due to work or other obligations (e.g., one-on-one case management, asynchronous vs synchronous video conferencing, smart phone applications) and examine how to culturally tailor these interventions for different at-risk populations. In sum, our findings support the need to translate evidence-based prenatal CBSM interventions into clinical practice to optimize the health of low-income mothers and their infants.

Practice and procedures

Depression risk

Depression risk status was determined using the Center for Epidemiologic Studies—Depression Scale (CES-D; Radloff, 1977). The CES-D is a 20-item, self-report measure assessing for affective and somatic symptoms of depression during the past week. Each item is rated on a 4-point scale, with higher scores (range = 0–60) reflecting greater symptoms of depression. A cut-off score of 16 was used in the Mothers & Babies Project to identify women at high risk for depression. The CES-D demonstrated good to excellent internal consistencies ($\alpha = 0.85–0.92$).

Perceived stress

Perceived stress was assessed in the SMART Moms Project using the 14-item version of the Perceived Stress Scale (PSS-14; Cohen & Williamson, 1988). The PSS-14 measures the degree to which situations in one's life over the past month were appraised as stressful, with higher scores (range = 0–56) reflecting higher stress levels. The PSS-14 demonstrated good internal consistency (α *range* = 0.74–0.75). For the Relaxation Day Pilot Project and the Mothers & Babies Project, women completed a one-item stress question where they were asked to rate their stress level on a 100-point visual analog scale (0 = *Not at all stressed*, 100 = *Extremely stressed*).

Pregnancy anxiety

Pregnancy anxiety was assessed using the 10-item Prenatal Anxiety Scale (PAS; Rini, Dunkel-Schetter, Wadhwa, & Sandman, 1999). The PAS assesses the frequency with which pregnant women worry or feel concerned about their health, their baby's health, labor and delivery, and caring for a baby. Each item is rated on a 4-point scale (range = 10–40), with higher scores reflecting greater levels of pregnancy anxiety. A median cut-off score of ≥ 17 was used in the SMART Moms Project to identify women with low vs high pregnancy anxiety. The PAS demonstrated good internal consistency in our study ($\alpha = 0.74$).

Salivary cortisol collection procedures

To provide saliva samples, women were given a saliva collection kit that included cryovials with saliva collection aids and instructions to collect saliva at home via passive drool several times on one collection day (e.g., immediately upon waking, 30, 45, and 60 min after waking, 12 p.m., 4 p.m., and 8 p.m. for SMART Moms Project). They were instructed to collect saliva on a weekday, making sure to abstain from behaviors known to interfere with the cortisol assay (e.g., brushing teeth) or to change cortisol concentrations (e.g., eating, exercising) for at least 60 min before sample collection. They were also provided a collection log to record their collection times and report any behaviors they believed may have interfered with saliva sampling. Adherence rates across all of our studies were high (85%–99%).

Salivary cortisol outcomes

After being thawed for biochemical analysis, saliva samples were centrifuged and salivary cortisol was analyzed using a time-resolved immunoassay with fluorescence detection. Intra- and interassay variability were both under 10%. Three cortisol summary scores were computed: the overall amount of cortisol secreted throughout the day (area under the curve, AUC), cortisol awakening response (CAR), and diurnal cortisol slope. Larger AUC numbers represent a greater amount of cortisol produced throughout the day. CAR measures the acute rise in cortisol typically seen after waking in the morning. Larger numbers represent a greater percent increase in cortisol from waking to 30 min after waking. Diurnal cortisol slope was estimated by calculating the change in cortisol from waking to 8 p.m. while taking into account individual collection time, with larger numbers representing a flatter diurnal cortisol slope (i.e., smaller decrease in cortisol levels across the day).

Mini-dictionary of terms

- ***Area under the curve (AUC).*** Represents the amount of cortisol produced throughout the day. Assessed by collecting several saliva samples throughout the day taking into account total collection time.
- ***Cognitive behavioral stress management.*** A psychosocial intervention (typically delivered in a group format) that consists of teaching active cognitive behavioral coping strategies (e.g., modifying maladaptive thoughts, improving assertiveness communication, reducing social isolation) and relaxation techniques (e.g., diaphragmatic breathing, progressive muscle relaxation) among stressed individuals to enhance their psychological functioning.
- ***Cortisol.*** A hormone released by the body's sympathetic nervous system in response to perceived threats in our environment. Cortisol levels rise during pregnancy to aid in fetal organ development. Altered patterns of cortisol have been associated with adverse maternal and infant health outcomes.
- ***Cortisol awakening response (CAR).*** Measures the acute rise in cortisol typically seen after waking in the morning and is hypothesized to prepare the body for the upcoming demands of the day. Larger numbers represent a greater percent increase in cortisol from waking to 30 min after waking.
- ***Diurnal cortisol.*** Represents the decline in cortisol levels typically observed from morning to bedtime within 1 day. Cortisol levels are typically replenished while sleeping at night with higher levels observed the next morning. Individuals that do not show this steep decline in cortisol throughout the day may be at greater risk for certain health problems (e.g., cardiovascular disease).
- ***Implementation fidelity.*** Degree to which a behavioral intervention was delivered as intended. Helps to better understand why an intervention was effective by assessing intervention delivery (what content was covered), intervention receipt (ensuring that participants received and understood content), and intervention enactment (participants' ability to perform intervention-related skills in relevant, real-life settings).
- ***Pregnancy anxiety.*** Persistent worries or fears experienced by some women during pregnancy related to their prenatal health, childbirth and delivery, the well-being of their baby, and future parenting.

Key facts

Key facts about cortisol during pregnancy

- Cortisol is a stress hormone that has been associated with a heightened risk for stress-related health complications during pregnancy and the postpartum period.
- Cortisol is the end-product of the hypothalamus-pituitary-adrenal (HPA) axis, is released by the adrenal glands, and is considered one of the main markers of the biological stress response.
- A dramatic change in cortisol secretion occurs during pregnancy with levels rising throughout pregnancy, playing an important role in fetal organ development, and returning to prepregnancy levels after childbirth.
- Higher concentrations of maternal cortisol during pregnancy have been related to adverse birth outcomes, including increased rates of preterm births, a higher need for infant resuscitation assistance at birth, and infant brain cell damage.
- A higher cortisol awakening response (CAR) during pregnancy has been associated with shorter gestational periods and a blunted or flat CAR with increased risk for postpartum depression.
- Flatter diurnal cortisol slopes (i.e., smaller decrease in cortisol levels across the day) during pregnancy have been associated with increased anxiety and impaired sleep in mothers and with low birthweight in infants.
- Altered cortisol patterns are particularly prevalent among low-income and ethnic minority mothers, which in turn have been associated with a greater risk for preterm birth.

Applications to other areas

In this chapter, we have reviewed our work examining the efficacy of prenatal CBSM interventions in regulating cortisol patterns among chronically stressed, low-income mothers and their infants. These CBSM interventions (i.e., SMART Moms Project, Mothers & Babies Project) are unique in that they provide manualized prenatal programs that have been effectively delivered in centrally located community-based hospitals to reach underserved low-income and ethnic minority mothers. These programs have the potential for wide applicability and can be readily delivered by prenatal health care professionals or paraprofessionals (e.g., social workers, OBGYNs, midwives, promotoras), as less than 25% of women report actively engaging in stress management during pregnancy and only 14% of pregnant women with mental health symptoms are referred for treatment by their health care provider (Birdee et al., 2014; Cox et al., 2016). Our results demonstrate that these CBSM interventions are effective in regulating perceived stress and cortisol patterns among low-income ethnic minority (African-American, Latina) mothers and those at risk for depression and pregnancy anxiety, which are the most common mental health symptoms experienced during pregnancy and postpartum (Dunkel Schetter & Tanner, 2012; Marcus & Heringhausen, 2009). They also show potential in regulating postpartum cortisol levels among infants of mothers who participate in CBSM during pregnancy. Therefore, these community-based CBSM interventions are needed to reach underserved women during pregnancy and provide them with the necessary skills and knowledge to facilitate stress management. Such interventions can also be extended to their partners and other family members involved in caring for their newborn (e.g., grandparents, siblings) to promote family-based stress management. Given the wide applicability of this work, future studies are needed to examine the short- and long-term effects of CBSM interventions in helping to reduce stress-related health complications and health disparities (e.g., postpartum depression, preterm births, impaired child cognitive development) in this population.

Summary points

- Prenatal stress can lead to significant long-term health problems for mothers and their infants.
- At greatest health risk are low-income women, who are prone to experiencing several psychosocial stressors and birth complications.
- One biological mechanism that has been identified in heightening a woman's risk for health complications during pregnancy and the postpartum period is the stress hormone cortisol.
- However, the best methods for regulating cortisol, in order to optimize maternal and infant health outcomes, have received little attention.
- Cognitive behavioral stress management (CBSM) is prominently recommended by the U.S. Preventive Services Task Force (USPSTF) to help prevent stress-related disorders during pregnancy given its efficacy in nonpregnant populations.
- However, few studies have systematically evaluated the effectiveness of CBSM interventions in regulating stress and cortisol levels among low-income pregnant women that are in most need of these programs.
- This chapter summarized our work examining the efficacy of prenatal CBSM interventions in regulating cortisol patterns among chronically stressed, low-income mothers and their infants.
- Our results show that prenatal CBSM interventions are effective in regulating perceived stress and cortisol patterns among low-income ethnic minority (African-American and Latina) mothers and those at risk for depression and pregnancy anxiety.
- Our results also show that CBSM is effective in regulating postpartum cortisol levels among infants of mothers who participate in CBSM during pregnancy.
- Our findings support the need to translate evidence-based prenatal CBSM interventions into clinical practice to optimize the health of low-income mothers and their infants.

References

Accortt, E. E., Freeman, M. P., & Allen, J. J. (2008). Women and major depressive disorder: Clinical perspectives on causal pathways. *Journal of Women's Health, 17*(10), 1583–1590. https://doi.org/10.1089/jwh.2007.0592.

Adam, E. K., Heissel, J. A., Zeiders, K. H., Richeson, J. A., Ross, E. C., Ehrlich, K. B., … Eccles, J. S. (2015). Developmental histories of perceived racial discrimination and diurnal cortisol profiles in adulthood: A 20-year prospective study. *Psychoneuroendocrinology, 62*, 279–291. https://doi.org/10.1016/j.psyneuen.2015.08.018.

Alhasanat, D., & Giurgescu, C. (2017). Acculturation and postpartum depressive symptoms among Hispanic women in the United States: Systematic review. *American Journal of Maternal Child Nursing, 42*(1), 21–28. https://doi.org/10.1097/NMC.0000000000000298.

American College of Obstetricians & Gynecologists (ACOG) Committee on Health Care for Underserved Women. (2006). Psychosocial risk factors: Perinatal screening and intervention. *Obstetrics & Gynecology, 108*(2), 469–477. https://doi.org/10.1097/00006250-200608000-00046.

Antoni, M. H. (2003). *Stress management intervention for women with breast Cancer*. Washington, DC: American Psychological Association.

Antoni, M. H. (2013). Psychosocial intervention effects on adaptation, disease course and biobehavioral processes in cancer. *Brain, Behavior, and Immunity, 30*(Suppl), S88–S98. https://doi.org/10.1016/j.bbi.2012.05.009.

Antoni, M. H., Cruess, D. G., Klimas, N., Carrico, A. W., Maher, K., Cruess, S., … Schneiderman, N. (2005). Increases in a marker of immune system reconstitution are predated by decreases in 24-h urinary cortisol output and depressed mood during a 10-week stress management intervention in symptomatic HIV-infected men. *Journal of Psychosomatic Research*, *58*(1), 3–13. https://doi.org/10.1016/j.jpsychores.2004. 05.010.

Beam, A. L., Fried, I., Palmer, N., Agniel, D., Brat, G., Fox, K., … Armstrong, J. (2020). Estimates of healthcare spending for preterm and low-birthweight infants in a commercially insured population: 2008–2016. *Journal of Perinatology*, *40*(7), 1091–1099. https://doi.org/10.1038/s41372-020-0635-z.

Beck, A. T., & Dozois, D. J. (2011). Cognitive therapy: Current status and future directions. *Annual Review of Medicine*, *62*, 397–409. https://doi.org/10.1146/annurev-med-052209-100032.

Birdee, G. S., Kemper, K. J., Rothman, R., & Gardiner, P. (2014). Use of complementary and alternative medicine during pregnancy and the postpartum period: an analysis of the National Health Interview Survey. *Journal of Women's Health*, *23*(10), 824–829. https://doi.org/10.1089/jwh.2013.4568.

Blumenshine, P., Egerter, S., Barclay, C. J., Cubbin, C., & Braveman, P. A. (2010). Socioeconomic disparities in adverse birth outcomes: A systematic review. *American Journal of Preventive Medicine*, *39*(3), 263–272. https://doi.org/10.1016/j.amepre.2010.05.012.

Breitenstein, S. M., Gross, D., Garvey, C. A., Hill, C., Fogg, L., & Resnick, B. (2010). Implementation fidelity in community-based interventions. *Research in Nursing & Health*, *33*(2), 164–173. https://doi.org/10.1002/nur.20373.

Buss, C., Entringer, S., Reyes, J. F., Chicz-DeMet, A., Sandman, C. A., Waffarn, F., & Wadhwa, P. D. (2009). The maternal cortisol awakening response in human pregnancy is associated with the length of gestation. *American Journal of Obstetrics and Gynecology*, *201*(4), 398e1–398e8. https://doi.org/10.1016/j.ajog.2009.06.063.

Clarke, T. C., Black, L. I., Stussman, B. J., Barnes, P. M., & Nahin, R. L. (2015). Trends in the use of complementary health approaches among adults: United States, 2002–2012. *National Health Statistics Reports*, *79*, 1–16.

Cohen, S., & Williamson, G. (1988). Perceived stress in a probability sample of the United States. In S. Spacapan, & S. Oskamp (Eds.), *The social psychology of health: Claremont symposium on applied social psychology*. Newbury Park, CA: Sage.

Cox, E. Q., Sowa, N. A., Meltzer-Brody, S. E., & Gaynes, B. N. (2016). The perinatal depression treatment cascade: Baby steps toward improving outcomes. *Journal of Clinical Psychiatry*, *77*(9), 1189–1200. https://doi.org/10.4088/JCP.15r10174.

Duan, H., Yuan, Y., Zhang, L., Qin, S., Zhang, K., Buchanan, T. W., & Wu, J. (2013). Chronic stress exposure decreases the cortisol awakening response in healthy young men. *Stress: The International Journal on the Biology of Stress*, *16*(6), 630–637. https://doi.org/10.3109/10253890.2013.840579.

Dunkel Schetter, C., Niles, A. N., Guardino, C. M., Khaled, M., & Kramer, M. S. (2016). Demographic, medical, and psychosocial predictors of pregnancy anxiety. *Paediatric and Perinatal Epidemiology*, *30*(5), 421–429. https://doi.org/10.1111/ppe.12300.

Dunkel Schetter, C., & Tanner, L. (2012). Anxiety, depression and stress in pregnancy: Implications for mothers, children, research, and practice. *Current Opinion in Psychiatry*, *25*(2), 141–148. https://doi.org/10.1097/YCO.0b013e3283503680.

Esch, T., & Stefano, G. B. (2010). The neurobiology of stress management. *Neuroendocrinology Letters*, *31*(1), 19–39.

Esperat, C., Feng, D., Zhang, Y., & Owen, D. (2007). Health behaviors of low-income pregnant minority women. *Western Journal of Nursing Research*, *29* (3), 284–300. https://doi.org/10.1177/0193945906295532.

Field, T., Diego, M., & Hernandez-Reif, M. (2006). Prenatal depression effects on the fetus and newborn: A review. *Infant Behavior & Development*, *29*(3), 445–455. https://doi.org/10.1016/j.infbeh.2006.03.003.

Fox, M., Entringer, S., Buss, C., DeHaene, J., & Wadhwa, P. D. (2015). Intergenerational transmission of the effects of acculturation on health in Hispanic Americans: A fetal programming perspective. *American Journal of Public Health*, *105*(Suppl 3), S409–S423. https://doi.org/10.2105/AJPH.2015.302571.

Freeman, M. P. (2019). Perinatal depression: Recommendations for prevention and the challenges of implementation. *JAMA*, *321*(6), 550–552. https://doi.org/10.1001/jama.2018.21247.

Fries, E., Dettenborn, L., & Kirschbaum, C. (2009). The cortisol awakening response (CAR): Facts and future directions. *International Journal of Psychophysiology*, *72*(1), 67–73. https://doi.org/10.1016/j.ijpsycho.2008.03.014.

Gennaro, S., O'Connor, C., McKay, E. A., Gibeau, A., Aviles, M., Hoying, J., & Melnyk, B. M. (2020). Perinatal anxiety and depression in minority women. *American Journal of Maternal Child Nursing*, *45*(3), 138–144. https://doi.org/10.1097/NMC.0000000000000611.

Glynn, L. M., Schetter, C. D., Chicz-DeMet, A., Hobel, C. J., & Sandman, C. A. (2007). Ethnic differences in adrenocorticotropic hormone, cortisol and corticotropin-releasing hormone during pregnancy. *Peptides*, *28*(6), 1155–1161. https://doi.org/10.1016/j.peptides.2007.04.005.

Goodman, J. H., & Tyer-Viola, L. (2010). Detection, treatment, and referral of perinatal depression and anxiety by obstetrical providers. *Journal of Women's Health*, *19*(3), 477–490. https://doi.org/10.1089/jwh.2008.1352.

Gunnar, M., & Quevedo, K. (2007). The neurobiology of stress and development. *Annual Review of Psychology*, *58*, 145–173. https://doi.org/10.1146/annurev.psych.58.110405.085605.

Hodyl, N. A., Aboustate, N., Bianco-Miotto, T., Roberts, C. T., Clifton, V. L., & Stark, M. J. (2017). Child neurodevelopmental outcomes following preterm and term birth: What can the placenta tell us? *Placenta*, *57*, 79–86. https://doi.org/10.1016/j.placenta. 2017.06.009.

Hofmann, S. G., Asnaani, A., Vonk, I. J., Sawyer, A. T., & Fang, A. (2012). The efficacy of cognitive behavioral therapy: A review of meta-analyses. *Cognitive Therapy and Research*, *36*(5), 427–440. https://doi.org/10.1007/s10608-012-9476-1.

Institute of Medicine. (2006). *Preterm birth: Causes, consequences, and prevention*. Washington, DC: National Academies Press.

Keenan, K., Gunthorpe, D., & Grace, D. (2007). Parsing the relations between SES and stress reactivity: Examining individual differences in neonatal stress response. *Infant Behavior & Development*, *30*, 134–145. https://doi.org/10.1016/j.infbeh.2006.08.001.

Kivlighan, K. T., DiPietro, J. A., Costigan, K. A., & Laudenslager, M. L. (2008). Diurnal rhythm of cortisol during late pregnancy: Associations with maternal psychological well-being and fetal growth. *Psychoneuroendocrinology*, *33*(9), 1225–1235. https://doi.org/10.1016/j.psyneuen.2008.06.008.

Kuhlman, K. R., Urizar, G. G., Jr., Robles, T. F., Yim, I. S., & Dunkel Schetter, C. (2019). Testing plausible biopsychosocial models in diverse community samples: Common pitfalls and strategies. *Psychoneuroendocrinology*, *107*, 191–200. https://doi.org/10.1016/j.psyneuen.2019.05.017.

Lara-Cinisomo, S., Clark, C. T., & Wood, J. (2018). Increasing diagnosis and treatment of perinatal depression in Latinas and African American women: Addressing stigma is not enough. *Women's Health Issues*, *28*(3), 201–204. https://doi.org/10.1016/j.whi.2018.01.003.

Larrabee Sonderlund, A., Schoenthaler, A., & Thilsing, T. (2021). The association between maternal experiences of interpersonal discrimination and adverse birth outcomes: A systematic review of the evidence. *International Journal of Environmental Research and Public Health*, *18*(4), 1465. https://doi.org/10.3390/ijerph18041465.

Lefmann, T., Combs-Orme, T., & Orme, J. G. (2017). Examining the inter-correlated effects of low income, life stress, and race on birth outcomes: A representative state study. *Social Work in Health Care*, *56*(6), 450–469. https://doi.org/10.1080/00981389.2017.1316811.

Lobel, M., Cannella, D. L., Graham, J. E., DeVincent, C., Schneider, J., & Meyer, B. A. (2008). Pregnancy-specific stress, prenatal health behaviors, and birth outcomes. *Health Psychology*, *27*(5), 604–615. https://doi.org/10.1037/a0013242.

Luecken, L. J., Lin, B., Coburn, S. S., MacKinnon, D. P., Gonzales, N. A., & Crnic, K. A. (2013). Prenatal stress, partner support, and infant cortisol reactivity in low-income Mexican American families. *Psychoneuroendocrinology*, *38*(12), 3092–3101. https://doi.org/10.1016/j.psyneuen.2013.09.006.

Marcus, S. M., & Heringhausen, J. E. (2009). Depression in childbearing women: When depression complicates pregnancy. *Primary Care*, *36*(1), 151–164. https://doi.org/10.1016/j.pop.2008.10.011.

Martin, J. A., Hamilton, B. E., & Osterman, M. (2020). Births in the United States, 2019. *NCHS Data Brief*, *387*, 1–8.

Mastorakos, G., & Ilias, I. (2003). Maternal and fetal hypothalamic-pituitary-adrenal axes during pregnancy and postpartum. *Annals of the New York Academy of Sciences*, *997*, 136–149. https://doi.org/10.1196/annals.1290.016.

Muñoz, R. F., Le, H.-N., Ippen, C. G., Diaz, M. A., Urizar, G. G., Jr., Soto, J., … Lieberman, A. F. (2007). Prevention of postpartum depression in low-income women: Development of the Mamás y Bebés/Mothers and Babies course. *Cognitive and Behavioral Practice*, *14*(1), 70–83. https://doi.org/10.1016/j.cbpra.2006.04.021.

Price, S. K., Corder-Mabe, J., & Austin, K. (2012). Perinatal depression screening and intervention: Enhancing health provider involvement. *Journal of Women's Health*, *21*(4), 447–455. https://doi.org/10.1089/jwh.2011.3172.

Radloff, L. S. (1977). The CES-D Scale: A self-report depression scale for research in the general population. *Applied Psychological Measurement*, *1*(3), 385–401. https://doi.org/10.1177/014662167700100306.

Reynolds, R. M., Labad, J., Buss, C., Ghaemmaghami, P., & Räikkönen, K. (2013). Transmitting biological effects of stress in utero: Implications for mother and offspring. *Psychoneuroendocrinology*, *38*(9), 1843–1849. https://doi.org/10.1016/j.psyneuen.2013.05.018.

Richter, J., Bittner, A., Petrowski, K., Junge-Hoffmeister, J., Bergmann, S., Joraschky, P., & Weidner, K. (2012). Effects of an early intervention on perceived stress and diurnal cortisol in pregnant women with elevated stress, anxiety, and depressive symptomatology. *Journal of Psychosomatic Obstetrics and Gynaecology*, *33*(4), 162–170. https://doi.org/10.3109/0167482X.2012.729111.

Rini, C. K., Dunkel-Schetter, C., Wadhwa, P. D., & Sandman, C. A. (1999). Psychological adaptation and birth outcomes: The role of personal resources, stress, and sociocultural context in pregnancy. *Health Psychology*, *18*(4), 333–345. https://doi.org/10.1037//0278-6133.18.4.333.

Sandman, C. A., Glynn, L. M., & Davis, E. P. (2013). Is there a viability-vulnerability tradeoff? Sex differences in fetal programming. *Journal of Psychosomatic Research*, *75*(4), 327–335. https://doi.org/10.1016/j.jpsychores.2013.07.009.

Scheyer, K., & Urizar, G. (2016). Altered stress patterns and increased risk for postpartum depression among low-income pregnant women. *Archives of Women's Mental Health*, *19*(2), 317–328. https://doi.org/10.1007/s00737-015-0563-7.

Siu, A. L., US Preventive Services Task Force (USPSTF), Bibbins-Domingo, K., Grossman, D. C., Baumann, L. C., Davidson, K. W., … Pignone, M. P. (2016). Screening for depression in adults: US Preventive Services Task Force Recommendation Statement. *JAMA*, *315*(4), 380–387. https://doi.org/10.1001/jama.2015.18392.

Sohr-Preston, S. L., & Scaramella, L. V. (2006). Implications of timing of maternal depressive symptoms for early cognitive and language development. *Clinical Child and Family Psychology Review*, *9*(1), 65–83. https://doi.org/10.1007/s10567-006-0004-2.

Talge, N. M., Neal, C., & Glover, V. (2007). Antenatal maternal stress and long-term effects on child neurodevelopment: How and why? *Journal of Child Psychology and Psychiatry*, *48*, 245–261. https://doi.org/10.1111/j.1469-7610.2006.01714.x.

Urizar, G. G., Jr., Caliboso, M., Gearhart, C., Yim, I. S., & Dunkel Schetter, C. (2019). Process evaluation of a stress management program for low-income pregnant women: The SMART Moms/Mamás LÍSTAS Project. *Health Education & Behavior*, *46*(6), 930–941. https://doi.org/10.1177/1090198119860559.

Urizar, G. G., Jr., Milazzo, M., Le, H. N., Delucchi, K., Sotelo, R., & Muñoz, R. F. (2004). Impact of stress reduction instructions on stress and cortisol levels during pregnancy. *Biological Psychology*, *67*(3), 275–282. https://doi.org/10.1016/j.biopsycho.2003.11.001.

Urizar, G. G., Jr., & Muñoz, R. F. (2011). Impact of a prenatal cognitive-behavioral stress management intervention on salivary cortisol levels in low-income mothers and their infants. *Psychoneuroendocrinology*, *36*(10), 1480–1494. https://doi.org/10.1016/j.psyneuen.2011.04.002.

Urizar, G. G., Jr., & Muñoz, R. F. (2021). Role of maternal depression on child development: A prospective analysis from pregnancy to early childhood. *Child Psychiatry and Human Development*. https://doi.org/10.1007/s10578-021-01138-1. Advance online publication.

Urizar, G. G., Jr., Yim, I. S., Rodriguez, A., & Schetter, C. D. (2019). The SMART Moms Program: A randomized trial of the impact of stress management on perceived stress and cortisol in low-income pregnant women. *Psychoneuroendocrinology*, *104*, 174–184. https://doi.org/10.1016/j.psyneuen.2019.02.022.

Urizar, G., & Kofman, Y. (2012). *The SMART moms (stress management and relaxation training for moms) course: A prenatal stress management program.*

Van den Bergh, B., van den Heuvel, M. I., Lahti, M., Braeken, M., de Rooij, S. R., Entringer, S., … Schwab, M. (2020). Prenatal developmental origins of behavior and mental health: The influence of maternal stress in pregnancy. *Neuroscience and Biobehavioral Reviews*, *117*, 26–64. https://doi.org/10.1016/j.neubiorev.2017.07.003.

Wesley, Y. (2006). Reduce stress: A stress reduction project for pregnant Black women. *Journal of Cultural Diversity*, *13*(4), 208–216.

Wilhelm, I., Born, J., Kudielka, B. M., Schlotz, W., & Wüst, S. (2007). Is the cortisol awakening rise a response to awakening? *Psychoneuroendocrinology*, *32*(4), 358–366. https://doi.org/10.1016/j.psyneuen.2007.01.008.

Yim, I. S., Tanner Stapleton, L. R., Guardino, C. M., Hahn-Holbrook, J., & Dunkel Schetter, C. (2015). Biological and psychosocial predictors of postpartum depression: Systematic review and call for integration. *Annual Review of Clinical Psychology*, *11*, 99–137. https://doi.org/10.1146/annurev-clinpsy-101414-020426.

Zijlmans, M. A., Riksen-Walraven, J. M., & de Weerth, C. (2015). Associations between maternal prenatal cortisol concentrations and child outcomes: A systematic review. *Neuroscience and Biobehavioral Reviews*, *53*, 1–24. https://doi.org/10.1016/j.neubiorev.2015.02.015.

Section B

Children and adolescents

Chapter 7

Adolescent acne excorie: Features and the use of cognitive behavioral therapy

Arezoo Moradi Tavallaei[a], Maryam Saeidi Sardabi[b], and Mahdi Razmara Ferezghi[c]
[a]*Faculty of Educational Science and Psychology, Department of Psychology, Ferdowsi University of Mashhad, Mashhad, Iran,* [b]*Faculty of Educational Science and Psychology, Department of Psychology, Shahid Chamran University of Ahvaz, Ahvaz, Iran,* [c]*Department of Dermatology, Imamreza Hospital of Mashhad, Mashhad University of Medical Sciences, Mashhad, Iran*

Abbreviations

AA alopecia areata
AE acne excorie
AV acne vulgaris
CBT cognitive-behavioral therapy
DSM diagnostic and statistical manual of mental disorders
ED excoriation disorder
SPS skin picking scale

Introduction

Acne and its features

Acne vulgaris (AV) is one of the chronic inflammatory dermatological diseases (Picardo, Eichenfield, & Tan, 2017), common in both young people and adults. The prevalence of AV in adolescents is stated as approximately 85% by some (Acer et al., 2019); based on a 2018 report, acne is described as more prevalent in those who are going through puberty, with a prevalence of almost 95% of this population (Skroza et al., 2018). This disease is characterized by comedones, papules, pustules, and nodules and it can appear on all parts of the body, including face, neck, back, etc. (Salman, Kurt, Topcuoglu, & Demircay, 2016).

Regarding the pathogenesis of acne, researchers focus on four factors: increase and alteration of sebum production, alteration of follicular skin keratinization that leads to comedones, colonization by *Propionibacterium acnes*, and inflammatory processes that involve innate and acquired immunity (Febyan, 2020). As mentioned, acne commonly develops during adolescence, a vulnerable time of significant hormonal changes, development, and emotional instability (Angus, 2021).

Psychological aspects of acne excorie

Psychodermatological disorders are generally categorized into two major groups: (1) cutaneous associations of psychiatric disorders, including acne excorie (AE) (also called excoriated acne), neurotic excoriations, skin manipulation associated with delusions of parasitosis, and unreasonable concerns about skin body image or odors that are inconsistent with dermatologist evaluation and (2) the large group of dermatologic disorders, such as rosacea, psoriasis, atopic dermatitis (AD), alopecia areata (AA), and acne, which are exacerbated by psychosocial stress and are comorbid with a wide range of psychiatric disorders, including major depressive disorder (Fried, Gupta, & Gupta, 2005). Another definition of psychodermatological disorders is skin diseases having a psychosomatic or behavioral aspect (Sharma, 2008).

As noted before, AE is a psychodermatological disorder in the first group having a major psychiatric dimension. Also, because of the special situation in the teenage years and the high prevalence of using social media during adolescence, individual appearance assumes more importance, and AE, with its effect on physical appearance, can affect teenagers

Handbook of Lifespan Cognitive Behavioral Therapy. https://doi.org/10.1016/B978-0-323-85757-4.00038-9

in both physical and psychological aspects because of its chronic and superficial features (Conforti et al., 2021). AE is diagnosed by acne lesions and scars from skin-picking behavior. Common psychological disorders play a major role in AE; as a common type of neurotic excoriation, it is mostly associated with mood and anxiety disorders (Fried et al., 2005). Kayiran, Karadag, and Jaffarany described psychological aspects of acne and mentioned that it can cause psychological disturbances in patients, such as anxiety, depression, anger, weak body perception, loss of self-confidence, social isolation, and activity restrictions (Mikhael, Hamed, Mansour, & Abdelrahman, 2019). Quality of life is one of the important aspects that is affected, being lowered in some ways by acne (Durai & Nair, 2015).

Based on a body of research, depression is another psychological factor that is mostly increased during acne flareups (Golchai, 2010; Uhlenhake, Yentzer, & Feldman, 2010). Acne patients experience depression at a rate about two to three times higher than the general population (Ohayon, 2007). Mild to moderate acne can lead to depression and suicidal ideation (Gupta, Gupta, & Watteel, 1998).

Anxiety is another important psychological factor that can affect both physical and psychological aspects of an individual's life. This factor, as with depression, has a relationship with AV and increases in those suffering from the disease (Samuels, Rosenthal, Lin, Chaudhari, & Natsuaki, 2020).

Because of the changes in appearance that acne can cause in individuals, another parameter that may change is self-esteem. A study by Delgard, Geiler, and Holm has shown that both male and female adolescents with acne have lower self-esteem than a control group (Dalgard, Gieler, Holm, Bjertness, & Hauser, 2008). Also, another study showed that if the acne lesions improve, self-esteem increases (Kouris et al., 2018).

Research conducted by Lukaviciute et al. (2020) reported that, compared to other skin diseases that involve the face, such as rosacea and perioral dermatitis, people suffering from acne experience more depression, anxiety, and suicidal ideation (Lukaviciute et al., 2020).

Stress is an important factor that can affect both physical and psychological aspects of an individual's life. Also, it can cause and increase acne lesions. Stress can be both an inciting and an exacerbating factor of acne and an outcome of the condition itself (Barr, Soutor, & Franklin, 2021).

Adolescents who experience more stress and suffer from acne pay more attention to their face and spend more time in front of the mirror, trying to remove acne nodules or other types of lesions in order to have smoother skin; these skin-picking behaviors, which are more common in stressed people, can lead to the type of AV called AE. One research study shows that AE is a common kind of AV in stressed patients (Aslan Kayiran, Karadag, & Jafferany, 2020).

Comorbidity with other psychiatric disorders has been reported in AV patients. Obsessive compulsive disorder, psychosis, and body dysmorphic syndrome are more commonly reported in patients having AV than in other persons. These patients spent about 2h a day concerned with their face (Marron et al., 2020). About 2% of patients with the AE lesion subtype have obsessive compulsive disorder and their time spent looking in the mirror is greater and is compulsive (Sarkar et al., 2016). Dermatologists consider AE to be a difficult condition to treat. Many dermatologists suspect that therapy may fail because of the patient inability to perform the necessary task for improvement: do not scratch (Burns, Breathnach, Cox, & Griffiths, 2010). Fig. 1 shows the defective cycle of comorbidity of psychological disturbance with AE and the importance of considering these aspects in treatment of AE.

Because of the occurrence of this disorder during puberty and the psychological vulnerability of adolescents, all of the problems mentioned can be experienced by teens more than other groups of patients, so in order to prevent future trauma, taking care of these patients in both the physical and psychological aspects is necessary.

Why is CBT useful in acne treatment?

Because of the aforementioned psychological dimensions of AE, this skin disease is put into the category of disorders called psychodermatological disorders. Also in the field of psychological disorder categorization, excoriation disorder (ED), compulsive skin picking, or AE when seen in patients with acne, has been categorized in the diagnostic and statistical manual of mental disorders, 5th edition (DSM-5) as an impulse control disorder that is grouped with obsessive-compulsive and related disorders (Toro-Martínez, 2014).

Acne and psychological disturbances cause an unhealthy cycle that may result in less ability to cope with the disease and loss of trust in treatments or compliance with treatments. This worsens the skin condition and increases the symptoms of depression and anxiety (Angus, 2021).

So, considering the psychological aspects of the disease can help patients suffering from acne in their personal and social life. Mind-body interactions should be considered when planning treatments for specific psychodermatological disorders (Sharma, 2008). As mentioned before, psychosocial stressors have been reported to precipitate cutaneous symptoms in 33%–98% of acne patients (Zhang et al., 2019), and many psychological disturbances are caused by acne, so using psychotherapy methods is helpful in reducing both acne lesions and psychological problems. Stress-alleviating methods

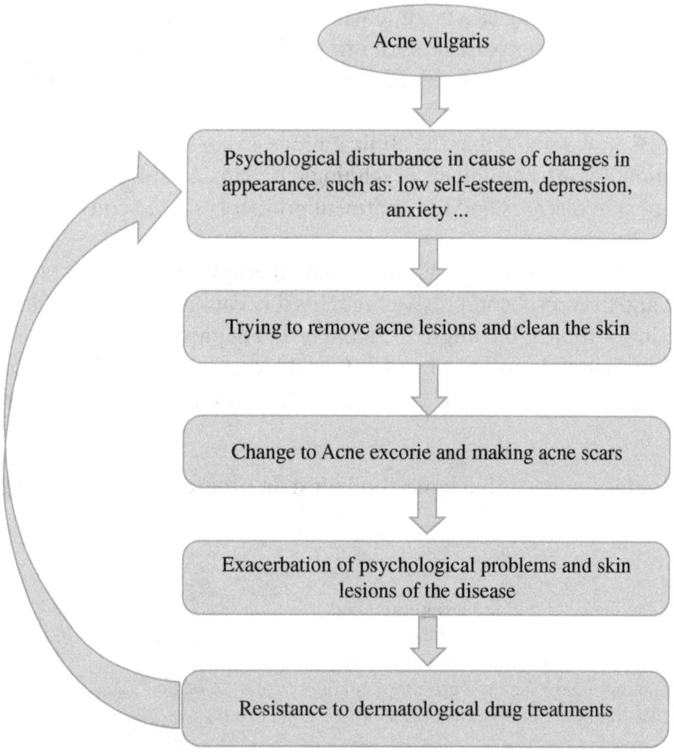

FIG. 1 The role of psychological problems in the process of AE. Disabling dermatological treatments and worsening prognosis of disease.

through cognitive-behavioral therapies (CBTs) should be considered (Griesemer, 1978; Shenefelt, 2018). CBT methods help to change dysfunctional thought patterns (cognitive) or actions (behavioral) that conflict with dermatological therapies (Miltenberger, Fuqua, & Woods, 1998). There is a paucity of data in behavioral treatment for AE in the literature and most of the previous studies are single-subject design or uncontrolled case series in this field. Thus the data are both lacking and contradictory.

Rosenbaum and Ayllon (1981) used habit reversal treatment for neurodermatitis. They taught awareness of the scratching behavior, developed a competing response practice of isometric exercise using fist clenching, and did symbolic rehearsal. After 6-month follow-ups, three patients out of four had significantly improved and healed (Schuck, Keijsers, & Rinck, 2011).

The picking component of AE was reported to be alleviated with CBT coupled with biofeedback, minocycline, and sertraline in a young adult woman (Davoudi, Omidi, Sehat, & Sepehrmanesh, 2017). This multimodal treatment approach is reported to be beneficial in resistant cases (Chatzikonstantinou et al., 2019).

The effectiveness of CBT has been studied in 25 patients with chronic skin picking and compared with a control group. Results show a greater decrease in skin picking after a 3-month follow up compared to the control group (Rosenbaum & Ayllon, 1981).

More recently, in a randomized trial involving 17 patients with skin picking, the severity of skin injury was measured at a 2-month follow-up assessment after four sessions of CBT. The treatment group showed significant improvement in severity of the lesions. It was suggested that CBT, even in a brief form, constitutes an adequate treatment option for skin-picking behavior (Griesemer, 1978).

In this chapter we introduce CBT as a useful method for better and faster treatment results in patients with AE in which they experience psychological disturbance alongside the dermatological disease.

Practice and procedures

CBT, as pioneered by Dr. Aaron Beck, is a short-term, goal-oriented psychotherapy treatment that helps patients understand and address dysfunctional thoughts or behaviors that lead to skin damage or psychological distress. By becoming aware of the negative and often unrealistic thoughts that awaken negative emotions, patients were able to begin engaging in healthier thinking patterns and consequently psychological impairments decreased and skin condition became better (Barr et al., 2021).

There are limited articles addressing CBT as a treatment of AE; many of these practices use CBT or other psychological treatment to heal psychological disturbances patients suffer from. But in this chapter we want to introduce CBT as a therapy that can help clinicians to treat people faster and reduce the chance of recurrence by controlling psychological aspects of this dermatological condition.

In a systemic review in 2019, it was noted that the clinical process of a series of skin diseases is the result of a complex and maybe reciprocal interaction among biological, psychological, and social factors, all of which can play a role in the occurrence and development of skin diseases and the treatment protocols should consider this multifactorial characteristic of skin diseases (Zhang et al., 2019).

In a clinical trial in Iran, researchers used CBT as an adjuvant therapy besides pharmacotherapy to reduce skin picking behaviors in patients with AE and to control compulsive thoughts that cause patients to pick at their acne comedones and so worsen their disease features. In addition, CBT improved anxiety and depression symptoms, compared to usual care. In this article, 32 adults with AE were randomly assigned to CBT or the control group. Both groups received similar standard medication. CBT was run in eight sessions during 2 months in the CBT group (Table 1). The self-reported skin picking

TABLE 1 Overview of the content of cognitive behavioral lifestyle program applied during 8 sessions in patients with acne excorie, Mashhad, Iran.

Session	Objectives
1	- Introduction of the therapist and the group members - Agreement on attendance, commitment, homework assignments, privacy, and buddy system - Providing information about the program and target - Providing information about the concepts of techniques - Understanding the rationale of cognitive behavior treatment
2	- Explaining the link between thoughts, feelings, and behaviors - Training about habitual behaviors - Making aware of the genesis of the skin picking habit and suggestion of new coping mechanisms to resolve conflicts more appropriately - Reveal and deal with feelings of frustration or anxiety - Describe the importance of cognition, beliefs, and attitudes, and their role in personality and behavior, constructive and positive beliefs
3	- Completing a thought record and setting realistic goals - Relaxation training to help reduce tension - Learning how to change undesirable beliefs with new ones - Illustrating attempts to control the urge to pick
4	- Understanding the concept of adaptation - Competing response training in which patients learn to perform movements that are incompatible with habitual behaviors - Exercises aimed at illustrating the limitations of control when aimed at private events, such as trying not to think of something, such as chocolate cake
5	- Contingency management consisting of contingent reinforcement for applying competing responses - As homework: They were suggested to practice with their friend about social challenges in interpersonal relationships and strategies for improving them with positive and constructive beliefs
6	- Generalization training to learn to control habitual behaviors in different everyday situations - An explanation about admission technique in cognitive therapy - Focus on changing the psychological function of the urge to pick from something threatening to just another verbal event. This involved cognitive exercises (repeating a thought over and over until it loses its meaning, contact with the present moment or mindfulness exercises, and "self as context" work)
7	- Evaluation of the lifestyle program and participants' progress - Students' challenges were discussed and they were questioned about their problems in implementing these techniques
8	- A comprehensive survey of the sessions and implications for everyday life - The final sessions involved reviewing the material from the earlier sessions, values work, and discussions regarding maintenance of decreases in picking - Giving confidence among participants to have healthy and successful personalities in the community and family

scale (SPS), clinical severity rating, and Beck anxiety and depression inventories were completed at the baseline and after a 2-month follow-up. Results of this clinical trial showed the clinical severity score of acne lesions in the CBT group was significantly reduced in pre- and posttreatment surveys, but in the control group there were no significant changes in the severity scale. Depression and anxiety, which had no significant differences at baseline in the two groups, were significantly reduced only in patients who were treated by CBT. Also, regarding treatment duration, patients who were treated by CBT in addition to routine dermatological treatment showed healing of lesions faster than those in the control group (healing period was about 5.70 ± 2.02 and 8.61 ± 2.94 months in the CBT and control groups, respectively) (Mashayekhi Goyonlo, Sardabi, Tavalaei, Khoshnevisan, & Razmara, 2020).

Jung and Hwang (2012) mentioned in their study that patients' bad habits can influence the outcome of acne treatment, and also that misunderstanding of skin care behaviors may lead to poor compliance. Thus it is crucial to understand patients' cognition and their common behavioral patterns of skin care in the treatment of acne. Based on this research, CBT can be combined with pharmacological treatment of acne. This treatment consists of several steps: determining patients' faulty cognition and behavior, educating them in realistic and appropriate targets, correct cognition and behavioral errors during CBT sessions, and getting feedback on results from patients throughout the treatment sessions. Then they will be treated for maintenance and preventing of recurrence. Patients reported more satisfaction with this combination of treatment methods. The combination can achieve a positive result in all types of acne and can reduce concerns about the recurrence or aggravation of acne (Jung & Hwang, 2012).

Research at the University of South Florida (2017) used a combination form of CBT and hypnosis to heal people with dermatological conditions. Results showed that psychocutaneous disorders such as pruritus (itching), acne, AE, psoriasis, trichotillomania, and urticaria were improved with this therapy (Shenefelt, 2018).

In a study in 2013, scientists said the objective physical severity of acne does not correlate with the patients' perception of severity and the impact on their life. Therefore it is crucial to focus on the patients' experience and perspective. Those patients who report a more harmful effect on their quality of life are in need of a personalized cognitive behavioral approach to supplement conventional medical therapy (Affleck, Chouliara, & Kielty, 2014).

Based on Angus (2021), the underlying principle is that the pathogenesis of acne is in relationship with psychological or psychiatric dimensions. Patients should be referred to psychological and psychiatric services if appropriate. Psychological, psychiatric, and medical interventions can occur in conjunction and it is probably optimal if they occur in parallel (Angus, 2021).

Conclusion

AE is a combined condition of psychology and dermatology, and it must be confronted with a combination of psychological and dermatological healing approaches. Excoriation is an important part of these diseases, which is a disruptive behavior and caused by vulnerable thoughts of trying to remove acne lesions. CBT with its special focus on disruptive thoughts and behaviors can be an effective treatment for patients who suffer from this illness.

Mini-dictionary of terms

Psychodermatological Disorder: A kind of dermatological diseases related to psychological problems.

Acne Excorie: A kind of dermatology diseases which have a bold psychological dimension.

Excoriation: A disruptive behavior in acne patients which caused by vulnerable thoughts.

Key facts

Some dermatological diseases have a psychiatric comorbidity or suffer from psychological problems.

Attention to psychological problems and comorbidity in the psychodermatological diseases are important and effective in treatment of patients.

Applications to other areas

The field of psychodermatology exists because of the importance of psychological aspects in most of the dermatological diseases. Psychological vulnerabilities may play a role in pathogenesis of the illness or be caused by the bad skin condition. For better healing and decreasing the risk of recurrence, most of the psychodermatological diseases, such as AA, psoriasis, AD, etc., can be treated by a combination of CBT and dermatological treatments.

Summary points

- Acne excorie has an important psychological component.
- Skin picking behaviors occur because of the thoughts that cause patients to clean skin of acne lesions.
- Using CBT, we can improve disruptive thoughts and behaviors, and then reduce the scars of acne.
- Using CBT, we can also improve people's compliance with the treatments and accelerate the healing process.
- CBT can improve psychological impairments caused by acne, such as depression, anxiety, and low self-esteem, and improve the quality of life of people who suffer from acne.
- CBT can also improve psychological impairments that caused the acne.

References

Acer, E., Altınoz, A. E., Erdogan, H. K., Agaoglu, E., Canakcı, N. Y., & Saracoglu, Z. N. (2019). Anxiety sensitivity and quality of life in patients with acne. *Annals of Dermatology*, *31*(4), 372–377.

Affleck, A., Chouliara, Z., & Kielty, S. (2014). Cognitive behavioral therapy for acne and other chronic skin disorders. *International Journal of Dermatology*, *53*(4), e246–e247.

Angus, J. (2021). Acne and psychodermatology. In *Psychodermatology in clinical practice* (pp. 267–282). Springer.

Aslan Kayiran, M., Karadag, A. S., & Jafferany, M. (2020). Psychodermatology of acne: Dermatologist's guide to inner side of acne and management approach. *Dermatologic Therapy*, *33*(6), e14150.

Barr, K. L., Soutor, C. A., & Franklin, J. M. M. (2021). Mind-body therapies. In *Integrative dermatology* (pp. 165–191). Springer.

Burns, T., Breathnach, S., Cox, N., & Griffiths, C. (2010). Psychodermatology and psychocutaneous disease. *Vol. 4. Rooks textbook of dermatology*. Oxford: Wiley-Blackwell.

Chatzikonstantinou, F., Miskedaki, A., Antoniou, C., Chatzikonstantinou, M., Chrousos, G., & Darviri, C. (2019). A novel cognitive stress management technique for acne vulgaris: A short report of a pilot experimental study. *International Journal of Dermatology*, *58*(2), 218–220.

Conforti, C., Giuffrida, R., Fadda, S., Fai, A., Romita, P., Zalaudek, I., & Dianzani, C. (2021). Topical dermocosmetics and acne vulgaris. *Dermatologic Therapy*, *34*(1), e14436.

Dalgard, F., Gieler, U., Holm, J.Ø., Bjertness, E., & Hauser, S. (2008). Self-esteem and body satisfaction among late adolescents with acne: Results from a population survey. *Journal of the American Academy of Dermatology*, *59*(5), 746–751.

Davoudi, M., Omidi, A., Sehat, M., & Sepehrmanesh, Z. (2017). The effects of acceptance and commitment therapy on man smokers' comorbid depression and anxiety symptoms and smoking cessation: A randomized controlled trial. *Addiction and Health*, *9*(3), 129.

Durai, P. C. T., & Nair, D. G. (2015). Acne vulgaris and quality of life among young adults in South India. *Indian Journal of Dermatology*, *60*(1), 33.

Febyan, W. K. (2020). Acne vulgaris in adults: A brief review on diagnosis and management. *International Journal of Research and Review*, *7*(5), 246–252.

Fried, R. G., Gupta, M. A., & Gupta, A. K. (2005). Depression and skin disease. *Dermatologic Clinics*, *23*(4), 657–664.

Golchai, J. (2010). Comparison of anxiety and depression in patients with acne vulgaris and healthy individuals. *Indian Journal of Dermatology*, *55* (4), 352.

Griesemer, R. D. (1978). *Emotionally triggered disease in a dermatologic practice. 8* (pp. 49–56). NJ: SLACK Incorporated Thorofare.

Gupta, M., Gupta, A., & Watteel, G. (1998). Perceived deprivation of social touch in psoriasis is associated with greater psychologic morbidity: An index of the stigma experience in dermatologic disorders. *Cutis*, *61*(6), 339–342.

Jung, J., & Hwang, E. J. (2012). Do patients with acne need cognitive behavioral therapy? An analysis of patient knowledge and behavior. *International Journal of Dermatology*, *51*(11), 1319–1324.

Kouris, A., Platsidaki, E., Christodoulou, C., Efstathiou, V., Markantoni, V., Armyra, K., … Kontochristopoulos, G. (2018). Patients' self-esteem before and after chemical peeling procedure. *Journal of Cosmetic and Laser Therapy*, *20*(4), 220–222.

Lukaviciute, L., Ganceviciene, R., Navickas, P., Navickas, A., Grigaitiene, J., & Zouboulis, C. C. (2020). Anxiety, depression, and suicidal ideation amongst patients with facial dermatoses (acne, rosacea, perioral dermatitis, and folliculitis) in Lithuania. *Dermatology*, *236*(4), 314–322.

Marron, S., Miranda-Sivelo, A., Tomas-Aragones, L., Rodriguez-Cerdeira, C., Tribo-Boixaro, M., Garcia-Bustinduy, M., … Diaz-Díaz, R. (2020). Body dysmorphic disorder in patients with acne: A multicentre study. *Journal of the European Academy of Dermatology and Venereology*, *34*(2), 370–376.

Mashayekhi Goyonlo, V., Sardabi, M. S., Tavalaei, A. M., Khoshnevisan, Z., & Razmara, M. (2020). Cognitive behavioral therapy as an adjuvant therapy in acne excoriee: A randomized controlled clinical trial. *The Journal of Dermatological Treatment*, 1–7. https://doi.org/10.1080/09546634.2020.1776207.

Mikhael, N. W., Hamed, A. M., Mansour, A. I., & Abdelrahman, E. S. (2019). Serum levels of brain-derived neurotrophic factor in patients with acne vulgaris. *Journal of Cosmetic Dermatology*, *18*(6), 1998–2003.

Miltenberger, R. G., Fuqua, R. W., & Woods, D. W. (1998). Applying behavior analysis to clinical problems: Review and analysis of habit reversal. *Journal of Applied Behavior Analysis*, *31*(3), 447–469.

Ohayon, M. M. (2007). Epidemiology of depression and its treatment in the general population. *Journal of Psychiatric Research*, *41*(3–4), 207–213.

Picardo, M., Eichenfield, L. F., & Tan, J. (2017). Acne and rosacea. *Dermatology and Therapy*, *7*(1), 43–52.

Rosenbaum, M. S., & Ayllon, T. (1981). The behavioral treatment of neurodermatitis through habit-reversal. *Behaviour Research and Therapy*, *19*(4), 313–318.

Salman, A., Kurt, E., Topcuoglu, V., & Demircay, Z. (2016). Social anxiety and quality of life in vitiligo and acne patients with facial involvement: A cross-sectional controlled study. *American Journal of Clinical Dermatology*, *17*(3), 305–311.

Samuels, D. V., Rosenthal, R., Lin, R., Chaudhari, S., & Natsuaki, M. N. (2020). Acne vulgaris and risk of depression and anxiety: A meta-analytic review. *Journal of the American Academy of Dermatology*, *83*(2), 532–541.

Sarkar, S., Patra, P., Mridha, K., Ghosh, S. K., Mukhopadhyay, A., & Thakurta, R. G. (2016). Personality disorders and its association with anxiety and depression among patients of severe acne: A cross-sectional study from eastern India. *Indian Journal of Psychiatry*, *58*(4), 378.

Schuck, K., Keijsers, G. P., & Rinck, M. (2011). The effects of brief cognitive-behaviour therapy for pathological skin picking: A randomized comparison to wait-list control. *Behaviour Research and Therapy*, *49*(1), 11–17.

Sharma, H. (2008). Psychogenic excoriation responding to fluoxetine: A case report. *Journal of the Indian Medical Association*, *106*(4), 245. 262–245, 262.

Shenefelt, P. D. (2018). Mindfulness-based cognitive hypnotherapy and skin disorders. *American Journal of Clinical Hypnosis*, *61*(1), 34–44.

Skroza, N., Tolino, E., Mambrin, A., Zuber, S., Balduzzi, V., Marchesiello, A., … Potenza, C. (2018). Adult acne versus adolescent acne: A retrospective study of 1,167 patients. *The Journal of Clinical and Aesthetic Dermatology*, *11*(1), 21.

Toro-Martínez, E. (2014). DSM-5: OCD and related disorders. *Vertex*, *25*(113), 63–67.

Uhlenhake, E., Yentzer, B. A., & Feldman, S. R. (2010). Acne vulgaris and depression: A retrospective examination. *Journal of Cosmetic Dermatology*, *9*(1), 59–63.

Zhang, X.-J., Wang, A.-P., Shi, T.-Y., Zhang, J., Xu, H., Wang, D.-Q., & Feng, L. (2019). The psychosocial adaptation of patients with skin disease: A scoping review. *BMC Public Health*, *19*(1), 1–15.

Chapter 8

Adolescent social anxiety: Bridging cognitive and social behavioral perspectives through a developmental lens

Anne C. Miers[a] and Carrie Masia Warner[b]
[a]*Unit Developmental and Educational Psychology, Institute of Psychology, Leiden University, Leiden, The Netherlands,* [b]*Psychology, Montclair State University, Montclair, NJ, United States*

Abbreviations

CBT cognitive behavioral therapy
PFC prefrontal cortex
SAD social anxiety disorder

Introduction

Social anxiety disorder (SAD), characterized by marked distress and avoidance in social situations and fear of negative evaluation (American Psychiatric Association, 2013), is the "prototypical adolescent disorder" (Rapee et al., 2019). Typical onset is in early to mid-teenage years with a prevalence of around 10% in Western samples (Merikangas et al., 2010; Rapee & Spence, 2004). SAD often persists from adolescence into adulthood (Pine, Cohen, Gurley, Brook, & Ma, 1998), leading to severe consequences in many aspects of life (Leigh & Clark, 2018). Social anxiety in adolescents manifests along a severity continuum (Rapee & Spence, 2004), with about 20% experiencing subclinical symptoms (Knappe, Sasagawa, & Creswell, 2015) like anticipatory worry prior to social events that dissipates once in the situation. These symptoms, although not usually diagnosable, can still be distressing and place youth at risk for worsening impairment. Unfortunately, social anxiety, at all levels, is largely unrecognized and untreated (Zarger & Rich, 2016).

In 2015, Blöte, Miers, Heyne, and Westenberg published a chapter about adolescent social anxiety in the school environment in which research on cognitive and social behavioral components were comprehensively reviewed. Hence, in the current chapter, we focus on empirical research published after 2015. The goals are threefold: (1) to review research pertaining to the relation between adolescent social anxiety and three factors: cognitive processes, social context and behavior, and neurodevelopmental findings regarding fear extinction, emphasizing how they interact; (2) to propose a new maintenance model of adolescent social anxiety integrating the three factors; and (3) to discuss the implications of the maintenance model for intervention. We start with a brief review of the available models of adolescent SAD and the extent to which they focus on both cognitive and social processes to provide a theoretical context.

Current models of adolescent social anxiety

The dominant model describing how social anxiety is maintained, despite regular exposure to benign social experiences, is the cognitive model of Clark and Wells (1995) (Clark, 2001). This framework was developed for adults and recently applied to adolescents by Leigh and Clark (2018). The cognitive model (see Fig. 1) describes how social situations activate a socially anxious person's fundamental beliefs about their social competencies and the importance of adhering to social rules. These assumptions lead to perceived social danger and threat, which in turn, activates a chain of cognitive, affective, and behavioral responses. Leigh and Clark (2018) concluded this model is supported by adolescent research demonstrating a positive association between social anxiety and cognitive processes such as, negative interpretation bias, self-focused attention, negative self-imagery, and pre- and postevent processing. However, as the majority of studies are correlational,

Handbook of Lifespan Cognitive Behavioral Therapy. https://doi.org/10.1016/B978-0-323-85757-4.00042-0

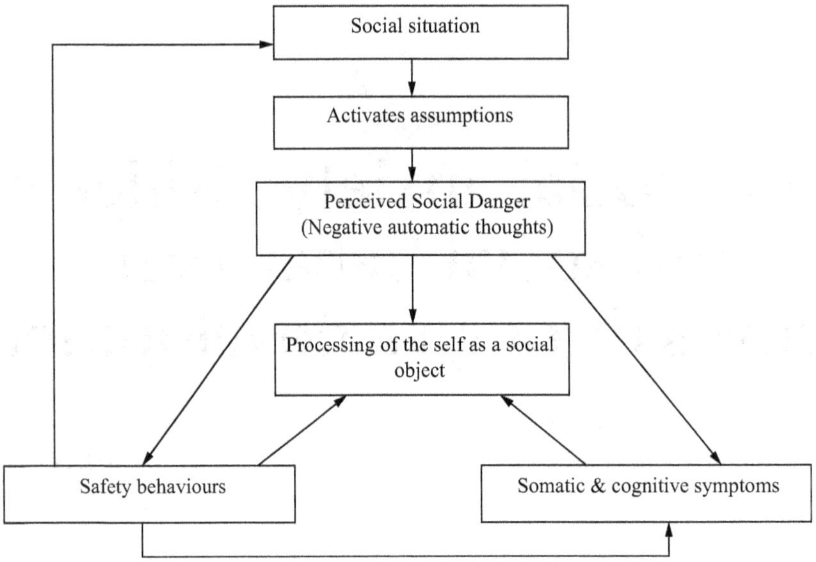

FIG. 1 The cognitive model of social anxiety disorder (Clark, 2001). This figure shows a model of thoughts and feelings that occur when a person with social anxiety disorder enters a social situation. *(From Clark, D. M. (2001). A cognitive perspective on social phobia. In W. R. Crozier, & L. E. Alden (Eds.),* International handbook of social anxiety: Concepts, research and interventions relating to the self and shyness (pp. 405–430). *New York: John Wiley & Sons Ltd, with permission.)*

they do not demonstrate empirical support for these processes *maintaining* social anxiety. Moreover, a simple downward extension of the adult model neglects the dynamic developmental influences of adolescence like the social behavioral influences of peer relationships and social skills.

Recently, Rapee et al. (2019) published a model of the influence of adolescent characteristics on the development of social-emotional disorders, including SAD. While this model delineates the *onset* of adolescent disorders, we include it because of its developmentally informed and comprehensive approach. The potential adolescent-specific factors are pubertal changes and accompanying hormones, self-concept, peer relationships, sleep changes, and emotional experience. The model draws attention to the broader social context and points to the importance of considering the interplay between factors and how they maintain social anxiety levels.

Cognitive processes and adolescent social anxiety

The cognitive variables we emphasize follow those reviewed by Leigh and Clark (2018) and are supported by two recent studies. First, in a longitudinal study investigating mediating pathways between childhood shyness and elevated social anxiety in adolescence, negative preevent processing (i.e., negative performance expectations) explained the association between shyness and later social anxiety (Blöte, Miers, Van den Bos, & Westenberg, 2019). While this social *self*-perception acted as a mediator, the negative social cognition regarding interpretation of *other* people's behavior (i.e., interpretation bias, defined as interpreting ambiguous social cues and signals from others as negative or threatening to the self) did not mediate the shyness-social anxiety link. Although this is not a direct test of maintenance, the authors suggested that negative self-perception in social situations, relative to overly negative interpretations of others, likely have a greater role in early stages of problematic social anxiety in late childhood and early adolescence.

Second, Chiu, Clark, and Leigh published a prospective study of cognitive processes and social anxiety in 11–14-year-olds. After accounting for baseline social anxiety levels, three cognitive processes uniquely predicted social anxiety 4–6 months later; in increasing strength, respectively: negative automatic thoughts, frequency of using safety behaviors, and self-focused attention. Postevent rumination was suggested to be indirectly associated with social anxiety via its influence on other cognitive processes such as negative automatic thoughts (Chiu, Clark, & Leigh, 2021). Indeed, Clark (2001) suggested that postevent rumination may connect social situations together through its negative influence on anticipatory processing about new social events.

Experimental studies of cognitive processes

Two experimental studies were published investigating the effect of manipulating mental self-imagery (Leigh, Chiu, & Clark, 2020), self-focused attention, and safety behaviors (Leigh, Chiu, & Clark, 2021) on social anxiety in nonclinical adolescents. In the first, high socially anxious adolescents engaged in two conversations with a naïve partner (Leigh et al., 2020). In one conversation, they were instructed to hold negative self-images in mind of feeling anxious during a social situation, and in the second, a benign image of feeling relaxed. Outcomes of anxiety and performance during conversations were collected from three sources: self-report, conversation partner, and independent assessors. The authors found a significant imagery effect on self-reported anxiety, whereby during the negative imagery condition, adolescents reported feeling and looking significantly more anxious relative to the benign condition. This study supports that forming a negative mental self-image increases state anxiety during social situations (Leigh et al., 2020).

In the second study, high and low socially anxious adolescents engaged in two conversations with naive partners with differing instructions (Leigh, Chiu, et al., 2021). One condition ("with") included instructions to use a specified set of safety behaviors and to focus on the self. The other ("without") did not require safety behaviors ("act naturally") and told participants to focus attention externally on the other person. Results revealed that participants in the "with" condition reported feeling significantly more anxious and believed to appear more anxious compared to the "without" condition. This higher self-reported anxiety was corroborated by reports from conversation partners. These detrimental effects occurred for all adolescents regardless of social anxiety level. This study provides evidence for safety behaviors and self-focused attention maintaining anxiety.

Social context and behavior and social anxiety

In the 2015 chapter, Blöte, Miers, Heyne, and Westenberg concluded that socially anxious adolescents, relative to nonanxious teenagers, have challenging social interactions with same-age peers and are less liked and accepted. These negative social outcomes are evident in first impression situations and within emerging and established relationships. Youth with social anxiety are seen as "different," a status that is not easily changed. The social behavioral factors implicated in the link between social anxiety and victimization or rejection were overt anxiety, poor social performance or being socially unskilled, and physical attractiveness. We recognize the ongoing debate regarding whether socially anxious adolescents' awkward social performance are reflective of true social skills deficits or are present but inhibited by anxiety. The goal of this section is not to resolve this debate, but rather to illustrate how the social context and cognitive processes are inextricably intertwined in relation to the presentation and maintenance of adolescent social anxiety.

Social behaviors and performance

It has been suggested that peer judgments of socially anxious adolescents are influenced by a general negative halo effect that involves an immediate negative impression (Blöte, Miers, Heyne, & Westenberg, 2015). This reaction may be driven by a specific characteristic, socially unskilled behavior, observed anxiety, physical unattractiveness, lack of expressiveness, or looking unfriendly. It is as yet unclear whether specific characteristics explain negative peer judgments. The two experimental studies conducted by Leigh and colleagues previously described, illustrate that awkward social performance may elicit short-term negative responses from others. In the first study, the manipulation of negative self-imagery in high socially anxious adolescents resulted in the spontaneous use of avoidant safety behaviors such as avoiding eye contact and not asking questions (Leigh et al., 2020). These avoidant behaviors significantly mediated the relation between negative imagery and ratings of poorer conversation quality. Thus, the action of engaging in such behaviors explained the association between participants' negative mental imagery and lower performance ratings from conversation partners and independent observers.

In the second study, low and high anxious adolescents were rated as more anxious and less likeable by interaction partners when actively engaging in self-focused attention and employing impression management safety behaviors (Leigh, Chiu, et al., 2021). In addition, conversations were rated more critically by uninvolved, independent observers. Hence, these studies show that certain social behaviors (e.g., avoiding eye contact, remaining quiet) do not come across well to others. As these studies did not track the behaviors of conversation partners, we do not know whether they showed negative social responses (e.g., disengaging from the interaction, showing disinterest, or ignoring). However, it is plausible that awkward social performance elicits short-term negative peer reactions during daily social interactions. These two studies demonstrate how social behavior and cognitive processes are intertwined as shown in Fig. 2: actively engaging in negative cognitive processes typical of social anxiety

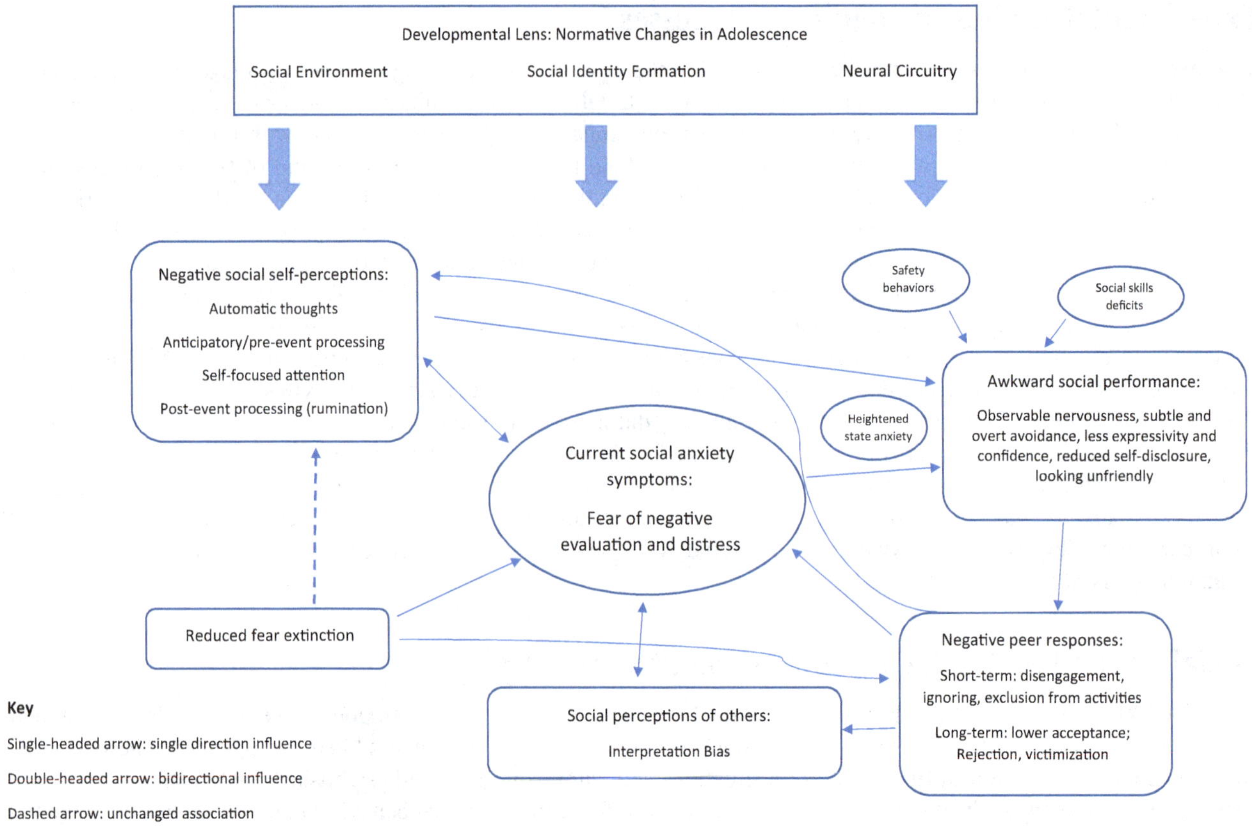

FIG. 2 Maintenance model of adolescent social anxiety. This figure shows pathways through which social anxiety is maintained in adolescents. These pathways include different factors including negative thoughts about the self, awkward behavior in social situations, and undesirable responses from other adolescents.

(e.g., self-focused attention) initiates awkward social performance, which in turn is responded to negatively by others, serving to reinforce anxiety in a feedback loop.

Peer interactions

Much research has documented long-term negative peer experiences, that is, difficulties in social functioning that are relatively stable such as rejection, low peer acceptance, and victimization. It is possible that short-term negative responses, over time, are associated with rejection, low peer acceptance, and victimization incidents, which in turn further make the social environment a threatening place and strengthen anxiety symptoms and their maintaining processes. Chiu, Clark, and Leigh (2021) published a systematic review and meta-analysis of prospective associations between peer interactions and social anxiety in adolescents aged 11–19 years. Peer rejection, or the degree to which an adolescent is disliked by peers, had a bidirectional relation with social anxiety. The strongest bidirectional association was between social anxiety and peer victimization, defined as repeated experiences of covert or overt aggressive acts from peers (Chiu et al., 2021). Peer acceptance, or the degree to which peers like to engage in social activities with the individual, was not associated with social anxiety.

How might victimization and rejection maintain social anxiety levels? Chiu et al. (2021) suggest a pathway via cognitive processes. That is, victimization and rejection may serve to confirm the socially anxious adolescents' negative self-perceptions about being unlikeable and socially unacceptable. In turn, this will make them feel anxious about future social situations and increase their fears of rejection from others. Similarly, and as proposed in the current model (Fig. 2), these long-term peer responses are likely to contribute to overly negative social perceptions of others, and in turn, reinforce social fears.

Neurodevelopmental processes: Fear extinction and social anxiety

Relatively recently, two studies were published both describing reduced fear extinction in adolescents (Baker, Den, Graham, & Richardson, 2014; Johnson & Casey, 2015). According to research in adolescent humans and rodents, fear extinction is impaired during adolescence (Baker et al., 2014), which was explained by an imbalance in the fear extinction system. In adult years, the crucial areas of the fear extinction circuit, namely the amygdala and prefrontal cortex (PFC), are fully matured and have strong connectivity, balancing fear inhibition and fear expression. In childhood years, the fear regulation system is in balance too, because subcortical structures such as the amygdala are still developing and connectivity between subcortical and cortical structures does not emerge until late adolescence. In contrast, during the adolescent years subcortical areas develop more quickly than cortical areas, creating an imbalance between the regions. As the subcortical structures are also hyperresponsive, this results in an excess of activity in these areas, which is left uncontrolled or uninhibited by the PFC. As stated in their paper "impairments in extinction during adolescence may be due to reduced prefrontal inhibitory control over excessive amygdala activation in the presence of fearful stimuli" (p. 93; Baker et al., 2014). As raised by Rapee et al. (2019), it is plausible that diminished fear extinction in the adolescent period would maintain anxious experiences and feelings.

A maintenance model of adolescent social anxiety

Based on the associations among constructs, we propose a *maintenance* model of adolescent social anxiety, depicted in Fig. 2. It is our aim to delineate the interacting processes that may explain how and why on a day-to-day basis, an adolescent experiencing anxiety about social situations and fear of negative evaluation, is unable to reduce these feelings, and instead, the social anxiety symptoms persist. We recognize that certain relations between different processes have been described previously (e.g., Chiu et al., 2021; Miers, Blöte, & Westenberg, 2011); here it is our intention to integrate the most important aspects into a combined model. We note that these interactions occur within a backdrop of several normative developmental changes. These include increased social demands and independent management of social relationships, and greater importance of, and sensitivity to, peers (Somerville, 2013); self-concept formation (Crocetti, Rubini, Branje, Koot, & Meeus, 2016); and changes in neural circuitry including improved monitoring of subcortical structures (e.g., amygdala) by the PFC (Caouette & Guyer, 2014). Together, they may explain the rise in social anxiety symptoms specifically, relative to other types, that occurs in adolescence (Westenberg, Drewes, Goedhart, Siebelink, & Treffers, 2004).

Integrating cognitive and social behavioral processes

We delineate the interactions between different processes beginning with negative social self-perceptions. An adolescent presented with an upcoming social situation, such as going to a friend's party or giving a presentation at school, is likely to think negatively about him or herself and his/her impending social performance. That is, adolescents will have negative expectations regarding their own social performance and predict negative social outcomes such as being ignored or rejected by peers. These negative self-perceptions are likely a manifestation of social beliefs that become activated and expressed before (anticipatory processing), during (self-focused attention), and after (postevent processing) the situation. As a result of these negative cognitions, adolescents may struggle to perform skillfully, or may behave in a way that they believe will prevent negative responses from peers, that is, using safety behaviors. The outcome of either is awkward social performance. This awkwardness is seen in, for example, less eye contact, using fewer gestures, less engagement and self-disclosure, appearing nervous, or inexpressiveness (Blöte, Poungjit, Miers, Van Beek, & Westenberg, 2015; Mesa, Le, & Beidel, 2015). Whether the same age-peers observing the socially anxious adolescent are new or familiar, the awkward social performance is likely to elicit negative responses immediately, such as an active disengagement from interacting, showing disinterest, and in the longer term, possibly lower acceptance and exclusion from different social activities. Immediate negative responses from peers will heighten the state anxiety experienced by the adolescent at that moment. Similarly, when the adolescent consistently experiences negative responses, these will be perceived as confirming both the preexisting fear of negative evaluation and social distress, as well as the negative social self-perceptions.

As reviewed earlier (Blöte, Miers, et al., 2015; Leigh & Clark, 2018), there is evidence to show negative peer responses to poor social performance; however, research to date has not tested the hypothesis proposed by the model that these negative peer responses reinforce existing social anxiety symptoms *and* negative social self-perceptions. In addition, the maintenance model proposes that, in social situations, negative responses from others worsen the social performance occurring in real time. Imagine a teenager giving a presentation in class who hardly looks at classmates to capture their attention and notices that they start talking among themselves. The teacher fails to ask students to listen and the socially anxious teenager becomes so fearful that he only manages to talk in a whisper, whereupon the class erupts into mayhem and the teacher becomes angry. Observing and measuring such iterative, small-scale, and subtle changes would add to our understanding of processes occurring *in the moment* during everyday social situations.

A second cognitive consequence of negative peer responses is the strengthening of negative thoughts about other people's behavior, that is, the interpretation bias. According to the current model, and alluded to in Miers et al. (2011), such interpretations may arise as a consequence of actual negative social experiences and therefore reflect a kernel-of-truth. Whether or not the interpretation bias is justified or indeed distorted, we suggest that unfavorable peer interactions will serve to heighten adolescents' tendency to perceive danger in social situations, thereby maintaining and potentially worsening anxiety from subclinical to clinical levels.

Fear extinction

A relatively new addition to adolescent maintenance theories is inefficient fear extinction. For an adolescent who has social fears, a process that might reduce anxiety during social events is learning to downregulate the learned fear. This may occur in the context of new information indicating that what was once dangerous is now a safe environment. We consider an adolescent who has experienced some bullying at school that causes the adolescent to become fearful of social situations with same-age peers. Upon moving to a new school, the adolescent (with the help of parents, neighborhood friends) has positive social experiences and is accepted. Such repeated social-safety information should allow the learned fear to be downregulated whereby the fear will diminish and eventually become extinct. However, when this process is delayed or impaired, positive social experiences of acceptance will not diminish anxiety during social events. If the adolescent, having moved to this new and encouraging school environment, still experiences anxious feelings in social situations, he or she may leave them prematurely. In and of itself, this lessens the opportunity to experience a decline in distress (and allow for new learning). Simultaneously, untimely departure from peer-group interactions may come across as disinterest and even haughtiness, potentially paving the way for negative peer responses and, in turn, maintaining anxiety. Finally, our framework suggests that limited fear extinction, even in the context of positive social experiences, precludes a change in negative cognitions to more adaptive thoughts.

Implications for interventions tailored to adolescents with social anxiety

Considerable evidence has accumulated to suggest that a standard CBT approach is not optimal for ameliorating social anxiety. That is, CBT, typically consisting of psychoeducation, relaxation, emotion identification, cognitive strategies, and gradual exposure, yields substantially lower response rates for SAD than other pediatric anxiety disorders (Ginsburg et al., 2011; Hudson et al., 2015). Ideally, these strategies should target the maintenance processes most applicable to adolescents with social anxiety, as specified in Fig. 2. Previous models have primarily emphasized strategies for altering problematic cognitive processes (Fig. 1). Adding to these prior advances, we recommend improving interventions through capitalizing on the adolescents' social context. Socially anxious adolescents exist in an interpersonally demanding environment that heightens anxiety, but also creates opportunities for innovative delivery of interventions. Integrating strategies within natural settings (e.g., schools) and involving peers may maximize efficacy and increase opportunities for positive outcomes. Such approaches have the potential to directly disrupt the negative feedback loop between social performance, negative peer reactions, and interfering cognitions.

Cognitive interventions

The preponderance of negative cognitive processes found in adolescents with social anxiety (Leigh & Clark, 2018; Spence & Rapee, 2016), as noted in Figs. 1 and 2, support using specific cognitive strategies to address them. Research on adults with SAD demonstrate that interventions addressing self-relevant cognitive factors proposed to maintain social anxiety (e.g., self-focused attention, worry and rumination before and after social events) are highly effective (Clark et al., 2003, 2006) and superior to traditional CBT (Rapee, Gaston, & Abbott, 2009). Leigh and Clark (2016) have developed a comprehensive, cognitive treatment for adolescents based on the adult version, Cognitive Therapy-SAD (CT-SAD) and grounded in the Clark and Wells (1995) framework in Fig. 1. A description of the adolescent version, CT-SAD-A, is beyond the scope of this chapter, but can be found in Leigh and Clark (2018). The main goal of CT-SAD-A is to correct problematic cognitive processes involving distorted negative self-images, self-focused attention, negative interpretation bias, and anticipatory worry and postevent processing. This individual treatment aims to help teenagers understand the harmful impact of self-focused attention and safety behaviors, train attention to be external and nonevaluative during social tasks, provide performance feedback using video recorded interactions (e.g., with adults available in the clinic) to correct negative interpretations, and conduct behavioral experiments to challenge negative predictions.

These strategies seem intuitive and initial efficacy findings are encouraging (Leigh, Creswell, et al., 2021). Yet, given the central role of peers and their evaluation during adolescence (Rapee et al., 2019), one-on-one treatment with an adult in a clinic setting may not be optimal. Interactions and behavioral experiments, as well as opinions from adults in therapeutic positions, may feel artificial, which likely causes socially anxious teenagers to discredit positive performance feedback, thereby limiting its therapeutic impact on social self-perceptions. In addition, even if cognitive changes do occur in these contrived settings, their transfer to the complex social worlds of teenagers is doubtful. To increase the developmental sensitivity and generalizability of cognitive interventions, we suggest conducting a portion of sessions with another teenager or two, or, possibly using videos of peers. For example, teenagers can witness actual negative responses from others when being overly self-focused and using safety behaviors and more positive ones when engaged in the social task at hand. We would expect that repeated observation of this kind would reveal that self-focused attention and safety behaviors make them look uninterested and awkward and lead to negative peer reactions. Incorporating peer evaluation, which is crucial to teenagers, increases treatment relevance and may facilitate more efficient and robust changes to cognitive distortions.

Social skills training

Socially anxious teenagers do not experience negative cognitions in a vacuum cut-off from their social world. Adjusting negative cognitions likely does not fully address the ineffectual social performance associated with social anxiety (Beidel, Turner, & Morris, 1999; Spence, Donovan, & Brechman-Toussaint, 1999). Impairments in verbal and nonverbal skills can be easily identified by peers, and may contribute to unfavorable social experiences and negative evaluation (Blöte, Miers, et al., 2015; Miers, Blöte, & Westenberg, 2010). Treatment programs that include social skills training, and not only social exposure, yield larger effect sizes (Masia Warner et al., 2016; Scaini, Belotti, Ogliari, & Battaglia, 2016), suggesting that a social behavioral ingredient may be critical.

Including skills that address impairments observed in socially anxious teenagers, Skills for Academic and Social Success (SASS), a school-based intervention (Masia Warner, Colognori, & Lynch, 2018) based on Social Effectiveness Therapy for Children (SET-C; Beidel, Turner, & Morris, 2000), contains four social skills groups (1) initiating conversations, (2) maintaining conversations and establishing friendships, (3) listening, remembering, and paying attention, and (4) assertiveness. A full description of these group training sessions is available in Masia Warner et al. (2018), but we highlight a few aspects here. Consistent with Leigh and Clark's discussion of the detrimental effect of self-focused attention on performance, SASS' session on listening, remembering and paying attention focuses on anxiety interference. It discusses how anxious thoughts and excessive self-monitoring and self-evaluation (e.g., What am I going to say next? That sounded so stupid) interfere with the ability to engage in conversations by compromising attentional resources needed to focus on social tasks. The therapist also explains that people who listen during conversations, as well as recall what others have said, appear more genuine and have more enjoyable and naturally flowing conversations. SASS employs direct practice to improve active listening and focus attention on the topic at hand (Masia Warner et al., 2018).

Another important aspect of skills training specifically for socially anxious adolescents is the intentional and repeated shaping of active and engaging nonverbal skills, such as speaking louder, increasing eye contact, and improving facial expressiveness and nonverbal gestures that communicate interest (Blöte, Miers, et al., 2015; Masia Warner et al., 2018). Because socially anxious adolescents often look unfriendly or unapproachable, all SASS skills sessions emphasize teaching students to become aware of these nonverbal behaviors and messages to others. Adolescents are reminded to smile, make eye contact, and to speak more often, clearly, and loudly throughout skills sessions. Shaping more socially-engaging behavior and eliminating safety behaviors may increase the likelihood of success in social situations and decrease the chances of negative peer responses, further encouraging adolescents to enter and approach situations they typically avoid and reducing harmful social self-perceptions that sustain social anxiety symptoms.

Exposure

Exposure to feared situations is central to the treatment of anxiety disorders (de Jong, Lommen, de Jong, & Nauta, 2019). Yet, as discussed earlier there is evidence to suggest that extinction-based exposure may be less effective in adolescents (Drysdale et al., 2014). In addition, given the brief nature of social interactions (e.g., asking a question in class, inviting someone to get together), it is often not feasible to keep adolescents in social situations for extended periods of time, making fear reduction even more difficult to achieve. While observed differences in fear extinction during adolescence may be seen as a vulnerability, preliminary evidence highlights several potential ways to strengthen exposure-based extinction learning in this developmental phase. For example, Golkar demonstrated stronger extinction and less fear recovery with repeated trials (Golkar, Bellander, & Öhman, 2013), as well as with social learning or opportunities to observe others handling anxiety-provoking situations (Golkar, Selbing, Flygare, Öhman, & Olsson, 2013). In addition, as pointed out by de Jong et al. (2019), context has been shown to influence the strength of learned associations (Bouton, 1993), and thus they recommend conducting exposure in multiple contexts to promote generalization of the learned, no fear response into adolescents' natural social world.

The extinction literature has several implications for the optimal use of exposure in adolescents. Because evaluation concerns are at the core of social anxiety and peers take on heightened importance during adolescence, implementing exposure in small groups of teenagers may be valuable. Not only can the group create a more clinically relevant setting for conducting exposures, but it also allows for observing, coaching, and encouraging others, which can support vicarious learning and potentially enhance fear reduction. In addition, when conducting exposures, it is essential to collaborate with schools possibly through involving school counselors or teachers. Not only do most common triggers of social anxiety routinely occur at school, but it is often the setting where social fears are initially acquired. Conducting exposures in everyday school situations affords the opportunity for repeated and frequent experiences in an authentic setting (Masia Warner et al., 2016). Schools also uniquely provide access to anxiety-provoking situations across various real-life contexts (e.g., classroom, school cafeteria, sports teams, clubs) and with multiple individuals (e.g., peers, teachers, coaches). Teachers can be enlisted to assist in exposure tasks (e.g., having a student enter a classroom late or answer questions in class; Ryan & Masia Warner, 2012). Finally, as schools are the main context where social relationships are established and interpersonal conflicts occur, school counselors are ideally positioned to coach adolescents in challenging situations that may arise (Fox, Masia Warner, & Drew, 2018). The natural environment inherent in schools can maximize exposure potency and generalization across social contexts (Coyle, Mahmud, Weeks, & Warner, 2020).

Peer interventions

The interventions discussed earlier have focused on changing cognitions or behaviors of adolescents that contribute to sustaining their social anxiety. However, highly central to our maintenance model for adolescent social anxiety is negative peer responses and experiences. The prominence of peers and adolescents' sensitivity to their evaluation is the hallmark of adolescence (Somerville, 2013). By nature of this developmental phase, it seems remiss to overlook interventions that leverage the use of peers and specifically adolescents' responsiveness to rewarding peer experiences. For example, Miers (2021) found that socially anxious young adults who received positive versus neutral feedback from peers after a speech reported significantly higher positive rumination in the week following and improved anticipatory thoughts about a hypothetical future speech. Certainly, performance feedback from other adolescents is a regular part of group interventions. However, its use might be strengthened by giving it greater emphasis and time, and making it more detailed and explicit. It is possible that experiencing positive peer responses in a controlled group setting initially will provide adolescents with the encouragement to take more social risk in the real world, and we know that this has to occur for benefits to be sustained.

To help bridge the therapeutic setting with the natural social context and promote the development of supportive peer relationships, SET-C and SASS include same-age peer assistants and social events (Masia Warner et al., 2018). SET-C, a clinic-based treatment for children with SAD includes 12 social events with outgoing children who are volunteers recruited through newspapers. SASS has the benefit of using school peers most of whom have been through the program, includes four 90-min social events (e.g., bowling, board games with pizza) held on weekends either in community settings or at school. The first social event is designed to be highly structured (e.g., bowling) with frequent encouragement and coaching of socially anxious adolescents and peer helpers to ensure positive social experiences. As events progress, they are intentionally more challenging, with less structure and direction, to mirror typical social experiences and peer reactions. Ensuring positive peer experiences early on may prepare adolescents to handle more neutral or even minor negative reactions without triggering a cycle of negative social self-perceptions. This resilience is essential for adolescence when navigating the challenging social landscape.

Conclusion

Our model provides a comprehensive account of processes that sustain social anxiety symptoms during the adolescent phase: an adolescent-specific maintenance model of social anxiety. It explicitly explains why adolescent social anxiety persists and does not "self-correct" as a result of regular exposure to social situations, like at school (Halldorsson & Creswell, 2017; Leigh & Clark, 2018). Moreover, we briefly describe a "developmental lens" in which to place these pathways: increased demands in the social environment, social identity formation, and changes in neural circuitry. We place emphasis on the social context of adolescents with social anxiety: what is different about their social behavior, social experiences, and social fear learning, relative to nonanxious teenagers, that influences social anxiety on a day-to-day basis? The model also aims to elucidate how the main factors interact to sustain social anxiety and generate new intervention avenues that capitalize on the social context and address processes central to maintaining social anxiety. We recommend leveraging peers and the natural social environment as a way to strengthen other approaches including cognitive strategies, exposure, and social skills training. In addition, explicitly incorporating positive peer feedback in therapeutic and real-life settings, particularly in the initial stages of intervention, may accelerate changes to the impervious negative cycle. School environments provide a rich context in which to involve peers and encourage social risk-taking, and allow for early detection and intervention, thus affording the chance to disrupt the implicated processes before they become ingrained and cause long-term impairment.

Practice and procedures

This chapter demonstrates how adolescent social anxiety and its maintenance can be researched using diverse methods. Social anxiety symptoms including negative cognitions are often assessed with self-report questionnaires. To assess social performance, it is useful to obtain measurements from adolescents as well as individuals in their social environment, ideally their peers. Different sources of information can be compared. Observer ratings of social skills, quality of interaction, verbal and nonverbal gestures, can be achieved with rating scales (e.g., the SPOSY; Blöte, Poungjit, et al., 2015). Short-term negative peer responses can be captured though recordings of social interactions and subsequent behavioral coding or independent observer ratings of behaviors in real time, such as in the classroom. Long-term negative responses have typically been assessed with sociometric nominations, whereby all pupils in a class are asked to nominate friends (e.g., Chiu et al., 2021). Finally, fear extinction can be studied using experiments in which a conditioned fear response is weakened by repeatedly exposing the participant to a conditioned stimulus (e.g., a group of age peers) without pairing it with the associated aversive consequence (e.g., peers laughing). A subsequent test of fear recovery can be likened to relapse after successfully completed exposure (Golkar, Bellander, & Öhman, 2013). Several of the intervention techniques and procedures discussed are available at www.oxcadatresources.com and in a book written specifically for school professionals to identify and treat social anxiety in schools (Masia Warner et al., 2018).

Mini-dictionary of terms

- **Negative social self-perceptions:** these are negatively valenced thoughts related to the social self and one's performance specifically in social situations. They can occur automatically and prior to, during, and after a social situation.
- **Awkward social performance:** this refers to the ways in which socially anxious adolescents behave in social situations and reflects impairments in verbal and nonverbal social skills.
- **Negative peer responses:** this term describes how socially anxious adolescents are responded to and treated by peers, both in the short- and long term.
- **Interpretation bias:** a cognitive bias defined by negative or threatening interpretation of ambiguous social behaviors in others.
- **Reduced fear extinction:** fear extinction is the process of unlearning an association between a stimulus and a feared response. This process is proposed to be compromised in adolescence.
- **Safety behaviors:** behaviors that socially anxious individuals use in social situations that they believe will reduce the likelihood of negative evaluation or help be socially effective (e.g., preparing what to say and monitoring what is said during an interaction). These behaviors are usually unhelpful and make the person socially ineffective.

Key facts of social anxiety disorder

- Social anxiety disorder is the second most common mental disorder in adolescence.
- Individuals with social anxiety disorder are more likely to underachieve in all education levels.
- Adolescent social anxiety disorder has the poorest treatment efficacy among anxiety disorders.
- Treatment utilization is very low, at around 15% of adolescents with social anxiety disorder.

Applications to other areas

In this chapter, we reviewed evidence for different factors that maintain social anxiety in adolescents, specifically those published after 2015. These factors are cognitive, social behavioral, and fear extinction processes. We integrated these factors into a new maintenance model of adolescent social anxiety. Unique to this model is the focus on the social context and how it interacts with cognitions and fear extinction to sustain social anxiety. Our framework is placed within a developmental context of normative changes, making it relevant to the developmental psychopathology field. This framework is likely relevant to other social-emotional disorders that arise in adolescence, such as generalized anxiety disorder and depression, due to the influence of similar factors such as social sensitivity and peer relationships (Rapee et al., 2019). As social anxiety often precedes the development of depression (Moitra, Herbert, & Forman, 2008), addressing factors identified in the model also has implications for preventing depression. In particular, using intervention strategies that aim to improve peer responses and functioning are likely to reduce loneliness and anhedonia. In addition, enhancing fear extinction for adolescents might generalize to multiple contexts and stimuli, such as those associated with specific phobias and generalized anxiety disorder. Our maintenance model and suggestions for intervention approaches are widely applicable to school psychology and school mental health in terms of targets that could be addressed in collaboration with school personnel. The emphasis of our framework on the social context and peers suggests that interventions for social anxiety in schools may be optimal.

Summary points

- This chapter proposes a comprehensive model of how social anxiety is maintained in adolescence.
- The adolescent-specific maintenance framework is an integration of research on cognitive, social behavioral, and fear extinction processes.
- Compared to previous social anxiety models, it emphasizes the importance of the social context and how it interacts with other processes.
- Inefficient fear extinction is included as an important factor that could sustain social anxiety.
- Interventions should target the maintenance processes identified and particularly aim to capitalize on the social context.
- We propose integrating intervention strategies within natural social contexts, such as schools, and involving peers where possible in cognitive interventions, social skills training, and exposures.

References

American Psychiatric Association. (2013). *Diagnostic and statistical manual of mental disorders (DSM-5®)*. American Psychiatric Publishing.

Baker, K. D., Den, M. L., Graham, B. M., & Richardson, R. (2014). A window of vulnerability: Impaired fear extinction in adolescence. *Neurobiology of Learning and Memory*, *113*, 90–100.

Beidel, D. C., Turner, S. M., & Morris, T. L. (1999). Psychopathology of childhood social phobia. *Journal of the American Academy of Child & Adolescent Psychiatry*, *38*(6), 643–650.

Beidel, D. C., Turner, S. M., & Morris, T. L. (2000). Behavioral treatment of childhood social phobia. *Journal of Consulting and Clinical Psychology*, *68*(6), 1072.

Blöte, A. W., Miers, A. C., Heyne, D. A., & Westenberg, P. M. (2015). Social anxiety and the school environment of adolescents. In K. Ranta, A. M. La Greca, L.-J. Garcia-Lopez, & M. Marttunen (Eds.), *Social anxiety and phobia in adolescents: Development, manifestation and intervention strategies* (pp. 151–181). Springer International Publishing AG.

Blöte, A. W., Miers, A. C., Van den Bos, E., & Westenberg, P. M. (2019). Negative social self-cognitions: How shyness may lead to social anxiety. *Journal of Applied Developmental Psychology*, *63*, 9–15.

Blöte, A. W., Poungjit, A., Miers, A. C., Van Beek, Y., & Westenberg, P. M. (2015). The Speech Performance Observation Scale for Youth (SPOSY): Assessing social performance characteristics related to social anxiety. *Journal of Experimental Psychopathology*, *6*(2), 168–184.

Bouton, M. E. (1993). Context, time, and memory retrieval in the interference paradigms of Pavlovian learning. *Psychological Bulletin*, *114*(1), 80.

Caouette, J. D., & Guyer, A. E. (2014). Gaining insight into adolescent vulnerability for social anxiety from developmental cognitive neuroscience. *Developmental Cognitive Neuroscience*, *8*, 65–76.

Chiu, K., Clark, D. M., & Leigh, E. (2021). Prospective associations between peer functioning and social anxiety in adolescents: A systematic review and meta-analysis. *Journal of Affective Disorders*, *279*, 650–661. https://doi.org/10.1016/j.jad.2020.10.055.

Chiu, K., Clark, D. M., & Leigh, E. (2021). Cognitive predictors of adolescent social anxiety. *Behaviour Research and Therapy*, *137*, 103801.

Clark, D. M. (2001). A cognitive perspective on social phobia. In W. R. Crozier, & L. E. Alden (Eds.), *International handbook of social anxiety: Concepts, research and interventions relating to the self and shyness*. New York: John Wiley & Sons Ltd.

Clark, D. M., Ehlers, A., Hackmann, A., McManus, F., Fennell, M., Grey, N., … Wild, J. (2006). Cognitive therapy versus exposure and applied relaxation in social phobia: A randomized controlled trial. *Journal of Consulting and Clinical Psychology*, *74*(3), 568.

Clark, D. M., Ehlers, A., McManus, F., Hackmann, A., Fennell, M., Campbell, H., … Louis, B. (2003). Cognitive therapy versus fluoxetine in generalized social phobia: A randomized placebo-controlled trial. *Journal of Consulting and Clinical Psychology*, *71*(6), 1058.

Clark, D. M., & Wells, A. (1995). A cognitive model of social phobia. In R. G. Heimberg, M. R. Liebowitz, D. A. Hope, & F. R. Schneier (Eds.), *Social phobia: Diagnosis, assessment, and treatment* (pp. 69–93). New York: Guilford Press.

Coyle, S., Mahmud, F., Weeks, C., & Warner, C. M. (2020). School-based treatment for children and adolescents with social anxiety disorder. In *Social skills across the life span* (pp. 237–254). Elsevier.

Crocetti, E., Rubini, M., Branje, S., Koot, H. M., & Meeus, W. (2016). Self-concept clarity in adolescents and parents: A six-wave longitudinal and multi-informant study on development and intergenerational transmission. *Journal of Personality*, *84*(5), 580–593.

de Jong, R., Lommen, M. J., de Jong, P. J., & Nauta, M. H. (2019). Using multiple contexts and retrieval cues in exposure-based therapy to prevent relapse in anxiety disorders. *Cognitive and Behavioral Practice*, *26*(1), 154–165.

Drysdale, A. T., Hartley, C. A., Pattwell, S. S., Ruberry, E. J., Somerville, L. H., Compton, S. N., … Walkup, J. T. (2014). Fear and anxiety from principle to practice: Implications for when to treat youth with anxiety disorders. *Biological Psychiatry*, *75*(11), e19–e20.

Fox, J. K., Masia Warner, C., & Drew, M. (2018). School-based intervention for adolescents with impairing social anxiety. In A. W. Leschied, D. H. Saklofske, & G. L. Flett (Eds.), *Handbook of school-based mental health promotion: An evidence-informed framework for implementation* (pp. 411–427). Springer International Publishing. https://doi.org/10.1007/978-3-319-89842-1_21.

Ginsburg, G. S., Sakolsky, D., Piacentini, J., Walkup, J. T., Coffey, K. A., Keeton, C. P., … March, J. (2011). Remission after acute treatment in children and adolescents with anxiety disorders: Findings from the CAMS. *Journal of Consulting and Clinical Psychology*, *79*(6), 806–813. https://doi.org/10.1037/a0025933.

Golkar, A., Bellander, M., & Öhman, A. (2013). Temporal properties of fear extinction—Does time matter? *Behavioral Neuroscience*, *127*(1), 59–69. https://doi.org/10.1037/a0030892.

Golkar, A., Selbing, I., Flygare, O., Öhman, A., & Olsson, A. (2013). Other people as means to a safe end: Vicarious extinction blocks the return of learned fear. *Psychological Science*, *24*(11), 2182–2190.

Halldorsson, B., & Creswell, C. (2017). Social anxiety in pre-adolescent children: What do we know about maintenance? *Behaviour Research and Therapy*, *99*, 19–36.

Hudson, J. L., Rapee, R. M., Lyneham, H. J., McLellan, L. F., Wuthrich, V. M., & Schniering, C. A. (2015). Comparing outcomes for children with different anxiety disorders following cognitive behavioural therapy. *Behaviour Research and Therapy*, *72*, 30–37.

Johnson, D. C., & Casey, B. J. (2015). Easy to remember, difficult to forget: The development of fear regulation. *Developmental Cognitive Neuroscience*, *11*, 42–55. https://doi.org/10.1016/j.dcn.2014.07.006.

Knappe, S., Sasagawa, S., & Creswell, C. (2015). Developmental epidemiology of social anxiety and social phobia in adolescents. In *Social anxiety and phobia in adolescents* (pp. 39–70). Springer.

Leigh, E., Chiu, K., & Clark, D. M. (2020). The effects of modifying mental imagery in adolescent social anxiety. *PLoS One*, *15*(4), e0230826.

Leigh, E., Chiu, K., & Clark, D. M. (2021). Self-focused attention and safety behaviours maintain social anxiety in adolescents: An experimental study. *PLoS One*, *16*(2), e0247703.

Leigh, E., & Clark, D. M. (2016). Cognitive therapy for social anxiety disorder in adolescents: A development case series. *Behavioural and Cognitive Psychotherpy*, *44*(1), 1–17. https://doi.org/10.1017/s1352465815000715.

Leigh, E., & Clark, D. M. (2018). Understanding social anxiety disorder in adolescents and improving treatment outcomes: Applying the cognitive model of Clark and Wells (1995). *Clinical Child and Family Psychology Review*, *21*(3), 388–414.

Leigh, E., Creswell, C., Stallard, P., Waite, P., Violato, M., Pearcey, S., … Clark, D. M. (2021). Delivering cognitive therapy for adolescent social anxiety disorder in NHS CAMHS: A clinical and cost analysis. *Behavioural and Cognitive Psychotherapy*, 1–13. https://doi.org/10.1017/S1352465821000035.

Masia Warner, C., Colognori, D., Brice, C., Herzig, K., Mufson, L., Lynch, C., … Moceri, D. C. (2016). Can school counselors deliver cognitive-behavioral treatment for social anxiety effectively? A randomized controlled trial. *Journal of Child Psychology and Psychiatry*, *57*(11), 1229–1238.

Masia Warner, C., Colognori, D., & Lynch, C. (2018). *Helping students overcome social anxiety: Skills for academic and social success (SASS)*. Guilford Publications.

Merikangas, K. R., He, J., Burstein, M., Swanson, S. A., Avenevoli, S., Cui, L., … Swendsen, J. (2010). Lifetime prevalence of mental disorders in US adolescents: Results from the National Comorbidity Survey Replication–Adolescent Supplement (NCS-A). *Journal of the American Academy of Child & Adolescent Psychiatry*, *49*(10), 980–989.

Mesa, F., Le, T.-A., & Beidel, D. C. (2015). Social skill-based treatment for social anxiety disorder in adolescents. In *Social anxiety and phobia in adolescents* (pp. 289–299). Springer.

Miers, A. C. (2021). An investigation into the influence of positive peer feedback on self-relevant cognitions in social anxiety. *Behaviour Change*, *38*, 193–207. https://doi.org/10.1017/bec.2021.8.

Miers, A. C., Blöte, A. W., & Westenberg, P. M. (2010). Peer perceptions of social skills in socially anxious and nonanxious adolescents. *Journal of Abnormal Child Psychology*, *38*(1), 33–41.

Miers, A. C., Blöte, A. W., & Westenberg, P. M. (2011). Negative social cognitions in socially anxious youth: Distorted reality or a kernel of truth? *Journal of Child and Family Studies*, *20*(2), 214–223.

Moitra, E., Herbert, J. D., & Forman, E. M. (2008). Behavioral avoidance mediates the relationship between anxiety and depressive symptoms among social anxiety disorder patients. *Journal of Anxiety Disorders*, *22*(7), 1205–1213.

Pine, D. S., Cohen, P., Gurley, D., Brook, J., & Ma, Y. (1998). The risk for early-adulthood anxiety and depressive disorders in adolescents with anxiety and depressive disorders. *Archives of General Psychiatry*, *55*(1), 56–64.

Rapee, R. M., Gaston, J. E., & Abbott, M. J. (2009). Testing the efficacy of theoretically derived improvements in the treatment of social phobia. *Journal of Consulting and Clinical Psychology*, *77*(2), 317.

Rapee, R. M., Oar, E. L., Johnco, C. J., Forbes, M. K., Fardouly, J., Magson, N. R., & Richardson, C. E. (2019). Adolescent development and risk for the onset of social-emotional disorders: A review and conceptual model. *Behaviour Research and Therapy*, *123*, 103501.

Rapee, R. M., & Spence, S. H. (2004). The etiology of social phobia: Empirical evidence and an initial model. *Clinical Psychology Review*, *24*(7), 737–767.

Ryan, J. L., & Masia Warner, C. (2012). Treating adolescents with social anxiety disorder in schools. *Child and Adolescent Psychiatric Clinics of North America*, *21*(1), 105.

Scaini, S., Belotti, R., Ogliari, A., & Battaglia, M. (2016). A comprehensive meta-analysis of cognitive-behavioral interventions for social anxiety disorder in children and adolescents. *Journal of Anxiety Disorders*, *42*, 105–112.

Somerville, L. H. (2013). The teenage brain: Sensitivity to social evaluation. *Current Directions in Psychological Science*, *22*(2), 121–127.

Spence, S. H., Donovan, C., & Brechman-Toussaint, M. (1999). Social skills, social outcomes, and cognitive features of childhood social phobia. *Journal of Abnormal Psychology*, *108*(2), 211.

Spence, S. H., & Rapee, R. M. (2016). The etiology of social anxiety disorder: An evidence-based model. *Behaviour Research and Therapy*, *86*, 50–67.

Westenberg, P. M., Drewes, M. J., Goedhart, A. W., Siebelink, B. M., & Treffers, P. D. (2004). A developmental analysis of self-reported fears in late childhood through mid-adolescence: Social-evaluative fears on the rise? *Journal of Child Psychology and Psychiatry*, *45*(3), 481–495.

Zarger, M. M., & Rich, B. A. (2016). Predictors of treatment utilization among adolescents with social anxiety disorder. *Children and Youth Services Review*, *71*, 191–198.

Chapter 9

Skin picking and hair pulling disorders in children and adolescents: The role of cognitive behavior therapy

Andrea Pozza

Department of Medical Sciences, Surgery and Neurosciences, University of Siena, Siena, Italy

Abbreviation

ACT Acceptance & Commitment Therapy

Introduction

Skin picking and hair pulling disorders: Main clinical and epidemiological features

Skin picking disorder is a psychopathological condition that consists of repetitive picking behaviors which result in significant tissue damage and are accompanied by significant subjective distress or functional impairment (American Psychiatric Association, 2013). Young individuals who show skin picking behaviors tend to pick small irregularities or blemishes or skin lesions such as scars, calluses, pimples or scabs resulting from previous picks (Snorrason, Smári, & Ólafsson, 2010). The most common picking areas are the most easily accessible ones, such as the face, arms, upper body, hands, legs, and cuticles (Roberts, O'Connor, & Bélanger, 2013). Epidemiological studies conducted in samples of late adolescents recruited from the adolescent general population suggest that more than 90% report occasional skin picking behaviors and around 4%–9% endorse significant impairment from recurrent, self-injurious skin picking (Bohne, Wilhelm, Keuthen, Baer, & Jenike, 2002; Calikusu, Kucukgoncu, Tecer, & Bestepe, 2012).

Clinical conceptualizations suggest that this condition includes two subtypes: an automatic one, that tends to occur outside of one's awareness, including situations where the individual picks his/her skin while engaged in a sedentary activity (e.g., reading/watching television); a focused subtype, performed to manage negative feelings or to correct perceived imperfections (e.g., Pozza, Albert, & Dèttore, 2020). Focused picking behaviors are generally triggered by negative emotions such as self-disgust, a feeling of guilt or shame, and they can generally produce a perceived sense of relief from such unpleasant emotional experiences in the short run.

Skin picking disorder generally tends to be more prevalent among females (Odlaug et al., 2013; Pozza, Giaquinta, & Dèttore, D., 2017). In a large sample of youth recruited from the general population, females reported significantly greater intensity and frequency of teasing behaviors than males and significantly higher levels of focused but not automatic skin picking (Pozza, Giaquinta, & Dèttore, 2016).

Although symptoms may first appear in various age groups, onset is often observed in adolescence coinciding with the onset of signs of puberty (Grant, Stein, Woods, & Keuthen, 2012; Odlaug & Grant, 2007); the average age of onset of the problem is reported to be between 12 and 16 years (Flessner & Woods, 2006; Wilhelm et al., 1999). According to Downing (2006), the onset in adolescence can be explained by the appearance of acne as a typical sign of body transformation during this developmental phase.

Hair pulling is a psychopathological condition characterized by recurrent pulling out of hair which generally determines noticeable hair loss (American Psychiatric Association, 2013). Individuals typically experience an increasing sense of tension prior to or when attempting to resist pulling and relief or pleasure when pulling. This repetitive behavior results in clinically significant distress or impairment and is not better accounted for by another mental health or medical condition

Handbook of Lifespan Cognitive Behavioral Therapy. https://doi.org/10.1016/B978-0-323-85757-4.00033-X

(American Psychiatric Association, 2013). As those who pick their skin, hair pullers typically tend to report increased levels of shame, self-blame, and frustration from pre- to postpulling, and lower levels of calmness after hair pulling episodes, while experiencing higher relief across the pulling cycle (Ghisi, Bottesi, Sica, Ouimet, & Sanavio, 2013).

The available evidence and theoretical models of hair pulling suggest that this condition may consist of the same subtypes, i.e., automatic and focused, as skin picking disorder, with similar phenomenological and functional characteristics (e.g., Flessner, Busch, Heideman, & Woods, 2008; Flessner, Woods, Franklin, Cashin, & Keuthen, 2008; Stein et al., 2010).

Skin picking and hair pulling disorders often present in comorbidity: in a study of university students Hajcak, Franklin, Simons, and Keuthen (2006) found that about 20% of the sample showed both conditions. There are several clinical features that these conditions have in common, such as age of onset, gender distribution, trigger situations, dysfunctional cognitions, difficulty in stopping repetitive behavior, presence of impulsivity and difficulty in emotional regulation, presence of these disorders in first-degree relatives, as well as the psychologically and socially disabling effects of the symptomatology such as shame and social disability experienced due to disfigurement of body parts (Wilhelm et al., 1999).

The specificity of skin picking and hair pulling in youth

Some researchers explored differences in the clinical characteristics of patients with early onset skin picking disorder (occurred within 10 years of age) and patients in whom the disorder had a later onset (after 10 years of age) (Odlaug & Grant, 2007). Forty-seven percent of patients reported that the disorder had had an early onset, an aspect that shows how frequent it is even in early developmental age. There were no differences between the two groups in the level of symptomatic severity or in the presence of specific psychiatric comorbidities. In the group with early-onset skin picking disorder, the period between the appearance of the first symptoms and the request for treatment tended to be longer than in the late-onset group (Odlaug & Grant, 2007). Another difference that emerged was that more frequent unconscious teasing behaviors were observed in the early onset group (Odlaug & Grant, 2007). In conclusion, the most important difference between early- and late-onset groups concerns the degree of awareness of the problem and the latency of the request for help, respectively lower and higher in the early onset group.

Cognitive behavioral therapy for skin picking and hair pulling disorders in youth

Habit reversal training

Habit reversal training and stimulus control (Azrin & Peterson, 1988) are the first-line behavioral techniques for skin picking and hair pulling disorders with any degree of symptomatic severity (Jones, Keuthen, & Greenberg, 2018). Habit reversal training consists of three distinct phases: (1) awareness training, (2) competitive response, and (3) social support. During the first phase, the clinician's goal is to help the young patient become more aware of skin picking/hair pulling behaviors, i.e., when, where, how they are activated and how they stop. The clinician and the young patient work together in trying to understand what this behavior consists of: the patient is invited to notice the moments in which the automatic behavior is triggered. Informal caregivers, i.e., parents, may be involved in this phase. In the next phase, in collaboration with the patient, the clinician identifies a competitive response, that is, a behavior that is incompatible with teasing. Such behavior, such as clenching both hands in a fist, must not attract the attention of people in public and must be maintained for a period of at least 60–90 s. Finally, during the last phase, the clinician and the patient identify a figure who provides emotional and social support to the latter during the exercises, stimulating him to implement the competitive response and positive reinforcements for having succeeded. During this process, family members (i.e., parents) may be contacted and involved.

In a preliminary phase, the therapist acts as a role model by showing the patient during the session the functional competitive behavior that is chosen in a way that is shared by both. Then the therapist and the patient identify a reference figure in his or her family environment who can be present in the initial phases during the situations most at risk of skin picking/hair pulling, reinforce the patient at home whenever she or he has succeeded in implementing the competitive behavior and encourage him or her to implement it when he or she has not implemented it correctly or not at all. The presence of the family member can be especially helpful in higher risk situations, such as at the mirror in the bathroom, at the television, or before taking a shower. In the early stages, as suggested by Deckersbach, Wilhelm, Keuthen, Baer, and Jenike (2002), it can be very useful to agree with the patient on an alarm signal that reminds them of the use of repetitive behaviors at certain times. It is equally important to agree on times when the patient can perform skin picking/hair pulling and to establish areas of the body or scalp where this is permitted. To this end, it may be effective to develop a hierarchy of teasing/pulling areas of the body/scalp and situations prior to skin picking/hair pulling ordered by the patient's perceived

TABLE 1 Situations which may trigger skin picking/hair pulling and competitive responses used in habit reversal training.

Situation	Examples of competitive actions
Watching television	For 2 min keep both hands busy by opening and closing your fists or hold on to an object such as the remote control, a rubber ball, or a rock
Reading a book	With your free hand hold another object such as a rubber ball or pencil or hold the book with both hands
Driving	Keep both hands on the steering wheel or radio buttons to change stations

level of difficulty (on a thermometer from 0 to 100) with respect to stopping the teasing/pulling behavior. Habit reversal exercises can be programmed by compiling a diary in which the therapist agrees with the times and schedules dedicated to homework. As the sessions progress, a generalization of the exercises in which the competitive action is carried out to similar places and contexts should be introduced. A list of competitive actions which may be used in the therapy is presented in Table 1. According to Deckersbach et al. (2002), habit reversal provides its greatest benefits to patients who need to become more aware of teasing/pulling, whereas it would be less effective for patients who use skin picking or hair pulling as a form of emotional regulation.

Stimulus control technique

This behavioral component can include the use of objects that limit skin picking/hair pulling in specific at-risk situations. For example, it may be helpful to teach the patient to use a pool cap every time he or she begins to undress to get in the shower or to use gloves or place patches on the ends of fingers while watching TV or studying. Stimulus control can also be graded with a hierarchy constructed according to a series of variables that play an important role in maintaining the problem. For example, weekly exercises can be planned with the therapist in which the patient is allowed to do skin picking/hair pulling but not with nails or only in certain areas of the body or only in a room with soft lights. The use of soft lighting can reduce the likelihood that the patient will notice skin imperfections or uneven or different hair/hair/eyebrows, which can be a trigger for the problematic behavior. At the same time, exercises can be agreed upon to prevent (or delay) skin picking/hair pulling by drawing up a list of distracting activities that immediately reduce the urge to scratch/tear, such as making a phone call to someone (if this is not a situation where the patient is at risk of automatic teasing), going for a walk, or cleaning one's room. It can be helpful if activities are planned with an attitude of doing something for ourselves, for our own well-being, or with an emphasis on personal values. These interventions are also intended to reduce the likelihood that the person will adopt other experiential avoidance strategies and control of internal states, such as brooding or rumination, which are often antecedent or concomitant to skin picking/hair pulling.

In order to avoid times when the person may experience feelings of emptiness, boredom, or depersonalization/derealization, triggers for body-focused behavior, it is helpful to compile a daily schedule of pleasurable activities spread throughout the day. As suggested by Deckersbach, Wilhelm, and Keuthen (2003), habit reversal can be integrated with relaxation training based for example on diaphragmatic breathing or autogenic training, which can prepare the patient for the next phases of therapy focused on emotional regulation.

The stimulus control technique is generally introduced at an early stage of the therapeutic path at the same time as habitat reversal training. Stimulus control refers to an intervention to modify the characteristics of the environment with the aim of reducing the salience of the behavioral antecedents of skin picking/hair pulling and therefore minimizing the reinforcing consequences of such behaviors (Azrin & Nunn, 1973). It is very important that competitive behavior can be emitted by the person in a contingent way with respect to the emergence of the impulse to tease (Miltenberger & Fuqua, 1985). Reducing the antecedents includes a series of interventions: (1) limiting the presence of behavioral triggers by removing for example from the surrounding environment those objects that can be used as reinforcement of skin picking/hair pulling, such as nail clippers or tweezers, or by covering mirrors, (2) provide/suggest alternative behaviors (e.g., handling an antistress object), (3) use measures that reduce the emission of dysfunctional behavior (e.g., wearing cotton gloves, hats or caps or cutting short nails), and (4) adopting strategies that increase the patient's awareness of his behavior (e.g., putting patches on his fingers or wearing noisy bracelets). Further examples of strategies that can limit the reinforcing effects of skin picking or hair pulling behaviors include the use of creams based on capsacin, a chemical compound responsible for the spiciness of peppers, with the aim of numbing the skin/hair area being teased/pulled or reducing the feeling of relief caused. A list of strategies used in therapy as a stimulus control.

Relaxation and generalization training: Two additional components of habit reversal training

Some authors (e.g., Deckersbach et al., 2002) consider two other components as integral parts of habit reversal training, namely a relaxation training aimed at reducing the state of tension (generally through autogenic training or diaphragmatic breathing) and generalization training of the habit reversal (which is applied to different contexts of the patient's daily life, starting initially from contexts in which it is easier for the patient to apply the principles of training). These two further components are inserted respectively between the first and second phase and after the third.

In a recent randomized trial (Schuck, Keijsers, & Rinck, 2011) with a wait-list control group of 35 university young students with skin picking disorder, the use of habit reversal training produced a large effect size and greater therapeutic results than the wait-list condition, at posttreatment and a three-month follow-up. However, in this study the behavioral technique was one of the components in a larger protocol that included also cognitive restructuring (Schuck et al., 2011).

Decoupling technique

A variant of habit reversal training introduced by Moritz and Rufer (2011) is the Decoupling technique. Habit reversal training and decoupling have in common the therapeutic characteristic of acting by creating interference at the motor level. The key difference between the two techniques is that while the first requires the subject to perform an antagonistic action with respect to the dysfunctional behavior, the second instead requires performing a movement that reproduces the typical movement of the disorder, precisely teasing skin or pulling hair, but modifying it.

The main of this approach is not preventing or blocking the behavioral response, instead introducing a similar but functional or harmless movement (Moritz & Rufer, 2011). In the decoupling technique, the final goal is modified; the subject is instructed, for example, to clasp his/her hands together or bring them to the ears or nose or hold them under the legs when sitting, instead of teasing. The new movement that is thus introduced should be carried out in an accelerated manner in order to prevent the usual motor sequence and consequently cause motor irritation as soon as the impulse to perform the dysfunctional gesture appears, which is often not performed with awareness. The state of irritation can help the patient to become aware and then use self-control strategies. Decoupling has recently been used with promising results in the treatment of trichotillomania (Moritz, Fricke, Treszl, & Wittekind, 2012; Moritz & Rufer, 2011) but further evaluations are needed in young clinical populations.

Cognitive strategies

Habit reversal training and stimulus control may be combined with cognitive techniques aimed at identifying dysfunctional beliefs that the individual has about himself/herself, others and world. Schuck et al. (2011) conducted a randomized controlled study on the effectiveness of a cognitive behavioral therapy with a predominant focus on cognitive aspects carried out for 5 weeks (6 sessions of 45 min in total) in comparison with a waiting list condition. The therapy includes the adaptation of a protocol already tested for trichotillomania (Van Minnen, Hoogduin, Keijsers, Hellenbrand, & Hendriks, 2003). During an initial phase psychoeducation is offered about the characteristics of skin picking/hair pulling and the maintenance cycles of the disorders, while subsequent sessions are focused on identifying dysfunctional beliefs and therapist-led restructuring interventions through descending arrow techniques and Socratic dialogue (Schuck et al., 2011). Cognitive restructuring is aimed at modifying beliefs specifically associated with skin picking disorder such as "I will never be able to stop as soon as I start teasing," "Teasing my skin/Pulling my hair gives me relief from all the worries I have in life," "When I'm sad I expect teasing my face/pull my hair to help me relax."

The cognitive intervention is combined with stimulus control techniques (e.g., planning activities to avoid situations at risk of picking episodes). The last session is dedicated to reviewing the techniques used, reinforcing the subjective experience of self-efficacy in the patient, and preventing relapse. Between one session and the next, homework is offered to propose exercises carried out in sessions with the therapist. The results of the study suggest that cognitive behavioral intervention with large effect sizes produces significant and superior changes to the waiting list on all outcome indices considered, self-reported and objective, the level of impulse to tease, the degree of functional deterioration, the distress caused by the symptoms, the ability to postpone picking/pulling, the degree of tension experienced before the teasing, and the extent of the lesions as shown by photographs of the skin areas taken before and after therapy (Schuck et al., 2011).

Psycho-physiological cognitive therapy

The Canadian group of Kieron O'Connor, Lavoie, and Schoendorff (2017) recently developed a new form of cognitive psychotherapy, an alternative approach to habit reversal training, the so-called Cognitive Psycho-Physiological Therapy, already studied as a treatment of Tourette's Syndrome. Unlike habit reversal training and other behavioral therapies for body-focused repetitive behaviors, this kind of treatment focuses on sensorimotor, physiological and emotional activation processes, which precede skin-teasing behaviors. The treatment lasts a total of 14 weeks and is divided into 10 weekly sessions conducted at office and then 1 month of practice at home.

The researchers published a recent preliminary open trial on 54 patients who reported repetitive behaviors focused on the body (O'Connor et al., 2017). The results showed large effect sizes and symptomatic improvement in 74% of the group with significant reductions in anxious-depressive symptoms and perfectionism as well as an increase in self-esteem and a maintenance of results at a 6-month follow-up (O'Connor et al., 2017). Therefore, this new therapy appears to be a promising strategy but needs further evaluation, particularly in children and adolescents, for whom data are not currently available.

Acceptance & Commitment Therapy (ACT)

Acceptance & Commitment Therapy (ACT; Hayes, Strosahl, & Wilson, 1999) is a third-generation therapeutic strategy that can be useful both with patients who show mainly automatic teasing behaviors and with those who have focused skin picking/hair pulling. The ACT therapeutic model is based on six processes, which consist of a series of specific exercises: Mindfulness, Self as Context, Defusion, Values, Committed Action, Acceptance. Each of the six therapeutic processes can play a very important role in the change process.

Mindfulness-based exercises are aimed at developing the ability to get in touch with the present moment, a capacity which can help the person both to acquire awareness of automatic teasing/pulling, as well as to recognize internal states that generate the impulse to tease/pull. Acceptance can allow the patient to learn emotional regulation strategies and at the same time reduce experiential avoidance or control/suppression strategies for unpleasant internal states. Defusion and the Self as Context allow the person to observe their internal states from the outside, to resize them as part of something broader, including the impulse to tease/pull and the mental events that precede it. The emphasis on Values and Committed Action can have a powerful motivating effect that helps the person to become more aware of the negative impact of the problematic behavior in their life, identify the reasons that can guide them in change and towards a life experience in line with his/her values, thus stimulating awareness of the dissonance between his/her values and teasing/pulling behaviors.

Twohig, Hayes, and Masuda (2006) evaluated the use of ACT exercises in a study on five university young students suffering from skin picking, who followed a treatment protocol with ACT inspired by the manual of Hayes et al. (1999), consisting of eight sessions every week of about 1 h each. The therapeutic path included two first sessions in which information on teasing/pulling behaviors was provided and a verbal contract was stipulated with each one on the duration, characteristics, and objectives of the path. The ACT intervention began by teaching the patient to distinguish between the urge to tease/pull (thoughts and body sensations that activate the need to tease) and the actual teasing/pulling behavior. In the next session, the young individual investigated all the strategies he/she has put in place so far to interrupt or avoid this dysfunctional behavior. The metaphor "The person in the hole" is introduced with the aim of demonstrating the ineffectiveness of the attempts made so far to neutralize the impulses. In 3–4 sessions, the therapist tries to highlight to the patient how any attempt to keep the impulse to tease/pull under complete control is unrealistic; this step is carried out through exercises aimed at illustrating the limits of the control strategies of emotions, thoughts, and bodily sensations. Thought suppression exercises are used ("Don't think about a blue rabbit") or the polygraph metaphor ("Imagine being attached to a polygraph and not having to let the world show any signs of discomfort or uncertainty when they ask you questions"). The therapeutic purpose of these exercises is to make the patient reflect on how unrealistic it is to try to control internal events such as thoughts or emotions, while it is more functional to control external events such as our behaviors.

Sessions 5–6 are therefore dedicated to transforming the impulses from something harmful and dangerous to simply verbal material, as if they were any other automatic thought that passes through our mind. This work is carried out with cognitive defusion exercises, such as, for example, repeating the thought for a long time uninterruptedly until it loses meaning. During this phase, metaphors may be used (e.g., the metaphor of passengers in the bus in which the patient represents the driver of a bus and passengers are the thoughts, which continually distract him/her by asking to take the directions they would like, taking him/her away from his/her own path). Mindfulness exercises may be applied as well (e.g., conscious breathing in the present moment) or exercises aimed at developing the self as a context (one of these is the metaphor of the chessboard where thoughts are the pawns on the chessboard, the patient's mind is something larger, it is the chessboard). The last two sessions are dedicated to reviewing the therapeutic material and the work carried out as a whole with the aim of promoting the maintenance of therapeutic results.

The results of the study by Twohig et al. (2006) demonstrated significant changes between the pre- and posttest both in a self-report measure of skin picking and in objective measures (photographs of the injured skin areas before and after the psychotherapy), and in measures of subjective feelings of anxiety and avoidance. However, these changes did not persist at follow-ups.

Fine and colleagues (2012) evaluated ACT therapy on two adolescents with trichotillomania. In this study, several modifications were made that were necessary to accommodate the protocol characteristics to their age. These modifications included (1) the involvement of parents in treatment sessions, (2) modifying the language used, (3) explaining several concepts more thoroughly, and (4) using age-appropriate examples when necessary. Overall, the results of this study indicated significant therapeutic outcomes including a reduction in the levels of subjective feelings of distress (Fine et al., 2012).

In summary, ACT may be considered a promising therapeutic for youth with skin picking and hair pulling disorders. It might be useful when it is combined with standard behavioral interventions, particularly for patients who endorse hair pulling as an emotion regulation strategy (i.e., focused subtype). However, while some preliminary has been carried out in adolescent, more data on children are necessary in the future.

Dialectical behavioral therapy

Dialectical behavioral therapy, conceived by the American psychologist Marsha Linehan (DBT; Linehan, 1993), is a form of structured treatment, based on teaching emotional regulation skills, which combines the elements of behavioral therapy with mindfulness and acceptance practices. There is currently no scientific evidence, based on efficacy studies, on the use of dialectical behavioral therapy for skin picking disorder.

However, techniques of DBT might be expected to be particularly suitable for those patients with emotional regulation difficulties and therefore with a predominantly focused subtype of skin picking disorder. For example, Welch and Kim (2012) used DBT as an enhancement to cognitive behavioral therapy in a 16-year-old girl diagnosed with trichotillomania. Results indicated a substantial decrease in the number of tears after 16 treatment sessions (Welch & Kim, 2012). DBT for adolescents requires several adaptations including the involvement of family members in multifamily skills groups, family therapy sessions, integrations of adolescent dialectical techniques, a reduction in treatment duration, and a simplification of terminology (MacPherson, Cheavens, & Fristad, 2013).

Fine and colleagues (2012) evaluated ACT therapy on two adolescents with trichotillomania. In this study, several modifications were made that were necessary to accommodate their age. These modifications included the involvement of parents in treatment sessions, modifying the language used, explaining several concepts more thoroughly, and using age-appropriate examples when necessary. Overall, the results of this study indicated significant therapeutic outcomes including a reduction in stress and distress (Fine et al., 2012). In summary, the components of ACT are an approach that is very useful when combined with standard behavioral interventions, particularly for patients using hair pulling as an emotion regulation strategy (i.e., focused behavior).

Mini-dictionary of terms

- Acceptance & Commitment Therapy = a third-generation therapy based on six processes and specific exercises: Mindfulness, Self as Context, Defusion, Values, Committed Action, and Acceptance.
- Dialectical behavioral therapy = a structured treatment, based on teaching emotional regulation skills, which combines the elements of behavioral therapy with mindfulness and acceptance practices.
- Habit reversal training = a technique based on three distinct phases: awareness training, competitive response, and social support.

Key facts

Skin picking and hair pulling disorders are psychopathological conditions that can be quite effectively treated through cognitive behavioral techniques in children and adolescents, such as habit reversal training, stimulus control technique, relaxation training, cognitive therapy, acceptance & commitment therapy, and dialectical behavior therapy.

Application to other areas

The contents of this chapter can be useful also for pediatric clinical practice, adolescent primary care, dermatological clinical practice, and psychiatric management of mental disorders.

Summary points

- Skin picking and hair pulling disorders often present in comorbidity and have in common several clinical features (e.g., age of onset, gender distribution, trigger situations, dysfunctional cognitions, difficulty in stopping repetitive behavior, presence of impulsivity and difficulty in emotional regulation).
- A psychotherapeutic approach aimed to target the vulnerability and maintenance factors common to both disorders is a useful strategy.
- The present chapter describes the use of cognitive behavioral therapy for skin picking and hair pulling disorders in youth as a multicomponent approach which includes behavioral interventions (e.g., habit reversal, stimulus control, relaxation training), cognitive strategies and third-wave approaches (ACT).

References

American Psychiatric Association. (2013). *Diagnostic and statistical manual of mental disorders* (5th ed.). Arlington, VA: American Psychiatric Association.

Azrin, N. H., & Nunn, R. G. (1973). Habit-reversal: A method of eliminating nervous habits and tics. *Behaviour Research and Therapy*, *11*, 619–628.

Azrin, N. H., & Peterson, A. L. (1988). Behavior therapy for Tourette's syndrome and tic disorders. In D. J. Cohen, R. D. Bruun, & J. L. Leckman (Eds.), *Tourette's syndrome and tic disorders: Clinical understanding and treatment* (pp. 238–255). New York, NY: John Wiley.

Bohne, A., Wilhelm, S., Keuthen, N. J., Baer, L., & Jenike, M. A. (2002). Skin picking in German students: Prevalence, phenomenology, and associated characteristics. *Behavior Modification*, *26*, 320–339.

Calikusu, C., Kucukgoncu, S., Tecer, Ö., & Bestepe, E. (2012). Skin picking in Turkish students: Prevalence, characteristics, and gender differences. *Behavior Modification*, *36*, 49–66.

Deckersbach, T., Wilhelm, S., Keuthen, N. J., Baer, L., & Jenike, M. A. (2002). Cognitive-behavior therapy for self-injurious skin picking: A case series. *Behavior Modification*, *26*(3), 361–377.

Deckersbach, T., Wilhelm, S., & Keuthen, N. (2003). Self-injurious skin picking: Clinical characteristics, assessment methods, and treatment modalities. *Brief Treatment & Crisis Intervention*, *3*(2), 249–260.

Downing, M. J. (2006). Relationship between excoriation practices, depression, anxiety, and obsessive-compulsive disorder among a college student sample. *Counseling and Clinical Psychology Journal*, *3*, 31–45.

Fine, K. M., Walther, M. R., Joseph, J. M., Robinson, J., Ricketts, E. J., Bowe, W. E., & Woods, D. W. (2012). Acceptance-enhanced behavior therapy for trichotillomania in adolescents. *Cognitive and Behavioral Practice*, *19*(3), 463–471. https://doi.org/10.1016/j.cbpra.2011.10.002.

Flessner, C. A., Busch, A. M., Heideman, P. W., & Woods, D. W. (2008). Acceptance-enhanced behavior therapy (AEBT) for trichotillomania and chronic skin picking: Exploring the effects of component sequencing. *Behavior Modification*, *32*, 579–594.

Flessner, C. A., Woods, D. W., Franklin, M. E., Cashin, S. E., & Keuthen, N. J. (2008). The Milwaukee inventory for subtypes of trichotillomania-adult version (MIST-A): Development of an instrument for the assessment of "focused" and "automatic" hair pulling. *Journal of Psychopathology and Behavioral Assessment*, *30*, 20–30.

Flessner, C. A., & Woods, D. W. (2006). Phenomenological characteristics, social problems, and the economic impact associated with chronic skin picking. *Behavior Modification*, *30*, 944–963.

Ghisi, M., Bottesi, G., Sica, C., Ouimet, A. J., & Sanavio, E. (2013). Prevalence, phenomenology and diagnostic criteria of hair-pulling in an Italian non-clinical sample: A preliminary study. *Journal of Obsessive-Compulsive and Related Disorders*, *2*, 22–29.

Grant, J. E., Stein, D. J., Woods, D. W., & Keuthen, N. J. (2012). *Trichotillomania, skin picking, and other body-focused repetitive behaviors*. Washington, DC: American Psychiatric Publishing.

Hajcak, G., Franklin, M. E., Simons, R. F., & Keuthen, N. J. (2006). Hairpulling and skin picking in relation to affective distress and obsessive-compulsive symptoms. *Journal of Psychopathology and Behavioral Assessment*, *28*(3), 177–185. https://doi.org/10.1007/s10862-005-9001-x.

Hayes, S. C., Strosahl, K. D., & Wilson, K. G. (1999). *Acceptance and commitment therapy: An experiential approach to behavior change*. New York, NY: Guilford Press.

Jones, G., Keuthen, N., & Greenberg, E. (2018). Assessment and treatment of trichotillomania (hair pulling disorder) and excoriation (skin picking) disorder. *Clinics in Dermatology*, *36*(6), 728–736.

Linehan, M. M. (1993). *Cognitive-behavioral treatment of borderline personality disorder*. New York, NY: The Guilford Press.

MacPherson, H., Cheavens, J., & Fristad, M. (2013). Dialectical behavior therapy for adolescents: Theory, treatment adaptations, and empirical outcomes. *Clinical Child and Family Psychology Review*, *16*, 59–80.

Miltenberger, R. G., & Fuqua, R. W. (1985). A comparison of contingent vs non-contingent competing response practice in the treatment of nervous habits. *Journal of Behavior Therapy and Experimental Psychiatry*, *16*, 195–200.

Moritz, S., & Rufer, M. (2011). Movement decoupling: A self-help intervention for the treatment of trichotillomania. *Journal of Behavior Therapy and Experimental Psychiatry*, *42*(1), 74–80. https://doi.org/10.1016/j.jbtep.2010.07.001.

Moritz, S., Fricke, S., Treszl, A., & Wittekind, C. E. (2012). Do it yourself! Evaluation of self-help habit reversal training versus decoupling in pathological skin picking: A pilot study. *Journal of Obsessive-Compulsive and Related Disorders*, *1*, 41–47.

O'Connor, K. P., Lavoie, M. E., & Schoendorff, B. (2017). *Managing tic and habit disorders: A cognitive psychophysiological treatment approach with acceptance strategies*. John Wiley & Sons.

Odlaug, B. L., & Grant, J. E. (2007). Childhood-onset pathologic skin picking: Clinical characteristics and psychiatric comorbidity. *Comprehensive Psychiatry, 48*, 388–393.

Odlaug, B. L., Lust, K., Schreiber, L. R. N., Christenson, G., Derbyshire, K., & Grant, J. E. (2013). Skin picking disorder in university students: Health correlates and gender differences. *General Hospital Psychiatry, 35*(2), 168–173.

Pozza, A., Albert, U., & Dèttore, D. (2020). Early maladaptive schemas as common and specific predictors of skin picking subtypes. *BMC Psychology, 8*, 1–11.

Pozza, A., Giaquinta, N., & Dèttore, D. (2016). Borderline, avoidant, sadistic personality traits and emotion dysregulation predict different pathological skin picking subtypes in a community sample. *Neuropsychiatric Disease and Treatment, 12*, 1861–1867.

Pozza, A., Giaquinta, N., & Dèttore, D. (2017). Impulsività e Alessitimia nei sottotipi di Skin Picking: uno studio su un ampio campione non-clinico. *Giornale Italiano di Psicologia, 44*, 917–938.

Roberts, S., O'Connor, K., & Bélanger, C. (2013). Emotion regulation and other psychological models for body-focused repetitive behaviors. *Clinical Psychology Review, 33*, 745–762.

Schuck, K., Keijsers, G. P., & Rinck, M. (2011). The effects of brief cognitive-behaviour therapy for pathological skin picking: A randomized comparison to wait-list control. *Behaviour Research and Therapy, 49*, 11–17.

Snorrason, Í., Smári, J., & Ólafsson, R. P. (2010). Emotion regulation in pathological skin picking: Findings from a non-treatment seeking sample. *Journal of Behavior Therapy and Experimental Psychiatry, 41*, 238–245.

Stein, D. J., Grant, J. E., Franklin, M. E., Keuthen, N., Lochner, C., Singer, H. S., & Woods, D. W. (2010). Trichotillomania (hair pulling disorder), skin picking disorder, and stereotypic movement disorder: Toward DSM-V. *Depression and Anxiety, 27*, 611–626.

Twohig, M. P., Hayes, S. C., & Masuda, A. (2006). A preliminary investigation of acceptance and commitment therapy as a treatment for chronic skin picking. *Behaviour Research and Therapy, 44*, 1513–1522.

Van Minnen, A., Hoogduin, K., Keijsers, G., Hellenbrand, I., & Hendriks, G. (2003). Treatment of trichotillomania with behavioral therapy or fluoxetine, a randomized, waiting list control study. *Archives of General Psychiatry, 60*, 517–522.

Welch, S., & Kim, J. (2012). DBT-enhanced cognitive behavioral therapy for adolescent trichotillomania: An adolescent case study. *Cognitive and Behavioral Practice, 19*, 183–193.

Wilhelm, S., Keuthen, N. J., Deckersbach, T., Engelhard, I. M., Forker, A. E., Baer, L., … Jenike, M. A. (1999). Self-injurious skin picking: Clinical characteristics and comorbidity. *Journal of Clinical Psychiatry, 60*, 454–459.

Further reading

Hallion, L. S., Park, J. M., & Keuthen, N. J. (2016). Treatment of an adult with excoriation (skin-picking) disorder. In E. A. Storch, & A. B. Lewin (Eds.), *Clinical handbook of obsessive-compulsive and related disorders* (pp. 273–287). Cham, Switzerland: Springer International Publishing.

Chapter 10

Cognitive behavioral therapy for child and adolescent hoarding disorder

Davíð R.M.A. Højgaard[a] and Gudmundur Skarphedinsson[b]

[a]*Department of Child and Adolescent Psychiatry, Aarhus University Hospital Psychiatry, Aarhus, Denmark* [b]*Faculty of Psychology, School of Health Sciences, University of Iceland, Reykjavík, Iceland*

Abbreviations

ADHD	attention deficit hyperactivity disorder
ASD	autism spectrum disorder
CBT	cognitive behavioral therapy
C-FOCI	Children's Florida Obsessive-Compulsive Scale
CHOCI	The Children's Obsessional Compulsive Inventory
CI	confidence interval
CSI	child savings inventory
CY-BOCS	Children's Yale-Brown Obsessive-Compulsive Scale
DIAMOND	Diagnostic Interview for Anxiety, Mood, and OCD and Related Neuropsychiatric Disorders
DSM-5	Diagnostic and Statistical Manual of Mental Disorders, 5th Edition
DSM-IV-TR	Diagnostic and Statistical Manual of Mental Disorders, 4th Edition, Text Revision
HAS	hoarding assessment scale
HD	hoarding disorder
HI	hoarding inventory
HRS-I	hoarding rating scale-interview
K-SADS-PL	the schedule for affective disorders and schizophrenia-present and lifetime version
OCD	obsessive-compulsive disorder
OCI-CV	obsessive-compulsive inventory-child version
OCPD	obsessive-compulsive personality disorder
SIHD	the structured interview for hoarding disorder
SI-R	saving inventory-revised for adults
SRI	serotonin reuptake inhibitors
UHSS	UCLA hoarding severity scale

Introduction

A very brief history of hoarding disorder

The first known reference to hoarding is in Dante's poem "The Inferno." It was written in the 14th century and described how hell consisted of nine circles that were ranked according to how bad they were. People who hoarded were placed in the fourth circle along with wasters. There they were supposed to crash heavy stones against each other, but the stones were thought to be a symbol of material wealth. Since then, hoarding has been referenced in many works, but it is not until the 1980s that compulsive hoarding was recognized by researchers as a phenomenon. In 2013, hoarding disorder (HD) was first classified as a distinct psychiatric disorder with the publication of DSM-5 (American Psychiatric Association, 2013), although it had already been theorized as an independent disorder in 1996 (Frost & Hartl, 1996; Zaboski et al., 2019).

Handbook of Lifespan Cognitive Behavioral Therapy. https://doi.org/10.1016/B978-0-323-85757-4.00037-7

Diagnosis of hoarding disorder in children and adolescents

Hoarding is thought to develop in childhood, although most people with HD first seek assistance from the health care system as adults. This has been confirmed in studies that have found symptom onset in childhood and a progressive worsening until adulthood (Grisham, Frost, Steketee, Kim, & Hood, 2006). It is therefore of utmost importance to identify symptoms of hoarding in children and adolescents in order to make the diagnosis as early as possible to avoid escalation of symptoms and a lengthy course of the disorder that may severely affect daily functioning and quality of life (Tolin et al., 2019).

Prior to the publication of DSM-5, hoarding was classified in DSM-IV-TR as one of the criteria for obsessive-compulsive personality disorder (OCPD) and defined as the inability to discard worn-out or worthless objects, even if no sentimental value is attached to them (American Psychiatric Association, 2013). Although not explicitly mentioned as a symptom of OCD in DSM-IV-TR, it stated that an OCD diagnosis should be considered in the case of severe hoarding. This could be particularly problematic if patients only displayed symptoms of hoarding but no other symptoms of OCD (Mataix-Cols et al., 2010).

In DSM-5, the following criteria must be fulfilled to in order to diagnose HD: (1) persistent difficulty in discarding or parting with possessions, regardless of their actual value; (2) perceived need to save items and distress associated with discarding them; (3) accumulation of possessions that clutter the active living area, and if these areas are uncluttered, it is because of the intervention of a third party; (4) the hoarding causes significant distress or impairment in one or more areas of functioning; (5) hoarding is not attributable to another medical condition; and (6) cannot be explained away by symptoms of another mental disorder, such as OCD. In addition to this, two specifiers are also defined: excessive acquisition and level of insight (good or fair insight, poor insight or delusional) (American Psychiatric Association, 2013). Excessive acquisition is seen in up to 85% of patients with HD (Frost, Tolin, Steketee, Fitch, & Selbo-Bruns, 2009), and reduced insight is commonly seen in children and adolescents with HD (Storch et al., 2007). HD in children and adolescents may be more episodic than in adult populations, and it is possible that hoarding-related clutter in this age group is less prevalent compared with adults, despite taking in account the intervention of their parents (Ivanov et al., 2021).

Even though HD is now seen as an independent disorder, hoarding can still be viewed as a part of another disorder such as OCD, as long as the hoarding symptoms are driven by obsessions. In such cases, an HD diagnosis is not given.

Prevalence of hoarding disorder

Knowledge about the prevalence of clinically significant hoarding symptoms in children and adolescents is currently extremely limited, and due to this there is considerable uncertainty about the prevalence of the disorder in this age group. A review has found the prevalence of HD to be 2.5% (CI 1.7%–3.6%) with similar rates for both males and females (Postlethwaite, Kellett, & Mataix-Cols, 2019). However, the review was focused on adults and was therefore not able to evaluate the influence of age on HD prevalence. A study that evaluated the prevalence rate among adolescents found it to be 2% (95% CI 1.6%–2.5%), based on a sample of 15-year-old twins ($N = 3974$), with higher prevalence in girls (Ivanov et al., 2013). By excluding the DSM-5 clutter criterion, the prevalence rate increased to 3.7% (95% CI 3.1%–4.3%) (Ivanov et al., 2013). Hoarding symptoms, without necessarily meeting the HD diagnostic criteria, have been found in almost 10% of a community sample of youths aged 6–17 years ($N = 16{,}718$) (Burton et al., 2016). The prevalence rate for younger children is even more uncertain, but a large study where parents were questioned about their children's hoarding behaviors found that 10% of the children showed some form of hoarding behavior, such as collecting or storing a large number of useless objects (Alvarenga et al., 2015). How many of these children would fulfill the diagnostic criteria for HD remains unknown.

Much more is known about the prevalence of hoarding symptoms in children and adolescents with OCD, but most studies have found it to be between 20% and 40% (Højgaard et al., 2019). However, a limitation to most of these studies is that hoarding symptoms are measured with an OCD-specific assessment, such as the Children's Yale-Brown Obsessive-Compulsive Scale (CY-BOCS), that may not catch non-OCD related hoarding symptoms.

The prevalence of HD in adults has been shown to increase linearly by 20% for every 5-year interval after the age of 15, with no gender differences (Cath, Nizar, Boomsma, & Mathews, 2017). There is thus a great need for research in the prevalence of HD in different populations and across the age span.

Age of onset for hoarding disorder

In a recent metaanalysis, the mean age of onset for HD was found to be 16.7 years of age (Zaboski et al., 2019). However, there is a possibility that children and adolescents diagnosed with OCD may be unaccounted for in these studies since hoarding was considered a part of OCD before the release of DSM-5. The true average age of onset could therefore be lower. An indication of this can be seen from a case series, where hoarding symptoms were found in all cases to start before the age of 8, and even before the age of 3 in one case (Plimpton, Frost, Abbey, & Dorer, 2009). An early HD onset is further supported by a study of children with OCD under the age of 10 where around 50% displayed hoarding symptoms (Frank et al., 2014), but caution should be taken as compulsive symptoms and hoarding symptoms are commonly seen in young children. Most studies of children and adolescents also lack a specific severity rating for the hoarding symptoms, making it difficult ascertain if their hoarding symptoms are clinically significant. When age of onset is examined retrospectively, the mean age of onset for hoarding symptoms has been shown to be around 13 years of age (Grisham et al., 2006). Age of onset studies applying properly validated assessment tools for HD that measure the severity of hoarding and are capable of distinguishing between HD and age-appropriate hoarding behavior in children are needed in order to reach a conclusion regarding the age of onset.

Comorbidity in hoarding disorder

HD and hoarding symptoms are commonly seen in concurrence with other psychiatric disorders, in particular OCD. However, due to the lack of systematic assessment of HD in children and adolescents, the rate of OCD in HD is uncertain. In adult studies where criteria for HD have been met, the OCD rate was found to vary between 37.5% and 56.7% (Frost, Steketee, & Grisham, 2004; Grisham, Brown, Savage, Steketee, & Barlow, 2007; Pertusa et al., 2008), but lower in community samples, where it was found to be below 18% (Frost, Steketee, & Tolin, 2011).

Studies that investigate comorbidity with other disorders are lacking, but most of the existing literature is based on samples of children and adolescents with OCD; this group, however, has been found to have a higher risk of comorbidity with tic disorders (Højgaard et al., 2019), autism symptoms (Boerema et al., 2019), and internalizing and externalizing symptoms (Storch et al., 2007), although results are somewhat inconsistent. Rates of attention deficit hyperactivity disorder (ADHD) have been observed to be elevated and to have a strong relationship with hoarding symptoms in childhood-onset OCD compared with general population samples (Sheppard et al., 2010). Furthermore, a study has found hoarding symptoms to correlate with ADHD and tic symptoms (Torres et al., 2016). Other studies have found hoarding symptoms to correlate with inattention in a mixed sample of patients with Gilles de la Tourette Disorder (Huisman-van Dijk, van de Schoot, Rijkeboer, Mathews, & Cath, 2016) and ADHD (Hacker et al., 2016), where both inattention and hyperactivity/impulsivity were distinctively associated with specific features of hoarding.

In a study on 204 children aged 7–13 with autism spectrum disorder (ASD) and comorbid anxiety or OCD symptoms, 34% had a moderate level of hoarding symptoms and 7% had severe or extreme levels of hoarding (Hacker et al., 2016). In younger children with ASD, between 4 and 7 years of age, the rate of hoarding symptoms has been shown to be 24.3% (Scahill et al., 2014).

Hoarding assessment for children and adolescents

In order to make a proper HD diagnosis and treatment planning, a proper assessment of hoarding symptoms needs to take place. The assessment must be based on validated instruments that take into account age-appropriate hoarding symptoms in children and adolescents (Hojgaard & Skarphedinsson, 2020). This is particularly important as collecting and saving behaviors can occur as age-appropriate behavior in children, and furthermore, children can have limited insight into their hoarding symptoms, thus further complicating assessment. Because children and adolescents usually live with their parents or under the supervision of adults, their ability to accumulate things is often hindered, masking the most noticeable symptom of HD and consequently limiting the impact on their daily functioning and quality of life. The DSM-5 takes this into account by stating that the criterion of accumulation does not need to be met as long as the living areas are uncluttered due to a third-party intervention, such as parents (American Psychiatric Association, 2013). In the assessment of children, it is also helpful to evaluate the nature of their attachment to things they have difficulty discarding as the nature of attachment may cause them severe distress if they feel they are damaging or hurting the item. Furthermore, a strong behavioral reaction to discarding is known to lead to negative behavioral reactions in children (Plimpton et al., 2009). Therefore, assessment of child and adolescent hoarding should not be performed with assessment tools developed for adults. Also, assessment tools developed for adults focus on clutter, but it is recommended that the assessment of children should rather focus on information processing deficits as well as the nature of their attachment to items (Plimpton et al., 2009; Storch et al., 2011). The information gained by this will also be valuable in the treatment of HD in this age group (Table 1)

Although many OCD-specific assessments such as the CY-BOCS include a checklist for hoarding symptoms, most of them are limited to few items and have no severity rating of the hoarding symptoms in isolation from the OCD symptoms. Similarly, assessments for other disorders do often not include hoarding assessment even though hoarding may be frequently seen in conjunction with those disorders (Mataix-Cols et al., 2010). Studies that investigate assessment of HD in children and adolescents are lacking, but part of the reason for this may be that no clinician-administered diagnostic tools are available for the assessment of HD in this group. Also, the DSM-5 version of the schedule for affective disorders

TABLE 1 Commonly expressed reasons for hoarding.

Qualities of living beings associated with object or items
Sentimental value
To avoid waste
The item/object may become useful in the future
The item has become a part of one's identity
The item has value
Important objects may accidentally get thrown out along with the item if discarded
Sometimes no reasons are given

and schizophrenia-present and lifetime version (K-SADS-PL) does not include a chapter on HD (Kaufman et al., 1997). However, HD will be included in the latest version of the child version of the Diagnostic Interview for Anxiety, Mood, and OCD and Related Neuropsychiatric Disorders (DIAMOND) (Tolin et al., 2018).

To date, the only HD-specific instrument available for children and adolescents is the child savings inventory (CSI) (Storch et al., 2011), but it includes 23 parent-rated items, is based on the saving inventory-revised for adults (SI-R) (Frost et al., 2004), and has four subscales: Discarding, Clutter, Acquisition, and Distress/Impairment (Storch, Muroff, et al., 2011). Each of the items is rated on a five-point Likert scale: (none, a little/minimal, some/moderate, most/much, almost all/completely), and the scale total score is calculated from all 23 items (Storch, Muroff, et al., 2011). The CSI subscales regarding discarding are aimed at measuring distress or interference associated with discarding items (e.g., "To what extent does your child have difficulty throwing things away that s/he does not need?"). The clutter subscale is intended to evaluate the clutter in the child's/adolescent's living areas and the distress that can be associated with it (e.g., "How much of your child's room (or where s/he sleeps, plays, etc.) is cluttered with possessions?"). The acquisition subscale measures the urge to acquire or purchase items (e.g., "How much control does your child have over his/her urges to acquire possessions that s/he does not need?"). The distress/impairment subscale measures distress and interference (e.g., "How distressing does your child find the task of throwing things away?"). This four-factor structure has, however, not been confirmed in a shortened 20-item version of the CSI (Soreni et al., 2018), but in this study, a three-factor solution based on 15 questions (CSI-15) was found (difficulty discarding, clutter, and distress/impairment). The main limitation of the CSI is that its psychometric properties have yet only been evaluated in child and adolescent OCD samples, which poses a limitation as HD may present differently in the absence of OCD or when exhibited with other comorbid disorders. Another limitation of the CSI is its lack of diagnostic utility as no DSM specific cut-off scores have been identified. Furthermore, the CSI is a parent-rated instrument only, but a clinician-rated instrument that comprises both the child's as well as parents' answers would make more accurate assessment and diagnosis of HD. Another instrument that assesses hoarding symptoms independently of other disorders is the compulsive hoarding in children: semistructured interview (Plimpton et al., 2009). This interview includes 27 questions focusing on the symptoms of hoarding behavior and the effect they may have on children, along with fluctuations in symptoms, ADHD and perfectionism, the child's own justifications for his/her behavior, and about the child's family history of hoarding symptoms and OCD.

There are several nonspecific hoarding assessments for children and adolescents available that measure hoarding symptoms in combination with symptoms of OCD, but as previously mentioned, many of these assessments do not assess all the domains central to hoarding and lack a severity rating of hoarding independent of other symptoms of OCD. One of these instruments is the CY-BOCS (Scahill et al., 1997), but it is a semistructured interview that includes a thorough checklist along with a severity measure of OCD. Hoarding is included in the checklist but is limited to only two items ("Fear of losing things" and "Difficulty throwing things away, saving bits of paper, string, etc."), although the clinician can add additional symptoms of hoarding in open text format (Scahill et al., 1997). The severity score is independent of the checklist and does therefore not provide any information of hoarding severity in isolation from other symptoms.

Other instruments that measure hoarding symptoms as part of OCD are available, but the compulsive inventory-child version (OCI-CV) (Foa et al., 2010), which is a self-report only, has more hoarding items than the CY-BOCS. The Children's Obsessional Compulsive Inventory (CHOCI) (Shafran et al., 2003) and the Children's Florida Obsessive-Compulsive Scale (C-FOCI) (Storch et al., 2009) are also well suited and commonly used self-reports designed to assess OCD symptoms and severity, including hoarding, but have the same limitations regarding HD as the other instruments.

Hoarding-specific assessments for use with adults or older adolescents

Currently, there are more assessment options for adults with HD. The structured interview for hoarding disorder (SIHD) is a good one, but it was designed based on the DSM-5 diagnostic criteria for HD (Nordsletten et al., 2013). Its strengths include the option to rule out hoarding behavior due to other disorders or medical conditions (Morris, Jaffee, Goodwin, & Franklin, 2016). The SIHD may be used to diagnose HD in adolescents 16 years of age and older (Mataix-Cols, Billotti, Fernández de la Cruz, & Nordsletten, 2013), but still needs to be adopted for use with younger adolescents and children.

There are also several interview-based assessments available for adults, such as the semistructured interviews: the hoarding rating scale-interview (HRS-I) (Tolin, Frost, & Steketee, 2010), and the UCLA hoarding severity scale (UHSS) (Saxena, Brody, Maidment, & Baxter, 2007). The interviews are similar, but the HRS-SR is better suited for use with adolescents as it refers only to the adolescent's own bedroom rather than the whole house or apartment, where the adolescent may have limited influence (Ivanov et al., 2013).

There are also several self-reports designed for adults that are worth mentioning, such as the hoarding scale (HS), but this is a 21-item self-report and was the first instrument designed to assess hoarding symptoms based on the contemporary theoretical model of hoarding (Frost & Gross, 1993). The HS has been adapted for use in children and adolescents under the name CSI (Storch, Muroff, et al., 2011). The SI-R is another self-report for adults that addresses some of the limitations of the HS (Frost et al., 2004). It measures hoarding on three subscales (1) excessive acquisition of purchased and free items, (2) saving and discarding behaviors, and (3) excessive clutter as a result of these behaviors (Frost et al., 2004). Finally, a short questionnaire named the hoarding assessment scale (HAS) is suitable for a quick assessment of hoarding symptoms (Schneider, Storch, Geffken, Lack, & Shytle, 2008).

The use of instruments designed for adults is best avoided when assessing children and young adolescents, but some of the aforementioned instruments for adults may be used for older adolescents, such as the SIHD.

Cognitive behavioral therapy for child and adolescent hoarding

Treatment for children and adolescents with hoarding problems

There is a lack of standardized treatment and systematic treatment studies for youth hoarding problems. One reason is that it is very rare that to meet children in clinics who have HD as their primary problem. However, it may potentially be beneficial to treat young people with hoarding problems early before they get more serious (Burton, Arnold, & Soreni, 2015).

In this section, evidence-based treatments for HD will be discussed. First, an overview of treatments for adult HD will be presented, as the only evidence of youth HD treatment is the downward extension of an adult CBT protocol, showing positive outcome but only in case studies.

Effectiveness of cognitive behavioral treatment

CBT for adult HD has proved to be effective. Tolin and colleagues (Tolin, Frost, Steketee, & Muroff, 2015) published a systematic review and metaanalysis by combining results from 12 HD samples. These patients were treated by recent CBT protocols (Muroff, Underwood, & Steketee, 2014; Tolin, Worden, Wootton, & Gilliam, 2017). Results showed positive outcome as large pre- to posteffect size was revealed for hoarding severity reduction ($g = 0.82$) and moderate effect size for hoarding-related impairment reduction ($g = 0.52$) (Tolin et al., 2015). Only one randomized controlled trial was included, which compared 12 weekly CBT sessions with waitlist control and showed moderate effect size (Steketee, Frost, Tolin, Rasmussen, & Brown, 2010).

The protocol by Tolin and colleagues lists seven interrelated HD treatment targets and strategies to address them. (1) Decision-making problems (e.g., deciding which items to discard and which items to sort) which may be caused by high levels of indecisiveness (Frost, Tolin, Steketee, & Oh, 2011; Samuels et al., 2002). A structured plan is proposed for regular sorting and discarding (in order to rehabilitate). It is also suggested to simplify the decision-making process so patients suffering may decide whether to donate, discard/recycle, or keep items. Thus, the focus is on minimizing and compensating for patients with potential executive functioning deficits. (2) Patients with HD frequently report maladaptive beliefs about different possessions and discarding. For instance, they report an overwhelming sense of responsibility for certain items (Frost, Hartl, Christian, & Williams, 1995), or they strongly fear that they will lose important information, or they feel a strong urge to prevent wastefulness or to maintain control over certain items by accumulating just in case that they are needed in the future. Cognitive restructuring is a recommended strategy to target these maladaptive ideas. For instance, one might recommend examining the evidence for and against a certain belief, and if necessary, to conduct behavioral experiments to test whether certain thoughts are accurate. (3) These patients often experience difficulties in regulating emotions, thus when items are discarded, they may react very emotionally as they are emotionally attached to the items and may react with sadness or depression when the item is discarded. Items may also be associated with a certain person, place, or past activity (Kellett & Holden, 2014; Steketee, Frost, & Kyrios, 2003). It is necessary to teach these patients about emotions, how they affect us, and about identifying these emotions properly. The protocol suggests emotional distress tolerance which includes strategies from acceptance and commitment therapy (Hayes, 2016). In addition, the patient is assisted in conducting exposure exercises to gradually confront these anxiety-provoking triggers. (4) Avoidance behavior is very common (e.g., avoiding discarding items as it evokes strong emotional reactions such as fear) and can be responsible for noncompliance and maintaining the HD symptoms. This needs to be addressed in each session. (5) Patients may also be impulsive which may be a problem that explains the excessive collecting and saving behavior. These behaviors may be enjoyable and provide a positive experience of emotions and cognitions (e.g., it is fun, and I feel that I am economical or thrifty) (Rasmussen, Steketee, Silverman, & Wilhelm, 2013). To address this, the protocol includes interventions used in substance abuse and compulsive buying treatments. It could include focusing on triggers for collecting items (e.g., specific situations, moods, and specific items). The patients are also encouraged to avoid these situations as they are associated with high-risk of collecting. Stimulus control is a critical component in refraining from collecting and saving items (challenging questions about usefulness and refrain temporarily). (6) Reduced motivation to change and poor insight may also impact the maintenance of HD (Frost, Tolin, & Maltby, 2010; Tolin, Fitch, Frost, & Steketee, 2010). The protocol suggests using motivational interviewing (MI) (Miller & Rollnick, 2013) in order to create ambivalence which may help motivate patients to modify their behavior. Direct confrontation and argumentation are not recommended when using MI. (7) Poor organizational skills are also common (Hartl et al., 2004; Woody, Kellman-McFarlane, & Welsted, 2014). For instance, items are not necessarily kept according to their categories (e.g., clothes in the wardrobe), thus the initial sorting and discarding work is much more difficult. Thus, it is important to teach and guide the patients to use organizational strategies (see Table 2).

Pharmacological treatment

Psychopharmacological treatments might also be considered for youths with HD. Only one case study was found that reported on effective treatment with fluoxetine and risperidone along with exposure treatment that resulted in remission for a 15-year-old girl with OCD and hoarding symptoms (Sockalingam & Zemans, 2007). In adult HD, studies are also very few. Only one metaanalysis exists, showing a mean response of 58% (95% CI 37%–76%) among 92 adult HD patients in 7 studies. Treatment consisted of serotonin reuptake inhibitors (SRI), venlafaxine, methylphenidate, and augmentation of SRI with quetiapine, minocycline, and naltrexone (Brakoulias, Eslick, & Starcevic, 2015). A more recent review shows that there are currently no future studies aiming at testing drugs in HD (Piacentino et al., 2019).

It is often necessary to consider appropriate pharmacological treatments for comorbid disorders, of which ADHD and OCD are likely the most common in youths. Thus, SRI treatment for OCD and stimulant medication for ADHD might have direct and indirect effects on HD. For instance, when a child with HD and ADHD is given treatment with stimulants, it might also have reduced attention deficits, thus increasing the child's ability to organize and sort different items (Pliszka & AACAP Work Group on Quality Issues, 2007; Rodriguez et al., 2013). It has been reported that adult HD is associated with attentional problems which may have an impact on the treatment of HD (Huisman-van Dijk et al., 2016).

For OCD and HD, SRI medication might reduce obsessional worries which trigger collecting and acquisition. However, the existing data on adult OCD with hoarding symptoms have shown poor outcome (Bloch, McGuire, Landeros-Weisenberger, Leckman, & Pittenger, 2010; Mataix-Cols, Rauch, Manzo, Jenike, & Baer, 1999). However, these studies did not report on specific hoarding symptom outcome.

TABLE 2 Key behaviors to target in CBT for hoarding disorder.

(1) Decision-making problems
(2) Maladaptive beliefs about different possessions and discarding
(3) Difficulties in regulating emotions when items are discarded
(4) Avoidance behavior (avoiding discarding items) is common
(5) Impulsivity is a problem that explains the excessive collecting and saving behavior
(6) Reduced motivation to change and poor insight may impact the maintenance of hoarding
(7) Poor organizational skills are common

Treatment for pediatric OCD with hoarding symptoms

The most extensive data come from general treatments of patients with OCD. In adult OCD, the data show that for patients with OCD and comorbid HD symptoms, the CBT outcome is in general poor compared to other symptoms of OCD (Bloch et al., 2014). In this metaanalysis, only two youth studies were included that showed similar results. The data on pediatric OCD are more positive, showing that youths with OCD and comorbid hoarding symptoms fare just as well as children with OCD without hoarding when treated with CBT (Højgaard et al., 2019; Masi et al., 2009; Olino et al., 2011; Rozenman et al., 2019; Storch, Rahman, et al., 2011).

Practical aspects of CBT for children with hoarding disorder

When working with young people with hoarding problems, it is important to know the similarities and the differences between young people and adults. Similarities include conducting a careful assessment of the acquisition, organization, and discarding patterns as well as other important issues. However, the clinician needs to adapt the language, methodology, and the process of the CBT to the appropriate age of the patient, for instance, specific age-related changes in how children relate and are attached to things. In addition, the organizational skills develop and improve with increasing age. Children are also without exception influenced by parents or other important family members. Thus, it is not as easy to accumulate or prevent others from knowing about their hoarding. This also means that others can intervene by removing things that are accumulated.

When working with children with hoarding, clinicians recommend that the following patterns are observed: (1) identify things that are associated with emotional attachment; (2) assess whether the child (and this is more common in younger children) believes that things have their own feelings, thus, by discarding them, they will become scared and sad; (3) that things may be important in order to remember specific events, for instance, toilet rolls which are empty may be related to a specific event in the life of the child (Whomsley, 2020) (see Table 3).

The recommended intervention is the E/RP, with the main focus of assisting the child to select and rank items according to importance and then slowly discarding them (Ale, Arnold, Whiteside, & Storch, 2014; Andersen & Thomsen, 2017; Gallo, Wilson, & Comer, 2013; McKay, 2016; Plimpton et al., 2009; Storch, Rahman, et al., 2011).

As mentioned earlier, no systematic treatment studies exist for pediatric HD. However, seven case studies were found reporting on treatment for eight individuals (Ale et al., 2014; Andersen & Thomsen, 2017; Gallo et al., 2013; McKay, 2016; Plimpton et al., 2009; Storch, Rahman, et al., 2011). The available studies show that the following CBT components are commonly applied: (1) parent-inclusion and giving parents adequate training to help their children, (2) E/RP, (3) contingency management (CM), and (4) cognitive training and strategies. All these components will be described below (see Table 4). A necessary initiating component is a thorough diagnostic and behavioral assessment.

TABLE 3 Treatment adaptations for children with HD.

(1) Identifying that things that are associated with emotional attachment
(2) Assessing beliefs of things having their own feelings (becoming scared and sad when discarded)
(3) Identifying whether things may be important in order to remember specific events

TABLE 4 CBT components in HD treatment for children.

(1) Parent-inclusion and adequate training for parents
(2) Exposure and response prevention
(3) Contingency management
(4) Cognitive training and strategies

(1) It may be essential to include parents in the treatment. In adult HD treatment, data suggest that more spouse (or other family member) assistance is associated with better treatment results (Muroff, Steketee, Bratiotis, & Ross, 2012; Tolin et al., 2015). All the available case studies on children with HD describe the significant role of parents (Ale et al., 2014; Andersen & Thomsen, 2017; Gallo et al., 2013; McKay, 2016; Plimpton et al., 2009; Storch, Rahman, et al., 2011). Parents may assist in treatment in many ways, such as helping with and facilitate homework and other treatment assignments (e.g., discarding and sorting tasks). Furthermore, they can be instructed to apply specific behavioral strategies included in the treatment. In addition, they can help and encourage their children to use the proper strategies between sessions. Parents may also need to use behavioral strategies to reduce treatment noncompliance or oppositional defiant behavior.

(2) Exposure to discarding possessions and not collecting new items is considered to be essential in HD treatment (Muroff et al., 2014; Steketee et al., 2010; Tolin et al., 2017). The case studies available apply this important strategy. What is necessary is to develop a hierarchy of item discarding and accumulation. The child is taught to rank these activities according to difficulty level. Then the child advances, in small incremental goals, from the current state to the end goal of having removed the clutter, and the child can discard without difficulties.

(3) Using contingency management may also be necessary. Specially to control potential oppositional and disruptive behavior (e.g., flare of anger or violence) and to motivate exposures, discarding and nonacquiring behavior (e.g., discarding or sorting). Children are rewarded for positive behaviors (e.g., participating in exposure or discarding) and response cost is applied for unacceptable behaviors (e.g., loss of token/privileges).

(4) Cognitive training and related strategies may also be useful. Cognitive restructuring would be applied to modify or reevaluate thoughts, responsible for hoarding behavior. Storch and colleagues (2011) describe a child with excessive hoarding of food items, who feared going hungry or starving. The child was taught to make statements about the availability of food. Then she was asked about her previous experiences of being hungry, for instance, periods when she was hungry for a short period of time but was given food eventually. She was also taught to evaluate the likelihood of being left without a chance to eat. As a part of cognitive strategies, children need to be taught problem-solving skills to solve problems and to identify alternative strategies. A common approach is to use didactic instruction modeling and role playing. For instance, by using the acronym RIBEYE to describe the problem-solving steps: (1) Relax, (2) Identify the problem, (3) Brainstorm potential solutions, (4) Evaluate solutions, (5) say Yes to the best solution, and (6) Encourage yourself for completion (Rohde, Feeny, & Robins, 2005).

Training children in decision-making may also be an important strategy to reduce and discard clutter. Children are instructed to sort items into smaller number of categories and according to relevance.

To conclude, we encourage practitioners and researchers to adapt a CBT adult HD protocol to pediatric HD and to collaborate across sites for feasibility and efficacy evaluation. The limited data suggest that pediatric HD is prevalent. As data on adults with HD suggest that early onset is common, early identification and treatment are necessary.

Conclusion

Hoarding symptoms often emerge in early adolescence. However, the symptoms of hoarding in this age group may not be as noticeable as clutter often is controlled by the parents or caretakers. Also, apart from OCD populations, the knowledge about hoarding symptoms is limited in this age group. Hoarding symptoms may also be milder in children and adolescents as they are known to become more severe with age, making early diagnosis and treatment when symptoms start to emerge important. The suggested CBT approach comprises four connected elements: (1) the use of parents as cotherapists, (2) exposure to discarding and refrain from collecting new items, (3) using contingency management for exposure and oppositional/disruptive behavior, and (4) cognitive training.

Applications to other areas

In this chapter, we discussed CBT for the treatment of HD in children and adolescents. As hoarding symptoms are often seen in conjunction with other psychiatric problems it may be of use to clinicians to screen and if positive to further assess by using some of the assessments described in this chapter. We do, however, advice against the use of instruments on other age-groups than they are designed for. As for the treatment principles described, they can be implemented in treatment of other disorders in case of comorbid hoarding symptoms. The general principles of CBT defined may also be adjusted and used for the treatment of other symptoms.

Practice and procedures

The following is an example of how a clinician could proceed in the evaluation of hoarding symptoms in a child/adolescent.

First, the diagnostic criteria should be evaluated, and a diagnosis given if they are fulfilled. In order to get a more detailed view of the hoarding symptoms, an assessment tool should be used, such as the CSI ore a more age-appropriate tool in case of older adolescents. In addition to the assessment, it is important to proceed with identifying things that may be associated with emotional attachment, explore whether the child believes that things have their own feelings, for example by asking if they believe they think the item may become scared or sad if thrown out. Then proceed with exploring if items at stake may be important in order to remember specific events that are important life events for the child. By following these steps, the clinician is laying a solid ground for proper treatment of hoarding disorder.

Summary points

- The average age of onset for hoarding disorder seems to be in early adolescence.
- There is an urgent need for properly validated assessment tools based on the DSM-5 criteria.
- Early identification and prevention of hoarding disorder is important to avoid worsening of symptoms and reduced functioning in adult life.
- A CBT strategy should comprise using parents as cotherapists, exposure to discarding and not collecting new items, using contingency management for exposure and oppositional/disruptive behavior, as well as cognitive training.
- Practitioners and researchers are encouraged to develop a CBT protocol for the treatment of child and adolescent hoarding disorder.

Key facts of hoarding disorder

- Hoarding was not recognized as a phenomenon of research until the 1980s.
- Hoarding disorder was first classified as a distinct psychiatric disorder in 2013 with the publication of DSM-5.
- Hoarding is believed to develop in childhood, although most people do not seek assistance from the health care system before they are adults.
- The prevalence of hoarding disorder is estimated to be around 2.5%.
- Hoarding disorder can be effectively treated with cognitive behavioral therapy (CBT).

Mini-dictionary of terms

Cognitive behavioral therapy (CBT) is a psychotherapeutic treatment method designed to treat problematic behavior and thought patterns that have a negative influence on the well-being of a person.

Cognitive restructuring is a psychological process of identifying and challenging thoughts that are irrational or maladaptive.

Comorbidity is the diagnosis of one or more additional disorders that co-occur with the main diagnosed disorder.

The Diagnostic and Statistical Manual of Mental Disorders (DSM) is a handbook of mental disorders that is designed to assist health care professionals in making accurate diagnoses.

Exposure is a therapy technique used in CBT to treat anxiety by exposing the person to the anxiety provoking stimuli and is based on the principle of respondent conditioning. Exposure will with time lead to a gradual decrease in the anxiety response associated with the exposed stimuli.

Maladaptive beliefs are beliefs that are inaccurate, rigid, and negatively biased and can play a role in the development of psychiatric disorders such as anxiety.

Prevalence is an epidemiological term that refers to the proportion of a particular population that is affected by a disorder at a specific time.

Psychological assessment is the gathering and combination of information from individuals used to evaluate behavior thoughts and other characteristics, than can be used to select treatment strategies and/or making a diagnosis.

A reward is an event, object or experience that is subjectively pleasurable or satisfying to the respondent and reinforces a specific behavior.

A self-report is a type of assessment tool where the respondent responds to questions concerning their own behavior or thoughts, without any assistance or interference.

References

Ale, C. M., Arnold, E. B., Whiteside, S. P. H., & Storch, E. A. (2014). Family-based behavioral treatment of pediatric compulsive hoarding: A case example. *Clinical Case Studies*, *13*(1), 9–21. https://doi.org/10.1177/1534650113504487.

Alvarenga, P. G., Cesar, R. C., Leckman, J. F., Moriyama, T. S., Torres, A. R., Bloch, M. H., … do Rosario, M. C. (2015). Obsessive-compulsive symptom dimensions in a population-based, cross-sectional sample of school-aged children. *Journal of Psychiatric Research*, *62*, 108–114. https://doi.org/10.1016/j.jpsychires.2015.01.018.

American Psychiatric Association. (2013). *Diagnostic and Statistical Manual of Mental Disorders, 5th Edition*. American Psychiatric Publishing. https://doi.org/10.1176/appi.books.9780890425596.893619.

Andersen, T. M. G., & Thomsen, P. H. (2017). Behandling af hamstringssyndrom hos en 12-årig dreng. *Ugeskrift for Laeger*, *179*(17), 1500–1502.

Bloch, M. H., Bartley, C. A., Zipperer, L., Jakubovski, E., Landeros-Weisenberger, A., Pittenger, C., & Leckman, J. F. (2014). Meta-analysis: Hoarding symptoms associated with poor treatment outcome in obsessive-compulsive disorder. *Molecular Psychiatry*, *19*(9), 1025–1030. https://doi.org/10.1038/mp.2014.50.

Bloch, M. H., McGuire, J., Landeros-Weisenberger, A., Leckman, J. F., & Pittenger, C. (2010). Meta-analysis of the dose-response relationship of SSRI in obsessive-compulsive disorder. *Molecular Psychiatry*, *15*(8), 850–855. https://doi.org/10.1038/mp.2009.50.

Boerema, Y. E., de Boer, M. M., van Balkom, A. J. L. M., Eikelenboom, M., Visser, H. A., & van Oppen, P. (2019). Obsessive compulsive disorder with and without hoarding symptoms: Characterizing differences. *Journal of Affective Disorders*, *246*(December 2018), 652–658. https://doi.org/10.1016/j.jad.2018.12.115.

Brakoulias, V., Eslick, G. D., & Starcevic, V. (2015). A meta-analysis of the response of pathological hoarding to pharmacotherapy. *Psychiatry Research*, *229*(1–2), 272–276. https://doi.org/10.1016/j.psychres.2015.07.019.

Burton, C. L., Arnold, P. D., & Soreni, N. (2015). Three reasons why studying hoarding in children and adolescents is important. *Journal of the Canadian Academy of Child and Adolescent Psychiatry = Journal de l'Academie Canadienne de Psychiatrie de l'enfant et de l'adolescent*, *24*(2), 128–130. http://www.ncbi.nlm.nih.gov/pubmed/26379725.

Burton, C. L., Crosbie, J., Dupuis, A., Mathews, C. A., Soreni, N., Schachar, R., & Arnold, P. D. (2016). Clinical correlates of hoarding with and without comorbid obsessive-compulsive symptoms in a community pediatric sample. *Journal of the American Academy of Child and Adolescent Psychiatry*, *55*(2), 114–121.e2. https://doi.org/10.1016/j.jaac.2015.11.014.

Cath, D. C., Nizar, K., Boomsma, D., & Mathews, C. A. (2017). Age-specific prevalence of hoarding and obsessive compulsive disorder: A population-based study. *The American Journal of Geriatric Psychiatry*, *25*(3), 245–255. https://doi.org/10.1016/j.jagp.2016.11.006.

Foa, E. B., Coles, M., Huppert, J. D., Pasupuleti, R. V., Franklin, M. E., & March, J. (2010). Development and validation of a child version of the obsessive compulsive inventory. *Behavior Therapy*, *41*(1), 121–132. https://doi.org/10.1016/j.beth.2009.02.001.

Frank, H., Stewart, E., Walther, M., Benito, K., Freeman, J., Conelea, C., … Garcia, A. (2014). Hoarding behavior among young children with obsessive-compulsive disorder. *Journal of Obsessive-Compulsive and Related Disorders*, *3*(1), 6–11. https://doi.org/10.1016/j.jocrd.2013.11.001.

Frost, R. O., & Gross, R. C. (1993). The hoarding of possessions. *Behaviour Research and Therapy*, *31*(4), 367–381. https://doi.org/10.1016/0005-7967(93)90094-B.

Frost, R. O., & Hartl, T. L. (1996). A cognitive-behavioral model of compulsive hoarding. *Behaviour Research and Therapy*, *34*(4), 341–350. https://doi.org/10.1016/0005-7967(95)00071-2.

Frost, R. O., Hartl, T. L., Christian, R., & Williams, N. (1995). The value of possessions in compulsive hoarding: Patterns of use and attachment. *Behaviour Research and Therapy*, *33*(8), 897–902. https://doi.org/10.1016/0005-7967(95)00043-W.

Frost, R. O., Steketee, G., & Grisham, J. (2004). Measurement of compulsive hoarding: Saving inventory-revised. *Behaviour Research and Therapy*, *42*(10), 1163–1182. https://doi.org/10.1016/j.brat.2003.07.006.

Frost, R. O., Steketee, G., & Tolin, D. F. (2011). Comorbidity in hoarding disorder. *Depression and Anxiety*, *28*(10), 876–884. https://doi.org/10.1002/da.20861.

Frost, R. O., Tolin, D. F., & Maltby, N. (2010). Insight-related challenges in the treatment of hoarding. *Cognitive and Behavioral Practice*, *17*(4), 404–413. https://doi.org/10.1016/j.cbpra.2009.07.004.

Frost, R. O., Tolin, D. F., Steketee, G., Fitch, K. E., & Selbo-Bruns, A. (2009). Excessive acquisition in hoarding. *Journal of Anxiety Disorders*, *23*(5), 632–639. https://doi.org/10.1016/j.janxdis.2009.01.013.

Frost, R. O., Tolin, D. F., Steketee, G., & Oh, M. (2011). Indecisiveness and hoarding. *International Journal of Cognitive Therapy*, *4*(3), 253–262. https://doi.org/10.1521/ijct.2011.4.3.253.

Gallo, K. P., Wilson, L. A. S., & Comer, J. S. (2013). Treating hoarding disorder in childhood: A case study. *Journal of Obsessive-Compulsive and Related Disorders*, *2*(1), 62–69. https://doi.org/10.1016/j.jocrd.2012.11.001.

Grisham, J. R., Brown, T. A., Savage, C. R., Steketee, G., & Barlow, D. H. (2007). Neuropsychological impairment associated with compulsive hoarding. *Behaviour Research and Therapy*, *45*(7), 1471–1483. https://doi.org/10.1016/j.brat.2006.12.008.

Grisham, J. R., Frost, R. O., Steketee, G., Kim, H. J., & Hood, S. (2006). Age of onset of compulsive hoarding. *Journal of Anxiety Disorders*, *20*(5), 675–686. https://doi.org/10.1016/j.janxdis.2005.07.004.

Hacker, L. E., Park, J. M., Timpano, K. R., Cavitt, M. A., Alvaro, J. L., Lewin, A. B., … Storch, E. A. (2016). Hoarding in children with ADHD. *Journal of Attention Disorders*, *20*(7), 617–626. https://doi.org/10.1177/1087054712455845.

Hartl, T. L., Frost, R. O., Allen, G. J., Deckersbach, T., Steketee, G., Duffany, S. R., & Savage, C. R. (2004). Actual and perceived memory deficits in individuals with compulsive hoarding. *Depression and Anxiety*, *20*(2), 59–69. https://doi.org/10.1002/da.20010.

Hayes, S. C. (2016). Acceptance and commitment therapy, relational frame theory, and the third wave of behavioral and cognitive therapies—Republished article. *Behavior Therapy*, *47*(6), 869–885. https://doi.org/10.1016/j.beth.2016.11.006.

Hojgaard, D. R. M. A., & Skarphedinsson, G. (2020). Assessment, treatment and the importance of early intervention of childhood hoarding. *Children Australia*, *45*(3), 145–152. https://doi.org/10.1017/cha.2020.17.

Højgaard, D. R. M. A., Skarphedinsson, G., Ivarsson, T., Weidle, B., Nissen, J. B., Hybel, K. A., … Thomsen, P. H. (2019). Hoarding in children and adolescents with obsessive-compulsive disorder: Prevalence, clinical correlates, and cognitive behavioral therapy outcome. *European Child & Adolescent Psychiatry*, *28*(8), 1097–1106. https://doi.org/10.1007/s00787-019-01276-x.

Huisman-van Dijk, H. M., van de Schoot, R., Rijkeboer, M. M., Mathews, C. A., & Cath, D. C. (2016). The relationship between tics, OC, ADHD and autism symptoms: A cross- disorder symptom analysis in Gilles de la Tourette syndrome patients and family-members. *Psychiatry Research*, *237*(7), 138–146. https://doi.org/10.1016/j.psychres.2016.01.051.

Ivanov, V. Z., Mataix-Cols, D., Serlachius, E., Brander, G., Elmquist, A., Enander, J., & Rück, C. (2021). The developmental origins of hoarding disorder in adolescence: a longitudinal clinical interview study following an epidemiological survey. *European Child & Adolescent Psychiatry*, *30*(3), 415–425. https://doi.org/10.1007/s00787-020-01527-2.

Ivanov, V. Z., Mataix-Cols, D., Serlachius, E., Lichtenstein, P., Anckarsäter, H., Chang, Z., … Rück, C. (2013). Prevalence, comorbidity and heritability of hoarding symptoms in adolescence: A population based twin study in 15-year olds. *PLoS One*, *8*(7). https://doi.org/10.1371/journal.pone.0069140, e69140.

Kaufman, J., Birmaher, B., Brent, D., Rao, U., Flynn, C., Moreci, P., … Ryan, N. (1997). Schedule for affective disorders and schizophrenia for school-age children-present and lifetime version (K-SADS-PL): Initial reliability and validity data. *Journal of the American Academy of Child & Adolescent Psychiatry*, *36*(7), 980–988. https://doi.org/10.1097/00004583-199707000-00021.

Kellett, S., & Holden, K. (2014). *Emotional attachment to objects in hoarding*. Oxford University Press. https://doi.org/10.1093/oxfordhb/9780199937783.013.013.

Masi, G., Millepiedi, S., Perugi, G., Pfanner, C., Berloffa, S., Pari, C., & Mucci, M. (2009). Pharmacotherapy in paediatric obsessive-compulsive disorder: A naturalistic, retrospective study. *CNS Drugs*, *23*(3), 241–252. https://doi.org/10.2165/00023210-200923030-00005.

Mataix-Cols, D., Rauch, S. L., Manzo, P. A., Jenike, M. A., & Baer, L. (1999). Use of factor-analyzed symptom dimensions to predict outcome with serotonin reuptake inhibitors and placebo in the treatment of obsessive-compulsive disorder. *The American Journal of Psychiatry*, *156*(9), 1409–1416. https://doi.org/10.1176/ajp.156.9.1409.

Mataix-Cols, D., Billotti, D., Fernández de la Cruz, L., & Nordsletten, A. E. (2013). The London field trial for hoarding disorder. *Psychological Medicine*, *43*(4), 837–847. https://doi.org/10.1017/S0033291712001560.

Mataix-Cols, D., Frost, R. O., Pertusa, A., Clark, L. A., Saxena, S., Leckman, J. F., … Wilhelm, S. (2010). Hoarding disorder: A new diagnosis for DSM-V? *Depression and Anxiety*, *27*(6), 556–572. https://doi.org/10.1002/da.20693.

McKay, D. (2016). Cognitive-behavioral treatment of hoarding in youth: A case illustration. *Journal of Clinical Psychology*, *72*(11), 1209–1218. https://doi.org/10.1002/jclp.22400.

Miller, W. R., & Rollnick, S. (2013). *Motivational interviewing: Preparing people for change* (3rd ed.). Guilford Press.

Morris, S. H., Jaffee, S. R., Goodwin, G. P., & Franklin, M. E. (2016). Hoarding in children and adolescents: A review. *Child Psychiatry and Human Development*, *47*(5), 740–750. https://doi.org/10.1007/s10578-015-0607-2.

Muroff, J., Steketee, G., Bratiotis, C., & Ross, A. (2012). Group cognitive and behavioral therapy and bibliotherapy for hoarding: A pilot trial. *Depression and Anxiety*, *29*(7), 597–604. https://doi.org/10.1002/da.21923.

Muroff, J., Underwood, P., & Steketee, G. (2014). In D. H. Barlow (Ed.), *Group treatment for hoarding disorder: Therapist guide (treatments that work)* (1st ed.). Oxford University Press. https://doi.org/10.1093/med:psych/9780199340965.001.0001.

Nordsletten, A. E., Fernández de la Cruz, L., Pertusa, A., Reichenberg, A., Hatch, S. L., & Mataix-Cols, D. (2013). The structured interview for hoarding disorder (SIHD): Development, usage and further validation. *Journal of Obsessive-Compulsive and Related Disorders*, *2*(3), 346–350. https://doi.org/10.1016/j.jocrd.2013.06.003.

Olino, T. M., Gillo, S., Rowe, D., Palermo, S., Nuhfer, E. C., Birmaher, B., & Gilbert, A. R. (2011). Evidence for successful implementation of exposure and response prevention in a naturalistic group format for pediatric OCD. *Depression and Anxiety*, *28*(September 2010), 342–348. https://doi.org/10.1002/da.20789.

Pertusa, A., Fullana, M. A., Singh, S., Alonso, P., Menchón, J. M., & Mataix-Cols, D. (2008). Compulsive hoarding: OCD symptom, distinct clinical syndrome, or both? *American Journal of Psychiatry*, *165*(10), 1289–1298. https://doi.org/10.1176/appi.ajp.2008.07111730.

Piacentino, D., Pasquini, M., Cappelletti, S., Chetoni, C., Sani, G., & Kotzalidis, G. D. (2019). Pharmacotherapy for hoarding disorder: How did the picture change since its excision from OCD? *Current Neuropharmacology*, *17*(8), 808–815. https://doi.org/10.2174/1570159X17666190124153048.

Plimpton, E. H., Frost, R. O., Abbey, B. C., & Dorer, W. (2009). Compulsive hoarding in children: Six case studies. *International Journal of Cognitive Therapy*, *2*(1), 88–104. https://doi.org/10.1521/ijct.2009.2.1.88.

Pliszka, S., & AACAP Work Group on Quality Issues. (2007). Practice parameter for the assessment and treatment of children and adolescents with attention-deficit/hyperactivity disorder. *Journal of the American Academy of Child and Adolescent Psychiatry*, *46*(7), 894–921. https://doi.org/10.1097/chi.0b013e318054e724.

Postlethwaite, A., Kellett, S., & Mataix-Cols, D. (2019). Prevalence of hoarding disorder: A systematic review and meta-analysis. *Journal of Affective Disorders*, *256*(January), 309–316. https://doi.org/10.1016/j.jad.2019.06.004.

Rasmussen, J., Steketee, G., Silverman, M., & Wilhelm, S. (2013). The relationship of hoarding symptoms to schizotypal personality and cognitive schemas in an OCD sample. *Journal of Cognitive Psychotherapy*, *27*(4), 384–396. https://doi.org/10.1891/0889-8391.27.4.384.

Rodriguez, C. I., Bender, J., Morrison, S., Mehendru, R., Tolin, D., & Simpson, H. B. (2013). Does extended release methylphenidate help adults with hoarding disorder?: A case series. *Journal of Clinical Psychopharmacology*, *33*(3), 444–447. https://doi.org/10.1097/JCP.0b013e318290115e.

Rohde, P., Feeny, N. C., & Robins, M. (2005). Characteristics and components of the TADS CBT approach. *Cognitive and Behavioral Practice*, *12*(2), 186–197. https://doi.org/10.1901/jaba.2005.12-186.

Rozenman, M., McGuire, J., Wu, M., Ricketts, E., Peris, T., O'Neill, J., … Piacentini, J. (2019). Hoarding symptoms in children and adolescents with obsessive-compulsive disorder: Clinical features and response to cognitive-behavioral therapy. *Journal of the American Academy of Child and Adolescent Psychiatry*. https://doi.org/10.1016/j.jaac.2019.01.017.

Samuels, J. F., Joseph Bienvenu, O., Riddle, M. A., Cullen, B. A. M., Grados, M. A., Liang, K. Y., … Nestadt, G. (2002). Hoarding in obsessive compulsive disorder: Results from a case-control study. *Behaviour Research and Therapy*, *40*(5), 517–528. https://doi.org/10.1016/S0005-7967(01)00026-2.

Saxena, S., Brody, A. L., Maidment, K. M., & Baxter, L. R. (2007). Paroxetine treatment of compulsive hoarding. *Journal of Psychiatric Research*, *41*(6), 481–487. https://doi.org/10.1016/j.jpsychires.2006.05.001.

Scahill, L., Dimitropoulos, A., McDougle, C. J., Aman, M. G., Feurer, I. D., McCracken, J. T., … Vitiello, B. (2014). Children's Yale-Brown Obsessive Compulsive Scale in autism spectrum disorder: Component structure and correlates of symptom checklist. *Journal of the American Academy of Child and Adolescent Psychiatry*, *53*(1). https://doi.org/10.1016/j.jaac.2013.09.018.

Scahill, L., Riddle, M. A., McSwiggin-Hardin, M., Ort, S. I., King, R. A., Goodman, W. K., … Leckman, J. F. (1997). Children's Yale-Brown Obsessive Compulsive Scale: Reliability and validity. *Journal of the American Academy of Child and Adolescent Psychiatry*, *36*(6), 844–852. https://doi.org/10.1097/00004583-199706000-00023.

Schneider, A. F., Storch, E. A., Geffken, G. R., Lack, C. W., & Shytle, R. D. (2008). Psychometric properties of the hoarding assessment scale in college students. *Illness, Crisis & Loss*, *16*(3), 227–236. https://doi.org/10.2190/IL.16.3.c.

Shafran, R., Frampton, I., Heyman, I., Reynolds, M., Teachman, B., & Rachman, S. (2003). The preliminary development of a new self-report measure for OCD in young people. *Journal of Adolescence*, *26*(1), 137–142. https://doi.org/10.1016/S0140-1971(02)00083-0.

Sheppard, B., Chavira, D., Azzam, A., Grados, M. A., Umaña, P., Garrido, H., & Mathews, C. A. (2010). ADHD prevalence and association with hoarding behaviors in childhood-onset OCD. *Depression and Anxiety*, *27*(7), 667–674. http://www.pubmedcentral.nih.gov/articlerender.fcgi?artid=2925836&tool=pmcentrez&rendertype=abstract.

Sockalingam, S., & Zemans, M. (2007). Compulsive hoarding associated with abortion. *Journal of the Canadian Academy of Child and Adolescent Psychiatry = Journal de l'Academie Canadienne de Psychiatrie de l'enfant et de l'adolescent*, *16*(4), 177–179. http://www.ncbi.nlm.nih.gov/pubmed/18392171.

Soreni, N., Cameron, D., Vorstenbosch, V., Duku, E., Rowa, K., Swinson, R., … McCabe, R. (2018). Psychometric evaluation of a revised scoring approach for the children's saving inventory in a Canadian sample of youth with obsessive–compulsive disorder. *Child Psychiatry and Human Development*, *49*(6), 966–973. https://doi.org/10.1007/s10578-018-0811-y.

Steketee, G., Frost, R. O., & Kyrios, M. (2003). Cognitive aspects of compulsive hoarding. *Cognitive Therapy and Research*, *27*(4), 463–479. https://doi.org/10.1023/A:1025428631552.

Steketee, G., Frost, R. O., Tolin, D. F., Rasmussen, J., & Brown, T. A. (2010). Waitlist-controlled trial of cognitive behavior therapy for hoarding disorder. *Depression and Anxiety*, *27*(5), 476–484. https://doi.org/10.1002/da.20673.

Storch, E. A., Khanna, M., Merlo, L. J., Loew, B. A., Franklin, M., Reid, J. M., … Murphy, T. K. (2009). Children's Florida Obsessive Compulsive Inventory: Psychometric properties and feasibility of a self-report measure of obsessive-compulsive symptoms in youth. *Child Psychiatry and Human Development*, *40*(3), 467–483. https://doi.org/10.1007/s10578-009-0138-9.

Storch, E. A., Lack, C. W., Merlo, L. J., Geffken, G. R., Jacob, M. L., Murphy, T. K., & Goodman, W. K. (2007). Clinical features of children and adolescents with obsessive-compulsive disorder and hoarding symptoms. *Comprehensive Psychiatry*, *48*(4), 313–318. https://doi.org/10.1016/j.comppsych.2007.03.001.

Storch, E. A., Muroff, J., Lewin, A. B., Geller, D., Ross, A., McCarthy, K., … Steketee, G. (2011). Development and preliminary psychometric evaluation of the Children's saving inventory. *Child Psychiatry and Human Development*, *42*(2), 166–182. https://doi.org/10.1007/s10578-010-0207-0.

Storch, E. A., Rahman, O., Park, J. M., Reid, J., Murphy, T. K., & Lewin, A. B. (2011). Compulsive hoarding in children. *Journal of Clinical Psychology*, *67*(5), 507–516. https://doi.org/10.1002/jclp.20794.

Tolin, D. F., Das, A., Hallion, L. S., Levy, H. C., Wootton, B. M., & Stevens, M. C. (2019). Quality of life in patients with hoarding disorder. *Journal of Obsessive-Compulsive and Related Disorders*, *21*, 55–59. https://doi.org/10.1016/j.jocrd.2018.12.003.

Tolin, D. F., Fitch, K. E., Frost, R. O., & Steketee, G. (2010). Family informants' perceptions of insight in compulsive hoarding. *Cognitive Therapy and Research*, *34*(1), 69–81. https://doi.org/10.1007/s10608-008-9217-7.

Tolin, D. F., Frost, R. O., & Steketee, G. (2010). A brief interview for assessing compulsive hoarding: The hoarding rating scale-interview. *Psychiatry Research*, *178*(1), 147–152. https://doi.org/10.1016/j.psychres.2009.05.001.

Tolin, D. F., Frost, R. O., Steketee, G., & Muroff, J. (2015). Cognitive behavioral therapy for hoarding disorder: A meta-analysis. *Depression and Anxiety*, *32*(3), 158–166. https://doi.org/10.1002/da.22327.

Tolin, D. F., Gilliam, C., Wootton, B. M., Bowe, W., Bragdon, L. B., Davis, E., … Hallion, L. S. (2018). Psychometric properties of a structured diagnostic interview for DSM-5 anxiety, mood, and obsessive-compulsive and related disorders. *Assessment*, *25*(1), 3–13. https://doi.org/10.1177/1073191116638410.

Tolin, D. F., Worden, B. L., Wootton, B. M., & Gilliam, C. M. (2017). *CBT for hoarding disorder: A group therapy program therapist's guide* (1st ed.). Wiley-Blackwell.

Torres, A. R., Fontenelle, L. F., Shavitt, R. G., Ferrão, Y. A., do Rosário, M. C., Storch, E. A., & Miguel, E. C. (2016). Comorbidity variation in patients with obsessive-compulsive disorder according to symptom dimensions: Results from a large multicentre clinical sample. *Journal of Affective Disorders*, *190*, 508–516. https://doi.org/10.1016/j.jad.2015.10.051.

Whomsley, S. R. C. C. (2020). An overview of hoarding difficulties in children and adolescents. *Children Australia*, *45*(3), 182–185. https://doi.org/10.1017/cha.2020.29.

Woody, S. R., Kellman-McFarlane, K., & Welsted, A. (2014). Review of cognitive performance in hoarding disorder. *Clinical Psychology Review*, *34*(4), 324–336. https://doi.org/10.1016/j.cpr.2014.04.002.

Zaboski, B. A., Merritt, O. A., Schrack, A. P., Gayle, C., Gonzalez, M., Guerrero, L. A., … Mathews, C. A. (2019). Hoarding: A meta-analysis of age of onset. *Depression and Anxiety*, *36*(6), 552–564. https://doi.org/10.1002/da.22896.

Chapter 11

School-based cognitive behavioral interventions for youth with anxiety

Bente Storm Mowatt Haugland[a], Gro Janne Wergeland[b], and Elisabeth Bakke Husabø[c]

[a]*Department of Clinical Psychology, Faculty of Psychology, University of Bergen, Bergen, Norway,* [b]*Department of Child and Adolescent Psychiatry, Division of Psychiatry, Haukeland, University Hospital, Bergen, Norway,* [c]*Regional Centre for Child and Youth Mental Health and Child Welfare, Norwegian Research Centre, Bergen, Norway*

Abbreviations

CAMHS	child and adolescent mental health services
CALIS	child anxiety life interference scale
CGI-S	clinical global impression scale severity
CK	Cool Kids
GCBT	group cognitive behavioral treatment
RCT	randomized control trial
SCAS	spence children anxiety scale
SMFQ	short moods and feelings questionnaire
STC	systematic text condensation
WL	wait list

Introduction

Youth anxiety—Prevalence, impairment, and long-term risk

Anxiety disorders are among the most common mental health problems of youth, with worldwide point prevalences of up to 8.5% for children and adolescents, and a lifetime prevalence rate up to 31.9% (Merikangas et al., 2010; Polanczyk, Salum, Sugaya, Caye, & Rohde, 2015; Vasileva, Graf, Reinelt, Petermann, & Petermann, 2021). In addition, many youth report subclinical levels of anxiety, with prevalence of 32% across countries (Balazs et al., 2013). Childhood onset of anxiety disorders is common (Vasileva et al., 2021), and prevalence rates tend to increase in adolescence relative to preadolescence (Ford, Goodman, & Meltzer, 2003). Anxiety disorders in youth rarely occur in isolation; they are both highly comorbid among themselves (Costello, Egger, Copeland, Erkanli, & Angold, 2011) and with other mental disorders (Essau, Lewinsohn, Lim, Ho, & Rohde, 2018).

Anxiety disorders severely impede development and wellbeing, and are associated with significant functional impairment and negative long-term consequences (Swan & Kendall, 2016). If left untreated, these disorders often continue into adulthood (Goodwin, Fergusson, & Horwood, 2004). Youth with anxiety disorders often experience difficulties in social and peer relationships, and poor functioning in school—developmentally important areas (de Lijster et al., 2018; Verduin & Kendall, 2008). Anxiety disorders represent a sizable societal burden with regard to healthcare costs, problems related to school absence, and time off work for parents (Bodden, Dirksen, & Bogels, 2008). These disorders also have a negative impact on youth's quality of life (Mendlowicz & Stein, 2000). Youth anxiety is furthermore a risk factor for later adult mental health problems, including other internalizing and externalizing disorders and substance abuse (Copeland, Angold, Shanahan, & Costello, 2014; Essau et al., 2018; Puleo, Conner, Benjamin, & Kendall, 2011). Thus, elevated levels of anxiety symptoms are associated with psychosocial impairment, impaired school performance, and increased risk for progression to clinical anxiety development of anxiety disorders (Angold, Costello, Farmer, Burns, & Erkanli, 1999; Swan & Kendall, 2016). Taken together, the negative impact of anxiety disorders on daily life and the risk for long-term negative consequences indicate a strong case for identification and early intervention for youth with anxiety.

Handbook of Lifespan Cognitive Behavioral Therapy. https://doi.org/10.1016/B978-0-323-85757-4.00044-4

Limited access and barriers to evidence-based treatment

Empirically supported and well-established treatments for anxiety disorders exist, with cognitive behavioral therapy (CBT) showing favorable outcomes across university clinics as well as in routine clinical care (Higa-McMillan, Francis, Rith-Najarian, & Chorpita, 2016; Wergeland, Riise, & Öst, 2021). Treatment gains have been maintained for up to 19 years posttreatment (Benjamin, Harrison, Settipani, Brodman, & Kendall, 2013). Unfortunately, only a minority of youth with anxiety disorders or elevated anxiety symptoms are identified and receive treatment (Chavira, Stein, Bailey, & Stein, 2004; Merikangas et al., 2011). Of those who receive treatment, even fewer (<3%) receive evidence-based treatment (Reardon, Harvey, & Creswell, 2020). Also, there are long delays from the onset of an anxiety disorder until initiation of treatment, ranging from 9 to 23 years (Thompson, Issakidis, & Hunt, 2008). Thus, a large proportion of youth go unnoticed and do not access mental health services.

Given the negative consequences of anxiety and the low percentage of youth with anxiety disorders or elevated anxiety symptoms who receive treatment, efforts are made to make interventions accessible in the everyday contexts of adolescents. Referrals of youth for treatment in child and adolescent mental health services (CAMHS) are typically initiated by their parents (Creswell, Waite, & Cooper, 2014). Accessing treatment in CAMHS requires that a parent notices and acknowledges that the youth experiences anxiety, that they schedule an appointment, and that the youth meet diagnostic criteria for an anxiety disorder. However, caregivers do not always recognize anxiety symptoms in their youth. Further, parents can be reluctant to seek help due to fear of stigma related to psychological problems. Also, they may not have knowledge about treatment options, or experience other barriers to treatment such as lack of transportation, high costs, complicated referral procedures, or limited service provision (Reardon, Harvey, Young, O'Brien, & Creswell, 2018; Salloum, Johnco, Lewin, McBride, & Storch, 2016). In addition, youth may themselves be hesitant to seek mental health treatment due to reasons such as lack of knowledge about mental health, preference for self-reliance, lack of accessibility of treatment, and concerns about confidentiality (Gulliver, Griffiths, & Christensen, 2010).

Prevention of anxiety

Prevention involves promoting psychosocial well-being and/or preventing symptoms to develop into full-blown, severe disorders (Bloom & Gullotta, 2014). Universal prevention is usually offered to the general public, or to a whole population group. Selective prevention, on the other hand, is given to individuals considered at higher risk than average of developing the target issue in focus, i.e., children of parents with anxiety disorders (Gullotta, 2015). Indicated prevention is usually given to individuals with elevated levels of the target issue, i.e., elevated anxiety symptoms. When a mental health disorder is at an early stage, indicated prevention may sometimes overlap with early treatment. Selective and indicated prevention may be combined and labeled targeted prevention. Preventive interventions are often implemented in settings not traditionally used for mental health services, such as schools or via internet.

Rationale for school-based interventions for youth anxiety

Schools are considered an ideal setting to offer prevention interventions. By delivering anxiety intervention in a school context we aim to increase access to interventions by reaching more youths with anxiety. Youth spend much of their daily lives in schools, and many anxiety-provoking situations occur in this setting. These include anxity-provoking situations related to academic performance, taking tests, or talking in front of others, and also situations where one is being evaluated (e.g., in sports), take part in social interactions, and require being separated from parents (Ginsburg, Becker, Drazdowski, & Tein, 2012; Masia Warner & Fox, 2012). According to school health nurses, anxiety is the most prevalent mental health problem among adolescents seeking help from school health services (Muggeo & Ginsburg, 2019). Thus, offering interventions in schools represent an opportunity to identify large number of youth at risk for or at an early stage of developing anxiety disorders (Lyon & Bruns, 2019). Administering interventions at school and during school hours may also facilitate training and generalizability, as a range of situations in the school setting can be used for practicing anxiety management skills (Beidas et al., 2012). Further, many of the barriers related to identification and access to treatment, as referred to above, may be overcome when anxiety interventions are provided in schools (Lyon & Bruns, 2019). Thus, offering help at school represents a way of bringing support to youth rather than youth and parents having to find help (Weeks, Hill, & Owen, 2017). This is an important step towards "democratizing access to services," with school-based interventions hopefully to a larger degree reach youth independent of parental income and/or knowledge of mental health problems.

Effects of school-based CBT for youth with anxiety

Several literature reviews and metaanalyses have looked at effects of anxiety prevention for children and adolescents (Christensen, Pallister, Smale, Hickie, & Calear, 2010; Lawrence, Rooke, & Creswell, 2017; Rasing, Creemers, Janssens, & Scholte, 2017; Teubert & Pinquart, 2011) (Table 1). Across metaanalyses, CBT has been found to be the most frequently used approach, with predominantly positive outcomes regarding reducing anxiety symptoms or preventing anxiety disorders (Christensen et al., 2010; Werner-Seidler et al., 2017). For metaanalyses specifically focusing on school-based indicated and targeted prevention, small to moderate effects have been found (Fisak et al., 2011; Werner-Seidler et al., 2017). Although one review has reported no evidence for the effectiveness of school-based targeted anxiety prevention (Caldwell et al., 2019), most reviews and metaanalyses show that school-based indicated and targeted CBT prevention for anxiety are effective (Hugh-Jones et al., 2020; Werner-Seidler et al., 2017).

Importantly, there is a dearth of research on prevention of anxiety among children and adolescents where long-term outcomes above 1-year are included (Fisak et al., 2011; Hugh-Jones et al., 2020). Knowledge about long-term outcomes is imperative when considering the effect of prevention interventions and is necessary to be able to conclude about the long-term value of anxiety-focused school-based CBT interventions.

Barriers against implementing school-based interventions

Mental health interventions tend to not be prioritized in schools, due to limited funding, staff capacity, and lack of national policy (Greenberg, 2010; Patalay et al., 2016). We believe school personnel, school managements, and policy makers need to be informed about the rational and advantages of school-based interventions. By reaching young people with effective interventions at an early stage we may improve their mental health, their quality of life, and subsequently also increase their learning capacity and academic progress.

Practice and procedures

In the following we describe the procedures, methods, and results from a Norwegian study successfully delivering indicated school-based CBT prevention/early intervention to adolescents with elevated anxiety symptoms. Also, experiences from the implementation of the interventions will be described. We believe experiences from this study may be relevant for implementation of school-based interventions for anxiety in other countries or cultures.

Implementing school-based CBT for adolescents with anxiety

CBT interventions were implemented in 18 junior-high schools in Norway (Haugland et al., 2017; Haugland et al., 2020). The intention was to offer CBT to adolescents with elevated levels of anxiety symptoms and thereby prevent future development of anxiety disorders. The aims were: (1) to examine the effect of indicated school-based CBT for adolescents with anxiety; (2) to compare two CBT program of different intensity; and (3) to examine whether potential effects were maintained 1-year after CBT. Furthermore, we wanted to examine the process of implementing school-based indicated prevention, e.g., how to recruit adolescent participants, and how to train and supervise providers in delivering CBT.

Adolescentes between 12 and 16 years were randomized to three conditions, all lasting 10 weeks; 10-session group cognitive behavioral treatment (GCBT), 5-session GCBT, or a delayed access wait list (WL). A total of 313 adolescents participated (mean 14.00 years, $SD = 0.84$, 84% girls). To participate the adolescents had to report mild to moderate levels of anxiety symptoms, i.e., a score of ≥ 25 on self-reported or parent reported anxiety on the Spence Children's Anxiety Scale (SCAS; Spence, 1998). Also, anxiety had to influence the youth's daily life, reported on the Child Aanxiety Life Interference Scale (CALIS; Lyneham et al., 2013). Having only a few exclusion criteria (i.e., having problems following group rules; demonstrating disruptive behavior; or having learning problems that would make it difficult to follow a manualized program) was important to ensure external validity. Only three adolescents were excluded from the interventions.

Multiple strategies were applied to recruit participants. Adolescents and parents were informed about the study at routine meetings with school nurses. School nurses gave presentations about anxiety and CBT interventions to teachers, parents, and youth. They also taught teachers how to identify adolescents with anxiety symptoms. Furthermore, the project was featured in local media, and through pamphlets and posters distributed at schools. Adolescents and parents were also informed about the project through a school survey on anxiety. Thus, self-referral as well as referrals from parents, teachers, and other health personnel were endorsed.

TABLE 1 Metaanalyses; effects of school-based preventive interventions for anxiety in youth.

Study	Studies included	% Female	Age range	Provider	Intervention type	Outcome	Level of prevention	Setting	ES	ES postintervention	ES f/u[a]	ES long-term f/u[b]
Caldwell et al. (2019)[c]	5	NR	12–18	MHP, school staff	CBT	sx	T	School	–	−0.22	–	−0.26
Fisak, Richard, and Mann (2011)[d]	17	35.8–100.0	3.9–16.3	MHP, lay providers	CBT, panic prevention, psychosocial debriefing	sx	T	School & parent-based	0.26	–	–	–
Lawrence et al. (2017)[c]	16	45.0–75.0	6–18	NR	CBT, biofeedback, play therapy	sx/dx	T	Any	–	−0.43/0.09	−0.46/0.17	−0.32/0.31
Rasing et al. (2017)	12	43.2–76.6	7–17	NR	CBT	sx	I	School & community	–	−0.17	−0.30	−0.19
Hugh-Jones, Beckett, Tumelty, and Mallikarjun (2020)	20	NR	7–17	MHP, computerized, researchers, school staff	CBT, play therapy, WM training, mindfulness, biofeedback	sx	I	School		−0.28	−0.35	−0.24
Teubert and Pinquart (2011)	36	27.0–100.0	3.9–17.05	Teachers, MHP	CBT, exercise, parenting skills training	sx/dx	T	Any	–	0.32/0.23	–	0.19/0.32[e]
Werner-Seidler, Perry, Calear, Newby, and Christensen (2017)	17	NR	7–22	School staff, MHP, researchers	CBT, social skills, wellbeing therapy	sx	T	School	–	0.22	–	–

Outcomes are self-reported and effect sizes are Hedges *g*. Effect sizes for diagnostic outcomes from Lawrence et al. (2017) are reported in risk ratios. All effect sizes are identical to how they are reported in the respective studies. Thus, both positive and negative ESs indicate desirable outcomes. NR = not reported; MHP = mental health provider; WM = working memory; sx = symptoms; dx = diagnosis; T = targeted; I = indicated.

[a] *Reflects from 3 to 6 months of follow-up.*

[b] *Reflects from 12 to 24 months of follow-up.*

[c] *ES are reported in standardized mean difference.*

[d] *ES are reported in weighted mean ES across postintervention and follow-up.*

[e] *Reflects from 1 to 24 months of follow-up.*

Reprinted from Husabø, E. B. (2021). *Implementation of indicated anxiety prevention in schools.* Thesis for the degree of PhD, University of Bergen Norway.

CBT programs

CBT interventions were delivered in groups with five to eight adolescents attending each group, receiving either the 5-session program (Vaag) or the 10-session program (Cool Kids). Vaag comprised weekly sessions of 45–90 min over the first 4 weeks, followed by a final session 5 weeks later (total 5.5 h). Session 2 is a joint youth-parent session. Included in the program is a self-help material (Raknes, 2010). Cool Kids (CK) comprises weekly 90-min sessions, plus two 90-min parents-only sessions (total 18 h). The adolescent group-based, school version of CK was applied (Rapee et al., 2006). CK includes workbooks for adolescents and parents. CK has previously shown efficacy as indicated, school-based prevention (Mifsud & Rapee, 2005).

Both programs include the two key components in anxiety-focused CBT, i.e., cognitive restructuring and exposure tasks. Also, both programs include in-session therapist led exposures and between-session homework exposure training. In Vaag, between sessions 4 and 5, participants performed exposure tasks at home, assisted by two short telephone calls or text messages from the group leaders. Thus, an important difference between the programs was the amount of therapist contact during the last half of the intervention. Information about the content and structure of the two programs can be found in Table 2.

TABLE 2 Content and structure of the Cool Kids and Vaag programs included in school-based anxiety indicated prevention in Norway.

Cool Kids (school version)				VÅG			
Session	Participants	Duration	Content	Session	Participants	Duration	Content
1	Adolescents	90 min	Psychoeducation. Anxiety, linking thoughts and feelings. Setting goals. Homework assignment	1	Adolescents	45 min	Psychoeducation. Anxiety, linking situations, thoughts and feelings. Setting goals. Homework assignment
2[a]	Adolescents	90 min	Cognitive restructuring ("realistic thinking") and self-rewards. Homework assignment	2	Adolescents & parents	60 min	Using self-help material to link situations, feelings and thoughts. Cognitive restructuring. Identifying avoidance. Helpful parenting. Homework assignment
3	Adolescents	90 min	Cognitive restructuring. Principles and application of exposure hierarchies (stepladders). Homework assignment	3	Adolescents	90 min	In-session exposure and behavioral experiments. Training plans. Homework assignment
4	Adolescents	90 min	Exposure hierarchy. Regulating anxiety by surfing emotions and worries. Homework assignment	4	Adolescents	90 min	In-session exposure and behavioral experiments. Training plans. Homework assignment
5[a]	Adolescents	90 min	Reviewing and revising exposure hierarchies (individual sessions 15–20 min)				
6	Adolescents	90 min	Simplifying cognitive restructuring. In-session exposure and behavioral experiments. Homework assignment		Adolescents	5–10 min × 2	Two telephone calls or text-messages, supporting the adolescents to follow the training plan

Continued

TABLE 2 Content and structure of the Cool Kids and Vaag programs included in school-based anxiety indicated prevention in Norway—cont'd

Cool Kids (school version)				VÅG			
Session	Participants	Duration	Content	Session	Participants	Duration	Content
7	Adolescents	90 min	In-session exposure and behavioral experiments. Homework assignment				
8	Adolescents	90 min	In-session exposure and behavioral experiments. Additional skills if needed to facilitate progress (e.g., problem solving, assertiveness). Homework assignment				
9	Adolescents	90 min	Troubleshooting exposure. In-session exposure. Homework assignment				
10	Adolescents	90 min	Reviewing goals. Positive and negative coping strategies. Future plans. Celebration	5	Adolescents	45 min	Review of progress so far. Future plans. Mutual support

[a] *Two psychoeducational parent session each 90 min. The first between session 1 and 3, and the second after session 5.*
Reprinted from Haugland, B. S. M., Raknes, S., Haaland, A. T., et al. (2017). School-based cognitive behavioral interventions for anxious youth: Study protocol for a randomized controlled trial. *Trials, 18*, 100. Originally published by BioMed Central.

Training and supervision

Each group session was administered by two group leaders. Group leaders were either health workers from community services (e.g., school nurses, community psychologists) or health workers from CAMHS. Thus, group leaders were mainly school personnel (i.e., school nurses) or mental health workers participating as part of their regular job. Group leaders were 93.8% women (mean age = 43.2 years, SD = 8.09, range 32–62), with an average of 6.7 years of experience working with youth (SD = 6.74, range 0–27 years). However, 83.9% had no prior CBT training and were considered novice CBT providers. The majority of the group leaders (75.0%) administered both GCBT programs.

Before conducting any groups, a 4-day skills-training workshop was given on basic CBT principles for anxiety, the content and procedures of the program manuals, and assessment procedures. During the implementation period, group leaders attended two additional 2-day workshops on youth anxiety, how to do exposure exercises, and troubleshooting any problems in delivering the interventions, recruiting adolescents, or collaborating with school personnel or others.

Regular supervision was considered necessary. Hence, supervision was mandatory for all group leaders. Experienced CBT therapists supervised the group leaders, with 3–10 h of supervision for each group, depending on the program and group leader's experience in delivering the interventions. Supervision was primarily administered face-to-face—and based on video-recordings of group sessions. To ensure high-quality supervision, instructions on duration, structure, and content of supervision were developed. Supervision was monitored by checklists completed by the supervisors after each session.

Effects on anxiety, depression, and sleep problems

Adolescents and their parents completed questionnaires at different time-points (before and after groups, after WL, and at 1-year follow up). Primary outcomes were changes in anxiety symptoms (SCAS) and impairment (CALIS) due to anxiety. Secondary outcomes were changes in depressive symptoms (Short Moods and Feelings Questionnaire; SMFQ) (Angold, Costello, Messer, & Pickles, 1995). Furthermore, group leaders rated all adolescents regarding overall severity of their anxiety symptoms on the Clinical Global Impression scale-Severity; CGI-S (Guy, 1976). Acceptable interrater reliability for the ratings was confirmed. Also, adolescents were assessed on sleep problems, as well as quality of life and self-efficacy. Finally, adolescents and parents rated their satisfaction with the interventions.

Both school-based GCBT programs had effect (Haugland et al., 2020). Adolescents, as well as their parents, reported significant lower youth anxiety after participating in GCBT compared to beforehand. The youth were less impaired due to anxiety after the GCBT. Also, parents reported reductions in the adolescents' depressive symptoms. Furthermore, severity of the adolescents' anxiety, rated by the group leaders, improved from before to after the GCBT. Similar improvements were not found in the WL group. Thus, school-based GCBT reduced anxiety, impairment, and depressive symptoms, with within-group effect sizes ranging from small to large, depending on program, informant, and measure (Table 3). Furthermore, outcomes were maintained or even improved at follow-up assessment one-year after completing the interventions. That is, the effect of the interventions lasted 1 year afterward, which is promising regarding the aim of preventing the development of anxiety disorders.

TABLE 3 Within and between effects (Cohen's *d*) for outcomes in a school-based anxiety indicated prevention in Norway.

	Within group effect sizes (*d*)			Between group effect sizes (*d*)
	Vaag	Cool Kids	WL	GCBT vs. WL
Anxiety symptoms (SCAS)				
Youth report				
Pre-post	0.41	0.62	0.31	0.34**
Pre-1-year follow-up	0.50	0.79	–	–
Parent report				
Pre-post	0.53	0.67	0.23	0.53***
Pre-1-year follow-up	0.81	0.90	–	–
Impairment from anxiety (CALIS)				
Youth report				
Pre-post	0.30	0.32	0.15	0.26*
Pre-1-year follow-up	0.32	0.42	–	–
Parent report				
Pre-post	0.61	0.55	0.21	0.51***
Pre-1-year follow-up	0.76	0.85	–	–
Depressive symptoms (SMFQ)				
Youth report				
Pre-post	0.17	0.28	0.06	0.23*
Pre-1-year follow-up	0.18	0.31	–	–
Parent report				
Pre-post	0.65	0.85	0.70	0.30**
Pre-1-year follow-up	0.81	1.00	–	–
Severity of anxiety (CGI-S)				
Pre-post	1.03	1.06	0.08	1.01***

CALIS, child anxiety life interference scale; CGI-S, clinical global impression-severity; SCAS, spence children anxiety scale; SMFQ, short mood and feeling questionnaire.
Interpretation of Cohen *d*: 0.2 = small, 0.5 = medium, and 0.8 = large.
*$P < 0.05$.
**$P < 0.01$.
***$P < 0.001$.

As anxiety and depression have been related to sleep problems in children and adolescents (Orchard, Gregory, Gradisar, & Reynolds, 2020), we examined insomnia, sleep onset latency and sleep duration in the adolescents. A high percentage confirmed sleep problems at preintervention, with 38.1% meeting criteria for insomnia, 34.8% reporting short sleep duration (<7 h), and 83.1% having long sleep onset latency (≥30 min) (Haugland et al., 2021). However, the anxiety-focused GCBT had no effect on the adolescents' self-reported sleep problems when compared to WL (Haugland et al., 2021). Due to the increased level of sleep problems found among the anxious youth, sleep problems should be assessed and techniques specifically targeting sleep problems included in future anxiety programs, to help adolescents with anxiety and cooccurring sleep problems.

Recruiting adolescents with anxiety disorders

Eighteen schools, all but one being public schools, signed up to deliver the school-based GCBT. One school terminated earlier than planned as the school management changed their prioritizations for the school health personnel. The remaining schools continued to recruit adolescents over a period of 2 years.

To examine more in depth the participants recruited to the school-based interventions, we assessed a subsample with a diagnostic interview. In the subsample ($n = 37$) a large majority (89.1%) met criteria for one or more mental health diagnosis. All had a primary anxiety diagnosis with a mean severity rating of 5.90 (SD = 1.30; range = 4–8). The most common primary diagnosis was generalized anxiety disorder which was found in about 30% of the youth. Of the youth having one or more anxiety diagnosis, only 12.1% had been in contact with community clinics during the last year (Husabo et al., 2020). Thus, by making indicated prevention available in schools, we may reach youth with clinical and impairing anxiety, who have not received treatment elsewhere.

The high number of adolescents in the subsample who met criteria for one or more diagnoses was surprising, considering that the school-based interventions were designed to enroll youth *at risk* for anxiety disorders (i.e., with elevated anxiety symptoms) and not those having already developed anxiety disorders. When implementing future school-based interventions one should prepare for this by tailoring the interventions to also help youth qualifying for a diagnosis or alternatively by creating well-functioning referral systems for youth in need of treatment rather than indicated interventions.

Treatment fidelity

Treatment fidelity in the delivery of interventions is about evaluating whether interventions are delivered as intended. Two important aspects of treatment fidelity are: *adherence*—the degree to which the core elements of a program are delivered, and *competence*—how skillfully the core elements of a program are delivered (Perepletchikova, Treat, & Kazdin, 2007). To date, very few have examined treatment fidelity in school-based CBT for anxiety. This is unfortunate as it makes it difficult to draw conclusions whether school-based CBT for anxiety is delivered according to manuals and with adequate competence. In the Norwegian school-based study all group sessions were videotaped. A team of seven clinical psychologists/ psychiatrist (28.6% male; M age = 48.70 years, range = 31–66 years; all Norwegian) coded the videotapes. The coders had clinical experience and high level of competence in CBT for anxiety. Two sessions from each group were coded for adherence and competence, representing 20% of the CK sessions and 40% of the Vaag sessions. The sessions chosen for assessing fidelity were selected to be comparable in duration and content across the two interventions. Coders were trained through scoring three videos reaching consensus ratings through discussions. Coders were certified by scoring four videos that were evaluated against established expert codes. Further, interrater reliability was established by expert coders double coding 23.9% of the videos (Husabo et al., 2021).

The results indicated that novice CBT providers delivered both CBT programs with good adherence and competence. Adherence and competence scores (ranging from 0 = no adherence/poor competence to 6 = thorough adherence/excellent competence) for each group (mean of the two rated sessions) ranged from 3.17 to 5.75 (mean = 4.41, SD = 0.56) for adherence and 2.75 to 5.88 (mean = 4.18, SD = 0.66) for competence. Treatment fidelity was somewhat higher in the brief 5-session program compared to in the 10-session program. This was interpreted as result of the longer program being more complex and allowing more flexibility, and therefore required somewhat more experience and skills from novice providers to deliver with fidelity. However, both programs were delivered with good treatment fidelity (Husabo et al., 2021).

User experiences: Participants and providers

Among adolescents randomized, 83.4% were defined as completers (attending 7 or more of the 10 sessions of CK or 4 or more of the 5 sessions of Vaag). There were no differences in retention between the two interventions. Both adolescents and parents gave positive evaluations after participating in the interventions, with high scores on a scale rating satisfaction, including items such as "I would recommend this program to a friend with similar problems." The mean rating of adolescents was 3.67 (SD = 0.90) and parent 3.80 (SD = 0.76) on a 5-point scale (1 = totally disagree, 5 = totally agree) where higher ratings indicated more satisfaction. Adolescents reported somewhat higher mean satisfaction for the longer program. The fact that so many completed the program, with high satisfaction among adolescents and parents, is an indication that the school-based CBT interventions were both feasible and acceptable.

We also wanted to know the group leaders' experiences from administering the GCBT, as this could indicate what factors they saw as helpful for the implementation of the interventions. This was especially important as most of the providers had limited experience in CBT. Seventeen of the 33 group leaders participated in three focus-group interviews (Husabo, Haugland, Wergeland, & Maeland, 2020). Between four and seven providers were included in each focus-group. At the time of the focus-group interviews the providers had delivered interventions for 2 years and administered between 2 to 8 groups each. Audio recordings of the interviews were transcribed verbatim. The focus-group interviews were analyzed qualitatively, applying a thematic cross-case analysis strategy (i.e., systematic text condensation, STC; Malterud, 2012). Four themes emerged, with the first (*Building competence and confidence through training, supervision, and collegial support*) referred to the providers experiencing that a supportive environment, receiving training, supervision, working in pairs, and participating in a network of providers contributed to an experience of competence. The second theme from the interviews (*The value of the initial assessment interview*) denoted how the assessment interview with adolescents and parents prior to inclusion in the GCBT served multiple purposes. The parents gained insight in their adolescent's problems, adolescents were motivated for active participation in the intervention, and providers received information they used to individualize the interventions to each of the participant. The third theme (*Experiencing that "it actually works," but different interventions may fit different adolescents*) denoted how providers gained trust in the effect of the interventions through observing changes in the functioning of the adolescents who participated in the GCBT. This motivated them to continue to recruit and deliver the interventions. With experience, the providers differentiated between the two programs, thinking that the short program matched better with the needs of some adolescents than others and vice versa. The final theme (*Crucial co-operation with the schools*) referred to the value of support from teachers. Providers highlighted the importance of including the interventions in the schools' routines and to build relations with teachers. The teachers helped facilitate implementation of the interventions. As teachers understood more about the nature of anxiety and the effect of the interventions, recruitment to the GCBT became easier (Husabo, Haugland, Wergeland, & Maeland, 2020).

Implementing school-based interventions for youth anxiety

According to Greenberg (2010) we need to increase our knowledge on implementation and sustainability of school-based interventions—to secure integration between educational policy and prevention, to understand factors influencing the integration of preventive programs with other ongoing activities in schools, and to continue developing and refining programs and models. To reach these goals, high level of collaboration between policy makers, researchers, mental health professionals, and educators is vital.

Close contact between providers of preventive interventions at different schools is recommended, regardless of what issue one is focusing on in school-based interventions. Regular contact among those responsible for implementing the interventions, e.g., the researcher or implementer, local coordinators, providers, and school managers is important. This may cause problems in recruitment, assessment, and delivery of interventions to be discovered and resolved at an early stage. Implementing school-based interventions may require that health personnel operate outside of their usual context. Success is dependent on all parties, researchers, health professionals, and educators, being respectful for the competence and skills of one another and having an open-mind and a willingness to learn from each other.

Evidence-based interventions may become ineffective and unsustainable if attention is not given to important factors related to the implementation process (Ogden et al., 2012). Thus, to ensure implementation quality in delivering school-based prevention for mental health problems, it is necessary to monitor the implementation process closely. This is probably the case for all school-based mental health interventions. Those who are involved in either the delivery or the decision to adopt interventions into the school (e.g., school principal or leader of school health services) need to have knowledge about common facilitators and barriers towards implementing school interventions. Furthermore, school leaders, teachers, parents, and school health services need to understand the rational for the prevention programs, why it is implemented and how it may be helpful for adolescents, for the school, and for school personnel. They need to understand that to help students learn, they may need to solve mental health issues such as elevated anxiety, that may otherwise interfere with school attendance and/or academic progress. Understanding the rational for the preventive interventions, how they work, and for whom they are intended is essential to achieve the support and collaboration the providers of school-based CBT need to deliver the intervention. This is probably also crucial for the interventions to be sustained over time.

Mini-dictionary of terms

- ***Exposure:*** a technique in anxiety-focused CBT involving repeated (often gradual) real or visualized confrontation with a feared situation, object, or event.
- ***Cognitive restructuring:*** a core component of CBT including a group of therapeutic techniques that help being aware of and challenging unhelpful thoughts, ideas, assumptions and thinking patterns, aiming to make these more balanced and accurate.
- ***Generalized anxiety:*** anxiety characterized by excessive, uncontrollable, and often irrational worry about many events or activities. Worries may focus on everyday matters such as health, finances, death, family, relationship, or school.
- ***Implementation:*** involves deliberate actions aimed at promoting the systematic use of interventions, methods, techniques, or procedures from research findings into routine practice.
- ***Indicated prevention:*** interventions offered to individuals at a very high-risk of developing the problem in focus. The risk can be indicated by the individuals having elevated levels of the target issue, i.e., elevated anxiety symptoms.
- ***Insomnia:*** a common sleep disorder characterized by sleep onset latency and/or difficulties staying asleep (waking up too early and not being able to sleep again). Associated with tiredness and sleepiness during daytime. Occurs at least three times per week and has been a problem for at least 1 month.
- ***School-based mental health interventions:*** interventions delivered in school facilities, often during school hours, either by school personnel (e.g., teachers), school health personnel (e.g., school nurses), or mental health professionals from community health services or specialized health/mental health services (e.g., psychologists, social workers).
- ***Sustainability:*** the extent to which newly implemented interventions or procedures are maintained over time.
- ***Sleep onset latency:*** the amount of time it takes from going from fully awake to sleeping.
- ***Targeted prevention:*** Includes both selective and indicated prevention, that is interventions targeting both individuals considered at higher risk than average of developing a disorder and individuals already demonstrating elevated levels of the target issue, i.e.. anxiety symptoms.
- ***Treatment fidelity:*** Whether the intervention is delivered as intended. The term may include adherence, i.e., the degree to which core elements of a program are delivered, competence, i.e., how skillfully core elements are delivered, differentiation, i.e., extent to which nonprescribed program content is avoided, and dosage, i.e., degree of the total program delivered.

Key facts

- Anxiety is among the most prevalent mental health disorders in youth.
- Only a small proportion of youth with anxiety is identified and offered help.
- School-based interventions reach youth who otherwise do not receive necessary mental health intervention.
- School-based anxiety-focused CBT has demonstrated effectiveness.
- Focus on implementation is needed for school-based interventions being delivered with quality over time.

Applications to other areas

We have reviewed findings related to school-based CBT for adolescents with anxiety. The experiences on how to implement school-based CBT as well as knowledge on facilitators of implementation are relevant when introducing mental health interventions into other everyday life contexts for adolescents (e.g., community health centers, youth clubs). Methods and findings from schools-based anxiety-focused interventions may be relevant also for school-based interventions targeting other mental health problems, e.g., depressive symptoms, sleep problems, eating problems, self-harm, externalizing symptoms. Common to all school-based interventions is that they need to fit with the context of the school-system. This means among others that they must be easily implemented in schools, be implemented with high fidelity by providers of the interventions, be acceptable to parents, youth, and teachers, and answer to a need experienced by policy makers, school-personnel, parents, and youth.

CBT applied in school-based prevention may be slightly different from CBT programs used in specialized treatment. However, investigating mechanisms of change, predictors, moderators, and mediators in preventive interventions may increase our understanding also of how to treat youth with anxiety disorders.

Summary points

- Anxiety disorders are among the most common mental health problems during adolescence and the costs and impairment associated with anxiety spans across several important domains, including the individual's social and emotional functioning, academic achievements and school attendance, family relationships, as well as societal costs.
- A large proportion of youth with anxiety do not access mental health services, do not receive evidence-based treatment (e.g., CBT), or experience long delays before entering treatment.
- By offering CBT as school-based indicated interventions, we may improve access to effective treatment.
- School-based indicated prevention for anxiety has demonstrated effectiveness with regard to reducing anxiety. This has been reported by several international reviews and was also found in a large trial including 313 adolescents from Norwegian junior high schools.
- To make school-based interventions successful and further develop this promising area within mental health research, we need to focus on implementation issues, that is how to recruit adolescents with anxiety, the training and supervision of providers of interventions, ensuring that interventions are delivered with high fidelity, as well as securing collaboration between providers of the interventions, school personnel, policy makers, and researchers.

References

Angold, A., Costello, E. J., Farmer, E. M. Z., Burns, B. J., & Erkanli, A. (1999). Impaired but undiagnosed. *Journal of the American Academy of Child & Adolescent Psychiatry*, *38*(2), 129–137. https://doi.org/10.1097/00004583-199902000-00011.

Angold, A., Costello, E. J., Messer, S. C., & Pickles, A. (1995). Development of a short questionnaire for use in epidemiological studies of depression in children and adolescents. *International Journal of Methods in Psychiatric Research*, *5*(4), 237–249.

Balazs, J., Miklosi, M., Kereszteny, A., Hoven, C. W., Carli, V., Wasserman, C., et al. (2013). Adolescent subthreshold-depression and anxiety: Psychopathology, functional impairment and increased suicide risk. *Journal of Child Psychology and Psychiatry*, *54*(6), 670–677. https://doi.org/10.1111/jcpp.12016.

Beidas, R. S., Mychailyszyn, M. P., Edmunds, J. M., Khanna, M. S., Downey, M. M., & Kendall, P. C. (2012). Training school mental health providers to deliver cognitive-behavioral therapy. *School Mental Health*, *4*(4), 197–206. https://doi.org/10.1007/s12310-012-9074-0.

Benjamin, C. L., Harrison, J. P., Settipani, C. A., Brodman, D. M., & Kendall, P. C. (2013, Oct). Anxiety and related outcomes in young adults 7 to 19 years after receiving treatment for child anxiety. *Journal of Consulting and Clinical Psychology*, *81*(5), 865–876. https://doi.org/10.1037/a0033048.

Bloom, M., & Gullotta, T. P. (2014). Definitions of primary prevention. In T. P. Gullotta, & M. Bloom (Eds.), *Encyclopedia of primary prevention and health promotion* (2nd ed., pp. 3–12). Kluwer Academic.

Bodden, D. H., Dirksen, C. D., & Bogels, S. M. (2008, May). Societal burden of clinically anxious youth referred for treatment: A cost-of-illness study. *Journal of Abnormal Child Psychology*, *36*(4), 487–497. https://doi.org/10.1007/s10802-007-9194-4.

Caldwell, D. M., Davies, S. R., Hetrick, S. E., Palmer, J. C., Caro, P., López-López, J. A., et al. (2019). School-based interventions to prevent anxiety and depression in children and young people: A systematic review and network meta-analysis. *The Lancet Psychiatry*, *6*(12), 1011–1020. https://doi.org/10.1016/s2215-0366(19)30403-1.

Chavira, D. A., Stein, M. B., Bailey, K., & Stein, M. T. (2004). Child anxiety in primary care: Prevalent but untreated. *Depression and Anxiety*, *20*(4), 155–164. https://doi.org/10.1002/da.20039.

Christensen, H., Pallister, E., Smale, S., Hickie, I. B., & Calear, A. L. (2010, Jun). Community-based prevention programs for anxiety and depression in youth: A systematic review. *Journal of Primary Prevention*, *31*(3), 139–170. https://doi.org/10.1007/s10935-010-0214-8.

Copeland, W. E., Angold, A., Shanahan, L., & Costello, E. J. (2014). Longitudinal patterns of anxiety from childhood to adulthood: The Great Smoky Mountains study. *Journal of the American Academy of Child & Adolescent Psychiatry*, *53*(1), 21–33. https://doi.org/10.1016/j.jaac.2013.09.017.

Costello, E. J., Egger, H. L., Copeland, W., Erkanli, A., & Angold, A. (2011). The developmental epidemiology of anxiety disorders: Phenomenology, prevalence, and comorbidity. In A. P. Field, & W. K. Silverman (Eds.), *Anxiety disorders in children and adolescents* (pp. 56–75). Cambridge University Press.

Creswell, C., Waite, P., & Cooper, P. J. (2014, Jul). Assessment and management of anxiety disorders in children and adolescents. *Archives of Disease in Childhood*, *99*(7), 674–678. https://doi.org/10.1136/archdischild-2013-303768.

de Lijster, J. M., Dieleman, G. C., Utens, E., Dierckx, B., Wierenga, M., Verhulst, F. C., et al. (2018). Social and academic functioning in adolescents with anxiety disorders: A systematic review. *Journal of Affective Disorders*, *230*, 108–117. https://doi.org/10.1016/j.jad.2018.01.008.

Essau, C. A., Lewinsohn, P. M., Lim, J. X., Ho, M. R., & Rohde, P. (2018). Incidence, recurrence and comorbidity of anxiety disorders in four major developmental stages. *Journal of Affective Disorders*, *228*, 248–253. https://doi.org/10.1016/j.jad.2017.12.014.

Fisak, B. J., Richard, D., & Mann, A. (2011). The prevention of child and adolescent anxiety: A meta-analytic review. *Prevention Science*, *12*, 255–268.

Ford, T., Goodman, R., & Meltzer, H. (2003). The British child and adolescent mental health survey 1999: The prevalence of DSM-IV disorders. *Journal of the American Academy of Child & Adolescent Psychiatry*, *42*(10), 1203–1211. https://doi.org/10.1097/00004583-200310000-00011.

Ginsburg, G. S., Becker, K. D., Drazdowski, T. K., & Tein, J.-Y. (2012). Treating anxiety disorders in inner city schools: Results from a pilot randomized controlled trial comparing CBT and usual care. *Child & Youth Care Forum*, *41*(1), 1–19. https://doi.org/10.1007/s10566-011-9156-4.

Goodwin, R. D., Fergusson, D. M., & Horwood, L. J. (2004). Early anxious/withdrawn behaviours predict later internalising disorders. *Journal of Child Psychology and Psychiatry*, *45*(4), 874–883. https://doi.org/10.1111/j.1469-7610.2004.00279.x.

Greenberg, M. T. (2010). School-based prevention: Current status and future challenges. *Effective Education*, *2*(1), 27–52. https://doi.org/10.1080/19415531003616862.

Gulliver, A., Griffiths, K. M., & Christensen, H. (2010). Perceived barriers and facilitators to mental health help-seeking in young people: A systematic review. *BMC Psychiatry*, *10*(113). http://www.biomedcentral.com/1471-244X/10/113.

Gullotta, T. P. (2015). Understanding primary prevention. In *Handbook of adolescent behavioral problems* (pp. 3–10). https://doi.org/10.1007/978-1-4899-7497-6_1.

Guy, W. (1976). *ECDEU assessment manual for psychopharmacology*. US Department of Health, Education, and Welfare Public Health Service Alcohol, Drug Abuse, and Mental Health Administration.

Haugland, B. S. M., Haaland, Å. T., Baste, V., Bjaastad, J. F., Hoffart, A., Rapee, R. M., et al. (2020). Effectiveness of brief and standard school-based cognitive-behavioral interventions for adolescents with anxiety: A randomized non-inferiority study. *Journal of the American Academy of Child & Adolescent Psychiatry*, *59*(4), 552–564. https://doi.org/10.1016/j.jaac.2019.12.003.

Haugland, B. S. M., Hysing, M., Baste, V., Wergeland, G. J., Rapee, R. M., Hoffart, A., et al. (2021). Sleep duration and insomnia in adolescents seeking treatment for anxiety in primary health care. *Frontiers in Psychology*, *12*(959). https://doi.org/10.3389/fpsyg.2021.638879.

Haugland, B. S. M., Hysing, M., Hoffart, A., Haaland, Å. T., Bjaastad, J. F., Wergeland, G. J., et al. (2021). Effect of early intervention for anxiety on sleep outcomes in adolescents: A randomized trial. *European Child & Adolescent Psychiatry*. https://doi.org/10.1007/s00787-021-01795-6.

Haugland, B. S. M., Raknes, S., Haaland, A. T., Wergeland, G. J., Bjaastad, J. F., Baste, V., et al. (2017). School-based cognitive behavioral interventions for anxious youth: Study protocol for a randomized controlled trial. *Trials*, *18*(100). https://doi.org/10.1186/s13063-017-1831-9.

Higa-McMillan, C. K., Francis, S. E., Rith-Najarian, L., & Chorpita, B. F. (2016). Evidence base update: 50 years of research on treatment for child and adolescent anxiety. *Journal of Clinical Child & Adolescent Psychology*, *45*(2), 91–113. https://doi.org/10.1080/15374416.2015.1046177.

Hugh-Jones, S., Beckett, S., Tumelty, E., & Mallikarjun, P. (2020). Indicated prevention interventions for anxiety in children and adolescents: a review and meta-analysis of school-based programs. *European Child & Adolescent Psychiatry*. https://doi.org/10.1007/s00787-020-01564-x.

Husabo, E., Haugland, B. S. M., McLeod, B. D., Ogden, T., Rapee, R. M., & Wergeland, G. J. (2020). Does school-based recruitment for anxiety interventions reach youth not otherwise identified? A comparison between a school-based sample and a clinical sample. *School Mental Health*. https://doi.org/10.1007/s12310-019-09357-7.

Husabo, E., Haugland, B. S. M., Wergeland, G. J., & Maeland, S. (2020). Providers' experiences with delivering school-based targeted prevention for adolescents with anxiety symptoms: A qualitative study. *School Mental Health*, *12*(4), 757–770. https://doi.org/10.1007/s12310-020-09382-x.

Husabo, E., Haugland, B. S. M., McLeod, B. D., Baste, V., Haaland, Å. T., Bjaastad, J. F., … Wergeland, G. J. (2021). Treatment fidelity in brief versus standard-length school-based interventions for youth with anxiety. *School Mental Health*. https://doi.org/10.1007/s12310-021-09458-2.

Lawrence, P. J., Rooke, S. M., & Creswell, C. (2017). Review: Prevention of anxiety among at-risk children and adolescents—A systematic review and meta-analysis. *Child and Adolescent Mental Health*, *22*(3), 118–130. https://doi.org/10.1111/camh.12226.

Lyneham, H. J., Sburlati, E. S., Abbott, M. J., Rapee, R. M., Hudson, J. L., Tolin, D. F., et al. (2013, Oct). Psychometric properties of the child anxiety life interference scale (CALIS). *Journal of Anxiety Disorders*, *27*(7), 711–719. https://doi.org/10.1016/j.janxdis.2013.09.008.

Lyon, A. R., & Bruns, E. J. (2019). From evidence to impact: Joining our best school mental health practices with our best implementation strategies. *School Mental Health*, *11*(1), 106–114. https://doi.org/10.1007/s12310-018-09306-w.

Malterud, K. (2012). Systematic text condensation: A strategy for qualitative analysis. *Scandinavian Journal of Public Health*, *40*(8), 795–805. https://doi.org/10.1177/1403494812465030.

Masia Warner, C., & Fox, J. K. (2012). Advances and challenges in school-based intervention for anxious and depressed youth: Identifying and addressing issues of sustainability. *School Mental Health*, *4*(4), 193–196. https://doi.org/10.1007/s12310-012-9087-8.

Mendlowicz, M. V., & Stein, M. B. (2000). Quality of life in individuals with anxiety disorders. *American Journal of Psychiatry*, *157*(5), 669–682. PM:10784456.

Merikangas, K. R., He, J. P., Burstein, M., Swanson, S. A., Avenevoli, S., Cui, L., et al. (2010). Lifetime prevalence of mental disorders in U.S. adolescents: Results from the national comorbidity survey replication—Adolescent Supplement (NCS-A). *Journal of the American Academy of Child & Adolescent Psychiatry*, *49*(10), 980–989. https://doi.org/10.1016/j.jaac.2010.05.017.

Merikangas, K. R., He, J. P., Burstein, M., Swendsen, J., Avenevoli, S., Case, B., et al. (2011). Service utilization for lifetime mental disorders in U.S. adolescents: Results of the national comorbidity survey-adolescent supplement (NCS-A). *Journal of the American Academy of Child & Adolescent Psychiatry*, *50*(1), 32–45. https://doi.org/10.1016/j.jaac.2010.10.006.

Mifsud, C., & Rapee, R. M. (2005). Early intervention for childhood anxiety in a school setting: Outcomes for an economically disadvantaged population. *Journal of the American Academy of Child & Adolescent Psychiatry*, *44*(10), 996–1004. https://doi.org/10.1097/01.chi.0000173294.13441.87.

Muggeo, M. A., & Ginsburg, G. S. (2019). School nurse perceptions of student anxiety. *Journal of School Nursing*, *35*(3), 163–168. https://doi.org/10.1177/1059840517752457.

Ogden, T., Bjornebekk, G., Kjobli, J., Patras, J., Christiansen, T., Taraldsen, K., et al. (2012). Measurement of implementation components ten years after a nationwide introduction of empirically supported programs—A pilot study. *Implementation Science*, *7*. https://doi.org/10.1186/1748-5908-7-49.

Orchard, F., Gregory, A. M., Gradisar, M., & Reynolds, S. (2020). Self-reported sleep patterns and quality amongst adolescents: Cross-sectional and prospective associations with anxiety and depression. *Journal of Child Psychology and Psychiatry*, *61*(10), 1126–1137. https://doi.org/10.1111/jcpp.13288.

Patalay, P., Giese, L., Stanković, M., Curtin, C., Moltrecht, B., & Gondek, D. (2016). Mental health provision in schools: Priority, facilitators and barriers in 10 European countries. *Child and Adolescent Mental Health, 21*(3), 139–147. https://doi.org/10.1111/camh.12160.

Perepletchikova, F., Treat, T. A., & Kazdin, A. E. (2007). Treatment integrity in psychotherapy research: Analysis of the studies and examination of the associated factors. *Journal of Consulting and Clinical Psychology, 75*(6), 829–841. https://doi.org/10.1037/0022-006X.75.6.829.

Polanczyk, G. V., Salum, G. A., Sugaya, L. S., Caye, A., & Rohde, L. A. (2015). Annual research review: A meta-analysis of the worldwide prevalence of mental disorders in children and adolescents. *Journal of Child Psychology and Psychiatry, 56*(3), 345–365. https://doi.org/10.1111/jcpp.12381.

Puleo, C. M., Conner, B. T., Benjamin, C. L., & Kendall, P. C. (2011, Jun). CBT for childhood anxiety and substance use at 7.4-year follow-up: A reassessment controlling for known predictors. *Journal of Anxiety Disorders, 25*(5), 690–696. https://doi.org/10.1016/j.janxdis.2011.03.005.

Raknes, S. (2010). *Psykologisk Førstehjelp. Ungdom [Psychological First Aid Kit. Teens]*. Oslo: Gyldendal Akademisk.

Rapee, R. M., Lyneham, H. J., Scniering, C. A., Wuthrich, V., Abbott, M. A., Hudson, J. L., et al. (2006). The cool kids child and adolescent anxiety program therapist manual. In *Centre for emotional health*. Sydney: Macquarie University.

Rasing, S. P. A., Creemers, D. H. M., Janssens, J., & Scholte, R. H. J. (2017). Depression and anxiety prevention based on cognitive behavioral therapy for at-risk adolescents: A meta-analytic review. *Frontiers in Psychology, 8*, 1066. https://doi.org/10.3389/fpsyg.2017.01066.

Reardon, T., Harvey, K., & Creswell, C. (2020). Seeking and accessing professional support for child anxiety in a community sample. *European Child & Adolescent Psychiatry, 29*(5), 649–664. https://doi.org/10.1007/s00787-019-01388-4.

Reardon, T., Harvey, K., Young, B., O'Brien, D., & Creswell, C. (2018). Barriers and facilitators to parents seeking and accessing professional support for anxiety disorders in children: Qualitative interview study. *European Child & Adolescent Psychiatry, 27*(8), 1023–1031. https://doi.org/10.1007/s00787-018-1107-2.

Salloum, A., Johnco, C., Lewin, A. B., McBride, N. M., & Storch, A. A. (2016). Barriers to access and participation in community mental health treatment for anxious children. *Journal of Affective Disorders, 196*, 54–61. https://doi.org/10.1016/j.jad.2016.02.026.

Spence, S. H. (1998). A measure of anxiety symptoms among children. *Behaviour Research and Therapy, 26*, 545–566.

Swan, A. J., & Kendall, P. C. (2016). Fear and missing out: Youth anxiety and functional outcomes. *Clinical Psychology: Science and Practice, 23*(4), 417–435. https://doi.org/10.1111/cpsp.12169.

Teubert, D., & Pinquart, M. (2011). A meta-analytic review on the prevention of symptoms of anxiety in children and adolescents. *Journal of Anxiety Disorders, 25*(8), 1046–1059. https://doi.org/10.1016/j.janxdis.2011.07.001.

Thompson, A., Issakidis, C., & Hunt, C. (2008). Delay to seek treatment for anxiety and mood disorders in an Australian clinical sample. *Behavior Change, 25*, 71–84. https://doi.org/10.1375/bech.25.2.71.

Vasileva, M., Graf, R. K., Reinelt, T., Petermann, U., & Petermann, F. (2021, Apr). Research review: A meta-analysis of the international prevalence and comorbidity of mental disorders in children between 1 and 7 years. *Journal of Child Psychology and Psychiatry, 62*(4), 372–381. https://doi.org/10.1111/jcpp.13261.

Verduin, T. L., & Kendall, P. C. (2008, May). Peer perceptions and liking of children with anxiety disorders. *Journal of Abnormal Child Psychology, 36*(4), 459–469. https://doi.org/10.1007/s10802-007-9192-6.

Weeks, C., Hill, V., & Owen, C. (2017). Changing thoughts, changing practice: Examining the delivery of a group CBT-based intervention in a school setting. *Educational Psychology in Practice, 33*(1), 1–15. https://doi.org/10.1080/02667363.2016.1217400. The possibility of bringing support rather than them finding it.

Wergeland, G. J. H., Riise, E. N., & Öst, L. G. (2021). Cognitive behavior therapy for internalizing disorders in children and adolescents in routine clinical care: A systematic review and meta-analysis. *Clinical Psychology Review, 83*, 101918. https://doi.org/10.1016/j.cpr.2020.101918.

Werner-Seidler, A., Perry, Y., Calear, A. L., Newby, J. M., & Christensen, H. (2017). School-based depression and anxiety prevention programs for young people: A systematic review and meta-analysis. *Clinical Psychology Review, 51*, 30–47. https://doi.org/10.1016/j.cpr.2016.10.005.

Chapter 12

Adolescents, Asperger syndrome, depression, and cognitive behavioral therapy

Jaiganesh Selvapandiyan
Department of Psychiatry, All India Institute of Medical Sciences, Vijayawada, Andhra Pradesh, India

Abbreviations

CBT cognitive behavior therapy
ToM Theory of Mind

Introduction

Asperger syndrome is a developmental disorder with features of autism-related symptoms in the backdrop of a well-functioning intellect. The autistic symptoms shared by persons having Asperger syndrome include deficits in social communication and restricted repetitive patterns of interests and behaviors. However, in stark contrast to most individuals suffering from autistic disorder, those with Asperger syndrome do not demonstrate significant deficits in cognitive functioning and language development. The twofold advantages resulting from the well-preserved intellect of individuals with Asperger syndrome are their superior adaptive functioning when compared to most individuals of autistic disorders and the greater trainability with cognition-based techniques in their later life.

Cognitive behavior therapy (CBT), a structured psychotherapy having a large empirical literature supporting its efficacy in treating depressive and anxiety disorders, has been adapted and modified for use in several other emotional disorders with disorder specific therapeutic models (Blackburn & Davidson, 1990; DeRubeis, Webb, Tang, & Beck, 2010; Hawton, Salkovkis, Kirk, & Clark, 1989; Hofmann, Asnaani, Vonk, Sawyer, & Fang, 2012). Research on the use of CBT in the Asperger syndrome population focuses mostly on comorbid anxiety disorder, anger management and social skills training, whereas research focusing on CBT usage in comorbid depressive disorders are limited and mostly case studies (Weston, Hodgekins, & Langdon, 2016). There is also notable lack of correspondence between the components of CBT utilized in the above trials, making the interpretation of results from these trials difficult. Hence, individual adaptation of CBT strategies with tailor-made approaches are recommended for effective outcome. Also, an experimental design following a single subject research is preferred to decipher the nuances of applying individual components of CBT in Asperger syndrome population (Betty, Stephenson, & Carter, 2015). The chapter goes on to discuss the features of CBT relevant to its application in adolescents suffering from Asperger syndrome with comorbid depressive disorder.

Depressive disorder in Asperger syndrome

Deficits in social communicative skills and nonverbal communicative skills predisposes an individual suffering from Asperger syndrome to increased emotional burden while performing everyday chores that require societal participation. These inherent social deficits make them experience increased levels of stress in both present and anticipated social activities. Hence, there is a great likelihood of them developing secondary depressive disorders when the perceived burden of social activity overwhelms their coping ability. In addition to social deficits, they also face problems related to adaptive functioning, which include inefficient life activities management, procrastination, impaired self-direction, and maladaptive problem solving. These, in turn, add to the risk of them developing a depressive disorder. Research also indicates the greater prevalence of depressive disorder in people suffering from Asperger's syndrome and high functioning autism

Handbook of Lifespan Cognitive Behavioral Therapy. https://doi.org/10.1016/B978-0-323-85757-4.00018-3

(Ghazuddin, Wieder-Mikhail, & Ghaziuddin, 1998; Green, Gilchrist, Burton, & Cox, 2000; Kim, Szatmari, Bryson, Streiner, & Wilson, 2000; Sterling, Dawson, Estes, & Greenson, 2008; Tantam, 2000; Tonge, Brereton, Gray, & Einfeld, 1999).

Adaptations to cognitive behavior therapy

The unique nature of Asperger syndrome's cognitive functioning demands several adaptations to the original cognitive behavioral protocol suggested by Beck (Beck, Rush, Shaw, & Emery, 1979). Since the main focus here is on depressive disorders arising in adolescents with Asperger syndrome, we will limit our discussion to this specific comorbidity.

Case conceptualization

CBT works on the premise that thought processes and behavioral patterns strongly influence a person's reaction to any specific situation. A cognitive case formulation tries to make sense of the individual's suffering by elaborating on the connections between their thought and behavioral and emotional domains (Tarrier, 2006). A person's thought processes in difficult situations are guided by their belief systems that are deeply ingrained within the mind. These belief systems get established in the person's mind from the individual's early life experiences. In addition to the cognitive processes, the behavioral repertoires too are closely linked to information processing and they get imprinted with the ensuing emotional state because of repeated presentations. The case formulation of a typical depressive disorder in an individual suffering from Asperger syndrome is presented in Fig. 1.

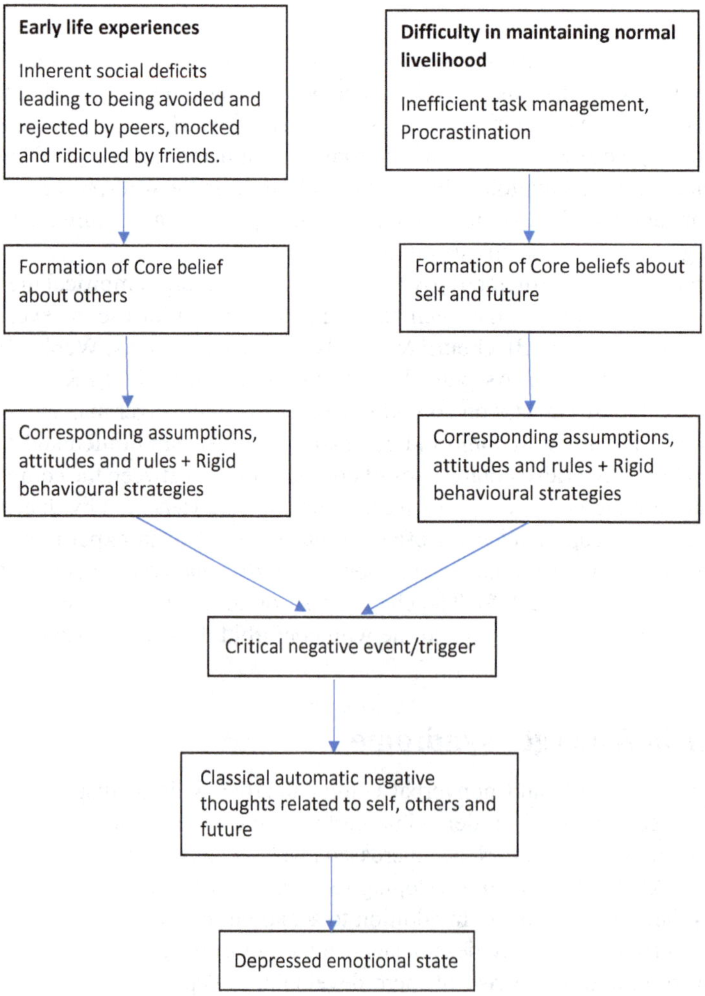

FIG. 1 Cognitive model of depression in Asperger's syndrome.

Therapeutic alliance

The hallmark of any standard CBT approach is the development of a strong and working therapeutic relationship. Needless to say, such a therapeutic alliance is even more important in a person suffering from Asperger syndrome who experiences inborn difficulties in building social bridges. However, a standard CBT approach usually recommends a collaborative therapeutic alliance wherein the goals of the treatment and the procedures done within the session are conjointly decided and implemented by the therapist and the client (Dozois & Bieling, 2010; Young, Rygh, Weinberger, & Beck, 2014). One has to be cautious while handling an Asperger syndrome patient in a collaborative sense, because they lack the motivation and intent to engage in a collaborative relationship that is mainly predicated on their processing impairments in nonsocial executive domain (Eisenmajer et al., 1996; Kleinhans, Akshoomoff, & Delis, 2005; Nyden, Gillberg, Hjelmquist, & Heiman, 1999; Ozonoff, South, & Miller, 2000; Pennington & Ozonoff, 1996). Therefore, a more directive approach is recommended in the backdrop of a strong therapeutic alliance.

Structure

CBT sessions are structured with clear and transparent steps implemented in every session. An agenda is set at the opening stage of the session, the current problems are prioritized and cognitive restructuring is done in the middle part and feedback is asked for and homework assignments are worked upon at closing the session (Beck, 2011). There is, however, a dilemma when it comes to structuring CBT sessions for Asperger syndrome patients. While there are recommendations to enforce structure to CBT sessions claiming that structured sessions help the patients to extrapolate the same to structure their lives, which is otherwise in a disorganized state (Anderson & Morris, 2006; Hare, 1997), considering their cognitive inflexibility, enforcing a structure in CBT sessions may result in premature discontinuation and attrition in therapy. Hence, the therapist has to be cautious in maintaining a balanced approach in structuring the sessions and have to exercise greater flexibility when the patient prefers unstructured sessions.

Setting goals

At the outset, we have two broad goals when we use CBT for an adolescent Asperger syndrome with comorbid depression. The first goal is to address the depressive symptomatology and the second is to improve the individual's competencies in the social sphere. When implementing the strategies, it has to be kept in mind that the goals should be broken down into multiple smaller chunks and goals that seem impossible should be dropped. Each factor described in Fig. 1 will become a platform for cognitive restructuring as we progress in the therapy from superficial automatic thoughts to deeper core beliefs.

Practice and procedures

Emotional education

Emotional education gains more importance in Asperger syndrome during the early phase of therapy. The interlinking between the thoughts, the emotional states, and maladaptive behavioral repertoires have to be clearly presented. Individuals suffering from Asperger syndrome also have deficits in Theory of Mind (ToM) which results in their failure to attribute the reason for others' behaviors to their respective emotional states, leading to self-referential or paranoid cognitions (Baron-Cohen & Jolliffe, 1997; Baron-Cohen, O'Riordan, Stone, Jones, & Plaisted, 1999; Blackshaw, Kinderman, Hare, & Hatton, 2001; Craig, Hatton, Craig, & Bentall, 2004; Heavey, Phillips, Baron-Cohen, & Rutter, 2000; Kleinman, Marciano, & Ault, 2001; Lombardo, Barnes, Wheelwright, & Baron-Cohen, 2007; Muris et al., 1999; Rutherford, Baron-Cohen, & Wheelwright, 2002). The personal experience of emotional states by individuals with Asperger syndrome is distorted in many ways and this could lead to misinterpretation of a particular emotional reaction of an Asperger syndrome individual by the therapist. (Example: An experience of mild boredom can lead to a sequence of exaggerated behavioral outbursts.) They may also face difficulties in verbalizing their emotional sufferings (Gaus, 2000; Holliday Willey, 2003). Hence, the therapist has to maintain an open mental state when evaluating the emotional reactions of a person suffering from Asperger syndrome.

Identifying thoughts

Identifying maladaptive automatic thoughts that have negative content and challenging them with evidence is the major work undertaken in cognitive restructuring. The most frequent variety of cognitive error identified in clinical work with individuals suffering from Asperger syndrome is "polarized thinking" (Anderson & Morris, 2006). The cognitive inflexibility of an individual with Asperger syndrome favors the rise and establishment of this type of cognitive distortion in them. As we are dealing with a disorder that faces more interpersonal conflicts, the arbitrary inference type is also identified as a more prevalent cognitive distortion in the cognitive repertoire of individuals with Asperger syndrome. Individuals with Asperger syndrome face great difficulty when asked to introspect. This results in challenges when they are asked to do a thought record exercise. Hence, the therapist has to be more alert to record the thought patterns that arise within the sessions as the client report a particular problem and put them into therapeutic testing.

Responding to thoughts

In evaluating and responding to cognitive errors, the therapist often uses the Socratic questioning approach with their clients (Greenberger & Padesky, 1995). In this approach, collecting evidence for and against the cognitions, looking for alternative explanations and looking for either the worst or the best outcomes are usually practiced. In the case of Asperger syndrome, if a therapist senses that Socratic questioning is not favorable in a specific real-life situation (Channon, Charman, Heap, Crawford, & Rios, 2001), then he/she should resort to a more directive approach. In a directive approach, the therapist will offer concrete opinions backed by logical evidence or proof. A directive approach should be considered only in the presence of a strong therapeutic alliance between the client and the therapist to minimize attrition.

Social skills training

Any CBT program in Asperger syndrome will be incomplete without explicitly addressing social deficits through social skills training (McAfee, 2001; Moyes, 2001). Social skill development is done through multiple methods within the framework of increasing prosocial polite behaviors and expanding social repertoires needed for everyday functioning. Both instrumental and relationship skills are taught. Instrumental skills like greeting with handshakes and turn taking are applicable in situations where people's behaviors are predictable and scripted. Again, relationship skills involving eye contact, body and head posture are essential in building interpersonal bonding. Paralinguistic aspects of communication such as voice volume and rate of speech should also be addressed to prepare individuals with Asperger syndrome have smooth conversations. Role-plays are important within the sessions to facilitate habilitation of newly learnt social skills.

Most often, interpersonal disputes with family members create stressful experiences for individuals suffering from Asperger syndrome. Interpersonal relationship modes training is helpful in addressing this area of dysfunction. In this strategy, the client is asked to identify significant family members with whom an attitude of peaceful communion is attempted even during nonconflict period.

Coping skills training

Individuals with Asperger syndrome tend to have poor problem-solving skills and inferior task management skills, which lead to nonprogression in their life goals. This should be addressed through time management training, assertiveness training, and graded task assignment to break procrastinations.

Adapting the techniques

The techniques used in CBT need to be attuned to the developmental characteristics of Asperger syndrome. This includes:

1. Extending the duration of the session to address slow learning ability of the client.
2. Limiting the use of metaphors to address the poor ability of the client in abstract thinking.
3. Using scriptures and depictions to address the difficulty of the client in expressing emotions through verbal means.
4. Using a pedantic style to address the overscrupulous character of the client.
5. Involving the client in each and every decision-making step to address their need for autonomy.

Mini-dictionary of terms

1. **Pervasive developmental disorder.** A group of disorders characterized by impaired development of social and communicative skills.
2. **Therapeutic alliance.** A cooperative working relationship between therapist and patient.

3. **Theory of Mind.** A set of intellectual abilities that helps us to perceive that others have an independent set of beliefs, attitudes, and assumptions that may differ from our own.
4. **Polarized thinking.** Perceiving a situation in only two categories.
5. **Arbitrary inference.** We tend to infer something as the truth without obvious evidence.
6. **Socratic questioning.** Inquiry into the logic of various abstractions.
7. **Habilitation.** The process of enhancing the level of functioning of an individual with a disability by providing appropriate training.
8. **Pedantic.** Giving too much attention to formal rules or small details.

Key facts of adolescent Asperger syndrome

- Asperger syndrome is a pervasive developmental disorder presenting unique challenges during adolescence.
- The well-preserved intellect in Asperger syndrome provides an opportunity for the therapist to teach new skills to an adolescent with Asperger syndrome.

Key facts of depressive disorder in Asperger syndrome

- The struggle faced by an individual with Asperger syndrome in social functioning contributes to excess stress and development of depressive disorder.
- Research reveals high incidence of depressive disorder in people suffering from Asperger syndrome.
- An untreated depressive disorder adds to the existing burden of an adolescent Asperger leading to even further deterioration in adaptive functioning.

Applications to other areas

We have reviewed several adaptations of CBT for application in the scenario of an adolescent suffering from Asperger syndrome with comorbid depressive disorder. We shall discuss further applicability of these techniques in other areas of therapeutics.

Of late, there has been an increase in the use of third-wave therapies for Asperger syndrome, in which meditative practices are directly used to train the individuals in the higher-order mindfulness techniques (de Bruin, Blom, Smit, van Steensel, & Bögels, 2015; Spek, van Ham, & Nyklíček, 2013). Asperger syndrome, being a developmental disorder, requires a multipronged approach wherein the above-described CBT strategies need to be implemented in a socio-cognitive framework before stepping into mindfulness meditative training (Selvapandiyan, 2019). The major difference between cognitively oriented programs and mindfulness strategies is in the method used to handle the cognitive pathologies. While adaptive manipulations of cognitions are utilized in the former program, in the latter program detached experiential techniques are utilized. It would be difficult for an individual experiencing intense depressive agony to shift from destructive ways of critical thinking to detached ways of mindful thinking. Therefore, mindfulness strategies have to be used as a part of holistic CBT paradigm, which is executed by priming the individual suffering from Asperger syndrome with developmentally oriented pragmatic cognitive methods (Selvapandiyan, 2019).

Recent research has identified multiple overlaps between Asperger syndrome and certain personality disorders (Dudas et al., 2017; Lugnegård, Hallerbäck, & Gillberg, 2012). CBT, being a common psychotherapy procedure for both Asperger syndrome and personality disorders, several adaptations recommended in this chapter for Asperger syndrome can be extrapolated to personality disorders (especially the techniques related to therapeutic alliance, therapy structure, and setting goals).

Summary points

- While adapting CBT for managing depressive disorder in Asperger syndrome, the developmental deficits/characteristics that are inherent to Asperger syndrome have to be taken into account.
- Building a strong therapeutic alliance is a challenge and, at the same time, an important prerequisite in the therapeutic process.
- The therapist needs to have more creativity and flexibility in implementing the therapeutic strategies.
- Social skills training and coping skills training should become an integral part of every CBT session.
- CBT should be viewed as a process rather than a one-stop solution, when we intend to treat an individual suffering from Asperger syndrome.

References

Anderson, S., & Morris, J. (2006). Cognitive behaviour therapy for people with Asperger syndrome. *Behavioural and Cognitive Psychotherapy, 34*, 293–303.

Baron-Cohen, S., & Jolliffe, T. (1997). Another advanced test of theory of mind: Evidence from very high functioning adults with autism or Asperger syndrome. *Journal of Child Psychology and Psychiatry, 38*, 813–822.

Baron-Cohen, S., O'Riordan, M., Stone, V., Jones, R., & Plaisted, K. (1999). Recognition of faux pas by normally developing children and children with autism and Asperger's disorder. *Journal of Autism and Developmental Disorders, 29*, 407–418.

Beck, J. S. (2011). *Cognitive behavior therapy: Basics and beyond* (2nd ed.). New York: Guilford Press.

Beck, A. T., Rush, A. J., Shaw, B. F., & Emery, G. (1979). *Cognitive therapy of depression*. New York: Guilford Press.

Betty, P. V. H., Stephenson, J., & Carter, M. (2015). Cognitive–behavioural approach for children with autism spectrum disorder: A literature review. *Journal of Intellectual & Developmental Disability, 40*, 213–229.

Blackburn, I. M., & Davidson, K. (1990). *Cognitive therapy for depression and anxiety: A practitioner's guide*. Oxford: Blackwell Scientific.

Blackshaw, A. J., Kinderman, P., Hare, D. J., & Hatton, C. (2001). Theory of mind, causal attribution and paranoia in Asperger syndrome. *Autism: The International Journal of Research and Practice, 5*, 147–163.

Channon, S., Charman, T., Heap, J., Crawford, S., & Rios, P. (2001). Real-life-type problem solving in Asperger's syndrome. *Journal of Autism and Developmental Disorders, 31*, 461–469.

Craig, J. S., Hatton, C., Craig, F. B., & Bentall, R. P. (2004). Persecutory beliefs, attributions and theory of mind: Comparison of clients with paranoid delusions, Asperger's syndrome and healthy controls. *Schizophrenia Research, 69*, 29–33.

de Bruin, E. I., Blom, R., Smit, F. M., van Steensel, F. J., & Bögels, S. M. (2015). MYmind: Mindfulness training for Youngsters with autism spectrum disorders and their parents. *Autism, 19*, 906–914.

DeRubeis, R. J., Webb, C. A., Tang, T. Z., & Beck, A. T. (2010). Cognitive therapy. In K. S. Dobson (Ed.), *Handbook of cognitive behavioral therapies* (pp. 277–317). New York: Guilford Press.

Dozois, D. J. A., & Bieling, P. J. (2010). Cognitive therapy for depression. In S. G. Hofmann, & M. A. Reinecke (Eds.), *Cognitive–behavioral therapy With adults: A guide to empirically-informed assessment and intervention* (pp. 1–18). Cambridge University Press.

Dudas, R. B., Lovejoy, C., Cassidy, S., Allison, C., Smith, P., & Baron-Cohen, S. (2017). The overlap between autistic spectrum conditions and borderline personality disorder. *PLoS One, 12*. https://doi.org/10.1371/journal.pone.0184447, e0184447.

Eisenmajer, R., Prior, M., Leekman, S., Wing, L., Gould, J., Welham, M., et al. (1996). Comparison of clinical symptoms in autism and Asperger's disorder. *Journal of the American Academy of Child and Adolescent Psychiatry, 35*, 1523–1531.

Gaus, V. (2000). "I feel like an alien": Individual psychotherapy for adults with Asperger's disorder using a cognitive behavioral approach. *NADD Bulletin, 3*, 62–65.

Ghazuddin, M., Wieder-Mikhail, W., & Ghaziuddin, N. (1998). Comorbidity of Asperger syndrome: A preliminary report. *Journal of Intellectual Disability Research, 42*, 279–283.

Green, J., Gilchrist, A., Burton, D., & Cox, A. (2000). Social and psychiatric functioning in adolescents with Asperger disorder compared with conduct disorder. *Journal of Autism and Developmental Disorders, 30*, 279–293.

Greenberger, D., & Padesky, C. A. (1995). *Mind over mood*. New York: Guilford Press.

Hare, D. J. (1997). The use of cognitive-behaviour therapy with people with Asperger syndrome: A case study. *Autism, 1*, 215–225.

Hawton, K., Salkovkis, P. M., Kirk, J., & Clark, D. M. (1989). *Cognitive behaviour therapy for psychiatric problems: A practical guide*. Oxford: Oxford University Press.

Heavey, L., Phillips, W., Baron-Cohen, S., & Rutter, M. (2000). The awkward moments test: A naturalistic measure of social understanding in autism. *Journal of Autism and Developmental Disorders, 30*, 225–236.

Hofmann, S. G., Asnaani, A., Vonk, I. J., Sawyer, A. T., & Fang, A. (2012). The efficacy of cognitive behavioral therapy: A review of meta-analyses. *Cognitive Therapy and Research, 36*, 427–440.

Holliday Willey, L. (2003). When the thunder roars. In L. Holliday Willey (Ed.), *Asperger syndrome in adolescence*. London: Jessica Kingsley.

Kim, K. A., Szatmari, P., Bryson, S. E., Streiner, D. L., & Wilson, F. (2000). The prevalence of anxiety and mood problems among children with autism and Asperger's disorder. *Autism, 4*, 117–132.

Kleinhans, N., Akshoomoff, N., & Delis, D. C. (2005). Executive functions in autism and Asperger's disorder: Flexibility, fluency, and inhibition. *Developmental Neuropsychology, 27*, 379–401.

Kleinman, J., Marciano, P., & Ault, R. (2001). Advanced theory of mind in high-functioning adults with autism. *Journal of Autism and Developmental Disorders, 31*, 29–36.

Lombardo, M. V., Barnes, J. L., Wheelwright, S. J., & Baron-Cohen, S. (2007). Self-referential cognition and empathy in autism. *PLoS One, 2*. https://doi.org/10.1371/journal.pone.0000883, e883.

Lugnegård, T., Hallerbäck, M. U., & Gillberg, C. (2012). Personality disorders and autism spectrum disorders: What are the connections? *Comprehensive Psychiatry, 53*, 333–340.

McAfee, J. (2001). *Navigating the social world: A curriculum for individuals with high-functioning autism and Asperger syndrome*. London: Jessica Kingsley.

Moyes, R. A. (2001). *Incorporating social goals in the classroom: A guide for teachers and parents of children with high-functioning autism and Asperger syndrome*. London: Jessica Kingsley.

Muris, P., Steerneman, P., Meesters, C., Merckelbach, H., Horselenberg, R., Van Den Hogan, T., et al. (1999). The TOM test: A new instrument for assessing theory of mind in normal children and children with pervasive developmental disorders. *Autism and Developmental Disorders*, *29*, 67–80.

Nyden, A., Gillberg, C., Hjelmquist, E., & Heiman, M. (1999). Executive function/attention deficits in boys with Asperger disorder, attention disorder and reading/writing disorder. *Autism*, *3*, 213–228.

Ozonoff, S., South, M., & Miller, J. (2000). DSM-IV defined Asperger disorder: Cognitive, behavioral and early history differentiation from high-functioning autism. *Autism*, *4*, 29–46.

Pennington, B. F., & Ozonoff, S. (1996). Executive functions and developmental psychopathology. *Journal of Child Psychology and Psychiatry*, *37*, 51–87.

Rutherford, M. D., Baron-Cohen, S., & Wheelwright, S. (2002). Reading the mind in the voice: A study with normal adults and adults with Asperger syndrome and high functioning autism. *Journal of Autism and Developmental Disorders*, *32*, 189–194.

Selvapandiyan, J. (2019). Improvising pragmatic cognitive-behavioral therapy for depressed adolescents with Asperger syndrome. *The Primary Care Companion for CNS Disorders*, *21*. https://doi.org/10.4088/PCC.18l02381.

Spek, A. A., van Ham, N. C., & Nyklíček, I. (2013). Mindfulness-based therapy in adults with an autism spectrum disorder: A randomized controlled trial. *Research in Developmental Disabilities*, *34*, 246–253.

Sterling, L., Dawson, G., Estes, A., & Greenson, J. (2008). Characteristics associated with presence of depressive symptoms in adults with autism spectrum disorder. *Journal of Autism and Developmental Disorders*, *38*, 1011–1018.

Tantam, D. (2000). Psychological disorder in adolescents and adults with Asperger disorder. *Autism*, *4*, 47–62.

Tarrier, N. (Ed.). (2006). *Case formulation in cognitive behaviour therapy: The treatment of challenging and complex cases* Routledge/Taylor & Francis Group.

Tonge, B., Brereton, A., Gray, K., & Einfeld, S. (1999). Behavioural and emotional disturbance in high-functioning autism and Asperger's disorder. *Autism*, *3*, 117–130.

Weston, L., Hodgekins, J., & Langdon, P. E. (2016). Effectiveness of cognitive behavioural therapy with people who have autistic spectrum disorders: A systematic review and meta-analysis. *Clinical Psychology Review*, *49*, 41–54.

Young, J. E., Rygh, J. L., Weinberger, A. D., & Beck, A. T. (2014). Cognitive therapy for depression. In D. H. Barlow (Ed.), *Clinical handbook of psychological disorders: A step-by-step treatment manual* (pp. 275–332). New York: The Guilford Press.

Chapter 13

Cognitive behavioral therapy for cannabis use disorder: A focus on adolescents

Alba González-Roz[a], Víctor Martínez-Loredo[b], Roberto Secades-Villa[a], and Sergio Fernández-Artamendi[c]
[a]*Addictive Behaviors Research Group, Department of Psychology, Faculty of Psychology, University of Oviedo, Oviedo, Principality of Asturias, Spain*
[b]*Department of Psychology and Sociology, University of Zaragoza, Teruel, Spain* [c]*Department of Psychology, Universidad Loyola Andalucía, Sevilla, Andalucía, Spain*

Abbreviations

ACRA	Community Reinforcement Approach for Adolescents
ADHD	attention-deficit/hyperactivity disorder
AUDIT	Alcohol Use Disorder Identification Test
CAST	Cannabis Abuse Screening Test
CBT	cognitive behavioral therapy
CBT-5	cognitive behavioral therapy—five sessions
CBT7	cognitive behavioral therapy—seven sessions
CM	contingency management
CPQ-A	Cannabis Problems Questionnaire for Adolescents
CPQ-A-S	Short Cannabis Problems Questionnaire for Adolescents
CRA	Community Reinforcement Approach
CSAT	the Center for Substance Abuse Treatment
CUD	cannabis use disorder
CUDIT	Cannabis Use Disorder Identification Test
CUDIT-R	Cannabis Use Disorder Identification Test-revised
CUPIT	Cannabis Use Problems Identification Test
CYT	the Cannabis Youth Treatment
DSM-IV	diagnostic and statistical manual of mental disorders, 4th edition
ESPAD	European School Survey Project on Alcohol and Other Drugs
MDFT	multidimensional family therapy
MET	motivational enhancement therapy
SBIRT	screening, brief intervention and referral to treatment
SDS	Severity Dependence Scale

Introduction

Prevalence of cannabis use in adolescents

Cannabis is the illicit drug most commonly used across the world, both among adults and adolescents. United States remains the region with the highest annual prevalence of cannabis use (8.8% among the population aged 15–64) (United Nations Office on Drugs and Crime, 2020). In Europe, the prevalence of use is about five times that of other substances. Around 15% (18.0 million) of young adults (aged 15–34) in Europe report using cannabis in the last year, with males being typically twice as likely to report use than females (European Monitoring Centre for Drugs and Drug Addiction, 2020a). Accordingly, cannabis-related disorders now play a major role in drug treatment admissions in high- and middle-income countries (European Monitoring Centre for Drugs and Drug Addiction, 2020a). Those entering treatment for problems related to cannabis are predominantly male, with a ratio of just under five males to each female.

Cannabis is also the most widely used drug among adolescents and young people. Globally, it is estimated that there were 11.6 million past-year users of cannabis among students aged 15–16 in 2018. This corresponds to an annual

Handbook of Lifespan Cognitive Behavioral Therapy. https://doi.org/10.1016/B978-0-323-85757-4.00013-4

prevalence of cannabis use of 4.7% among this age group, a rate that is higher than the rate among the general population aged 15–64 (3.9%) (United Nations Office on Drugs and Crime, 2020).

Cannabis was the most widely used illicit drug in all European School Survey Project on Alcohol and Other Drugs (ESPAD) countries (European Monitoring Centre for Drugs and Drug Addiction, 2020b). On average, 16% of students had used cannabis at least once in their lifetime, 7.1% had used cannabis in the last 30 days, and 2.4% reported that they had first used cannabis at age 13 or younger (European Monitoring Centre for Drugs and Drug Addiction, 2020b). Boys reported cannabis use to a larger extent than girls (18% vs 13%). Using the Cannabis Abuse Screening Test (CAST) (Legleye, Karila, Beck, & Reynaud, 2007), 4.0% of students in the total ESPAD population (4.7% of the boys and 3.3% of the girls) were classified as high-risk cannabis users. This corresponds to an average proportion of 35% among students who reported cannabis use in the last year (European Monitoring Centre for Drugs and Drug Addiction, 2020b). Although these epidemiological studies have consistently shown that cannabis use is higher among boys than girls, this gap seems to be slightly narrowing (Secades-Villa & Fernández Artamendi, 2017), and transition from experimental cannabis use to development of cannabis use disorder (CUD) seems faster in girls than in boys (Khan et al., 2013).

Consequences of cannabis use in adolescence

In recent years, the chemical profiles of newly engineered cannabis strains may be becoming more harmful as technology and production methods improve. This increase in potency poses higher risk of cannabis use, mostly among adolescents, a particularly vulnerable period for drugs use, including cannabis, as the brain is still developing (ElSohly et al., 2016).

The risk of developing dependence on cannabis among those who have ever used the drug (even once) has been estimated at 9% (Lopez-Quintero et al., 2011), rate that rises to 17% among users who start using cannabis in adolescence (Anthony, 2006). Early onset and regular cannabis use have shown to be significant risk factors for multiple health problems (Copeland, Rooke, & Swift, 2013; Hall, 2006). Thus, evidence indicates that adolescence is a sensitive period during which cannabis use may result in adverse neurocognitive effects that appear to show a level of permanency into adulthood (Blest-Hopley, Colizzi, Giampietro, & Bhattacharyya, 2020). Adolescent cannabis use is associated with an increased risk for psychosis later in life (Kiburi, Molebatsi, Ntlantsana, & Lynskey, 2021), and depression and suicidality in young adulthood (Gobbi et al., 2019). Cannabis use in this population is also a risk factor for violence in youths (Dellazizzo et al., 2020). Additionally, early cannabis use increases risks for other illegal and legal substance use (Secades-Villa, Garcia-Rodríguez, Jin, Wang, & Blanco, 2015; Weinberger, Zhu, Lee, Xu, & Goodwin, 2021).

In sum, this body of research supports the link between cannabis use and numerous mental health disorders in adolescence, and underscores the importance of developing innovative and specific prevention, assessment, and treatment strategies aimed at curtailing cannabis use and ameliorating some of these problems in this population.

Assessment of cannabis misuse and CUD in adolescence

Objectives of clinical assessment

The clinical assessment pursues a comprehensive evaluation of the pattern of cannabis use (frequency, quantity, and level of severity), and other domains involved in the onset and maintenance of this behavior. The functional analysis is the fundamental assessment method for CUD. In doing this, it is important to thoroughly identify the specific characteristics of any situation where individuals use cannabis, including both internal (e.g., physical and emotional states, thoughts) and external triggers. The assessment of dispositional variables (i.e., physical condition, use of other substances, family environment, neighborhood, socioeconomic status, etc.) as well as motivating operations that may have a more generalized impact on cannabis use is of great interest. Verbal rules (i.e., expectancies and cannabis-related cognitions) associated with cannabis use and its potential consequences are an important component of the functional analysis as well. Regarding the consequences, it is possible that individuals use cannabis in different situations aimed at obtaining different experiences or benefits and, therefore, all functions of cannabis use associated with specific contexts should be identified.

Building and enhancing the therapeutic alliance with the adolescent/family

Low or nontherapeutic alliance is associated with greater cannabis use (Berry, Gregg, Lobban, & Barrowclough, 2016). Brief motivational interventions such as the motivational enhancement therapy (MET; Miller, Zweben, DiClemente, & Rychtarik, 1992) are useful strategies to gather relevant information at the initial stages of treatment, enhance motivation, and elicit the clients' commitment to change. Brief motivational interventions in primary care contexts have shown positive effects on managing ambivalence, cessation efforts, and negative consequences associated with substance use (D'Amico et al., 2018; Kells, Burke, Parker, Jonestrask, & Shrier, 2019). Motivational interview may be also useful to frame the client's current behavior into extended molar patterns connected with their goals and personal values. In the context of discounted value of temporally distant goals compared to immediate availability of cannabis, this may help to discuss the overall utility of the client's behavior (Ladd, Murphy, & Borsari, 2021). Importantly, most of the available treatments also include motivational interviewing strategies during the first sessions. Of the available protocols, the brief intervention and referral to treatment (SBIRT) represents a feasible approach to target motivation to change in cannabis users with beneficial outcomes in high school and community settings (Maslowsky, Capell, Moberg, & Brown, 2017; Mitchell, Gryczynski, O'Grady, & Schwartz, 2013).

Assessment and diagnostic tools

As already mentioned, the assessment in cannabis users should consider an exhaustive screening of different areas, including, but not limited to, the motivation state of change, frequency, and quantity of cannabis use, associated negative psychosocial consequences as well as physical problems and mental-health co-occurrent problems and disorders (Fernández-Artamendi, Martínez-Loredo, & López-Núñez, 2021). Relatedly, the rates of tobacco and cannabis co-use are markedly high in the adolescent population (Conway et al., 2018), and self-registers of tobacco cigarettes and joints may assist with treatment planning and encourage less motivated groups to cease using tobacco. Continued monitoring of expired air carbon monoxide can be regarded as useful for both assessment and educational purposes on tobacco and cannabis effects (Goldstein, Gans, Ripley-Moffitt, Kotsen, & Bars, 2018). Overall, no single questionnaire can be considered as a gold standard for evaluating all aspects related to cannabis use. Accordingly, this section will offer an overview of the main assessment and screening measures of cannabis-related problems and severity of CUD. The clinical utility and scoring system are presented in Table 3 (see the "Practice and procedures" section).

Screening of CUD

One of the most used and efficient screening tools for CUD in youths (aged 14–22) is the CAST (Legleye et al., 2007; Legleye, Piontek, & Kraus, 2011). It comprises six questions that explore either lifetime or past 12 months' frequency of nonrecreational use, memory problems, motivation to reduce/stop using cannabis, unsuccessful quit attempts, and cannabis use-associated problems. Although the items are answered using a Likert-type scale (never, rarely, sometimes, quite often, and very often), users' responses are usually dichotomized in the so-called "binary" version with scores ranging between 0 and 6.

Based on the Alcohol Use Disorder Identification Test (AUDIT), Adamson and Sellman (2003) developed a 10-item questionnaire to screen adults for cannabis abuse or dependence (DSM-IV) within the past 6 months: The Cannabis Use Disorder Identification Test or CUDIT. Seven years later, an updated version (CUDIT-R; Adamson et al. (2010)) was developed, maintaining four of the original items, with four additional ones. Mimetically with the AUDIT, the CUDIT-R assesses the following domains: frequency of use, cannabis-related problems (abuse), dependence, and psychological features.

Despite both the CUDIT and CUDIT-R being developed for the general population, some studies have informed of adequate psychometric properties for adolescents and young adults (Annaheim, Rehm, & Gmel, 2008). Notwithstanding, a more recent version for youth covering cannabis use motives, leisure time, and work/school impairment seems preferable when assessing more severe problematic cannabis users (Annaheim, Scotto, & Gmel, 2010).

Another questionnaire, the Severity Dependence Scale (SDS) was initially developed as a short screener for substance use dependence in adults. In 2006, Martin, Copeland, Gates, and Gilmour (2006) validated the five-item SDS in an adolescent sample (aged 14–18) of cannabis users. The SDS covers past 3 months and explores, in a four-point scale, perceived control over cannabis use, concerns about current use, and difficulties in cutting down on cannabis.

Other screening measures of cannabis-related problems

Two other widely used questionnaires to assess problems associated with cannabis use are the Cannabis Problem Questionnaire for Adolescents (CPQ-A; Martin, Copeland, Gilmour, Gates, & Swift, 2006) and the Cannabis Use Problems Identification Test (CUPIT; Bashford, Flett, & Copeland, 2010). The original version of the CPQ-A comprised 50 items, but the most widely used version is shorter (27 items) (Martin, Copeland, Gilmour, et al., 2006). The CPQ-A evaluates cannabis-related problems and psychosocial and mental health problems in the last 3 months. A shorter version of 12 items is also available (CPQ-A-S; Proudfoot, Vogl, Swift, Martin, & Copeland, 2010), which may offer more clinically useful information for treatment planning and evaluation, despite having weaker psychometric properties compared to the CAST.

The CUPIT (Bashford et al., 2010) assesses at-risk adolescents from the general population. The total score allows to discriminate between individuals with cannabis abuse, dependence, and nonproblem use. Of note is that the validation study offers two cut-offs; the optimal cut-off maximizing both sensitivity (87%) and specificity (79%) was ≥ 20, which also offered a positive predictive value of 98% at 1 year. Nonetheless, the authors recommend a cut-off of ≥ 12 in order to reduce the number of "diagnostic orphans" (individuals reporting some problems but not reaching the established diagnostic threshold).

Empirical support on psychological treatments for cannabis misuse and CUD in adolescents

Most effective treatment options available for cannabis misuse and CUD include cognitive behavioral therapy (CBT), MET, family-based therapies, the Adolescent Community Reinforcement Approach (ACRA), and contingency management (CM) (Dennis et al., 2004; Hoch et al., 2012; Hogue, Henderson, Becker, & Knight, 2018; Hogue, Henderson, Ozechowski, & Robbins, 2014; Winters et al., 2021). In particular, psychological interventions based on cognitive behavioral models have shown considerable empirical support for the treatment of CUD (Fadus et al., 2019; Tanner-Smith, Wilson, & Lipsey, 2013) and other related problems such as hazardous drinking (Kay-Lambkin, Baker, Lewin, & Carr, 2009) or coexisting mental health problems (Hendriks, van der Schee, & Blanken, 2012). In practice, motivational interviewing and CM are usually integrated into CBT protocols, as there is evidence suggesting significant booster effects on abstinence as a result of adding CM to CBT/MET (Carroll et al., 2006).

One of the most significant research efforts to improve treatment effectiveness for cannabis users is the initiative launched by the Center for Substance Abuse Treatment (CSAT), the Cannabis Youth Treatment study (CYT; Dennis et al., 2002, 2004). This project was developed in response to the high number of adolescent cannabis users and the lack of short-term outpatient treatments targeting cannabis use in this population. The CYT study examined the cost-effectiveness of five outpatient therapies for adolescent cannabis users, including Multidimensional Family Therapy (MDFT), ACRA, and the combination of MET and CBT in different lengths. Results revealed that the combination of five sessions of MET+CBT delivered over 6 weeks and ACRA were the most cost-effective in terms of sustained reductions in substance use and involvement in illegal activities (Dennis et al., 2004; Ramchand, Griffin, Suttorp, Harris, & Morral, 2011).

Treatment rationale and intervention components in CBT

In the CBT model, the therapist provides patients with the appropriate knowledge on their addictive behavior, as well as with the necessary training to deal with high-risk situations triggering substance use. Although different CBT protocols are available, core strategies include the following: (1) providing psychoeducation on the factors associated with cannabis use; (2) training in effective coping skills to overcome difficult personal life events, including mood-management strategies and relaxation techniques; (3) assisting with family problems and school difficulties; and (4) training in relapse prevention strategies to effectively identify internal (e.g., emotional related) and external (e.g., peer substance use) high-risk situations linked to cannabis use.

The MET/CBT5 therapy integrates several of the former components and has been manualized by the CSAT. The MET/CBT5 protocol (Web, Seudder, Kaminer, & Kadden, 2002) consists of five sessions and is delivered once per week. The first two sessions are held individually and seek to enhance motivation to change through assessments of reasons to change, prior cessation attempts, and perceived self-efficacy. Three additional group-based sessions with a maximum of six people are then delivered to train individuals in effective refusal skills, increase the engagement in nonsubstance use activities, and develop a healthy social network. As an improved alternative, the CBT7 is intended to follow this former protocol and facilitate sustained abstinence through mood-management training and relapse prevention (Dennis et al., 2002).

Of the available CBT approaches for cannabis use in adolescents, the ACRA is considered one of the most effective (Aguinaldo et al., 2019). Its core strategies and several recommendations for their successful implementation are presented in subsequent sections.

The Adolescent Community Reinforcement Approach

Community Reinforcement Approach (CRA) was developed for adults with alcohol use problems in the 70s, but was latter adapted to adolescents (Godley et al., 2001). CRA attempts to rearrange contingencies through increases in nonsubstance use behaviors. With this purpose, ACRA (the acronym that CRA takes for adolescents) uses basic CBT-based principles, including operant techniques and skill-training activities to help adolescents and caregivers in dealing with substance use. ACRA can be implemented either in groups or individually and in a variety of contexts, such as residential and outpatient treatments and school settings (Godley, Smith, Meyers, & Godley, 2016). The treatment is flexible, but the recommended treatment length is 12–14 weeks. Although it offers 19 procedures, the therapist can go through each depending on the client's needs. The number of sessions is a suggestion, but the authors recommend implementing a minimum of ten 1-h sessions with the adolescent, two 1-h sessions with the client's caregiver, and two 1.5-h sessions with both.

The main goals of ACRA include (Godley et al., 2001, 2016): (1) building a positive relationship between the adolescent and the therapist; (2) promoting engagement in positive social activities; (3) promoting positive peer relationships; (4) improving the relationship with the family; and (5) facilitating abstinence from cannabis, alcohol and other drugs. With the caregivers, ACRA pretends to promote their participation in the treatment process, help the adolescent reduce or eliminate substance-related problems, and provide information to the caregivers about effective parenting practices. The updated version of the ACRA was published in 2016 and covers the following themes. The suggestive flow of sessions and main therapeutic procedures are summarized in Table 1.

TABLE 1 Main topics covered in the ACRA treatment.

Format (caregivers/ client)	Themes/strategies
Client only sessions	• Overview of ACRA • Discussion on client's relevant life areas and treatment goals • Homework assignment
	• Rapport building (therapeutic motivational style) • Review of homework completion • Revision of treatment goals • Exploring the quality of social relationships • Functional analysis of substance use and prosocial behavior
	• Communication skills • Problem-solving skills • Happiness scales • Goals of counseling • Schedule of prosocial activities • Training in drink/drug refusal skills • Systematic encouragement • Relapse prevention • Anger management • Alcohol and drug testing • Other ACRA procedures if needed (couple relationship skills, job seeking skills, medication adherence, sobriety sampling)
Caregivers-only session	• Rapport building • Rationale of ACRA • Training in effective parenting • Training in communications skills • Training in problem solving
Caregiver + the adolescent sessions	• Reinforcement of therapy attendance • Examining positive things, the family and the client appreciation for each other • Problem solving and communication styles (training in effective communication styles)

This table provides the structure and content of sessions in ACRA. *Note*: *ACRA*, The Adolescent Community Reinforcement Approach.

Functional analysis of substance use behavior

The functional analysis should serve as a guide for the semistructure interview where the therapist will explain the purpose of the assessment and gather information to identify stimuli within the three-term contingency: internal/external triggers (conditioned and/or discriminative stimuli, motivating operations, and verbal rules), specific descriptions of substance use behaviors, and short-term positive and long-term negative consequences (O'Brien, Haynes, & Kaholokula, 2016). The functional analysis is performed individually during the first two sessions, although it should be continuously revised along the therapy to track the client's changes. As a practical recommendation, the manualized ACRA treatment (Godley et al., 2016) notes that it is important to use several techniques that help build an adequate rapport before starting the functional analysis so as to create a comfortable environment. To facilitate the analysis, the therapist can suggest to the client to describe a typical situation of cannabis use and then ask specific questions. Once all the relevant information have been gathered, the therapist should analyze the functional relationship between all the identified elements, provide a summary of the information, and praise the adolescent for participating. Table 2 shows an example of the functional analysis form and Fig. 1 shows a schema with the main elements comprising the functional analysis of substance use.

Functional analysis of prosocial behavior

A strength of ACRA is that it focuses on prosocial or substance-free activities linked to client's own life goals or values. As recommended by the authors of the manualized ACRA treatment (Godley et al., 2016), the therapist can first focus on one typical activity that the adolescent has enjoyed in the last 6 months. It is possible that clients are unable to identify any positive substance-free activity. In this case, it is useful to provide them with a list of different activities available in the community, tailored to clients' age. Once the functional analysis of different prosocial activities is examined, the therapist should summarize the information and explain to the client how they will use this information to promote alternative behaviors during treatment. Additionally, the client and therapist should work together on a homework assignment based on

TABLE 2 **Example of a completed functional analysis of substance use behavior record sheet.**

A		B	C	
What external things (who, where, when, what) trigger your behavior?	What internal things (thoughts and emotional or physical feelings) trigger your behavior?	Using behavior (what, how much, how long)	What happens or what do you feel in the short term?	What happens or do you feel in the long term?
– My friends John, Bob, Kelly and Jeff	– Thinking, getting high and having fun	– Marijuana – 2–3 blunts – 3–4 h	– I feel fine being with friends in the park talking and laughing	– Problems with my parents
– Bob's house – The park	– Getting relaxed	– Marijuana – 2–3 blunts – 3–4 h	– It's cool to be out of my house	– I forget things, dry mouth and eyes, problems to focus on things – I feel less like doing things – Doing homework gets harder
– The evening – Weekends around 12 p.m.	– Restlessness – Bored, overwhelmed			
– Being alone	– Bored, overwhelmed	– Marijuana – 1 blunt – Until I go to sleep (1 h)	– I feel relaxed and enjoy "my time"	
– My bedroom				
– Nighttime				

This table provides an example on how to register the A[antecedents]-B[behavior]-C[consequences] chain of different situations related to cannabis use during the interview. *Notes: A*, antecedents; *B*, behaviors; *C*, consequences.

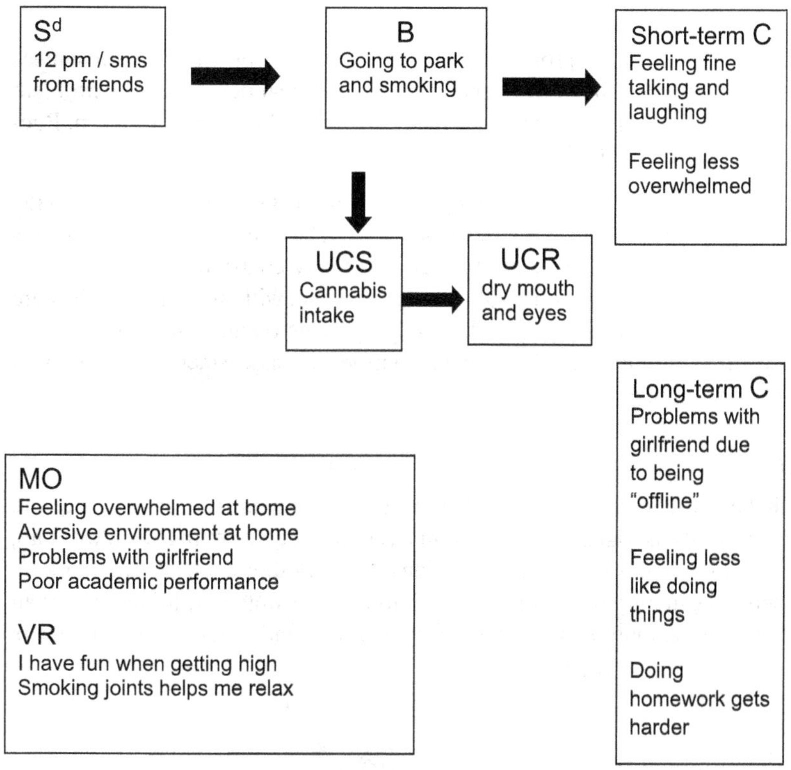

FIG. 1 Example of functional analysis of smoking behavior in the park with friends. This figure illustrates a functional analysis of smoking behavior based on the information provided in Table 2. *B*, behavior; *C*, consequence; *MO*, motivating operations; S^d, discriminative stimuli; *UCR*, unconditioned response; *UCS*, unconditioned stimulus; *VR*, verbal rules.

the identified activities. In doing this, it is necessary to plan when, how, for how long and how frequent will the adolescent engage in each behavior during the following weeks. It is important to envisage any obstacle or short-term negative consequence (Godley, Smith, Passetti, & Subramaniam, 2014).

Drug refusal skills

The objective of training in refusal skills is to provide clients with adequate skills to avoid using drugs when offered and maintain their abstinence. This procedure can be used either when clients are still using any drug and want to refuse future opportunities to use them, or when they have relapsed (Godley et al., 2016; Meyers, Roozen, & Smith, 2011).

Training in how to effectively refuse peer drug offerings intends also to promote social support by identifying people in the clients' networks who could be supportive in high-risk relapse situations. In addition to reinforcing existing refusal skills and promoting their generalizability by identifying when those skills are used, the six refusal skills that are taught are: to say "no" assertively, to display assertive body language, to suggest alternative activities, to change the subject of the conversation, to directly address the issue with whomever persists in offering the drug, and to leave the situation if any of the previous techniques does not work. During the training, it is recommended that the clinician role-plays not only the role of a person offering drugs, but also the client in order to give the latter the opportunity to observe their nonassertive performance in an easier way.

Communication skills

The objective of this procedure is to train the client in effective communication skills to improve the quality of the interactions between the clients and their networks (Meyers et al., 2011). This module is used in client-only sessions and in caregiver-only and family sessions. To foster the generalizability of skills trained in session, it is important to use real-life situations where communication problems played a major role in the past (Godley et al., 2016).

After discussing with the clients why it is important to communicate positively (i.e., to increase the odds of getting what the client really wants in a conflictive situation), the therapist should teach three communication elements (an understanding statement, a partial responsibility statement, and an offer to help), practice them multiple times via role playing and exchange the roles (Godley et al., 2014).

Problem solving

Problem solving is aimed at training adolescents on how to solve problems without using cannabis or other substances (Meyers et al., 2011). This procedure is individualized and consists of three main steps: (1) a clear definition of a problem situation, (2) brainstorming to get several alternatives, and (3) using pros and cons to encounter the optimal solution. Recommendations for its effective implementation are (Godley et al., 2016):

- Framing the problem-solving technique: Introducing the rationale for working on problem solving should be done in the context of the boy/girl life problems. It will be important to choose current problems that she or he would like to address in the session and envisage any issues (skills deficits/social or personal barriers) that may contribute.
- The therapist should encourage the client to consider as many potential solutions as possible, without judging if they are adequate at first. Two principles guide this phase: going for quantity and avoid judgments on any solution.
- Training in problem solving should end with the assignment of the implementation, the assessment of the chosen solution, and continued monitoring of the outcome attained.

Caregiver-adolescent relationship skills

This procedure seeks to work collaboratively with the family or a significant other to enhance their motivation to participate in the therapy and educate on effective communication skills together with a family version of goal setting (Godley et al., 2014). During weeks 3–5, the therapist trains the caretaker in positive communications skills (Godley et al., 2016). Working in role-playing several real situations helps to discuss the quality of current interactions and illustrate how to implement an understanding statement into a conversation, the self-attribution of partial responsibility, and offering help (Meyers et al., 2011).

Job seeking skills

This module seeks to involve the adolescent in prosocial activities. The therapist trains the adolescent in a realistic and systematic approach to finding a job. This training is conducted individually and there is no specific number of sessions (Godley et al., 2016). Job seeking skills should consider existing skills and previous work experience. Procedural steps are as follow: exploring which areas are of interest, making a list of job contacts, role-playing to practice social skills on how to search for a job, asking for an application, and doing an interview.

Anger management

This is an optional module in ACRA that seeks to train adolescents on how to identify anger and recognize early psychophysiological signs triggering this emotion (Godley et al., 2016). Using functional analysis, the adolescent tries to identify antecedents and consequences related to the problem behavior and plan rewarding activities that may help to deal with anger. The adolescent is encouraged to take time to cool down, implement problem-solving skills, or use assertive communication skills whenever he/she feels anger-related signs (Godley et al., 2014).

Relapse prevention

One essential part of ACRA treatment, and other CBT protocols as well, includes training in relapse prevention skills (Meyers et al., 2011). Once the adolescent gets sober, it is important to work on relapse prevention through the functional analysis of relapse situations. Once the external and internal triggers (early warning signs of relapse) are identified, it is easy to modify the behavior early in the chain through the implementation of effective coping skills. With this purpose, the therapist reviews with the patient the difference between lapse (slip or substance use in a single occasion or for a short period of time) and relapse (return to using drugs in a regular basis) and works in a collaborative way with the client to identify personal high-risk situations related to continued cannabis use; CBT strategies that were previously trained (e.g., assertive communication, drug refusal skills) are incorporated in this plan that specifies a schedule for the implementation of specific activities/strategies upon necessity (Godley et al., 2016).

Summary and conclusions

CUD is a highly prevalent condition in the youth population that is also accompanied by other co-occurring behaviors, such as tobacco use, alcohol-related problems, and mental health disorders. Currently, there is no medication approved for CUD; however, several psychosocial treatments are available for the adolescent population. CBT, and particularly the ACRA treatment, are among the most effective therapies to date. ACRA produces significant reductions on cannabis use and positively impacts other areas, such as internalizing and externalizing behaviors. In this therapy, the adolescent and his/her caregivers work in collaboration to facilitate substance use abstinence. During 12/14 weeks, the therapist covers the most relevant variables related to substance use maintenance (poor social skills, problem-solving strategies, family problems). In doing this, the functional analysis of behavior is one of the core assessment and treatment procedures that are individualized to each person. Other assessment measures with sound psychometric properties for the adolescent population are the CUDT-R, the CAST, and the CUPIT.

Practice and procedures

See Table 3.

Mini-dictionary of terms

- **Functional analysis of behavior.** Procedure aimed at establishing the contingency relationships between the elements comprising a behavioral stream, based on Pavlovian and operant conditioning.
- **Cannabis use disorder.** The term "cannabis use disorder" is the label used by the DSM when an individual meets 2 out of the 11 specific criteria in the previous 12 months.
- **Stages of change.** A fundamental component of the Transtheoretical Model of Change proposed by Prochaska and DiClemente. It describes the readiness to change of a given person regarding a specific health-related behavior.
- **Motivational enhancement therapy.** The motivational enhancement therapy was designed by Miller and Rollnick and has as a main goal to help clients to solve their ambivalences about engaging in treatment and quitting.
- **Refusal skills.** This is essentially a coping strategy that prevents youth from relapsing back to substance use once they have reduced or abstained from drugs.

Key facts of cognitive behavioral therapy implementation

- The functional analysis of cannabis use and substance-free behaviors are key elements of both the assessment and treatment.
- Motivational enhancement strategies need to be implemented during the initial sessions.
- CBT is among the most effective interventions for promoting enduring abstinence and reducing psychosocial and individual risk factors related to cannabis problems.
- The optimal intervention for reducing cannabis use is likely to be CBT+MET+CM.
- Evidence suggests that CUD interventions should consist of a minimum of 14 sessions.

Applications to other areas

Timely, this chapter surges in the context of increasing cannabis use rates in the adolescent population. This review work provides readers with the fundamentals of CBT implementation procedures, and clinical recommendations for the assessment and treatment processes. Its scope is directed to both researchers and clinicians, as well as to students at different levels. The assessment and treatment procedures described herein can be used to address both cannabis misuse and CUD. CBT is among the most effective therapy for cannabis in the adolescent population. It has provided evidence on its effectiveness for both reducing cannabis use and promoting abstinence in both adolescents and adults (Hoch et al., 2012). Numerous studies have also shown positive outcomes in other substance use disorders and nonsubstance use domains, such as internalizing (e.g., depression, anxiety, prevention of psychosis, and externalizing (e.g., ADHD, delinquency) problems. This is clinically relevant as it is estimated that up to 31% of the adolescents aged 10–19 also present at least one co-occurrent mental health disorder (Silva et al., 2020).

Summary points

- The cannabis use screening should consider different areas, including the motivation state of change, frequency, and quantity of cannabis use, associated negative consequences and mental-health comorbidity.
- Adding tobacco to cannabis joints is a common practice in adolescents, and both conditions should be monitored through treatment.

TABLE 3 Screening and assessment tools for cannabis misuse and cannabis use disorder in adolescents.

Measuring tool	Clinical utility	Psychometric properties	Scoring
CAST (Legleye, 2018; Legleye et al., 2007)	Problematic patterns of cannabis use	Reliability: $\alpha = 0.81$ Association with DSM-5 CUD scores: $r = 0.69$	Total score: 0–24 Cut-off suggesting moderate DSM-5 CUD: ≥5 Sensitivity: 86.6% Specificity: 78.2% PPV (cut-off ≥5): 43.2% NPV (cut-off ≥5): 96.8% Cut-off suggesting severe DSM-5 CUD: ≥8 Sensitivity: 86% Specificity: 86.7% PPV (cut-off ≥8): 27.3% NPV (cut-off ≥8): 99.1%
CPQ-A (Martin, Copeland, Gilmour, et al., 2006)	Cannabis-related problems	Three factors with reliability ranging 0.72–0.88 Association with DSM-IV CUD scores: $r = 0.80$	Total score: 0–27 Cut-off suggesting dependence (Spanish version by Fernández-Artamendi et al., 2012): ≥5.5 Specificity = 86.7% Sensitivity = 80.4% Cut-off suggesting abuse (Spanish version by Fernández-Artamendi et al., 2012): ≥4.5 Specificity = 76.3% Sensitivity = 76.5%
CPQ-A-S (Proudfoot et al., 2010)	Cannabis-related problems	Reliability: $\alpha = 0.73$	Total score: 0–12
CUDIT-R (Adamson et al., 2010)	CUD	Reliability: $\alpha = 0.91$	Total score: 0–32 Cut-off suggesting current CUD: ≥13 Sensitivity: 91% Specificity: 90% PPV (cut-off ≥13): 96% NPV (cut-off ≥13): 80%
CUPIT (Bashford et al., 2010)	Cannabis related problems	Two factors: Reliability "impaired control": $\alpha = 0.92$ Reliability "problems": $\alpha = 0.83$	Total score: 0–82 Cut-off suggesting DSM-IV CUD or abuse: ≥12 Sensitivity: 98% Specificity: 35% 1-year PPV: 95.1%
SDS (Martin, Copeland, Gates, et al., 2006)	Dependence severity	Reliability: $\alpha = 0.83$ Test-retest: ICC = 0.88 Association with number of DSM-IV cannabis dependence symptoms: $r = 0.76$	Total score: 0–15 Cut-off suggesting current CUD: ≥4 Specificity = 94.3% Sensitivity = 65.1%

This table provides a set of assessment tools for cannabis in the adolescent population. It informs on their psychometric properties, their clinical utility, and the scoring system. *Note: CAST*, Cannabis Abuse Screening Test; *DSM*, Diagnostic and Statistical Manual of Mental Disorders; *CPQ-A*, Cannabis Problems Questionnaire-Adolescents; *CPQ-A-S*, the shortened version of the Cannabis Problems Questionnaire for Adolescents; *CUD*, cannabis use disorder; *CUDIT-R*, Cannabis Use Disorder Identification Test-Revised; *CUPIT*, Cannabis Use Problems Identification Test; *ICC*, intra-class correlation; *NPV*, negative predictive value; *PPV*, positive predictive value; *SDS*, Severity of Dependence Scale.

- CBT is effective in reducing cannabis use and facilitating abstinence.
- ACRA is a CBT approach that seeks to enhance abstinence from cannabis, alcohol and other substances while enhancing family and social support.
- Adding MET and CM to CBT enhances abstinence outcomes.

References

Adamson, S. J., Kay-Lambkin, F. J., Baker, A. L., Lewin, T. J., Thornton, L., Kelly, B. J., & Sellman, J. D. (2010). An improved brief measure of cannabis misuse: The Cannabis Use Disorders Identification Test-Revised (CUDIT-R). *Drug and Alcohol Dependence*, *110*, 137–143.

Adamson, S. J., & Sellman, J. D. (2003). A prototype screening instrument for cannabis use disorder: The Cannabis Use Disorders Identification Test (CUDIT) in an alcohol-dependent clinical sample. *Drug and Alcohol Review*, *22*, 309–315.

Aguinaldo, L. D., Squeglia, L. M., Gray, K. M., Coronado, C., Lees, B., Tomko, R. L., & Jacobus, J. (2019). Behavioral treatments for adolescent cannabis use disorder: A rationale for cognitive retraining. *Current Addiction Reports*, *6*, 437–442.

Annaheim, B., Rehm, J., & Gmel, G. (2008). How to screen for problematic cannabis use in population surveys: An evaluation of the Cannabis Use Disorders Identification Test (CUDIT) in a Swiss sample of adolescents and young adults. *European Addiction Research*, *14*, 190–197.

Annaheim, B., Scotto, T. J., & Gmel, G. (2010). Revising the Cannabis Use Disorders Identification Test (CUDIT) by means of Item Response Theory. *International Journal of Methods in Psychiatric Research*, *19*, 142–155.

Anthony, J. C. (2006). The epidemiology of cannabis dependence. In R. A. Roffman, & R. S. Stephens (Eds.), *Cannabis dependence: Its nature, consequences and treatment* (pp. 58–105). Cambridge: Cambridge University Press.

Bashford, J., Flett, R., & Copeland, J. (2010). The Cannabis Use Problems Identification Test (CUPIT): Development, reliability, concurrent and predictive validity among adolescents and adults. *Addiction*, *105*, 615–625.

Berry, K., Gregg, L., Lobban, F., & Barrowclough, C. (2016). Therapeutic alliance in psychological therapy for people with recent onset psychosis who use cannabis. *Comprehensive Psychiatry*, *67*, 73–80.

Blest-Hopley, G., Colizzi, M., Giampietro, V., & Bhattacharyya, S. (2020). Is the adolescent brain at greater vulnerability to the effects of cannabis? A narrative review of the evidence. *Frontiers in Psychiatry*, *26*(11), 859.

Carroll, K. M., Easton, C. J., Nich, C., Hunkele, K. A., Neavins, T. M., Sinha, R., … Rounsaville, B. J. (2006). The use of contingency management and motivational/skills-building therapy to treat young adults with marijuana dependence. *Journal of Consulting and Clinical Psychology*, *74*, 955–966.

Conway, K. P., Green, V. R., Kasza, K. A., Silveira, M. L., Borek, N., Kimmel, H. L., … Compton, W. M. (2018). Co-occurrence of tobacco product use, substance use, and mental health problems among youth: Findings from wave 1 (2013–2014) of the population assessment of tobacco and health (PATH) study. *Addictive Behaviors*, *76*, 208–217.

Copeland, J., Rooke, S., & Swift, W. (2013). Changes in cannabis use among young people: Impact on mental health. *Current Opinion in Psychiatry*, *26*, 325–329.

D'Amico, E. J., Parast, L., Shadel, W. G., Meredith, L. S., Seelam, R., & Stein, B. D. (2018). Brief motivational interviewing intervention to reduce alcohol and marijuana use for at-risk adolescents in primary care. *Journal of Consulting and Clinical Psychology*, *86*, 775–786.

Dellazizzo, L., Potvin, S., Dou, B. Y., Beaudoin, M., Luigi, M., Giguère, C.É., & Dumais, A. (2020). Association between the use of cannabis and physical violence in youths: A meta-analytical investigation. *American Journal of Psychiatry*, *177*, 619–626.

Dennis, M., Godley, S. H., Diamond, G., Tims, F. M., Babor, T., Donaldson, J., … Funk, R. (2004). The Cannabis Youth Treatment (CYT) Study: Main findings from two randomized trials. *Journal of Substance Abuse Treatment*, *27*, 197–213.

Dennis, M., Titus, J. C., Diamond, G., Donaldson, J., Godley, S. H., Tims, F. M., … Steering Committee, C. Y. T. (2002). The Cannabis Youth Treatment (CYT) experiment: Rationale, study design and analysis plans. *Addiction*, *97*, 16–34.

ElSohly, M. A., Mehmedic, Z., Foster, S., Gon, C., Chandra, S., & Church, J. C. (2016). Changes in cannabis potency over the last 2 decades (1995–2014): Analysis of current data in the United States. *Biological Psychiatry*, *79*, 613–619.

European Monitoring Centre for Drugs and Drug Addiction. (2020a). *European drug report 2020: Trends and developments*. Lisbon: EMCDDA.

European Monitoring Centre for Drugs and Drug Addiction. (2020b). *ESPAD report 2019. Results from the European School Survey Project on Alcohol and Other Drugs*. Lisbon: EMCDDA.

Fadus, M. C., Squeglia, L. M., Valadez, E. A., Tomko, R. L., Bryant, B. E., & Gray, K. M. (2019). Adolescent substance use disorder treatment: An update on evidence-based strategies. *Current Psychiatry Reports*, *21*, 96.

Fernández-Artamendi, S., Fernández-Hermida, J. R., García-Cueto, E., Secades-Villa, R., García-Fernández, G., & Barrial-Barbén, S. (2012). Spanish adaptation and validation of the adolescent-Cannabis Problems Questionnaire (CPQ-A). *Adicciones*, *24*, 41–49.

Fernández-Artamendi, S., Martínez-Loredo, V., & López-Núñez, C. (2021). Sex differences in comorbidity between substance use and mental health in adolescents: Two sides of the same coin. *Psicothema*, *33*, 36–43.

Gobbi, G., Atkin, T., Zytynski, T., Wang, S., Askari, S., Boruff, J., … Mayo, N. (2019). Association of cannabis use in adolescence and risk of depression, anxiety, and suicidality in young adulthood: A systematic review and meta-analysis. *JAMA Psychiatry*, *76*, 426–434.

Godley, S. H., Meyers, R. J., Smith, J. E., Karvinen, T., Titus, J. C., Godley, M. D., … Kelberg, P. (2001). *The Adolescent Community Reinforcement Approach for adolescent cannabis users*. Rockville, MD: Substance Abuse and Mental Health Services Administration.

Godley, S. H., Smith, J. E., Meyers, R. J., & Godley, M. (2016). *The Adolescent Community Reinforcement Approach: A clinical guide for treating substance use disorders*. Normal, IL: Chestnut Health Systems.

Godley, S. H., Smith, J. E., Passetti, L. L., & Subramaniam, G. (2014). The Adolescent Community Reinforcement Approach (A-CRA) as a model paradigm for the management of adolescents with substance use disorders and co-ocurring psychiatric disorders. *Substance Abuse*, *35*, 352–363.

Goldstein, A. O., Gans, S. P., Ripley-Moffitt, C., Kotsen, C., & Bars, M. (2018). Use of expired air carbon monoxide testing in clinical tobacco treatment settings. *Chest*, *153*, 554–562.

Hall, W. (2006). The mental health risks of adolescent cannabis use. *PLoS Medicine*, *3*(2), e39.

Hendriks, V., van der Schee, E., & Blanken, P. (2012). Matching adolescents with a cannabis use disorder to multidimensional family therapy or cognitive behavioral therapy: Treatment effect moderators in a randomized controlled trial. *Drug and Alcohol Dependence*, *125*, 119–126.

Hoch, E., Noack, R., Henker, J., Pixa, A., Höfler, M., Behrendt, S., … Wittchen, H. U. (2012). Efficacy of a targeted cognitive-behavioral treatment program for cannabis use disorders (CANDIS *). *European Neuropsychopharmacology*, *22*, 267–280.

Hogue, A., Henderson, C. E., Becker, S. J., & Knight, D. K. (2018). Evidence base on outpatient behavioral treatments for adolescent substance use, 2014–2017: Outcomes, treatment delivery, and promising horizons. *Journal of Clinical Child and Adolescent Psychology*, *47*, 499–526.

Hogue, A., Henderson, C. E., Ozechowski, T. J., & Robbins, M. S. (2014). Evidence base on outpatient behavioral treatments for adolescent substance use: Updates and recommendations 2007–2013. *Journal of Clinical Child and Adolescent Psychology*, *43*, 695–720.

Kay-Lambkin, F. J., Baker, A. L., Lewin, T. J., & Carr, V. J. (2009). Computer-based psychological treatment for comorbid depression and problematic alcohol and/or cannabis use: A randomized controlled trial of clinical efficacy. *Addiction*, *104*, 378–388.

Kells, M., Burke, P. J., Parker, S., Jonestrask, C., & Shrier, L. A. (2019). Engaging youth (adolescents and young adults) to change frequent marijuana use: Motivational Enhancement Therapy (MET) in primary care. *Journal of Pediatric Nursing*, *49*, 24–30.

Khan, S. S., Secades-Villa, R., Okuda, M., Wang, S., Pérez-Fuentes, G., Kerridge, B. T., & Blanco, C. (2013). Gender differences in cannabis use disorders: Results from the National Epidemiologic Survey of Alcohol and Related Conditions. *Drug and Alcohol Dependence*, *130*, 101–108.

Kiburi, S. K., Molebatsi, K., Ntlantsana, V., & Lynskey, M. T. (2021). Cannabis use in adolescence and risk of psychosis: Are there factors that moderate this relationship? A systematic review and meta-analysis. *Substance Abuse*, *22*, 1–25.

Ladd, B. O., Murphy, J. G., & Borsari, B. (2021). Integration of motivational interviewing and behavioral economic theories to enhance brief alcohol interventions: Rationale and preliminary examination of client language. *Experimental and Clinical Psychopharmacology*, *29*, 90–98.

Legleye, S. (2018). The Cannabis Abuse Screening Test and the DSM-5 in the general population: Optimal thresholds and underlying common structure using multiple factor analysis. *International Journal of Methods in Psychiatric Research*, *27*, e1597.

Legleye, S., Karila, L., Beck, F., & Reynaud, M. (2007). Validation of the CAST, a general population Cannabis Abuse Screening Test. *Journal of Substance Use*, *12*, 233–242.

Legleye, S., Piontek, D., & Kraus, L. (2011). Psychometric properties of the Cannabis Abuse Screening Test (CAST) in a French sample of adolescents. *Drug and Alcohol Dependence*, *113*, 229–235.

Lopez-Quintero, C., Pérez de los Cobos, J., Hasin, D. S., Okuda, M., Wang, S., Grant, B. F., & Blanco, C. (2011). Probability and predictors of transition from first use to dependence on nicotine, alcohol, cannabis, and cocaine: Results of the national epidemiologic survey on Alcohol and Related Conditions (NESARC). *Drug and Alcohol Dependence*, *115*, 120–130.

Martin, G., Copeland, J., Gates, P., & Gilmour, S. (2006). The Severity of Dependence Scale (SDS) in an adolescent population of cannabis users: Reliability, validity and diagnostic cut-off. *Drug and Alcohol Dependence*, *83*, 90–93.

Martin, G., Copeland, J., Gilmour, S., Gates, P., & Swift, W. (2006). The Adolescent Cannabis Problems Questionnaire (CPQ-A): Psychometric properties. *Addictive Behaviors*, *31*, 2238–2248.

Maslowsky, J., Capell, J. W., Moberg, D. P., & Brown, R. L. (2017). Universal school-based implementation of screening brief intervention and referral to treatment to reduce and prevent alcohol, marijuana, tobacco, and other drug use: Process and feasibility. *Substance Abuse: Research and Treatment*, *11*, 1–10.

Meyers, R. J., Roozen, H. G., & Smith, J. E. (2011). The community reinforcement approach: An update of the evidence. *Alcohol Research and Health*, *33*, 380–388.

Miller, W. R., Zweben, A., DiClemente, C. C., & Rychtarik, R. G. (1992). *Motivational enhancement therapy manual. A clinical research guide for therapists treating individuals with alcohol abuse and dependence*. Rockville, MD: National Institute on Alcohol Abuse and Alcoholism.

Mitchell, S. G., Gryczynski, J., O'Grady, K. E., & Schwartz, R. P. (2013). SBIRT for adolescent drug and alcohol use: Current status and future directions. *Journal of Substance Abuse Treatment*, *44*, 463–472.

O'Brien, W., Haynes, S. N., & Kaholokula, J. K. (2016). Behavioral assessment and the functional analysis. In C. M. Nezu, & A. M. Nezu (Eds.), *The Oxford handbook of cognitive and behavioral therapies*. Oxford, UK: Oxford University Press.

Proudfoot, H., Vogl, L., Swift, W., Martin, G., & Copeland, J. (2010). Development of a short cannabis problems questionnaire for adolescents in the community. *Addictive Behaviors*, *35*, 734–737.

Ramchand, R., Griffin, B. A., Suttorp, M., Harris, K. M., & Morral, A. (2011). Using a cross-study design to assess the efficacy of motivational enhancement therapy-cognitive behavioral therapy 5 (MET/CBT5) in treating adolescents with cannabis-related disorders. *Journal of Studies on Alcohol and Drugs*, *72*, 380–389.

Secades-Villa, R., & Fernández Artamendi, S. (2017). Gender differences in cannabis use disorders. In V. R. Preedy (Ed.), *The handbook of cannabis and related pathologies* (pp. 131–137). London: Academic Press.

Secades-Villa, R., Garcia-Rodríguez, O., Jin, C. J., Wang, S., & Blanco, C. (2015). Probability and predictors of the cannabis gateway effect: A national study. *International Journal of Drug Policy*, *26*, 135–142.

Silva, S. A., Silva, S. U., Ronca, D. B., Gonçalves, V. S. S., Dutra, E. S., & Carvalho, K. M. B. (2020). Common mental disorders prevalence in adolescents: A systematic review and meta-analyses. *PLoS One*, *15*, e0232007.

Tanner-Smith, E. E., Wilson, S. J., & Lipsey, M. W. (2013). The comparative effectiveness of outpatient treatment for adolescent substance abuse: A meta-analysis. *Journal of Substance Abuse Treatment*, *44*, 145–158.

United Nations Office on Drugs and Crime. (2020). *World drug report 2020*. Vienna: United Nations publication.

Web, C., Seudder, M., Kaminer, Y., & Kadden, R. (2002). *CYT cannabis youth treatment series volume 2*. Rockville, MD: U.S. Department of Health and Human Services, Substance Abuse and Mental Health Services Administration Center for Substance Abuse Treatment.

Weinberger, A. H., Zhu, J., Lee, J., Xu, S., & Goodwin, R. D. (2021). Cannabis use and the onset of cigarette and e-cigarette use: A prospective, longitudinal study among youth in the United States. *Nicotine & Tobacco Research*, *23*, 609–613.

Winters, K. C., Mader, J., Budney, A. J., Stanger, C., Knapp, A. A., & Walker, D. D. (2021). Interventions for cannabis use disorder. *Current Opinion in Psychology*, *38*, 67–74.

Chapter 14

Anxiety in youth with asthma and cognitive behavioral therapy

Gemma Sicouri
Black Dog Institute and School of Psychology, University of New South Wales, Sydney, NSW, Australia

Abbreviations

ADIS-C/P	Anxiety Disorders Interview Schedule for children and parents
CBT	cognitive behavioral therapy
HPA	hypothalamopituitary axis
RCT	randomized control trial

Introduction

Asthma is a serious respiratory disease, characterized by a reversible inflammatory condition of airways, with symptoms including coughing, wheezing, and difficulty breathing. It is the most common chronic illness in youth and affects around 1 in 9 children and adolescents (Akinbami, Simon, & Rossen, 2016). Children and adolescents (also referred to as "youth") with asthma have been shown to be at increased risk of psychological disorders, and anxiety disorders in particular (Dudeney, Sharpe, Jaffe, Jones, & Hunt, 2017). Anxiety in youth with asthma is associated with significant emotional and financial cost to individuals, their families, and wider society. Yet, there is limited understanding of the factors associated with the development and maintenance of anxiety disorders in youth with asthma—and this, unfortunately, has translated into a lack of evidence-based treatment options for this population. This chapter provides an overview of the current research on anxiety in youth with asthma and includes practical information about how to adapt cognitive behavioral therapy (CBT) for youth anxiety to account for an individual's asthma status. We conclude the chapter by highlighting the need for further research to improve outcomes for this population.

Asthma in youth

Asthma is the most common chronic illness in youth worldwide (World Health Organisation, 2021). In the United States, estimates show that asthma affects over 5 million children and adolescents (under the age of 18 years), of which 44% have experienced one or more asthma exacerbations (also known as asthma attacks) in 2019 (Centers for Disease Control and Prevention, 2019). There are large disparities in terms of prevalence rates and asthma outcomes across population groups (see Canino, McQuaid, & Rand, 2009). Youth asthma is also extremely diverse in terms of intrusiveness of symptoms and treatment burden; however, with proper asthma management, most children and adolescents can live free of symptoms and without significant functional impairment. Yet, despite the availability of effective treatments, asthma morbidity is a major public health issue. It is one of the most common causes of school absenteeism and accounts for over $5.92 billion in direct healthcare costs annually in the United States (Perry, Braileanu, Palmer, & Stevens, 2019). Understanding the factors that influence asthma morbidity—including psychological factors, such as anxiety—is crucial to guide treatments designed to reduce the impact of both disorders on the lives of children, their families, and society.

Causes of asthma

Asthma is thought to be caused by an interaction of genetic and environmental risk factors, although the strength of evidence varies across studied factors (Clawson et al., 2019). Inherited attributes (e.g., sex, family heredity), prenatal and

Handbook of Lifespan Cognitive Behavioral Therapy. https://doi.org/10.1016/B978-0-323-85757-4.00005-5

postnatal events (e.g., maternal smoking, maternal obesity), and early childhood exposures (e.g., bronchiolitis, poorer maternal psychosocial functioning) may contribute to the development of asthma. Other risk factors include indoor and outdoor allergens, air pollution, being allergic, and increased frequency of respiratory infections (Australian Institute of Health and Welfare, 2020).

Asthma severity and control

Two important concepts which guide the treatment of asthma in youth are a child's asthma control and asthma severity. Asthma control is the extent to which the features of asthma have been reduced or removed. It comprises of two components: the pattern of asthma symptoms and how a child responds to medical intervention. Well-controlled asthma means that the asthma symptoms are infrequent, there is no limitation of normal activities and there are low risks of asthma exacerbations. Asthma severity means how much treatment is needed to keep asthma well-controlled. It can be categorized into intermittent or persistent asthma, and persistent asthma can be categorized into either mild, moderate, or severe. Asthma exacerbations (also known as asthma attacks) are acute asthma events which involve a restriction of airways and worsening asthma symptoms. Asthma exacerbations are triggered by environmental factors (such as tobacco smoke or indoor allergens), poor asthma control (perhaps due to poor medical adherence or the absence of a written asthma management plan), and psychological states (such as anxiety or stress). Asthma symptoms, particularly asthma exacerbations, are scary for children and their families and highlight the need for a good understanding of a child's asthma severity and control. A thorough assessment of a child's asthma symptoms, medication use, lung functioning, and functional impairment is thus required to determine a child's asthma severity and control.

Asthma and anxiety in youth

Youth with asthma experience significant psychological difficulties, and notably higher rates of anxiety disorders (Lu et al., 2012; Pinquart & Shen, 2011). A meta-analysis of 15 cross-sectional studies found that approximately one in five children and adolescents with asthma have an anxiety disorder (Dudeney et al., 2017), which is three times as high as the prevalence of anxiety disorders in children without asthma (Lawrence et al., 2015). Anxiety disorders tend to occur at a higher rate compared to other psychological disorders comorbid with asthma, such as behavioral problems or depression (Pinquart & Shen, 2011; Vila, Nollet-Clemençon, de Blic, Mouren-Simeoni, & Scheinmann, 2000). Further, children with asthma experience higher levels of anxiety symptoms compared to children with other chronic illnesses (e.g., diabetes or cancer; Pinquart & Shen, 2011). Thus, there seems to be a degree of specificity in the asthma-anxiety relationship.

Children with asthma experience a range of anxiety disorders. Generalized anxiety disorder and social anxiety disorder are three times higher for children with asthma compared to those without asthma. Separation anxiety disorder, panic disorder, and agoraphobia are twice as high (Dudeney et al., 2017). The spread of anxiety disorders in children with asthma is different to that seen in adults, which tends to show a stronger association between panic disorder and asthma (Hasler et al., 2005; Katon, Richardson, Lozano, & McCauley, 2004). This comorbidity is likely due to the overlap in physical symptoms of asthma and panic disorder (Kessler et al., 2005).

Impact of anxiety disorders in youth with asthma

Anxiety disorders in youth with asthma has a significant impact on a young person's functioning and development, including poorer health outcomes, such as increased asthma symptoms, increased functional impairment, greater hospital utilization and emergency department use, poorer asthma control, and reduced quality of life (Richardson et al., 2006). There are several pathways by which asthma and anxiety might influence each other (Goodwin, Bandiera, Steinberg, Ortega, & Feldman, 2012). A child with asthma may impact the development of anxiety due to increased stress, impairment, or discomfort. Conversely, the presence of an anxiety disorder might exacerbate asthma symptoms due to emotions, thoughts, and behaviors (e.g., avoidance of taking medications due to fears about the side effects). The presence of an anxiety disorder may also impact on the socio-cognitive development of children as they enter adolescence. For example, anxiety in the context of asthma may reduce a child's individuation from their parents, which may interfere with peer relationships and reduce a child's self-confidence (Katon et al., 2004). Thus, youth with this dual diagnosis are at greater risk of a range of difficulties associated with both asthma and anxiety.

A CBT framework

Cognitive behavioral models of youth anxiety highlight the interaction between a person's cognitive processes, emotional experiences, and behavioral responses in the development and maintenance of anxiety disorders in youth (Hudson & Rapee, 2004). In contrast to adults, these models also emphasize the role and influence of family factors in youth anxiety disorders. We will use a CBT framework to consider the development and maintenance of anxiety disorders in youth with asthma.

Internal factors

Biological factors

Biological changes associated with the presence of an anxiety disorder may increase vulnerability to asthma onset and disease course. For example, Peters and Fritz (2010) propose that stress or anxiety could activate the autonomous nervous system or hypothalamopituitary adrenal axis (HPA), resulting in bronchorestrictions (Peters & Fritz, 2010). Anxiety has also been implicated via its impact on the neuroendocrine systems' effects on lung inflammation and the brainstem's respiratory sensor system. Katon et al. (2004) highlight several biological processes and physiological processes which help to explain asthma-anxiety comorbidity—including oversensitivity of the neural circuits involved in fear responding to experiences of breathlessness, carbon dioxide hypersensitivity, and higher levels of airway resistance. Asthma medications may also change biological processes that affect anxiety (Clawson et al., 2019).

Cognitive bias

Cognitive bias plays an important role in the maintenance and development of anxiety in youth and may also account for higher levels of anxiety in youth with asthma. The experience of asthma may generate fearful or catastrophic beliefs about asthma symptoms, which, in turn, may provoke panic attacks (Carr, Lehrer, Rausch, & Hochron, 1994; Park, Sawyer, & Glaun, 1996).

Park et al. (1996) developed a theoretical model adapted from a cognitive model of panic disorder (see Fig. 1). This model proposes that the basis of anxiety in those with asthma is the threat from symptoms of asthma itself. In panic disorder, people misinterpret normal bodily sensations (e.g., shortness of breath) as catastrophic or potentially life threatening, which, in turn, leads to an increase in anxiety. People might engage in avoidance, escape, or other behaviors to feel safe, which ultimately serves to maintain or worsen their anxiety symptoms. Thus, anxiety in a child with asthma increases when they misinterpret normal bodily sensations (e.g., breathlessness) as indicative of an asthma attack. If a child responds to these normal bodily sensations in the same way as they would if they were asthma symptoms (e.g., take quick relief medication), it is unlikely their symptoms would subside. Understandably, this would further increase anxiety in a child with asthma as their interpretation of the situation as potentially life threatening seems more likely. Subsequently, they would

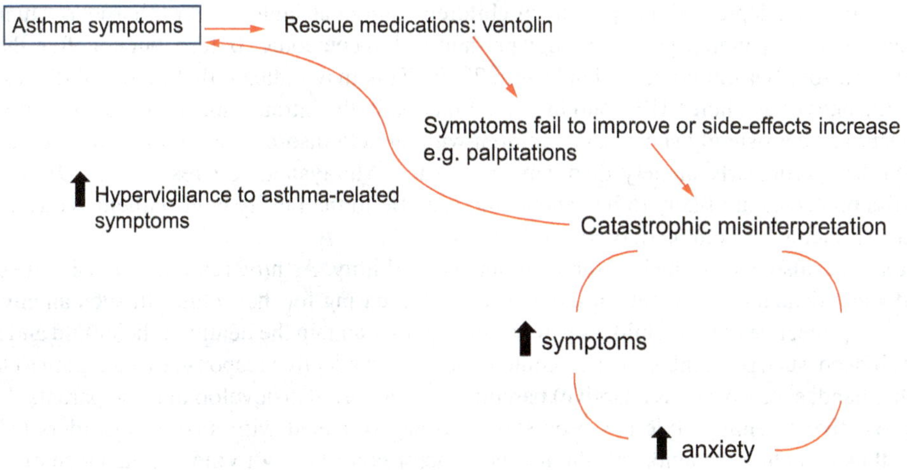

FIG. 1 A cognitive model of asthma and anxiety. *(Adapted from Park, S. J., Sawyer, S. M., & Glaun, D. E. (1996). Childhood asthma complicated by anxiety: An application of cognitive behavioural therapy.* Journal of Paediatrics & Child Health, 32*(2), 183–187.)*

attempt to engage in the same maladaptive coping behaviors as people with anxiety (e.g., avoidance of activities), which might maintain anxiety in the long run.

A handful of studies have been conducted to determine if children with asthma and anxiety display a cognitive bias toward asthma-specific stimuli. One study found that children with asthma demonstrated greater attentional biases toward asthma stimuli compared to children without asthma (Lowther, Newman, Sharp, & McMurray, 2016). However, other research showed that these biases were similar to those shown in children with anxiety more generally (Dudeney et al., 2017). Our research showed that children with asthma interpreted ambiguous information related to asthma as more threatening and engaged in more avoidant responding compared to children without asthma (Sicouri et al., 2017b). However, youth with comorbid asthma and anxiety demonstrated similar threat interpretation biases and avoidant responding to general threat scenarios as youth with anxiety only. Overall, these findings indicate that CBT interventions adapted for children with anxiety and asthma should target both general and asthma-specific cognitive biases.

Learning and behavioral mechanisms

Learning and behavioral mechanisms could also explain asthma and anxiety comorbidity in children. Pairing previous neutral stimuli (e.g., taking part in an activity) may become associated with negative asthma experiences (e.g., having an asthma exacerbation) through classical conditioning and result in conditioned anxiety (e.g., anxiety about taking part in an activity due to an experience of having an asthma exacerbation). Further avoidance and escape behaviors may inadvertently reinforce these behaviors via operant conditioning. Behaviors, such as avoidance (e.g., reducing or not taking part in activities), escape (e.g., quick relief medication use), or safety behaviors (e.g., medication overuse) in response to anxiety may serve to maintain and exacerbate their anxiety. Further, youth with asthma are generally encouraged to avoid potential asthma triggers; however, those with anxiety this may generalize avoidance to unrelated activities (e.g., school) to reduce their fears about potential asthma exacerbations.

Environmental factors

Asthma events

Anxiety could be secondary to asthma events, such as living with the stress of living with a chronic health condition (Goodwin et al., 2012). Experiencing a life-threatening asthma episode could lead to severe anxiety, such as posttraumatic stress disorder (Kean, Kelsay, Wamboldt, & Wamboldt, 2006).

Family and parent factors

Family factors, such as parent psychopathology and parenting behaviors, have also been implicated in asthma outcomes (Goodwin et al., 2012; Kaugars, Klinnert, & Bender, 2004). In terms of parent psychopathology, parents of children with asthma are more anxious and depressed that parents of children without asthma (Easter, Sharpe, & Hunt, 2015) and some longitudinal studies show that parent psychopathology predicts asthma outcomes before, but not after, the birth of the child (Cookson, Granell, Joinson, Ben-Shlomo, & Henderson, 2009). This may indicate that maternal distress influences child asthma through influences of an altered HPA and immune function in the intrauterine environment. Similarly, research on children with anxiety has consistently shown that children with anxiety disorders are more likely to have a parent with a mental health disorder, particularly anxiety disorders (Lawrence, Murayama, & Creswell, 2019). The mechanisms by which this transmission occurs are likely to be genetic and specific to the family environment. Thus, parental psychopathology may be a risk factor for both anxiety and asthma in youth.

Parenting factors may also play a role in asthma-anxiety comorbidity. Asthma requires intensive medical management and places considerable demands on parents or carers involved in caring for their child. In such an environment, parents may respond by being protective of their child, as parents attempt to maintain the health of their child and regulate their own exposure to stress. Indeed, such parental protection could be seen as an adaptive response to managing a child with a chronic illness. On the other hand, excessive (less adaptive) parental control may also develop in these parents. In children without asthma, an overprotective parenting style has been shown to be associated with anxiety disorders (McLeod, Wood, & Weisz, 2007) and thus parenting behaviors might also be a factor contributing to the maintenance of anxiety in children with and without asthma (Sicouri et al., 2017c). There is also some evidence that a parent may provide the child with information relevant to a specific or generalized threat or the value of avoidant coping (Sicouri et al., 2017b).

Practice and procedures

In this section, we provide a brief overview of the practice and procedures for treating anxiety in youth with asthma. First, we outline treatments for youth with asthma and youth with anxiety disorders. Second, we briefly review studies that have evaluated CBT to treat anxiety disorders in youth with asthma. Finally, we describe the main assessment and treatment components of a CBT program for this population.

Treatment for youth with asthma

The goal of treatment for youth asthma is to have well-controlled asthma using minimal medical intervention. Whereas long-term controller medication, such as corticosteroids, is used daily to prevent symptoms; quick relief medication is used to treat acute asthma symptoms and asthma exacerbations. To help children and families manage medication for their child's asthma, health providers usually complete an asthma action plan. This provides information about the severity of a child's asthma, asthma triggers, graded instructions for what do to in different settings and what medications to take at different levels of asthma severity. Other nonpharmacological approaches to managing asthma include parent or youth smoking cessation, avoidance of asthma triggers, physical exercise, breathing exercises, and a healthy diet. Evidence for each of these approaches is mixed and medical intervention is still recommended as the first line of treatment for asthma.

CBT treatment for anxiety disorders in youth

CBT is the most widely evaluated treatment (Creswell, Waite, & Hudson, 2020) for children and adolescents with anxiety, and it is routinely recommended as the frontline treatment (e.g., NICE CG31 (NICE, 2005) and NICE CG159 (NICE, 2014)). In a recent Cochrane review, James, Reardon, Soler, James, and Creswell (2020) evaluated the findings of 87 randomized control trials (RCTs) of CBT for anxiety disorders in children and adolescents and concluded that CBT was "probably more effective in the short-term that waiting lists/no treatment." Studies reported a 50% remission rate (i.e., free of a primary anxiety disorder diagnosis) in children who completed treatment (James et al., 2020). These studies have usually investigated CBT for multiple anxiety disorders; however, there is some evidence that disorder-specific treatments may have potential in improving treatment outcomes for youth (e.g., social anxiety; Ingul, Aune, & Nordahl, 2014).

Empirical research on CBT for anxiety disorders in youth with asthma

Given the efficacy of CBT for youth with anxiety, it seems likely that CBT would also be a useful intervention for anxiety disorders in youth with asthma. However, only four uncontrolled trials have examined the question of whether CBT is effective for anxiety disorders in youth with asthma (Marriage & Henderson, 2012; Papneja & Manassis, 2006; Park et al., 1996; Sicouri et al., 2017a). One study involved a retrospective file review of cases of childhood asthma where a recent worsening in function had been reported but where respiratory testing did not confirm a worsening in condition (Park et al., 1996). Twenty-three cases of children aged 10–12 years old were identified and the researchers determined that these children had developed an increasing fear associated with their asthma with an increasing awareness of its potential life-threatening nature. The researchers used a CBT model based on panic disorder to treat these children and found that treatment resulted in fewer hospitalizations. This study, while preliminary, is encouraging for the potential application of CBT that specifically targets asthma-related fears.

Papneja and Manassis (2006) compared the treatment response to CBT of 36 children (8–12 years old) with asthma and anxiety to that of 36 children with anxiety disorders alone. The results of this study suggested that, although both groups improved significantly, children with asthma improved significantly less compared to children with anxiety alone in terms of anxiety symptoms and anxiety diagnostic severity.

One potential reason for this finding is that there was no indication that CBT was adapted to account for the asthma status of the children in the study. If the role of specific fears related to asthma was not considered, it is likely that the treatment may not have been optimal.

Another study investigated the effect of CBT on anxiety and asthma-related quality of life in 10 children (aged 7–16 years) with asthma and anxiety (Marriage & Henderson, 2012). The treatment was adapted to account for the child's asthma status, and included breathing and relaxation exercises and basic cognitive restructuring.

Children reported fewer anxiety symptoms and higher quality of life at posttreatment and at 3-month follow-up. However, the study did not include diagnostic measures of child anxiety and so it is unknown whether the intervention achieved clinically meaningful change for anxiety.

Building on this research, we developed a CBT treatment for children with anxiety disorders which was adapted to account for children's asthma status (Sicouri et al., 2017a). The aim of the program was to (1) target factors that have been found to contribute to the maintenance and development of anxiety in children with asthma and (2) to provide good education and promote a problem-focused coping style on the management of asthma. While larger RCTs are needed to determine if this CBT is an effective intervention for children with anxiety and asthma comorbidity, the outcomes of this small pilot study provided preliminary proof-of-concept evidence for the efficacy of CBT for this population. This treatment is described below.

CBT treatment for anxiety disorders in youth with asthma

CBT uses a variety of cognitive and behavioral treatment components, such as modifying anxiety-related thought content and exposure to result in changes in anxiety symptoms. Some CBT programs include parents and target parental behaviors that might exacerbate or maintain anxiety (e.g., reassurance provision or allowing avoidance). The program described below is an adapted CBT program for children with anxiety disorders or elevated anxiety symptoms which is tailored to account for a child's asthma status. This program can be delivered in individual or group format, and it typically runs for eight sessions (see Table 1 for a CBT program). Parents are typically included in treatment for children aged 7–12 years. Treatment components are designed to address the key factors that research suggest maintain problematic anxiety in the context of an asthma diagnosis.

Identification and assessment

Assessment of asthma

Before starting treatment, it is important to discuss the child's asthma history and current functioning with the child's medical practitioner. Obtaining the child's general medical and developmental history, including, information about the child's asthma control, frequency, and severity of asthma symptoms, and exacerbations, medications, asthma triggers, medication adherence, and a child's perceptions of symptoms are important components of this discussion. In addition, it is important to obtain a copy of the child's asthma action plan which provides clear information and guidelines for managing the child's asthma. It may also be worthwhile obtaining the medical practitioner's view regarding the child's psychological functioning and ability and motivation to engage in psychological treatment.

It is also important to assess the child's asthma functioning with the child and family. This can be done with the use of semistructured interviews, such as the Family Asthma Management Assessment System (McQuaid, Walders, Kopel, Fritz, & Klinnert, 2005). This assesses key areas in relation to the child's asthma self-management, such as asthma knowledge and medication adherence.

TABLE 1 CBT program for anxiety disorders in youth with asthma.

Session	Treatment component
Session 1	Psychoeducation (1) Information about asthma, information about anxiety and the flight-or-fight response, identifying bodily sensations
Session 2	Psychoeducation (2) The relationship between asthma and anxiety, telling apart asthma and anxiety symptoms. Monitoring worry. Deep breathing exercises *Parents only*: Parenting a child with asthma and encouraging nonanxious behaviors
Session 3	The relationship between thoughts, feelings, and behaviors
Session 4	Cognitive restructuring (detective thinking)
Session 5	Graded exposure with planning and pacing (stepladders)
Session 6	Reviewing stepladders
Session 7	Problem-solving and assertiveness skills
Session 8	Living well with asthma (asthma management) Relapse prevention

Several valid measures can also be used as part of a comprehensive asthma assessment. Asthma control can be assessed with the use of questionnaires administered to the child and/or parent, such as the Asthma Control Questionnaire (Juniper, Gruffydd-Jones, Ward, & Svensson, 2010), the Asthma Therapy Assessment Questionnaire (Skinner et al., 2004), and the Asthma Control Test (Nathan et al., 2004)/Childhood Asthma Control Test (Liu et al., 2007). Asthma triggers can be assessed with the use of the Asthma Trigger Inventory (Wood et al., 2006). The Pediatric Quality of Life Questionnaire (Juniper et al., 1996) is a reliable and valid interviewer-administered scale that assesses daily function that children with asthma report as most troublesome. A carer quality of life assessment questionnaire (Juniper, O'Byrne, Guyatt, Ferrie, & King, 1999) may also be useful to determine how much a child's asthma has impacted carers in terms of emotional functioning and activity limitation.

Assessment of anxiety

The gold-standard method for assessing anxiety disorders in children and adolescents is a structured diagnostic interview, such as the Anxiety Disorders Interview Schedule for children and parents (ADIS-C/P; Silverman & Albano, 1996). This provides a comprehensive assessment of anxiety (including symptoms, severity, and interference) using information from both the parent/carer and the child. However, while standardized assessments such as the ADIS-C/P are commonly used in research trials, they are time-consuming and rarely used in clinical settings. Diagnostic interviews are typically supplemented with psychometrically reliable and valid questionnaire measures of symptoms (e.g., Spence Children's Anxiety Scale; Spence, 1998) and the impact of anxiety symptoms on the child's life at home, outside the home, and the impact on the wider family (e.g., Child Anxiety Life Interference Scale, Lyneham et al., 2013). Use of multiple sources (e.g., parent, young person, teacher), rather than relying on only one reporter, can lead to a richer and potentially more accurate understanding of the child's symptoms (Reardon et al., 2019). Behavioral observations can also provide useful assessment data on levels and features of a child's anxiety, as well as the family's response to the child.

A thorough and competent assessment also requires a broader assessment of the onset, development, and context of anxiety symptoms, as well as information regarding the child's medical (other than asthma) history, developmental history, school and social history, and family psychiatric history. Given the possible role of the family in maintaining anxiety disorders, a competent assessment should also consider the broader family factors which may be maintaining the disorder, including the level of family functioning, the parent-child relationship, and parenting behaviors theorized to play a role in youth anxiety.

Finally, it is essential that a comprehensive assessment should also include appropriate consideration of risk of low mood, self-harm and suicide, as well as factors that increase the risk of suicidal ideation and behavior, such as being bullied or alcohol and drug problems (Reijntjes, Kamphuis, Prinzie, & Telch, 2010; Robinson, Sareen, Cox, & Bolton, 2011).

Case conceptualization

Consistent with a CBT approach, a case conceptualization which maps out the relationship between an individual's thoughts, emotions, and behaviors, as well as considering broader factors, such as the role of parents in maintaining anxiety, needs to be developed. Critical to this conceptualization is understanding the relationship between asthma and anxiety. For example, anxiety in children with asthma may be specific to asthma (e.g., anxiety about symptoms, triggers, or treatment), or it may be unrelated and more generalized.

Psychoeducation

Information about asthma

Clinicians should provide an overview of what asthma is, its symptoms, and how to assess severity. Common triggers for asthma exacerbations can be discussed and asking parents and children what they do to manage triggers. It is also important to discuss aspects of asthma management in relation to a child's individual asthma action plan.

The relationship between asthma and anxiety

The rationale for learning about anxiety management in the context of a child's asthma diagnosis should be provided. Clinicians can highlight that when people have asthma, they may develop fears about their illness and feel they are not in control of their symptoms. These catastrophic thoughts may stop them doing things they want to do, or rely on others for help, which then impacts their confidence in themselves. Children with decreased self-efficacy are less able to learn important skills to manage their asthma which may lead to decreased medication adherence and poorer asthma management skills. This results in poorer asthma outcomes, including more school days missed. It is important to point out that not all children with asthma have anxiety, but research suggests that children are three times as likely to have an anxiety disorder than children without asthma. Worry may be about asthma, or it may be about other things like the dark, giving a speech in class or making mistakes. Therefore, anxiety management skills are important for asthma management.

Information about anxiety

Information about what anxiety is, including the flight and fight response, should be provided. Common physical symptoms of anxiety are discussed, including increased heart rate, trembling, and increased breathing rate. For people with asthma, it is important to highlight that the increased breathing rate is the most important symptom to be aware of. Changes in breathing rate might lead to over-breathing or hyperventilating which may constrict airways, worsening shortness of breath. This means that it may become harder to breathe, leading to a cycle of hyperventilation and anxiety, which may then trigger an asthma exacerbation. For example, in this cycle, the flight-or-fight response can be triggered by a worrying thought, changes in the body, or being in a stressful situation. This may lead to an increased rate of breathing, as well as other physical symptoms, such as increased heart rate. If the person begins to breathe in quickly through the mouth, any shortness of breath may be worsened, which may then be interpreted as "the worst" is about to happen then and there, for example, the person will have an asthma attack. This leads to an anxiety asthma cycle, shown in Fig. 2. It may be helpful to introduce deep breathing techniques as a skill to help bring breathing rates back to normal to help stop the anxiety asthma cycle.

Telling apart asthma and anxiety

Clinicians should provide information on how to distinguish asthma from anxiety symptoms. Children with asthma may be hypersensitive to changes in breathing and their body. Physical symptoms brought on by the flight or fight response may be easily confused with the symptoms of an asthma exacerbation. Research has shown that there are three unique symptoms that occur only during an asthma attack but not during the flight or fight responses; these are: wheezing, coughing, and mucus production. By observing these symptoms, children and their parents can tell apart an asthma attack from heightened anxiety and help them respond to the symptoms accordingly.

Monitoring worry

A child can monitor their anxiety using a worry scale which rates the intensity of their emotions on a scale of 0 (very relaxed) to 10 (extremely worried). Children are encouraged to record situations in which they feel anxious, particularly in relation to their asthma. For example, children are asked to provide information on (1) the situation; (2) the thoughts experienced by asking "What do I think will happen?" "What could go wrong?" or "What bad thing am I expecting?"; (3) the intensity of worry/fear using the worry scale; and (4) behavior (what she does). In doing this exercise, it is helpful to point out the relationship between thoughts and asthma behaviors (such as taking quick relief asthma medication).

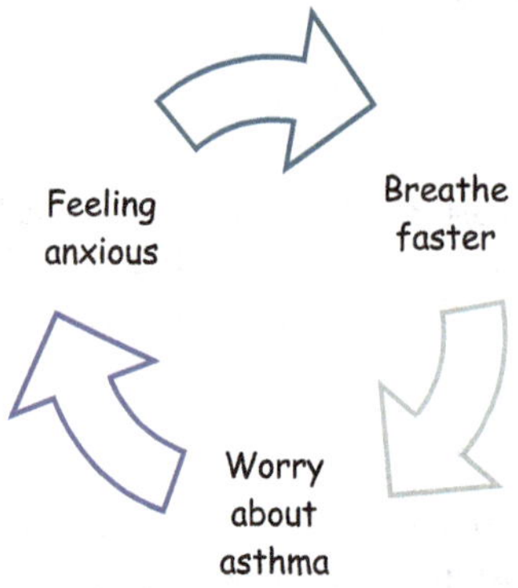

FIG. 2 The asthma-anxiety cycle.

Cognitive restructuring (detective thinking)

Cognitive restructuring, also known as detective thinking, involves teaching children to think more realistically by examining evidence for thoughts to work out if the thoughts are true or not. Clinicians should discuss the impact of thinking errors on feelings and behaviors; namely that people who are anxious typically overestimate how likely it is than an unpleasant event will happen, or they overestimate how bad the consequences will be if the even does happen. Detective thinking involves examining whether the thoughts are true or not by asking questions as "What has happened in the past?" "What else could happen?" "What are the facts?" "What might someone else think or feel in the situation?" "What would you think about a friend in this situation?" and "How would you handle it if the worst did happen?" Finally, based on all the evidence, the child is asked to judge how likely is it really that the bad thing will happen?" Children are then asked to come up with a more realistic and helpful thought about the situation and to provide a worry rating both before and after detective thinking. Parents can also be encouraged to come up with some examples of detective thinking and share their own worries. An example of an asthma-related detective thinking worksheet is shown in Fig. 3.

Gradual exposure with planning and pacing

To overcome fears, it is important that they are faced in a gradual way. This is known as gradual exposure. The goal of exposure is to reduce physiological arousal and reverse unhelpful behaviors, which may be maintaining anxiety (e.g., avoidance). Undertaking exposures can also provide evidence for cognitive restructuring tasks and undertaking cognitive restructuring tasks prior to completing exposure tasks may be helpful. Exposure tasks are usually designed with the use of a "stepladder," which begins with small fears and increases to bigger fears. The steps are designed to be manageable and frequently repeated. Rewarding the child for completing each step is encouraged to increase a child's motivation to complete the stepladder. For children with asthma, what about exposures to potentially dangerous situations? Asthma is a serious respiratory disease, and so it is natural that children and parents feel anxious and act cautiously. While these concerns are normal and avoidance of asthma triggers is important, being overly cautious can also be a problem. Excessive avoidance of activities does not encourage children to learn that they can manage on their own and reduce their self-efficacy and confidence.

Planning and pacing

The key to facing asthma-related fears is to remember that there are limits to children's activity levels, so clinicians need to help children and parents realize the fine balance between facing their fears and staying within safety limits. Planning and

EVENT *My breathing will speed up after playing ball games with friends*

THOUGHT *I'm going to have an asthma attack* (Worry rating: 10/10)

EVIDENCE

Alternative outcomes (facts, how likely, other possibilities, other people, past)

I don't get asthma attacks every time I play ball and run around

I could be getting out of breath from too much exercise

I might feel out of breath, but I may not cough or wheeze

Realistic consequences (how bad, disaster or not, how can I cope, how long will it last)

If I am feeling out of breath I can rest for a while to see if my breathing calms down

REALISTIC THOUGHT *Even though I might feel out of breath, I may not cough and wheeze. I can rest and see if my breathing slows down* (Worry rating 3/10)

FIG. 3 Example of a child's detective thinking worksheet.

pacing are two important principles to remember when standing up to asthma-related fears. Planning means thinking ahead about how children will carry the activities out and making sure time for rest breaks is scheduled. Pacing means taking the time for activities, rather than rushing them. Planning and pacing are important because it is important to do things slowly rather than all at once. Being aware of children's limits, taking small steps and taking rest are all important concepts to keep in mind when designing stepladders for children to face asthma-related fears.

A good example to use for planning and pacing is exercise avoidance. Children with asthma may fear and avoid physical exercise as it can bring on breathless sensations and may trigger an asthma attack. However, regular exercise can improve muscle strength and lung function, which leads to improved health and fitness and confidence in managing shortness of breath during activities. Therefore, people with asthma should do regular exercise, but they need to do it gradually and to plan plenty of rest breaks in between. An example of using planning and pacing in a stepladder is shown in Fig. 4.

Modifying parent factors

Providing parents with information about their potential role in maintaining their child's anxiety may be helpful. Research shows that parents of children with asthma and/or anxiety tend to be more overprotective and stressed than parents of children without asthma and/or anxiety. While this is normal and understandable—particularly when parenting a child with asthma, sometimes parents need to be aware of their own behavior to help their child face their fears and learn skills to manage their asthma independently, especially as their child enters adolescence. Parents can encourage their child's brave behavior by using rewards for nonanxious behavior and independence (e.g., through attention and praise), modeling non-anxious behaviors, and challenging their own thoughts and fears about their child tackling asthma-related fears. Parents are encouraged to talk openly to their child about their asthma and the risks, as well working with their child to develop a self-management asthma plan, if it is developmentally appropriate.

GOAL: To do exercise regularly so I can be fit enough to try out for the soccer team

HIGH STEPS

Try out for the soccer team

Practice dribbling the ball for 10 mins and rest and jog for 5 mins

Practice dribbling the ball for 5 mins and rest and jog for 5 mins

Practice passing the ball for 10 mins and rest and jog for 5 mins

MIDDLE STEPS

Practice passing the ball for 5 mins and rest and jog for 5 mins

Walk for 10 minutes and rest and jog for 5 minutes

Walk for 5 minutes and rest and jog for 5 minutes

Walk for 5 minutes and rest and jog for 2 minutes

LOW STEPS

Walk for 15 minutes and rest

Walk for 10 minutes and rest

Walk for 5 minutes and rest

SETBACKS: When the child was at Step 7, they caught a cold and had an asthma attack. They took a break from the stepladder to get some rest. When they felt better, they started practicing their stepladder from Step 5 to make sure their body was up to doing exercise again.

FIG. 4 Example of a child's stepladder with planning and pacing.

General skills training

General skills training may also be relevant. Problem-solving is a skill used to help children to generate and choose adaptive responses rather than limited or avoidant responses. Problem-solving may be particularly helpful for children with asthma to increase their autonomy in managing their asthma and medication. Assertiveness skills might also increase a child's ability to deal confidently with other people (e.g., eye contact), or to deal with potential teasing and bullying (e.g., ignoring).

Living well with asthma (asthma management)

Finally, it is important to emphasize that children with asthma can live a normal healthy life if they successfully manage their condition. Good asthma management requires knowledge about asthma, adherence to treatment regime (e.g., medication adherence, managing triggers), a healthy lifestyle (e.g., balanced diet, regular exercise), positive beliefs, social skills, family and social support, problem-solving skills, and access to good healthcare. Clinicians can encourage children and their parents to use their asthma management skills to manage and understand their asthma, ensure adherence to treatment, and encouraging skills to foster independence in managing their asthma and anxiety.

Summary and future research directions

Overall, there is very limited evidence on the efficacy of CBT for anxiety disorders in youth with asthma. RCTs with larger sample sizes are needed to confirm its efficacy. Nevertheless, clinicians can easily adapt CBT for youth anxiety disorders—known to be highly efficacious—for youth with asthma. The main adaptations relate to psychoeducation about asthma and its relationship with anxiety, targeting cognitions that might be maintaining anxiety in the context of asthma, and incorporating planning and pacing into gradual exposure tasks. Given the high prevalence of anxiety in youth with asthma, practitioners working with youth with asthma are encouraged to routinely assess for anxiety disorders. Prospective designs would be helpful to assess the developmental trajectory of anxiety and asthma status in children as they relate to particular risk factors (e.g., cognitive bias) to inform interventions which are designed to target these factors.

Mini-dictionary of terms

Asthma: a chronic respiratory condition characterized by a reversible inflammatory condition of airways, with symptoms including coughing, wheezing, and difficulty breathing.

Asthma action plan: a set of instructions for managing a person's asthma.

Asthma exacerbation: also known as an "asthma attack." This usually means having symptoms (e.g., wheezing, chest tightness, and breathing difficulties) severe enough to need treatment in the emergency department or admission to hospital. Sometimes people use asthma exacerbation to describe someone experiencing asthma symptoms.

Asthma control: the extent to which asthma features have been reduced or removed. Well-controlled asthma means asthma symptoms are infrequent (e.g., twice per week) and there is no or limited functional impairment (e.g., limitation of normal activities).

Asthma morbidity: frequency of acute asthma exacerbations, emergency department visits and hospitalizations due to asthma.

Asthma severity: how much treatment is needed to keep asthma well-controlled. Severe asthma is asthma which remains uncontrolled despite treatable factors having been addressed and maximum inhaled treatment is being taken.

Asthma triggers: situations or activities that bring on asthma symptoms.

Cognitive restructuring: also known as "detective thinking." Identifying unhelpful thoughts and learning to modify them.

Gradual exposure: facing fears in a step-by-step way.

Planning and pacing: planning means thinking ahead about how to carry out activities. Pacing means taking the time for activities, rather than rushing them.

Key facts

- Asthma is a serious respiratory disease with symptoms including coughing, wheezing, and difficulty breathing. It is the most common chronic illness in youth.
- Youth with asthma experience higher rates of anxiety disorders. Youth with both disorders experience significant impairment in functioning and experience higher rates of both anxiety and asthma symptoms.

- Cognitive behavioral models explain the development and maintenance of anxiety disorders in youth with asthma. Biological, cognitive, and environmental factors play a role in understanding how anxiety and asthma comorbidity develops and is exacerbated.
- CBT can be modified to effectively treat anxiety disorders in children with asthma. Treatment components include psychoeducation, cognitive restructuring, gradual exposure with planning and pacing and modifying parent behaviors. The primary goal of treatment is anxiety diagnosis remission and to improve asthma related quality of life.

Applications to other areas

In this chapter, we have described a CBT treatment for anxiety disorders in youth with asthma. This treatment is unique in that it has been tailored for youth with asthma and targeted factors that have been shown to be associated with anxiety in youth with asthma. The treatment is based on CBT for youth with anxiety disorders and also has applicability to anxiety disorders in youth with other chronic illnesses, such as inflammatory bowel disease (see Chapter 20). Anxiety disorders are seen in a variety of chronic health conditions in youth (Pinquart & Shen, 2011) and thus CBT can be potentially tailored to treat anxiety in youth with other chronic health conditions.

It also seems likely that factors which confer vulnerability for anxiety disorders in youth with asthma may also confer vulnerability for youth with other chronic illnesses, such as hypervigilance to symptoms and parenting factors—which could be targeted as part of a tailored CBT treatment (e.g., Catarozoli et al., 2019).

More broadly, this chapter is relevant to CBT treatments for youth in general. Adaptations to CBT programs that have been developed for adults to account for a child's developmental stage and age are relevant considerations. This includes modifications around the language, as well as the decision about whether to include parents in treatment.

Summary points

- This chapter focused on anxiety disorders in youth with asthma.
- Anxiety disorders are highly prevalent in youth with asthma.
- Factors which might confer vulnerability for anxiety in youth with asthma include both internal (e.g., genetics, cognitive factors) and environmental factors (e.g., parenting behaviors).
- There are very few evidence-based treatments for this population. This chapter described a CBT approach which was adapted to treat anxiety disorders in youth with asthma.
- Further research is needed to better understand the factors that confer vulnerability for anxiety disorders in youth with asthma and to examine the efficacy of CBT for this population.

References

Akinbami, L. J., Simon, A. E., & Rossen, L. M. (2016). Changing trends in asthma prevalence among children. *Pediatrics*, *137*(1), 1–7. https://doi.org/10.1542/peds.2015-2354.

Australian Institute of Health and Welfare. (2020). *AIHW—Reports-chronic-repiratory-conditions—Asthma—Contents—Asthma*. Retrieved 27 July 2021, from https://www.aihw.gov.au/reports/chronic-respiratory-conditions/asthma/contents/asthma.

Canino, G., McQuaid, E. L., & Rand, C. S. (2009). Addressing asthma health disparities: A multilevel challenge. *Journal of Allergy and Clinical Immunology*, *123*(6), 1209–1217.

Carr, R. E., Lehrer, P. M., Rausch, L. L., & Hochron, S. M. (1994). Anxiety sensitivity and panic attacks in an asthmatic population. *Behaviour Research and Therapy*, *32*(4), 411–418. https://doi.org/10.1016/0005-7967(94)90004-3.

Catarozoli, C., Brodzinsky, L., Salley, C. G., Miller, S. P., Lois, B. H., & Carpenter, J. L. (2019). Necessary adaptations to CBT with pediatric patients. In R. D. Friedberg, & J. K. Paternostro (Eds.), *Handbook of cognitive behavioral therapy for pediatric medical conditions* (pp. 103–117). Cham: Springer International Publishing.

Centers for Disease Control and Prevention. (2019). *CDC—Asthma—Most recent asthma data*. Retrieved 27 July 2021, from https://www.cdc.gov/asthma/most_recent_national_asthma_data.htm.

Clawson, A. H., Ruppe, N., Nwankwo, C., Blair, A., Baudino, M., & Mehdi, N. (2019). Cognitive behavioral therapy for youth with asthma: Anxiety as an example. In R. D. Friedberg, & J. K. Paternostro (Eds.), *Handbook of cognitive behavioral therapy for pediatric medical conditions* (pp. 345–367). Cham: Springer International Publishing.

Cookson, H., Granell, R., Joinson, C., Ben-Shlomo, Y., & Henderson, A. J. (2009). Mothers' anxiety during pregnancy is associated with asthma in their children. *Journal of Allergy and Clinical Immunology*, *123*(4), 847–853. e811 https://doi.org/10.1016/j.jaci.2009.01.042.

Creswell, C., Waite, P., & Hudson, J. (2020). Practitioner review: Anxiety disorders in children and young people–assessment and treatment. *Journal of Child Psychology and Psychiatry*, *61*(6), 628–643.

Dudeney, J., Sharpe, L., Jaffe, A., Jones, E. B., & Hunt, C. (2017). Anxiety in youth with asthma: A meta-analysis. *Pediatric Pulmonology*, *52*(9), 1121–1129.

Easter, G., Sharpe, L., & Hunt, C. J. (2015). Systematic review and meta-analysis of anxious and depressive symptoms in caregivers of children with asthma. *Journal of Pediatric Psychology*, *40*(7), 623–632. https://doi.org/10.1093/jpepsy/jsv012.

Goodwin, R. D., Bandiera, F. C., Steinberg, D., Ortega, A. N., & Feldman, J. M. (2012). Asthma and mental health among youth: Etiology, current knowledge and future directions. *Expert Review of Respiratory Medicine*, *6*(4), 397–406. https://doi.org/10.1586/ers.12.34.

Hasler, G., Gergen, P. J., Kleinbaum, D. G., Ajdacic, V., Gamma, A., Eich, D., … Angst, J. (2005). Asthma and panic in young adults. *American Journal of Respiratory and Critical Care Medicine*, *171*(11), 1224–1230. https://doi.org/10.1164/rccm.200412-1669OC.

Hudson, J. L., & Rapee, R. M. (2004). From anxious temperament to disorder: An etiological model. In R. G. Heimberg, C. L. Turk, & D. S. Mennin (Eds.), *Generalized anxiety disorder: Advances in research and practice* (pp. 51–74). New York: Guilford Press.

Ingul, J. M., Aune, T., & Nordahl, H. M. (2014). A randomized controlled trial of individual cognitive therapy, group cognitive behaviour therapy and attentional placebo for adolescent social phobia. *Psychotherapy and Psychosomatics*, *83*(1), 54–61.

James, A. C., Reardon, T., Soler, A., James, G., & Creswell, C. (2020). Cognitive behavioural therapy for anxiety disorders in children and adolescents. *Cochrane Database of Systematic Reviews*, *11*.

Juniper, E. F., Gruffydd-Jones, K., Ward, S., & Svensson, K. (2010). Asthma control questionnaire in children: Validation, measurement properties, interpretation. *European Respiratory Journal*, *36*(6), 1410–1416. https://doi.org/10.1183/09031936.00117509.

Juniper, E. F., Guyatt, G. H., Feeny, D. H., Ferrie, P. J., Griffith, L. E., & Townsend, M. (1996). Measuring quality of life in children with asthma. *Quality of Life Research*, *5*(1), 35–46. https://doi.org/10.1007/bf00435967.

Juniper, E. F., O'Byrne, P. M., Guyatt, G. H., Ferrie, P. J., & King, D. R. (1999). Development and validation of a questionnaire to measure asthma control. *European Respiratory Journal*, *14*(4), 902–907. https://doi.org/10.1034/j.1399-3003.1999.14d29.x.

Katon, W. J., Richardson, L., Lozano, P., & McCauley, E. (2004). The relationship of asthma and anxiety disorders. *Psychosomatic Medicine*, *66*(3), 349–355. https://doi.org/10.1097/01.psy.0000126202.89941.ea.

Kaugars, A. S., Klinnert, M. D., & Bender, B. G. (2004). Family influences on pediatric asthma. *Journal of Pediatric Psychology*, *29*(7), 475–491. https://doi.org/10.1093/jpepsy/jsh051.

Kean, E. M., Kelsay, K., Wamboldt, F., & Wamboldt, M. Z. (2006). Posttraumatic stress in adolescents with asthma and their parents. *Journal of the American Academy of Child and Adolescent Psychiatry*, *45*(1), 78–86. https://doi.org/10.1097/01.chi.0000186400.67346.02.

Kessler, R. C., Berglund, P., Demler, O., Jin, R., Merikangas, K. R., & Walters, E. E. (2005). Lifetime prevalence and age-of-onset distributions of DSM-IV disorders in the national comorbidity survey replication. *Archives of General Psychiatry*, *62*(6), 593–602. https://doi.org/10.1001/archpsyc.62.6.593.

Lawrence, D., Johnson, S., Hafekost, J., Boterhoven de Haan, K., Sawyer, M., Ainley, J., & Zubrick, S. R. (2015). *The mental health of children and adolescents: Report on the second Australian child and adolescent survey of mental health and wellbeing*. Canberra, ACT, Australia: Department of Health.

Lawrence, P. J., Murayama, K., & Creswell, C. (2019). Systematic review and meta-analysis: Anxiety and depressive disorders in offspring of parents with anxiety disorders. *Journal of the American Academy of Child & Adolescent Psychiatry*, *58*(1), 46–60. https://doi.org/10.1016/j.jaac.2018.07.898.

Liu, A. H., Zeiger, R., Sorkness, C., Mahr, T., Ostrom, N., Burgess, S., … Manjunath, R. (2007). Development and cross-sectional validation of the Childhood Asthma Control Test. *Journal of Allergy and Clinical Immunology*, *119*(4), 817–825. https://doi.org/10.1016/j.jaci.2006.12.662.

Lowther, H., Newman, E., Sharp, K., & McMurray, A. (2016). Attentional bias to respiratory- and anxiety-related threat in children with asthma. *Cognition and Emotion*, *30*(5), 953–967. https://doi.org/10.1080/02699931.2015.1036842.

Lu, Y., Mak, K. K., Van Bever, H. P., Ng, T. P., Mak, A., & Ho, R. C. M. (2012). Prevalence of anxiety and depressive symptoms in adolescents with asthma: A meta-analysis and meta-regression. *Pediatric Allergy and Immunology*, *23*(8), 707–715.

Lyneham, H. J., Sburlati, E. S., Abbott, M. J., Rapee, R. M., Hudson, J. L., Tolin, D. F., & Carlson, S. E. (2013). Psychometric properties of the Child Anxiety Life Interference Scale (CALIS). *Journal of Anxiety Disorders*, *27*(7), 711–719. https://doi.org/10.1016/j.janxdis.2013.09.008.

Marriage, D., & Henderson, J. (2012). Cognitive behaviour therapy for anxiety in children with asthma. *Nursing Children and Young People*, *24*(9), 30–34.

McLeod, B. D., Wood, J. J., & Weisz, J. R. (2007). Examining the association between parenting and childhood anxiety: A meta-analysis. *Clinical Psychology Review*, *27*(2), 155–172. https://doi.org/10.1016/j.cpr.2006.09.002.

McQuaid, E. L., Walders, N., Kopel, S. J., Fritz, G. K., & Klinnert, M. D. (2005). Pediatric asthma management in the family context: The family asthma management system scale. *Journal of Pediatric Psychology*, *30*(6), 492–502. https://doi.org/10.1093/jpepsy/jsi074.

Nathan, R. A., Sorkness, C. A., Kosinski, M., Schatz, M., Li, J. T., Marcus, P., … Pendergraft, T. B. (2004). Development of the asthma control test: A survey for assessing asthma control. *Journal of Allergy and Clinical Immunology*, *113*(1), 59–65. https://doi.org/10.1016/j.jaci.2003.09.008.

NICE. (2005). *Obsessive-compulsive disorder and body dysmorphic disorder: Treatment. (CG31). NICE guidance*. Retrieved from https://www.nice.org.uk/guidance/cg31/chapter/1-Guidance.

NICE. (2014). *Social anxiety disorder: Recognition, assessment and treatment (CG159). NICE guidance*. Retrieved from https://www.nice.org.uk/guidance/cg159.

Papneja, T., & Manassis, K. (2006). Characterization and treatment response of anxious children with asthma. *Canadian Journal of Psychiatry*, *51*(6), 393–396.

Park, S. J., Sawyer, S. M., & Glaun, D. E. (1996). Childhood asthma complicated by anxiety: An application of cognitive behavioural therapy. *Journal of Paediatrics and Child Health*, *32*(2), 183–187.

Perry, R., Braileanu, G., Palmer, T., & Stevens, P. (2019). The economic burden of pediatric asthma in the United States: Literature review of current evidence. *Pharmacoeconomics*, *37*(2), 155–167. https://doi.org/10.1007/s40273-018-0726-2.

Peters, T. E., & Fritz, G. K. (2010). Psychological considerations of the child with asthma. *Child and Adolescent Psychiatric Clinics of North America*, *19*(2), 319–333. ix https://doi.org/10.1016/j.chc.2010.01.006.

Pinquart, M., & Shen, Y. (2011). Anxiety in children and adolescents with chronic physical illnesses: A meta-analysis. *Acta Paediatrica*, *100*(8), 1069–1076.

Reardon, T., Creswell, C., Lester, K. J., Arendt, K., Blatter-Meunier, J., Bögels, S. M., … Herren, C. (2019). The utility of the SCAS-C/P to detect specific anxiety disorders among clinically anxious children. *Psychological Assessment*, *31*(8), 1006.

Reijntjes, A., Kamphuis, J. H., Prinzie, P., & Telch, M. J. (2010). Peer victimization and internalizing problems in children: A meta-analysis of longitudinal studies. *Child Abuse & Neglect*, *34*(4), 244–252. https://doi.org/10.1016/j.chiabu.2009.07.009.

Richardson, L. P., Lozano, P., Russo, J., McCauley, E., Bush, T., & Katon, W. (2006). Asthma symptom burden: Relationship to asthma severity and anxiety and depression symptoms. *Pediatrics*, *118*(3), 1042–1051. https://doi.org/10.1542/peds.2006-0249.

Robinson, J., Sareen, J., Cox, B. J., & Bolton, J. M. (2011). Role of self-medication in the development of comorbid anxiety and substance use disorders: A longitudinal investigation. *Archives of General Psychiatry*, *68*(8), 800–807. https://doi.org/10.1001/archgenpsychiatry.2011.75.

Sicouri, G., Sharpe, L., Hudson, J. L., Dudeney, J., Jaffe, A., & Hunt, C. (2017a). A case series evaluation of a pilot group cognitive behavioural treatment for children with asthma and anxiety. *Behaviour Change*, *34*(1), 35–47. https://doi.org/10.1017/bec.2017.3.

Sicouri, G., Sharpe, L., Hudson, J. L., Dudeney, J., Jaffe, A., Selvadurai, H., & Hunt, C. (2017c). Parent-child interactions in children with asthma and anxiety. *Behaviour Research and Therapy*, *97*, 242–251. https://doi.org/10.1016/j.brat.2017.08.010.

Sicouri, G., Sharpe, L., Hudson, J. L., Dudeney, J., Jaffe, A., Selvadurai, H., … Hunt, C. (2017b). Threat interpretation and parental influences for children with asthma and anxiety. *Behaviour Research and Therapy*, *89*, 14–23. https://doi.org/10.1016/j.brat.2016.11.004.

Silverman, W. K., & Albano, A. M. (1996). *The anxiety disorders interview schedule for DSM-IV—Child and parent versions*. San Antonio. TX: Psychological Corporation.

Skinner, E. A., Diette, G. B., Algatt-Bergstrom, P. J., Nguyen, T. T., Clark, R. D., Markson, L. E., & Wu, A. W. (2004). The Asthma Therapy Assessment Questionnaire (ATAQ) for children and adolescents. *Disease Management*, *7*(4), 305–313. https://doi.org/10.1089/dis.2004.7.305.

Spence, S. H. (1998). A measure of anxiety symptoms among children. *Behaviour Research and Therapy*, *36*(5), 545–566. https://doi.org/10.1016/S0005-7967(98)00034-5.

Vila, G., Nollet-Clemençon, C., de Blic, J., Mouren-Simeoni, M. C., & Scheinmann, P. (2000). Prevalence of DSM IV anxiety and affective disorders in a pediatric population of asthmatic children and adolescents. *Journal of Affective Disorders*, *58*(3), 223–231. https://doi.org/10.1016/S0165-0327(99)00110-X.

Wood, B. L., Cheah, P. A., Lim, J., Ritz, T., Miller, B. D., Stern, T., & Ballow, M. (2006). Reliability and validity of the Asthma Trigger Inventory applied to a pediatric population. *Journal of Pediatric Psychology*, *32*(5), 552–560. https://doi.org/10.1093/jpepsy/jsl043.

World Health Organisation. (2021). *WHO—Newsroom-fact-sheets–detail–asthma*. Retrieved 27 July 2021, from https://www.who.int/news-room/fact-sheets/detail/asthma.

Chapter 15

Cognitive-behavioral therapy adapted for youth with comorbid anxiety and autism spectrum disorder

Kristen H. Erps, Emily R. Jellinek, Lindsey N. Landry, Andrew G. Guzick, Sophie C. Schneider, and Eric A. Storch
Department of Psychiatry and Behavioral Sciences, Baylor College of Medicine, Houston, TX, United States

Abbreviations

ASD autism spectrum disorder
CBT cognitive-behavioral therapy
ERP exposure and response prevention

Introduction to autism spectrum disorder

Autism spectrum disorder (ASD) is a developmental disability marked by restricted, repetitive behaviors and interests, as well as deficits in social interaction and communication (American Psychiatric Association, 2013). Approximately 1 in 44 children meet criteria for ASD, and boys are 4.2 times more likely to receive this diagnosis than girls (Maenner et al., 2021). The symptom presentation across genders vary, which may impact the likelihood of receiving an ASD diagnosis (Halladay et al., 2015). Females with ASD compared to males with ASD on average have better social skills (Head, Mcgillivray, & Stokes, 2014), less repetitive, stereotyped behaviors (Mandy et al., 2012), and may have lower average intellectual ability (Banach et al., 2009; Fombonne, 2005; Rivet & Matson, 2011).

ASD occurs in all racial, ethnic, and socioeconomic groups, though its prevalence is reported to increase along with socioeconomic status (SES) among Black, White, and Hispanic youth in the United States, likely due to increased access to diagnostic services among individuals from higher SES backgrounds (Durkin et al., 2017). In countries with universal access to diagnoses, health, and educational services, this overrepresentation of ASD in high SES families is not seen (Delobel-Ayoub et al., 2015). Instead, lower SES is associated with higher risk for ASD in these countries (Rai et al., 2012).

ASD and anxiety

It is common for youth with ASD to have cooccurring psychiatric disorders; however, the prevalence of comorbid diagnoses varies across studies (Hossain et al., 2020). One systematic review of individuals with ASD found that 94% reported a cooccurring psychiatric disorder (Richa, Fahed, Khoury, & Mishara, 2014), with other studies demonstrating a positive association between age and number of comorbid psychiatric diagnoses (De Giorgi et al., 2019). Psychiatric comorbidities that occur with relative commonality include anxiety disorders, obsessive-compulsive disorder (OCD), attention-deficit/hyperactivity disorder (ADHD), conduct disorders, mood disorders (e.g., depression and bipolar disorder), suicidal ideation and attempts, sleep disorders, and psychotic disorders (Hossain et al., 2020). Anxiety is one of the most common cooccurring psychiatric disorders for individuals with ASD, with prevalence ranging from 30% to 50% across systematic reviews and meta-analyses (Van Steensel, Bögels, & Perrin, 2011). Although prevalence estimates often range from 30% to 50%, some outlier studies have seen higher or lower prevalence estimates (Hossain et al., 2020). Lower prevalence estimates often occur in studies that do not use diagnostic interviews or ASD-specific adjustments (Van Steensel et al., 2011). A recent meta-analysis examined the prevalence of various anxiety disorders for individuals with ASD (Hollocks, Lerh, Magiati, Meiser-Stedman, & Brugha, 2019). The overall lifetime prevalence for any comorbid anxiety disorders was 42%. The lifetime prevalence of specific anxiety disorders varied for individuals with ASD, with social

Handbook of Lifespan Cognitive Behavioral Therapy. https://doi.org/10.1016/B978-0-323-85757-4.00002-X

anxiety disorder (29%), obsessive-compulsive disorder (OCD; 22%), generalized anxiety disorder (26%), and panic/agoraphobia (28%) the most common. Youth with ASD often also experience atypical phobias not usually seen in typically developing youth, such as fears of other speaking foreign languages or of certain household appliances, and may also have a higher prevalence of certain phobias (e.g., fears of dark, storms, crowds) than typical youth (Evans, Canavera, Kleinpeter, Maccubbin, & Taga, 2005).

Individuals with ASD and comorbid anxiety disorders often have a more complex clinical presentation which may impact accurate detection, diagnosis, and treatment. This may be due in part to the overlapping symptomatology for ASD and anxiety disorders which is further exacerbated by characteristics of ASD. For example, while both individuals with ASD and social anxiety disorder may present with deficits in social interaction and communication, the social-emotional deficits characteristic of ASD may make differential assessment and diagnosis more difficult (Kleberg et al., 2017; White, Oswald, Ollendick, & Scahill, 2009). Similarly, youth with OCD and ASD both experience repetitive thoughts and behaviors, and distinguishing between certain types of symptoms (e.g., ordering/arranging rituals in OCD done to relieve distress vs lining up behaviors in youth with ASD they may enjoy). In addition, individuals with ASD and cooccurring anxiety may present with elevated levels of impairment, as anxiety can exacerbate symptoms of ASD related to social avoidance, speech and language, and disruptive/externalizing behavioral concerns (Wood & Gadow, 2010).

Cognitive-behavioral therapy for anxiety

For typically developing children and adolescents with anxiety disorders, exposure-based cognitive-behavioral therapy (CBT) is an effective treatment (Higa-Mcmillan, Francis, Rith-Najarian, & Chorpita, 2016; McGuire et al., 2015). CBT has robust empirical support for its effectiveness for use with youth with anxiety (Banneyer, Bonin, Price, Goodman, & Storch, 2018; Wang et al., 2017). Further, researchers have demonstrated the effectiveness of CBT compared to other treatment modalities for anxiety in children with ASD. Children with anxiety disorders that receive exposure-based CBT are more likely to achieve full recovery and have higher response rates of remission in anxiety diagnoses compared to alternative treatments (see James, James, Cowdrey, Soler, & Choke, 2015; Silk et al., 2018).

Components of CBT for anxiety

Overview

CBT for typically developing children with anxiety disorders usually consists of 12–18 one-hour sessions that focus on anxiety management strategies followed by exposures (Ale, Mccarthy, Rothschild, & Whiteside, 2015; Walkup et al., 2008; Whiteside et al., 2020). Often these CBT sessions are delivered to the child only, with parents being included minimally. However, there are family CBT protocols and parent-led CBT protocols that include parents during treatment sessions or during exposures that have been effective (e.g., Lebowitz, Marin, Martino, Shimshoni, & Silverman, 2020; Rudy, Zavrou, Johnco, Storch, & Lewin, 2017).

Exposure

Exposure is a well-established, key component of CBT for youth with anxiety disorders (also termed exposure and response prevention (ERP) when used with OCD due to a focus on preventing ritualistic and/or safety behaviors) (Ale et al., 2015; Peris et al., 2017; Whiteside et al., 2020). Theoretically, exposure decreases the negative reinforcement associated with avoiding anxiety provoking stimuli, as avoidance results in decreased anxiety levels initially but does not provide opportunities for corrective learning about feared situations, and thus, anxiety is maintained or worsened in the long term. Through repeated exposures, the feared trigger elicits less distress, and the child learns that the feared outcome does not happen, or they can cope with it if it does (Craske, Treanor, Conway, Zbozinek, & Vervliet, 2014). Exposure aims to promote habituation, which is the decrease of anxiety in the presence of an anxiety provoking or fearful stimuli while minimizing anxiety-reducing behaviors (Benito & Walther, 2015). In application, exposure involves exposing youth to anxiety provoking stimuli and having the child refrain from engaging in rituals, avoidance, or escape behaviors.

For example, a child with OCD who has contamination fears with associated cleaning rituals would be gradually exposed to fear evoking stimuli without compensatory rituals by using a fear hierarchy that starts with low anxiety provoking situations (e.g., shaking hands with a family member without washing) and then working up to something more anxiety provoking (e.g., touching the ground without washing). Exposure is an integral part of CBT for anxiety; CBT interventions that include exposures are significantly more robust than those without it, as exposure practice results in more effective treatment (Peris et al., 2017; Tiwari, Kendall, Hoff, Harrison, & Fizur, 2013). Involving parents in exposures may be a beneficial way to increase the focus on exposure in treatment. Including parents and training them to be "exposure coaches" may help facilitate habituation to feared stimuli outside of sessions although future research is needed to provide clarity on the involvement of parents in anxiety treatments (Maric, van Steensel, & Bögels, 2018; Rudy et al., 2017).

Affect recognition and cognitive appraisal

Affect recognition and cognitive reappraisal are common, well-established treatment components of CBT for anxiety (Whiteside et al., 2020). CBT protocols for anxiety typically begin with psychoeducation about anxiety, where information is provided to help understand anxiety such as the prevalence, symptoms, and cognitive-behavioral framework of anxiety. Next, children are often taught affect recognition in order to help them label and recognize when they are feeling anxious or scared and somatic responses to anxiety (Silverman, 1996). After children are able to identify anxious feelings, cognitive reappraisal is used in CBT for anxiety to help children challenge and modify anxious thoughts into alternative, more helpful thoughts that will decrease anxiety (James et al., 2015).

Reinforcement and modeling

When doing exposures in CBT for anxiety, contingent reinforcement is an evidence-based practice that is critical to the treatment in order to promote engaging in anxiety confronting behaviors (Higa-Mcmillan et al., 2016). Research has indicated that receiving a reward after exposures predict better outcomes following the exposure tasks (Tiwari et al., 2013). Reinforcers for completing exposure tasks and engaging in the target behavior include therapist praise/rewards, self-reward/self-praise, and tangible rewards (Higa-Mcmillan et al., 2016, 2017). In addition to the child confronting their own fearful stimuli, evidence supports the effectiveness of modeling as a well-established CBT component (Higa-Mcmillan et al., 2016). Modeling refers to when therapists, peers, or parents engage in nonfearful responses to anxiety (e.g., approach behaviors). This component of CBT is useful to help promote vicarious learning and encourage the imitation of nonfearful responses to anxiety.

Adaptation of CBT for youth with ASD

Individuals with ASD and cooccurring anxiety present with unique needs which may impact the effectiveness of typical CBT interventions. Adaptations to CBT that have shown to improve treatment efficacy include the use of reward contingencies leveraging special interests, significant family involvement, and the use of communication and other teaching strategies (e.g., including social skills, longer sessions) that are beneficial for individuals with ASD (Wood et al., 2020).

Contingent reinforcement

With exposure to feared situations a major component of anxiety therapy, individuals with ASD and anxiety may benefit from additional motivation to face their fears. Contingent reinforcement systems provide individuals with the opportunity to earn rewards by engaging in brave behavior tasks (i.e., exposures) and facing their fears (Pittig, Hengen, Bublatzky, & Alpers, 2018). By providing this specific reinforcement, individuals will be more motivated and more likely to attempt these exposure practices in later sessions (Pittig & Dehler, 2019). For individuals with ASD and comorbid anxiety, additional reinforcement within therapy can improve treatment motivation and continued brave behavior practice.

Rewards should be identified collaboratively with a child to ensure they are personally salient. Rewards can be tangible (i.e., physical or material), such as toys or money, or intangible (i.e., nonmaterial), such as praise or time doing a preferred activity (Kazdin, 2013). By recognizing the various reward options, clinicians can work with clients to determine a preferred reward and ensure that the reinforcement is meaningful and motivating (Lee, Yu, Martin, & Martin, 2010). For individuals with ASD, preferred rewards are particularly effective as this population often develops restricted interests (Danial & Wood, 2013). Clinicians can build upon a client's specific, restricted interests by incorporating aligning rewards into treatment. For example, for children with restricted interests related to trains, clinicians can incorporate trains as a preferred reward (e.g., receiving a toy train, watching train videos).

Though rewards are effective in maintaining motivation and continued brave practices, there are a few considerations when integrating rewards. First, rewards should be received soon after the child completes a brave task to strengthen the association between the practice and the reward (Piazza, Roane, & Karsten, 2011). Second, rewards should be exclusively contingent on the child practicing the brave behaviors (i.e., the reward should not be readily accessible without any contingency). This ensures that children stay motivated to earn the reward by engaging in a specific behavior (Piazza et al., 2011). Third, there should be variability in rewards (Wine & Wilder, 2009). By creating a reward list with multiple options, this can maintain the child's engagement to continue bravery practices of increasing difficulty.

Family involvement

When working with youth with anxiety, it is important to consider how family systems have contributed to the maintenance and persistence of anxious behaviors. It is common for family members to provide reassurance and environmental accommodations to reduce anxiety in youth, such as allowing avoidance of feared situations or attending to ritualistic behaviors, which can ultimately hinder anxiety treatment (Storch et al., 2008). This is seen both universally among families of anxious youth with and without ASD (Frank et al., 2020; Storch, Lewin, et al., 2015; Storch, Zavrou, et al., 2015).

For youth with ASD and anxiety, involving caregivers in treatment can buffer any communication or behavioral difficulties that may be present. Youth with ASD who are young or low-functioning may particularly benefit from caregiver involvement as they can aid in hierarchy creation, goal-setting, and choosing appropriate and effective rewards, though some level of family involvement is indicated across childhood and adolescence. It is important for clinicians to recognize the role that family members and caregivers play in both youth's daily routines and anxiety management. Recent studies support this, highlighting that involvement of family members in anxiety treatment leads to decreased anxiety symptoms compared to those whose families are not involved (Salloum et al., 2018; Storch et al., 2020).

Communication and teaching strategies

Individuals with ASD may have communication and behavioral challenges that negatively impact treatment. Basic teaching and behavioral strategies, such as prompting, modeling, and the aforementioned contingent reinforcement, can be effective supplements to exposure-based interventions with this population. Prompting is an effective way to aid youth in engaging in a task, but it is important for clinicians to ensure that they are providing the right amount of aid. Clinicians should aim to provide just enough aid to steer the youth into practicing on their own, but not too much as that may take away from their learning. In addition to prompting, modeling can be an effective way to engage youth in exposure practices and teaching new strategies. When modeling, it may be necessary to provide visualizations (e.g., pictures, videos), as individuals with ASD may have difficulty observing and imitating an in-vivo model. A virtual model can increase engagement and also effectiveness of this learning strategy.

In addition to incorporating behavioral strategies into treatment, it may also be appropriate to include social skills or social communication exposure practices. Individuals with ASD often display communication deficits, such as literal interpretations, difficulty seeing the "big picture," and black-and-white (i.e., concrete) thinking. Perspective-taking techniques such as the use of visualized language, or visuals (objects, drawing) that represent verbal content, can be an effective way to promote social interactions within the context of CBT (Ekman & Hiltunen, 2015). The presence of social deficits can increase the presence of anxiety symptoms (Wood et al., 2009). For example, an individual with ASD and social anxiety may have limited social skills which have perpetuated their anxiety. In this case, the individual may be worried both about how to engage socially and what the peer response will be. To effectively treat the social anxiety, the clinician should first ensure that the client has the social skills necessary to successfully engage in the exposure practice tasks. Therefore, addressing social deficits through direct teaching and modeling within the context of therapy can better prepare the client for exposure tasks related to social anxiety and increase the effectiveness of exposure-based interventions for individuals with ASD.

Efficacy research

Recent research demonstrates the effectiveness of adapted CBT protocols for youth with ASD. Many of these treatment protocols include adaptations described in this chapter such as social communication modules, increased family involvement, reward systems, and extended session times (Hunsche & Kerns, 2019).

One of the most well-studied CBT protocols is an adapted modular exposure-based CBT protocol, Behavioral Interventions for Anxiety in Children with Autism (BIACA). Building on a number of smaller trials (e.g., Storch et al., 2013; Storch, Lewin, et al., 2015; Storch, Zavrou, et al., 2015; Wood et al., 2009), Wood et al. (2020) examined the use of BIACA for children ages 7–13 with ASD and anxiety in comparison to Coping Cat and treatment as usual. While both Coping Cat and BIACA were beneficial in reducing anxiety, the adapted CBT program for children showed greater effectiveness compared to Coping CAT and treatment as usual in reducing anxiety *and* improving social communication functioning (Wood et al., 2020).

Indeed, many adapted CBT protocols for this population have proven to have strong efficacy when compared to waitlist controls (Reaven, Blakeley-Smith, Culhane-Shelburne, & Hepburn, 2012;; Wood et al., 2009, 2015) and treatment as usual (Storch et al., 2013; Storch, Lewin, et al., 2015; Storch, Zavrou, et al., 2015; Wood et al., 2020). Most trials are focused on children (e.g., Reaven et al., 2012; Storch et al., 2013; Wood et al., 2009; Wood et al., 2020), yet there is emerging research on CBT protocols for use with adolescents (e.g., Storch, Lewin, et al., 2015; Storch, Zavrou, et al., 2015; Wood et al., 2015). In a review of the effectiveness of the 10 CBT programs for anxiety in children and adolescents with ASD by Hunsche and Kerns (2019), two programs were found to be probably efficacious treatment and five were found as possibly efficacious treatment. It is important to note that using adapted CBT protocol may also simultaneously decrease other ASD-related symptoms that contribute to anxiety (emotional dysregulation symptoms, social communication deficits, and adaptive functioning deficits) in children with ASD (e.g., Wood et al., 2021) and children with ASD *and* anxiety disorders (e.g., Drahota, Wood, Sze, & Van Dyke, 2011; Storch, Lewin, et al., 2015; Storch, Zavrou, et al., 2015; Wood et al., 2020).

Taken together, their findings suggest that there are many CBT-based programs that are effective in reducing anxiety symptoms in children and adolescents with ASD and cooccurring anxiety (Hunsche & Kerns, 2019). Future research is needed to evaluate the effectiveness of such CBT programs compared to other treatments and to continue refining CBT programs for ASD and anxiety to increase efficacy.

Practice and procedures

When treatment planning for anxiety in youth with ASD, there are general components of CBT that should be incorporated. These include psychoeducation, fear hierarchy creation, in-session exposure, and check-in and planning of exposure "homework" (i.e., at-home graded exposure practices). Throughout treatment, clinicians should assess how best to involve family and/or caregivers, need for supplemental modules for social skills and communication, and any behavior management strategies. In addition, incorporating reward systems and youth's specific interests into therapeutic discussions or exercises at the onset of treatment can aid in rapport building and increase youth buy-in. It is important to note that individual needs will vary greatly by client, and specific needs should be assessed and addressed in ongoing collaboration with the family. Clinicians should remain flexible as they apply CBT for youth with ASD and anxiety and be open to considering additional modifications as necessary. An example treatment plan is provided in Table 1 and accommodations in Table 2.

TABLE 1 Example CBT treatment plan for youth with ASD.

Visit	Session objectives	With who?
1	Psychoeducation about anxiety and ASD	Family
	Recognizing emotions	Child
	Set goals	Family
2	Teach and practice relaxation skills	Family
	Introduce the worry scale	Child
	Introduce cognitive restructuring	Parent
3	Teach and practice realistic, helpful thoughts	Child
	Introduce the rationale of exposure	Parent
4	Explain fighting fear by facing fear	Child
	Make a fears and worries list	Family
	Create the first exposure hierarchy	Family

Continued

TABLE 1 Example CBT treatment plan for youth with ASD—cont'd

Visit	Session objectives	With who?
5	Review stepladders progress and problem-solve	Family
	Practice identifying helpful and unhelpful thoughts	Child
	Identify parenting traps that may maintain anxiety	Parent
6–9+	Review helpful thinking or relaxation skills	Child
	Review stepladders progress and problem-solve	Family
	Complete in-session exposure(s)	Family
	Create another exposure hierarchy when ready	Family
Final	Review skills learned and progress	Child
	Discuss maintenance and set-backs	Parent
	Celebration	Family

This table provides a breakdown of objectives and audience (i.e., Parent, Child, Family) for each CBT session. *Note.* Beginning with session 2, the first task for all sessions should include reviewing homework and providing praise and rewards for task completion.

TABLE 2 Overview of adaptations to CBT components for individuals with ASD.

Accommodations for ASD	Indicated when
Family involvement	Clients need more external support
Contingent reinforcement	Clients have little motivation to engage with materials or practice exposure tasks
Behavioral teaching strategies	Clients are noncompliant, lack interest and motivation, or display disruptive behaviors
Social and adaptive skill modules	Clients present with social difficulties, adaptive difficulties, or social anxiety
Parent training modules	Clients engage in disruptive, noncompliant, or aggressive behaviors
	When current parenting practices are reinforcing and maintaining the client's anxiety

This table shows various adaptations to CBT when working with individuals with ASD and when they should be implemented.

Mini-dictionary of terms

Exposure. A core CBT skill where youth are gradually exposed to increasing fearful stimuli so youth can learn that feared outcomes will not happen, or if they do, that they can cope with them. Exposure also allows youth to practice anxiety management skills in vivo. An important component of exposure often involves response prevention, which occurs when youth are made to refrain from compulsions, avoidance, safety, or escape behaviors that would usually engage in to help minimize distress.

Contingent reinforcement. Providing reinforcement (praise, tangible rewards, tokens) that is dependent on engaging (or not engaging in) in a specific behavior. It is important for therapists, parents, or other caretakers to make sure that reinforcers are only given if the contingency is met (if the child attempts an exposure task). If the contingency is not met, then the child should not receive the reinforcer.

Fear hierarchy. An exposure treatment plan that lists specific situations an individual will face, starting with low anxiety situations and gradually increasing in anxiety until the individual conquers their fear.

Visualized language. Visuals (drawings, objects, images) that are used to help convey language. In therapy, this may include using a visual for fear ratings (e.g., a scale from 1 to 10 with faces that range from most relaxed to most worried) or a visual schedule (e.g., pictures representing the session plan).

Prompting. Assistance or a cue that is provided to help an individual engage in a behavior. Prompting may be verbal (narrating directions of what the child needs to do), full physical (hand-over-hand guidance to assist the child in completing a task), partial physical (helping direct the child's body toward the task), or a gesture (pointing to what needs to be done).

Modeling. Demonstration of a desired behavior to promote imitation. A therapist, parent, or other caregiver can demonstrate the task that is requested of the child (e.g., complete an exposure task) and demonstrate a low-anxiety response.

Cognitive reappraisal. Reinterpreting an emotion eliciting situation to alter its meaning (the thoughts feelings and emotions associated with it). This involves recognizing unhelpful thoughts and changing those thoughts to more helpful ones. In exposure therapy with children, therapists can prompt the child to identify a simple, helpful thought for them to think as they attempt an exposure task (e.g., "I can do this!" or "This is not too bad!").

Key facts of adapting CBT for youth with ASD

- Providing rewards that are contingent on engaging in brave behavior tasks and related to specific interests leads to increased motivation for youth with ASD.
- Involving family members in therapy allows the provider to address how the family system has maintained anxiety symptoms through accommodation and avoidance.
- Caregivers are often well-equipped to support goal-setting, provision of rewards, and behavioral management within the context of CBT.
- Behavioral teaching strategies, such as prompting and modeling, promote engagement and understanding of exposure-based practices.
- Including modules that address social and adaptive skill deficits can promote engagement in CBT and also decrease other, nonanxiety-related behaviors that are present in youth with ASD.
- Adaptations generally maintain the core elements of CBT for typically developing youth (exposure therapy, psychoeducation, cognitive skills).

Application to other areas

The adaptations described in this chapter to CBT with ERP can be used not only with children with ASD, but they are also applicable to adults with ASD with developmental modifications. Clinicians should consider modifying the material as seen fit, so that it is age appropriate for the individual being seen. For example, for some adults with ASD, instead of including family members, it may be helpful to include partners or spouses in treatment. Subsequently, the adaptations made to CBT to suit youth with ASD may be useful for individuals with other developmental disabilities, or social communication disorders.

Presently, the current effectiveness research has evaluated CBT for individuals with high cognitive functioning (e.g., Storch, Lewin, et al., 2015, Storch, Zavrou, et al., 2015; Wood et al., 2020). However, aspects of the CBT protocols for individuals with ASD may be applicable for use with lower cognitively functioning individuals (IQ $<$ 70). Psychoeducation, cognitive reappraisal, and ERP practices with reinforcement have been shown to reduce anxiety in individuals with ASD and Intellectual Disabilities (Moskowitz et al., 2017). Modifications such as including behavioral intervention plans and positive behavior supports within CBT protocols may be useful within this population (Moskowitz et al., 2017). Clinical judgment and flexibility should be used when considering what modifications may be needed for CBT to be effective with other populations.

Summary points

- Anxiety disorders are the most common psychiatric comorbidity for individuals with ASD.
- Exposure-based CBT is effective for treating anxiety in typically developing youth, but modification to treatment increases appropriateness for youth with ASD.
- CBT adaptations for individuals with ASD include creating contingency reward systems, inclusion of family in therapy sessions, and a focus on social communication with perspective-taking techniques.
- Recent studies support the efficacy of CBT for youth with ASD and provide some support for the value of these adaptations beyond traditional CBT.

- CBT has led to a reduction in other, nonanxiety-related behaviors in youth with ASD.
- Therefore, adapting CBT for youth with ASD and anxiety is effective in decreasing both anxiety symptoms and other, nonanxiety concerns.

References

Ale, C. M., Mccarthy, D. M., Rothschild, L. M., & Whiteside, S. P. H. (2015). Components of cognitive behavioral therapy related to outcome in childhood anxiety disorders. *Clinical Child and Family Psychology Review*, *18*(3), 240–251. https://doi.org/10.1007/s10567-015-0184-8.

American Psychiatric Association. (2013). *Diagnostic and statistical manual of mental disorders* (5th ed.). Washington, DC: Author.

Banach, R., Thompson, A., Szatmari, P., Goldberg, J., Tuff, L., Zwaigenbaum, L., & Mahoney, W. (2009). Brief report: Relationship between non-verbal IQ and gender in autism. *Journal of Autism and Developmental Disorders*, *39*(1), 188–193. https://doi.org/10.1007/s10803-008-0612-4.

Banneyer, K. N., Bonin, L., Price, K., Goodman, W. K., & Storch, E. A. (2018). Cognitive behavioral therapy for childhood anxiety disorders: A review of recent advances. *Current Psychiatry Reports*, *20*(8). https://doi.org/10.1007/s11920-018-0924-9.

Benito, K. G., & Walther, M. (2015). Therapeutic process during exposure: Habituation model. *Journal of Obsessive-Compulsive and Related Disorders*, *6*, 147–157. https://doi.org/10.1016/j.jocrd.2015.01.006.

Craske, M. G., Treanor, M., Conway, C. C., Zbozinek, T., & Vervliet, B. (2014). Maximizing exposure therapy: An inhibitory learning approach. *Behaviour Research and Therapy*, *58*, 10–23.

Danial, T. J., & Wood, J. J. (2013). Cognitive behavioral therapy for children with autism: Review and considerations for future research. *Journal of Developmental & Behavioral Pediatrics*, *34*(9), 702–715. https://doi.org/10.1097/DBP.0b013e31829f676c.

De Giorgi, R., De Crescenzo, F., D'Alò, G. L., Rizzo Pesci, N., Di Franco, V., Sandini, C., & Armando, M. (2019). Prevalence of non-affective psychoses in individuals with autism spectrum disorders: A systematic review. *Journal of Clinical Medicine*, *8*(9), 1304. https://doi.org/10.3390/jcm8091304.

Delobel-Ayoub, M., Ehlinger, V., Klapouszczak, D., Maffre, T., Raynaud, J.-P., Delpierre, C., & Arnaud, C. (2015). Socioeconomic disparities and prevalence of autism spectrum disorders and intellectual disability. *PLoS One*, *10*(11). https://doi.org/10.1371/journal.pone.0141964, e0141964.

Drahota, A., Wood, J. J., Sze, K. M., & Van Dyke, M. (2011). Effects of cognitive behavioral therapy on daily living skills in children with high-functioning autism and concurrent anxiety disorders. *Journal of Autism and Developmental Disorders*, *41*(3), 257–265.

Durkin, M. S., Maenner, M. J., Baio, J., Christensen, D., Daniels, J., Fitzgerald, R., … Yeargin-Allsopp, M. (2017). Autism spectrum disorder among US children (2002–2010): Socioeconomic, racial, and ethnic disparities. *American Journal of Public Health*, *107*(11), 1818–1826. https://doi.org/10.2105/ajph.2017.304032.

Ekman, E., & Hiltunen, A. J. (2015). Modified CBT using visualization for autism spectrum disorder (ASD), anxiety and avoidance behavior—A quasi-experimental open pilot study. *Scandinavian Journal of Psychology*, *56*(6), 641–648.

Evans, D. W., Canavera, K., Kleinpeter, F. L., Maccubbin, E., & Taga, K. (2005). The fears, phobias and anxieties of children with autism spectrum disorders and down syndrome: Comparisons with developmentally and chronologically age matched children. *Child Psychiatry and Human Development*, *36*, 3–26. https://doi.org/10.1007/s10578-004-3619-x.

Fombonne, E. (2005). The changing epidemiology of autism. *Journal of Applied Research in Intellectual Disabilities*, *18*(4), 281–294. https://doi.org/10.1111/j.1468-3148.2005.00266.x.

Frank, H. E., Kagan, E. R., Storch, E. A., Wood, J. J., Kerns, C. M., Lewin, A. B., … Kendall, P. C. (2020). Accommodation of anxiety in youth with autism spectrum disorder: Results from the TAASD study. *Journal of Clinical Child & Adolescent Psychology*, 1–11. https://doi.org/10.1080/15374416.2020.1759075.

Halladay, A. K., Bishop, S., Constantino, J. N., Daniels, A. M., Koenig, K., Palmer, K., … Szatmari, P. (2015). Sex and gender differences in autism spectrum disorder: Summarizing evidence gaps and identifying emerging areas of priority. *Molecular Autism*, *6*(1). https://doi.org/10.1186/s13229-015-0019-y.

Head, A. M., Mcgillivray, J. A., & Stokes, M. A. (2014). Gender differences in emotionality and sociability in children with autism spectrum disorders. *Molecular Autism*, *5*(1), 19. https://doi.org/10.1186/2040-2392-5-19.

Higa-Mcmillan, C. K., Francis, S. E., Rith-Najarian, L., & Chorpita, B. F. (2016). Evidence base update: 50 years of research on treatment for child and adolescent anxiety. *Journal of Clinical Child & Adolescent Psychology*, *45*(2), 91–113. https://doi.org/10.1080/15374416.2015.1046177.

Hollocks, M. J., Lerh, J. W., Magiati, I., Meiser-Stedman, R., & Brugha, T. S. (2019). Anxiety and depression in adults with autism spectrum disorder: A systematic review and meta-analysis. *Psychological Medicine*, *49*(4), 559–572. https://doi.org/10.1017/s0033291718002283.

Hossain, M. M., Khan, N., Sultana, A., Ma, P., McKyer, E. L. J., Ahmed, H. U., & Purohit, N. (2020). Prevalence of comorbid psychiatric disorders among people with autism spectrum disorder: An umbrella review of systematic reviews and meta-analyses. *Psychiatry Research*, *287*. https://doi.org/10.1016/j.psychres.2020.112922, 112922.

Hunsche, M. C., & Kerns, C. M. (2019). Update on the effectiveness of psychotherapy for anxiety disorders in children and adolescents with ASD. *Bulletin of the Menninger Clinic*, *83*(3), 326–352. https://doi.org/10.1521/bumc.2019.83.3.326.

James, A. C., James, G., Cowdrey, F. A., Soler, A., & Choke, A. (2015). Cognitive behavioural therapy for anxiety disorders in children and adolescents. *Cochrane Database of Systematic Reviews*. https://doi.org/10.1002/14651858.cd004690.pub4.

Kazdin, A. E. (2013). *Behavior modification in applied settings*. Waveland Press.

Kleberg, J. L., Högström, J., Nord, M., Bölte, S., Serlachius, E., & Falck-Ytter, T. (2017). Autistic traits and symptoms of social anxiety are differentially related to attention to others' eyes in social anxiety disorder. *Journal of Autism and Developmental Disorders*, *47*(12), 3814–3821. https://doi.org/10.1007/s10803-016-2978-z.

Lebowitz, E. R., Marin, C., Martino, A., Shimshoni, Y., & Silverman, W. K. (2020). Parent-based treatment as efficacious as cognitive-behavioral therapy for childhood anxiety: A randomized noninferiority study of supportive parenting for anxious childhood emotions. *Journal of the American Academy of Child & Adolescent Psychiatry*, *59*(3), 362–372.

Lee, M. S. H., Yu, C. T., Martin, T. L., & Martin, G. L. (2010). On the relation between reinforcer efficacy and preference. *Journal of Applied Behavior Analysis*, *43*(1), 95–100. https://doi.org/10.1901/jaba.2010.43-95.

Maenner, M. J., Shaw, K. A., Bakian, A. V., Bilder, D. A., Durkin, M. S., Esler, A., … Cogswell, M. E. (2021). Prevalence and characteristics of autism spectrum disorder among children aged 8 years—Autism and developmental disabilities monitoring network, 11 sites, United States, 2018. *MMWR Surveillance Summaries (Washington, D.C.: 2002)*, *70*(11), 1–16. https://doi.org/10.15585/mmwr.ss7011a1.

Mandy, W., Chilvers, R., Chowdhury, U., Salter, G., Seigal, A., & Skuse, D. (2012). Sex differences in autism spectrum disorder: Evidence from a large sample of children and adolescents. *Journal of Autism and Developmental Disorders*, *42*(7), 1304–1313. https://doi.org/10.1007/s10803-011-1356-0.

Maric, M., van Steensel, F. J., & Bögels, S. M. (2018). Parental involvement in CBT for anxiety-disordered youth revisited: Family CBT outperforms child CBT in the long term for children with comorbid ADHD symptoms. *Journal of Attention Disorders*, *22*(5), 506–514.

McGuire, J. F., Piacentini, J., Lewin, A. B., Brennan, E. A., Murphy, T. K., & Storch, E. A. (2015). A meta-analysis of cognitive behavior therapy and medication for child obsessive–compulsive disorder: Moderators of treatment efficacy, response, and remission. *Depression and Anxiety*, *32*(8), 580–593.

Moskowitz, L. J., Walsh, C. E., Mulder, E., Mclaughlin, D. M., Hajcak, G., Carr, E. G., & Zarcone, J. R. (2017). Intervention for anxiety and problem behavior in children with autism spectrum disorder and intellectual disability. *Journal of Autism and Developmental Disorders*, *47*(12), 3930–3948. https://doi.org/10.1007/s10803-017-3070-z.

Peris, T. S., Caporino, N. E., O'Rourke, S., Kendall, P. C., Walkup, J. T., Albano, A. M., … Compton, S. N. (2017). Therapist-reported features of exposure tasks that predict differential treatment outcomes for youth with anxiety. *Journal of the American Academy of Child & Adolescent Psychiatry*, *56*(12), 1043–1052. https://doi.org/10.1016/j.jaac.2017.10.001.

Piazza, C. C., Roane, H. S., & Karsten, A. (2011). Identifying and enhancing the effectiveness of positive reinforcement. In W. W. Fisher, C. C. Piazza, & H. S. Roane (Eds.), *Handbook of applied behavior analysis* (pp. 151–164). The Guilford Press.

Pittig, A., & Dehler, J. (2019). Same fear responses, less avoidance: Rewards competing with aversive outcomes do not buffer fear acquisition, but attenuate avoidance to accelerate subsequent fear extinction. *Behaviour Research and Therapy*, *112*, 1–11. https://doi.org/10.1016/j.brat.2018.11.003.

Pittig, A., Hengen, K., Bublatzky, F., & Alpers, G. (2018). Social and monetary incentives counteract fear-driven avoidance: Evidence from approach-avoidance decisions. *Journal of Behavior Therapy and Experimental Psychiatry*, *60*, 69.

Rai, D., Lewis, G., Lundberg, M., Araya, R., Svensson, A., Dalman, C., … Magnusson, C. (2012). Parental socioeconomic status and risk of offspring autism spectrum disorders in a Swedish population-based study. *Journal of the American Academy of Child and Adolescent Psychiatry*, *51*(5), 467–476. e6 https://doi.org/10.1016/j.jaac.2012.02.012.

Reaven, J., Blakeley-Smith, A., Culhane-Shelburne, K., & Hepburn, S. (2012). Group cognitive behavior therapy for children with high-functioning autism spectrum disorders and anxiety: A randomized trial. *Journal of Child Psychology and Psychiatry*, *53*(4), 410–419. https://doi.org/10.1111/j.1469-7610.2011.02486.x.

Richa, S., Fahed, M., Khoury, E., & Mishara, B. (2014). Suicide in autism spectrum disorders. *Archives of Suicide Research: Official Journal of the International Academy for Suicide Research*, *18*(4), 327–339. https://doi.org/10.1080/13811118.2013.824834.

Rivet, T. T., & Matson, J. L. (2011). Review of gender differences in core symptomatology in autism spectrum disorders. *Research in Autism Spectrum Disorders*, *5*(3), 957–976. https://doi.org/10.1016/j.rasd.2010.12.003.

Rudy, B. M., Zavrou, S., Johnco, C., Storch, E. A., & Lewin, A. B. (2017). Parent-led exposure therapy: A pilot study of a brief behavioral treatment for anxiety in young children. *Journal of Child and Family Studies*, *26*(9), 2475–2484.

Salloum, A., Andel, R., Lewin, A. B., Johnco, C., McBride, N. M., & Storch, E. A. (2018). Family accommodation as a predictor of cognitive-behavioral treatment outcome for childhood anxiety. *Families in Society: The Journal of Contemporary Social Services*, *99*(1), 45–55. https://doi.org/10.1177/1044389418756326.

Silk, J. S., Tan, P. Z., Ladouceur, C. D., Meller, S., Siegle, G. J., Mcmakin, D. L., … Ryan, N. D. (2018). A randomized clinical trial comparing individual cognitive behavioral therapy and child-centered therapy for child anxiety disorders. *Journal of Clinical Child & Adolescent Psychology*, *47*(4), 542–554. https://doi.org/10.1080/15374416.2016.1138408.

Silverman, W. K. (1996). *Vol. 1. Anxiety disorders interview schedule for DSM-IV: Parent interview schedule*. Oxford University Press.

Storch, E. A., Arnold, E. B., Lewin, A. B., Nadeau, J. M., Jones, A. M., De Nadai, A. S., … Murphy, T. K. (2013). The effect of cognitive-behavioral therapy versus treatment as usual for anxiety in children with autism spectrum disorders: A randomized, controlled trial. *Journal of the American Academy of Child & Adolescent Psychiatry*, *52*(2), 132–142. e132 https://doi.org/10.1016/j.jaac.2012.11.007.

Storch, E. A., Lewin, A. B., Collier, A. B., Arnold, E., De Nadai, A. S., Dane, B. F., … Murphy, T. K. (2015). A randomized controlled trial of cognitive-behavioral therapy versus treatment as usual for adolescents with autism spectrum disorders and comorbid anxiety. *Depression and Anxiety*, *32*(3), 174–181. https://doi.org/10.1002/da.22332.

Storch, E. A., Merlo, L. J., Larson, M. J., Marien, W. E., Geffken, G. R., Jacob, M. L., … Murphy, T. K. (2008). Clinical features associated with treatment-resistant pediatric obsessive-compulsive disorder. *Comprehensive Psychiatry*, *49*(1), 35–42. https://doi.org/10.1016/j.comppsych.2007.06.009.

Storch, E. A., Schneider, S. C., De Nadai, A. S., Selles, R. R., McBride, N. M., Grebe, S. C., … Lewin, A. B. (2020). A pilot study of family-based exposure-focused treatment for youth with autism spectrum disorder and anxiety. *Child Psychiatry and Human Development*, *51*(2), 209. https://doi.org/10.1007/s10578-019-00923-3.

Storch, E. A., Zavrou, S., Collier, A. B., Ung, D., Arnold, E. B., Mutch, P. J., … Murphy, T. K. (2015). Preliminary study of family accommodation in youth with autism spectrum disorders and anxiety: Incidence, clinical correlates, and behavioral treatment response. *Journal of Anxiety Disorders*, *34*, 94–99. https://doi.org/10.1016/j.janxdis.2015.06.007.

Tiwari, S., Kendall, P. C., Hoff, A. L., Harrison, J. P., & Fizur, P. (2013). Characteristics of exposure sessions as predictors of treatment response in anxious youth. *Journal of Clinical Child & Adolescent Psychology*, *42*, 34–43.

Van Steensel, F. J. A., Bögels, S. M., & Perrin, S. (2011). Anxiety disorders in children and adolescents with autistic spectrum disorders: A meta-analysis. *Clinical Child and Family Psychology Review*, *14*(3), 302–317.

Walkup, J. T., Albano, A. M., Piacentini, J., Birmaher, B., Compton, S. N., Sherrill, J. T., & Kendall, P. C. (2008). Cognitive behavioral therapy, sertraline, or a combination in childhood anxiety. *New England Journal of Medicine*, *359*(26), 2753–2766.

Wang, Z., Whiteside, S. P. H., Sim, L., Farah, W., Morrow, A. S., Alsawas, M., … Murad, M. H. (2017). Comparative effectiveness and safety of cognitive behavioral therapy and pharmacotherapy for childhood anxiety disorders. *JAMA Pediatrics*, *171*(11), 1049. https://doi.org/10.1001/jamapediatrics.2017.3036.

White, S. W., Oswald, D., Ollendick, T., & Scahill, L. (2009). Anxiety in children and adolescents with autism spectrum disorders. *Clinical Psychology Review*, *29*(3), 216–229. https://doi.org/10.1016/j.cpr.2009.01.003.

Whiteside, S. P. H., Sim, L. A., Morrow, A. S., Farah, W. H., Hilliker, D. R., Murad, M. H., & Wang, Z. (2020). A meta-analysis to guide the enhancement of CBT for childhood anxiety: Exposure over anxiety management. *Clinical Child and Family Psychology Review*, *23*(1), 102–121. https://doi.org/10.1007/s10567-019-00303-2.

Wine, B., & Wilder, D. A. (2009). The effects of varied versus constant high-, medium-, and low-preference stimuli on performance. *Journal of Applied Behavior Analysis*, *42*(2), 321–326. https://doi.org/10.1901/jaba.2009.42-321.

Wood, J. J., Drahota, A., Sze, K., Har, K., Chiu, A., & Langer, D. A. (2009). Cognitive behavioral therapy for anxiety in children with autism spectrum disorders: A randomized, controlled trial. *Journal of Child Psychology and Psychiatry*, *50*(3), 224–234. https://doi.org/10.1111/j.1469-7610.2008.01948.x.

Wood, J. J., Ehrenreich-May, J., Alessandri, M., Fujii, C., Renno, P., Laugeson, E., … Storch, E. A. (2015). Cognitive behavioral therapy for early adolescents with autism spectrum disorders and clinical anxiety: A randomized controlled trial. *Behavior Therapy*, *46*(1), 7–19. https://doi.org/10.1016/j.beth.2014.01.002.

Wood, J. J., & Gadow, K. D. (2010). Exploring the nature and function of anxiety in youth with autism spectrum disorders. *Clinical Psychological Science and Practice*, *17*, 281–292. https://doi.org/10.1111/j.1468-2850.2010.01220.x.

Wood, J. J., Kendall, P. C., Wood, K. S., Kerns, C. M., Seltzer, M., Small, B. J., … Storch, E. A. (2020). Cognitive behavioral treatments for anxiety in children with autism spectrum disorder. *JAMA Psychiatry*, *77*(5), 474. https://doi.org/10.1001/jamapsychiatry.2019.4160.

Wood, J. J., Sze Wood, K., Chuen Cho, A., Rosenau, K. A., Cornejo Guevara, M., Galán, C., … Hellemann, G. (2021). Modular cognitive behavioral therapy for autism-related symptoms in children: A randomized controlled trial. *Journal of Consulting and Clinical Psychology*, *89*(2), 110.

Chapter 16

Cognitive-behavioral therapy for children with autism and anxiety

Amanda R. Johnson, Samara Wolpe, Ingrid S. Tien, Virginia Sklar Muscatello, and Jeffrey J. Wood
Department of Education, University of California, Los Angeles, Los Angeles, CA, United States

Abbreviations

CC Coping Cat
EBT evidence-based treatments
EDI emotion dysregulation inventory
ER emotion regulation
FYF facing your fears
ID intellectual disability
OCD obsessive-compulsive disorder
PRT pivotal response treatment
RCT randomized control trial
TAU treatment as usual

Introduction

Many individuals with autism similarly struggle with anxiety and emotion dysregulation in conjunction with other core autism-related behavior challenges (e.g., social interaction and communication struggles). Children with both autism and anxiety symptoms seem to experience elevated levels of distress above and beyond the challenges associated with autism alone (Smith, Ollendick, & White, 2019). When examining the ways in which autism-related characteristics intersect with symptoms of anxiety, it is crucial to bear in mind the heterogeneity of clinical needs in this population. Varied presentations of emotion dysregulation include internalizing (e.g., anxiety or depression) or externalizing (e.g., disruptive behavior) symptoms (Cai, Richdale, Uljarević, Dissanayake, & Samson, 2018). Due to the common manifestation of anxiety symptoms in children with autism, there is a heightened need for efficacious interventions to target these symptoms.

The increased presence of anxiety among school-aged children with autism can provide additional challenges due to potential functional impairment above and beyond the presence of autism characteristics (Mazefsky, Yu, White, Siegel, & Pilkonis, 2018; Storch et al., 2012; White et al., 2014). For example, school-aged children with both autism and anxiety often exhibit traditional childhood fears (e.g., separation anxiety, generalized anxiety) as well as specific fears and anxieties (e.g., specific sounds, minor changes in routine, germs, mealtime rituals; Kerns et al., 2014). With amplified rates of specific fears and anxieties in conjunction with autism-related characteristics, these children often have greater difficulties with adaptive functioning, social skills, loneliness, potential exhibition of self-injurious behaviors, family conflict, and behavioral challenges that manifest in varying contexts (e.g., with peers, in academic settings).

Cognitive-behavioral therapy (CBT) is an established treatment used to address an array of symptoms (e.g., anxiety, mood disorders) in typically developing (TD) adults and children (García-Escalera, Chorot, Valiente, Reales, & Sandín, 2016). CBT has also been increasingly used for verbal children with autism to address co-occurring mental health needs related to emotion dysregulation, with research indicating CBT as a promising treatment for reducing anxiety symptoms in this population (Storch et al., 2013, 2015; Wood et al., 2009, 2015, 2020). Despite this emergent research, evaluation of the overall body of evidence examining the effect of CBT on symptoms of anxiety in children with autism is necessary due to the preliminary nature of existing studies.

Handbook of Lifespan Cognitive Behavioral Therapy. https://doi.org/10.1016/B978-0-323-85757-4.00007-9

Therefore, the goal of this review is twofold: (1) to provide a summary explaining the heterogeneous and complex nature of autism and anxiety and (2) to showcase the investigation of existing CBT research for anxiety in children with autism. Upon reviewing recent studies, future directions for this field of research are discussed.

Anxiety and autism

Children with autism have an increased likelihood of anxiety disorders, as up to 80% of children with autism meet the criteria of a clinical diagnosis of anxiety (Simonoff et al., 2008), in comparison to the 7.1% of children nationwide who have diagnosed anxiety (Ghandour et al., 2018). Children with autism as young as five have been found to be struggling with anxiety (Keen, Adams, Simpson, den Houting, & Roberts, 2019). Anxiety, not autism-related concerns, are often the most cited reason for children visiting clinical care, leading to speculation if anxiety is better categorized as an autism-related trait or a separate co-occurring disorder (Kerns & Kendall, 2012). High levels of generalized worry and fear of specific situations have been noted since the first clinical descriptions of autism (Kanner, 1943). The most common types of anxiety diagnosed among children with autism are specific phobias, obsessive-compulsive disorder, and social anxiety disorder (Reaven et al., 2009; White et al., 2014). Despite the increased number of children with autism meeting criteria for an anxiety disorder diagnosis, clinicians often have a difficult time identifying specific anxiety disorders in individuals with autism (MacNeil, Lopes, & Minnes, 2009). It can be difficult to meaningfully separate anxiety from other areas of mental health need and psychosocial functioning in children with autism (e.g., Wood & Gadow, 2010) given the overlapping underlying causes (e.g., common genes, the stress associated with having autism, executive dysfunction) and the interdependency of symptom areas. For instance, social skills related to autism can reduce social confidence and increase self-consciousness, and the fight-or-flight component of anxiety can heighten aggression and defiance.

Despite the high rates of comorbidity, there is considerable heterogeneity in the presentation of anxiety symptoms in children with autism. In a semistructured interview assessment, Kerns et al. (2014) explored different presentations of anxiety in youth on the autism spectrum, defined by the similarity to meet the typical anxiety disorder symptoms described in the Diagnostic and Statistical Manual of Mental Disorders—Fifth Edition (DSM-5; American Psychiatric Association, 2013). Of participants, 63% were found to have significant anxiety leading to impairment of quality of life, with many anxiety symptoms resembling those described in the DSM. Some children's anxiety symptoms were correlated with characteristics of autism, including difficulties with social interactions and communication. Despite the varied and heterogeneous phenotypes of individuals with autism, negative thoughts were found to be a manifestation of both forms of anxiety and autism.

As previously mentioned, clinicians frequently confuse anxiety behaviors with autism-related challenges, as a clinician's focus on autism may overshadow anxiety struggles (Mason & Scior, 2004). This potential "diagnostic overshadowing" can lead to children's anxiety symptoms failing to receive adequate clinical treatment. Lacking treatment for a child's anxiety may have detrimental effects on both a child's short-term distress as well as their long-term well-being. Delaying treatment can lead to the continued reinforcement of negative coping strategies and avoidance behaviors, allowing the anxiety to build (White et al., 2014; Wood et al., 2009).

Effects of anxiety in children with autism

Parent reports indicate anxiety symptoms in children with autism have a deleterious effect on their child's wellbeing and family functioning (den Houting, Adams, Roberts, & Keen, 2018). Additionally, children with autism and co-occurring anxiety report more self-injurious behavior, depressive symptoms, and parental stress compared with children with autism who did not experience clinically significant anxiety. In the same study, the authors found that children with autism and anxiety experience more autism-related symptoms than children with autism but no anxiety, suggesting that anxiety and autism are intertwined, and both require adequate treatment to prevent negative effects on a child's overall wellbeing.

Qualitative studies on both children with autism and their parents suggest that anxiety is associated with a lower quality of life, impacting children across home, social, and school settings (Adams & Emerson, 2021). Anxiety disorders in childhood are also a significant predictor of anxiety disorders in adulthood, often associated with negative outcomes such as substance abuse and increased risk of suicidal behavior (Allen, Benningfield, & Blackford, 2020). Considering that individuals with autism already have lower quality of life outcomes across multiple domains in adulthood (Howlin & Magiati, 2017), addressing comorbid anxiety disorders in this population may be important to improve overall well-being across the life course.

Untreated anxiety in this population can also lead to poorer educational performance and subsequent underemployment after graduation (Perihan et al., 2020). School, an already stressful place for many TD children, can present a variety of additional anxiety-related challenges for students with autism. Parents of school-age children with autism and co-occurring anxiety reported that their children's anxiety negatively affected their concentration, learning, behavior, and child's ability to perform to the best of their capabilities while in school (Adams & Emerson, 2021). An investigation into the effects of anxiety in children with autism between 10 and 12 years old found that 57% of children reported difficulty completing schoolwork due to anxiety, 87% of parents reported that the child's classroom performance was affected by anxiety, and 43% of those parents claimed children experienced a large impact on classroom performance as a result of the presentation of anxiety symptoms (den Houting et al., 2018). Several characteristics of school were reported by parents to heighten anxiety in their children with autism, with common triggers including overwhelming or negative sensory experiences, academic expectations, social demands, and general uncertainty (Simpson, Adams, Alston-Knox, Heussler, & Keen, 2019).

Cognitive-behavioral therapy and autism

Psychotherapy is a commonly used mental health service for youth with autism (Zablotsky et al., 2015). CBT for anxiety problems typically involves teaching cognitive skills as well as helping clients directly face the situations that provoke anxiety. In this process, called in vivo exposure, the client learns that their feared situation is not as catastrophic as the client might have thought, but also reminds the client of the positive mechanisms they have to address their anxiety symptoms. For children with autism and concurrent anxiety, the general CBT approach of challenging irrational, fearful beliefs and developing rational beliefs has been utilized in various manual-based programs, although the approach has varied widely from program to program via the level of emphasis placed on in vivo exposure relative to other treatment elements (i.e., cartooning, role playing).

In a metaanalysis of 23 studies of CBT in children with autism and co-occurring anxiety conducted by Perihan et al. (2020), there was a moderate effect size in the reduction of anxiety symptoms. CBT protocols that emphasized parental involvement outperformed treatments that focused solely on the child in eventual reduction of anxiety symptoms and parent-reported negative externalizing behaviors. Furthermore, Perihan and colleagues found that treatments with a duration shorter than 12 weeks yielded smaller effects than treatments lasting 12 weeks or longer, suggesting that treatment type (i.e., group, family, or individual) and duration of treatment are important moderators in the efficacy of CBT for children with autism and anxiety.

Variations of CBT for children with autism and anxiety

The gold standard for demonstrating a treatment's effectiveness is through the evaluation of randomized control trials (RCT). In an RCT, children are randomly assigned to one of two (or more) treatment conditions, specifically children assigned to a therapy condition are compared with those who are not (or those in a comparison treatment condition). Several small RCTs have investigated the potential benefits of a CBT for school-aged children with autism and interfering anxiety with varying rates of success at reducing both anxiety and autism symptomatology (Reaven, Blakeley-Smith, Culhane-Shelburne, & Hepburn, 2012; Storch et al., 2013, 2015; Wood et al., 2009, 2015, see Table 1). There are similarities among many of the core elements within evaluated CBT programs. Elements typically include a structured (and measurable), goal-oriented approach that is time-limited and utilizes a combination of psychoeducation, cognitive skills training (e.g., reframing), exposure therapy (i.e., facing fears), and positive reinforcement. However, the designs of these programs are varied; such as group vs. individual, modular designs, or individual components.

Some researchers in the autism field have started to emphasize the importance of a personalized approach to CBT, herein referred to as *adapted CBT* (cf. Wood et al., 2020). Adapted CBT is a modular approach that provides an adjusted intervention to children with autism and interfering anxiety with modifications guided by each child's clinical needs. While standard-of-practice CBT [e.g., Coping Cat (CC); Kendall & Hedtke, 2006] is a comprehensive CBT treatment developed for TD children with anxiety disorders. In a three-site study by Wood and colleagues comparing standard-of-practice CBT and adapted CBT with treatment as usual (TAU), CBT was found to be efficacious for youth with autism and maladaptive and interfering anxiety, with an adapted CBT approach exhibiting further advantages on standardized outcome measures (2020). Appropriate modifications of treatment for individuals may make CBT accessible and effective for a wider range of children with autism ranging in verbal ability and living skills (Spain & Happé, 2020).

TABLE 1 RCTs of efficacious CBT programs for youth with autism and anxiety.

Study authors & year	Symptoms targeted	Design	Age range	Outcome measures
Wood et al. (2009)	Anxiety + ASD	RCT (Wait-list control)	7–11	ADIS-C/P, CGI, MASC
Reaven et al. (2009)	Anxiety + ASD	RCT (Wait-list control, pilot)	8–14	ADOS, SCQ, WASI, SCARED
Drahota, Wood, Sze, and Van Dyke (2011)	Anxiety + ASD	RCT (Wait-list control)	7–11	Children's daily living skills and related parental intrusiveness
Storch et al. (2013)	Anxiety + ASD	RCT (TAU control)	7–11	PARS, ADIS-C/P, CGI-Severity & Improvement
Wood, Fujii, Renno, and Van Dyke (2014)	Anxiety + ASD	RCT (TAU control)	7–11	Observed Social Communication-Related Autism Symptom Severity
McConachie et al. (2014)	Anxiety + ASD	RCT (Wait-list control)	9–13	ADOS, SCQ, WASI, SCARED
Ehrenreich-May et al. (2014)	Anxiety + ASD	RCT	11–14	ADIS-IV-C/P, PARS, CGI-S, RCADS
Hepburn, Blakeley-Smith, Wolff, and Reaven (2016)	Anxiety + ASD	RCT (Wait-list control, pilot telehealth)	7–19	SCARED, PSOC
Storch et al. (2015)	Anxiety + ASD	RCT (TAU control)	11–16	ADIS-IV-C/P, PARS, CGI-Severity & Improvement, SACA, CIS-P, CBCL, MASC-Parent, SRS, RCADS
Wood et al. (2015)	Anxiety + ASD	RCT (Wait-list control)	11–15	ADIS-IV-C/P, PARS, CGI-S, MASC-Parent, RCADS
Wood et al. (2020)	Anxiety + ASD	RCT (TAU control)	7–13	PARS, CBCL, SRS-2, CAIS
Wood et al. (2021)	Anxiety + ASD	RCT (social-skills intervention control)	6–13	POPE, YTP

Note. Measure abbreviations are as follows: Anxiety Disorders Interview Schedule (ADIS), Autism Diagnostic Observation Schedule (ADOS), Child Anxiety Impact Scale (CAIS), Child Behavior Checklist (CBCL), Clinical Global Impression (CGI), Columbia Impairment Scale (CIS), Multidimensional Anxiety Scale for Children (MASC), Parenting Sense of Competence (PSOC), Pediatric Anxiety Rating Scale (PARS), Playground Observation of Peer Engagement (POPE), Revised Children's Anxiety and Depression Scale (RCADS), Screen for Anxiety and Related Emotional Disorders in Children (SCARED), Service Assessment for Children & Adolescents (SACA), Social Communication Questionnaire (SCQ), Social-Responsiveness Scale-2 (SRS-2), Wechsler Abbreviated Scales of Intelligence (WASI), Youth Top Problems Scale (YTP).

Narrative review of research on CBT for anxiety in children with autism (practice and procedures)

Group-based CBT for anxiety in children with autism. Several studies set the stage for further examination into the applicability of CBT for children with autism and co-occurring anxiety with promising preliminary results. In a brief intervention study, Sofronoff, Attwood, and Hinton (2005) implemented a 6-week group CBT intervention for children ages 10–12 with the diagnosis of Asperger Syndrome (now within the larger DSM-5 Autism Diagnostic Criteria) and parent-reported anxiety, with an emphasis on identifying feelings, project work, and group discussion. Researchers saw a reduction in reported anxiety symptoms both immediately post intervention and at 3-month follow-up among CBT participants.

In a further examination of the benefits of a similar group therapy setting for children with autism, McConachie et al. (2014) disseminated a brief group CBT based on the "Exploring Feelings" manual (Attwood, 2004) to 32 participants, which had the same core elements as the study mentioned above. One group received immediate group therapy, with the other group receiving delayed therapy. Attendance was very high (91%), and qualitative interviews of parents and children indicated that both found the process helpful, with parents in particular reporting that the group had a positive effect on their child's anxiety. However, while 76% of the participating children with autism in the immediate treatment group achieved a reduction in their anxiety symptoms compared with only 33% in the delayed treatment group, there were no participants who improved significantly enough to test out of the criteria for at least one anxiety disorder.

Another seminal early exploration of group CBT for anxiety in children with autism was conducted by Reaven et al. (2009). Participants included verbally fluent children with autism who had an IQ above 70. Participants expressed significant anxiety symptoms as defined by parent report scores exceeding clinical cutoffs in the Screen for Child Anxiety Related Emotional Disorders (SCARED; Birmaher et al., 1999). Lastly, children could not be simultaneously receiving any other psychological therapies explicitly targeting anxiety, although all other interventions (both behavioral and pharmacological) were acceptable. Thirty-three families participated in the study, with 31 families included in the final sample. Families were accepted into the program based in order of expressed interest, with early responders placed in the treatment group and later responders placed on the waitlist.

The 12-week treatment was divided into two 6-week intervention blocks, each targeting different skills, using the unpublished CBT Manual Fighting Worry and Facing Fears. CBT sessions consisted of large group work as well as parental psychoeducation about anxiety and CBT, defining and differentiating between "adaptive" and "excessive" levels of parental protection, and encouraging parents to allow their children to experience levels of anxiety through exposure to certain fears to allow them the opportunity to implement learned coping skills. The first 6 weeks provided an introduction into anxiety symptoms and focused on educating participants on the physical symptoms of anxiety, teaching them to recognize their own symptoms of anxiety and to then implement physical calming techniques (e.g., deep breathing, soothing self-talk). Children and parents worked together to identify frequently occurring anxiety-provoking situations that interfere with everyday functioning. In the second intervention block, children engaged in graded exposure tasks (slowly building up to facing fears in both imagined and real situations) for the fears they identified in the first section. Since traditional CBT protocols for children with anxiety require skills that are often difficult for children with autism—for instance, good linguistic and abstract thinking skills—modifications were made to address the participants' learning requirements. There was an enhanced focus on visual approaches, video modeling, parent participation, multiple choice instead of fill-in-the-blank answers to questions, and increased opportunities for repetition.

At the end of the 12 weeks, children in the active treatment group displayed a decrease in anxiety symptom severity relative to the waitlist group as measured by parent report, although not by child report. The authors theorized that the lack of improvement in the child report measure may have been due to the children's increased ability to recognize their own anxiety symptoms, which could cause over-reporting in the posttreatment stage. A less optimistic interpretation of these results is that parents are overestimating the improvement in their children's anxiety due to the placebo effect given their potentially increased investment in the intervention. This study, while lacking some of the rigors of later RCTs examining similar protocols, provided promising initial support for the application of modified CBT for anxiety in children with autism.

Reaven et al. (2012) provide further evidence of the efficacy of CBT in children with autism in the context of group therapy sessions. The authors used the facing your fears (FYF) protocol (Reaven, Blakeley-Smith, Nichols, & Hepburn, 2011), the same procedure as described above, a family-focused group CBT program developed specifically for children with autism. Fifty children with autism and anxiety were randomized either to FYF or TAU for 12 weeks, during which the treatment group underwent 12 group sessions, each 1.5 h long. Reaven et al. (2012) found that participants who underwent the FYF intervention had significantly fewer anxiety diagnoses postintervention compared with the participants in the TAU group and experienced a comparative increase in meeting criteria for positive treatment response. The reductions in anxiety symptoms observed posttreatment remained relatively constant at 3- and 6-month follow-up.

The FYF intervention has also been examined in children with autism in a telehealth context. Hepburn et al. (2016) evaluated the feasibility of disseminating FYF over a virtual platform. While fidelity proved excellent and improvements in anxiety were observed in the sample of children with autism and anxiety disorders throughout the treatment, some families in the sample struggled with the technology, a fact which significantly affected the quality of some of the treatment sessions. However, this study took place in 2016 when, likely, fewer participants would be familiar with telehealth and receiving virtual healthcare compared to 2021.

In addition to the benefits of reducing anxiety symptoms in children with autism, there may be social and emotional benefits to group therapies like FYF beyond the clinical targets of the intervention. Rapee, Wignall, Hudson, and Schniering (2000) hypothesized that group therapies and interventions may provide a rare opportunity for socialization with peers struggling with similar challenges, potentially reducing social isolation, and promoting a sense of normalization for children whose social interactions are traditionally both challenging and isolating. For this reason, among others, social skills interventions often take on the form of group coaching (e.g., the UCLA PEERS®; Laugeson, Frankel, Gantman, Dillon, & Mogil, 2012).

Individual and family CBT for anxiety in children with autism. A study by McNally Keehn, Lincoln, Brown, and Chavira (2013) examined the effect of a modified version of the CC program, a cognitive-behavioral treatment protocol developed for TD children with anxiety disorders, in children with autism and anxiety (Kendall & Hedtke, 2006). In addition to tracking changes in measures of anxiety symptoms, they assessed subsequent changes in anxiety diagnoses, as well as parent-reported anxiety. The study consisted of a 16-week, manualized CBT intervention, with the primary goal of empowering children with the tools to both recognize signs of anxious arousal and equip them with the tools to manage these anxious reactions. The authors found that participants in the treatment group experienced significant reduction in both anxiety symptoms and anxiety diagnoses postintervention compared with the waitlist-control group, with the gains from the intervention maintained at follow-up. Although CBT was previously demonstrated to improve anxiety symptoms in neurotypical populations, this study successfully demonstrated its usefulness for children with autism.

Another study of the CC program examined the potential moderating effect of autism symptomology in the outcome of unmodified family and individual CBT treatments for TD children with anxiety (Puleo & Kendall, 2011). In the first of two studies, the authors found that children with moderate levels of autism traits as determined by the social responsiveness scale (SRS; Constantino & Gruber, 2012) predicted reduced responsiveness to individual CBT (ICBT), but not to family CBT (FCBT). Puleo and Kendall also found that parent-reported SRS scores were more sensitive predictors of treatment outcomes than observational codes. These findings echo recommendations from other experts for parental involvement in CBT for children with autism, given the frequency and variety of difficulties in communication and self-regulation in this population (e.g., Hassenfeldt, Lorenzi, & Scarpa, 2015). In a related study, Puleo and Kendall (2011) found that completion of at-home exposures was higher in FCBT. The authors hypothesized that this reflects the increased benefit of allowing parents to act as facilitators for children with autism whose communication skills may otherwise be barriers to engagement in the treatment. Furthermore, the benefit of FBCT promoting increased at-home exposure tasks may be particularly salient to children with autism who are more impacted, thereby providing more accessible opportunities to practice these skills. Hassenfeldt et al. (2015) further emphasize the efficacy of parental involvement in CBT interventions for children with autism, reporting positive effects for parents as well as children. Parents of children with autism who participate in the child's intervention may experience improvements in mental health and levels of stress (Hassenfeldt et al., 2015).

Modular CBT for children with autism and anxiety. Despite the apparent efficacy of CBT in reducing anxiety in children with autism, its effects may not be as robust when compared to CBT's effect on anxiety in TD children. In an analysis of 48 studies, Hofmann, Asnaani, Vonk, Sawyer, and Fang (2012) found that the effects of CBT for TD children were larger than the effects of CBT for children with autism. However, this disparity may have been due to a lack of individual customization of CBT programs. As a result, several CBT programs have been adapted to target core characteristics of autism with the goal of improving the efficacy of CBT in this population.

The rationale for adapted CBT programs for youth with autism is multifaceted. The allocation of specific practices relevant to each child's symptom profile may promote greater clinical gains due to the accrued benefits of individualized treatment. There are also contextual factors specific to autism that may cause anxiety (e.g., social communication challenges, sensory overstimulation, etc.) which should be addressed specifically in any treatment targeting anxiety in this population (Wood et al., 2009). Youth with autism may also benefit from more parental involvement in treatment than is typical in standard of practice CBT (Puleo & Kendall, 2011).

Several CBT protocols have been successfully modified while keeping the core components of CBT intact, demonstrating that modified or modular CBTs are successful at encompassing the effective components of CBT for youth with autism (Reaven et al., 2011; Sofronoff et al., 2005; Wood et al., 2009). Modular CBT consists of the guided combination of individually scripted techniques that are explicitly matched to the child's individual strengths and needs (Chorpita, Taylor, Francis, Moffitt, & Austin, 2004). The modular format of adapted CBT allows for components of treatment to be apportioned based on need giving an opportunity for a personalized treatment program for supporting the reduction of a child's anxiety and autism-related struggles. An example of an adapted CBT treatment approach is Behavioral Interventions for Anxiety in Children with Autism (BIACA; Wood et al., 2009, 2015, 2020).

Wood et al. (2009) implemented an initial RCT testing BIACA against a waitlist-control condition for children with autism and anxiety. The participants consisted of 40 children (7–11 years old) with co-occurring autism and anxiety, accompanied by their primary caregiver. Participants were randomly assigned to 16 sessions of CBT or a 3-month waitlist (36 completed treatment or waitlist). The intervention program consisted of: (a) 16 weekly 90-min sessions (time in session split evenly between children and parents) to facilitate parent engagement, (b) a modular format guided by an algorithm to personalize treatment because of the heterogeneous presentation of autism and anxiety across all participants, (c) disruptive behavior is addressed with incentive-based practices, (d) inclusion of modules on social engagement skills, (e) special interests are incorporated into treatment to facilitate greater child engagement. Independent evaluators blind to treatment conditions conducted structured diagnostic interviews, while parents and children completed anxiety symptom checklists at baseline and posttreatment/postwaitlist. Youth randomized to BIACA had primary outcomes comparable to those found in CBT for TD anxious youth, with BIACA demonstrating superiority to waitlist in response rate (76.5% vs. 8.7%), clinical severity ($d = 2.5$), and SRS scores ($d = 0.75$; Wood et al., 2009).

Subsequently, Storch et al. (2013) replicated the initial RCT of BIACA by randomizing a sample of 45 children with co-occurring autism and anxiety to either BIACA or TAU. At the conclusion of the 16 weeks, 75% of the children randomized to BIACA met the criteria for "treatment responders," compared with only 14% of participants randomized to the TAU condition. These gains were largely maintained at 3-month follow-up, suggesting that BIACA is a durable, effective treatment for anxiety in youth with autism.

In a second replication study, Storch et al. (2015) randomized 31 youth (11–16 years old) with autism and clinically significant anxiety to BIACA or TAU. Participants randomized to the TAU condition were able to seek any treatment of their choosing (or choose to forgo treatment) for 16 weeks. Overall, 68.8% of participants responded to treatment, which was superior to TAU and on par with past results among youth with autism. These replications are pivotal in the evaluation of treatment programs to ensure generalizability and durability of the initial promising results. Presently, BIACA is the only CBT program for autism that meets criteria as "efficacious" based on American Psychological Association Division 53 Criteria (Southman-Gerow & Prinstein, 2014), based on its multiple successful replications by independent research groups and use of fully randomized designs with active control groups and randomization at the youth level. These studies underscore that CBT is a promising treatment for youth with co-occurring autism and anxiety.

Similar findings emerged in a second study of BIACA for anxiety in early adolescents with autism (Wood et al., 2015), in which 33 youth with autism (11–15 years old) were randomized to 16 sessions of BIACA or a waitlist. BIACA significantly outperformed the waitlist group in reduction of anxiety symptoms on the PARS. The group assigned to the BIACA condition experienced significantly greater reductions in clinician-rated anxiety severity relative to the waitlist arm ($d = 0.74$); 79% of the BIACA group versus 28.6% of the waitlist group were treatment responders.

Most recently, Wood et al. (2020) conducted a three-site RCT which evaluated the efficacy of two versions of CBT (BIACA and the standard-of-practice CBT for children with anxiety, CC) for anxiety in youth with autism (2020). This study was the first to compare an adapted CBT program to an active CBT comparison condition for children with autism. Participants included a volunteer sample of 167 children with autism and maladaptive and interfering anxiety (7–13 years old) who were randomized to BIACA, CC, or TAU. In this study, the primary outcome measure was the PARS, which was rated by an independent evaluator who was unaware of which treatment condition the participants were assigned to. The PARS was selected as a primary outcome measure in line with other studies examining similar protocols and following measurement best practice (LeCavalier et al., 2015).

BIACA outperformed CC and TAU on the primary outcome measure and several secondary outcome measures. This study is especially relevant as it is the first major test of an adapted CBT protocol in comparison to another established, manualized evidence-based, standard-of-practice CBT program. Furthermore, BIACA outperformed the other treatment conditions across each of the three sites, highlighting both the specificity and wide applicability of its effects as well as the importance of designing CBT programs specifically for youth with autism. Adapted CBT not only appears to be an efficacious treatment for youth with autism and maladaptive and interfering anxiety, but also demonstrates additional advantages due to its adapted and personalized nature.

Discussion and future directions

CBT is a treatment modality with a long-standing history of success in treating a wide range of anxiety disorders and emotion dysregulation in TD children and adults. The results of this review suggest that CBT may also be an efficacious treatment for youth with autism and maladaptive, interfering anxiety. While many of the studies reviewed were high quality RCTs, they were limited by several factors. For example, several studies used small sample sizes, relatively unrepresentative samples, and randomization at the group level, limiting generalizability of study results. Despite this, most studies used reliable and valid outcome measures administered by independent evaluators blind to treatment conditions, offering a certain level of confidence in the promise of CBT in this population—particularly when supported by the findings of the small number of more stringent RCTs in this group of clinical trials. Thus, several possible future directions for this field are explored.

Services, implementation, and dissemination. As the practices of both individual and family CBT have been found to be well-established, evidence-based interventions (EBIs) when implemented in university settings, an imperative next step to investigate the efficacy of CBT for reducing anxiety in individuals with autism will be to test the effectiveness of CBT in community settings. Stuart and colleagues investigated the effects of the 2008 Mental Health Parity and Addiction Equity Act on service receipt and out of pocket costs for children with autism in their families, discovering that ⅔ of children in the sample used outpatient mental health services (Stuart et al., 2017). Additionally, CBT may also be a cost-effective alternative when compared to other treatment options for anxiety (e.g., medication) for individuals with autism (Storch et al., 2015). Therefore, the preponderance of children with autism seeking outpatient mental health services combined with the cost-effectiveness of this treatment alternative emphasizes the potential reach and serviceability of CBT treatments for the substantial number of children with autism and co-occurring anxiety.

Most children with autism receive community-based services. A study by Zablotsky et al. (2015) found that 61.7% of children with autism in their sample received only community-based services, with that percentage rising to 72.2% when examining children with autism and ID (intellectual disability). Therefore, it is imperative that providers are trained in using evidence-based, efficacious practices to best support their clients. Similarly, it is vital that these EBIs are validated in community-based care settings to ensure generalizability in this heterogeneous population.

Despite the demonstrable importance of connecting community-based care with EBIs, there is a significant research-to-practice gap. Pickard, Meza, Drahota, and Brikho (2018) examined service use and community-based in for-profit care for children with autism in Southern California to capture what services children are receiving in service delivery systems traditionally underrepresented by autism research. Despite a high number of providers endorsing using practices with "established evidence" (76.3%), an even higher number endorsed using practices without established evidence (79.1%), or with undetermined evidence (78.9%). Clearly, there is a need for dissemination and implementation support for efficacious practices like CBT in community service settings for children with autism.

Conclusion

Despite the promising results reviewed, researchers should continue investigating CBT as treatment for children and adolescents with autism. Although further research with more stringent methodologies is necessary, extant findings are encouraging. Researchers are encouraged to further develop CBT protocols for affected youth while using assessments of core autism symptoms as well as emotional and behavioral functioning as key outcome measures. The continued implementation of CBT treatments for this population is recommended and further development of CBT for children with autism and co-occurring anxiety is a hopeful direction for intervention research in this field.

Mini-dictionary of terms

- ***Autism.*** Autism, or autism spectrum disorder, refers to a heterogeneous condition characterized by difficulty with social skills, nonverbal communication, speech, and repetitive behaviors.
- ***Maladaptive and interfering anxiety.*** Experiencing an anxiety response in unwarranted situations, leading to a set of dysfunctional strategies that lead to increased anxiety responses and can interfere with social relationships, daily functioning, or quality of life.
- ***Adaptive functioning.*** How well an individual can handle common situations in life, such as self-care, communication skills, or social situations.
- ***Adapted cognitive-behavioral therapy.*** CBT that has been altered to accommodate specific learning needs or differences.
- ***Modular cognitive-behavioral therapy.*** CBT that contains flexible modules that clinicians can select based on their client's individual needs to better target specific problems.

Key facts

- Children with autism have an increased likelihood of anxiety disorders, as up to 80% of children with autism meet the criteria of a clinical diagnosis of anxiety (Simonoff et al., 2008), in comparison to the 7.1% of children nationwide who have diagnosed anxiety (Ghandour et al., 2018).
- Anxiety, not autism-related concerns, is often the most cited reason for children visiting clinical care.
- The most common types of anxiety diagnosed among children with autism are specific phobias, obsessive-compulsive disorder, and social anxiety disorder.
- Despite the increased number of children with autism meeting criteria for an anxiety disorder diagnosis, clinicians often have a difficult time identifying specific anxiety disorders in individuals with autism.
- There is considerable heterogeneity in the presentation of anxiety symptoms in children with autism.

- Parent reports indicate anxiety symptoms in children with autism have a deleterious effect on their child's wellbeing and family functioning.
- Psychotherapy is a commonly used mental health service for youth with autism.
- CBT for anxiety problems typically involves teaching cognitive skills as well as helping clients directly face the situations that provoke anxiety.
- The gold standard for demonstrating a treatment's effectiveness is through the evaluation of RCT.

Applications to other areas

Expanding CBT to children with intellectual disability. Additionally, future research should investigate how CBT treatments can be broadened to address previously untargeted subpopulations and symptoms within autism. For example, many studies presented have included verbal individuals with autism and could be expanded to meet the needs of individuals with minimal language or those with an intellectual disability. Individuals with autism and an ID also experience comorbid psychopathologies (e.g., anxiety) at high rates (Brereton, Tonge, & Einfeld, 2006). However, treatments for these comorbidities have, at present, largely focused on populations of individuals on the autism spectrum with comparatively lower support needs (Bradley, Henderson, & Monfore, 2004).

Intensity of treatment. The optimal intensity of CBT treatments should also be investigated. Other autism treatments, such as applied behavior analysis (ABA) programs and pivotal response treatment (PRT; Koegel & Koegel, 2006), are consistently delivered over the course of many months or years and in multiple settings (i.e., in a child's home, school, and community). Future research should retain the relative efficiency of CBT's outpatient treatment model and examine the effectiveness of more intensive CBT programs over the course of a longer time frame designed to meet the needs of participants in their daily lives and to make an impact on their overall symptom severity across the emotional, behavioral, and social domains. While length and intensity of treatment is often determined by the clinician to meet the needs of individual clients, general guidelines or suggestions would be helpful for clinicians to be able to make well-informed suggestions for their heterogeneous, neurodivergent clients.

Emotional dysregulation and autism. Although originally not considered a "core" feature of autism, individuals on the spectrum often experience problems with emotion dysregulation. This difficulty can be attributed to heightened emotional responsiveness to stimuli coupled with poor emotional control (Mazefsky et al., 2013). Specifically, individuals with autism often fail to utilize strategies such as situation selection, situation modification, attentional deployment, cognitive change, and response modulation (i.e., five areas of emotion regulation, ER: Gross, 2008). Furthermore, when there is some element of ER at play, it is typically less effective than the ER exhibited by TD individuals (Samson et al., 2014). These ER difficulties are often compounded by co-occurring psychiatric conditions in individuals on the spectrum such as anxiety, mood and attentional disorders, and schizophrenia, which promote further emotion dysregulation (Rosen, Mazefsky, Vasa, & Lerner, 2018).

Summary points

- Children with autism frequently experience clinically significant anxiety.
- This anxiety can present in complex ways, sometimes eluding detection.
- CBT has proven efficacious in reducing anxiety symptoms for children with autism.
- Although all formats of CBT are useful in reducing anxiety, parental/familial involvement is recommended.
- CBT can also be effective in improving general ER in this population.
- Although CBT is a promising treatment, there are limitations both in the research and practicality of the treatment, suggesting the need for further research into the use of CBT for anxiety in children with autism.
- There is an ongoing need to test the effectiveness of CBT for children with autism in community settings, and, if justified, to disseminate CBT to practitioners in these settings.

References

Adams, D., & Emerson, L. M. (2021). The impact of anxiety in children on the autism spectrum. *Journal of Autism and Developmental Disorders, 51*(6), 1909–1920.

Allen, K. B., Benningfield, M., & Blackford, J. U. (2020). Childhood anxiety—If we know so much, why are we doing so little? *JAMA Psychiatry, 77*(9), 887–888.

American Psychiatric Association. (2013). *Diagnostic and statistical manual of mental disorders* (5th ed.). Washington, DC: American Psychiatric Association.

Attwood, T. (2004). Cognitive behaviour therapy for children and adults with Asperger's syndrome. *Behaviour Change*, *21*(3), 147–161.

Birmaher, B., Brent, D. A., Chiappetta, L., Bridge, J., Monga, S., & Baugher, M. (1999). Psychometric properties of the screen for child anxiety related emotional disorders (SCARED): A replication study. *Journal of the American Academy of Child & Adolescent Psychiatry*, *38*(10), 1230–1236.

Bradley, R., Henderson, K., & Monfore, D. A. M. (2004). A national perspective on children with emotional disorders. *Behavioral Disorders*, *29*(3), 211–223.

Brereton, A. V., Tonge, B. J., & Einfeld, S. L. (2006). Psychopathology in children and adolescents with autism compared to young people with intellectual disability. *Journal of Autism and Developmental Disorders*, *36*(7), 863–870.

Cai, R. Y., Richdale, A. L., Uljarević, M., Dissanayake, C., & Samson, A. C. (2018). Emotion regulation in autism spectrum disorder: Where we are and where we need to go. *Autism Research*, *11*(7), 962–978.

Chorpita, B. F., Taylor, A. A., Francis, S. E., Moffitt, C., & Austin, A. A. (2004). Efficacy of modular cognitive behavior therapy for childhood anxiety disorders. *Behavior Therapy*, *35*(2), 263–287.

Constantino, J. N., & Gruber, C. P. (2012). *Social responsiveness scale—second edition (SRS-2)*. Western Psychological Services.

den Houting, J., Adams, D., Roberts, J., & Keen, D. (2018). Exploring anxiety symptomatology in school-aged autistic children using an autism-specific assessment. *Research in Autism Spectrum Disorders*, *50*(1), 73–82.

Drahota, A., Wood, J. J., Sze, K. M., & Van Dyke, M. (2011). Effects of cognitive behavioral therapy on daily living skills in children with high-functioning autism and concurrent anxiety disorders. *Journal of Autism and Developmental Disorders*, *41*(3), 257–265.

Ehrenreich-May, J., Storch, E. A., Queen, A. H., Hernandez Rodriguez, J., Ghilain, C. S., Alessandri, M., et al. (2014). An open trial of cognitive-behavioral therapy for anxiety disorders in adolescents with autism spectrum disorders. *Focus on Autism and Other Developmental Disabilities*, *29*(3), 145–155.

García-Escalera, J., Chorot, P., Valiente, R. M., Reales, J. M., & Sandín, B. (2016). Efficacy of transdiagnostic cognitive-behavioral therapy for anxiety and depression in adults, children and adolescents: A meta-analysis. *Revista de Psicopatología y Psicología Clínica*, *21*(3), 147–175.

Ghandour, R. M., Jones, J. R., Lebrun-Harris, L. A., Minnaert, J., Blumberg, S. J., Fields, J., et al. (2018). The design and implementation of the 2016 National Survey of Children's Health. *Maternal and Child Health Journal*, *22*(8), 1093–1102.

Gross, J. J. (2008). Emotion regulation. In *Vol. 3. Handbook of emotions* (pp. 497–513). New York, NY: The Guilford Press.

Hassenfeldt, T. A., Lorenzi, J., & Scarpa, A. (2015). A review of parent training in child interventions: Applications to cognitive-behavioral therapy for children with high-functioning autism. *Review Journal of Autism and Developmental Disorders*, *2*(1), 79–90.

Hepburn, S. L., Blakeley-Smith, A., Wolff, B., & Reaven, J. A. (2016). Telehealth delivery of cognitive-behavioral intervention to youth with autism spectrum disorder and anxiety: A pilot study. *Autism*, *20*(2), 207–218.

Hofmann, S. G., Asnaani, A., Vonk, I. J., Sawyer, A. T., & Fang, A. (2012). The efficacy of cognitive behavioral therapy: A review of meta-analyses. *Cognitive Therapy and Research*, *36*(5), 427–440.

Howlin, P., & Magiati, I. (2017). Autism spectrum disorder: Outcomes in adulthood. *Current Opinion in Psychiatry*, *30*(2), 69–76.

Kanner, L. (1943). Autistic disturbances of affective contact. *Nervous Child*, *2*(3), 217–250.

Keen, D., Adams, D., Simpson, K., den Houting, J., & Roberts, J. (2019). Anxiety-related symptomatology in young children on the autism spectrum. *Autism*, *23*(2), 350–358.

Kendall, P. C., & Hedtke, K. A. (2006). *Cognitive-behavioral therapy for anxious children: Therapist manual*. Workbook Publishing.

Kerns, C. M., & Kendall, P. C. (2012). The presentation and classification of anxiety in autism spectrum disorder. *Clinical Psychology: Science and Practice*, *19*(4), 323.

Kerns, C. M., Kendall, P. C., Berry, L., Souders, M. C., Franklin, M. E., Schultz, R. T., et al. (2014). Traditional and atypical presentations of anxiety in youth with autism spectrum disorder. *Journal of Autism and Developmental Disorders*, *44*(11), 2851–2861.

Koegel, R. L., & Koegel, L. K. (2006). *Pivotal response treatments for autism: Communication, social, & academic development*. Paul H Brookes Publishing.

Laugeson, E. A., Frankel, F., Gantman, A., Dillon, A. R., & Mogil, C. (2012). Evidence-based social skills training for adolescents with autism spectrum disorders: The UCLA PEERS® program. *Journal of Autism and Developmental Disorders*, *42*(6), 1025–1036.

LeCavalier, L., Wood, J. J., Halladay, A. K., Jones, N. E., Aman, M. G., Cook, E. H., et al. (2015). Measuring anxiety as a treatment endpoint in youth with autism spectrum disorder. *Journal of Autism and Developmental Disorders*, *44*(5), 1128–1143.

MacNeil, B. M., Lopes, V. A., & Minnes, P. M. (2009). Anxiety in children and adolescents with autism spectrum disorders. *Research in Autism Spectrum Disorders*, *3*(1), 1–21.

Mason, J., & Scior, K. (2004). 'Diagnostic overshadowing' amongst clinicians working with people with intellectual disabilities in the UK. *Journal of Applied Research in Intellectual Disabilities*, *17*(2), 85–90.

Mazefsky, C. A., Herrington, J., Siegel, M., Scarpa, A., Maddox, B. B., Scahill, L., et al. (2013). The role of emotion regulation in autism spectrum disorder. *Journal of the American Academy of Child & Adolescent Psychiatry*, *52*(7), 679–688.

Mazefsky, C. A., Yu, L., White, S. W., Siegel, M., & Pilkonis, P. A. (2018). The emotion dysregulation inventory: Psychometric properties and item response theory calibration in an autism spectrum disorder sample. *Autism Research*, *11*(6), 928–941.

McConachie, H., McLaughlin, E., Grahame, V., Taylor, H., Honey, E., Tavernor, L., et al. (2014). Group therapy for anxiety in children with autism spectrum disorder. *Autism*, *18*(6), 723–732.

McNally Keehn, R. H., Lincoln, A. J., Brown, M. Z., & Chavira, D. A. (2013). The Coping Cat program for children with anxiety and autism spectrum disorder: A pilot randomized controlled trial. *Journal of Autism and Developmental Disorders*, *43*(1), 57–67.

Perihan, C., Burke, M., Bowman-Perrott, L., Bicer, A., Gallup, J., Thompson, J., et al. (2020). Effects of cognitive behavioral therapy for reducing anxiety in children with high functioning ASD: A systematic review and meta-analysis. *Journal of Autism and Developmental Disorders*, *50*(6), 1958–1972.

Pickard, K., Meza, R., Drahota, A., & Brikho, B. (2018). They're doing what? A brief paper on service use and attitudes in ASD community-based agencies. *Journal of Mental Health Research in Intellectual Disabilities*, *11*(2), 111–123.

Puleo, C. M., & Kendall, P. C. (2011). Anxiety disorders in typically developing youth: Autism spectrum symptoms as a predictor of cognitive-behavioral treatment. *Journal of Autism and Developmental Disorders*, *41*(3), 275–286.

Rapee, R. M., Wignall, A., Hudson, J. L., & Schniering, C. A. (2000). *Treating anxious children and adolescents: An evidence-based approach*. New Harbinger Publications.

Reaven, J., Blakeley-Smith, A., Culhane-Shelburne, K., & Hepburn, S. (2012). Group cognitive behavior therapy for children with high-functioning autism spectrum disorders and anxiety: A randomized trial. *Journal of Child Psychology and Psychiatry*, *53*(4), 410–419.

Reaven, J., Blakeley-Smith, A., Nichols, S., & Hepburn, S. (2011). *Facing your fears: Group therapy for managing anxiety in children with high-functioning autism spectrum disorders*. Paul Brookes Publishing.

Reaven, J. A., Blakeley-Smith, A., Nichols, S., Dasari, M., Flanigan, E., & Hepburn, S. (2009). Cognitive-behavioral group treatment for anxiety symptoms in children with high-functioning autism spectrum disorders: A pilot study. *Focus on Autism and Other Developmental Disabilities*, *24*(1), 27–37.

Rosen, T. E., Mazefsky, C. A., Vasa, R. A., & Lerner, M. D. (2018). Co-occurring psychiatric conditions in autism spectrum disorder. *International Review of Psychiatry*, *30*(1), 40–61.

Samson, A. C., Phillips, J. M., Parker, K. J., Shah, S., Gross, J. J., & Hardan, A. Y. (2014). Emotion dysregulation and the core features of autism spectrum disorder. *Journal of Autism and Developmental Disorders*, *44*(7), 1766–1772.

Simonoff, E., Pickles, A., Charman, T., Chandler, S., Loucas, T., & Baird, G. (2008). Psychiatric disorders in children with autism spectrum disorders: Prevalence, comorbidity, and associated factors in a population-derived sample. *Journal of the American Academy of Child & Adolescent Psychiatry*, *47*(8), 921–929.

Simpson, K., Adams, D., Alston-Knox, C., Heussler, H. S., & Keen, D. (2019). Exploring the sensory profiles of children on the autism spectrum using the short sensory profile-2 (SSP-2). *Journal of Autism and Developmental Disorders*, *49*(5), 2069–2079.

Smith, I. C., Ollendick, T. H., & White, S. W. (2019). Anxiety moderates the influence of ASD severity on quality of life in adults with ASD. *Research in Autism Spectrum Disorders*, *62*(1), 39–47.

Sofronoff, K., Attwood, T., & Hinton, S. (2005). A randomised controlled trial of a CBT intervention for anxiety in children with Asperger syndrome. *Journal of Child Psychology and Psychiatry*, *46*(11), 1152–1160.

Southman-Gerow, M. A., & Prinstein, M. J. (2014). Evidence base updates: The evolution of psycholoigcal treatments for children and adolescents. *Journal of Clinical Child & Adolescent Psychology*, *43*(1), 1–6.

Spain, D., & Happé, F. (2020). How to optimise cognitive behaviour therapy (CBT) for people with autism spectrum disorders (ASD): A Delphi study. *Journal of Rational-Emotive & Cognitive-Behavior Therapy*, *38*(2), 184–208.

Storch, E. A., Arnold, E. B., Lewin, A. B., Nadeau, J. M., Jones, A. M., De Nadai, A. S., et al. (2013). The effect of cognitive-behavioral therapy versus treatment as usual for anxiety in children with autism spectrum disorders: A randomized, controlled trial. *Journal of the American Academy of Child & Adolescent Psychiatry*, *52*(2), 132–142.

Storch, E. A., Larson, M. J., Ehrenreich-May, J., Arnold, E. B., Jones, A. M., Renno, P., et al. (2012). Peer victimization in youth with autism spectrum disorders and co-occurring anxiety: Relations with psychopathology and loneliness. *Journal of Developmental and Physical Disabilities*, *24*(6), 575–590.

Storch, E. A., Lewin, A. B., Collier, A. B., Arnold, E., De Nadai, A. S., Dane, B. F., et al. (2015). A randomized controlled trial of cognitive-behavioral therapy versus treatment as usual for adolescents with autism spectrum disorders and comorbid anxiety. *Depression and Anxiety*, *32*(3), 174–181.

Stuart, E. A., McGinty, E. E., Kalb, L., Huskamp, H. A., Busch, S. H., Gibson, T. B., et al. (2017). Increased service use among children with autism spectrum disorder associated with mental health parity law. *Health Affairs*, *36*(2), 337–345.

White, S. W., Mazefsky, C. A., Dichter, G. S., Chiu, P. H., Richey, J. A., & Ollendick, T. H. (2014). Social-cognitive, physiological, and neural mechanisms underlying emotion regulation impairments: Understanding anxiety in autism spectrum disorder. *International Journal of Developmental Neuroscience*, *39*(1), 22–36.

Wood, J. J., Drahota, A., Sze, K., Har, K., Chiu, A., & Langer, D. A. (2009). Cognitive behavioral therapy for anxiety in children with autism spectrum disorders: A randomized, controlled trial. *Journal of Child Psychology and Psychiatry*, *50*(3), 224–234.

Wood, J. J., Ehrenreich-May, J., Alessandri, M., Fujii, C., Renno, P., Laugeson, E., et al. (2015). Cognitive behavioral therapy for early adolescents with autism spectrum disorders and clinical anxiety: A randomized, controlled trial. *Behavior Therapy*, *46*(1), 7–19.

Wood, J. J., Fujii, C., Renno, P., & Van Dyke, M. (2014). Impact of cognitive behavioral therapy on observed autism symptom severity during school recess: A preliminary randomized, controlled trial. *Journal of Autism and Developmental Disorders*, *44*(9), 2264–2276.

Wood, J. J., & Gadow, K. D. (2010). Exploring the nature and function of anxiety in youth with autism spectrum disorders. *Clinical Psychology: Research and Practice*, *17*(4), 281–292.

Wood, J. J., Kendall, P. C., Wood, K. S., Kerns, C. M., Seltzer, M., Small, B. J., et al. (2020). Cognitive behavioral treatments for anxiety in children with autism spectrum disorder: A randomized clinical trial. *JAMA Psychiatry*, *77*(5), 474–483.

Wood, J. J., Sze Wood, K., Chuen Cho, A., Rosenau, K. A., Cornejo Guevara, M., Galán, C., et al. (2021). Modular cognitive behavioral therapy for autism-related symptoms in children: A randomized controlled trial. *Journal of Consulting and Clinical Psychology*, *89*(2), 110.

Zablotsky, B., Pringle, B. A., Colpe, L. J., Kogan, M. D., Rice, C., & Blumberg, S. J. (2015). Service and treatment use among children diagnosed with autism spectrum disorders. *Journal of Developmental and Behavioral Pediatrics*, *36*(2), 98–105.

Chapter 17

The use of dialectical behavior therapy in childhood and adolescent eating disorders

Rebecca C. Kamody[a,b,c] and E. Thomaseo Burton[d,e]

[a]*Child Study Center, Yale School of Medicine, New Haven, CT, United States,* [b]*Department of Pediatrics, Yale School of Medicine, New Haven, CT, United States,* [c]*Department of Psychiatry, Yale School of Medicine, New Haven, CT, United States,* [d]*Department of Pediatrics, College of Medicine, University of Tennessee Health Science Center, Memphis, TN, United States,* [e]*Children's Foundation Research Institute, Le Bonheur Children's Hospital, Memphis, TN, United States*

Abbreviations

AN	anorexia nervosa
ARIFD	avoidant/restrictive food intake disorder
BED	binge-eating disorder
BN	bulimia nervosa
DBT	dialectical behavior therapy
DBT-A	dialectical behavior therapy for adolescents
FBT	family-based treatment
LOC eating	loss-of-control eating
OBE	objective binge episode
OSFED	other specified feeding and eating disorders
RO DBT	radically open dialectical behavior therapy

Introduction

Eating disorders encompass a group of serious psychiatric disorders characterized by persistent and abnormal eating patterns. As a diagnostic category of the *DSM-5*, feeding and eating disorders share significant overlap regarding eating behavior, but differ in a number of important ways (American Psychiatric Association, 2013; World Health Organization, 2019). Presentations can range from intense focus on body shape and restriction of nutritional intake, as seen in anorexia nervosa (AN), to discrete episodes of eating large quantities of food with an accompanying sense of loss-of-control, as seen in binge-eating disorder (BED), and with subsequent compensatory behaviors as seen in bulimia nervosa (BN). Regardless of specific diagnosis, eating disorders impair physical, social, and psychological functioning.

The prevalence of eating disorders has risen over the years and the global burden of disordered eating behaviors represents significant public health concern (Santomauro et al., 2021). Individuals who engage in disordered eating are more likely to experience medical complications (e.g., malnutrition, gastrointestinal dysfunction, cardiometabolic disease) as well as psychiatric comorbidities (e.g., anxiety, depression, substance use/abuse, and suicidality; (Treasure, Duarte, & Schmidt, 2020). While at the time of writing of this chapter, the long-term effects of the coronavirus disease 2019 (COVID-19) on individuals with eating disorders is unknown, the greater susceptibility to illness, psychological distress associated with uncertainty and social isolation, and decreased access to medical and mental health care has the potential to exacerbate prevalence and severity of disordered eating behaviors (Fernández-Aranda et al., 2020), including among youth (Graell et al., 2020). Further, the increased use of social network sites and video streaming platforms during the pandemic lockdown has been shown to result in increased drive of thinness, body image concerns, and disordered eating behavior risk (Vall-Roqué, Andrés, & Saldaña, 2021). As such, it is more critical than ever to identify effective approaches to targeting disordered eating during the childhood and adolescent developmental stages to prevent further escalation into protracted presentations.

Handbook of Lifespan Cognitive Behavioral Therapy. https://doi.org/10.1016/B978-0-323-85757-4.00039-0

While eating disorders affect individuals of all ages, childhood and adolescence is a particularly sensitive period for onset (Favaro, Caregaro, Tenconi, Bosello, & Santonastaso, 2009). Eating behaviors developed in childhood track into adulthood, and researchers have uncovered distinct trajectories that may predict later risk for disordered eating (Herle et al., 2020). Although youth may experience clinically significant distress related to disordered eating behaviors, they often do not meet frequency or duration criteria (Mitchison et al., 2020). Eating disorder nosology has expanded to classify such cases as *other specified feeding and eating disorders* (OSFED), which subsumes atypical AN, subthreshold BN, subthreshold BED, purging disorder, and night eating syndrome (American Psychiatric Association, 2013). Individuals engaging in subthreshold disordered eating behaviors are still susceptible to physical, psychological, and social sequelae associated with a diagnosis of an eating disorder (Micali et al., 2015).

Treatment for eating disorders in youth

Only about 20% of individuals with eating disorders seek treatment, and the numbers are lower for youth with subthreshold presentations (Forrest, Smith, & Swanson, 2017; Micali et al., 2015). Such findings highlight cultural differences that may influence clinical presentation, and emphasize the importance of efficacious and evidence-based prevention and intervention efforts for populations not meeting standard diagnostic thresholds. Current evidence supports the use of psychological modalities such as cognitive behavioral therapy (CBT), interpersonal therapy (IPT), and family-based treatment (FBT) for treatment of eating disorders among children and adolescents (Hay, 2020). Despite this accrual of evidence, however, the majority of youth engaging in these treatments fail to reach symptom remission, and maintenance is low among those who do (Wilson, Grilo, & Vitousek, 2007). Dialectical behavior therapy (DBT; Linehan, 1993a, 1993b) which was originally developed as a treatment for adults diagnosed with borderline personality disorder, has shown significant promise as an alternate treatment option for youth exhibiting disordered eating behaviors (Reilly et al., 2020). Traditional DBT encompasses 4 modules, shown in Table 1.

DBT has been adapted to treat youth with transdiagnostic emotional dysregulation (Miller, Rathus, & Linehan, 2006). As a comprehensive evidenced-based practice, DBT for adolescents (DBT-A) includes numerous therapeutic aspects, including individual and family therapy, telephone coaching, consult team for providers, and DBT skills training groups (Miller et al., 2006). Included in the DBT-A adaptations to skills includes the additional skill module of "Walking the Middle Path," which focuses on the parent-child relationship in treatment and addressing dialectical dilemmas inevitably encountered (see Table 2; Rathus, Campbell, Miller, & Smith, 2015).

TABLE 1 Four core skills of DBT.

Skill module	Focus in traditional DBT	Application to eating disorder treatment
Mindfulness	Learning to practice of nonjudgmental awareness and acceptance of present thoughts, feelings, and experiences	Application of nonjudgmental observation to foods, body image, and eating practices
Distress tolerance	Learning to manage difficult or impossible to change situations without making things worse	Use of "crisis" coping skills to resist urge to engage in disordered eating behavior when emotionally overwhelmed
Emotion regulation	Learning to change emotions in order to cope with the current situation	Regulating emotions and reducing emotional vulnerability to reduce urges to use disordered eating as means of coping
Interpersonal effectiveness	Learning to ask for what you want and need from others while preserving self-respect and the relationship	Learning to effectively navigate interpersonal stressors to prevent urge to engage in disordered eating triggered by maladaptive interpersonal interactions

TABLE 2 Walking the middle path skill module.

Skill module	Focus in traditional DBT-A	Application to eating disorder treatment
Walking the middle path	Targets skill deficits in the parent-teen relationship and within families with emphasis on validation	Emphasis on family involvement in child's eating disorder treatment; balancing dialectic of empowering parental support and adolescent autonomy with making change

The skills of DBT-A are generalizable to a number of adolescent psychiatric presentations, including AN, BN, and BED (Fischer & Peterson, 2015; Kamody, Thurston, Pluhar, Han, & Burton, 2019; Salbach-Andrae, Bohnekamp, Pfeiffer, Lehmkuhl, & Miller, 2008). Despite evidence highlighting the effectiveness of DBT-A for youth with emotional dysregulation and DBT for adults with eating disorders, the use of DBT in youth with eating disorders is a relatively emerging area of study. Of note, the American Academy of Child and Adolescent Psychiatry has identified the use of DBT to treat children and adolescents with eating disorders as a psychosocial intervention that necessitates future study (Lock & La Via, 2015). Though this call was published more than 5 years ago, further study continues to be needed.

Theoretical model for DBT for loss-of-control eating

Rooted in the affect regulation model, DBT was created to build adaptive skills in individuals who engage in maladaptive coping strategies, and has been adapted for dysregulated eating (Safer, Telch, & Chen, 2009). For example, a young person may engage in binge-eating behaviors in response to interpersonal challenges or restrict their eating to cope with traumatic experiences. Both binge-eating specifically (that may occur in the context of BED, BN, and AN-binge purge type), and loss-of-control (LOC) eating more generally, can be conceptualized using the affect regulation model of DBT (see Fig. 1; Safer et al., 2009). According to this model, bingeing and LOC eating occur as an attempt at regulating one's emotions (Safer et al., 2009; Safer, Couturier, & Lock, 2007). Research has found negative affect and emotion regulation difficulties are related to binge eating (Dingemans, Danner, & Parks, 2017), and difficulties identifying emotions and inadequate emotion regulation strategies found to strengthen the relation between emotion regulation difficulties and binge eating (Whiteside et al., 2007). A review of the literature suggests evidence that binge eating is triggered by negative emotions, and using food to cope can provide short-term relief (Leehr et al., 2015). Though eating may provide immediate relief from negative emotions, research suggests that after engaging in overeating, negative affect actually increases (Haedt-Matt & Keel, 2011),

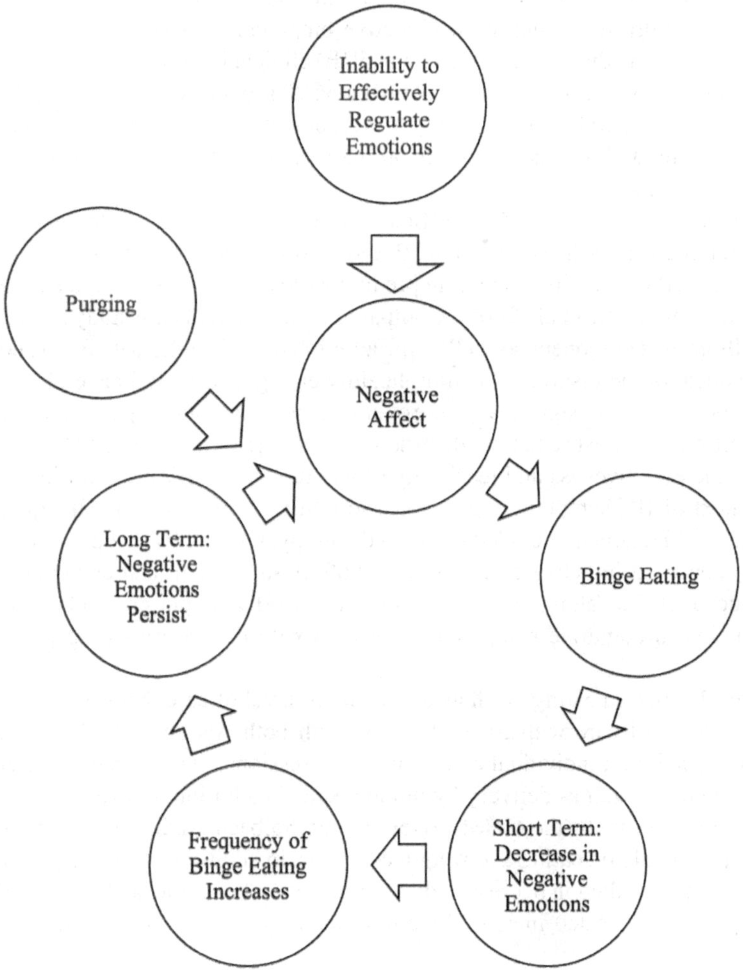

FIG. 1 Application of DBT Affect Regulation Model to Binge Eating (with application to binge/purge cycle).

thus perpetuating the cycle of dysregulated emotions leading to disordered eating. This theoretical understanding of LOC eating supports the affect regulation model to explain LOC disordered eating, and highlights the particular relevance of DBT in the management of BED, BN, and OSFED presentations.

Studies evaluating use of DBT for disordered eating in youth

Elements and mechanisms of change within DBT (Lynch, Chapman, Rosenthal, Kuo, & Linehan, 2006) can be integrated into eating disorder treatments in a number of ways, all with the goal of increasing youth's ability to regulate their affect in adaptive ways to replace disordered eating. In a recent review of the literature Reilly and colleagues distinguished between the use of DBT as a primary treatment, versus the use of DBT skills adjunctive to another eating disorder treatment protocol (Reilly et al., 2020). Of the studies reviewed, the authors distinguished between those that utilize a more comprehensive DBT model (individual sessions, skills training, family sessions, etc.) and those that includes DBT skills only modalities. For the sake of this chapter, both approaches will be acknowledged, in addition to those that integrate DBT skills with other evidence-based modalities for eating disorders.

Studies utilizing DBT as the primary treatment include AN, BN, BED, and OSFED presentations. It is important to note that diagnostic terminology related to eating disorders was updated in *DSM-5*, and research published before 2013 used the *eating disorder not otherwise specified* (EDNOS) diagnosis to encompass eating disorder presentations not meeting full diagnostic criteria for AN or BN. For the sake of consistency in the present chapter, we instead use the most up-to-date diagnostic term of OSFED.

The first case study to document the application of traditional DBT to treat a youth with a disordered eating presentation included an adolescent with a diagnosis of BED (Safer et al., 2007). The adolescent worked with an individual provider on an outpatient basis, and attended 21 weeks of outpatient individual therapy that focused on learning DBT skills (see Table 1) in individual sessions (as opposed to group), and included the review of a DBT diary card, phone coaching, and family sessions. This individual outpatient approach resulted in drastic reduction of objective binge episodes (OBEs), as well as decreases in disordered eating based on the structured eating disorder examination (EDE) clinical interview, which were maintained at follow up. At baseline, the subject of this case study had reported 22 OBEs on 20 separate days. At 3 months posttreatment, the client reported only one OBE (Safer et al., 2007). In this study, developmental adaptations for adolescence included incorporating family sessions in treatment, and included sessions on interpersonal effectiveness, particularly related to effectively achieving objectives with parents.

Similarly at the outpatient level of a care, one case series examined the efficacy of a 25-week DBT treatment that included both individual therapy and group skills training (Salbach-Andrae et al., 2008). Twelve adolescent females with AN and BN participated, and the use of this outpatient DBT modality resulted in pre to posttreatment reductions in eating disorder related behaviors and psychopathology (Salbach-Andrae et al., 2008). Compared to the original case study above, this adaption of traditional DBT-A included the skills group component, as well as the addition of a "Dealing with Food and Body Image" skills module, that included psychoeducation on disordered eating, healthy eating, and body image disturbances. This protocol resulted in high retention (over 90%), with significant improvements to functional impairment and disordered eating pathology based on a validated self-report measure (eating disorder-inventory [EDI], second edition). In another a pilot study at the outpatient level of care, the effectiveness and feasibility of an outpatient DBT program for 10 adolescent females endorsing bingeing (in the context of BN symptom expression and OSFED), nonsuicidal self-injury (NSSI), and suicidal ideation (Fischer & Peterson, 2015). Treatment included individual therapy, DBT skills group training, a DBT therapist consultation group, and access to coaching in between sessions. At both posttreatment and at 6-month follow-up, participants were found to have reduced OBEs, along with reductions in NSSI and purging behaviors (Fischer & Peterson, 2015), in addition to a reduction in cognitive symptoms associated with disordered eating based on the EDE.

Comprehensive DBT at higher levels of care for disordered eating, such as the inpatient level of care, have also been evaluated. Implementation of DBT at the inpatient level of care with 31 adolescents with both AN and BN (Salbach, Klinkowski, Pfeiffer, Lehmkuhl, & Korte, 2007) included both individual and group DBT sessions, that included shorter duration and increased frequency of the skill curriculum, as well as delivery by nursing staff in addition to mental health providers. Similar to the case series at the outpatient level of care by the same lead researcher (Salbach-Andrae et al., 2008), this inpatient adaptation included a group focused specifically on eating and weight concerns, in addition to standard DBT skills sessions. The treatment at the inpatient level of care resulted in decreased disordered eating behaviors and decreased use of maladaptive weight loss strategies. Findings also suggested increased weight among youth presenting with AN (Salbach et al., 2007).

As noted above, in addition to studies examining implementation of more comprehensive DBT modalities, there has also been exploration of DBT skills only and adjunctive DBT skills interventions. With regard to skills only interventions, interventions may be adapted to use the skills only approach as preventative approaches to subthreshold disordered eating presentations. In this realm, DBT skills adapted interventions similarly demonstrate promise. In one open pilot trial for youth presenting with binge eating and emotional overeating behaviors (Kamody et al., 2019), DBT skills were adapted and streamlined to be presented in a culturally responsive manner to underserved and underrepresented youth in the Mid-south United States (see Pluhar, Kamody, Sanchez, Thurston, & Burton, 2018) for detailed intervention description). Of the 15-youth completing the condensed 10-week DBT skills group intervention, there was significant reductions in both self and parent report of emotional overeating behaviors (i.e., eating in response to emotional triggers) and in OBEs. Examination of qualitative data from exit interviews suggested acceptance-based skills (e.g., distress tolerance skill module) to be most helpful to youth with subthreshold binge eating in this culturally-responsive DBT skills only intervention (Kamody, Thurston, & Burton, 2020). Post hoc analyses of the pilot trial data suggested that youths' health locus of control and perception of their physical appearance impacted both initial engagement and outcomes of the intervention (Kamody, Thurston, & Burton, 2021). The importance of self-perception and body image found in this skills only intervention further supports the benefits of the incorporation of psychoeducation modules on food and body image integrated into the more comprehensive DBT treatments described above (Salbach et al., 2007; Salbach-Andrae et al., 2008).

In addition to work in DBT skills only intervention, there is the also the field of research integrating DBT skills into other comprehensive eating disorder treatment modalities. These have most notably been as adjunctive intervention to FBT (Lock & le Grange, 2005), and have been applied for youth with AN, BN, and OSFED. The rationale for this adjunctive integration is often to increase individual skill knowledge and self-regulation, while utilizing strong evidence-based approaches focused on parental and environmental supports (i.e., FBT). Thus, while optimizing and supporting environmental intervention to promote immediate gains and success in treatment, youth may also work toward building their self-regulation skills including both acceptance and change-based DBT coping skills.

One study evaluated the integration of DBT skills with FBT at the outpatient level of care, and suggests promise (Peterson, Van Diest, Mara, & Matthews, 2019). Among 18 adolescents enrolled in an FBT for AN, atypical AN, or OSFED, this study integrated weekly DBT skills into outpatient treatment. After completing the 6-month treatment, youth demonstrated reduction in overall eating disorder pathology (based on self-report on the EDE questionnaire [EDE-Q]), as well as an increase in adaptive coping skills and techniques. Another study integrated the FBT approach at the intensive outpatient program (IOP) level of care with DBT skills training (Johnston, O'Gara, Koman, Baker, & Anderson, 2015). Within the treatment, youth and families participated in treatment 3–4 h per week for 7–8 weeks, which included more traditional FBT sessions, with inclusion of both adolescent and parent DBT skills group weekly. At the conclusion of treatment, youth demonstrated overall significant reduction in eating disorder pathology based on self-report measures (EDE-Q), as well as significant weight gain, with these gains maintained by close to two-thirds (64%) of youth at 1-year follow up (Johnston et al., 2015).

The integration of DBT skills adjunctive to FBT has also been evaluated at the partial hospitalization level of care for youth with BN (Murray et al., 2015). Within this study, treatment included the integration of individual, family, parent-only, and multifamily components. The integration of these evidence-based approaches included utilizing FBT-styled family sessions, with multifamily DBT skills groups (in which all involved family members learned skills), 6 days a week (3–10 h per day) for an average of approximately 77 days. Based on self-report assessments (EDE-Q), youth evidenced significant reductions in overall eating disorder symptoms, as well as BN specific symptoms (including reduction in binge episodes and self-induced vomiting). In a second study at the PHP level of care (Pennell, Webb, Agar, Federici, & Couturier, 2019), implementation of a DBT-informed program for youth with a range of disordered eating presentations (including AN-restricting type, AN-binge/purge type, BN, ARFID, and OSFED) integrating FBT was examined through retrospective hospital record review. Treatment occurred 5 days a week for 7–10 h, which included DBT skills lessons daily, along with family meals following an FBT format. Based on retrospective medical record review, youth engaged in this DBT skills-informed treatment experienced significant increase in weight at discharge, and reduction in restricting and binge-purge behaviors. In addition, nearly 80% remained out of the hospital without readmission at the follow up 2 years later (Pennell et al., 2019).

Theoretical model for DBT for over-control eating

Another application of DBT for eating disorders is the development and adaptation of radically open dialectical behavior therapy (RO DBT; Lynch et al., 2013). As compared to traditional DBT which focuses on targeting conditions of emotion dysregulation (e.g., binge eating, LOC eating, bingeing/purging), RO DBT was developed to conversely focus on

TABLE 3 Three core features of RO DBT promoting psychological health.

	Feature of psychological health in RO DBT	Application to eating disorder treatment
Receptivity and openness	Being open to new experiences and challenging automatic reaction to feedback to promote learning	Increasing comfort with new experiences with others; increasing comfort with less control of eating
Flexible control	Learning to flexibly adapt to changes in the environment	Increasing comfort with more intuitive/less restrictive eating (i.e., increasing comfort with less control of intake)
Intimacy and social-connectedness	Survival requires connecting with others in order to form bonds	Social connectedness decreases emotional loneliness, reducing strength of restriction as way of coping

maladaptive patterns and disorders of over-control, including disordered eating associated with restriction, such as AN (Lynch et al., 2013). Similar to traditional DBT, RO DBT is a comprehensive evidence-based treatment, that focuses on three key features (see Table 3).

RO DBT is based on a foundational assumption that emotional expression is needed to help individuals communicate and connect with others (Lynch, Hempel, & Dunkley, 2015). As such, the theory behind the treatment suggests that psychological experiences, including cognition, emotions, and behaviors, are social signals to others to communicate. Therefore, those with maladaptive over-control may struggle to accurately express their emotions, and subsequently have barriers in connecting with others (Lynch et al., 2015). This leads to experiences of emotional loneliness, leading to feelings of being unsafe and using maladaptive behaviors to cope (e.g., restriction). For individuals with maladaptive over-control eating, their restrictive eating perpetuates a cycle of distancing themselves from others, and increasing feelings of emotional loneliness (see Fig. 2; Lynch et al., 2013).

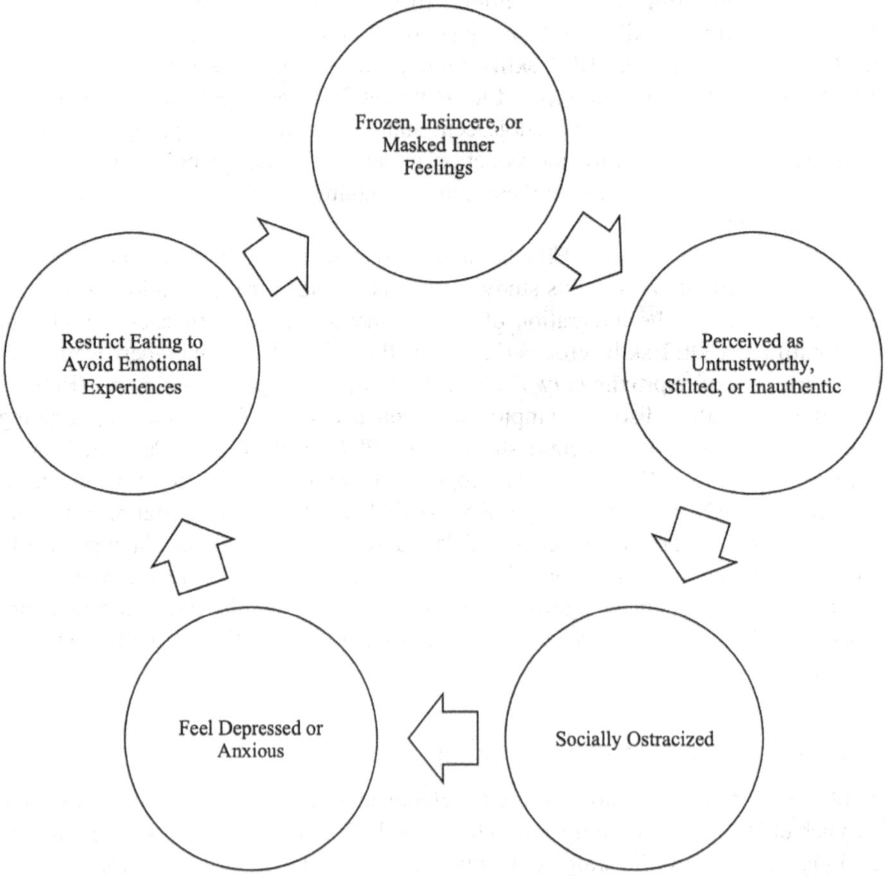

FIG. 2 Application of RO DBT inhibited expression model for restrictive eating.

Studies evaluating use of RO DBT for disordered eating

When originally piloted in adults with AN, comprehensive RO DBT was found to result in significant improvements in weight gain, decreases in disordered eating, and decreases in eating disorder psychopathology (Lynch et al., 2013). Only recently, however, has RO DBT been piloted in a singular study with adolescents (Baudinet et al., 2020). In the pilot study, 131 adolescents presenting with restrictive disordered eating presentations at an intensive eating disorder day treatment program received 2.5 h of RO DBT treatment incorporated in two sessions each week for an average of 13.4 weeks (as part of their intensive day treatment program). Among youth within the program, eating disorder symptoms were significantly negatively associated with social connectedness and positively associated with cognitive inflexibility (consistent with theoretical underpinnings of over-control model of RO DBT). After the course of RO DBT treatment, significant improvements were found in drive for thinness (based on the EDI, third edition), social connectedness, and emotional expressiveness (Baudinet et al., 2020).

Bringing together the DBT for eating disorders in youth research

Taken together, though limited, the studies detailed above suggest that both comprehensive DBT modalities for youth, as well as DBT skills only and DBT skills adjunctive interventions evidence promise as an approach to treating youth with eating disorders. In total, only 10 intervention studies have evaluated and been presented here (with additional publications based on the same line of research; e.g., Kamody et al., 2020, 2021). Four of these studies examined more comprehensive traditional DBT applications for youth with disordered eating (Fischer & Peterson, 2015; Safer et al., 2007; Salbach et al., 2007; Salbach-Andrae et al., 2008), one evaluated more comprehensive RO DBT applications in youth with restrictive eating (Baudinet et al., 2020), one examined DBT skills only for subthreshold disordered eating (Kamody et al., 2019), and four examined the use of DBT skills as adjunctive to another gold-standard eating disorder treatment (FBT) to increase adaptive coping and self-regulation (Johnston et al., 2015; Murray et al., 2015; Pennell et al., 2019; Peterson et al., 2019). This limited research encompasses a wide range of heterogeneous disordered eating presentations at varying levels of severity and impairment (ranging from restrictive to dysregulated presentations). Across these presentations, DBT suggests promise; however, much more rigorous and extensive research is needed. The limited research, and lack of controlled clinical trials, limits the conclusions that can be made about the treatment's efficacy. Given the promise of this treatment to both prevent and treat emerging and full-threshold eating disorder presentations, however there are a number of directions for the research to pursue (see Table 4).

Practice and procedures

With regard to evaluation of treatment studies, both qualitative and quantitative approaches have been used to assess measures of acceptability, feasibility, and effectiveness of DBT interventions for youth with eating disorders. Though limited in overall number of studies, as described in this chapter there have been a range of the types of studies involving naturalistic and "real world" populations with eating disorders, with the data collection and evidence thus far including case studies (Safer et al., 2007), case series (Salbach-Andrae et al., 2008), and pilot trials (Fischer & Peterson, 2015; Kamody et al., 2019). As such, qualitative data has primarily focused on interviews assessing patient and family perspectives on their experiences in DBT interventions (e.g., Kamody et al., 2020).

Of the studies examined in this chapter, the quantitative and self-report measures include both structured diagnostic interviews (specifically, the EDE; Cooper & Fairburn, 1987), as well as self-report questionnaires, including the EDE-questionnaire (Fairburn & Beglin, 1994), and the eating disorder inventory (Garner, Olmstead, & Polivy, 1983; both second and third editions (EDI-II, EDI-3)). Both the EDE-Q and EDI are validated assessments that use self-report scales to assess eating disorder pathology, as well as frequency data to gain insight into the frequency of occurrence of specific eating disorder-related behaviors (e.g., OBE, purging). Other developmentally adapted reports of disordered eating presentations used in DBT intervention studies include the emotional eating scale (EES) for children and adolescents, to assess use of eating as a way to soothe or regulate emotions (Tanofsky-Kraff et al., 2007).

Finally, in addition to these measures of eating disorder pathology and frequency of disordered eating behaviors through interview and questionnaire report, the promise and success of DBT interventions has thus far included review of hospital charts to gain additional information on weight gain and anthropometric data, along with data on feasibility through assessment of retention in treatment studies, as well as readmission to higher levels of care (e.g., inpatient hospitalizations; Pennell et al., 2019).

TABLE 4 Future directions for research in DBT in childhood and adolescent eating and feeding disorders.

Category	Research needs
Methodology of treatment studies	• Use of randomized controlled trials to examine efficacy of DBT for childhood and adolescent eating disorders compared to other treatment modalities • Examination of implementation of DBT for childhood and adolescent eating disorders at varying levels of care (e.g., inpatient, partial hospitalization, intensive outpatient, in-home, outpatient) through controlled trials • Examining developmental patterns and trajectories of recovery of youth engaged in the DBT for eating disorder treatment • Use of ecological momentary assessment to assess changes in disordered eating pathology among youth in addition to self and parent report • Examination of psychopharmacological treatments in conjunctive with DBT curricula to treat protracted disordered eating presentations
Prevention	• Assessment and prevention studies using tiered models to introduce DBT skills to manage disordered eating prior to progression to full-criteria eating disorder presentations • Identify risk and protective factors for child and adolescent outcomes in DBT treatment for eating disorders • Use of dissemination and implementation science framework to develop effective implementation in alternative settings for youth (e.g., primary care provider offices, schools, etc.)
Theory	• Examine the affect regulation model of disordered eating (see Fig. 1) through developmental lens to understand impact of different stages of development on the progression of disordered eating in youth • Examination of RO DBT inhibited expression model (see Fig. 2) through developmental lens to better understand impact on childhood and adolescent development on the RO DBT approach to restrictive eating
Correlates	• Identifying how other aspects of cultural identity (e.g., race, ethnicity, gender, socioeconomic status, etc.) impact outcomes in DBT interventions • Exploring role of aspects of cultural identity on engagement in DBT intervention through different modalities and differing levels of care

Mini-dictionary of terms

- Anorexia nervosa (AN): *DSM-5* diagnosis characterized by low body weight, intense fear of gaining weight, and disturbance in how body shape/weight are experienced (includes both restricting type and binge/purge type).
- Bulimia nervosa (BN): *DSM-5* diagnosis characterized by recurrent episodes of binge eating and recurrent associated compensatory behaviors (not when in the context of AN binge-purge type).
- Binge-eating disorder (BED): *DSM-5* diagnosis characterized recurrent episodes of binge eating without associated compensatory behaviors, that includes experience of loss of control while eating, and distress about binge eating.
- Dialectical behavior therapy (DBT): evidence-based treatment for affect dysregulation aimed at decreasing problem behaviors and replacing them with more skillful behaviors, such as mindfulness, distress tolerance, emotion regulation, and interpersonal effectiveness.
- Dialectical behavior therapy for Adolescents: evidence-based adaptation of DBT for adolescents that includes navigating parent-child dialectical dilemmas and conflict.
- Radically open dialectical behavior therapy (RO DBT): intervention developed from traditional DBT that is focused on treating psychiatric presentations indicative of maladaptive over control (e.g., AN), as opposed to emotional dysregulation.

Key facts

Key facts of eating disorders in children and adolescents

- Research suggests earlier average age of onset for disordered eating than previously believed, with onset in adolescence most common.

- Prior to development of full criteria eating disorders based on the *DSM-5*, youth often display subthreshold disordered eating earlier in development.
- Research suggests that early intervention and prevention of disordered eating and eating disorders can prevent development of more protracted presentations.

Key facts of DBT

- DBT is an evidence-based treatment with robust support for use in treatment for diagnostic presentations of emotional and behavioral dysregulation.
- Comprehensive DBT for adolescents includes multiple components, including individual sessions, family work, intersession coaching, therapist consult team, and DBT skills training via groups.
- DBT skills, which focus on change-based skills and acceptance-based skills (including mindfulness, distress tolerance, emotion regulation, and interpersonal effectiveness), hold promise of being effective in increasing more effective coping adjunctive to other evidence-based treatments.

Key facts of DBT for eating disorders

- Use of DBT as a treatment for eating disorders is based on the affect regulation model.
- DBT as an effective intervention modality for treating eating disorders demonstrates promise across disordered eating presentations, including those related to restriction (e.g., AN), and those indicative of dysregulation (e.g., BN, BED).
- DBT skills themselves may support prevention of disordered eating progression prior to presentations becoming more severe and protracted.

Key facts of DBT for childhood and adolescent eating disorders

- Limited research has examined comprehensive DBT as a treatment for eating disorders in youth.
- The research available suggests promise, and has been identified by the psychiatric community as warranting further study.
- Research suggests that in youth, DBT treatment for disordered eating may also serve to reduce other behaviors of concern through this same modality (e.g., self-harm).
- Both DBT skills only intervention and DBT skills adjunctive to FBT treatment suggest promise as treatment to support prevention and intervention of disordered eating in youth.

Application to other areas

DBT, an extension of CBT, was originally developed to aid individuals in replacing self-destructive behaviors with more adaptive coping strategies. In this chapter, we reviewed the burgeoning application of DBT in the treatment of eating disorders among children and adolescents. The majority of work in this realm has focused on intervention. However, considering that disordered eating patterns may present as early as 2 years of age (Herle et al., 2020), more work around prevention may be merited. Of course, the core skills of mindfulness, distress tolerance, emotion regulation, and interpersonal effectiveness may be abstract for young children, and the middle path skill module more applicable to adolescent-parent dialectical dilemmas. Nevertheless, considering the efficacy of FBT in treating eating disorders in youth (Lock & le Grange, 2005), and the promising findings related to adjunctive DBT skills to FBT treatment for adolescents reviewed in this chapter (Johnston et al., 2015; Murray et al., 2015; Pennell et al., 2019; Peterson et al., 2019), DBT may hold promise as an adjunctive therapy for younger children with emerging disordered eating presentations.

In FBT, parental caregivers are recruited to aid in treatment by taking control of their child's eating disorder. Gaining skills to live in the present moment, develop healthy ways to manage stress, regulate intense emotions, and engage with others in an effective way has strong potential to support caregivers as they support their child. This may include the incorporation of the DBT for children (DBT-C) curriculum, an emerging evidence-based developmental adaptation for dysregulated presentations in younger youth (Perepletchikova et al., 2017).

Summary points

- This chapter focuses on the use and DBT to treat children and adolescents with eating disorders and subthreshold disordered eating.
- DBT is an evidence-based therapeutic approach to treating presentations of transdiagnostic emotion regulation.
- DBT has been developmentally adapted for adolescents (DBT-A), and more recently for children (DBT-C).
- DBT has evidenced promise as both a primary treatment and DBT skills as an adjunctive treatment for the treatment and prevention of eating disorders.
- The research thus far, while promising, is limited. More extensive research, including the use of controlled trials, is needed to establish DBT as an efficacious treatment for eating disorders in youth.

References

American Psychiatric Association. (2013). *Diagnostic and statistical manual of mental disorders (DSM-5®)*. American Psychiatric Pub.

Baudinet, J., Simic, M., Griffiths, H., Donnelly, C., Stewart, C., & Goddard, E. (2020). Targeting maladaptive overcontrol with radically open dialectical behaviour therapy in a day programme for adolescents with restrictive eating disorders: An uncontrolled case series. *Journal of Eating Disorders, 8*(1), 68. https://doi.org/10.1186/s40337-020-00338-9.

Cooper, Z., & Fairburn, C. (1987). The eating disorder examination: A semi-structured interview for the assessment of the specific psychopathology of eating disorders. *International Journal of Eating Disorders, 6*(1), 1–8. https://doi.org/10.1002/1098-108X(198701)6:1<1::AID-EAT2260060102>3.0.CO;2-9.

Dingemans, A., Danner, U., & Parks, M. (2017). Emotion regulation in binge eating disorder: A review. *Nutrients, 9*(11), 1274. https://doi.org/10.3390/nu9111274.

Fairburn, C. G., & Beglin, S. J. (1994). Assessment of eating disorders: Interview or self-report questionnaire? *International Journal of Eating Disorders, 16*, 363–370. https://doi.org/10.1002/1098-108X(199412)16:4<363::AID-EAT2260160405>3.0.CO;2-#.

Favaro, A., Caregaro, L., Tenconi, E., Bosello, R., & Santonastaso, P. (2009). Time trends in age at onset of anorexia nervosa and bulimia nervosa. *The Journal of Clinical Psychiatry, 70*(12), 1715–1721.

Fernández-Aranda, F., Casas, M., Claes, L., Bryan, D. C., Favaro, A., Granero, R., et al. (2020). COVID-19 and implications for eating disorders. *European Eating Disorders Review, 28*(3), 239.

Fischer, S., & Peterson, C. (2015). Dialectical behavior therapy for adolescent binge eating, purging, suicidal behavior, and non-suicidal self-injury: A pilot study. *Psychotherapy, 52*, 78–92. https://doi.org/10.1037/a0036065.

Forrest, L. N., Smith, A. R., & Swanson, S. A. (2017). Characteristics of seeking treatment among US adolescents with eating disorders. *International Journal of Eating Disorders, 50*(7), 826–833.

Garner, D. M., Olmstead, M. P., & Polivy, J. (1983). Development and validation of a multidimensional eating disorder inventory for anorexia nervosa and bulimia. *International Journal of Eating Disorders, 2*(2), 15–34. https://doi.org/10.1002/1098-108X(198321)2:2<15::AID-EAT2260020203>3.0.CO;2-6.

Graell, M., Morón-Nozaleda, M. G., Camarneiro, R., Villaseñor, Á., Yáñez, S., Muñoz, R., et al. (2020). Children and adolescents with eating disorders during COVID-19 confinement: Difficulties and future challenges (doi:10.1002/erv.2763). *European Eating Disorders Review, 28*(6), 864–870. https://doi.org/10.1002/erv.2763.

Haedt-Matt, A. A., & Keel, P. K. (2011). Revisiting the affect regulation model of binge eating: A meta-analysis of studies using ecological momentary assessment. *Psychological Bulletin, 137*(4), 660–681. https://doi.org/10.1037/a0023660.

Hay, P. (2020). Current approach to eating disorders: A clinical update. *Internal Medicine Journal, 50*(1), 24–29.

Herle, M., De Stavola, B., Hübel, C., Abdulkadir, M., Ferreira, D. S., Loos, R. J., et al. (2020). A longitudinal study of eating behaviours in childhood and later eating disorder behaviours and diagnoses. *The British Journal of Psychiatry, 216*(2), 113–119.

Johnston, J. A. Y., O'Gara, J. S. X., Koman, S. L., Baker, C. W., & Anderson, D. A. (2015). A pilot study of Maudsley family therapy with group dialectical behavior therapy skills training in an intensive outpatient program for adolescent eating disorders. *Journal of Clinical Psychology, 71*(6), 527–543. https://doi.org/10.1002/jclp.22176.

Kamody, R. C., Thurston, I. B., & Burton, E. T. (2020). Acceptance-based skill acquisition and cognitive reappraisal in a culturally responsive treatment for binge eating in adolescence. *Eating Disorders, 28*(2), 184–201. https://doi.org/10.1080/10640266.2020.1731055.

Kamody, R. C., Thurston, I. B., & Burton, E. T. (2021). Adolescent engagement in a binge-eating behavioral health intervention: Influence of perceptions of physical appearance and locus of control. *Children (Basel, Switzerland), 8*(2), 102. https://doi.org/10.3390/children8020102.

Kamody, R. C., Thurston, I. B., Pluhar, E. I., Han, J. C., & Burton, E. T. (2019). Implementing a condensed dialectical behavior therapy skills group for binge-eating behaviors in adolescents. *Eating and Weight Disorders, 24*, 367–372. https://doi.org/10.1007/s40519-018-0580-4.

Leehr, E. J., Krohmer, K., Schag, K., Dresler, T., Zipfel, S., & Giel, K. E. (2015). Emotion regulation model in binge eating disorder and obesity—A systematic review. *Neuroscience & Biobehavioral Reviews, 49*, 125–134. https://doi.org/10.1016/j.neubiorev.2014.12.008.

Linehan, M. M. (1993a). *Cogntive-behavioral treatment of borderline personality disorder*. Guilford Press.

Linehan, M. M. (1993b). *Skills training manual for treating borderline personality disorder*. Guilford Press.

Lock, J., & La Via, M. C. (2015). Practice parameter for the assessment and treatment of children and adolescents with eating disorders. *Journal of the American Academy of Child & Adolescent Psychiatry, 54*(5), 412–425. https://doi.org/10.1016/j.jaac.2015.01.018.

Lock, J., & le Grange, D. (2005). Family-based treatment of eating disorders. *The International Journal of Eating Disorders*, *37*(Suppl), S64–S67, discussion S87-69. https://doi.org/10.1002/eat.20122.

Lynch, T. R., Chapman, A. L., Rosenthal, M. Z., Kuo, J. R., & Linehan, M. M. (2006). Mechanisms of change in dialectical behavior therapy: Theoretical and empirical observations. *Journal of Clinical Psychology*, *62*(4), 459–480. https://doi.org/10.1002/jclp.20243.

Lynch, T. R., Gray, K. L. H., Hempel, R. J., Titley, M., Chen, E. Y., & O'Mahen, H. A. (2013). Radically open-dialectical behavior therapy for adult anorexia nervosa: Feasibility and outcomes from an inpatient program. *BMC Psychiatry*, *13*(1), 293. https://doi.org/10.1186/1471-244X-13-293.

Lynch, T. R., Hempel, R. J., & Dunkley, C. (2015). Radically open-dialectical behavior therapy for disorders of over-control: Signaling matters. *American Journal of Psychotherapy*, *69*(2), 141–162. https://doi.org/10.1176/appi.psychotherapy.2015.69.2.141.

Micali, N., Solmi, F., Horton, N. J., Crosby, R. D., Eddy, K. T., Calzo, J. P., et al. (2015). Adolescent eating disorders predict psychiatric, high-risk behaviors and weight outcomes in young adulthood. *Journal of the American Academy of Child & Adolescent Psychiatry*, *54*(8), 652–659. e651.

Miller, A. L., Rathus, J. H., & Linehan, M. M. (2006). *Dialectical behavior therapy with suicidal adolescents*. The Guilford Press.

Mitchison, D., Mond, J., Bussey, K., Griffiths, S., Trompeter, N., Lonergan, A., et al. (2020). DSM-5 full syndrome, other specified, and unspecified eating disorders in Australian adolescents: Prevalence and clinical significance. *Psychological Medicine*, *50*(6), 981–990.

Murray, S. B., Anderson, L. K., Cusack, A., Nakamura, T., Rockwell, R., Griffiths, S., et al. (2015). Integrating family-based treatment and dialectical behavior therapy for adolescent bulimia nervosa: Preliminary outcomes of an open pilot trial. *Eating Disorders: The Journal of Treatment & Prevention*, *23*, 336–344. https://doi.org/10.1080/10640266.2015.1044345.

Pennell, A., Webb, C., Agar, P., Federici, A., & Couturier, J. (2019). Implementation of dialectical behavior therapy in a day hospital setting for adolescents with eating disorders. *Journal of the Canadian Academy of Child and Adolescent Psychiatry*, *28*, 21–29. https://www.ncbi.nlm.nih.gov/pubmed/31001348.

Perepletchikova, F., Nathanson, D., Axelrod, S. R., Merrill, C., Walker, A., Grossman, M., et al. (2017). Randomized clinical trial of dialectical behavior therapy for preadolescent children with disruptive mood dysregulation disorder: Feasibility and outcomes. *Journal of the American Academy of Child & Adolescent Psychiatry*, *56*(10), 832–840. https://doi.org/10.1016/j.jaac.2017.07.789.

Peterson, C. M., Van Diest, A. M. K., Mara, C. A., & Matthews, A. (2019). Dialectical behavioral therapy skills group as an adjunct to family-based therapy in adolescents with restrictive eating disorders. *Eating Disorders: The Journal of Treatment & Prevention*, 1–13. https://doi.org/10.1080/10640266.2019.1568101 (epub ahead of print).

Pluhar, E. I., Kamody, R. C., Sanchez, J., Thurston, I. B., & Burton, E. T. (2018). Description of an intervention to treat binge-eating behaviors among adolescents: Applying the template for intervention descriptions and replication. *International Journal of Eating Disorders*, *51*, 1128–1133. https://doi.org/10.1002/eat.22954.

Rathus, J., Campbell, B., Miller, A., & Smith, H. (2015). Treatment acceptability study of walking the middle path, a new DBT skills module for adolescents and their families. *American Journal of Psychotherapy*, *69*(2), 163–178.

Reilly, E. E., Orloff, N. C., Luo, T., Berner, L. A., Brown, T. A., Claudat, K., et al. (2020). Dialectical behavioral therapy for the treatment of adolescent eating disorders: A review of existing work and proposed future directions. *Eating Disorders*, *28*(2), 122–141.

Safer, D. L., Couturier, J. L., & Lock, J. (2007). Dialectical behavior therapy modified for adolescent binge eating disorder: A case report. *Cognitive and Behavioral Practice*, *14*(2), 157–167. https://doi.org/10.1016/j.cbpra.2006.06.001.

Safer, D. L., Telch, C. F., & Chen, E. Y. (2009). *Dialectical behavior therapy for binge eating and bulimia*. The Guilford Press.

Salbach, H., Klinkowski, N., Pfeiffer, E., Lehmkuhl, U., & Korte, A. (2007). Dialectical behavior therapy for adolescents with anorexia and bulimia nervosa (DBT-AN/BN)—A pilot study. *Praxis der Kinderpsychologie und Kinderpsychiatrie*, *56*(2), 91–108. https://doi.org/10.13109/prkk.2007.56.2.91 (Dialektisch-behaviorale Therapie für jugendliche Patientinnen mit Anorexia und Bulimia nervosa (DBT-AN/BN) - eine Pilotstudie.).

Salbach-Andrae, H., Bohnekamp, I., Pfeiffer, E., Lehmkuhl, U., & Miller, A. L. (2008). Dialectical behavior therapy of anorexia and bulimia nervosa among adolescents: A case series. *Cognitive and Behavioral Practice*, *15*, 415–425. https://doi.org/10.1016/j.cbpra.2008.04.001.

Santomauro, D. F., Melen, S., Mitchison, D., Vos, T., Whiteford, H., & Ferrari, A. J. (2021). The hidden burden of eating disorders: An extension of estimates from the global burden of disease study 2019. *The Lancet Psychiatry*, *8*, 320–328.

Tanofsky-Kraff, M., Theim, K. R., Yanovski, S. Z., Bassett, A. M., Burns, N. P., Ranzenhofer, L. M., et al. (2007). Validation of the emotional eating scale adapted for use in children and adolescents (EES-C). *International Journal of Eating Disorders*, *40*, 232–240. https://doi.org/10.1002/eat.20362.

Treasure, J., Duarte, T. A., & Schmidt, U. (2020). Eating disorders. *The Lancet*, *395*, 899–911.

Vall-Roqué, H., Andrés, A., & Saldaña, C. (2021). The impact of COVID-19 lockdown on social network sites use, body image disturbances and self-esteem among adolescent and young women. *Progress in Neuro-Psychopharmacology & Biological Psychiatry*, *110*. https://doi.org/10.1016/j.pnpbp.2021.110293, 110293.

Whiteside, U., Chen, E., Neighbors, C., Hunter, D., Lo, T., & Larimer, M. (2007). Difficulties regulating emotions: Do binge eaters have fewer strategies to modulate and tolerate negative affect? *Eating Behaviors*, *8*(2), 162–169. https://doi.org/10.1016/j.eatbeh.2006.04.001.

Wilson, G. T., Grilo, C. M., & Vitousek, K. M. (2007). Psychological treatment of eating disorders. *American Psychologist*, *62*(3), 199.

World Health Organization. (2019). *ICD11: International statistical classification of diseases and related health problems (11th Revision)*. World Health Organization Publication.

Chapter 18

CBT interventions for pediatric Type 1 diabetes mellitus

Omer Shaked[a,b] and Daniel Hamiel[c,d,e]

[a]*Pediatric Endocrine and Diabetes Unit, Sheba Medical Center, Edmond and Lily Safra Children's Hospital, Ramat-Gan, Israel,* [b]*The Louis and Gabi Weisfeld School of Social Work, Bar-Ilan University, Ramat-Gan, Israel,* [c]*Baruch Ivcher School of Psychology, IDC Herzliya, Herzliya, Israel* [d]*Cohen-Harris Resilience Center, Tel-Aviv, Israel,* [e]*Tel-Aviv-Brüll Community Mental Health Center, Clalit Health Services, Tel-Aviv, Israel*

Abbreviations

ACT acceptance and commitment therapy
BGAT blood glucose awareness training
CBT cognitive behavioral therapy
DKA diabetic ketoacidosis
DBT dialectical behavior therapy
FOH fear of hypoglycemia
HbA1C glycated hemoglobin
MBCT mindfulness-based cognitive therapy
PTSD posttraumatic stress disorder
SH severe hypoglycemia
T1DM Type 1 diabetes mellitus

Introduction

Type 1 diabetes mellitus (T1DM) is one of the most common endocrine disorders diagnosed in childhood or adolescence (Maahs, West, Lawrence, & Mayer-Davis, 2010). Currently, T1DM is an incurable chronic illness with no known preventive measures and is caused by the autoimmune destruction of beta cells in the pancreas, resulting in insulin deficiency and hyperglycemia. Insulin injections or a continuous subcutaneous insulin pump in combination with glucose self-monitoring, diet, and exercise are required to control glucose concentrations. Failure to take an adequate amount of insulin can result in hyperglycemia and diabetic ketoacidosis (DKA), which is a potentially life-threatening event, while an excess of insulin can lead to hypoglycemia with neuroglycogenic symptoms such as shakiness, nausea, sweating, irritability, fatigue, excessive hunger, weakness, seizure, and coma. Long-term microvascular complications of diabetes such as retinopathy, nephropathy, and neuropathy can be prevented or delayed by intensive diabetes management (American Diabetes Association, 2016; Chiang et al., 2018). In the case of childhood chronic disease, optimal disease management involves integrating the illness into the family's daily routine (Civita & Dobkin, 2004). At younger ages, parent-child collaboration requires more parental sensitivity and attentiveness to the child's needs, and should be accompanied by support for the child's autonomy, manifested by showing an understanding of the child's preferences and providing choices regarding health management (Allen, Hauser, Eickholt, Bell, & O'Connor, 1994). Family members (parents or other members of the family) may not contribute equally to coping efforts or may not share all coping responsibilities, depending on the illness demands.

Support for children and adolescents with T1DM may be crucial. Chronic diseases in general, and T1DM in particular, are associated with several psychiatric disorders. Children and adolescents with T1DM are at risk for developing anxiety and mood disorders (Dybdal et al., 2017). In addition, detailed meal planning, precision in food portions, and constant monitoring of carbohydrate intake (which are essential to the management of T1DM) may place adolescent females at higher levels of concern regarding food and can lead to the emergence of eating disorders (Toni et al., 2017). Furthermore,

Handbook of Lifespan Cognitive Behavioral Therapy. https://doi.org/10.1016/B978-0-323-85757-4.00016-X

as insulin therapy often causes severe hypoglycemia (SH), T1DM is associated with a high prevalence of posttraumatic stress disorder (PTSD) symptoms (Al Hayek, Robert, Braham, Issa, & Al Sabaan, 2015). Since children and adolescents with chronic poor metabolic control are more likely to have underlying psychosocial problems or psychiatric disorders than children with good metabolic control (Delamater, de Wit, McDarby, Malik, & Acerini, 2014), psychological support is highly significant.

Cognitive behavioral therapy

Cognitive behavioral therapy (CBT) is an effective intervention designed to provide help for the mentioned psychological problems. The approach was developed by Beck and Elis, who claimed that psychopathology is the result of distorted thoughts and beliefs that manifest in distorted emotions and consequently in maladaptive behaviors (Beck, 1997).

Classical (or second-wave) CBT would address the content of thoughts and beliefs of the adolescents. It is derived from the conception that the source of suffering is not necessarily the agonizing objective reality to which one is exposed, as much as one's cognitions and interpretations of reality that may lead to subjective painful reactions. To prevent such emotional reactions, CBT therapy therefore aims for a replacement of extreme thoughts and beliefs by more moderate and realistic thoughts and beliefs. T1DM onset diagnosis, for example, may be accompanied by the thought: "My life is over, I'm a sick boy, I am disabled," which may be followed by a sense of extreme sadness and withdrawal from life. CBT therapy may therefore strive to identify that thought, provide the skills to recognize that particular disturbing thought, and replace it (or reframe it) with a less extreme and therefore less disturbing thought, that may not produce such sadness: for example, "Things are going to be different, more difficult and challenging now. I will have to take care of my eating and take care of my glucose levels. It is going to be tough, but I can handle it."

Third-wave CBT therapy is less confrontational than classical CBT. Third-wave CBT therapy does not attempt to directly challenge and replace the content of thinking. Based on contextual concepts derived from the context in which the thoughts and emotions appear in the person's mind and not related to the content of the thoughts, the third-wave approach emphasizes topics such as nonjudgmental awareness of the moment (mindfulness), acceptance of emotions, and questions about the effectiveness of thoughts in advancing one's goals and values.

New models and intervention approaches include acceptance and commitment therapy (ACT), dialectical behavior therapy (DBT), mindfulness-based cognitive therapy (MBCT), functional analytic psychotherapy, metacognitive therapy, and several others (Hayes & Hofmann, 2018). The common factor in all methods of third-wave CBT is the rejection of the classical CBT main tendency to refute and repair destructive thoughts. Moreover, third-wave CBT stresses that the source of suffering is the agonizing attempt to solve and get rid of unpleasant emotions and feel better by continued efforts to repair thoughts and beliefs. Such attempts are reflected by constant thinking activity, which is the actual cause that fuels more painful emotional reactions. Third-wave CBT therapy would encourage the patient to identify and accept his own thoughts and feelings while enhancing the ability to freely choose how to react to them. The unpleasant feelings in the third-wave conceptualization aren't something that the person has to get rid of and immediately relieve, but rather a sign of being challenged and a starting point for personal development. Third-wave CBT would address the thought of an adolescent with T1DM "I cannot take these insulin injections anymore" by reminding them that that thought is not a fact of life, but a reflection of an acceptable and understandable present frustrated state of mind, and would challenge the thought at the same time. This state of mind should not automatically determine our behavior, not now and not later.

The problem with this thought, according to the third-wave CBT, isn't the content of the thought but its continued existence in the mind. The appearance of the thought is quite normal and understandable. On the other hand, letting such a thought stay in the patient's mind with its terminal meaning for a long time might cause tremendous pressure on the patient and may lead to dysfunction in dealing with diabetes. Third-wave CBT would argue that this state of mind is a reflection of a mental event that has crossed the mind of the adolescent and that the adolescent can freely decide if and how to relate to it. In this case, the understanding that the thought does not promote the adolescent's goals but hurts them, can help them not to cooperate with it. Working in this way may decrease the stress caused by the accompanied imagination of possible medical consequences (for not injecting insulin).

The current review will therefore discuss applications of the various CBT interventions among children and adolescents with T1DM. Since a common differentiation between children (ages 0–14) and adolescents (ages 15–18) exists (International Diabetes Federation, 2019) regarding demographic variables and medical care, and since parental care is prominent at younger ages, a distinction between these age groups has also been applied here. The following paragraphs will present classical and third-wave CBT therapeutic interventions for parents and children regarding these two age groups, addressing psychological issues that commonly characterize coping with T1DM: diagnosis, chronicity, and dangers.

Interventions upon pediatric T1DM diagnosis

When a child is more dependent on their parents, the parent's reaction to the diagnosis of T1DM and to the treatment may have a crucial influence. Parents often experience grief, loss of a "healthy child," and loss of spontaneity in life. These experiences may later develop into depression symptoms. If the onset involved drastic deterioration in the child's physical health and significant loss of weight, they may feel guilt because of late diagnosis. In cases in which intensive care for the child was required and/or loss of consciousness occurred, the event may also be experienced as traumatic by the parents, who could later develop anxiety symptoms (Anderson & Bracket, 2007; Schiffrin, 2001) and posttraumatic stress symptoms in the form of reexperiencing the event (i.e., flashback or nightmares), avoidance behaviors (i.e., decreased physical activities or socializing activities), and hyperarousal (i.e., tension and difficulty concentrating) and impairment in the overall level of functioning (Landolt et al., 2002). Often parents have more extensive worries and fundamental uncertainty regarding the future.

During infancy and the first years of life the child is too young and premature developmentally to participate in meaningful intervention. Then, the parents may require CBT intervention setting of their own (Snoek & Skinner, 2002). Interventions for the parents differ from family interventions. Since T1DM pervades nearly all of the children's activities during the helpless ages, including play, meals, sleep, and social interactions, CBT intervention for parents would promote adaptation (Mackey et al., 2016). To reconstruct life, the intervention is designed to improve identifying age-appropriate expectations regarding diabetes management, use of breathing as a stress reduction practice, applying a strategy of problem-solving, and incorporate behavioral parenting skills that target eating, sleep, constructing social support networks, parental self-care and general child's safety. The intervention also addresses the parent's anxious or depressive attitudes towards diabetes onset through the acquisition of positive thinking skills.

At preschool and younger ages, the child's emotions may be entangled with his parents'; many classical CBT interventions are designed for the entire family, that is for the children and their parents together. These interventions aim to increase the entire family sense of competence after such a troubling event. In addition, during a time of major life adjustments that follow T1DM onset, which may rapidly undermine the family's confidence, CBT intervention would target the sense of agency and capacity. CBT interventions upon T1DM diagnosis would therefore focus on establishing familial strong communications, improving social problem-solving skills, and through applying recognition of the associations between thoughts, feelings, and actions (the main working tool in CBT), assist in building a guided self-dialogue among all family members for efficient stress management (Ambrosino et al., 2008; Grey et al., 2009).

At older ages, adolescents' struggle with T1DM diagnosis may involve loss of sense of normalcy, loss of independence, and loss of healthy self-image. This struggle could translate into social anxiety, depression symptoms, or sometimes behavior that represents a rebellion and noncompliance with the medical treatment. They may also feel frightened after being hospitalized (or even in an intensive care unit) and perhaps develop anxiety symptoms and sometimes PTSD symptoms (Schiffrin, 2001; Snoek & Skinner, 2002). Since adolescents have yet to construct a differentiated self-identity, they may also absorb their parent's sentiments on top of their own struggle when diagnosed with T1DM.

Classical CBT interventions with families of adolescents focus on practical problem-solving skills to cope with the weakening effect of sadness and worries that may follow the diagnosis. In other words, the problem-solving approach provides families with a pragmatic and more specific day-to-day manageable perspective on diabetes rather than an extensively generalized self-defeating approach, through realistic behavioral goals and a structured coping vocabulary that can replace a nonproductive anxious and melancholic approach. Along with the acquisition of a structured coping vocabulary, the intervention aims to promote diagnosis acceptance and ways that parents can support acceptance, and share their feelings about having diabetes (Carpenter, Price, Cohen, Shoe, & Pendley, 2014). These interventions provide the families with the rationale for using a structured problem-solving dialogue to address diabetes at a sensitive time. The intervention provides practice for family communications to establish productive dialectical patterns, and provide practice for family acquisition of adapting skills for favorable lifestyle changes, to ease the acceptance of T1DM diagnosis and to ease the time required for the family to become accustomed to living with T1DM. Classical CBT will therefore help families to challenge their negative thoughts and replace them with more rational ones in order to recover from the unpleasant emotions. Therapists will attempt to direct patients to an alternative interpretation of the disease—that although the child is now different in some way from other children, the conclusion of never having a normal life is exaggerated and that in most areas they can continue to function like any other family.

Third-wave CBT interventions upon diagnosis recognize that children and adolescents may become identified with a conceptualization of the self as ill, or adopt illness as their identity, the consequences of categorizing essential thoughts such as "I am sick" as deterministic and oppressive (as opposed to a more descriptive self-conceptualization as coping with the medical condition and physiological restrictions that constitute "diabetes"). This kind of fusion with the generalized thought "I am sick" often held by children and adolescents upon diagnosis may turn into a rigid representation of reality, which makes it too hard for them to accept T1DM as part of life. Lack of acceptance can later be translated into denial mechanisms and avoidance behaviors in order to distance themselves from an unbearable self-concept (Hadlandsmyth, White, Nesin, & Greco, 2013). The fusion with a generalized thought structure can be common among chronic patients; however, T1DM compels a great deal of daily reminders of the illness due to the strict treatment requirements, and thus has the potential to invoke even more frequent occasions for the generalization mechanism to be activated.

While classical CBT intervention addresses the generalized identification of patients with the illness through reframing of thought content, third-wave CBT would address their reactions to such thoughts. Upon T1DM onset, ACT intervention would promote the perspective that, while a disease like diabetes is now an integral part of life, there is a lot of unnecessary suffering involved here (Lindholm-Olinder, Fischier, Fries, et al., 2015). The source of unnecessary suffering according to ACT is the desire to find a solution to the unpleasant thoughts and the feelings derived from them. For example, someone who is diagnosed with diabetes and is regularly holding the thought "I will always be different and never live a normal life like other children" will probably suffer from the desire to find a complete solution to the disturbing thoughts and feelings derived from them, a solution that of course does not exist. In this case he or she will continue to be preoccupied with thoughts and focus on a sense of victimhood or try to deny the disease and not treat it properly in order to feel like everyone else.

In ACT interventions, the child or the adolescent learns to accept and contain the unpleasant emotion and not to argue with the thought that stands behind it. The content of the thought isn't the target of therapy but rather how not to be controlled by it. In ACT there is no attempt to refute the child's thought but only to accept that an unpleasant emotion is a part of life and that there is no point in trying to control thoughts and emotions. Using ACT therapy teaches not to be managed by the thought and not to add to the necessary suffering. The children can learn in therapy to recognize what is important for them and to give up the focus on trying to solve what cannot be solved and thus minimize the suffering.

ACT interventions would include exercises to decrease attempts to control thoughts and feelings, and identify behaviors aimed to control or avoid emotional pain that may reinforce its existence. The ACT framework also includes mindfulness exercises that aim to promote acceptance and contact with the present moment, breathing exercises to decrease anxiety, and mindfulness exercises to notice the ongoing process of thinking and experience it as just thinking. Finally, ACT interventions emphasize the importance of personally identified values and goals (which could also include goals specifically related to diabetes care). Values and goals are supposed to guide one to explore the thought, checking whether it promotes or distances one from them (Amsberg, Wijk, & Livheim, 2018; Hadlandsmyth et al., 2013).

CBT interventions for chronicity of pediatric T1DM

A core issue for many T1DM patients is having to cope with the meaning underlying the chronicity of the disease: in other words, the understanding that T1DM is uncurable. Knowing that no matter what you do the condition will not improve may cause discouragement and noncompliance with medical treatment. In addition, the ongoing and never-ending medical precautions required to prevent hyperglycemia and hypoglycemia may be experienced as a heavy burden by parents and children alike. At the same time, there can be no relief from the daily medical demands since the health consequences for not complying with the treatment can be fatal over the long run. Therefore it is not surprising that, because of the never-ending and frustrating burden, many T1DM patients and their parents are at risk for stress, anxiety, and depression (Buchberger et al., 2016).

An acute manifestation of this stressful and depressing uncurable chronic state would therefore be exhaustion and a feeling of burnout. At younger ages, a premature child has no capacity to endure Sisyphean daily tasks on his own and is indeed dependent on the parents' encouragement and support for coping with the chronicity of T1DM. Accordingly, classical CBT interventions will focus on teaching the parents emotional regulation to decrease their exhaustion. Teaching the parents emotional regulation reflects the concept of "load sharing," wherein close relationships and stronger emotional regulation among one family member can enhance another's goal-directed behaviors. For example, parents with efficient emotion regulation may be better able to support their distressed or frustrated child, in times the child's dysregulation interferes with his scaffolding of diabetes management. Conversely, in families where both the youth and parents experience poor emotional regulation, there may be no one able to "pick up the slack" concerning the complex diabetes-related tasks, and families may experience a cycle in which low parent and youth emotional regulation amplifies family tension. This stress may then weaken parent and youth emotional regulation further, pushing family members to the boundaries of their capacity, creating greater barriers to collaborative family diabetes management (Bauer, Hilliard, Albright, et al., 2020).

During adolescence, the chronicity of T1DM can manifest in familial conflicts. Adolescence has traditionally been known as a challenging time, as adolescents are trying to establish autonomy and challenge authority (Commissariat,

Kenowitz, Trast, Heptulla, & Gonzalez, 2016). Adolescents tend to engage in rebellious and pleasure-seeking behavior, value privacy, have a greater awareness of self-image, and above all, are more influenced by peer interactions in forming their identity (Hamilton & Daneman, 2002). Adolescents claim independence from their parents and begin to relate more to their peers (Cheung, Cureton, & Canham, 2006). Thus peer acceptance becomes especially important in developing social connections and friendships, and, by extension, a positive sense of self. The self-conceptualization that begins in early adolescence is shaped through experiences and comparisons with others and influenced by peers' expectations and perceptions (Bong & Skaalvik, 2003).

Adolescents with T1DM experience all of the previously mentioned challenges while also needing to manage their illness. Managing T1DM often involves greater parental involvement and oversight, causing adolescents with the illness to differ from their peers (Helgeson, Snyder, Escobar, Siminerio, & Becker, 2007). Peers can potentially exert a negative influence on diabetes management, particularly if concern over peer reactions becomes more valuable to adolescents with the illness than does managing their diabetes (Wysocki & Greco, 2006). To conclude, adolescents with T1DM may rebel against their parents within a process of individuation, and as part of their rebellion, adolescents may reject values and customs that represent their elders when T1DM care is identified with the parents.

Rebelling against parents therefore might include rejection of the T1DM treatment régime since it is opposing youth spontaneity, impulsiveness, and sense of freedom. Classical CBT interventions with families of adolescents would focus on problem-solving skills due to the potential of conflicts. The problem-solving approach provides families with a clear and replicable communication structure through current behavioral goals rather than sorting out past issues, and permits families to have a more beneficial collaboration (Carpenter et al., 2014). These interventions provide the families with the rationale for using a formal and structured problem-solving dialogue without being captivated by emotionally laden controlling issues that may increase the potential for family conflicts. Structured problem-solving aims for the moderation of thought mechanisms and the reduction of extreme approaches with realistic views. CBT practice may include specific, short-term goals that can be accomplished within the next week; choosing an incentive; brainstorming solutions; and determining whether solutions are working. For practice, each step can also be introduced through nondiabetes case examples, and after successful experience the dialogue can be applied to a relevant diabetes-related problem. These interventions provide practice for family communications, establishing productive communication patterns, identifying negative interactions experienced by youth as parent's annoying constant "nagging" that may lead to increased resistance. Aspects of adaptive family cooperation also include flexibility of thinking to identify alternative practical solutions (Bauer et al., 2020), which can provide families with the skills to find a path through reoccurring dead-end communication patterns that tend to result in tension and frustration.

Exhaustion and feeling burnout, which may be prevalent among parents for children with T1DM, are central topics that are addressed in third-wave CBT interventions. Parents who are destined to routines of glucose management are at risk for detachment and apathetic automated response. ACT intervention would target automated behaviors and promote awareness, to increase psychological vitality and consequently decrease exhaustion and burnout (Byrne, Ghráda, O'Mahony, & Brennan, 2021). Increasing parent's awareness and vitality through ACT interventions is practiced within structured mindfulness exercises (Sairanen, Lappalainen, Lappalainen, & Hiltunen, 2020). Practicing mindfulness in a population that copes with chronic medical states aims at enhancing the ability to reflect on choices regarding medical treatment (Ludwig & Kabat-Zinn, 2008). ACT interventions therefore place significant attention on choice. Practicing mindfulness by focusing nonjudgmentally on the present moment is the answer to the default automatic response related to the past and the future of the unbearable chronic characteristic of T1DM. Parents and children learn how to move from stress reaction to stress response under choice. Mindfulness makes choices possible. By focusing on awareness, the intervention addresses feelings of unwilling entrapment in routines and consequently has the potential to increase the persistence of parental involvement in the children's medical treatment through valued actions rather than automatically executing unpleasant tasks such as pricking and endless measuring (Byrne et al., 2021). Mindfulness exercise for parents of children with T1DM aims at enhancing receptive attention to and awareness of current experiences and the present reality (Van Gampelaere, Luyckx, Goethals, et al., 2020). By enhancing awareness only of this very specific present experience, rather than the entire never-ending exhausting medical régime, ACT interventions promote the parents' optimism and capacity to support their child, who is trapped in a Sisyphean reality (de Bruin et al., 2014). Here, the role of awareness is in promoting the capacity of making distinctions between the present, the past, and the future, between the event taking place now and the impact of frustrating events of the past and those that are expected to happen in the future. In a situation where there is no present awareness, there is a tendency to automatically react to the past and to the predicted future. In such a situation, it will be almost impossible to relate to what is happening now and respond with an appropriate corresponding response. Awareness can be regarded therefore as a necessary condition to decrease ineffective automatic reactions and to allow for a more appropriate and responsible response that matches the present and the person's values and goals.

Among adolescents, the aspiration for individuated self-determination is a fundamental part of the developmental phase. As mentioned earlier, adolescents are trying to establish autonomy and challenge authority (Commissariat et al., 2016), are more influenced by peer interactions in forming their identity (Hamilton & Daneman, 2002), claim independence from their parents, and begin to relate more to their peers (Cheung et al., 2006). When self-determination includes rejection of values that represents their elders, it may also include rejection of the T1DM treatment régime since it opposes youth's sense of spontaneity under the perception of parental control. In other words, taking care of diabetes because of being told to may increase detachment and resentment. When the most important factor in the life of adolescents is the desire to be similar and equal to their peers, having to stop so many times to measure and stabilize their glucose level can make them feel different, and even may cause a sense of disability that they might attempt to deny. As a result, they might act against the need of the moment to address their medical diabetes care. Their behavior in this case might not fit the responsibility that is already required of them at this stage of their lives. Instead, it could reflect their wish to rebel and be equal to their peers. It will therefore be against the direction of some of their own values and wishes—to be healthy and independently take care of themselves.

ACT intervention for T1DM is designed to focus on personal responsibility not as opposing significant others, but as an internally motivating choice according to personal values and wishes, to which adolescents may be more receptive (Hadlandsmyth et al., 2013).

Accordingly, ACT intervention emphasizes the understanding of the adolescent's point of view over criticisms of being irresponsible, while giving up the attempt to change their approach. From an understanding of their point of view, the desire to be independent and the desire to be an equal member of their peer group, it will be possible to understand why T1DM treatment is so difficult for them and to identify with them. When adolescents feel understood, there is a much better chance that they will be willing to choose an independent way of taking responsibility, a way that also takes into account their age needs. In the same way, they too will be able to free themselves from the natural desire of their age to automatically rebel against parents and decrease self-defeating behaviors on an issue as important to them as their health.

Interventions for fear of hypoglycemia

Hypoglycemia (plasma glucose level <3.9 mmol/L [<70 mg/dL]) is the major limiting factor in the glycemic management of T1DM (Chamberlain et al., 2017). SH is defined by the patient's urgent need for the assistance of another person (American Diabetes Association, 2016). A ranging prevalence between 7.8% (Bulsara, Holman, Davis, & Jones, 2004), 7.5% (McCoy et al., 2012), and 20% (Gilsanz, Karter, Beeri, Quesenberry, & Whitmer, 2018) of reported SH among T1DM patients was found, which makes SH a primary medical issue. The SH is accompanied with various symptoms—such as trembling, sweating, warmness, nausea, drowsiness, dizziness, confusion, tiredness, difficulty speaking, shivering, headache, inability to concentrate, blurred vision, weakness, hunger, and pounding heart (Hepburn et al., 1991). It should be emphasized that SH is considered a most adverse event, since it may put patients at risk for injury and death (Seaquist, Anderson, Childs, et al., 2013). This particular feature of SHis also accompanied by the understanding that one's life is in danger, and therefore matches the criteria for a traumatic event (American Psychiatric Association, 2013), and thus may also result in PTSD.

A probable result of SH is an intense emotional reaction termed *fear of hypoglycemia* (FOH) (Gonder-Frederick, Schmidt, Vajda, et al., 2011). This fear holds crucial importance since it jeopardizes medical treatment due to its role in essential disagreements between patients and attending medical staff regarding coping with diabetes: patients and their families may try to avoid in any way the disturbing symptoms and the sensation of danger that can result from low blood-glucose level while the attending medical staff does not take into consideration the unpleasant symptoms and focuses more on the preferable glucose level (Böhmea, Bertin, Cosson, & Chevalier., 2013). In addition, while attempting to avoid such ill experiences, numerous patients omit insulin to circumvent low blood-glucose levels (Martyn-Nemeth, Schwarz Farabi, Mihailescu, Nemeth, & Quinn, 2016). Therefore the previously mentioned doctor-patient differences may consequently cause patients and their families to resist the recommended medical treatment (Böhmea et al., 2013).

One way to refuse medical treatment may take the form of insulin omission. Insulin omission may result in DKA and hospitalizations; in addition, insulin omission may put patients at risk for hyperglycemia and glycated hemoglobin (HbA1C), which in turn results in long-term complications (Morris et al., 1997). To conclude, FOH is a major psychological factor to be considered in T1DM treatment.

Parents with FOH are often worrying about their child having a low blood glucose while asleep and their child not recognizing a low glucose (Van Name et al., 2018). Parental fear of hypoglycemia may interfere with the achievement of glycemic treatment targets, as some families may aim for higher blood glucose levels for their child to avoid low glucose levels. Since the fear can be intense, children may absorb and internalize their parent's attitude and classical CBT interventions are therefore designed for the family as a whole. The intervention strives for a change in both cognitions and behaviors via reframing of catastrophic thinking and relaxation training. Relaxation training includes breathing exercises to balance anxiety level, since anxiety may distort the parent's and the child's catastrophic interpretations of a current medical state. The reframing of thought content is achieved first via medical psychoeducation. The cognition of loss of control that threatens the parents and the child is addressed by blood glucose awareness training (BGAT). The training improves the ability to detect and interpret blood-glucose levels. The underlying informative mechanism is aiming to reduce the sense of uncertainty associated with hypoglycemic episodes and increase confidence in recognizing and anticipating hypoglycemia through a better medical understanding and to increase the accuracy of recognizing symptoms such as mood changes, cognitive difficulties, and bodily functions (Driscoll, Raymond, & Naranjo, 2016).

Third-wave CBT intervention would not strive to reframe one's thought content nor directly aim to decrease one's fear or any other unpleasant emotion. Third-wave CBT promotes acceptance of fear as a natural emotional response when living with the dangers of diabetes and aims to help the patient not to be controlled by the wish to feel good or automatically react to reduce the fear, no matter what the cost. It encourages giving up the wish to stay occupied with the threatening thoughts in order to solve them. It encourages patients' inquiry into whether catastrophic thought promotes and protects them or harms them. It accepts cognitions that may naturally accompany such dangers, and aims to increase resilience to that fear (Aalders et al., 2018).

The fear of hypoglycemia may drive parents to focus more intensely on threatening cues. A third-wave CBT intervention accordingly is designed to promote parent's receptive attention to a wider perspective on the current medical situation, since a fear reaction narrows the scope of attention (Van Gampelaere et al., 2019). That is, low glucose level can initiate negative cognitions (e.g., "my child is going to lose consciousness") and consequent negative emotional feelings (e.g., fear). When such negative cognitions are experienced as real and as accurate evaluations of the medical event, the parents and the child may feel compelled to avoidance behaviors (i.e., insulin omission). Third-wave CBT intervention can help the family to acquire defusion techniques for these cases. Defusion techniques are derived from a principle that was addressed previously, and are based on the observation of thoughts as mental events that cross our mind, as cognitive experiences but not necessarily as an objective truth. The techniques are based on the understanding that a current thought and emotion rely more on personal history and past experience and on future expectations, than on what is actually happening in reality right now, at the moment. Practicing defusion is the practice of tolerance to distressing thoughts without automatically engaging in their content and in action (and more particularly, avoidance actions). Taken together, defusion techniques are a practice of increasing awareness to distressing internal experiences with suspension of automatic responses (Bogusch & O'Brien, 2019).

Adolescents with T1DM, who fuse subjective experiences with reality, might attempt to control their emotions via avoidance strategies. Consequently, they might react inflexibly and display poorer adherence and poorer disease management. ACT intervention explicitly focuses on personal autonomy, to which adolescents may be particularly receptive. In an ACT framework, mindfulness exercises aim to promote acceptance, contact with the present moment, and cognitive defusion. Mindfulness sessions may begin with exercises noticing breath, sounds, and the associated physical sensations: sensation of skin on clothes or the feeling of an arm against a chair. Mindfulness exercises can then progress to exercises noticing the ongoing stream of thinking. In mindfulness training the trainee learns to ground awareness within the physical dimension, that is, the present physical state and bodily sensations, and gives up refueling harmful thoughts and feelings that are fused in past and future contexts. This process makes it possible to respond to the current reality in the most appropriate way and in accordance with the trainee's values. Ultimately, mindfulness exercises would be applied directly to notice diabetes-adherence behaviors such as mindful eating, exercising, blood glucose monitoring, and mindfully administering insulin injections. Adolescents would be encouraged to notice the thoughts, emotions, and physical sensations that occurred prior to, during, and after these activities, to simply observe each thought as it occurred and to be willing to experience each emotion and physical sensation that emerges. Through mindfulness practice, the adolescent learns to experience the thought as less compelling, and as a consequence to experience their autonomy by not being subordinated to their thoughts (Hadlandsmyth et al., 2013).

CBT intervention for T1DM and eating disorders among adolescents

The prevalence of eating disorders among adolescents with Type 1 diabetes is twice that of the general population (Clery, Stahl, Ismail, Treasure, & Kan, 2017). During T1DM onset, as a result of insulin deficiency, adolescents with T1DM often see their weight drastically decreasing, and because of insulin treatments they regain weight rapidly. This weight fluctuation can exacerbate discomfort with body image and may lead to disturbed eating behaviors (Coleman & Caswell, 2020). Furthermore, withholding or reducing the recommended insulin treatment offers individuals with T1DM a tempting way to achieve instant weight loss (Kınık, Gönüllü, Vatansever, & Karakaya, 2017). The damage to body image can also damage the sense of self-value, and may result in attempts to reach correction with the help of body control. This can be achieved by extreme sports or weight loss, accompanied with the temptation to lose weight easily with the help of a reduction in insulin injection, which could be the cause of the increase in the incidence of eating disorders in diabetic adolescents (Coleman & Caswell, 2020). Although it is not currently acknowledged as a formal diagnosis, the term "diabolimia" stands for this very distinguishable kind of eating disorder that signifies insulin omission, more frequently found among female adolescents, who attempt to gain control over their body (Clery et al., 2017). Since the long-term and the short-term potential consequences of eating disorders among diabetic adolescents are destructive, T1DM has been referred to as "the world's most dangerous eating disorder" (Torjesen, 2019).

Classical CBT intervention for adolescents with T1DM would address attitudes and personality characteristics related to them, which are commonly associated with the development of eating disorders. CBT intervention would therefore target characteristics such as perfectionism, low self-esteem ("I have to be perfect (or healthy) in order to be precious"), and interpersonal difficulties that may be underlying mechanisms ("I have to get constant assurance in order to feel appreciated and loved") (Fairburn, 2008). Such characteristics are being addressed within classical CBT intervention combined with motivational interviewing techniques conducted by mental health practitioners and also by nurses (Ismail et al., 2010). A nurse, who can professionally monitor medical adherence while providing empathic care, can facilitate new behaviors along cognitive changes, through successful management of medical treatment. The aim of the intervention would be to enable adolescents to identify and modify unhelpful diabetes-specific cognitions and behaviors which may contribute to suboptimal glycemic control. For example, a change in the cognitive distortion "overgeneralization" (which can be manifested by beliefs in the form of "I'm worthless") can be achieved through sessions dedicated to better behavioral self-care that would successfully improve glycemic control, during a continuous striving for strength and good health, which may promote an alternative approach (for example, "when I take care of myself I can feel like a healthy person").

In the same manner, more cognitive distortions in the form of "dichotomous thinking," "negative filter," "mind reading," and "magnification," that can result in setting unhelpful rigid eating rules. For example, when there is a hold on the cognitive distortion of "mind reading," the adolescent can say: "Everyone perceives me as sick" and therefore decide that he\she should show everyone how good he looks and therefore try to lose weight or do extreme sports. This can be addressed by behavioral experiments and activities which can promote examination of advantages and disadvantages of different types of coping (Ismail et al., 2010).

Third-wave CBT would target an adolescent's elevated perfectionism and flawed self-image. ACT intervention is designed to address the tendency to "clamp down" on eating and regulating strict dietary rules, the tendency to give up palatable foods, and avoiding insulin treatments to reduce the impact of eating on weight. ACT intervention would focus on the unique emotional which activate eating behaviors and cognitions that precede eating or the result of eating, strive to increase awareness to stream of thinking in the moment and cope more effectively using acceptance through mindfulness exercise. The goal will be first to identify thoughts such as: "I have to look great," "I look awful," "I am sick," "I am fat," ... and first examine the effect they have on the adolescent's feeling and behavior. The adolescents personal values would be clarified at this point and applied to first understand how they are challenged by this kind of thoughts and then encourage engagement in healthy behaviors, even when eating is experienced as disturbing. Instead of trying to challenge the cognitive distortion involved in these thoughts, thoughts will be defined as supportive or hurtful of values and will be addressed accordingly with the help of the defusion, mindfulness, and other techniques of the third-wave CBT. Acceptance interventions also encourage a more compassionate and less punitive approach. In one of the studies of ACT and eating disorders in T1DM (Merwin, Moskovich, & Babyak, 2021) between sessions, a mobile-phone app was designed to assist adolescents to cope with diabetes-specific feelings and routines such as feeling overwhelmed in taking insulin at each meal, and creating mindfulness specific coping skills presented at times of heightened emotional distress in the form of practice watching "in vivo" thoughts and emotions as waves.

Efficiency of CBT interventions for pediatric T1DM

Several randomized control trial studies (RCTs) examined the effectiveness of classical second-wave CBT interventions in stabilizing blood-glucose levels among youth with T1DM, well-being and mental health. A meta-analysis conducted by Pillay, Armstrong, and Butalia (2015) included 22 studies of underaged (<18) T1DM population and compared them with nine studies among adults (>18) with T1DM. Mean age of the youth participants ranged from 9.7 to 15.4 years and the mean HbA1c ranged from 7.4% to 15.7% at baseline. Pillay et al. reported that adults (MD −0.28 [CI, −0.57 to 0.01] percentage points) appeared to benefit from HbA1c decrease more than youths at the end of the intervention (MD −0.12 [CI, −0.43 to 0.19] percentage points). However, the effectiveness of the interventions at the 12-month follow-up appeared higher for a youth decrease in HbA1c (MD, −0.52 [CI, −1.04 to 0.00] percentage points) than for adults (MD, −0.14 [CI, −1.28 to 1.00] percentage points). A more recent study conducted by Wei et al. (2018) randomized 85 youth aged 11–16 years to receive either 6 weekly sessions of individual CBT interventions or individual supportive counseling, with two follow-up sessions at 6 and 12 months. Wei et al. reported that HbA1c in the supportive counseling group increased from 8.4 [6.4–12.5]% to 9.3 [6.5–13.9]%; $P = .001$), while the CBT group maintained 8.7 [6.4–14]% and 8.9 [6.8–13.9]%). Wei et al. also reported that more patients who have undergone CBT individual interventions showed an improved or maintained HbA1c levels after 24 months (62.5% vs. 35.5%, $P = .032$(.

A study conducted by Ellis, Carcone, and Slatcher (2019) examined the efficiency of third-wave CBT compared with classical CBT in stabilizing blood-glucose levels among youth with T1DM. Ellis et al. reported that among 48 participants aged 16–20 years of age that were randomly assigned to MBSR (Mindfulness Based Stress Reduction) group intervention, cognitive-behavioral stress management group intervention (CBSM), or a diabetes support group, MBSR was found to reduce self-reported stress at end of treatment ($P = .03$, $d =$!0.49) and 3-month follow-up ($P = .01$, $d =$!0.67), while no significant difference in HbA1c improvement was found between the MBSR and the CBSM groups. Nevertheless, only diabetes support group participants had improved glycemic control at the end of treatment ($P = .01$, $d =$!0.62). Wei et al., however, stated that the small sample size, which was designed as a pilot test of the efficacy of MBSR, had lack of adequate power to detect group differences.

Practice and procedures

Interventions begin with two to three sessions of assessment, creating a working alliance, and gaining understanding of the individual and the family's backgrounds. History of previous family member's illness and trauma, along with patterns of the family's communications and the patient's self-image, are particularly relevant for designing a proper intervention. Furthermore, since core beliefs of the patient concerning their illness, and the patient's relationships with these beliefs, are the grounds for the intervention, they need to fully be understood. This information can be achieved directly through the interview but also by monitoring. At this point patients (and parents if relevant) are guided to identify their automatic thoughts related to the disease, thoughts that usually emerge in moments when they are emotionally overwhelmed. The therapist collects these thoughts to address them in the next step. This first step of therapy also includes psychoeducation about the possible psychological effects of T1DM on life.

The classic second-wave CBT interventions then proceed to four to five sessions of cognitive restructuring—identifying cognitive mistakes and nonfunctional assumptions that are related to attitudes toward T1DM and learning how to reinterpret events and interactions. These sessions are dedicated to facilitating the patient's everyday diabetes experiences regarding the relevant issue (i.e., adjustment to diagnosis; chronicity; traumatic stress; or eating disorder) and the practice of thoughts reframing. At the same therapeutic stage, there will also be work on emotional regulation to enable the new cognitive attitude.

In the rest of the sessions, the intention will be to use the new cognitive restructuring to change some negative behavior that was affected by old cognitions (poor compliance with treatment, concealment, closure, and social isolation). When the patient is young, the work is carried out along with establishing healthy familial communication.

Third-wave CBT is similar in its early stages. Here, too, the first stage of the protocol is intake and psychological education. Here, too, it follows by monitoring of thoughts and attitudes concerning the disease and in general, but the sequel is different. After about four sessions designed for intake, psychological education, and monitoring, most of the sessions that follow (about six) are devoted to the following topics: (A) strengthening the ability to be in the present moment and deal with the disease in the present tense and nonjudgmentally. This stage will include mindfulness exercises; (B) emotional regulation; (C) identify the values or areas of interest that are important to the child from which he was moving away because of negative attitudes toward the disease and take new actions according to these identified values and areas of interest; (D) learn to defuse and move away from the content of the thoughts to enable new behavior concerning the disease; and (E) develop self-compassion to replace self-criticism usually found in patients with T1DM.

Both second- and third-wave CBT interventions are concluded with two to three sessions of summary. The summary facilitates the patient's reflections on the process, personal growth, and relapse prevention through rehearsing all stages of the intervention. It is also recommended to set follow-up sessions several weeks after the intervention, according to the patient's needs.

Applications to other areas

Three of the central issues discussed here have relevance to psychosocial interventions in various chronic medical conditions. CBT interventions upon illness onset; interventions for chronicity of illnesses; and interventions for traumatic aspects of illnesses should be applied in various medical domains and can be endorsed by clinicians with appropriate semantical adjustments. In addition, the distinction between the therapeutic work of CBT in children and adolescents which is discussed there, alongside the different impact that parents have on children at different ages, shouldn't be different in other chronic medical conditions. Both the parent's nonefficient coping model factor found during childhood with its application to the children's adherence to treatment, and the rebellion factor found during adolescence alongside the adolescent's attempts to differentiate coping from that of their parents, should be found to be chief factors in other chronic medical conditions in pediatric medicine.

Summary

In this chapter we addressed CBT principles to deal with emotional difficulties and psychopathology that may occur in children and adolescent patients as well as with their parents, due to T1DM. Common to CBT in its different types is the focus on the role of the patient's extreme thoughts and beliefs in creating suffering. We introduced the therapeutic method in classical CBT, which is mainly the placement of less extreme and more balanced alternative thinking, with the use of a problem-solving approach and self-regulation techniques.

We introduced the method of the third-wave of CBT and especially the ACT model, which see the patient's main source of suffering not in the content of the negative thinking but in trying to solve the emotional difficulties by arguing with the extreme thoughts or by avoiding events that evoke the unpleasant thoughts and emotions. In the third-wave CBT the goal is to accept the unpleasant emotions that accompany thoughts, identify the ineffectiveness of thoughts in advancing a person to achieve his goals and fulfill his values, and give up cooperation with the extreme thought (either by stop arguing with it or by stop avoidance). For this purpose, various techniques are offered that help the patient to regulate emotions, understanding the essence of thoughts as mental events that appear in mind and not as reality, expanding the focus of attention, identifying values, and being committed to them and mindfulness.

We focus our discussion on the working process of CBT in T1DM in four central issues: interventions upon pediatric T1DM diagnosis, interventions for chronicity of pediatric T1DM, interventions for fear of hypoglycemia, and for eating disorders that are related to T1DM. In all of these subjects, we have demonstrated how the principles of classical and third-wave CBT, emphasizing the role of the impact of thoughts over emotions, help maintain coping with T1DM.

We made the distinction between the therapeutic work of CBT in children and adolescents and discussed the different impact that parents have on children at different ages. The main factor in childhood was found to be more related to adherence to the parenting nonefficient coping model while in adolescent it was more a rebellion and an attempt to differentiate their coping from that of their parents. We also demonstrated how the interventions are applicable to parents as well as to children and presented studies that showed the effectiveness of CBT treatments of both types in the different problems of adaptation associated with T1DM.

Mini-dictionary of terms

T1DM: The most common endocrine disorder diagnosed in childhood. T1DM is an incurable chronic illness with no known preventive measures and is caused by the autoimmune destruction of beta cells in the pancreas.

CBT: An effective intervention, claiming that psychopathology is the result of distorted thoughts and beliefs that manifest in distorted emotions and consequently in maladaptive behaviors. CBT therapy therefore aims for a replacement of extreme thoughts and beliefs by more moderate and realistic thoughts and beliefs.

Hypoglycemia: Medical event in which plasma glucose level drops to <3.9 mmol/L (<70 mg/dL). SH is defined by the patient's urgent need for the assistance of another person. It is a life-endangering event.

Fear of hypoglycemia: Intense emotional reaction to hypoglycemia, accompanied by the understanding that one's life is in danger; it therefore matches the criteria for a traumatic event and thus may also result in posttraumatic stress symptoms.

Diabolimia: A diabetes-specific eating disorder, more frequently found among female adolescents, who attempt to gain control over their body by withholding or reducing the recommended insulin treatment, attempting to achieve instant weight loss.

Summary points

Classical CBT interventions for T1DM focus on replacing extreme thoughts with a more balanced thinking, problem-solving approach, along with self-regulation techniques.

In third-wave CBT, the goal is to accept unpleasant emotions that accompany thoughts, while identifying the ineffectiveness of thoughts fulfilling goals and values.

Four central issues are the target of CBT interventions for T1DM: interventions upon diagnosis, interventions for chronicity of illness, interventions for PTSD, and eating disorders.

CBT interventions require a distinction between the therapeutic work of CBT in children and adolescents, and parent's impact on their children.

The main factor in childhood is more related to adherence to the parenting nonefficient coping model.

The main factor in adolescence is more related to rebellion and the attempt to differentiate coping from that of the parents.

References

Aalders, J., Hartman, E., Nefs, G., Nieuwesteeg, A., Hendrieckx, C., Aanstoot, H. J., et al. (2018). Mindfulness and fear of hypoglycaemia in parents of children with Type 1 diabetes: Results from Diabetes MILES Youth – The Netherlands. *Diabetic Medicine*, *35*(5), 650–657.

Al Hayek, A. A., Robert, A. A., Braham, R. B., Issa, B. A., & Al Sabaan, F. S. (2015). Predictive risk factors for fear of hypoglycemia and anxiety-related emotional disorders among adolescents with Type 1 diabetes. *Medical Principles and Practice*, *24*, 222–230. https://doi.org/10.1159/000375306.

Allen, J. P., Hauser, S. T., Eickholt, C., Bell, K. L., & O'Connor, T. G. (1994). Autonomy and relatedness in family interactions as predictors of expressions of negative adolescent affect. *Journal of Research on Adolescence*, *4*, 535–552.

Ambrosino, J. M., Fennie, K., Whittemore, R., Jaser, S., Dowd, M. F., & Grey, M. (2008). Short-term effects of coping skills training in school-age children with Type 1 diabetes. *Pediatric Diabetes*, *9*, 74–82. https://doi.org/10.1111/j.1399-5448.2007.00356.x.

American Diabetes Association. (2016). In W. T. Cefalu (Ed.), *Standards of medical care in diabetes*. USA: American Diabetes Association.

American Psychiatric Association. (2013). *Diagnostic and statistical manual of mental disorders* (5th). Arlington, VA: Author.

Amsberg, S., Wijk, I., & Livheim, F. (2018). Acceptance and commitment therapy (ACT) for adult Type 1 diabetes management: Study protocol for a randomised controlled trial. *BMJ Open*, *8*. https://doi.org/10.1136/bmjopen-2018-022234, e022234.

Anderson, B., & Bracket, J. (2007). Diabetes during childhood. In F. J. Snoek, & T. C. Skinner (Eds.), *Psychology in diabetes care*. NY: John Wiley & Sons, Ltd.

Bauer, K. W., Hilliard, M. E., Albright, D., et al. (2020). The role of parent self-regulation in youth Type 1 diabetes management. *Current Diabetes Reports*, *20*, 37. https://doi.org/10.1007/s11892-020-01321-z.

Beck, A. T. (1997). The past and future of cognitive therapy. *The Journal of Psychotherapy Practice and Research*, *6*(4), 276–284.

Bogusch, L. M., & O'Brien, W. H. (2019). The effects of mindfulness-based interventions on diabetes-related distress, quality of life, and metabolic control among persons with diabetes: A meta-analytic review. *Behavioral Medicine*, *45*(1), 19–29.

Böhmea, P. E., Bertin, E., Cosson, N., & Chevalier. (2013). Fear of hypoglycaemia in patients with Type 1 diabetes: Do patients and diabetologists feel the same way? *Diabetes & Metabolism*, *39*, 63–70. https://doi.org/10.1016/j.diabet.2012.10.006.

Bong, M., & Skaalvik, E. M. (2003). Academic self-concept and self-efficacy: How different are they really? *Educational Psychology Review*, *15*, 1–40. https://doi.org/10.1023/A:1021302408382.

Buchberger, B., Huppertz, H., Krabbe, L., Lux, B., Mattivi, J. T., & Siafarikas, A. (2016). Symptoms of depression and anxiety in youth with Type 1 diabetes: A systematic review and meta-analysis. *Psychoneuroendocrinology*, *70*, 70–84.

Bulsara, M. K., Holman, C. D. A. J., Davis, E. A., & Jones, T. W. (2004). The impact of a decade of changing treatment on rates of severe hypoglycemia in a population-based cohort of children with type 1 diabetes. *Diabetes Care*, *27*(10), 2293–2298.

Byrne, G., Ghráda, Á. N., O'Mahony, T., & Brennan, E. (2021). A systematic review of the use of acceptance and commitment therapy in supporting parents. *Psychology and Psychotherapy: Theory, Research and Practice*. https://doi.org/10.1111/papt.12282, e12282.

Carpenter, J. L., Price, J. E. W., Cohen, M. J., Shoe, K. M., & Pendley, J. S. (2014). Multifamily group problem-solving intervention for adherence challenges in pediatric insulin-dependent diabetes. *Clinical Practice in Pediatric Psychology*, *2*(2), 101–115. https://doi.org/10.1037/cpp0000059.

Chamberlain, J. J., Kalyani, R. R., Leal, S., Rhinehart, A. S., Shubrook, J. H., Skolnik, N., et al. (2017). Treatment of type 1 diabetes: Synopsis of the 2017 American Diabetes Association Standards of Medical Care in Diabetes. *Annals of Internal Medicine*, *167*(7), 493–498.

Cheung, R., Cureton, V. Y., & Canham, D. L. (2006). Quality of life in adolescents with Type 1 diabetes who participate in Diabetes Camp. *The Journal of School Nursing*, *22*(1), 53–58. https://doi.org/10.1177/10598405060220010901.

Chiang, J. L., Maahs, D. M., Garvey, K. C., Hood, K. K., Laffel, L. M., Weinzimer, S. A., et al. (2018). Type 1 diabetes in children and adolescents: A position statement by the American Diabetes Association. *Diabetes Care*, *41*(9), 2026–2044.

Civita, M. D., & Dobkin, P. L. (2004). Pediatric adherence as a multidimensional and dynamic construct, involving a triadic partnership. *Journal of Pediatric Psychology*, *29*(3), 157–169. https://doi.org/10.1093/jpepsy/jsh018.

Clery, P., Stahl, D., Ismail, K., Treasure, J., & Kan, C. (2017). Systematic review and meta-analysis of the efficacy of interventions for people with Type 1 diabetes mellitus and disordered eating. *Diabetic Medisin*, *34*(12), 1667–1675. https://doi.org/10.1111/dme.13509.

Coleman, S. E., & Caswell, N. (2020). Diabetes and eating disorders: An exploration of 'Diabulimia'. *BMC Psychology*, *8*, 101. https://doi.org/10.1186/s40359-020-00468-4.

Commissariat, P. V., Kenowitz, J. R., Trast, J., Heptulla, R. A., & Gonzalez, J. S. (2016). Developing a personal and social identity with Type 1 diabetes during adolescence: A hypothesis generative study. *Qualitative Health Research*, *26*(5), 672–684. https://doi.org/10.1177/1049732316628835.

de Bruin, E. I., Zijlstra, B. J., Geurtzen, N., van Zundert, R. M., van de Weijer-Bergsma, E., Hartman, E. E., et al. (2014). Mindful parenting assessed further: Psychometric properties of the Dutch version of the Interpersonal Mindfulness in Parenting Scale (IM-P). *Mindfulness*, *5*(2), 200–212.

Delamater, A. M., de Wit, M., McDarby, V., Malik, J., & Acerini, C. L. (2014). Psychological care of children and adolescents with Type 1 diabetes. *Pediatric Diabetes*, *15*, 232–244.

Driscoll, K. A., Raymond, J., & Naranjo, D. (2016). Fear of hypoglycemia in children and adolescents and their parents with Type 1 diabetes. *Current Diabetes Reports*, *16*, 77. https://doi.org/10.1007/s11892-016-0762-2.

Dybdal, D., Tolstrup, J. S., Sildorf, S. M., Boisen, K. A., Svensson, J., Skovgaard, A. M., et al. (2017). Increasing risk of psychiatric morbidity after childhood onset type 1 diabetes: A population-based cohort study. *Diabetologia*. https://doi.org/10.1007/s00125-017-4517-7.

Ellis, D. A., Carcone, A. I., & Slatcher, R. (2019). Efficacy of mindfulness-based stress reduction in emerging adults with poorly controlled, type 1 diabetes: A pilot randomized controlled trial. *Pediatric Diabetes*, *20*, 226–234. https://doi.org/10.1111/pedi.12807.

Fairburn, C. G. (2008). *Cognitive behavior therapy and eating disorders*. Guilford Press.

Gilsanz, P., Karter, A. J., Beeri, M. S., Quesenberry, C. P., Jr., & Whitmer, R. A. (2018). The bidirectional association between depression and severe hypoglycemic and hyperglycemic events in type 1 diabetes. *Diabetes Care*, *41*(3), 446–452.

Gonder-Frederick, L. A., Schmidt, K. M., Vajda, K. A., et al. (2011). Psychometric properties of the hypoglycemia fear survey-ii for adults with Type 1 diabetes. *Diabetes Care*, *34*(4), 801–806. https://doi.org/10.2337/dc10-1343.

Grey, M., Whittemore, R., Jaser, S., Ambrosino, J., Lindemann, E., Liberti, L., et al. (2009). Effects of coping skills training in school-age children with type 1 diabetes. *Research in Nursing & Health*, *32*, 405–418. https://doi.org/10.1002/nur.20336.

Hadlandsmyth, K., White, K. S., Nesin, A. E., & Greco, L. A. (2013). Proposing an acceptance and commitment therapy intervention to promote improved diabetes management in adolescents: A treatment conceptualization. *International Journal of Behavioral Consultation and Therapy*, *7*(4), 12–15. https://doi.org/10.1037/h0100960.

Hamilton, J., & Daneman, D. (2002). Deteriorating diabetes control during adolescence: Physiological or psychological? *Journal of Pediatric Endocrinology and Metabolism*, *2002*(15), 115–126.

Hayes, S. C., & Hofmann, S. G. (2018). *Process-based CBT: The science and core clinical competencies of cognitive behavioral therapy*. Oakland, CA: New Harbinger.

Helgeson, V. S., Snyder, P. R., Escobar, O., Siminerio, L., & Becker, D. (2007). Comparison of adolescents with and without diabetes on indices of psychosocial functioning for three years. *Journal of Pediatric Psychology*, *32*(7), 794–806. https://doi.org/10.1093/jpepsy/jsm020.

Hepburn, D. A., Deary, I. J., Frier, B. M., Patrick, A. W., Quinn, J. D., & Fisher, B. M. (1991). Symptoms of acute insulin-induced hypoglycemia in humans with and without IDDM: Factor-analysis approach. *Diabetes Care*, *14*(11), 949–957.

International Diabetes Federation. (2019). *IDF diabetes atlas* (9th ed.). Brussels, Belgium: International Diabetes Federation.

Ismail, K., Maissi, E., Thomas, S., Chalder, T., Schmidt, U., & Bartlett, J. (2010). A randomised controlled trial of cognitive behaviour therapy and motivational interviewing for people with Type 1 diabetes mellitus with persistent sub-optimal glycaemic control: a diabetes and psychological therapies (ADaPT) study. *Health Technology Assessment*, *14*(22).

Kınık, M. F., Gönüllü, F. V., Vatansever, Z., & Karakaya, I. (2017). Diabulimia, a Type I diabetes mellitus-specific eating disorder. *Turkish Archives of Pediatrics/Türk Pediatri Arşivi*, *52*(1), 46.

Landolt, M. A., Ribi, K., Laimbacher, J., Vollrath, M., Gnehm, H. E., & Sennhauser, F. H. (2002). Brief report: Posttraumatic stress disorder in parents of children with newly diagnosed Type 1 diabetes. *Journal of Pediatric Psychology*, *27*(7), 647–652. https://doi.org/10.1093/jpepsy/27.7.647.

Lindholm-Olinder, A., Fischier, J., Fries, J., et al. (2015). A randomised wait-list controlled clinical trial of the effects of acceptance and commitment therapy in patients with Type 1 diabetes: A study protocol. *BMC Nursing*, *14*, 61. https://doi.org/10.1186/s12912-015-0101-y.

Ludwig, D. S., & Kabat-Zinn, J. (2008). Mindfulness in medicine. *Journal of the American Medical Association*, *300*, 1350–1352.

Maahs, D. M., West, N. A., Lawrence, J. M., & Mayer-Davis, E. J. (2010). Epidemiology of Type 1 diabetes. *Endocrinology and Metabolism Clinics of North America*, *39*(3), 481–497. https://doi.org/10.1016/j.ecl.2010.05.011.

Mackey, E. R., Herbert, L., Monaghan, M., Cogen, F., Wang, J., & Streisand, R. (2016). The feasibility of a pilot intervention for parents of young children newly diagnosed with type 1 diabetes. *Clinical Practice in Pediatric Psychology*, *4*(1), 35.

Martyn-Nemeth, P., Schwarz Farabi, S., Mihailescu, D., Nemeth, J., & Quinn, L. (2016). Fear of hypoglycemia in adults with Type 1 diabetes: Impact of therapeutic advances and strategies for prevention—A review. *Journal of Diabetes and its Complications*, *30*(1), 167–177. https://doi.org/10.1016/j.jdiacomp.2015.09.003.

McCoy, R. G., Van Houten, H. K., Ziegenfuss, J. Y., Shah, N. D., Wermers, R. A., & Smith, S. A. (2012). Increased mortality of patients with diabetes reporting severe hypoglycemia. *Diabetes Care*, *35*(9), 1897–1901.

Merwin, R. M., Moskovich, A. A., & Babyak, M. (2021). An open trial of app-assisted acceptance and commitment therapy (iACT) for eating disorders in Type 1 diabetes. *Journal of Eating Disorders*, *9*, 6. https://doi.org/10.1186/s40337-020-00357-6.

Morris, A. D., Boyle, D. I., McMahon, A. D., Greene, S. A., Mac Donald, T. M., & Newton, R. W. (1997). Adherence to insulin treatment, glycaemic control, and ketoacidosis in insulin-dependent diabetes mellitus. The DARTS/MEMO Collaboration. Diabetes audit and research in Tayside Scotland. Medicines Monitoring Unit. *Lancet (London, England)*, *350*(9090), 1505–1510.

Pillay, J., Armstrong, M. J., & Butalia, S. (2015). Behavioral programs for Type 1 diabetes mellitus: A systematic review and meta-analysis. *Annals of Internal Medicine*, *163*, 836–847. https://doi.org/10.7326/M15-1399.

Sairanen, E., Lappalainen, R., Lappalainen, P., & Hiltunen, A. (2020). Mediators of change in online acceptance and commitment therapy for psychological symptoms of parents of children with chronic conditions: An investigation of change processes. *Journal of Contextual Behavioral Science*, *15*, 123–130. https://doi.org/10.1016/j.jcbs.2019.11.010.

Schiffrin, A. (2001). Psychosocial issues in pediatric diabetes. *Current Diabetes Reports*, *1*, 33–40.

Seaquist, E. R., Anderson, J., Childs, B., et al. (2013). Hypoglycemia and diabetes: A report of a workgroup of the American diabetes association and the endocrine society. *The Journal of Clinical Endocrinology and Metabolism*, *98*(5), 1845–1859. https://doi.org/10.1210/jc.2012-4127.

Snoek, F. J., & Skinner, T. C. (2002). Psychological counselling in problematic diabetes: Does it help? *Diabetic Medicine*, *19*(4), 265–273.

Toni, G., Berioli, M. G., Cerquiglini, L., Ceccarini, G., Grohmann, U., Principi, N., et al. (2017). Eating disorders and disordered eating symptoms in adolescents with Type 1 diabetes. *Nutrients*, *9*, 906. https://doi.org/10.3390/nu9080906.

Torjesen, I. (2019). Diabulimia: The world's most dangerous eating disorder. *BMJ*, *364*, l982. https://doi.org/10.1136/bmj.l982.

Van Gampelaere, C., Luyckx, K., Goethals, E. R., et al. (2020). Parental stress, anxiety and trait mindfulness: associations with parent–child mealtime interactions in children with Type 1 diabetes. *Journal of Behavioral Medicine*, *43*, 448–459. https://doi.org/10.1007/s10865-020-00144-3.

Van Gampelaere, C., Luyckx, K., Van Ryckeghem, D. M., van der Straaten, S., Laridaen, J., Goethals, E. R., et al. (2019). Mindfulness, worries, and parenting in parents of children with type 1 diabetes. *Journal of Pediatric Psychology*, *44*(4), 499–508.

Van Name, M. A., Hilliard, M. E., Boyle, C. T., Miller, K. M., DeSalvo, D. J., Anderson, B. J., et al. (2018). Nighttime is the worst time: Parental fear of hypoglycemia in young children with Type 1 diabetes. *Pediatric Diabetes*, *19*, 114–120. https://doi.org/10.1111/pedi.12525.

Wei, C., Allen, R. J., Tallis, P. M., Ryan, F. J., Hunt, L. P., Shield, J. P. H., et al. (2018). Cognitive behavioural therapy stabilises glycaemic control in adolescents with Type 1 diabetes—Outcomes from a randomised control trial. *Pediatric Diabetes*, *19*, 106–113. https://doi.org/10.1111/pedi.12519.

Wysocki, T., & Greco, P. (2006). Social support and diabetes management in childhood and adolescence: Influence of parents and friends. *Current Diabetes Reports*, *6*, 117–122. https://doi.org/10.1007/s11892-006-0022-y.

Chapter 19

Empirically supported cognitive-behavioral interventions for adolescent anorexia and bulimia nervosa

Madelyn Johnson[a], Kathryn M. Huryk[a], Sasha Gorrell[a], and Daniel Le Grange[a,b]

[a]*Department of Psychiatry and Behavioral Sciences, University of California, San Francisco, CA, United States* [b]*Department of Psychiatry, The University of Chicago (Emeritus), Chicago, IL, United States*

Abbreviations

AAN	atypical anorexia nervosa
ACT	acceptance and commitment therapy
AFT	adolescent-focused therapy
AN	anorexia nervosa
ARFID	avoidant restrictive food intake disorder
ASFT	acceptance-based separated family treatment
BN	bulimia nervosa
CBT-A	CBT-E for adolescents
CBT-E	cognitive-behavioral therapy-enhanced for eating disorders
DBT	dialectical behavioral therapy
EBW	expected body weight
ED	eating disorders
EDE	eating disorder examination
EFT	emotional-focused therapy
EOT	end of treatment
FBT	family-based treatment
FT	family therapy
HLC	higher level of care
IFT	intensive family therapy
IPC	intensive parental coaching
NICE	The National Institute for Health Care Services
OSFED	other specified feeding or eating disorders
PDT	psychodynamic therapy
PFT	parent-focused therapy
RCT	randomized controlled trial
RO-DBT	radically open DBT
SFT	systemic family therapy
SPT	supportive psychotherapy
UP	unified protocol

Introduction

Eating disorders (ED) are pernicious and deadly psychiatric disorders that significantly impair physical health and psychosocial functioning (American Psychiatric Association [APA], 2013). In particular, restrictive EDs such as anorexia nervosa (AN) and bulimia nervosa (BN) are associated with increased mortality and high treatment costs (van Hoeken & Hoek, 2020), as well as significant global disease burden (Erskine, Whiteford, & Pike, 2016). Features of AN include markedly

Handbook of Lifespan Cognitive Behavioral Therapy. https://doi.org/10.1016/B978-0-323-85757-4.00020-1

TABLE 1 Clinical differences across diagnoses.

	Anorexia nervosa (AN)	Atypical anorexia nervosa (AAN)	Bulimia nervosa (BN)
Weight	Significant low body weight in the context of age, sex, developmental trajectory and physical health OR persistent behavior interfering with weight gain	Lose a large amount of weight, but are not considered significantly underweight in the context of age, sex, developmental trajectory and physical health	No weight criteria; often present at normal weight in the context of age, sex, developmental trajectory and physical health
Eating disorder symptoms	Inadequate food intake; may include binge eating and purging; maladaptive exercise	Inadequate food intake; may include binge eating and purging; maladaptive exercise	Cycles of binge eating and compensatory behaviors (e.g., self-induced vomiting; exercise; laxative use)
	Body image distortion and dissatisfaction; lack of recognition of low weight	Body image distortion and dissatisfaction; lack of recognition of weight loss	Body image distortion and dissatisfaction
	Undue influence of body weight or shape on self-evaluation	Undue influence of body weight or shape on self-evaluation; similar or more severe than AN	Undue influence of body weight or shape on self-evaluation
	Intense fear of gaining weight	Intense fear of gaining weight	Behaviors are intended to promote weight loss or prevent weight gain
Medical complications	Compromises all physical systems, secondary to malnutrition (e.g., bradycardia; hypotension; amenorrhea)	Same as for AN	Same as for AN/AAN with additional risk related to self-induced vomiting and other compensatory behaviors

low body weight, intense fear of weight gain, body image disturbance, and weight loss behaviors (Table 1) (APA, 2013). Among youth and adolescents, there may also be a failure to meet expected body weight based on an individual's growth history, or failure to maintain a normal developmental trajectory (e.g., growing in height but not weight). Individuals with BN similarly experience significant body image disturbance, with behavior that is characterized by recurrent binge-eating episodes (i.e., consuming an unusually large amount of food in a short amount of time with the experience of a lack of control) and inappropriate compensatory behaviors that are intended to influence weight or shape (e.g., self-induced vomiting, laxative, diuretic, diet-pill or other medication misuse, fasting, or excessive exercise).

Although lower weight has traditionally been equated with greater illness severity in AN, patients with atypical AN (AAN; a disorder classified under the umbrella diagnosis of other specified feeding or eating disorders [OSFED]), experience ED psychopathology and significant weight loss but are not underweight relative to norms for age and height (APA, 2013).

Patients with AAN comprise nearly a third of inpatient ED programs (Sawyer, Whitelaw, Le Grange, Yeo, & Hughes, 2016; Whitelaw, Gilbertson, Lee, & Sawyer, 2014; Whitelaw, Lee, Gilbertson, & Sawyer, 2018) and may have greater weight loss, and demonstrate equal or greater ED pathology and medical acuity than those with AN (Garber et al., 2019; Sawyer et al., 2016). Other individuals diagnosed with OSFED may present with specific ED behaviors (e.g., night eating) that do not otherwise meet full criteria for another eating and feeding disorder. Avoidant restrictive food intake disorder (ARFID) is characterized by a failure to meet nutritional or energy needs and is associated with nutritional deficiency, weight loss, and/or impairment of psychosocial functioning; unlike other restrictive EDs, ARFID does not feature body image disturbances. For the purposes of this text, we focus primarily on interventions for AN, AAN, and BN, given the breadth of the current evidence-base for the treatment of these disorders among youth and adolescents.

Early identification and intervention of EDs is crucial, as even subclinical cases can lead to long-term and serious physiological and psychological issues (Le Grange & Loeb, 2007). Earlier diagnosis and treatment of EDs is associated with improved treatment outcomes and a better long-term prognosis for ED recovery (Treasure & Russell, 2011). The National Institute for Health Care Services (NICE, 2017) and the Canadian practice guidelines for the treatment of children and adolescents with EDs (Couturier et al., 2020) support family-based treatment (FBT) as the recommended first-line treatment approach for adolescent AN and BN. These guidelines further express support for cognitive-behavioral therapy enhanced (CBT-E) to treat adolescent AN and BN for individuals for whom FBT is unacceptable, contraindicated (e.g., if no caretaker is able to participate in the treatment process), or ineffective (Table 2).

TABLE 2 Features of the treatment approach in FBT and CBT-E for adolescent eating disorders.

Feature of the treatment approach	FBT	CBT-E
Role of the family	Parents are considered the primary resource in facilitating progress in treatment	Routine involvement of parents, but limited
	Siblings participate in treatment to the extent indicated	
Role of the therapist	Expert guide to parents in supporting the adolescent in recovery	Collaborative relationship with adolescent who is an active participant in the process of change
Structure and foci	*Three phases*	*Three phases*
	Foci of phases:	Foci of phases:
	(1) Parents charged with restoring nutrition and weight restoration; also charged with interrupting cycle of ED behaviors	(1) Creation of personal formulation of disorder; increasing adolescent motivation for change and engagement in treatment
	(2) Gradually restore independence in eating and exercise to adolescent	(2) If willing, patients are helped to regain weight and address ED psychopathology and maintaining processes; emphasis on addressing shape and weight concern
	(3) Address typical developmental concerns that have been impacted by ED; identifying features relevant to maintain progress and preventing relapse	(3) Maintaining progress and correcting setbacks
Typical format	Primarily outpatient; ~ 20, 50–60 min sessions	Primarily outpatient; ~30–40, 50 min sessions
	Sessions are weekly at the start of treatment and bi-weekly (or further spaced) as indicated in later stages of treatment	Patient attends 2 sessions in preparation for treatment and subsequent sessions are generally weekly
	Session two consists of a family meal	
	Parents and youth participate in every session	Parents attend one session alone near the start of treatment, followed by 4–12, 15 min sessions with family together over the course of treatment

Note: FBT, family-based treatment; CBT-E, cognitive-behavioral therapy enhanced (for eating disorders); ED, eating disorder.

In the following text, we describe cognitive-behavioral and behavioral interventions for adolescent AN and BN, with a specific focus on the empirical evidence that supports CBT-E and FBT. We briefly explore other treatment options that have demonstrated a preliminary evidence base, offer recommendations for when a particular treatment option might be indicated, and conclude with suggestions for future directions in the clinical care of adolescents with restrictive EDs.

Cognitive-behavioral therapy for eating disorders

CBT-E was developed as a transdiagnostic treatment for all EDs. Although currently less empirically evaluated than FBT for the treatment of adolescents, CBT-E has emerged as a compelling and promising treatment approach for restrictive EDs (Dalle Grave, Calugi, Doll, & Fairburn, 2013). CBT-E is primarily conceived of as an individual form of therapy, with limited parental involvement. This approach aims to engage patients in creating change and taking ownership of their recovery by enabling them with skills to manage their symptoms (Dalle Grave, Eckhardt, Calugi, & Le Grange, 2019). Accordingly, an initial goal in CBT-E is to engage the adolescent in treatment and actively involve them in the process of change. Once the treatment has begun, the therapist aids the adolescent in increasing understanding of their ED psychopathology, and in beginning to disrupt these maladaptive thoughts and behaviors. During each session, the adolescent reviews the "tasks" that they have agreed to work on in between sessions; the actions taken by the adolescent between sessions in redirecting their own thoughts and behavior directly relate to the rate of change that is made during the course of therapy. Throughout, the adolescent is empowered to become actively involved in their own treatment planning and formulation, including in identifying strategies for relapse prevention (Dalle Grave et al., 2019).

Below, we describe the typical format and structure for CBT-E for adolescents with AN, AAN, and BN, the role of the family, and how the treatment approach might vary slightly according to diagnosis. We also summarize the current evidence base on CBT-E for restrictive EDs, along with recent work investigating its effectiveness when compared with FBT, and factors that may mediate or moderate treatment outcome.

Phases and format

Even before beginning treatment, the individualistic nature of CBT-E invites the adolescent, rather than their parents, to decide to take control of the ED, including when to begin treatment, what issues to focus on, and how they will approach addressing them. Once initiated, CBT-E includes three main phases of treatment. In the first phase of CBT-E, the therapist joins with the adolescent in formulating a shared conceptualization of the factors that maintain the ED, setting treatment goals, and in deciding when and how to challenge ED cognitions (Fig. 1) (Dalle Grave et al., 2019; Hutton & Morrison, 2013). Collaborative work between the therapist and adolescent encourages age-appropriate autonomy (Stacey et al., 2011), specifically in asking the patient to think critically about their own treatment goals and to practice making more informed decisions, which may lead to a reduction in symptoms (Stacey et al., 2011). Since the adolescent is actively involved in treatment and is deciding for themselves what they will or will not do, they will never be prescribed something that they do not willingly agree to (Dalle Grave et al., 2019). The first phase places an emphasis on establishing a pattern of regular eating and engaging the patient in daily self-monitoring of eating and related thoughts and behaviors.

The main focus of the next phase of CBT-E is to address ED cognitions, with particular emphasis on concerns around shape and weight, and to manage appropriate eating without compensatory behaviors. The adolescent is encouraged to examine and disrupt the mechanisms that are maintaining their ED. This phase of treatment includes psychoeducation about EDs more generally (e.g., the impact of restriction on risk for binge eating, the role of body checking and avoidance behaviors in maintaining dissatisfaction, etc.), as well as about ED-related factors which may be particularly relevant for a given patient (e.g., the emotion-regulatory function of exercise behavior), both of which will assist them in addressing their ED in the context of treatment (Dalle Grave et al., 2019). There is a continued emphasis on using self-monitoring records to help patients better understand their ED and to track how their behaviors are changing.

The final goal of CBT-E is to create lasting change for the adolescent, including by working to increase awareness of triggers for ED cognitions and by developing strategies for resisting engagement in ED behaviors. This final phase collectively addresses any signs of the ED reactivating and guides the adolescent in planning for how to prevent relapse (Dalle Grave et al., 2019; Fairburn, 2008).

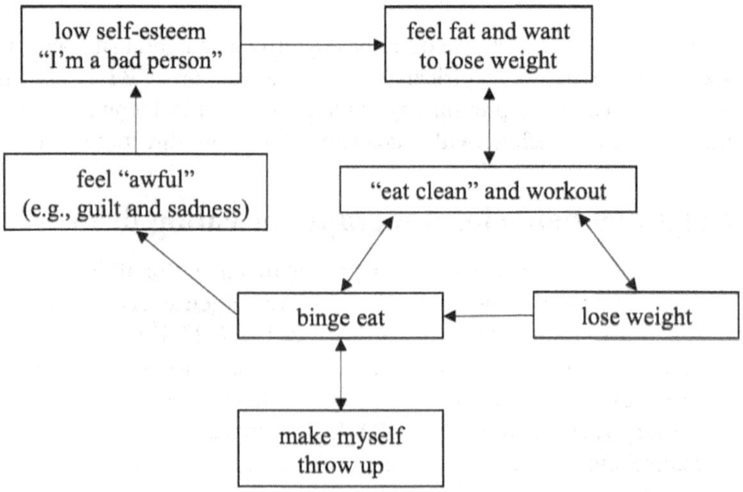

FIG. 1 Sample CBT case formulation. This is a visual representation of a case of other specified feeding or eating disorder (OSFED) in a young person's own words. Over the course of CBT-A, the patient and their therapist collaboratively develop a working case formulation, which can be reviewed and revised throughout treatment.

Role of the family, and diagnostic considerations

Distinct from CBT-E for adults, parents are included as a means of support for the adolescent when developmentally appropriate, although less so than in FBT. Although CBT-E is centrally focused on engaging the adolescent in treatment, parents and other caretaking family members are invited to attend a single one-hour assessment session toward the start of the treatment process. This session may be used to help identify family factors that will potentially hinder the adolescent's treatment success, and to provide general psychoeducation and guidance around EDs and the process of recovery. This parental session is conducted without the presence of the adolescent; however, as treatment progresses, parents are asked to attend a small portion of subsequent follow-up sessions to keep the family and adolescent apprised of progress or hindrances that may be related to the family system and structure of support.

CBT-E for AN

The CBT-E therapist and adolescent devote the initial weeks in treatment to discussing and processing reasons for and against making changes, including in understanding the importance of weight gain (Dalle Grave & Calugi, 2020; Fairburn, 2008). In the context of AN, the discussions around willingness to change allows the adolescent themselves to decide to regain weight rather than having this decision forced upon them (Dalle Grave et al., 2019). Once the adolescent has decided to make the changes necessary to recover from AN, weight gain is a primary focus followed by addressing low self-esteem, perfectionism, and interpersonal deficits that may present in the context of the disorder (Dalle Grave & Calugi, 2020). Monitoring of nutrition and weight gain continue throughout CBT-E for AN until the weight goals are met (Dalle Grave et al., 2019). If an adolescent is considerably underweight (e.g., <90% of their treatment goal weight), treatment may extend beyond the course indicated for an adolescent with less weight to gain.

CBT-E for BN

CBT-E to treat adolescent BN differs slightly in its focus. Specifically, emphasis on weight gain may be less important than it is for AN, so time can be devoted to addressing the style of eating (e.g., regularity of eating, and how much food and how quickly food is consumed), urges to binge eat, purge, or other compensatory behaviors, and understanding bodily cues (e.g., hunger and fullness) (Dalle Grave & Calugi, 2020; Fairburn, 2008). As the adolescent advances through treatment, they continue to review their own progress and are encouraged to address any barriers to change, including mechanisms that maintain patterns of binge eating and compensatory behaviors (e.g., shape and weight concerns or mood). Just as in CBT-E for AN, during the final phase of treatment, the adolescent may examine the development of the ED and the origins of the body image disturbances, and plan and prepare for relapse prevention (Dalle Grave & Calugi, 2020; Fairburn, 2008).

Empirical support for CBT-E in adolescents

Although the efficacy of CBT-E to treat adult EDs is well established (Byrne, Fursland, Allen, & Watson, 2011; Poulsen et al., 2014), there is less empirical evidence for CBT-E when adapted for adolescents. To date, there have been no randomized controlled trials (RCTs) that examine CBT-E's utility in treating adolescent AN. For treatment of adolescent BN, two RCTs have compared CBT-E to either FBT (Le Grange, Lock, Agras, Bryson, & Jo, 2015) or psychodynamic therapy (PDT; Stefini et al., 2017). In the first of these BN studies, although FBT was more effective at promoting abstinence from binge-eating and purging behaviors by end of treatment (EOT) and 6-month follow-up, there were no differences found between the two treatments in symptom report by the 12-month follow-up (Le Grange et al., 2015). When comparing CBT-E to PDT, the two treatments had comparable remission rates (33% and 31%, respectively). Both treatments also demonstrated significant improvement in secondary outcomes (e.g., stable or decreased frequency in binge eating and purging, respectively) which remained steady for both groups at 12-month follow-up (Stefini et al., 2017).

A recent nonrandomized effectiveness trial showed that FBT was more efficient in achieving weight restoration in underweight adolescents with restrictive EDs when compared to CBT-E (Le Grange et al., 2020). However, CBT-E and FBT had similar weight outcomes at 6- and 12-month follow-up and an equal reduction in other ED symptomology and cognitions at EOT. Together, these findings underscore CBT-E's feasibility as a treatment option for adolescents with restrictive EDs (Le Grange et al., 2020).

Mediators and moderators for CBT-E

Due to the limited evidence base on CBT-E in adolescent EDs, there are few identified moderators of treatment. In the above-mentioned effectiveness trial, the sample was divided into a "lower weight" and "higher weight" cohort (Le Grange et al., 2020). In the lower weight cohort only, there were four identified treatment effect moderators of weight gain at EOT: depression, age, prior hospitalization for a comorbid psychiatric disorder, and family status (Le Grange et al., 2020). Patients with higher levels of depression achieved significantly less weight gain when treated with CBT-E and older participants gained weight more quickly in CBT-E than those who were younger (Le Grange et al., 2020). This is likely related to the autonomous structure that CBT-E relies upon, such that older patients may be able to better manage the responsibilities of participating in CBT-E. In this study, previous psychiatric hospitalizations were associated with weight loss in CBT-E and for both CBT-E and FBT, patients living with their family of origin experienced less weight gain than those living in reconstituted families, a difference that was only significant for those receiving CBT-E. Taken together, these studies show very promising support for the use of CBT-E in treating adolescents with BN and further suggest certain clinical and demographic features that warrant further inquiry based on their potential moderating effect on treatment outcomes.

Family-based treatment

Following publication of the first RCT to examine family therapy (FT) in the treatment of AN and BN (Russell, Szmukler, Dare, & Eisler, 1987), the field of ED treatment experienced a steady evolution over the course of subsequent decades. Historically, parents and families had been erroneously blamed for causing their children's EDs, and they were thus often excluded from their care, by design. This promising trial of FT initiated a paradigm shift regarding caregiver involvement in treatment, later prompting the development of manualized FBT (Lock & Le Grange, 2012). Early research supported the premise that parents were active and powerful contributors to their children's health (Eisler et al., 2000; Le Grange, Eisler, Dare, & Russell, 1992), and FBT has since been rigorously tested over the past nearly three decades (Rienecke, 2017). With the recognition that families are capable of guiding their loved ones through ED recovery, parents and caregivers are now empowered to become welcomed members of the treatment team, with defined roles.

Principles of FBT

FBT is grounded in several tenets that serve to guide the clinician, patient, and family through the treatment of AN and BN (Table 3). As outlined in the published treatment manual, FBT takes an agnostic stance toward the cause of the disorder, acknowledging that there are multiple interacting risk factors that contribute developing an ED, including complex epigenetic and neurobiological pathways (Lock & Le Grange, 2012). FBT acknowledges that science cannot yet explain the precise etiology of each individual's ED, and posits that the family's energy is best spent taking action to save their child, rather than searching for elusive answers as to how or why the ED developed.

Instead, FBT clinicians help families prioritize weight restoration and renourishment, temporarily deferring focus on other issues and activities. Families are provided with psychoeducation on the importance of addressing malnourishment, which can otherwise lead to medical instability and maintenance of the ED. Caregivers are charged with the task of ensuring that their child eats adequately and regularly to prevent the ED from further harming the body and disrupting adolescent development. Additionally, FBT emphasizes externalization of the ED, wherein the patient is viewed as a person with an illness, who cannot be blamed for the disorder's development or symptoms. Learning to externalize the illness can help families empathize and connect with the healthy part of their loved one. Likewise, when family members attribute challenging aspects of the ED to the illness itself, they may reduce feelings of blame toward the adolescent.

Because FBT is expressly focused on empowering parents, the therapist takes on the role of consultant and collaborator while respecting parents' expertise on their child and family. This nonauthoritarian therapeutic stance allows caregivers to discover and practice methods that work best within their family system. Although the therapist will guide and assist the family, caregivers are responsible for renourishing their child. The therapist's energy is directed toward helping parents rebuild confidence in their ability to feed their child by emphasizing their strengths and unique position to enact change. Rather than telling the parents precisely how they should go about doing this, the therapist supports them in using their own skills and resources to bring about recovery (Rienecke, 2017).

TABLE 3 Tenets of FBT.

Tenet	Description
Illness agnosticism	No assumptions are made about the causes of the ED, instead it is viewed as a complex, multifaceted illness. There is no blame placed on the adolescent or the family members for the development of the ED
	Treatment is centered on prioritizing what needs to be done to help the adolescent recover as quickly as possible, rather than on identifying what factors caused the ED
Externalizing the illness	Separating the child and the ED is stressed throughout treatment. All participants in treatment are encouraged to understand that the child is not in control of their ED, they did not choose to develop their ED, and they cannot choose to stop the ED
	By externalizing the illness, parental criticism (i.e., expressed emotion) is reduced, which is known to have a negative impact on treatment outcomes
Nonauthoritarian therapeutic stance	The therapist assumes a consultant role, acknowledging that the caregivers are experts on their own children and family
	The therapist guides the family through recovery by assisting the parents in finding the best way to refeed their child and restore weight
Parental empowerment	FBT charges the parents with an active recovery role. The energy of the clinician is focused on empowering the parents to feel confident as they interact with their child and the ED, especially during the refeeding process
	The therapist will not provide specific meal plans or directions for recovery, which allows the parents to do what they feel is best for their child and family as they build confidence in their own abilities
Primary focus of symptom reduction	Weight restoration and medical stability is the primary initial focus of treatment. This prevents any further physical damage to the adolescent and keeps the family focused on restoring their physical health

Note: FBT, family-based treatment; ED, eating disorder.

FBT for AN

FBT was initially manualized and tested for the treatment of adolescent AN, and much of the subsequent FBT literature has focused on AN specifically. Here we describe the phases and format of FBT for AN and AAN, as well as the roles of the family and therapist. We then review the empirical support for FBT for AN and its variations in the treatment of this disorder.

Phases and format

There are three phases of FBT that guide the family through treatment for AN. The first phase is focused on restoring the weight and nutrition of the ill adolescent, and caregivers are tasked with supporting their child to eat regularly and prevent behaviors that interfere with weight gain (Lock & Le Grange, 2012). Following an assessment, the therapist begins meeting with the patient and their family on a weekly basis. Each session is spent with the family as a whole, although the therapist first meets briefly with the adolescent to weigh them, discuss the past week, and provide support. During the first session, the therapist begins to engage the whole family in the therapy by taking a history of the ED and how it has impacted their family. Employing the principles of FBT, the therapist charges parents with renourishment. In the second session, the therapist guides the family through a meal together to empower the parents as they navigate nourishing their child at home (Lock & Le Grange, 2012). Throughout the rest of phase one, the clinician assists caregivers in developing and honing the skills and strategies that enable them to re-feed their child. The parents will continue to take full responsibility of feeding their child as weight restoration and symptom reduction progresses (Lock & Le Grange, 2012).

As the adolescent makes strides with weight gain and decreased engagement in ED behaviors, they can begin to regain some independence over their eating and exercise. Phase two of FBT is focused on this monitored transition of control over weight-related behaviors back to the adolescent, and through steady progress they regain a developmentally-appropriate level of autonomy (Lock & Le Grange, 2012). The family will enter phase three when the adolescent is no longer engaging in ED behaviors and is able to maintain a healthy body weight for their expected growth based on their personal development history. An example of a weight graph that might typically be evidenced in FBT can be found in Fig. 2. During

FIG. 2 Sample weight trajectory of a teen who recovered from anorexia nervosa. This graph illustrates the weight trajectory of a cisgender female who developed anorexia nervosa and "fell off" her growth curve at 14-years-old. Following a brief hospitalization for medical stabilization, her parents pursued family-based treatment and helped her return to her prior developmental trajectory.

the third and final phase of FBT, the therapist helps the family address remaining issues related to normal adolescent development that may have been interrupted by the ED, and psychiatric comorbidities if present (Lock & Le Grange, 2012). The therapist also guides the family through treatment termination and relapse prevention planning.

Role of the family and therapist

Grounded in the belief that parents and families are the most valuable agents of change in ED recovery, the FBT therapist strives to actively involve each member of the family in treatment (Lock & Le Grange, 2012). During the initial assessment and throughout FBT, the therapist seeks to understand who is involved in the young person's life, and how they can each be engaged in the process of restoring health. Adults in caregiving roles are charged with helping their child to gain weight and normalize eating behaviors, while siblings and other family members provide support outside of mealtimes. Depending on the family's unique structure and culture, FBT might include not only parents, but also step-parents, grandparents, siblings, cousins and other important people. Research on FBT suggests that it is equally effective for a variety of family structures (e.g., separated, single-parent, grandparent-headed household) as it is for "intact" married couples (Lock, Agras, Bryson, & Kraemer, 2005; Lock, Couturier, Bryson, & Agras, 2006). Studies have also shown that among family members, mothers attend FBT sessions most frequently (Hughes, Burton, Le Grange, & Sawyer, 2018) and they typically spend the most time actively caring for their ill child (Rhind et al., 2016). That said, the therapist should strongly encourage all available parents to participate in treatment, as greater FBT session attendance by fathers has been shown to predict better treatment outcomes, in terms of both physical and cognitive recovery (Hughes et al., 2018). Although there is much yet to be learned about the involvement of siblings, qualitative research has found that FBT can help siblings to better understand the ED and learn how to support their family members (van Langenberg et al., 2018).

In FBT, the therapist acts as a guide to support the family through recovery while respecting the parents' strength and expertise on their own child. This allows the parents to reposition themselves as an authority figure over the ED (Lock & Le Grange, 2012). In order to effectively move through the phases of FBT, parents are often supported by the therapist to employ behavioral-modification strategies with their child that help to both discourage behaviors that derive from the illness (e.g., cutting food into many small pieces), as well as to encourage adaptive behaviors that support the path to recovery (e.g., effectively using coping skills to manage ED-related distress [Fig. 3]). From a theoretical perspective, principles of behaviorism include contingency management (Skinner, 1966), whereby parents may need to establish real-world consequences when behavioral goals are not met. For example, if a child is unable to finish breakfast within an agreed-upon time frame, parents may decide that the child should not attend school that day until the meal is finished. For many parents, rewarding their child for "good" behavior may feel regressive or developmentally inappropriate, particularly for older adolescents. For other parents, it is possible they have never had to use contingencies for "bad" behavior before. In these instances, the therapist can help parents to draw from their own expert knowledge of their child, and identify what rewards or negative consequences will be most likely to effectively shape their child's behavior to align less with the illness, and more toward recovery.

In reducing blame on the adolescent for their illness (i.e., agnosticism), the therapist is able to guide parents toward noticing what their child is doing well—both in the context of the ED, and in other areas of their adolescent life. In this way, parents are better able to reduce their own frustration with the entrenched ED behavior, and to empathically engage in more positive parenting practices. Drawing from principles of parent management training (e.g., Kazdin, 1997), parents may be guided to learn how to give simple and effective directions to their child that confer warmth, but also firmness (e.g., "I know this is really hard for you, but we expect you to finish that last bite on your plate.") In this example, adolescents who are struggling with completing a meal may display heightened emotion, including yelling or throwing things. Rather than engaging in the emotional outburst with the child, a more effective parenting approach may be to instead, diffuse the emotion of the moment. This may be orchestrated by intentionally not attending to the outburst and/or working to support the child in more effectively tolerating their distress (e.g., introducing an adaptive distraction).

Empirical support for FBT for AN

To date, six RCTs have examined the efficacy of FBT among adolescents with AN. In each of these RCTs, measured outcomes for the adolescent include both weight gain (e.g., reaching the expected body weight [EBW]) and cognitive recovery indices, which are assessed through the eating disorder examination interview (EDE; Fairburn, Cooper, & O'Connor, 2014). The first RCT of manualized FBT for AN compared this family approach to adolescent-focused therapy (AFT),

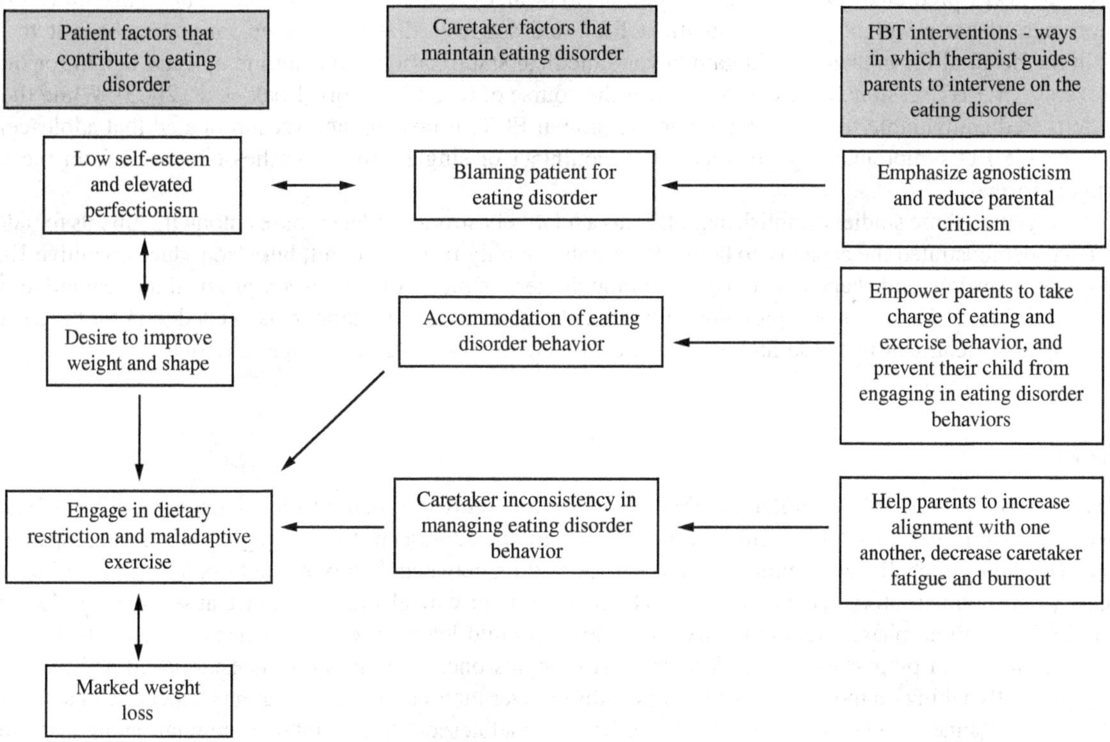

FIG. 3 Sample FBT case formulation. This is a visual representation of a case of anorexia nervosa (AN). Over the course of FBT, the therapist works to collaboratively guide caretakers to effectively support their child's recovery. As caretaker factors that maintain the eating disorder evolve over the course of treatment, the therapist may adjust the intervention strategies relative to this initial formulation.

a form of individual therapy rooted in psychodynamic theory. Compared to those who received AFT, a significantly greater proportion of those who received FBT were in full remission (i.e., >95% EBW and EDE within one standard deviation of community norms) at both 6- and 12-month follow-up (Lock et al., 2010). Systemic family therapy (SFT) has a primary focus on the family system and the relationships within a given system, and would not address the ED (i.e., the normalization of eating and weight), unless it is specifically raised by the family (Agras et al., 2014). When compared to SFT, patients randomized to FBT gained weight more quickly and required less time spent in the hospital for medical stabilization (Agras et al., 2014).

In an effort to continuously improve clinical outcomes for the treatment of adolescent EDs, two RCTs were conducted that compared FBT in differing forms: conjoint FT and separated FT (Eisler et al., 2000; Le Grange et al., 1992). Conjoint FT resembles the current format of standard FBT, where everyone in the family meets together for treatment sessions. Separated FT divides the session such that the same therapist first meets with the adolescent alone and then meets with the parents without the child present. In both RCTs, the separated format demonstrated significantly improved weight gain and treatment outcomes (Eisler et al., 2000; Le Grange et al., 1992). Through these trials, it was discovered that adolescents from more critical families fared better in a separated rather than conjoint format, findings that drove the development of parent-focused therapy (PFT). PFT is a separated version of FBT, where parents meet on their own with the clinician while the adolescent is directly monitored by another medical professional, who weighs the patient and provides brief supportive counseling at each session. PFT may be recommended for a variety of reasons, including the presence of high expressed emotion or parental criticism, which has been shown to negatively impact treatment outcomes in FBT (Eisler et al., 2000; Le Grange et al., 1992). When compared with FBT, remission rates were higher in PFT at EOT (44 v 22%) but did not differ at 6- or 12-month follow-up (Le Grange et al., 2016). These findings suggested that FBT and PFT were comparable, although secondary analyses showed that PFT was more effective at reducing criticism in families with high expressed emotion (Allan, Le Grange, Sawyer, McLean, & Hughes, 2018).

Varying lengths of hospitalization prior to beginning outpatient FBT have been studied, as well as varying lengths of FBT, to determine the most efficient course of treatment. An RCT compared a brief (for medical stability) with a longer (to achieve weight restoration, i.e., 90% EBW) hospitalization before beginning outpatient FBT (Madden et al., 2015). Results demonstrated that irrespective of hospitalization length, early weight gain by session 4 of outpatient FBT (e.g., >1.8 kg) predicted remission and higher %EBW at EOT and 12-month follow-up. This study highlights the importance of early weight gain in outpatient FBT and suggests that there is no long-term benefit to keeping patients in the hospital for greater weight gain beyond medical stabilization. To examine optimal length for outpatient FBT, 10 versus 20 FBT sessions were compared over the course of 6 or 12 months (Lock et al., 2005). While the shorter version delivered comparable results to the longer version at EOT, follow-up analyses indicated that adolescents with higher obsessive ED symptomatology or those with nonintact or single-parent families benefited from the standard (i.e., longer) format.

Taken altogether, these studies establish that FBT has a relatively strong evidence-base among treatments for adolescent AN. FBT has demonstrated the capacity to help adolescents not only restore weight, but also reduce cognitive ED symptomatology (e.g., weight and shape concerns), even though there is often a lag between physical and cognitive recovery (Lock & Le Grange, 2019). Further exploration of treatment mediators and moderators is needed to tailor the treatment to maximize positive treatment response and boost clinical outcomes (Lock & Le Grange, 2019).

FBT for BN

The phases of FBT for BN follow similarly to that for AN, only modified to better fit the diagnosis. FBT-BN begins with phase one, which aims to re-establish regular eating and prevent the adolescent from engaging in purging, especially post-mealtime. This helps to challenge common BN patterns of cycling between dietary restriction, losing control over eating, and inappropriate compensatory behaviors. Like FBT for AN, a family meal still takes place at session two (Le Grange & Lock, 2007). Throughout phase one, the family will help their child learn to feel safe when eating foods that may have become associated with binge-eating episodes. Phase two begins once regular eating is established and the adolescent can begin gradually taking on more control and responsibility over their eating, while parents supervise in a developmentally appropriate manner (Le Grange & Lock, 2007). Once the adolescent has maintained regular eating and minimized compensatory behaviors, phase three addresses familial and adolescent issues separate from BN, as well as focusing on termination of treatment and relapse prevention (Le Grange & Lock, 2007).

Empirical support for FBT for BN

Research on psychological treatments for youth with BN is limited. Of the four RCTs published to date, three have examined the efficacy of FT or FBT for adolescent BN and have compared these family approaches to various individual treatments, including self-guided CBT, supportive psychotherapy (SPT), and CBT-E for adolescents (CBT-A). Although those who participated in self-guided CBT had a greater reduction in binge eating frequency 6 months into treatment, by EOT the FT and self-guided CBT groups showed comparable levels of reduction in binge eating, purging behaviors, and ED cognitions (Schmidt et al., 2007). FBT outperformed SPT, both in reducing episodes of both binge eating and purging and in improving ED psychopathology (Le Grange, Crosby, Rathouz, & Leventhal, 2007). Compared to CBT-A, FBT has demonstrated greater efficacy in promoting abstinence from binge eating and purging, at both EOT and 6-months posttreatment follow-up (Le Grange et al., 2015). However, in this same study there were no significant differences between FBT and CBT-A at 12-months posttreatment follow-up. Taken together, these findings, while limited, provide compelling evidence of the efficacy for FBT for adolescents with BN.

Mediators and moderators for FBT

Some specific patient and family characteristics have been shown to impact the course and outcomes of FBT, primarily in AN. These include early weight gain, expressed emotion, and cognitive rigidity.

Early weight gain

For adolescents with AN, early weight gain is a robust predictor of overall response to treatment in FBT. Specifically, those who gain approximately 2–3 kg by the fourth session are most likely be in remission at EOT (Doyle, Le Grange, Loeb, Doyle, & Crosby, 2010; Le Grange et al., 2014; Madden et al., 2015). Studies have also shown that increased parental self-efficacy (i.e., degree to which parents feel they are able to effectively intervene on the disorder) predicts better FBT outcomes (Byrne, Accurso, Arnow, Lock, & Le Grange, 2015; Robinson, Strahan, Girz, Wilson, & Boachie, 2013). To this end, the FBT model can be modified for those who do not gain adequate weight within the first month of treatment.

This adaptive, stepped-care model involves adding three sessions of intensive parental coaching (IPC) to standard FBT with the goal of improving parental self-efficacy and subsequently, weight gain (Lock et al., 2015). IPC begins with a family session (when indicated) to reinvigorate the family's motivation to save their child from the crisis of malnutrition. The second session is held with parents only, during which the therapist helps them to identify barriers to renourishment. During the third IPC session, the FBT therapist coaches parents through a second family meal. Pilot research has demonstrated that this adaptation to FBT can improve rates of weight restoration and remission in early "nonresponders" (Lock et al., 2015).

Expressed emotion

High expressed emotion, in the form of parental criticism, hostility or emotional overinvolvement, has been shown to negatively impact FBT outcomes (Eisler et al., 2000; Le Grange et al., 1992; Rienecke, Accurso, Lock, & Le Grange, 2016). In these studies, the presence of parental criticism, specifically, predicted lower levels of improvement in ED psychopathology at EOT (Rienecke et al., 2016). Additionally, maternal hostility was significantly associated with poorer family functioning and lower body weight at EOT in those who received FBT, compared to those who received AFT (Lock et al., 2010).

Addressing negative emotions, modifying parental criticism, and strengthening both the parental alignment and therapeutic alliance earlier in treatment may promote FBT's effectiveness (Hughes, Sawyer, Accurso, Singh, & Le Grange, 2019; Le Grange et al., 2016). As described above, families with high expressed emotion may achieve better outcomes in a separated or parent-focused format, rather than the traditional conjoint approach to FBT (Allan et al., 2018; Eisler et al., 2000; Le Grange et al., 1992). Emotional-focused therapy (EFT) may be used as an adjunctive approach with standard FBT for families who are particularly challenged in communicating effectively (Robinson, Dolhanty, & Greenberg, 2015). EFT seeks to create a feeling of trust, mutual respect, and strong communication within families, which may help to facilitate the process of FBT.

Cognitive rigidity

Cognitive rigidity (e.g., inflexibility of thoughts or habits around food or eating) is an identified moderator of adolescent ED treatment (Lock & Le Grange, 2019) and in two studies, those with AN with heightened ED-related obsessions and compulsions fared better in FBT than in AFT or PFT (Le Grange et al., 2012, 2016). Those reporting severe ED-related obsessionality required lengthier treatment to achieve similar outcomes to those without (Lock et al., 2005). Taken together, those with pronounced food and eating-related obsessions and compulsions may require longer treatment and may have better outcomes in a conjoint family-based approach, rather than in individual or separated-format approaches.

Choosing a treatment

Factors to consider

When selecting a treatment for adolescents with AN or BN, careful consideration of the known mediators and moderators of each approach may help to inform treatment selection. For example, although FBT may be particularly efficient at restoring weight in underweight adolescents (Le Grange et al., 2020), families with high expressed emotion may benefit from a separated format of FBT (i.e., PFT) or another treatment type altogether (e.g., CBT-E), to reduce conflict and criticism that may negatively impact clinical outcomes (Rienecke et al., 2016).

When FT is unacceptable, contraindicated, or ineffective (e.g., when there is a history of abuse or other issues that prevent parents from being able to engage in FBT), taking a CBT-E approach may be preferred (NICE, 2017). However, for adolescents who are younger, have higher levels of depression, or previous hospitalizations for comorbid disorders, CBT-E may not be the best treatment option and another treatment type may be recommended (e.g., FBT; Le Grange et al., 2020). These factors should be weighed against available treatment options when treatment planning.

Higher level of care

Outpatient treatment is appropriate for most adolescents with ED, and it is strongly recommended that youth receive care in the least intensive treatment environment possible (Couturier et al., 2020; NICE, 2017). However, many patients will require a higher level of care (HLC) at some point in their illness to address medical instability, acute suicidality, complex psychiatric comorbidities, or the failure of outpatient treatment (SAHM, 2015). HLC settings include inpatient medical or psychiatric hospitals, day treatment programs (i.e., partial hospitalization and intensive outpatient), and residential treatment centers. Although HLC treatment is sometimes necessary, many such programs are not evidence-based and few have reported on their outcomes (Anderson et al., 2017; Couturier et al., 2020). There is emerging evidence that both FBT (e.g., Huryk et al., 2020; Rienecke & Richmond, 2018) and CBT-E (Dalle Grave, Conti, Sartirana, Sermattei, & Calugi, 2021) can be adapted for youth in HLC settings, which may improve treatment effectiveness and mitigate the risks associated with transitions in care.

Outpatient FBT has also been scaled up into a 5-day form of intensive family therapy (IFT), which has been shown to reduce ED symptoms and risk for hospitalization (Eisler et al., 2016; Marzola et al., 2015; Rockwell, Boutelle, Trunko, Jacobs, & Kaye, 2011). Based on the same principles as traditional FBT, the goal of IFT is to provide families with a concentrated dose of treatment to help them build knowledge and skills that will bring about rapid and lasting symptom change. IFT is geared toward families who need to bolster their work in standard FBT, those who are stepping down from HLC settings, as well as those who do not have local access to specialized ED treatment. IFT programming can be implemented with a single family or in a multifamily format, wherein groups of families share resources and mutual support.

Future directions

An explicit focus of both FBT and CBT-E for AN and BN is to quickly restore the adolescent back to a path of normal development, absent of their ED. Although early intervention yields remission for many youth who receive treatment, it is still critical that future research and implementation efforts continue to increase positive prognosis for a greater number of patients. In particular, specialized treatments such as FBT yield improvement in weight outcomes at the conclusion of a standard course of treatment, but for many, there is a latency in improvement in cognitive ED symptomology (Lock & Le Grange, 2019). Less is known currently about potential mediating and moderating factors that may specifically predict for whom, and why a particular treatment approach (e.g., FBT vs. CBT-E) may be more or less effective in fully resolving the cognitive symptoms associated with the ED. Future adequately powered trials might also examine the comparative efficacy of a specific treatment approach, in samples with greater diversity of transdiagnostic eating presentations (e.g., ARFID; pediatric obesity) and demographic features (i.e., examining the intersectionality of race, gender and ethnicity).

Perhaps more important than improving effectiveness of these varied approaches is the urgent need for dissemination and accessibility of evidence-based treatment for youth with EDs. The need for increased access to evidence-based ED treatments continues to grow; however, the accessibility of trained providers remains limited, and provider training options are financially burdensome. There have been some recent strides taken to allow for these necessities to be more accessible and cost effective. Web-based trainings may be a practical alternative to in-person trainings. Additional research is needed to evaluate the comparative efficacy of in-person versus web-based training, including potential trade-offs between quality and accessibility.

Expansion in treatment formats

Due to the limited number of specialized providers, finding a clinician nearby for treatment can be extremely difficult, especially for families that live in rural areas, away from academic institutions, where providers are often clustered. Telehealth may be a feasible and potentially more affordable option for families that cannot access in-person therapeutic services. The efficacy of telehealth has been examined in FBT, with promising indications that a telehealth approach may yield positive clinical outcomes (Anderson, Byrne, Goodyear, Reichel, & Le Grange, 2015). However, there have been no studies to date examining the efficacy of CBT-E in a telehealth delivery format and across treatment modalities, and future efforts should be made to continue to investigate the feasibility and efficacy of virtual delivery.

Self-guided options may be more cost-effective than standard therapy while still achieving promising clinical outcomes. Specifically, some work indicates that self-guided CBT offers comparable treatment effects to FT for adolescents with BN (Schmidt et al., 2007). When a self-guided approach was compared to FBT for adolescent AN, while the sample size was small, outcomes were comparable to previous RCTs involving FBT (Lock, Darcy, Fitzpatrick, Vierhile, & Sadeh-Sharvit, 2017). More rigorous exploration of self-guided options is necessary, but current evidence is promising.

Other evidence-based treatment approaches

"Third wave" psychotherapy treatment modalities have emerged as an evolution and extension of traditional CBT approaches (Hayes, 2004). Although FBT and standard CBT-E remain the current first-line recommended approaches for treating adolescents with restrictive EDs (Couturier et al., 2020; NICE, 2017) several third wave CBT treatments bear promise, and warrant further study. In particular, research over the past two decades highlights the use of dialectical behavioral therapy (DBT) (e.g., Wisniewski & Kelly, 2003), acceptance and commitment therapy (ACT) (e.g., Merwin, Zucker, & Timko, 2013), and the unified protocol (UP) (Thompson-Brenner et al., 2021) in treating adolescents with EDs. Other adjunctive treatments to FBT or CBT-E that have shown preliminary signs of efficacy among adolescents with restrictive EDs include the use of mirror exposure to specifically target body image concerns (Klimek, Wei, & Blashill, 2020) and other specific types of exposure intervention (Butler & Heimberg, 2020).

Each of the three third wave approaches identified just above have been conceptualized as an adaptation of existing treatment approaches, specific for adolescents with EDs, and have included the support of caretakers, as appropriate. In the case of DBT, a recent uncontrolled case series of Radically Open DBT (RO-DBT) demonstrated preliminary efficacy in targeting maladaptive overcontrol among adolescents with restrictive EDs (Baudinet et al., 2020). Similarly, among youth with EDs, an ACT approach is hypothesized to increase cognitive and behavioral flexibility on the part of the adolescent, and to reduce experiential avoidance and illness accommodation both on the part of the patient and their caretaker (Merwin et al., 2013; Timko, Zucker, Herbert, Rodriguez, & Merwin, 2015). In a recent trial of an integration of ACT with a separated family approach (i.e., acceptance-based separated family treatment; ASFT), results demonstrated preliminary feasibility, acceptability, and efficacy for reducing ED pathology and promoting weight gain in adolescents with AN (Timko et al., 2015). Finally, taking a transdiagnostic approach to target both ED and comorbid psychopathology, a recent report of the use of the UP in United States-based residential ED treatment programs indicated favorable outcomes on ED symptom severity, especially among those with greater baseline experiential avoidance, and improved depression, mindfulness, and anxiety sensitivity (Thompson-Brenner et al., 2021).

Summary and conclusions

The last decade has witnessed advances in the development of FBT for adolescent EDs, with emerging evidence of the efficacy of CBT-E. Taken together, there is a robust body of evidence supporting the efficacy of interventions that emphasize family involvement, along with factors that may support a more individual approach, with family support. Further research is needed to investigate the precise nature of optimal family involvement, and specifically to determine for whom, and under what conditions certain types of treatment might be most effective. Finally, the dissemination of specialty treatments such as FBT and CBT-E is critical to the advancement of evidence-based treatment of ED in adolescents, toward which all efforts ought to be made.

Practice and procedures

Mental health and medical providers should routinely assess for ED symptomatology as a part of standard practice. General ED assessment guidelines include evaluating signs and symptoms of an ED across the physical, behavioral, and cognitive-emotional domains, and when indicated, should include the use of one of several widely available empirically-based measures designed to assess and diagnose EDs. It is also important to assess for the presence of comorbid psychiatric symptoms, both secondary to malnutrition, but also any symptoms or diagnoses that were premorbid to the onset of ED symptoms. Although weight restoration may alleviate some mood symptoms for some youth, others may require more intensive services to address co-occurring psychiatric presentations. During assessment, it may also be useful to gather a thorough family history, which may be beneficial when considering how, when, and which family members can become involved in treatment. Regular assessment of weight, vital signs, and laboratory values are also critical both at entry to, and throughout the course of treatment to monitor the adolescent's physical health. Due to the complexity of the ED diagnosis, and the severity and medical acuity of these disorders, an interdisciplinary approach to treatment is typically recommended. In most cases, adolescents with EDs benefit from a cohesive team that coordinates services among mental health providers, pediatric or adolescent medicine providers, and nutritional experts who collectively support the adolescent and family through recovery. This interdisciplinary coordination of care can support parents as they navigate weight restoration and disrupting ED behaviors, as well as build confidence in their skills and ability to overcome the ED. When communication from providers varies, it can unnecessarily complicate and undermine patient and parental efforts. In contrast, when families receive consistent communication from a coordinated team, both patient and parental efforts may be considerably well-supported within the process of recovery.

Mini-dictionary of terms

Adolescent-focused therapy: an individual psychotherapy for adolescents with anorexia nervosa.

Anorexia nervosa: an eating disorder characterized by low weight, body image disturbances, and fear of weight gain or behaviors that interfere with weight gain.

Atypical anorexia nervosa: an eating disorder characterized by significant weight loss in an individual who may appear to be at a normal of higher weight, body image disturbances, and fear of weight gain or behaviors that interfere with weight gain.

Avoidant restrictive food intake disorder: an eating disorder characterized by disturbances in eating that result in failure to meet nutritional needs, driven by low appetite, avoidance of foods due to color, texture, or smell, or concern that eating will lead to choking, vomiting, or another negative consequence.

Bulimia nervosa: an eating disorder characterized by episodes of binge eating and inappropriate compensator behaviors and body image disturbances.

Cognitive-behavioral therapy enhanced for eating disorders: a transdiagnostic cognitive-behavioral treatment that focuses on disrupting mechanisms that maintain eating disorders.

Emotional-focused therapy: a therapeutic approach that focuses on emotions, promoting trust, mutual respect, and communication.

Expected body weight: an estimated weight based on one's height, age, and historical growth and development.

Family-based treatment: a family-based behavioral intervention that focuses on parent-led renourishment and disruption of eating disorder behaviors.

Family therapy: a psychotherapy that focuses on family systems, communication and conflict.

Intensive parental coaching: an adaption to family-based treatment that aims to improves parental self-efficacy when the adolescent has not gained weight early in treatment.

Other specified feeding or eating disorder: a category of eating disorders that do not fit more specific criteria, but cause significant distress or impairment in psychosocial functioning.

Parent-focused therapy: a separated version of family-based treatment, where the parents meet with the clinician alone and the child is followed by another medical professional.

Supportive psychotherapy: an individual form of psychotherapy that is nondirective and focuses on supporting patients though self-expression within a therapeutic relationship.
Systemic family therapy: a therapeutic approach that focuses on the family system and the relationships within the system.

Key facts

- FBT is the leading evidence-based intervention for adolescent AN.
- FBT is a recommended treatment for adolescent BN.
- CBT-E has shown promise in the treatment of adolescent EDs.
- Family involvement is recommended when treating adolescent EDs.
- Future research is needed to indicate for whom and under which conditions certain types of treatment for adolescent EDs might be indicated.

Applications to other areas

This chapter has reviewed empirically supported cognitive-behavioral and behavioral interventions for adolescent ED, including CBT-E and FBT. On the whole, this literature indicates that families are often assets in the treatment of restrictive EDs among youth, some of the most severe of the psychiatric disorders. The history of professionals harmfully misattributing blame for the disorder to parents of children with EDs underscores the necessity of using scientifically-validated approaches to treat psychopathology. Additionally, evidence continues to support the positive impact that parental involvement can have on treatment outcomes (Hughes et al., 2018). Furthermore, each of the evidence-based treatments discussed in this chapter emphasizes the value of promoting early behavioral change to facilitate cognitive and emotional improvements. Given the risks associated with EDs, as well as the complexities of comorbid presentations, clinicians who follow evidence-based practices in working with other populations should routinely screen for EDs. In other settings (e.g., primary care or school counseling), brief empirically-based ED screening measures should be employed to facilitate earlier intervention and better treatment outcomes. Although specialized treatment is often indicated, parents and clinicians can support youth at-risk for EDs by encouraging regular, adequate eating to support normal adolescent growth and development. On the other hand, it is also critical that those in targeted ED treatment also receive subsequent or simultaneous care that can address co-occurring symptoms of depression, anxiety, and other issues that may complicate recovery.

Summary points

- Adolescent EDs, such as AN, AAN and BN, are serious illnesses that require early intervention, and families should be involved in treatment whenever possible.
- FBT currently has the strongest evidence base in the treatment of adolescent EDs, and this intervention can be modified and adapted to improve outcomes.
- CBT-E has also shown promise in the treatment of adolescent EDs, and it is currently recommended when FBT is not feasible or effective.
- Additional research is needed to directly compare the efficacy of FBT and CBT-E, as well as to elucidate conditions under which one may be indicated over the other.
- Outpatient treatment is appropriate for most adolescents with EDs, and there is an urgent need to improve the dissemination and accessibility of evidence-based interventions.
- Hospital and intensive care settings can incorporate the empirically supported principles of FBT and CBT-E.
- More research is needed on potential alternatives to FBT and CBT-E, such as third wave CBT approaches.

Funding

Dr. Gorrell is supported by the National Institutes of Mental Health (K23MH126201).

Conflicts of interest

Dr. Le Grange receives royalties from Guilford Press and Routledge Press for books on family-based treatment and is a co-owner of the Training Institute for Child and Adolescent Eating Disorders which trains professionals in family-based treatment.

On behalf of all identified authors, we wish to confirm that there are no other known conflicts of interest associated with this publication and there has been no significant financial support for this work that could have influenced its outcome.

References

Agras, W. S., Lock, J., Brandt, H., Bryson, S. W., Dodge, E., Halmi, K. A., ... Woodside, B. (2014). Comparison of 2 family therapies for adolescent anorexia nervosa: A randomized parallel trial. *JAMA Psychiatry*, *71*(11), 1279–1286. https://doi.org/10.1001/jamapsychiatry.2014.1025.

Allan, E., Le Grange, D., Sawyer, S. M., McLean, L. A., & Hughes, E. K. (2018). Parental expressed emotion during two forms of family-based treatment for adolescent anorexia nervosa. *European Eating Disorders Review*, *26*, 46–52. https://doi.org/10.1002/erv.2564.

Anderson, K. E., Byrne, C., Goodyear, A., Reichel, R., & Le Grange, D. (2015). Telemedicine of family-based treatment for adolescent anorexia nervosa: A protocol of a treatment development study. *Journal of Eating Disorders*, *3*, 25. https://doi.org/10.1186/s40337-015-0063-1.

Anderson, L. K., Reilly, E. E., Berner, L., Wierenga, C. E., Jones, M. D., Brown, T. A., ... Cusack, A. (2017). Treating eating disorders at higher levels of care: Overview and challenges. *Current Psychiatry Reports*, *19*, 48. https://doi.org/10.1007/s11920-017-0796-4.

APA. (2013). *Diagnostic and statistical manual of mental disorders* (5th ed.). American Psychiatric Association.

Baudinet, J., Simic, M., Griffiths, H., Donnelly, C., Stewart, C., & Goddard, E. (2020). Targeting maladaptive overcontrol with radically open dialectical behaviour therapy in a day programme for adolescents with restrictive eating disorders: An uncontrolled case series. *Journal of Eating Disorders*, *8*(1), 1–13. https://doi.org/10.1186/s40337-020-00338-9.

Butler, R. M., & Heimberg, R. G. (2020). Exposure therapy for eating disorders: A systematic review. *Clinical Psychology Review*, *78*. https://doi.org/10.1016/j.cpr.2020.101851.

Byrne, C. E., Accurso, E. C., Arnow, K. D., Lock, J., & Le Grange, D. (2015). An exploratory examination of patient and parental self-efficacy as predictors of weight gain in adolescents with anorexia nervosa. *International Journal of Eating Disorders*, *48*(7), 883–888. https://doi.org/10.1002/eat.22376.

Byrne, S. M., Fursland, A., Allen, K. L., & Watson, H. (2011). The effectiveness of enhanced cognitive behavioural therapy for eating disorders: An open trial. *Behaviour Research and Therapy*, *49*(4), 219–226. https://doi.org/10.1016/j.brat.2011.01.006.

Couturier, J., Isserlin, L., Norris, M., Spettigue, W., Brouwers, M., Kimber, M., McVey, G., Webb, C., Findlay, S., Bhatnagar, N., Snelgrove, N., Ritsma, A., Preskow, W., Miller, C., Coelho, J., Boachie, A., Stienegger, C., Loewen, R., Loewen, T., ... Pilon, D. (2020). Canadian practice guidelines for the treatment of children and adolescents with eating disorders. *Journal of Eating Disorders*, *8*, 4. https://doi.org/10.1186/s40337-020-0277-8.

Dalle Grave, R., & Calugi, S. (2020). *Cognitive behavior therapy for adolescents with eating disorders*. Guilford Press.

Dalle Grave, R., Calugi, S., Doll, H. A., & Fairburn, C. G. (2013). Enhanced cognitive behaviour therapy for adolescents with anorexia nervosa: An alternative to family therapy? *Behaviour Research and Therapy*, *51*(1), R9–R12. https://doi.org/10.1016/j.brat.2012.09.008.

Dalle Grave, R., Conti, M., Sartirana, M., Sermattei, S., & Calugi, S. (2021). Enhanced cognitive behaviour therapy for adolescents with eating disorders: A systematic review of current status and future perspectives. *Italian Journal of Eating Disorders and Obesity*, *3*(29). https://doi.org/10.32044/ijedo.2021.01.

Dalle Grave, R., Eckhardt, S., Calugi, S., & Le Grange, D. (2019). A conceptual comparison of family-based treatment and enhanced cognitive behavior therapy in the treatment of adolescents with eating disorders. *Journal of Eating Disorders*, *7*, 42. https://doi.org/10.1186/s40337-019-0275-x.

Doyle, P. M., Le Grange, D., Loeb, K., Doyle, A. C., & Crosby, R. D. (2010). Early response to family-based treatment for adolescent anorexia nervosa. *International Journal of Eating Disorders*, *43*(7), 659–662. https://doi.org/10.1002/eat.20764.

Eisler, I., Dare, C., Hodes, M., Russell, G., Dodge, E., & Le Grange, D. (2000). Family therapy for adolescent anorexia nervosa: the results of a controlled comparison of two family interventions. *Journal of Child Psychology and Psychiatry*, *41*(6), 727–736.

Eisler, I., Simic, M., Hodsoll, J., Asen, E., Berelowitz, M., Connan, F., ... Landau, S. (2016). A pragmatic randomised multi-centre trial of multifamily and single-family therapy for adolescent anorexia nervosa. *BMC Psychiatry*, *16*(1). https://doi.org/10.1186/s12888-016-1129-6.

Erskine, H. E., Whiteford, H. A., & Pike, K. M. (2016). The global burden of eating disorders. *Current Opinion in Psychiatry*, *29*(6), 346–353. https://doi.org/10.1097/yco.0000000000000276.

Fairburn, C. G. (2008). *Cognitive behavior therapy and eating disorders*. Guilford Press.

Fairburn, C. G., Cooper, Z., & O'Connor, M. (2014). Eating disorder examination (17.0D). In C. G. Fairburn (Ed.), *Cognitive behavior therapy and eating disorders*. New York: Guilford Press.

Garber, A. K., Cheng, J., Accurso, E. C., Adams, S. H., Buckelew, S. M., Kapphahn, C. J., ... Golden, N. H. (2019). Weight loss and illness severity in adolescents with atypical anorexia nervosa. *Pediatrics*, *144*(6). https://doi.org/10.1542/peds.2019-2339.

Hayes, S. C. (2004). Acceptance and commitment therapy, relational frame theory, and the third wave of behavioral and cognitive therapies. *Behavior Therapy*, *35*(4), 639–665. https://doi.org/10.1016/S0005-7894(04)80013-3.

Hughes, E. K., Burton, C., Le Grange, D., & Sawyer, S. M. (2018). The participation of mothers, fathers, and siblings in family-based treatment for adolescent anorexia nervosa. *Journal of Clinical Child and Adolescent Psychology*, *47*, S456–S466. https://doi.org/10.1080/15374416.2017.1390756.

Hughes, E. K., Sawyer, S. M., Accurso, E. C., Singh, S., & Le Grange, D. (2019). Predictors of early response in conjoint and separated models of family-based treatment for adolescent anorexia nervosa. *European Eating Disorders Review*, *27*(3), 283–294. https://doi.org/10.1002/erv.2668.

Huryk, K. M., Casasnovas, A. F., Feehan, M., Paseka, K., Gazzola, P., & Loeb, K. L. (2020). Lower rates of readmission following integration of family-based treatment in a higher level of care. *Eating Disorders*, 1–8. https://doi.org/10.1080/10640266.2020.1823173.

Hutton, P., & Morrison, A. P. (2013). Collaborative empiricism in cognitive therapy for psychosis: A practice guide. *Cognitive and Behavioral Practice*, *20* (4), 429–444. https://doi.org/10.1016/j.cbpra.2012.08.003.

Kazdin, A. E. (1997). Parent management training: Evidence, outcomes, and issues. *Journal of the American Academy of Child & Adolescent Psychiatry*, *36*(10), 1349–1356. https://doi.org/10.1097/00004583-199710000-00016.

Klimek, P., Wei, B., & Blashill, A. J. (2020). Exploring moderators of mirror exposure on pre- to post changes in body image outcomes: Systematic review and meta-analysis. *Eating Disorders*, 1–22. https://doi.org/10.1080/10640266.2020.1791665.

Le Grange, D., Crosby, R. D., Rathouz, P. J., & Leventhal, B. (2007). A randomized controlled comparison of family-based treatment and supportive psychotherapy for adolescent bulimia nervosa. *Archives of General Psychiatry*, *64*(9), 1049–1056. https://doi.org/10.1001/archpsyc.64.9.1049.

Le Grange, D., Eckhardt, S., Dalle Grave, R., Crosby, R. D., Peterson, C. B., Keery, H., … Martell, C. (2020). Enhanced cognitive-behavior therapy and family-based treatment for adolescents with an eating disorder: A non-randomized effectiveness trial. *Psychological Medicine*, 1–11. https://doi.org/10.1017/S0033291720004407.

Le Grange, D., Eisler, I., Dare, C., & Russell, G. F. M. (1992). Evaluation of family treatments in adolescent anorexia nervosa: A pilot study. *International Journal of Eating Disorders*, *12*, 347–357. https://doi.org/10.1002/1098-108X(199212)12:4<347::AID-EAT2260120402>3.0.CO;2-W.

Le Grange, D., Hughes, E. K., Court, A., Yeo, M., Crosby, R. D., & Sawyer, S. M. (2014). Randomized clinical trial of parent-focused treatment and family-based treatment for adolescent anorexia nervosa. *Journal of the American Academy of Child and Adolescent Psychiatry*, *55*(8), 683–692. https://doi.org/10.1016/j.jaac.2016.05.007.

Le Grange, D., Hughes, E. K., Court, A., Yeo, M., Crosby, R. D., & Sawyer, S. M. (2016). Randomized clinical trial of parent-focused treatment and family-based treatment for adolescent anorexia nervosa. *Journal of the American Academy of Child and Adolescent Psychiatry*, *55*(8), 683–692.

Le Grange, D., & Lock, J. (2007). *Treating bulimia in adolescents: A family-based approach*. The Guilford Press.

Le Grange, D., Lock, J., Agras, W. S., Bryson, S., & Jo, B. (2015). Randomized clinical trial of family-based treatment and cognitive-behavioral therapy for adolescent bulimia nervosa. *Journal of the American Academy of Child and Adolescent Psychiatry*, *54*(11), 886–894. https://doi.org/10.1016/j.jaac.2015.08.008.

Le Grange, D., Lock, J., Agras, W. S., Moye, A., Bryson, S. W., Jo, B., & Kraemer, H. C. (2012). Moderators and mediators of remission in family-based treatment and adolescent focused therapy for anorexia nervosa. *Behaviour Research and Therapy*, *50*(2), 85–92. https://doi.org/10.1016/j.brat.2011.11.003.

Le Grange, D., & Loeb, K. (2007). Early identification and treatment of eating disorders: Prodrome to syndrome. *Early Intervention in Psychiatry*, *1*, 27–39. https://doi.org/10.1111/j.1751-7893.2007.00007.x.

Lock, J., Agras, S. W., Bryson, S., & Kraemer, H. C. (2005). A comparison of short- and long-term family therapy for adolescent anorexia nervosa. *Journal of the American Academy of Child and Adolescent Psychiatry*, *44*(7), 632–639. https://doi.org/10.1097/01.chi.0000161647.82775.0a.

Lock, J., Couturier, J., Bryson, S., & Agras, S. (2006). Predictors of dropout and remission in family therapy for adolescent anorexia nervosa in a randomized clinical trial. *The International Journal of Eating Disorders*, *39*(8), 639–647. https://doi.org/10.1002/eat.20328.

Lock, J., Darcy, A., Fitzpatrick, K. K., Vierhile, M., & Sadeh-Sharvit, S. (2017). Parental guided self-help family based treatment for adolescents with anorexia nervosa: A feasibility study. *International Journal of Eating Disorders*, *50*(9), 1104–1108. https://doi.org/10.1002/eat.22733.

Lock, J., & Le Grange, D. (2012). *Treatment manual for anorexia nervosa: A family-based approach* (2nd ed.). Guilford Press.

Lock, J., & Le Grange, D. (2019). Family-based treatment: Where are we and where should we be going to improve recovery in child and adolescent eating disorders. *The International Journal of Eating Disorders*, *52*(4), 481–487. https://doi.org/10.1002/eat.22980.

Lock, J., Le Grange, D., Agras, W. S., Fitzpatrick, K. K., Jo, B., Accurso, E., … Stainer, M. (2015). Can adaptive treatment improve outcomes in family-based therapy for adolescents with anorexia nervosa? Feasibility and treatment effects of a multi-site treatment study. *Behaviour Research and Therapy*, *73*, 90–95. https://doi.org/10.1016/j.brat.2015.07.015.

Lock, J., Le Grange, D., Agras, W. S., Moye, A., Bryson, S. W., & Jo, B. (2010). Randomized clinical trial comparing family-based treatment with adolescent-focused individual therapy for adolescents with anorexia nervosa. *Archives of General Psychiatry*, *67*(10), 1025–1032. https://doi.org/10.1001/archgenpsychiatry.2010.128.

Madden, S., Miskovic-Wheatley, J., Wallis, A., Kohn, M., Lock, J., Le Grange, D., … Touyz, S. (2015). A randomized controlled trial of in-patient treatment for anorexia nervosa in medically unstable adolescents. *Psychological Medicine*, *45*(2), 415–427. https://doi.org/10.1017/S0033291714001573.

Marzola, E., Knatz, S., Murray, S. B., Rockwell, R., Boutelle, K., Eisler, I., & Kaye, W. H. (2015). Short-term intensive family therapy for adolescent eating disorders: 30-month outcome. *European Eating Disorders Review*, *23*(3), 210–218. https://doi.org/10.1002/erv.2353.

Merwin, R. M., Zucker, N. L., & Timko, C. A. (2013). A pilot study of an acceptance-based separated family treatment for adolescent anorexia nervosa. *Cognitive and Behavioral Practice*, *20*(4), 485–500. https://doi.org/10.1016/j.cbpra.2012.11.001.

National Institute for Health and Care Excellence. (2017). *Eating disorders (NICE guideline 69)*.

Poulsen, S., Lunn, S., Daniel, S. I., Folke, S., Mathiesen, B. B., Katznelson, H., & Fairburn, C. G. (2014). A randomized controlled trial of psychoanalytic psychotherapy or cognitive-behavioral therapy for bulimia nervosa. *The American Journal of Psychiatry*, *171*(1), 109–116. https://doi.org/10.1176/appi.ajp.2013.12121511.

Rhind, C., Salerno, L., Hibbs, R., Micali, N., Schmidt, U., Gowers, S., … Treasure, J. (2016). The objective and subjective caregiving burden and caregiving behaviours of parents of adolescents with anorexia nervosa. *European Eating Disorders Review*, *24*, 310–319. https://doi.org/10.1002/erv.2442.

Rienecke, R. D. (2017). Family-based treatment of eating disorders in adolescents: Current insights. *Adolescent Health, Medicine and Therapeutics*, *8*, 69–79. https://doi.org/10.2147/AHMT.S115775.

Rienecke, R. D., Accurso, E. C., Lock, J., & Le Grange, D. (2016). Expressed emotion, family functioning, and treatment outcome for adolescents with anorexia nervosa. *European Eating Disorders Review*, *24*(1), 43–51. https://doi.org/10.1002/erv.2389.

Rienecke, R. D., & Richmond, R. L. (2018). Three-month follow-up in a family-based partial hospitalization program. *Eating Disorders*, *26*, 278–289. https://doi.org/10.1080/10640266.2017.1388665.

Robinson, A. L., Dolhanty, J., & Greenberg, L. (2015). Emotion-focused family therapy for eating disorders in children and adolescents. *Clinical Psychology & Psychotherapy*, *22*(1), 75–82.

Robinson, A. L., Strahan, E., Girz, L., Wilson, A., & Boachie, A. (2013). 'I know I can help you': Parental self-efficacy predicts adolescent outcomes in family-based therapy for eating disorders. *European Eating Disorders Review*, *21*, 108–114. https://doi.org/10.1002/erv.2180.

Rockwell, R., Boutelle, K., Trunko, M., Jacobs, M., & Kaye, W. (2011). An innovative short-term, intensive, family-based treatment for adolescent anorexia nervosa: Case series. *European Eating Disorders Review*, *19*(4), 362–367. https://doi.org/10.1002/erv.1094.

Russell, G. F., Szmukler, G. I., Dare, C., & Eisler, I. (1987). An evaluation of family therapy in anorexia nervosa and bulimia nervosa. *Archives of General Psychiatry*, *44*(12), 1047. https://doi.org/10.1001/archpsyc.1987.01800240021004.

Sawyer, S. M., Whitelaw, M., Le Grange, D., Yeo, M., & Hughes, E. K. (2016). Physical and psychological morbidity in adolescents with atypical anorexia nervosa. *Pediatrics*, *137*(4). https://doi.org/10.1542/peds.2015-4080, e20154080.

Schmidt, U., Lee, S., Beecham, J., Perkins, S., Treasure, J., Yi, I., … Eisler, I. (2007). A randomized controlled trial of family therapy and cognitive behavior therapy guided self-care for adolescents with bulimia nervosa and related disorders. *The American Journal of Psychiatry*, *164*(4), 591–598. https://doi.org/10.1176/ajp.2007.164.4.591.

Skinner, B. F. (1966). Contingencies of reinforcement in the design of a culture. *Behavioral Science*, *11*(3), 159–166. https://doi.org/10.1002/bs.3830110302.

Society for Adolescent Health and Medicine. (2015). Position paper of the Society for Adolescent Health and Medicine: Medical management of restrictive eating disorders in adolescents and young adults. *Journal of Adolescent Health*, *56*(1), 121–125. https://doi.org/10.1016/j.jadohealth.2014.10.259.

Stacey, D., Légaré, F., Col, N. F., Bennett, C. L., Barry, M. J., Eden, K. B., … Wu, J. H. C. (2011). Decision aids for people facing health treatment or screening decisions. *Cochrane Database of Systematic Reviews*. https://doi.org/10.1002/14651858.cd001431.pub4.

Stefini, A., Salzer, S., Reich, G., Horn, H., Winkelmann, K., Bents, H., … Kronmüller, K. T. (2017). Cognitive-behavioral and psychodynamic therapy in female adolescents with bulimia nervosa: A randomized controlled trial. *Journal of the American Academy of Child and Adolescent Psychiatry*, *56*(4), 329–335. https://doi.org/10.1016/j.jaac.2017.01.019.

Thompson-Brenner, H., Singh, S., Gardner, T., Brooks, G. E., Smith, M. T., Lowe, M. R., & Boswell, J. F. (2021). The renfrew unified treatment for eating disorders and comorbidity: Long-term effects of an evidence-based practice implementation in residential treatment. *Frontiers in Psychiatry*, *12*, 226. https://doi.org/10.3389/fpsyt.2021.641601.

Timko, C. A., Zucker, N. L., Herbert, J. D., Rodriguez, D., & Merwin, R. M. (2015). An open trial of acceptance-based separated family treatment (ASFT) for adolescents with anorexia nervosa. *Behaviour Research and Therapy*, *69*, 63–74. https://doi.org/10.1016/j.brat.2015.03.011.

Treasure, J., & Russell, G. (2011). The case for early intervention in anorexia nervosa: Theoretical exploration of maintaining factors. *The British Journal of Psychiatry*, *199*(1), 5–7. https://doi.org/10.1192/bjp.bp.110.087585.

van Hoeken, D., & Hoek, H. W. (2020). Review of the burden of eating disorders: Mortality, disability, costs, quality of life, and family burden. *Current Opinion in Psychiatry*, *33*(6), 521–527. https://doi.org/10.1097/YCO.0000000000000641.

van Langenberg, T., Duncan, R. E., Allen, J. S., Sawyer, S. M., Le Grange, D., & Hughes, E. K. (2018). "They don't really get heard": A qualitative study of sibling involvement across two forms of family-based treatment for adolescent anorexia nervosa. *Eating Disorders*, *26*(4), 373–387. https://doi.org/10.1080/10640266.2018.1453632.

Whitelaw, M., Gilbertson, H., Lee, K. J., & Sawyer, S. M. (2014). Restrictive eating disorders among adolescent inpatients. *Pediatrics*, *134*(3), e758–e764. https://doi.org/10.1542/peds.2014-0070.

Whitelaw, M., Lee, K. J., Gilbertson, H., & Sawyer, S. M. (2018). Predictors of complications in anorexia nervosa and atypical anorexia nervosa: Degree of underweight or extent and recency of weight loss? *The Journal of Adolescent Health*, *63*(6), 717–723. https://doi.org/10.1016/j.jadohealth.2018.08.019.

Wisniewski, L., & Kelly, E. (2003). The application of dialectical behavior therapy to the treatment of eating disorders. *Cognitive and Behavioral Practice*, *10*(2), 131–138. https://doi.org/10.1016/S1077-7229(03)80021-4.

Chapter 20

Inflammatory bowel disease and cognitive behavioral therapy (CBT) in the young

L. Stapersma[a,b], J.C. Escher[c], and E.M.W.J. Utens[b,d]

[a]*Department of Child and Adolescent Psychiatry/Psychology, Erasmus Medical Centre Sophia's Children Hospital, Rotterdam, The Netherlands,* [b]*Levvel, Center for Child Psychiatry/Department of Child and Adolescent Psychiatry, Academic Medical Center, Amsterdam, The Netherlands,* [c]*Department of Pediatric Gastroenterology, Erasmus Medical Centre Sophia's Children Hospital, Rotterdam, The Netherlands,* [d]*Research Institute of Child Development and Education, University of Amsterdam, Amsterdam, The Netherlands*

Abbreviations

CAU	care-as-usual
CBT	cognitive behavioral therapy
CD	Crohn's disease
HRQOL	health-related quality of life
IBD	inflammatory bowel disease
IBD-U	inflammatory bowel disease unclassified
RCT	randomized controlled trial
SNDT	supportive nondirective therapy
TRF	tumor necrosis factor
UC	ulcerative colitis

Inflammatory bowel disease in youth

Inflammatory bowel disease

Inflammatory bowel disease (IBD) is a chronic disease, characterized by inflammation of the gastrointestinal tract. IBD encompasses two types: Crohn's disease (CD) and ulcerative colitis (UC). IBD-unclassified (IBD-U) is used for colitis that cannot (yet) be classified as either CD or UC. The disease course fluctuates from periods of active disease (relapses, with increased symptoms) and periods of clinical remission (no symptoms present). CD and UC share several symptoms, such as abdominal pain, (bloody) diarrhea, anemia, and systemic symptoms, such as fatigue, lack of appetite, and weight loss (Griffiths, 2004; Rosen, Dhawan, & Saeed, 2015). The main difference between CD and UC is the location and type of the inflammation (see Fig. 1). In CD, any part of the gastrointestinal tract from mouth to anus can be affected, whereas in UC, inflammation often is located at the colon or rectum. Furthermore, the onset of CD is more insidious than that of UC, and CD often presents itself in "skip lesions," that is, some parts of the intestines are affected, whereas others are not (Sauer & Kugathasan, 2009; Szigethy, McLafferty, & Goyal, 2011). IBD can be considered when typical symptoms are present and abnormalities in blood and stool are apparent. The diagnosis can only be confirmed after extensive endoscopy of both the upper and lower parts of the intestinal tract with multiple mucosal biopsies. The etiology of IBD is complex, with several factors playing a role. A genetic susceptibility often is present and then combined with a dysregulated immunological response to the bacterial gut flora. Environmental factors such as infection, certain foods, smoking, and psychological stress can further trigger the dysregulation (Griffiths, 2004; Szigethy et al., 2011).

IBD in children and adolescents

Approximately 35% of patients receive the diagnosis of IBD before they are 25 years of age (Bequet et al., 2017; Rosen et al., 2015). Children and adolescents (*hereafter referred to as youth*) often present with malnutrition, growth failure, or

Handbook of Lifespan Cognitive Behavioral Therapy. https://doi.org/10.1016/B978-0-323-85757-4.00015-8

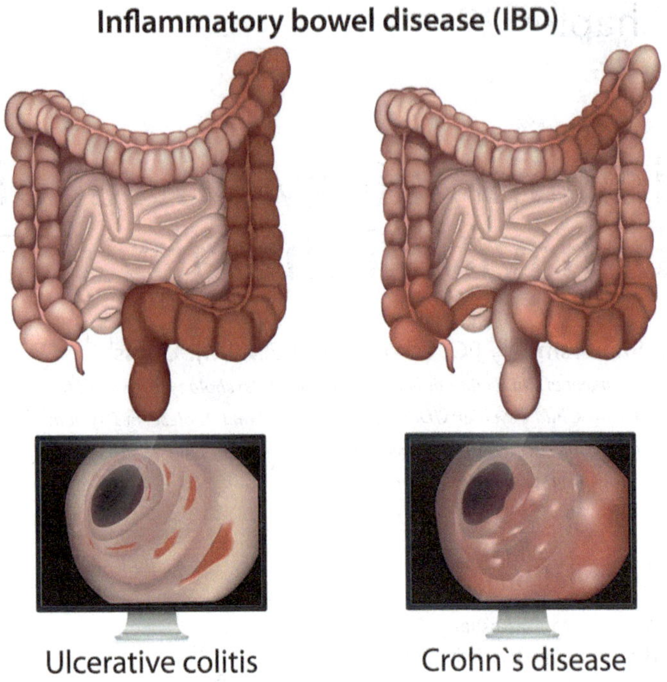

FIG. 1 Differences in location of inflammation between Crohn's disease (CD) and ulcerative colitis (UC). CD and UC are the two main types of inflammatory bowel disease (IBD). Both are characterized by inflammation in the gastrointestinal tract. The location and type of the inflammation, however, differs.

delayed puberty (Sauer & Kugathasan, 2009). Worldwide incidence rates of IBD range from 11.4 (Asia/the Middle East, Oceania) to 13.9 (North America) to 23 (Europe) per 100,000 person-years. Incidence rates are rising, which most likely can be attributed to the rising incidence of CD (Benchimol et al., 2011; Sykora et al., 2018).

Medical treatment

The medical treatment for IBD is focused on the suppression of the inflammation and aimed at the induction and maintenance of clinical remission, since IBD cannot be cured. Preferably treatment also aims to heal to inflamed mucosal (Atreya & Neurath, 2017). First-line treatment for children with active disease is exclusive enteral nutrition. This is less invasive than pharmacological treatment, but requires adherence to a strict liquid diet for 8 weeks. If enteral nutrition fails, corticosteroids are often used for induction of remission in moderate-to-severe IBD. After remission is established, immunomodulators are used to maintain remission (Gomollon et al., 2017; Magro et al., 2017). In refractory IBD, biologicals such as infliximab or adalimumab (anti-TRF-treatments) or vedolizumab (antiintegrin) can be used. Surgery is indicated when noninflammatory strictures are present in CD, or when all treatments fail in UC.

Psychological aspects of IBD in youth

Psychological problems

The biopsychosocial model implies that psychological and social factors are likely to be interrelated with disease symptoms and functional outcomes (Drossman, 1998), see Fig. 2 for a possible conceptualization of interrelationship between IBD and psychosocial factors. Due to the unpredictable course of the disease and the chronic nature of IBD, adolescents are at risk to experience psychosocial problems (Mackner et al., 2013; Ross, Strachan, Russell, & Wilson, 2011). The psychological problems most studied in IBD patients are anxiety and depression. The importance of anxiety and depression in IBD is also stressed by the mutual effects anxiety/depression and IBD can have on each other.

In adults, a metaanalysis showed that patients with IBD have a high risk for having anxiety and/or depression, consisting of either subclinical anxiety/depression or clinical anxiety or depressive disorders (Neuendorf, Harding, Stello, Hanes, & Wahbeh, 2016). In youth, several studies indicate that patients often experience anxiety/depression (Greenley et al., 2010; Ross et al., 2011). Somewhat in contrast, a metaanalysis from our group showed that anxiety/depression were prevalent, although not more prevalent than in healthy populations (Stapersma, van den Brink, Szigethy, Escher, & Utens, 2018).

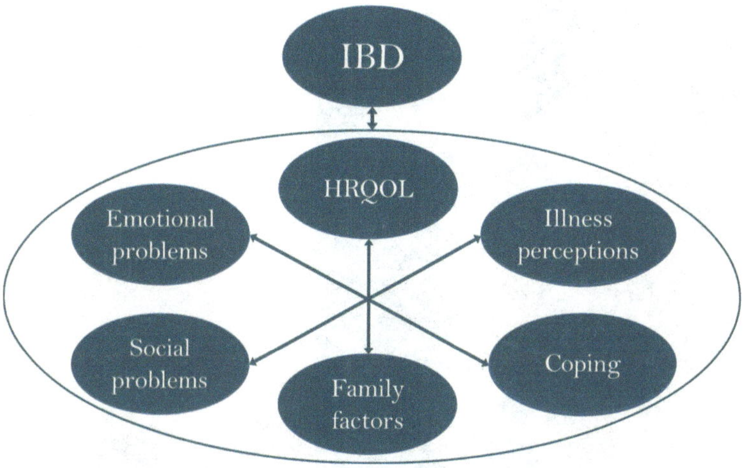

FIG. 2 Conceptualization of interrelationships between IBD and psychosocial factors. In pediatric IBD, several psychosocial factors play a role in the disease outcome of patients. This model displays a conceptualization of potential interrelationships between IBD and psychosocial factors. Evidence has been found for several interrelationships between factors. This conceptualization as a whole, however, has not been substantiated, but rather is a visual aid.

However, these results should be interpreted with caution since across studies varying instruments and cutoffs were used to assess symptoms. Further, only a few studies investigated anxiety/depressive disorders.

Bidirectional relationship inflammation and anxiety/depression in IBD

In IBD, the relationship between inflammation and anxiety/depression seems bidirectional (Peppas et al., 2021). Studies in youth are scarce, but some evidence for this bidirectionality has been found in adults. Anxiety and depression are associated with clinical relapse (Bernstein et al., 2010) or higher clinical disease activity (Marrie, Graff, Fisk, Patten, & Bernstein, 2021) during follow-up. More specifically, depression has a negative impact on disease course (Alexakis, Kumar, Saxena, & Pollok, 2017) and is related to a shorter time to relapse when compared to anxiety (Mikocka-Walus, Pittet, Rossel, von Kanel, & Swiss, 2016). Evidence of the reverse was found as well: clinical disease activity impacts anxiety and depression (Mikocka-Walus, Knowles, Keefer, & Graff, 2016; Panara et al., 2014). Recently, two studies provided even more evidence for the bidirectionality between inflammation and anxiety/depression. Clinical disease activity predicted change in perceived stress, where perceived stress predicted change in clinical disease activity (Sexton et al., 2017). In addition, clinical disease activity gave a sixfold risk on anxiety 2 years later, and baseline anxiety and depression were associated with several indicators of increased clinical disease activity (Gracie, Guthrie, Hamlin, & Ford, 2018). In youth, clinical disease activity was associated with anxiety and depression (Brooks et al., 2016).

Brain-gut axis

The common combination of IBD inflammation with psychological problems (anxiety, depression, stress) can be explained by the brain-gut axis (see Fig. 3). The hypothesized brain-gut axis proposes communication between the brain and the gut, via interactions between the automatic nervous system, the central nervous system, the stress system (hypothalamic-pituitary-adrenal [HPA] axis), the corticotropin-releasing factor system, and the intestinal response. The communication seems bidirectional; simply stated IBD increases stress and stress increases IBD (Bonaz & Bernstein, 2013; Peppas et al., 2021). Evidence comes from animal studies showing that stress induces reactive inflammation in colitis models (Mawdsley & Rampton, 2005). In human, elevated levels of inflammatory markers have been found in patients with depression (Dowlati et al., 2010; Rosenblat, Cha, Mansur, & McIntyre, 2014) and anxiety (Hou & Baldwin, 2012; Vogelzangs, Beekman, de Jonge, & Penninx, 2013). A vicious circle arises, in which inflammation and anxiety/depression negatively influence each other (Bernstein, 2017; Peppas et al., 2021).

Other mental health problems

Next to anxiety and depression, health-related quality of life (HRQOL) is studied extensively in youth with IBD. Several studies have shown that youth with IBD have lower HRQOL than healthy controls (Brooks et al., 2019; Ross et al., 2011). Whether youth with IBD report lower HRQOL than youth with other somatic disease is unclear (Knowles et al., 2018). Other types of psychological problems than anxiety and depression in youth with IBD have been studied to a very limited extent. A few studies investigated externalizing problems in youth with IBD; none reported differences with normative data (Greenley et al., 2010). Recently, a population-based analysis was made in patients 5–24 years of age. The prevalence of

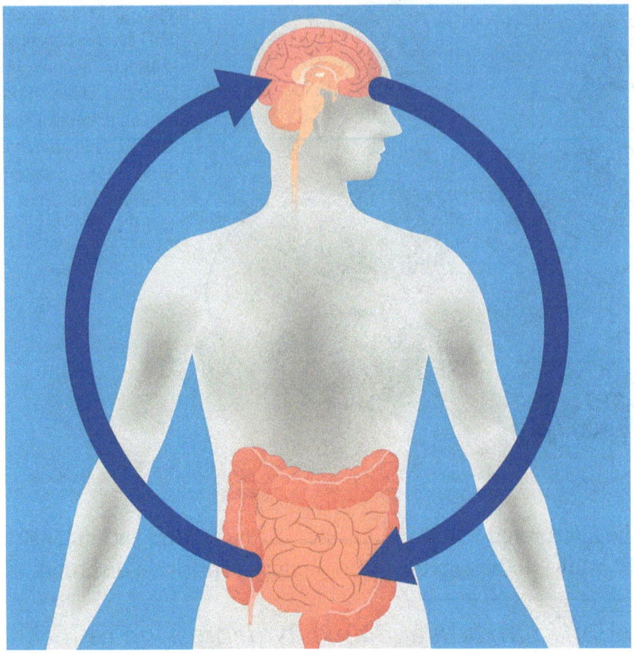

FIG. 3 Brain-gut axis. The brain-gut axis proposes an explanation between inflammation and anxiety/depression, involving interactions between several nervous and stress systems, thereby facilitating communication between the brain and the gut.

any psychiatric disorder was 21.6%. This prevalence was mainly explained by anxiety and depression. Of all individual psychiatric disorders that the authors counted in IBD patients, less than 5% was other than anxiety and depression. The risk of psychiatric disorders was higher in females and among older patients between 15 and 24 years (Thavamani, Umapathi, Khatana, & Gulati, 2019).

Other psychological factors

Besides the above-described psychological problems and HRQOL, other psychological factors are important for youth with IBD.

Coping. First, coping has been investigated in some studies in youth with IBD. It is unclear whether youth with IBD use more positive or negative coping styles, compared to healthy controls. Nevertheless, coping seems to be positively related to psychological outcomes in these patients (McCombie, Mulder, & Gearry, 2013a, 2013b). Maladaptive or passive coping was related to more anxiety and depression (Xanthopoulos, 2006), while adaptive coping was associated with higher HRQOL (van der Zaag-Loonen, Grootenhuis, Last, & Derkx, 2004).

Illness perceptions. Second, illness perceptions influence adjustment to the IBD. In adults, unfavorable illness perceptions have been shown to negatively affect outcomes and adjustment (Jordan, Sin, Fear, & Chalder, 2016; Tribbick et al., 2017). Although scarcely studied in children and adolescents, illness perceptions are suggested to be associated with psychological problems (Brooks et al., 2016; Stapersma et al., 2019).

Social problems. Third, being diagnosed with or having IBD can have a large impact on social functioning, especially in adolescence, a life phase in which social activities are of great importance for social development. Although studies based on parent-report did not show differences in social functioning between youth with IBD and healthy controls, studies based on self-report did (Greenley et al., 2010). More specific, in youth, IBD was associated with worse social competence and more social problems (Mackner, Vannatta, & Crandall, 2012). It has been shown that youth with IBD have a delayed (psycho)social development, i.e., they less often have side-jobs, were less frequently going out, were older when falling in love for the first time, and the transition from family dependence to close peer friendships was delayed (Hummel et al., 2013).

Treatment adherence. Last, since IBD is a chronic disease, treatment adherence is very important as well. Nonadherence can lead to unnecessary escalation of medical treatment. Rates for nonadherence in adolescents have been shown to range from 2% to 93%, depending on definition of adherence, used methodology, and type of medication (Spekhorst, Hummel, Benninga, van Rheenen, & Kindermann, 2015). Recently, Knowles and Alex (2020) reported that 65%–90% of adolescents adhere perfectly. Often, in adolescence, nonadherence is unintended, i.e., they forget to take their medication, sleep in during weekends, are not at home, or find their regimen quite complex. However, these so-called barriers interact with psychological factors, such as anxiety/depression, family dysfunction, and maladaptive coping.

Family factors

Not only the youth itself, but also the larger family system is affected by the IBD of the patient. Although family members, especially parents, have an important role in the management of the disease, literature has largely focused on the medical and psychological functioning of the identified patient. However, several family factors have been examined in youth with IBD, almost exclusively in cross-sectional studies. Recently, a comprehensive review has been published with a summary of the findings (Cushman, Shih, & Reed, 2020). First, Cushman et al. (2020) found that parental stress, i.e., the stress associated with the care for a child with an illness was high in parents of youth with IBD, and often comparable to the stress level of parents of youth with other chronic illnesses. Higher parental stress is associated with more psychosocial problems and lower HRQOL in the child with IBD. Their review showed that the association between parental stress and disease factors is less clear.

Second, the review showed that a substantial amount (approximately half) of parents experiences psychosocial problems (anxiety and depressive symptoms, stress). Results on how psychosocial functioning of fathers of IBD youth differs from fathers in healthy populations were mixed. Mothers of youth with IBD reported significantly higher levels of psychological problems than mothers of healthy controls and of normative samples. In general, mothers of children with IBD report comparable rates of psychological problems with mothers of youth with other chronic illnesses, which are higher than in healthy controls. Cushman et al. (2020) found in multiple studies that parental psychosocial functioning was related to child HRQOL, psychosocial functioning, and disease factors (i.e., disease activity and pain).

Third, family functioning refers to the dynamics within a family, including, for example, communication, problem solving, and affective availability. Although not extensively studied, several aspects of family functioning appeared to be negatively affected in families with a pediatric IBD patient. Worse family functioning is associated with worse child outcomes, either psychosocial or disease-related (i.e., disease activity and medication adherence) as well (Cushman et al., 2020). In sum, it is important to consider family factors in youth with IBD, since these have been shown to be related to disease outcomes.

Cognitive behavioral therapy in youth with IBD

CBT and anxiety and depression

To date, only a few handful of studies tested the effect of CBT on psychological outcomes in youth with IBD. Nearly all focused on either anxiety, depression, or both as outcome. In a small pilot study, nine patients improved in their anxiety after following a CBT protocol that included discussing illness concerns (Reigada et al., 2013). In a small randomized controlled trial (RCT), Szigethy and colleagues found that subclinical depression was effectively treated with a disease-specific CBT protocol (PASCET-PI; Primary and Secondary Control Enhancement Therapy for Physical Illness, see Fig. 4), showing better outcomes than care-as-usual (CAU, standard medical care), also at 12-month follow-up (Szigethy et al., 2007; Thompson et al., 2012). The authors later confirmed the effectiveness of CBT in improving clinical depression in a larger RCT, but then supportive nondirective therapy (SNDT) showed similar results (Szigethy et al., 2014). In contrast, a three-session social learning CBT tested in a large RCT appeared to be not better in improving subclinical anxiety and depression than a time and attention control (Levy et al., 2016). In addition, a disease-specific CBT protocol adapted to target both anxiety and depression, improved these outcomes, but did not outperformed the CAU control group (Stapersma et al., 2018, 2019). Recently, also a study was planned to test the effectiveness of mindfulness-based CBT in youth with IBD (Ewais et al., 2020), although no results are published yet. Concluding, overall, promising results of CBT have been found, although not all studies found significant differences between the youth in CBT groups and controls.

CBT and disease activity/disease course

Three cohorts investigated the effect of CBT on disease activity. Szigethy et al. (2014) showed a greater reduction in disease activity in the CBT group compared to SNDT. On the other hand, Levy et al. (2016) and van den Brink et al. (2019) did not find an additional effect of CBT on reduction of flares or time to relapse. However, exploratory analyses suggested that

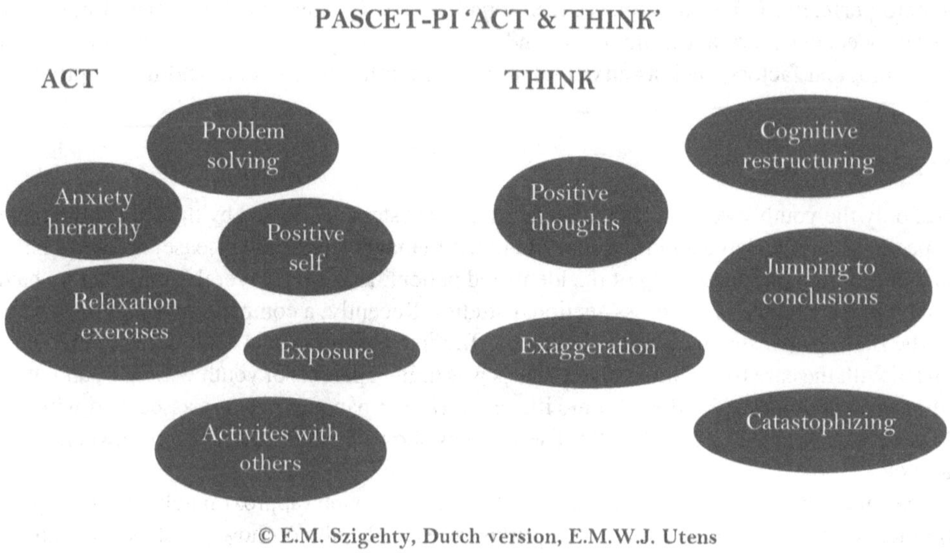

FIG. 4 Overview of the PASCET-PI. The PASCET-PI (Primary and Secondary Control Enhancement Therapy for Physical Illness) is a cognitive behavioral therapy (CBT)-protocol, with disease-specific elements (see Practice and Procedures for more details). The protocol entails behavioral interventions (summarized using the acronym ACT) and cognitive interventions (summarized with the acronym THINK).

younger children and youth with more clinical disease activity may benefit from CBT in improving their disease course. In sum, mixed results have been found with respect to the effect of CBT on disease activity.

CBT and other psychosocial outcomes

Very few studies have tested the effects of CBT on outcomes other than anxiety and depression. A few studies focused on HRQOL, even less on coping, and the results were mixed (Levy et al., 2016; Stapersma, van den Brink, van der Ende, Szigethy, et al., 2019; Szigethy et al., 2014). Nevertheless, CBT may be effective to improve HRQOL (Levy et al., 2016; Szigethy et al., 2014).

Evidence from adults with IBD indicates that CBT has been effective in improving outcomes such as pain, fatigue, and IBD relapses (McCombie et al., 2013a, 2013b). Recently, a metaanalysis found that in adults, psychological therapies can improve anxiety, depression, and HRQOL, although with small effect sizes (Davis, Bolin, Crane, & Crandell, 2020). Most importantly, it seems that psychological therapies should be targeted at patients that are most likely to benefit. Unfortunately, only few studies to date adhered to this advice. Patients with psychological problems or higher disease activity are most likely to benefit from CBT.

In pediatric IBD, only two studies have been performed to improve parental outcomes by CBT. The study of Levy et al. (2016) tested a three-session social learning CBT and reported less parental maladaptive responses to children's symptoms. Furthermore, Greenley et al. (2015) tested a problem-solving skills training for families, but did not included parental outcomes. In other pediatric chronic illnesses, however, it has been found that CBT can improve parenting behavior, and problem-solving therapy can improve parental behavior and mental health (Law, Fisher, Eccleston, & Palermo, 2019).

Clinical implications

With respect to psychological problems in youth with IBD, pediatric gastroenterologists should be aware of the coexistence of anxiety and depression with IBD, and even more of the mutual effects the disease and the psychological problems can have. Until now, it is not completely clear yet to which extent anxiety/depression and IBD impact each other. Clinicians and future studies should take into account the fluctuating course of the disease, when trying to understand anxiety and depression in pediatric IBD. To be able to target treatment, systematic screening on anxiety and depression of these patients is of the utmost importance. Furthermore, other psychological factors should also be considered, when youth with IBD experience psychosocial problems. Coping, illness perceptions, and family factors all can play an important role.

Regarding psychological treatment for youth with IBD, it is crucial to gain more insight into which patients benefit the most from what type of psychological intervention. It seems important to attune patients to the psychological intervention most suited for them, based on their disease activity or psychological problems. If patients experience clinical psychological problems, psychotherapy can enhance the medical disease management as well. For subclinical psychological problems, monitoring of these problems may be sufficient (Stapersma, van den Brink, van der Ende, et al., 2018; Stapersma, van den Brink, van der Ende, Szigethy, et al., 2019). Attention for these subclinical problems may be helpful, for example, because knowing a professional keeps track of your psychological problems already can be stress relieving. For clinicians working with youth with IBD, it can support the psychological treatment to monitor the IBD disease activity between sessions. Tailoring the treatment to the disease course and relapses should also be considered. Especially when patients are hospitalized, psychological treatment may be provided via telephone, videoconferencing, or in the hospital.

Practice and procedures

In youth with IBD, a disease-specific CBT protocol has been developed by E.M. Szigethy, the Primary and Secondary Control Enhancement Therapy for Physical Illness (PASCET-PI). She and her colleagues adapted the original PASCET (Weisz, Southam-Gerow, Gordis, & Connor-Smith, 2003) to a protocol tailored to treat depression in patients with IBD (Szigethy et al., 2007). The most important modifications were to focus on an appropriate locus of control, to integrate the illness narrative (see Fig. 5) into the treatment goals, to apply CBT skills to illness-related problems (i.e., problem solving focusing on medication adherence) and to incorporate psychoeducation about the relationship between depression and IBD. In addition, in the protocol, flexible delivery of the sessions was advocated (e.g., via phone or during medical visits). Next, E.M.W.J. Utens expanded the protocol to focus on anxiety as well, by adding an anxiety hierarchy and anxiety-focused relaxation, see, for example, van den Brink et al. (2016).

Mini-dictionary of terms

- *Inflammatory bowel disease*: a chronic disease characterized by relapsing inflammation of the gastrointestinal tract, with two main types Crohn's disease and ulcerative colitis.
- *Disease activity*: the level of activity, i.e., the level of inflammation in the gastrointestinal tract caused by the IBD. Terms used for in what state IBD is are active versus inactive.
- *Disease course*: the fluctuation and course of IBD, encompassing, e.g., relapses of inflammation and the levels of disease activity in a patient.
- *Anxiety*: an emotion characterized by feelings of tension and restlessness, worrying thoughts, and physical symptoms, such as raised blood pressure or palpitations.
- *Depression*: feelings of sadness, loss of interest, with cognitive distortions (negative thinking, difficulty concentrating), and somatic symptoms such as problems with sleeping and eating.
- *Brain-gut axis*: hypothesized mechanism with interactions between the autonomic nervous system, the central nervous system, the stress system (hypothalamic-pituitary-adrenal axis), the corticotropin-releasing factor system, and the intestinal response, facilitating communication between the brain and the gut.
- *Coping*: cognitive and behavioral strategies of an individual to deal with stress or demanding situations.
- *Illness perceptions*: representations someone has about the illness, its treatment, and the consequences of the illness.

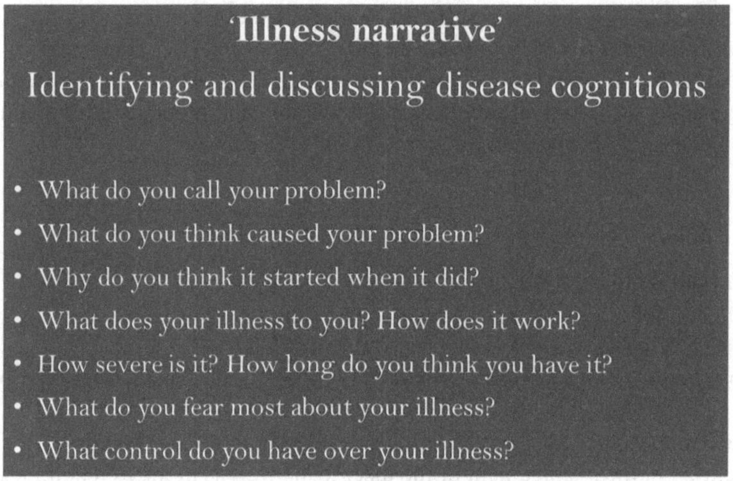

© E.M. Szigehty, Dutch version, E.M.W.J. Utens

FIG. 5 Illness narrative. The illness narrative is the most central adaptation of the PASCET-PI, in which the disease cognitions a patient has are discussed extensively and are monitored during the treatment phase.

- *Family factors*: a broad range of factors having to do with family (of a pediatric patient), such as the interactions or communication between family members, problems of parents, or coping styles used by the family and its members.
- *Cognitive behavioral therapy*: short evidence-based form of psychotherapy, focusing on symptoms in the present or future, in which interactions between thoughts (cognitions), actions (behavior) and feelings (emotions) are being targeted.

Key facts of IBD

- In IBD, the intestinal mucosa is affected by inflammation, leading to a variety of symptoms, such as abdominal pain, bloody stools, and fatigue.
- Patients with IBD also can have fistulas, i.e., connections that are formed from the gastrointestinal tract to other parts of organs or the surface of a nearby skin.
- Treatment of IBD has shifted from symptom control to mucosal healing. First, for induction of remission nutritional therapy (in CD), mesalamines (in UC) or corticosteroids (in both CD and UC) are first-line therapy. For maintenance of remission, immunomodulators can be used. Surgery is used for complications such as stricture of the bowel or fistulizing disease in CD or can be curative in UC.
- In children, treatment considerations are optimization of nutrition, achievement of normal pubertal development, and facilitation of emotional and social development.
- The process of transition to adult care is important to manage in the context of pediatric chronic illnesses. In IBD, this is important since disease management shifts from parents to the adolescent or young adult.

Applications to other areas

The field of pediatric psychology focuses on psychological factors and treatment in youth with somatic illnesses. With respect to application the other areas, CBT in other pediatric conditions should be considered. In general, psychological treatment has been shown to be able to improve psychological problems in youth with a chronic somatic illness (Bennett, Shafran, Coughtrey, Walker, & Heyman, 2015; Sansom-Daly, Peate, Wakefield, Bryant, & Cohn, 2012). However, not in all somatic illnesses, the evidence for the effectiveness of psychotherapy is firmly established. A recent Cochrane review indicated that the quality of RCTs in pediatric patients varied strongly (Thabrew et al., 2018). Therapies designed to treat anxiety and depression, especially those based on CBT principles, are suggested to be effective in treating subclinical psychological symptoms in children with a chronic somatic illness. When targeted at other psychological problems, psychological therapies did not perform better than control conditions (Thabrew, Stasiak, Hetrick, Donkin, et al., 2018). With respect to e-health interventions for youth with chronic somatic illnesses, these authors report in a follow-up review, that e-health may be promising (Thabrew et al., 2018). However, studies were often of low quality, and there is room for further development.

For youth with a chronic somatic illness, it seems important that standard generic CBT-protocols should be adapted to be used for pediatric patients. Behavioral and cognitive techniques can be applied to pediatric populations with somatic illnesses. However, specific aspects of the somatic disease should receive sufficient attention and being integrated in the psychotherapeutic process. In general, having a chronic illness severely impacts healthy psychosocial development. This impact should be addressed by therapists. More specifically, in auto-immune disorders (such as IBD, asthma, or rheumatic diseases), the mutual impact that disease symptoms and psychological problems can have on each other should be considered. Furthermore, disease management by the patient and his/her family is important to discuss.

Summary points

- Having IBD increases the risk for psychological problems, specifically anxiety and depression.
- Even in quiescent disease, youth with IBD can experience anxiety or depressive symptoms.
- Systematic and repeated screening for anxiety and depression is important in youth with IBD.
- CBT for youth with IBD is promising, but should be optimized: which patients have to be treated with what type of psychotherapy?
- Studies on psychotherapy in IBD should consider the erratic disease course when assessing treatment outcomes.
- The family system should be received more attention in the management of the disease, as well as in the psychological treatment.

References

Alexakis, C., Kumar, S., Saxena, S., & Pollok, R. (2017). Systematic review and meta-analysis: The impact of a depressive state on disease course in adult inflammatory bowel disease. *Alimentary Pharmacology & Therapeutics*, *46*(3), 225–235.

Atreya, R., & Neurath, M. F. (2017). Current and future targets for mucosal healing in inflammatory bowel disease. *Visceral Medicine*, *33*(1), 82–88. https://doi.org/10.1159/000458006.

Benchimol, E. I., Fortinsky, K. J., Gozdyra, P., Van den Heuvel, M., Van Limbergen, J., & Griffiths, A. M. (2011). Epidemiology of pediatric inflammatory bowel disease: A systematic review of international trends. *Inflammatory Bowel Diseases*, *17*(1), 423–439. https://doi.org/10.1002/ibd.21349.

Bennett, S., Shafran, R., Coughtrey, A., Walker, S., & Heyman, I. (2015). Psychological interventions for mental health disorders in children with chronic physical illness: A systematic review. *Archives of Disease in Childhood*, *100*(4), 308–316.

Bequet, E., Sarter, H., Fumery, M., Vasseur, F., Armengol-Debeir, L., Pariente, B., … EPIMAD Group. (2017). Incidence and phenotype at diagnosis of very-early-onset compared with later-onset paediatric inflammatory bowel disease: A population-based study [1988–2011]. *Journal of Crohn's and Colitis*, *11*(5), 519–526.

Bernstein, C. N. (2017). The brain-gut axis and stress in inflammatory bowel disease. *Gastroenterology Clinics of North America*, *46*(4), 839–846.

Bernstein, C. N., Singh, S., Graff, L. A., Walker, J. R., Miller, N., & Cheang, M. (2010). A prospective population-based study of triggers of symptomatic flares in IBD. *American Journal of Gastroenterology*, *105*(9), 1994–2002. https://doi.org/10.1038/ajg.2010.140.

Bonaz, B. L., & Bernstein, C. N. (2013). Brain-gut interactions in inflammatory bowel disease. *Gastroenterology*, *144*(1), 36–49. https://doi.org/10.1053/j.gastro.2012.10.003.

Brooks, A. J., Norman, P., Peach, E. J., Ryder, A., Scott, A. J., Narula, P., … Rowse, G. (2019). Prospective study of psychological morbidity and illness perceptions in young people with inflammatory bowel disease. *Journal of Crohn's and Colitis*, *13*(8), 1003–1011.

Brooks, A. J., Rowse, G., Ryder, A., Peach, E. J., Corfe, B. M., & Lobo, A. J. (2016). Systematic review: Psychological morbidity in young people with inflammatory bowel disease—Risk factors and impacts. *Alimentary Pharmacology & Therapeutics*, *44*(1), 3–15.

Cushman, G., Shih, S., & Reed, B. (2020). Parent and family functioning in pediatric inflammatory bowel disease. *Children (Basel, Switzerland)*, *7*(10), 188. https://doi.org/10.3390/children7100188.

Davis, S. P., Bolin, L. P., Crane, P. B., & Crandell, J. (2020). Non-pharmacological interventions for anxiety and depression in adults with inflammatory bowel disease: A systematic review and meta-analysis. *Frontiers in Psychology*, *11*, 538741.

Dowlati, Y., Herrmann, N., Swardfager, W., Liu, H., Sham, L., Reim, E. K., & Lanctot, K. L. (2010). A meta-analysis of cytokines in major depression. *Biological Psychiatry*, *67*(5), 446–457.

Drossman, D. A. (1998). Presidential address: Gastrointestinal illness and the biopsychosocial model. *Psychosomatic Medicine*, *60*(3), 258–267.

Ewais, T., Begun, J., Kenny, M., Headey, A., Tefay, M., & Kisely, S. (2020). Mindfulness-based cognitive therapy experiences in youth with inflammatory bowel disease and depression: Findings from a mixed methods qualitative study. *BMJ Open*, *10*(11), e041140.

Gomollon, F., Dignass, A., Annese, V., Tilg, H., Van Assche, G., Lindsay, J. O., … Ecco. (2017). 3rd European evidence-based consensus on the diagnosis and management of crohn's disease 2016. Part 1. Diagnosis and medical management. *Journal of Crohn's and Colitis*, *11*(1), 3–25.

Gracie, D. J., Guthrie, E. A., Hamlin, P. J., & Ford, A. C. (2018). Bi-directionality of brain-gut interactions in patients with inflammatory bowel disease. *Gastroenterology*, *154*(6), 1635–1646. e1633. doi:S0016-5085(18)30068-4 [pii] https://doi.org/10.1053/j.gastro.2018.01.027.

Greenley, R. N., Gumidyala, A. P., Nguyen, E., Plevinsky, J. M., Poulopoulos, N., Thomason, M. M., … Kahn, S. A. (2015). Can you teach a teen new tricks? Problem solving skills training improves oral medication adherence in pediatric patients with inflammatory bowel disease participating in a randomized trial. *Inflammatory Bowel Diseases*, *21*(11), 2649–2657.

Greenley, R. N., Hommel, K. A., Nebel, J., Raboin, T., Li, S. H., Simpson, P., & Mackner, L. (2010). A meta-analytic review of the psychosocial adjustment of youth with inflammatory bowel disease. *Journal of Pediatric Psychology*, *35*(8), 857–869. doi:jsp120 [pii] 10.1093/jpepsy/jsp120.

Griffiths, A. M. (2004). Specificities of inflammatory bowel disease in childhood. *Best Practice & Research. Clinical Gastroenterology*, *18*(3), 509–523.

Hou, R., & Baldwin, D. S. (2012). A neuroimmunological perspective on anxiety disorders. *Human Psychopharmacology: Clinical and Experimental*, *27*(1), 6–14.

Hummel, T. Z., Tak, E., Maurice-Stam, H., Benninga, M. A., Kindermann, A., & Grootenhuis, M. A. (2013). Psychosocial developmental trajectory of adolescents with inflammatory bowel disease. *Journal of Pediatric Gastroenterology and Nutrition*, *57*(2), 219–224.

Jordan, C., Sin, J., Fear, N. T., & Chalder, T. (2016). A systematic review of the psychological correlates of adjustment outcomes in adults with inflammatory bowel disease. *Clinical Psychology Review*, *47*, 28–40. https://doi.org/10.1016/j.cpr.2016.06.001.

Knowles, S. R., & Alex, G. (2020). Medication adherence across the life span in inflammatory bowel disease: Implications and recommendations for nurses and other health providers. *Gastroenterology Nursing*, *43*(1), 76–88. https://doi.org/10.1097/sga.0000000000000467.

Knowles, S. R., Graff, L. A., Wilding, H., Hewitt, C., Keefer, L., & Mikocka-Walus, A. (2018). Quality of life in inflammatory bowel disease: A systematic review and meta-analyses—Part I. *Inflammatory Bowel Diseases*, *24*(4), 742–751.

Law, E., Fisher, E., Eccleston, C., & Palermo, T. M. (2019). Psychological interventions for parents of children and adolescents with chronic illness. *Cochrane Database of Systematic Reviews*, *3*, CD009660.

Levy, R. L., van Tilburg, M. A., Langer, S. L., Romano, J. M., Walker, L. S., Mancl, L. A., … Whitehead, W. E. (2016). Effects of a cognitive behavioral therapy intervention trial to improve disease outcomes in children with inflammatory bowel disease. *Inflammatory Bowel Diseases*, *22*(9), 2134–2148.

Mackner, L. M., Greenley, R. N., Szigethy, E., Herzer, M., Deer, K., & Hommel, K. A. (2013). Psychosocial issues in pediatric inflammatory bowel disease: Report of the north american society for pediatric gastroenterology, hepatology, and nutrition. *Journal of Pediatric Gastroenterology and Nutrition*, *56*(4), 449–458. https://doi.org/10.1097/MPG.0b013e3182841263.

Mackner, L. M., Vannatta, K., & Crandall, W. V. (2012). Gender differences in the social functioning of adolescents with inflammatory bowel disease. *Journal of Clinical Psychology in Medical Settings*, *19*(3), 270–276. https://doi.org/10.1007/s10880-011-9292-x.

Magro, F., Gionchetti, P., Eliakim, R., Ardizzone, S., Armuzzi, A., Barreiro-de Acosta, M., … Colitis, O. (2017). Third European evidence-based consensus on diagnosis and management of ulcerative colitis. Part 1. Definitions, diagnosis, extra-intestinal manifestations, pregnancy, cancer surveillance, surgery, and ileo-anal pouch disorders. *Journal of Crohn's and Colitis*, *11*(6), 649–670.

Marrie, R. A., Graff, L. A., Fisk, J. D., Patten, S. B., & Bernstein, C. N. (2021). The relationship between symptoms of depression and anxiety and disease activity in IBD over time. *Inflammatory Bowel Diseases*, *27*(8), 1285–1293. https://doi.org/10.1093/ibd/izaa349.

Mawdsley, J. E., & Rampton, D. S. (2005). Psychological stress in IBD: New insights into pathogenic and therapeutic implications. *Gut*, *54*(10), 1481–1491.

McCombie, A. M., Mulder, R. T., & Gearry, R. B. (2013a). How IBD patients cope with IBD: A systematic review. *Journal of Crohn's and Colitis*, *7*(2), 89–106.

McCombie, A. M., Mulder, R. T., & Gearry, R. B. (2013b). Psychotherapy for inflammatory bowel disease: A review and update. *Journal of Crohn's and Colitis*, *7*(12), 935–949. doi:S1873-9946(13)00069-X [pii] 10.1016/j.crohns.2013.02.004.

Mikocka-Walus, A., Knowles, S. R., Keefer, L., & Graff, L. (2016). Controversies revisited: A systematic review of the comorbidity of depression and anxiety with inflammatory bowel diseases. *Inflammatory Bowel Diseases*, *22*(3), 752–762.

Mikocka-Walus, A., Pittet, V., Rossel, J. B., von Kanel, R., & Swiss, I. B. D. C. S. G. (2016). Symptoms of depression and anxiety are independently associated with clinical recurrence of inflammatory bowel disease. *Clinical Gastroenterology and Hepatology*, *14*(6), 829–835 e821. doi:S1542-3565(16)00047-1 [pii] https://doi.org/10.1016/j.cgh.2015.12.045.

Neuendorf, R., Harding, A., Stello, N., Hanes, D., & Wahbeh, H. (2016). Depression and anxiety in patients with inflammatory bowel disease: A systematic review. *Journal of Psychosomatic Research*, *87*, 70–80.

Panara, A. J., Yarur, A. J., Rieders, B., Proksell, S., Deshpande, A. R., Abreu, M. T., & Sussman, D. A. (2014). The incidence and risk factors for developing depression after being diagnosed with inflammatory bowel disease: A cohort study. *Alimentary Pharmacology and Therapeutics*, *39*(8), 802–810. https://doi.org/10.1111/apt.12669.

Peppas, S., Pansieri, C., Piovani, D., Danese, S., Peyrin-Biroulet, L., Tsantes, A. G., … Bonovas, S. (2021). The brain-gut axis: Psychological functioning and inflammatory bowel diseases. *Journal of Clinical Medicine*, *10*(3).

Reigada, L. C., Benkov, K. J., Bruzzese, J. M., Hoogendoorn, C., Szigethy, E., Briggie, A., … Warner, C. M. (2013). Integrating illness concerns into cognitive behavioral therapy for children and adolescents with inflammatory bowel disease and co-occurring anxiety. *Journal for Specialists in Pediatric Nursing*, *18*(2), 133–143. https://doi.org/10.1111/jspn.12019.

Rosen, M. J., Dhawan, A., & Saeed, S. A. (2015). Inflammatory bowel disease in children and adolescents. *JAMA Pediatrics*, *169*(11), 1053–1060.

Rosenblat, J. D., Cha, D. S., Mansur, R. B., & McIntyre, R. S. (2014). Inflamed moods: A review of the interactions between inflammation and mood disorders. *Progress in Neuro-Psychopharmacology and Biological Psychiatry*, *53*, 23–34. https://doi.org/10.1016/j.pnpbp.2014.01.013.

Ross, S. C., Strachan, J., Russell, R. K., & Wilson, S. L. (2011). Psychosocial functioning and health-related quality of life in paediatric inflammatory bowel disease. *Journal of Pediatric Gastroenterology and Nutrition*, *53*(5), 480–488. https://doi.org/10.1097/MPG.0b013e31822f2c32.

Sansom-Daly, U. M., Peate, M., Wakefield, C. E., Bryant, R. A., & Cohn, R. J. (2012). A systematic review of psychological interventions for adolescents and young adults living with chronic illness. *Health Psychology*, *31*(3), 380–393. https://doi.org/10.1037/a0025977.

Sauer, C. G., & Kugathasan, S. (2009). Pediatric inflammatory bowel disease: Highlighting pediatric differences in IBD. *Gastroenterology Clinics of North America*, *38*(4), 611–628.

Sexton, K. A., Walker, J. R., Graff, L. A., Bernstein, M. T., Beatie, B., Miller, N., … Targownik, L. E. (2017). Evidence of bidirectional associations between perceived stress and symptom activity: A prospective longitudinal investigation in inflammatory bowel disease. *Inflammatory Bowel Diseases*, *23*(3), 473–483.

Spekhorst, L. M., Hummel, T. Z., Benninga, M. A., van Rheenen, P. F., & Kindermann, A. (2015). Adherence to oral maintenance treatment in adolescents with inflammatory bowel disease: A systematic review. *Journal of Pediatric Gastroenterology and Nutrition*, *62*(2), 264–270. https://doi.org/10.1097/MPG.0000000000000924.

Stapersma, L., van den Brink, G., Szigethy, E. M., Escher, J. C., & Utens, E. (2018). Systematic review with meta-analysis: Anxiety and depression in children and adolescents with inflammatory bowel disease. *Alimentary Pharmacology & Therapeutics*, *48*(5), 496–506. https://doi.org/10.1111/apt.14865.

Stapersma, L., van den Brink, G., van der Ende, J., Bodelier, A. G., van Wering, H. M., Hurkmans, P. C. W. M., … Utens, E. M. W. J. (2019). Illness perceptions and depression are associated with health-related quality of life in youth with inflammatory bowel disease. *International Journal of Behavioral Medicine*. https://doi.org/10.1007/s12529-019-09791-6.

Stapersma, L., van den Brink, G., van der Ende, J., Szigethy, E. M., Beukers, R., Korpershoek, T. A., … Utens, E. (2018). Effectiveness of disease-specific cognitive behavioral therapy on anxiety, depression, and quality of life in youth with inflammatory bowel disease: A randomized controlled trial. *Journal of Pediatric Psychology*, *43*(9), 967–980. doi:5025052 [pii] 10.1093/jpepsy/jsy029.

Stapersma, L., van den Brink, G., van der Ende, J., Szigethy, E. M., Groeneweg, M., de Bruijne, F. H., … Utens, E. M. W. J. (2019). Psychological outcomes of a cognitive behavioral therapy for youth with inflammatory bowel disease: Results of the HAPPY-IBD randomized controlled trial at 6- and 12-month follow-up. *Journal of Clinical Psychology in Medical Settings*, 1–17.

Sykora, J., Pomahacova, R., Kreslova, M., Cvalinova, D., Stych, P., & Schwarz, J. (2018). Current global trends in the incidence of pediatric-onset inflammatory bowel disease. *World Journal of Gastroenterology*, *24*(25), 2741–2763.

Szigethy, E., Bujoreanu, S. I., Youk, A. O., Weisz, J., Benhayon, D., Fairclough, D., … DeMaso, D. R. (2014). Randomized efficacy trial of two psychotherapies for depression in youth with inflammatory bowel disease. *Journal of the American Academy of Child and Adolescent Psychiatry*, *53*(7), 726–735. doi:S0890-8567(14)00297-4 [pii] 10.1016/j.jaac.2014.04.014.

Szigethy, E., Kenney, E., Carpenter, J., Hardy, D. M., Fairclough, D., Bousvaros, A., … DeMaso, D. R. (2007). Cognitive-behavioral therapy for adolescents with inflammatory bowel disease and subsyndromal depression. *Journal of the American Academy of Child and Adolescent Psychiatry*, *46* (10), 1290–1298. https://doi.org/10.1097/chi.0b013e3180f6341f S0890-8567(09)61847-5 [pii].

Szigethy, E., McLafferty, L., & Goyal, A. (2011). Inflammatory bowel disease. *Pediatric Clinics of North America*, *58*(4), 903–920. x–xi. doi:S0031-3955 (11)00066-6 [pii] https://doi.org/10.1016/j.pcl.2011.06.007.

Thabrew, H., Stasiak, K., Hetrick, S. E., Donkin, L., Huss, J. H., Highlander, A., … Merry, S. N. (2018). Psychological therapies for anxiety and depression in children and adolescents with long-term physical conditions. *Cochrane Database of Systematic Reviews*, *12*(12), CD012488.

Thabrew, H., Stasiak, K., Hetrick, S. E., Wong, S., Huss, J. H., & Merry, S. N. (2018). E-Health interventions for anxiety and depression in children and adolescents with long-term physical conditions. *Cochrane Database of Systematic Reviews*, *8*(8), CD012489.

Thavamani, A., Umapathi, K. K., Khatana, J., & Gulati, R. (2019). Burden of psychiatric disorders among pediatric and young adults with inflammatory bowel disease: A population-based analysis. *Pediatric Gastroenterology, Hepatology & Nutrition*, *22*(6), 527–535.

Thompson, R. D., Craig, A., Crawford, E. A., Fairclough, D., Gonzalez-Heydrich, J., Bousvaros, A., … Szigethy, E. (2012). Longitudinal results of cognitive behavioral treatment for youths with inflammatory bowel disease and depressive symptoms. *Journal of Clinical Psychology in Medical Settings*, *19*(3), 329–337. https://doi.org/10.1007/s10880-012-9301-8.

Tribbick, D., Salzberg, M., Connell, W., Macrae, F., Kamm, M., Bates, G., … Knowles, S. (2017). Differences across illness perceptions in inflammatory bowel disease and their relationships to psychological distress and quality of life. *Gastroenterology Nursing*, *40*(4), 291–299.

van den Brink, G., Stapersma, L., Bom, A. S., Rizopolous, D., van der Woude, C. J., Stuyt, R. J. L., … Escher, J. C. (2019). Effect of cognitive behavioral therapy on clinical disease course in adolescents and young adults with inflammatory bowel disease and subclinical anxiety and/or depression: Results of a randomized trial. *Inflammatory Bowel Diseases*, *25*(12), 1945–1956. https://doi.org/10.1093/ibd/izz073.

van den Brink, G., Stapersma, L., El Marroun, H., Henrichs, J., Szigethy, E. M., Utens, E. M., & Escher, J. C. (2016). Effectiveness of disease-specific cognitive-behavioural therapy on depression, anxiety, quality of life and the clinical course of disease in adolescents with inflammatory bowel disease: Study protocol of a multicentre randomised controlled trial (HAPPY-IBD). *BMJ Open Gastroenterology*, *3*(1). https://doi.org/10.1136/bmjgast-2015-000071 bmjgast-2015-000071 [pii], e000071.

van der Zaag-Loonen, H. J., Grootenhuis, M. A., Last, B. F., & Derkx, H. H. F. (2004). Coping strategies and quality of life of adolescents with inflammatory bowel disease. *Quality of Life Research*, *13*(5), 1011–1019. https://doi.org/10.1023/B:QURE.0000025598.89003.0c.

Vogelzangs, N., Beekman, A. T. F., de Jonge, P., & Penninx, B. W. J. H. (2013). Anxiety disorders and inflammation in a large adult cohort. *Translational Psychiatry*, *3*(4). https://doi.org/10.1038/tp.2013.27, e249.

Weisz, J. R., Southam-Gerow, M. A., Gordis, E. B., & Connor-Smith, J. (2003). Primary and secondary control enhancement training for youth depression: Applying the deployment-focused model of treatment development and testing. In *Evidence-based psychotherapies for children and adolescents* (pp. 165–182). New York, NY, US: The Guilford Press.

Xanthopoulos, M. S. (2006). *The relationships among psychological distress, stress, disease symptom activity, and coping in adolescents diagnosed with Crohn's disease* (Doctor of Philosophy). Philadelphia: Drexel University. Dissertation Abstracts International Section B: The Sciences and Engineering database.

Chapter 21

Pediatric obsessive-compulsive disorder, CBT, and fear extinction learning

Abigail E. Candelari[a], Daniel A. Geller[b], and Eric A. Storch[a]
[a]*Department of Psychiatry and Behavioral Sciences, Baylor College of Medicine, Houston, TX, United States* [b]*Massachusetts General Hospital Department of Psychiatry, Harvard Medical School, Boston, MA, United States*

Abbreviations

CR conditioned response
CS conditioned stimulus
ERP exposure and response prevention
OCD obsessive-compulsive disorder
SUDs subjective units of distress
US unconditioned stimulus

Obsessive-compulsive disorder (OCD) affects approximately 1%–2% of children and adolescents (Rapoport et al., 2000; Zohar, 1997), and results in significant functional impairment in multiple domains of life (Piacentini, Bergman, Keller, & McCracken, 2003). Childhood onset OCD disrupts social functioning, impairs family functioning, reduces likelihood of academic success, and if left untreated, leads to sustained functional impairment in adulthood (Rapaport, Clary, Fayyad, & Endicott, 2005).

Pediatric OCD is characterized by obsessions, which are intrusive and recurrent thoughts, images, or doubts which trigger negative affect, such as disgust, fear, or distress. Compulsions (rituals) in response to obsessions are designed to neutralize or alleviate the negative affective state. Rituals may be observable or unobservable to others. Observable behaviors may include repetitive washing, evening up behaviors, strict daily rituals, or checking; unobservable compulsions are mental acts such as counting, praying, replaying, or replacing an intrusive thought with a favorable thought.

Functional impairment is exacerbated with co-occurring diagnoses, which are common (Storch et al., 2008). Disorders that commonly co-occur with OCD include anxiety disorders (Farrell, Barrett, & Piacentini, 2006), Tourette's Syndrome and other tic disorders (Storch et al., 2008), attention-deficit hyperactivity disorder (ADHD) (Geller, 2006), disruptive behavior disorders (Farrell et al., 2006), and major depressive disorder (MDD) (Storch et al., 2012). Other serious psychiatric problems, such as substance abuse and self-harm behaviors, occur at higher rates among youth with OCD (Piacentini et al., 2003).

Psychosocial factors that influence onset and maintenance of pediatric OCD

There appears to continuity in the phenotypic presentation of OCD in children and adults; however, variable insight, familial patterns, comorbidities, some differences in phenotypic expression, and developmental progression of symptoms differentiate children from adults with OCD (Geller, Homayoun, & Johnson, 2021). Insight is variable among youth, although most have good to excellent insight (Selles et al., 2018). However, even among children who have moderate insight, ability to articulate cognitive and emotional states may be lacking.

Symptom presentation is generally the same across youth and adults. At times, certain symptoms are less complex in youth versus adults (e.g., sexual obsessions). Further, ritualistic and superstitious behaviors are common in early childhood (Luke et al., 2021), and certain symptoms, including perfectionism, are inadvertently reinforced by parents and teachers, making identification of OCD among youth challenging. These developmental influences on the expression childhood OCD contribute to underdiagnosis, underrecognition, and misdiagnosis (Geller, 2006).

Handbook of Lifespan Cognitive Behavioral Therapy. https://doi.org/10.1016/B978-0-323-85757-4.00028-6

Examination of OCD symptom constellations by age group has revealed that symmetry and ordering, and "just-right" rituals are more prevalent in pediatric OCD (Labad et al., 2008). Some children report a vague feeling of unease or discomfort until certain actions are performed repeatedly, without specific obsessions driving rituals. The absence of concrete cognitive obsessions may make ERP less successful because habituation to vague, but intense feelings of discomfort will not be able to utilize cognitive strategies (Geller et al., 2021). Examination of OCD symptom content has also revealed that sexual and religious/morality obsessions may be over-represented among adolescents (Geller et al., 2001), highlighting developmental influences in the phenotypic expression of OCD during a time marked by navigating autonomy in decisions about moral and/or religious beliefs. Pediatric OCD symptom content is generally stable but may transform over time (Wu et al., 2019). Changes in OCD symptom presentation should be considered in the context of appropriate developmental changes, including stages of attachment, navigating increased autonomy, and moral development.

Behavioral factors

OCD encompasses behavioral, cognitive, genetic, and biological risk and maintenance factors. From a behavioral perspective, children and adolescents demonstrate learned responses to anxiety, which are maintained by negative reinforcement (i.e., distress reduction) (Lewin, Wu, McGuire, & Storch, 2014). Children with OCD engage in compulsions to alleviate anxiety or distress related to obsessions. Once the compulsion or ritual is completed, distress is alleviated, thereby negatively reinforcing the child or adolescent for ritual engagement and increasing the chances of further rituals. Similarly, avoidance of fear-provoking situations relieves anxiety in the short term, but in the long term, fears are strengthened through negative reinforcement (Franklin, Freeman and March, 2018).

Cognitive factors

Emotion dysregulation, anxiety sensitivity, perfectionism, cognitive biases, rigidity, and sensitivity to disgust each influence expression of OCD. Cognitive distortions refer to patterns of thinking in which a child interprets a situation as catastrophic or fuses worst-case scenario thoughts with truth, and these types of cognitive misappraisals are common among children with OCD as they attempt to make sense of intrusive thoughts.

Although the content of thoughts do not differ between people with and without an OCD diagnosis (Rachman, 2007), research shows that cognitive misinterpretations such as thought–action fusion, believing that "normal" people do not have intrusive thoughts, intolerance of anxiety/uncertainty, and the belief that the distress associated with obsessions is uncontrollable, are common among people with OCD.

Anxiety sensitivity, emotion regulation, and family factors

Individuals who experience anxiety in response to the common physiological symptoms accompanying negative emotions tend to have more emotion dysregulation difficulties (Campbell-Sills, Barlow, Brown, & Hofmann, 2006). Excessive disgust reactions also predict problems with emotion regulation (Cisler, Reardon, Williams, & Lohr, 2007), and is especially significant in contamination symptoms (Cisler, Olatunji, & Lohr, 2009). There is a growing body of evidence suggesting that sensitivity for disgust is related to OCD symptom severity, particularly for contamination-themed obsessions (Cervin, Perrin, Olsson, Claesdotter-Knutsson, & Lindvall, 2021). This association has been found among clinical and nonclinical samples of youth, and among clinical samples, reductions in disgust sensitivity during OCD treatment are correlated with reductions in OCD symptoms (Olatunji, Armstrong, & Elwood, 2017).

In addition to passing on genetic vulnerabilities for anxiety and OCD, parents play a role in children's symptomatology through their behaviors, including but not limited to accommodation, modeling of anxiety, and emotion socialization. As symptoms progress, parents commonly become intimately involved in their child's rituals. This involves verbal reassurance or engaging in "verbal checking," engaging in verbal rituals, and permitting children to avoid feared stimuli (Storch et al., 2007). These behaviors are done to help alleviate the child's distress by providing relief in the immediate moment; however, these accommodations unintentionally reinforce the cycle of obsessions and compulsions. Other specific parenting behaviors thought to play a role in reinforcing OCD-related beliefs include modeling of anxious behaviors (e.g., encouraging avoidance of "contaminated" household objects) and modeling expressions of disgust (Mathieu, Conlon, Waters, McKenzie, & Farrell, 2020). Parents of children with OCD may impose rules more rigidly and are more likely to engage in excessive overprotection, criticism, and intrusive parental oversight (Collins & Coles, 2018; Salkovskis, Shafran, Rachman, & Freeston, 1999). Parents are also thought to reinforce inflated sense of responsibility in some childhood OCD cases (Mathieu et al., 2020).

Cognitive-behavioral therapy for pediatric OCD

For children and adolescents with OCD, ERP alone or in combination with pharmacotherapy is the gold-standard treatment (Weisz et al., 2012). The American Academy of Child and Adolescent Psychiatry practice parameters recommend pharmacotherapy with ERP for individuals with moderate-to-severe OCD, and cognitive-behavioral therapy with ERP for younger children and for patients with mild, moderate, and severe OCD (Geller & March, 2012). Specifically, ERP has demonstrated superiority in attention-control conditions (Freeman et al., 2014; Piacentini et al., 2011), and in some studies, SSRI medications (Pediatric OCD Treatment Study Team, 2004; Storch et al., 2013).

In the cognitive-behavioral model, mechanisms of fear acquisition and extinction play an important conceptual role in explaining symptom development, maintenance and treatment of OCD. Exposure and response prevention (ERP) is a type of cognitive-behavioral treatment (Craske & Barlow, 2007). While protocols vary some, ERP approaches include psychoeducation, hierarchy development, and substantial involvement in ERP. Metaanalyses examining cognitive therapy treatment outcomes in over 100 trials designated exposures as the first-line approach for anxiety in children and adolescents (Higa-McMillan, Francis, Rith-Najarian, & Chorpita, 2016), however it is used to a lesser extent for OCD. Fear reduction is the central goal of exposure therapy. Exposure and response prevention involves systematic and gradual exposure to feared situations, without presence of safety behaviors/rituals. Metaanalytic investigations have designated exposures as a first-line approach for fear-based disorders in children, adolescents, and adults (Higa-McMillan et al., 2016; McGuire et al., 2015; Öst, Riise, Wergeland, Hansen, & Kvale, 2016).

Mechanisms implicated in CBT for fear-based psychiatric disorders

Despite several decades of research demonstrating the efficacy of ERP, mechanisms implicated in fear reduction in CBT are commonly misunderstood and commonly misrepresented in the clinical literature. This is unfortunate, given that a strong theoretical understanding of mechanisms underlying change in ERP is critical to clinicians' successful delivering treatment (Abramowitz, 2013). In the following sections, we discuss the historical context in which systematic desensitization and habituation, which are considered precursors to modern theory, are outlined. Next, fear extinction learning, fear conditioning, threat generalization, and inhibitory learning are summarized, and clinical implications and future directions are discussed.

Historical context

Reciprocal inhibition

Reciprocal inhibition became the predominant theory behind exposure therapy during the rise of Behaviorism. This theory posits that the pairing of an aversive stimulus with a response that is physiologically incompatible with fear will prevent the fear response, thereby making it neutral in effect (Wolpe, 1968). Rooted in Pavlovian era conditioning principles, Wolpe tested reciprocal inhibition in cats that had undergone a fear-conditioning process. By administering electric shocks to the cats while in their cages, they eventually grew to fear the cage; yet eventually, when food was given to the cats while in their cages, they demonstrated a reduced fear response (Wolpe, 1954). Wolpe was the first to use systematic desensitization to treat phobias. In the first stage of his 1960s experiments, Wolpe trained patients in deep muscle relaxation, which was thought to be physiologically incompatible with fear. After a thorough discussion of feared situations, the therapist and patient ordered items on a fear hierarchy from most to least anxiety-provoking.

Systematic desensitization stressed a gradual approach to the hierarchy, in which fear reduction was accomplished with lower items on the hierarchy first, thereby enabling higher items to provoke less anxiety than they had at the start of treatment. Finally, patients were asked to vividly imagine the feared stimulus and then use progressive muscle relaxation taught in the first stage of treatment. As this process continued, therapists would move to the next hierarchy item when the patient indicated experiencing no distress on the current item. Wolpe reported that 90% of cases were "much improved" on self-report measures of responses to feared stimuli after termination of treatment (Wolpe, 1954). Reciprocal inhibition was thought to underlie systematic desensitization during this era; however, later work in the 1970s demonstrated fear reduction in studies with exposures, but no relaxation component (Dawson & McMurray, 1978; Gillan & Rachman, 1974; Waters, McDonald, & Koresko, 1972).

Habituation

The predominant rationale guiding exposures in CBT has been Emotion Processing Theory (Foa & Kozak, 1986), which emphasizes self-reported reduction of fear via subjective units of distress (SUDs; Wolpe, Brady, Serber, Agras, & Liberman, 1973) ratings as the key measures of therapeutic exposure. This theory proposes that our memory of fear involves beliefs about the feared stimulus, associated response, and the meaning of the two together (referred to as a *fear structure*; Foa & Kozak, 1986). During exposure therapy, these structures are activated and presented with information that is incompatible. This incompatible information is what enables changes in the patient's meaning of the feared stimulus (namely, reductions in the perceived threat). A graded fear hierarchy and SUDS ratings were emphasized in this approach, giving patients autonomy in the overall treatment course. Foa et al. (2005) suggested that exposures be terminated when at least a 50% reduction in SUDs is achieved.

Habituation, which refers to decrement in a natural anxiety response, emerged as the next predominant theory of fear reduction in exposures in Emotion Processing Theory (Mathews, 1971). Foa and Kozak (1986) hypothesized that within-session habituation to physiological arousal enables integration of newly learned meaning of the feared stimulus, and this process then enables between-session habituation to occur. As such, displays of within- and between-session habituation were considered evidence of successful integration of new meaning, and were considered to be predictors of treatment outcome. In their review of the exposure therapy literature, Craske et al. (2008) argued that many studies do not assess between-session habituation, and among the studies that did include assessment of between-session habituation, not all found it to predict treatment response.

In addition, while within-session habituation of fear and physiological symptoms appears to occur in treatment, there is limited evidence that it predicts long-term treatment outcomes (Craske et al., 2008). Habituation leads to clear reductions in CR during exposure to a stimulus; however, it notably does not involve new learning. Fear extinction learning, rather than habituation, is now widely thought to provide a better explanation of fear reduction in exposure.

Mechanisms implicated in extinction learning

Fear conditioning

Exposure therapy is rooted in Pavlovian conditioning principles. *Fear conditioning* (also rereferred to as *threat conditioning* or *fear acquisition*) is the process in which a neutral stimulus (conditioned stimulus or CS) is repeatedly paired with an aversive stimulus (unconditioned stimulus, US), it will eventually elicit a conditioned response (CR) even in the absence of the aversive stimulus. This is referred to as the CS-US association. For example, the presentation of a neutral bell sound paired with an electric shock will eventually elicit a fear response when the sound is presented without the shock.

In fear extinction, repeated presentation of the conditioned stimulus without the unconditioned stimulus will lead to an eventual decline in the conditioned response to the CS-US association (McGuire, Orr, Wu, et al., 2016). When encountering a CS and experiencing the associated psychophysiological changes (e.g., racing thoughts, chills or hot flashes, rapid heart rate, respiration, etc.), youth with fear-based psychiatric disorders commonly engage in behaviors such as avoidance, rituals/compulsions, or safety behaviors, to quickly eliminate alleviate the distress that they experience during the conditioned response. Among youth with OCD, the quick reduction in unpleasant emotions following the safety behavior inadvertently reinforces the behavior, thereby promoting use of avoidance/rituals in the future and maintaining the fear.

Threat generalization is another term relevant in anxiety. It occurs when a conditioned response to a threat generalizes to stimuli that share similar characteristics to the CS (Glenn et al., 2012). Threat generalization is clinically observed across multiple anxiety-based psychiatric disorders. For example, if a child is stung by an insect, their perception of threat may generalize to other insects or animals with similar characteristics (e.g., size, appearance). Similarly, youth with OCD have been found to generalize perceived threat across different contaminated stimuli, even if some stimuli are ranked in the lower range of the hierarchy. Threat generalization involves increasingly large fear responses as stimuli more closely approximate the CS previously paired with the US, and increasingly less extreme as the stimuli becomes less similar to the CS (Dunsmoor & Paz, 2015).

Some youth have a more generalized and/or sustained fear response. *Contextual conditioning*, which posits that the contextual background in which specific threat cues are encountered play an important role in the acquisition and extinction of CRs, has been used to explain this phenomenon (McGuire, Orr, Wu, et al., 2016). For example, if a child is bitten by a dog while visiting a friend on one occasion, the child may experience fear going back to the friend's house even if the dog has been removed from the house. Fear responses to cues that provided the context within which the threat CS and/or US were presented are the result of context conditioning (Geller et al., 2017).

Extinction learning

Extinction learning refers to the process of reduction in response to a threat CS during repeated exposures of the threat cue, in the absence of the US and the safety behaviors (e.g., rituals/compulsions, avoidance). This process does not eradicate the CS-US association formed during fear conditioning. Instead, a new association between the CS and no US is formed, and this new "CS-no US" association competes for expression with the originally learned CS-US association (Myers & Davis, 2007). Over repeated exposures to the threat cue, the *new* CS-no US association becomes stronger, and eventually extinguishes the CR previously generated by the CS-US association (Geller et al., 2017). To illustrate, a child who was attacked by a dog might be gradually exposed to other dogs without any negative consequences, in order to establish a new learned association. Exposure therapy would typically introduce a variety of dogs, beginning with less fear-inducing (such as a smaller, or calm dog) to more fear-inducing (such as a larger, louder, or more active dog). As the newly learned association is strengthened through repetition, the previously learned association and the CR generated by the association becomes inhibited. In other words, the association "not all dogs attack me" becomes engrained and replaces the previously learned association. Importantly, when youth engage in compulsions, rituals, or safety behaviors during the exposure, these behaviors prevent the new learning that is necessary to establish a competing threat-inhibitory association from taking place (Geller et al., 2021).

Although the terms continue to be erroneously used interchangeably in the clinical literature, there are important conceptual distinctions between the terms "habituation" and "extinction learning." *Habituation* refers to the decrease of a natural response that is instinctively elicited by a US, while extinction learning refers to the decrease of a conditioned (acquired) response. Extinction retention (also known as *extinction recall* in the literature) refers to the extent to which an inhibitory response previously learned during fear extinction is retained over time (McGuire & Storch, 2019). In ERP, extinction learning focuses on the ability to acquire and strengthen the CS-no US association during nonreinforced exposures, whereas extinction retention measures the ability to retain the learned association (e.g., between therapy sessions).

Threat conditioning and extinction in youth with anxiety

Mixed evidence for clinical and nonclinical group differences in extinction learning has been reported. Although some studies found extinction learning across clinical and nonclinical groups without significant group differences (Shechner et al., 2015), other studies found that youth with anxiety disorder diagnoses exhibited deficits in extinction learning (Waters, Henry, & Neumann, 2009), particularly on self-report measures.

Poorer extinction retention has been observed among subclinical and clinically anxious youth (Craske et al., 2008). Anxious youth exhibited differences in neural activation during extinction retention testing (Britton et al., 2014), suggesting that extinction retention may be influenced by individual differences in neural architecture activated during extinction. The literature to date where extinction learning deficits were observed has predominantly characterized these as deficits in inhibitory learning (i.e., difficulty inhibiting the CR, produced by the initial CS-US pairings, following extinction trials).

Threat conditioning and extinction in youth with OCD

Research on threat conditioning and extinction learning among youth with OCD is limited to a small number of studies. In an examination of conditioned fear and extinction in youth with OCD, (McGuire, Orr, Wu, et al. (2016) administered a differential fear-conditioning task and found no group differences during fear acquisition, with both groups exhibiting differential fear conditioning of a skin conductance response (SCR) when exposed to the conditioned stimuli, compared to the nonconditioned stimuli. This research suggests that children with OCD acquire a conditioned threat response similarly to nonclinical controls, however youth with OCD appear to experience a different pattern of extinction learning: youth with OCD exhibited a persistent fear to the CS relative to the US during the extinction phase, providing further evidence for a deficit in inhibitory learning.

One study examining fear acquisition and extinction learning using SCR revealed that youth with OCD exhibited a stronger orienting SCR to initial presentations of stimuli prior to CBT initiation, larger SCRs overall and larger SCR to a visual CS+ during fear acquisition, suggesting impaired inhibitory learning (Geller et al., 2017) in which the initial CR to the CS is more likely to persist throughout extinction (McGuire, Orr, Essoe, et al., 2016).

Geller et al. (2019) found that CBT responders and nonresponders acquired their conditioned fear SCRs similarly; however, differences were observed during the extinction phase between CBT responders and nonresponders. Treatment responder or nonresponder status interacted with the conditioned stimulus type, showing that CBT responders differentiated the stimulus paired with (CS+) and without (CS−) the unconditioned stimulus correctly, whereas the CBT nonresponders experienced difficulty with this. Results suggest that differential fear extinction appears to be an underlying clinical correlate of response to treatment in pediatric OCD and may be an important factor for future research.

Implications for CBT and future directions

Threat conditioning and extinction appear to be influenced by several individual factors, including (but not limited to) developmental stage (e.g., early childhood vs adolescence), age, gender, attention biases, contingency awareness, and psychopathology in parents and children. Future studies should continue to take these factors consideration to better understand their relationship with CBT treatment outcomes.

Notable methodological discrepancies exist between psychophysiological and self-report measures assessing threat conditioning and extinction in youth. Patterns in the observed discrepancies suggest that youth may report fear to be "extinguished" via a self-report measure, while continuing to show physiological signs (e.g., SCR) of fear response. On the other hand, some children may not become consciously aware that their fear has diminished in absence of psychophysiological reactivity, given that psychophysiological awareness develops over time. This group of youth may continue to endorse anxiety on self-report measures. For these individuals, a focus on emotion recognition strategies alongside ERP may help to strengthen awareness between self-reported thoughts and somatic feelings (physiological arousal). While discrepancies across measures are relatively common in the childhood psychopathology literature, future work should carefully examine their methodological designs to improve this weakness.

Children with OCD have demonstrated impaired ability to discriminate threats during differential cued tasks and generalization tasks in several studies and have demonstrated poor awareness of the CS-US contingency (Glenn et al., 2012; Kadosh et al., 2015; McGuire, Orr, Essoe, et al., 2016), even after ruling-out attention deficit hyperactivity disorder (ADHD) and controlling for other psychiatric disorders. Given that threat discrimination and contingency awareness are important components of extinction learning, there may be some benefit to improving threat discrimination early on in CBT to facilitate extinction learning in later exposures.

Given its role in cognitive interpretations and memory, attention can be a critical aspect of threat acquisition. Clinicians should address inattention or poor threat discrimination as they are able to during CBT, as youth who have impaired threat perception may not fully benefit from exposures if they are unable to identify the correct target stimulus. Interventions treating children with OCD and other fear-based psychiatric disorders should consider incorporating these constructs in future work. Adapting attention bias modification tasks to teach accurate threat discrimination in CBT may be a clinically advantageous future direction.

Greater pretreatment psychophysiological extinction learning predicted greater reductions in child-rated anxiety after group CBT in one study (Waters & Pine, 2016), however, youths' self-reported extinction and extinction retention (via SUDs ratings; Wolpe et al., 1973) has yielded mixed evidence as a predictor of CBT treatment outcome in CBT (Kircanski & Peris, 2015). Taken together, evidence suggests that deficits in fear extinction learning in youth are attributable to impaired *inhibitory learning*, or difficulty inhibiting the initial CS-US association with the new CS-no US association learned during extinction. Illness severity and anxiety sensitivity are associated with differential conditioning, and these constructs may be important to examine in future studies on threat conditioning and extinction learning among youth with OCD. Finally, methods should be strengthened in future studies examining youth who display discrepancies in fear outcomes between their psychophysiological data and self-report measures, as these youth may be at heightened risk for nonresponse to CBT.

Practice and procedures

Children are often reliable reporters of their own *overall* anxiety symptoms; however, children's reports of each individual symptom are more likely to vary from parent reports. Taken together, it is preferable to conduct assessments with multiple reporters whenever possible. Unstructured clinical interviews, semistructured interviews, parent and child self-report measures, clinician rating scales, and teacher-report measures are clinically useful in the assessment process with children with OCD (see Merlo, Chaix, Yang, Lynch, & Råstam, 2005 for a review).

The Children's Yale-Brown Obsessive–Compulsive Scale (CY-BOCS; Scahill et al., 1997) is a clinician-rated semistructured interview which assesses OCD symptoms, along with clinician ratings of severity of obsessions and compulsions and a global OCD severity score. The CY-BOCS has demonstrated excellent internal consistency (Storch et al., 2004), good interrater reliability (Scahill et al., 1997), strong convergent and discriminant validity (Scahill et al., 1997; Storch et al.,

2004), and adequate treatment sensitivity (Pediatric OCD Treatment Study Team, 2004). Child self-report measures should also be utilized. The Children's Florida Obsessive–Compulsive Inventory (C-FOCI; Storch et al., 2009) and the Obsessive–Compulsive Inventory-Child Version (OCI-CV; Foa et al., 2010) are two popular measures. Data gleaned from child- or parent-report measures and the CY-BOCS should be interpreted alongside clinical impressions from a diagnostic clinical interview.

Core modules of ERP include psychoeducation, cognitive training, mapping OCD symptoms, exposure and response prevention (ERP), relapse prevention, and generalization of the skills learned in treatment. These goals do not differ for youth and adult patients, however, ERP with pediatric patients typically includes involvement of parents.

First, parents and children are provided with psychoeducation, in which they learn about the nature of OCD and anxiety, their child's symptoms, and how the symptoms manifest for their child. Importantly, psychoeducation emphasizes the ability to identify symptoms, label them as OCD, externalize the symptoms, and then challenge the intrusive thoughts or urges through a variety of cognitive strategies. "Externalizing" OCD by "bossing back" intrusive thoughts and Socratic thinking are examples of cognitive techniques commonly used in ERP with children.

ERP sessions generally involve interval review, determination of goals, introduction of new information, in-session therapist-assisted exposure practice, assigning ERP homework to be completed during the week, and progress monitoring. During in-session exposure practice, the patient is purposefully and systematically exposed to feared situations, *without* engaging in compulsions, to facilitate new learning. Exposures are conducted in a graded manner; for example, a child with contamination fears may start by touching surfaces perceived as contaminated, such as a counter, and eventually habituate to touching more difficult items such as a toilet seat, garbage can, etc. The final component of ERP is relapse prevention, in which the child and family work with the therapist to learn strategies for maintaining treatment gains.

For clinicians who are interested in developing mastery of ERP, *OCD in Children and Adolescents* by John March, M. D., and Karen Mulle, BSN, MTS, MSW, is a commonly used treatment manual. We also recommend the following handbooks for clinicians and families: *Obsessive-Compulsive Disorders: A Complete Guide to Getting Well and Staying Well* by Fred Penzell, Ph.D., *Talking Back to OCD: The Program That Helps Kids and Teens Say "No Way" and Parents Say "Way to Go"* by John March, M.D., and *Freeing Your Child From Obsessive-Compulsive Disorder* by Tamar Chansky, Ph.D. Many other useful resources can be found on the International OCD Foundation's website (https://iocdf.org/).

Mini-dictionary of terms

Obsessive-compulsive disorder (OCD). A psychiatric disorder characterized by intrusive, distressing thoughts (obsessions) and associated behaviors (compulsions) designed to neutralize the obsession.

Obsession. Refers to an intrusive thought, urge, or idea, often accompanied by doubts, disgust, anxiety, etc. in patients with OCD.

Compulsion. Refers to an intentional behavior, often performed in a ritualistic way, which obeys OCD. Compulsions are designed to quickly reduce the distress caused by the obsession.

Exposure. After multiple prolonged **exposures** to the perceived threat, anxiety in response to the feared stimulus decreases.

Exposure plus response prevention (ERP). Exposure plus response prevention states that successful exposures must involve **refraining from engaging in compulsions**.

Habituation. Refers to the neurobehavioral response after successive trials (exposures) in which anxiety symptoms decrease.

Extinction retention (or extinction recall). Refers to the extent to which an inhibitory response previously learned during fear extinction is retained over time.

Extinction learning. Refers to the decrease of a conditioned (acquired) response.

Subjective units of distress (SUDs). Refers to patients' subjective ratings (usually from 1 to 10) indicating their level of distress/anxiety.

Fear hierarchy. A list of phobic stimuli, ranked from least to most anxiety-provoking.

Graded exposure. In contrast to the method called flooding, **graded exposure** begins low on the patient's fear hierarchy, and gradually works up to the more fear-inducing stimuli.

Differential reinforcement. Refers to turning attention **toward mutually pleasant interactions, and away from any potentially negative interactions**. For example, playing checkers while waiting for SUDs to come down during an exposure.

Generalization. Refers to teaching patients to apply strategies learned in treatment to targets not specifically addressed on a hierarchy.

Relapse prevention. Refers to the last stage in treatment in which the patient learns how to anticipate OCD triggers and practice skillful coping in advance, to prevent OCD relapse posttreatment.

Key facts of fear extinction learning

- ERP involves psychoeducation, fear hierarchy development, and graded exposures to feared stimuli to achieve symptom reduction.
- Fear extinction learning refers to the process of building newly learned associations, which compete with and eventually replace the original learned fear response.
- Repeated exposures to a threat cue in the absence of safety behaviors results in new learning and fear extinction.
- Deficits in fear extinction are attributable to difficulty inhibiting the initial association with the new association learned during extinction. Future work should examine how inhibitory learning deficits, anxiety sensitivity, and differential conditioning affect OCD symptom presentation among youths.

Application to other areas

In this chapter on ERP for pediatric OCD, we reviewed the theoretical underpinnings and historical context for using ERP. Numerous studies have shown that fear extinction learning is what leads to OCD symptom reduction in ERP. Extinction learning involves building new learned associations, which eventually replace the originally learned fear response. However, not all children with OCD respond to ERP, and research suggests that this may be due to differential fear conditioning. Factors including developmental stage, gender, attention biases, contingency awareness, and comorbid psychopathologies appear to influence threat conditioning and extinction. Future work should examine how inhibitory learning deficits, anxiety sensitivity, and differential conditioning affect OCD presentation and treatment response among pediatric populations.

Summary points

- This chapter outlines the etiology, risk, and maintenance factors for pediatric OCD, with a particular focus on assessment, ERP treatment, and the theoretical principles behind ERP.
- Clinicians should consider expected developmental stages when assessing for pathological vs developmentally appropriate symptoms. Clinicians should also incorporate data from multiple reporters in their evaluations.
- ERP is the gold-standard treatment for OCD, and as such, treatment for youth with OCD should focus mainly on exposure and response prevention.
- Parents are heavily involved in treatment for pediatric OCD and should have the opportunity to learn the theoretical underpinnings of ERP. Clinicians should engage parents in as much psychoeducation as needed before proceeding with exposures.
- A strong theoretical understanding of ERP includes understanding of habituation, fear conditioning, contextual conditioning, fear generalization, and fear extinction learning, as well as inhibitory learning principles.

References

Abramowitz, J. S. (2013). The practice of exposure therapy: Relevance of cognitive-behavioral theory and extinction theory. *Behavior Therapy*, *44*(4), 548–558.

Britton, W. B., Lepp, N. E., Niles, H. F., Rocha, T., Fisher, N. E., & Gold, J. S. (2014). A randomized controlled pilot trial of classroom-based mindfulness meditation compared to an active control condition in sixth-grade children. *Journal of School Psychology*, *52*(3), 263–278.

Campbell-Sills, L., Barlow, D. H., Brown, T. A., & Hofmann, S. G. (2006). Acceptability and suppression of negative emotion in anxiety and mood disorders. *Emotion*, *6*(4), 587.

Cervin, M., Perrin, S., Olsson, E., Claesdotter-Knutsson, E., & Lindvall, M. (2021). Involvement of fear, incompleteness, and disgust during symptoms of pediatric obsessive–compulsive disorder. *European Child & Adolescent Psychiatry*, *30*(2), 271.

Cisler, J. M., Olatunji, B. O., & Lohr, J. M. (2009). Disgust sensitivity and emotion regulation potentiate the effect of disgust propensity on spider fear, blood injection-injury fear, and contamination fear. *Journal of Behavior Therapy and Experimental Psychiatry*, *40*(2), 219–229.

Cisler, J. M., Reardon, J. M., Williams, N. L., & Lohr, J. M. (2007). Anxiety sensitivity and disgust sensitivity interact to predict contamination fears. *Personality and Individual Differences*, *42*(6), 935–946.

Collins, L. M., & Coles, M. E. (2018). A preliminary investigation of pathways to inflated responsibility beliefs in children with obsessive compulsive disorder. *Behavioural and Cognitive Psychotherapy*, *46*(3), 374–379.

Craske, M. G., & Barlow, D. H. (2007). *Mastery of your anxiety and panic: Therapist guide. Vol. 2*. Oxford University Press.

Craske, M. G., Waters, A. M., Bergman, R. L., Naliboff, B., Lipp, O. V., Negoro, H., & Ornitz, E. M. (2008). Is aversive learning a marker of risk for anxiety disorders in children? *Behaviour Research and Therapy*, *46*(8), 954–967.

Dawson, R. W., & McMurray, N. E. (1978). Desensitization without hierarchical presentation and concomitant relaxation. *Australian Journal of Psychology*, *30*(2), 119–132.

Dunsmoor, J. E., & Paz, R. (2015). Fear generalization and anxiety: Behavioral and neural mechanisms. *Biological Psychiatry*, *78*(5), 336–343.

Farrell, L., Barrett, P., & Piacentini, J. (2006). Obsessive–compulsive disorder across the developmental trajectory: Clinical correlates in children, adolescents and adults. *Behaviour Change*, *23*(2), 103–120.

Foa, E. B., Coles, M., Huppert, J. D., Pasupuleti, R. V., Franklin, M. E., & March, J. (2010). Development and validation of a child version of the obsessive compulsive inventory. *Behavior Therapy*, *41*(1), 121–132.

Foa, E. B., & Kozak, M. J. (1986). Emotional processing of fear: Exposure to corrective information. *Psychological Bulletin*, *99*(1), 20.

Foa, E. B., Liebowitz, M. R., Kozak, M. J., Davies, S., Campeas, R., Franklin, M. E., ... Tu, X. (2005). Randomized, placebo-controlled trial of exposure and ritual prevention, clomipramine, and their combination in the treatment of obsessive-compulsive disorder. *American Journal of Psychiatry*, *162*(1), 151–161. https://doi.org/10.1176/appi.ajp.162.1.151.

Franklin, M. E., Freeman, J. B., & March, J. S. (2018). *Treating OCD in children and adolescents: A cognitive-behavioral approach*. Guilford Publications.

Freeman, J., Garcia, A., Frank, H., Benito, K., Conelea, C., Walther, M., & Edmunds, J. (2014). Evidence base update for psychosocial treatments for pediatric obsessive-compulsive disorder. *Journal of Clinical Child and Adolescent Psychology*, *43*(1), 7–26. https://doi.org/10.1080/15374416.2013.804386.

Geller, D. A. (2006). Obsessive-compulsive and spectrum disorders in children and adolescents. *Psychiatric Clinics*, *29*(2), 353–370.

Geller, D. A., Biederman, J., Faraone, S., Agranat, A., Cradock, K., Hagermoser, L., ... Coffey, B. J. (2001). Developmental aspects of obsessive compulsive disorder: Findings in children, adolescents, and adults. *Journal of Nervous and Mental Disease*, *189*(7), 471–477.

Geller, D. A., Homayoun, S., & Johnson, G. (2021). Developmental considerations in obsessive compulsive disorder: Comparing pediatric and adult-onset cases. *Frontiers in Psychiatry*, *12*, 678538.

Geller, D. A., & March, J. (2012). Practice parameter for the assessment and treatment of children and adolescents with obsessive-compulsive disorder. *Journal of the American Academy of Child & Adolescent Psychiatry*, *51*, 98–113. https://doi.org/10.1016/j.jaac.2011.09.019.

Geller, D. A., McGuire, J. F., Orr, S. P., Pine, D. S., Britton, J. C., Small, B. J., ... Storch, E. A. (2017). Fear conditioning and extinction in pediatric obsessive compulsive disorder. *Annals of Clinical Psychiatry*, *29*(1), 17.

Geller, D. A., McGuire, J. F., Orr, S. P., Small, B. J., Murphy, T. K., Trainor, K., ... Storch, E. A. (2019). Fear extinction learning as a predictor of response to cognitive behavioral therapy for pediatric obsessive compulsive disorder. *Journal of Anxiety Disorders*, *64*, 1–8. https://doi.org/10.1016/j.janxdis.2019.02.005.

Gillan, P., & Rachman, S. J. (1974). An experimental investigation of desensitization in phobic patients. *British Journal of Psychiatry*, *124*(581), 392–401.

Glenn, C. R., Klein, D. N., Lissek, S., Britton, J. C., Pine, D. S., & Hajcak, G. (2012). The development of fear learning and generalization in 8–13 year-olds. *Developmental Psychobiology*, *54*(7), 675–684.

Higa-McMillan, C. K., Francis, S. E., Rith-Najarian, L., & Chorpita, B. F. (2016). Evidence base update: 50 years of research on treatment for child and adolescent anxiety. *Journal of Clinical Child & Adolescent Psychology*, *45*(2), 91–113.

Kadosh, K. C., Haddad, A. D., Heathcote, L. C., Murphy, R. A., Pine, D. S., & Lau, J. Y. (2015). High trait anxiety during adolescence interferes with discriminatory context learning. *Neurobiology of Learning and Memory*, *123*, 50–57.

Kircanski, K., & Peris, T. S. (2015). Exposure and response prevention process predicts treatment outcome in youth with OCD. *Journal of Abnormal Child Psychology*, *43*(3), 543–552.

Labad, J., Menchon, J. M., Alonso, P., Segalas, C., Jimenez, S., Jaurrieta, N., ... Vallejo, J. (2008). Gender differences in obsessive–compulsive symptom dimensions. *Depression and Anxiety*, *25*(10), 832–838.

Lewin, A. B., Wu, M. S., McGuire, J. F., & Storch, E. A. (2014). Cognitive behavior therapy for obsessive-compulsive and related disorders. *Psychiatric Clinics*, *37*(3), 415–445.

Luke, A. K., Ankney, R., Wilton, E. P., Gladstone, T. R., Berlin, K. S., & Flessner, C. A. (2021). Developmental trajectories of pediatric obsessive–compulsive symptoms. *Research on Child and Adolescent Psychopathology*, *49*(12), 1635–1648. https://doi.org/10.1007/s10802-020-00742-4.

Mathews, A. M. (1971). Psychophysiological approaches to the investigation of desensitization and related procedures. *Psychological Bulletin*, *76*(2), 73.

Mathieu, S. L., Conlon, E. G., Waters, A. M., McKenzie, M. L., & Farrell, L. J. (2020). Inflated responsibility beliefs in paediatric OCD: Exploring the role of parental rearing and child age. *Child Psychiatry & Human Development*, *51*(4), 552–562.

McGuire, J. F., Orr, S. P., Essoe, J. K.-Y., Mccracken, J. T., Storch, E. A., & Piacentini, J. (2016). Extinction learning in childhood anxiety disorders, obsessive compulsive disorder and post-traumatic stress disorder: Implications for treatment. *Expert Review of Neurotherapeutics*, *16*, 1155–1174. https://doi.org/10.1080/14737175.2016.1199276.

McGuire, J. F., Orr, S. P., Wu, M. S., Lewin, A. B., Small, B. J., Phares, V., ... Storch, E. A. (2016). Fear conditioning and extinction in youth with obsessive-compulsive disorder. *Depression and Anxiety*, *33*(3), 229–237.

McGuire, J. F., Piacentini, J., Lewin, A. B., Brennan, E. A., Murphy, T. K., & Storch, E. A. (2015). A meta-analysis of cognitive behavior therapy and medication for child obsessive compulsive disorder: Moderators of treatment efficacy, response, and remission. *Depression and Anxiety*, *32*(8), 580–593. https://doi.org/10.1002/da.22389.

McGuire, J. F., & Storch, E. A. (2019). An inhibitory learning approach to cognitive-behavioral therapy for children and adolescents. *Cognitive and Behavioral Practice*, *26*(1), 214–224.

Merlo, J., Chaix, B., Yang, M., Lynch, J., & Råstam, L. (2005). A brief conceptual tutorial of multilevel analysis in social epidemiology: Linking the statistical concept of clustering to the idea of contextual phenomenon. *Journal of Epidemiology & Community Health*, *59*(6), 443–449.

Myers, K. M., & Davis, M. (2007). Mechanisms of fear extinction. *Molecular Psychiatry*, *12*(2), 120–150.

Olatunji, B. O., Armstrong, T., & Elwood, L. (2017). Is disgust proneness associated with anxiety and related disorders? A qualitative review and meta-analysis of group comparison and correlational studies. *Perspectives on Psychological Science*, *12*(4), 613–648.

Öst, L. G., Riise, E. N., Wergeland, G. J., Hansen, B., & Kvale, G. (2016). Cognitive behavioral and pharmacological treatments of OCD in children: A systematic review and meta-analysis. *Journal of Anxiety Disorders*, *43*, 58–69.

Pediatric OCD Treatment Study Team. (2004). Cognitive-behavior therapy, sertraline, and their combination for children and adolescents with obsessive-compulsive disorder: The Pediatric OCD Treatment Study (POTS) randomized controlled trial. *Journal of the American Medical Association*, *292*, 1969–1976.

Piacentini, J., Bergman, R. L., Chang, S., Langley, A., Peris, T., Wood, J. J., & McCracken, J. (2011). Controlled comparison of family cognitive behavioral therapy and psychoeducation/relaxation training for child obsessive-compulsive disorder. *Journal of the American Academy of Child & Adolescent Psychiatry*, *50*(11), 1149–1161.

Piacentini, J., Bergman, R. L., Keller, M., & McCracken, J. (2003). Functional impairment in children and adolescents with obsessive-compulsive disorder. *Journal of Child and Adolescent Psychopharmacology*, *13*(2, Suppl. 1), 61–69.

Rachman, S. (2007). Unwanted intrusive images in obsessive compulsive disorders. *Journal of Behavior Therapy and Experimental Psychiatry*, *38*(4), 402–410.

Rapaport, M. H., Clary, C., Fayyad, R., & Endicott, J. (2005). Quality-of-life impairment in depressive and anxiety disorders. *American Journal of Psychiatry*, *162*(6), 1171–1178.

Rapoport, J. L., Inoff-Germain, G., Weissman, M. M., Greenwald, S., Narrow, W. E., Jensen, P. S., … Canino, G. (2000). Childhood obsessive–compulsive disorder in the NIMH MECA Study: Parent versus child identification of cases. *Journal of Anxiety Disorders*, *14*(6), 535–548.

Salkovskis, P., Shafran, R., Rachman, S., & Freeston, M. H. (1999). Multiple pathways to inflated responsibility beliefs in obsessional problems: Possible origins and implications for therapy and research. *Behaviour Research and Therapy*, *37*(11), 1055–1072.

Scahill, L., Riddle, M. A., McSwiggin-Hardin, M., Ort, S. I., King, R. A., Goodman, W. K., … Leckman, J. F. (1997). Children's Yale-Brown obsessive compulsive scale: Reliability and validity. *Journal of the American Academy of Child & Adolescent Psychiatry*, *36*(6), 844–852.

Selles, R. R., Højgaard, D. R., Ivarsson, T., Thomsen, P. H., McBride, N., Storch, E. A., … Stewart, S. E. (2018). Symptom insight in pediatric obsessive-compulsive disorder: Outcomes of an international aggregated cross-sectional sample. *Journal of the American Academy of Child and Adolescent Psychiatry*, *57*(8), 615.

Shechner, T., Britton, J. C., Ronkin, E. G., Jarcho, J. M., Mash, J. A., Michalska, K. J., … Pine, D. S. (2015). Fear conditioning and extinction in anxious and nonanxious youth and adults: examining a novel developmentally appropriate fear-conditioning task. *Depression and Anxiety*, *32*(4), 277–288.

Storch, E. A., Bussing, R., Small, B. J., Geffken, G. R., McNamara, J. P., Rahman, O., … Murphy, T. K. (2013). Randomized, placebo-controlled trial of cognitive-behavioral therapy alone or combined with sertraline in the treatment of pediatric obsessive–compulsive disorder. *Behaviour Research and Therapy*, *51*(12), 823–829.

Storch, E. A., Geffken, G. R., Merlo, L. J., Jacob, M. L., Murphy, T. K., Goodman, W. K., … Grabill, K. (2007). Family accommodation in pediatric obsessive–compulsive disorder. *Journal of Clinical Child and Adolescent Psychology*, *36*(2), 207–216.

Storch, E. A., Khanna, M., Merlo, L. J., Loew, B. A., Franklin, M., Reid, J. M., … Murphy, T. K. (2009). Children's Florida obsessive compulsive inventory: Psychometric properties and feasibility of a self-report measure of obsessive–compulsive symptoms in youth. *Child Psychiatry and Human Development*, *40*(3), 467–483.

Storch, E. A., Lewin, A. B., Larson, M. J., Geffken, G. R., Murphy, T. K., & Geller, D. A. (2012). Depression in youth with obsessive-compulsive disorder: Clinical phenomenology and correlates. *Psychiatry Research*, *196*(1), 83–89.

Storch, E. A., Merlo, L. J., Larson, M. J., Geffken, G. R., Lehmkuhl, H. D., Jacob, M. L., … Goodman, W. K. (2008). Impact of comorbidity on cognitive-behavioral therapy response in pediatric obsessive-compulsive disorder. *Journal of the American Academy of Child & Adolescent Psychiatry*, *47*(5), 583–592.

Storch, E. A., Murphy, T. K., Geffken, G. R., Soto, O., Sajid, M., Allen, P., … Goodman, W. K. (2004). Psychometric evaluation of the Children's Yale–Brown obsessive-compulsive scale. *Psychiatry Research*, *129*(1), 91–98.

Waters, A. M., Henry, J., & Neumann, D. L. (2009). Aversive Pavlovian conditioning in childhood anxiety disorders: Impaired response inhibition and resistance to extinction. *Journal of Abnormal Psychology*, *118*(2), 311.

Waters, W. F., McDonald, D. G., & Koresko, R. L. (1972). Psychophysiological responses during analogue systematic desensitization and non-relaxation control procedures. *Behaviour Research and Therapy*, *10*(4), 381–393.

Waters, A. M., & Pine, D. S. (2016). Evaluating differences in Pavlovian fear acquisition and extinction as predictors of outcome from cognitive behavioural therapy for anxious children. *Journal of Child Psychology and Psychiatry*, *57*(7), 869–876.

Weisz, J. R., Chorpita, B. F., Palinkas, L. A., Schoenwald, S. K., Miranda, J., Bearman, S. K., … Research Network on Youth Mental Health. (2012). Testing standard and modular designs for psychotherapy treating depression, anxiety, and conduct problems in youth: A randomized effectiveness trial. *Archives of General Psychiatry*, *69*(3), 274–282.

Wolpe, J. (1954). Reciprocal inhibition as the main basis of psychotherapeutic effects. *AMA Archives of Neurology & Psychiatry*, *72*(2), 205–226.

Wolpe, J. (1968). Psychotherapy by reciprocal inhibition. *Conditional Reflex: A Pavlovian Journal of Research & Therapy*, *3*(4), 234–240.

Wolpe, J., Brady, J. P., Serber, M., Agras, W. S., & Liberman, R. P. (1973). The current status of systematic densitization. *American Journal of Psychiatry*, *130*(9), 961–965.

Wu, M. S., Geller, D. A., Schneider, S. C., Small, B. J., Murphy, T. K., Wilhelm, S., & Storch, E. A. (2019). Comorbid psychopathology and the clinical profile of family accommodation in pediatric OCD. *Child Psychiatry & Human Development*, *50*(5), 717–726.

Zohar, A. H. (1997). The epidemiology of obsessive-compulsive disorder in children and adolescents. *Child and Adolescent Psychiatric Clinics of North America*, *8*(3), 445–460.

Chapter 22

Parent-child interactions and cognitive-behavioral therapy: A focus on anxious children

Daniëlle Van der Giessen and Francisca J.A. Van Steensel
Department of Child Development and Education, University of Amsterdam, Amsterdam, Netherlands

Abbreviations

AD	anxiety disorder
CBT	cognitive-behavioral therapy
CCBT	child-focused cognitive-behavioral therapy
FCBT	family-focused cognitive-behavioral therapy
PCBT	parent-focused cognitive-behavioral therapy
PCI	parent-child interaction

Introduction

Anxiety disorders (ADs) are highly prevalent throughout childhood and adolescence (Polanczyk, Salum, Sugaya, Caye, & Rohde, 2015). Environmental factors are more strongly related to children's AD than genetic factors (Eley et al., 2015), and extensive research shows that parenting is related to the development and maintenance of child anxiety (Möller, Nikolić, Majdandžić, & Bögels, 2016; Percy, Creswell, Garner, O'Brien, & Murray, 2016; van der Bruggen, Stams, & Bögels, 2008). Therefore, parents are often involved in cognitive-behavioral therapy (CBT) to enhance treatment efficacy. While CBT is an efficacious and effective treatment for children's AD (e.g., Reynolds, Wilson, Austin, & Hooper, 2012), there is empirical inconsistency whether parental inclusion in CBT is more effective than child focused CBT (Carnes, Matthewson, & Boer, 2019; Manassis et al., 2014).

Surprisingly and intriguingly, most treatment studies only evaluated the effects of CBT on children's AD (e.g., Manassis et al., 2014), and on parenting and family functioning using questionnaire data (e.g., Jongerden & Bögels, 2015). Only a handful of treatment studies examined effects of CBT on parent-child interactions using observational data. Yet, there is growing evidence that parent-child interaction patterns are more closely related to child anxiety when using observational paradigms than when using self- and parent reports (McLeod, Wood, & Weisz, 2007), and observational measures of parent-child behavior have been suggested to be particularly sensitive to change following interventions (Aspland & Gardner, 2003). Overall, observations of parent-child interactions can provide insight into which real-life interaction patterns and outcomes of interest change after CBT, and whether parental involvement leads to different changes in interaction patterns.

This review first discusses *observational studies* demonstrating which behavioral, cognitive, and emotional features of parent-child interactions are related to child anxiety (see Fig. 1). Second, existing observational studies examining changes in parent-child interactions after CBT are discussed. Finally, implications for future research and clinical practice are presented.

Parent-child interactions and anxiety disorders

Theoretical models of childhood anxiety emphasize the importance of behavioral, cognitive, and emotional characteristics of reciprocal parent-child interactions in the etiology and maintenance of children's ADs (Chorpita & Barlow, 1998;

Handbook of Lifespan Cognitive Behavioral Therapy. https://doi.org/10.1016/B978-0-323-85757-4.00029-8

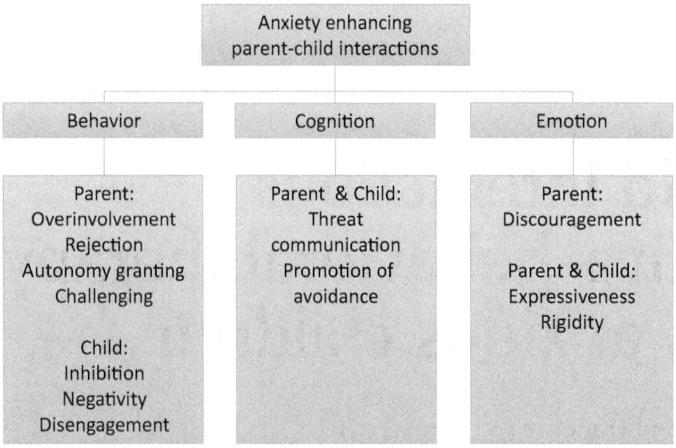

FIG. 1 Dimensions and subdimensions of anxiety enhancing parent-child interactions.

Creswell, Murray, Stacey, & Cooper, 2011; Hudson & Rapee, 2004). We now briefly define and explain behaviors, cognitions, and emotions during parent-child interactions that are thought to be related to childhood anxiety, and shortly review recent observational studies providing empirical evidence for this link.

Behavior during parent-child interactions

Several parenting behaviors have been thought to play a role in the development of childhood anxiety. First, *parental overinvolvement* is a broad parenting behavior dimension consisting of several subcomponents (e.g., overcontrol, overprotection, intrusiveness, psychological control), and it refers to excessive restrictive, intrusive, and overprotective parenting and discouragement of children's independence (Möller et al., 2016). This is thought to increase children's threat perception, to reduce children's perceived control over threat, to diminish children's opportunities for exposure to novelty or potentially difficult situations, and to limit children's opportunities to explore their environment and develop age-appropriate behavior (Chorpita & Barlow, 1998; van der Bruggen et al., 2008). Parents that *encourage and grant* children's *autonomy* and independence might enhance children's feelings of control over their environment, thereby potentially preventing or reducing children's anxiety (Chorpita & Barlow, 1998). Second, *parental rejection* is defined as hostile, negative, low in warmth, critical and disapproving behavior, which might lead to children's negative self-view, undermining effective emotion regulation, thereby increasing child anxiety (Chorpita & Barlow, 1998; McLeod et al., 2007). Third, *challenging behavior* refers to parental encouragement that push children's physical limits (e.g., rough-and-tumble play, tickling, competing), as well their socio-emotional limits (e.g., teasing, encouraging competition and performance) (Majdandžić, de Vente, Colonnesi, & Bögels, 2018). Through this challenging parenting behavior children learn to be braver in unfamiliar situations, to explore the world and new situations, which might foster children's confidence and coping skills, thereby potentially preventing the development of anxiety.

Empirical observational evidence shows strongest and quite consistent associations between more parental overinvolvement and more child anxiety (Asbrand, Hudson, Schmitz, & Tuschen-Caffier, 2017; McLeod et al., 2007; van der Bruggen et al., 2008). Importantly, associations were found in longitudinal studies (Hudson, Murayama, Meteyard, Morris, & Dodd, 2019; Lewis-Morrarty et al., 2012), in father-child and mother-child interactions (Rork & Morris, 2009; Teetsel, Ginsburg, & Drake, 2014), in experimental studies when manipulating maternal control (Thirlwall & Creswell, 2010), and associations appeared to be stronger for older children (van der Bruggen et al., 2008; Waite & Creswell, 2015). More overprotection toward toddlers during low stress tasks was also linked to *less* child anxiety 4 years later (Majdandžić et al., 2018), suggesting the potential benefits of sensitively displayed protective behavior which might depend on the operationalization of overprotection, and children's age (Möller et al., 2016). Furthermore, more observed parental autonomy support, especially during discussion tasks (McLeod et al., 2007; van der Bruggen et al., 2008), and more challenging behavior, particularly of fathers, during physical play tasks with young children (Majdandžić et al., 2018; Majdandžić, Möller, De Vente, Bögels, & Van Den Boom, 2014), was related to *less* child anxiety, suggesting the potential benefits of these parenting behaviors. Evidence for the link between observed parental rejection and childhood anxiety is more mixed, with studies showing no effects (Asbrand et al., 2017; Bögels, Bamelis, & van der Bruggen, 2008), small effects (McLeod et al., 2007), and different effects for subdimensions of rejection (Waite & Creswell, 2015). Finally, although it is thought that parental anxiety also affects parenting behavior, as anxious parents tend to underestimate their child's abilities and view new situations as threating themselves, observational evidence is somewhat mixed (e.g., Murray et al., 2012; Teetsel et al., 2014; van der Bruggen et al., 2008).

Though children's behavior is also thought to play a role in shaping anxiety enhancing parenting behaviors (Hudson & Rapee, 2004), only few observational studies examine child behavior during interactions. One study showed that at first anxious parents tend to withdraw from parent-child interactions, letting the child deal with potentially stressful tasks on their own, but when the child expresses negative affect parents tend to react with excessive levels of control toward the child (Woodruff-Borden, Morrow, Bourland, & Cambron, 2002). In line with this, anxious parents of anxious children showed reduced tolerance of their child's negative emotions which might explain that higher levels of intrusive, fearful, and negative parental behavior during the interactions were also found (Creswell, Apetroaia, Murray, & Cooper, 2013). Finally, when mothers of behaviorally inhibited preschoolers showed more overinvolvement at age 4, these children reported more anxiety symptoms at age 12 (Hudson et al., 2019). Together, results seem to underscore that anxiety in children is likely to stem from the transactional nature of parent-child interactions.

Cognitions during parent-child interactions

Another risk factor for child anxiety is how parents communicate their own (anxious) cognitions about the environment and about the child. *Threat communication or verbalizations* refer to situations where parents implicitly or explicitly communicate to their child negative information that indicates potential threat or promotes avoidance of feared situations (Percy et al., 2016). Frequent and excessive threat communication may enhance children's anxious cognitions, such as bias toward threat information and an underestimation of coping abilities, leading to increased fear in children. This might also reinforce parents' threat communication, which can ultimately increase children's chance to develop and maintain an AD (Creswell et al., 2011).

A systematic review shows that observational evidence is strongest for the role of parents' verbal and nonverbal fear communication in the development of anxious cognitions of children (Emerson, Ogielda, & Rowse, 2019). More specifically, children exposed to fearful verbal and nonverbal threat communication of parents while talking about an upcoming spelling test, reported poorer performance-related anxious cognitions compared to children exposed to calm and nonthreat parental communication, and this effect was stronger for fathers than mothers (Burstein & Ginsburg, 2010). Also, when mothers were observed during a 5-min speech planning period to display more verbal and nonverbal threat and anxiety toward their child, this was associated with children's lower self-evaluations after the speech, indicating that children underestimate their coping abilities (Becker & Ginsburg, 2011). Finally, high trait anxious mothers who received negative information about an unknown animal told more negative stories to their children, which produced higher levels of children's fear beliefs (Muris, van Zwol, Huijding, & Mayer, 2010).

Another systematic observational review shows that fear relevant verbal communication of parents and children is associated with children's anxious responses and symptomatology (Percy et al., 2016). Specially, anxious children have been found to interpret ambiguous scenarios in a threatening manner and chose avoidant responses, and after family discussions these avoidance responses increased suggesting the parents reinforced through communication children's anxious responses (Barrett, Rapee, Dadds, & Ryan, 1996). Related, children started threat-related discussions about scenario's more often than mothers, and more verbal expansion of mothers on threat-related discussions about the scenarios was related to more child anxiety (Hane & Barrios, 2011). Finally, socially anxious mothers have also been found to be less encouraging and more likely to communicate potential threat when discussing the experience of starting school with their children, and this communication style was related with higher levels of internalizing problems, particularly for behaviorally inhibited children (Murray et al., 2014).

Emotions during parent-child interactions

Parent-child emotion communication is also thought to play an important role in childhood anxiety. Parent-child *emotion communication* is a very broad construct and it refers to the way parents and children express, encourage, and discuss emotions, and to the ability of dyads to flexibly switch between emotions, and it is thought to be a key mechanism in children's emotional socialization and socio-emotional functioning (Morris, Silk, Steinberg, Myers, & Robinson, 2007; Van der Giessen & Bögels, 2018). Parent-child conversations which are dismissing, unsupportive, discouraging, negative, and emotionally rigid do not provide children with ample opportunities to learn to recognize, understand, express, and manage emotional situations. This may accentuate children's emotion regulation vulnerabilities, which places them at greater risk for the development of anxiety (Morris et al., 2007; Suveg, Morelen, Brewer, & Thomassin, 2010).

Emotional expressiveness, that is the amount of positive and negative emotions, of both parents and children during interactions is related to child anxiety. Parents of AD children expressed fewer positive emotions during emotional discussions (Hudson, Comer, & Kendall, 2008; Suveg, Zeman, Flannery-Schroeder, & Cassano, 2005), used more negative emotion words during a cognitive and a social task (Hosey & Woodruff-Borden, 2012), and showed less positive and more negative affect during emotion discussions with boys particularly (Suveg et al., 2008), than parents of nonanxious children. A longitudinal study shows that parents' aggressive affect, but not dysphoric or positive affect, during a positive discussion predicted higher levels of anxiety symptoms in adolescents 2 years later (Schwartz et al., 2012). AD children also tend to display fewer positive emotions during emotionally challenging interactions with their parents than nonanxious children (Hudson et al., 2008; Suveg et al., 2008). Importantly, evidence supports differential reciprocal emotion patterns between anxious and nonanxious parent-child dyads (Schrock & Woodruff-Borden, 2010; Williams, Kertz, Schrock, & Woodruff-Borden, 2012). Nonetheless, emotional expressivity of parent-child dyads is not in all studies related child anxiety (Brumariu & Kerns, 2015; Suveg et al., 2005; Van der Giessen et al., 2015; Van der Giessen & Bögels, 2018).

Parents of AD children have also been found to discourage emotional discussions and these children show more emotion regulation difficulties during interactions. Mothers of AD children have been found discourage children's emotion discussions (Suveg et al., 2005), and to invalidate children's feelings, show less warmth and interest, and elaborate less about these emotional events than mothers of nonanxious children (Brumariu & Kerns, 2015). Also, a lack of supportive maternal emotions during emotional discussions longitudinally predicted adolescents' anxiety symptoms, particularly for girls who showed a more helpless pattern of parasympathetic regulation when watching sad video's (Hastings, Klimes-Dougan, Kendziora, Brand, & Zahn-Waxler, 2014). During conflict interactions both fathers and mothers of AD youth showed more emotion dismissing and less emotion encouragement (Hurrell, Houwing, & Hudson, 2017). During discussions of children's past emotions, both fathers and mothers of AD children engaged in few explanatory discussions of emotions, but only fathers of AD children were more likely to discourage angry emotion discussions than fathers of nonanxious children (Suveg et al., 2008). Importantly, AD children also have been found to be less engaged in the conversation with parents, to express affect with greater intensity, to show less congruence in affect and behavior (Brumariu & Kerns, 2015), to have greater difficulty with regulating their emotions during the conversation (Hurrell et al., 2017), and to show fewer problem-solving emotion regulation strategies (Suveg et al., 2008) than non-AD children.

Emotional rigidity, that is the inability to flexibly switch between emotions, to get stuck in emotions, and to have smaller emotional repertoire (Van der Giessen & Bögels, 2018), is also found to be associated with child anxiety. More emotional rigidity of parent-child dyads was concurrently related to more internalizing problems in high risk children in kindergarten (Hollenstein, Granic, Stoolmiller, & Snyder, 2004), and predicted more anxiety symptoms of adolescents 5 years later (Van der Giessen et al., 2015). Parent-child dyads with an AD child, and particularly when mothers also had an AD, showed less emotional flexibility during conflict interactions than parent-child dyads with nonanxious children (Van der Giessen & Bögels, 2018).

Summary of parent-child interactions and anxiety disorders

Extensive *observational studies* show that parent-child interactions characterized by overinvolvement, rejection, unchallenging behavior, negative threat verbalizations, negative affect, emotion dismissing, and emotional rigidity are to some degree related to (clinical) anxiety in childhood. However, most observational studies (a) examined mother-child interactions, (b) relied on cross-sectional designs, (c) focused mainly on parents' behavior, cognitions, and emotions during interactions, (d) used dissimilar coding systems and assessment contexts, (e) did not investigate bidirectional or transactional communication patterns, (f) did not examine parental anxiety, (g) did not study effects of children's age and gender, (h) differently conceptualized cognitive, emotional, and behavioral features of interactions, and (i) did not examine these features of interactions in tandem. Future longitudinal studies should rigorously assess the reciprocal nature of parent-child interactions in anxious and nonanxious parents and children, to more thoroughly understand how behavior, cognitions, and emotions during interactions together contribute to development and maintenance of childhood anxiety.

CBT for childhood anxiety

While CBT is effective for treating children's AD with almost 60% of the children being free from (one of) their anxiety disorders after treatment (James, Reardon, Soler, James, & Creswell, 2020), CBT is not effective for everyone. It is important to unravel who profits from CBT and who does not, and to understand the factors related to treatment response. Therefore, first treatment studies are reviewed that examined whether features of parent-child interactions changed after

CBT and were linked to treatment outcome. Next, parental involvement in CBT is discussed and how this possibly relates to changes in parent-child interactions.

Changes in parent-child interactions after CBT

Table 1 provides an overview of studies examining parent-child interactions in relation to CBT. Overall, evidence shows that parental behavior during interactions can change after treatment: maternal encouragement of bravery increased (Silk et al., 2013), and decreases in rejection were found in parent-child interactions (Van der Giessen et al., 2019). For subdimensions of overinvolvement results were mixed, with studies showing after CBT *decreases* in intrusiveness (Wood et al., 2009), *increases* in intrusiveness (Hiller et al., 2016), and *increases* in psychological control (Van der Giessen et al., 2019). Additionally, Hiller et al. (2016) showed no significant changes for other parental behaviors (e.g., encouragement, overprotection) during interactions.

Changes in parental behaviors were inconsistently related to treatment outcome. More maternal encouragement of bravery prior to treatment was associated with better treatment response (Silk et al., 2013). For older children (10–13 years), but not for younger children (6–9 years), who followed Family focused CBT, decreases in parental intrusiveness were related to decreases in child anxiety (Wood et al., 2009). In contrast, no associations between parents' behavior (overinvolvement, expression of fearfulness, rejection, psychological control) and treatment outcome were also evident (Creswell et al., 2008; Van der Giessen et al., 2019). Moreover, it is unclear whether the child's anxiety decreased (more) because parents' behaviors changed, or whether parents' behavior changed as result of children's decreases in anxiety.

Several issued need to be mentioned when interpreting these abovementioned findings. First, only a handful of observational studies examined changes in parent-child interactions after CBT, and these studies exclusively focused on changes in parental behaviors while child behavior as well as parent and child cognitions and emotions were largely neglected. For example, none of the studies included assessments of child behavior (e.g., asking for help, seeking reassurance) during interactions, and did not examine how children's behavior might elicit parental behavior and/or how children's behavior during interactions might have changed after treatment.

Second, studies have predominantly focused on interactions of mother-child dyads. One notable exception is the study of Van der Giessen et al. (2019) which not only examined mother-child and father-child interactions but also investigated mother-father-child interactions. Results showed that psychological control increased from pre- to posttreatment in triadic interaction for mothers (particularly in child focused CBT), but increases were less strong for fathers (particularly in the family focused CBT). Although replication is needed, such findings stress the importance of studying family interactions beyond mother-child interactions.

Parental involvement in CBT

Since parents' behavior during parent-child interactions can change (see earlier), we need to know how parents were involved in CBT to understand which treatment factors may facilitate these changes. Before discussing this issue, we need to discuss the growing evidence for involving parents in treatment. Although some individual studies support that the involvement of parents in the child's treatment is beneficial, most (meta-analytical) reviews do not find evidence for family-focused (FCBT) or parent-focused CBT (PCBT) outperforming child CBT (CCBT) (e.g., Reynolds et al., 2012). One complicating factor may be that there is large heterogeneity in the way and extent to which parents are involved in children's treatment (Kreuze, Pijnenborg, de Jonge, & Nauta, 2018; Manassis et al., 2014). Evidence indicates that a strong emphasis on contingency management or transfer of control in treatment is beneficial for children's long-term anxiety remission rates (Kreuze et al., 2018), and high parental involvement is beneficial in the long term for secondary treatment outcomes (e.g., depressive symptoms, social competence) as well (Kreuze et al., 2018).

As parental involvement seems beneficial for the maintenance of the treatment effects, for further improvements, or to prevent relapse, it underlines the need to look more closely how exactly parental involvement is defined in CBT studies that examine changes in parent-child interactions, to be able to understand which type or amount of parental involvement is needed to elicit change.

Parental involvement in CBT studies examining changes in parent-child interactions

For the selected CBT studies (see Table 1), the type of parental involvement in CBT is specified in Table 2. Parental involvement varies a lot across studies from fully parent guided treatment (e.g., Creswell et al., 2008; Hiller et al., 2016), to child-focused treatment with minimal parental involvement (e.g., Silk et al., 2013; CCBT in Van der Giessen

TABLE 1 Overview of selected CBT studies using observational measures to examine whether parent-child interactions have changed after CBT, and/or how these variables are related to treatment outcome.

Authors	Participants	CBT treatment	Observational task	Observational measures	Findings
Creswell, Willetts, Murray, Singhal, and Cooper (2008)	22 children aged 6–12 years with a primary AD, and their mothers	Individual parent-focused CBT	During the fourth treatment session, a 5 min speech task was administered The child was asked to give a 3–5 min presentation about a chosen topic	During the speech the (verbal and nonverbal) *behaviors of mothers* were coded: (1) overinvolvement and (2) mother's expression of fearfulness	Maternal expression of fear, but not overinvolvement, was associated with children's anxiety level Being diagnosis-free after treatment was not related to mothers' parenting behaviors
Hiller et al. (2016)	High anxious parents (46 mothers, 3 fathers, and 10 couples)of 60 children (7–12 years) with an AD	Parents were randomized to follow either (1) Brief Guided Parent Delivered CBT (BPD-CBT) or (2) BPD-CBT with addition of strategies to promote parental tolerance of children's emotions (TCNE)	Pretreatment a speech task in which the child was asked to prepare and give a speech was administered. Parent and child were given 5 min alone to prepare this. The parents were asked to give a brief introduction prior to the child's speech The task was also administered posttreatment but then in the presence of a stranger who observed and rated the speech	During the preparation moment of 5 min the following *parent behaviors* were coded: (1) expressed anxiety, (2) overprotection, (3) intrusiveness, (4) encouragement, (5) warmth, and (6) quality of relationship	Posttreatment an increase in intrusiveness was observed Little evidence was found for differences between conditions (BPD-CBT vs TCNE) for either child outcome or changes in parent behaviors during parent-child interactions. One exception was for the variable "quality of relationship" (which seemed to have improved slightly in the TCNE condition, but decreased slightly in the BPD-CBT condition)
Silk et al. (2013)	47 children aged 9–13 years, and their primary caregiver (94% mother)	Children were randomized to either (1) individual child-focused CBT or (2) Child-Centered Therapy (CCT)	Pre- and posttreatment children were instructed to give a speech in front of a camera. They could choose from three topics (topics differed at pre- and posttreatment assessment). Children were given the option to do a second speech	Maternal encouragement of bravery *(parent behavior)* to complete the task and to do the optional speech was coded	Maternal encouragement of bravery was increased after CBT, but not after CCT. In addition, more maternal encouragement of bravery prior to treatment, was associated with better treatment response
Van der Giessen, Colonnesi, and Bögels (2019)	128 AD children aged 8–18 years, their mothers and fathers	Children were randomized to either (1) Child CBT (CCBT) or (2) Family CBT (FCBT)	Children participated in random order in two dyadic interactions: a father-child conflict discussion and a mother-child conflict discussion. They also took part in a triadic conflict discussion and a triadic anxiety discussion (in random order) All tasks lasted around 5 min	*Parents (verbal and nonverbal) behavior* on two dimensions was coded: (1) rejection (e.g., criticism, hostility, frustration) versus warmth, and (2) psychological control (e.g., intrusiveness, guilt induction)	A decrease in rejection was found from pre- to posttreatment (no difference between FCBT and CCBT). Psychological control did not change from pre- to posttreatment in the dyadic interactions, but increased in the triadic interactions (depending on type of treatment, gender of the parent, and type of task). Parental rejection or psychological control was not found to be related to treatment effectiveness

TABLE 1 Overview of selected CBT studies using observational measures to examine whether parent-child interactions have changed after CBT, and/or how these variables are related to treatment outcome—cont'd

Authors	Participants	CBT treatment	Observational task	Observational measures	Findings
Wood, McLeod, Piacentini, and Sigman (2009)	35 children with an AD aged 6–13 years	Children were randomized to either (1) Child CBT (CCBT) or (2) Family CBT (FCBT)	Pre- and posttreatment children were instructed to put on a belt (with the rationale that it would contain a heart-rate monitor). They were told that they could to it themselves but that their parent could help when needed. The task was designed to be difficult and to elicit parental assistance	A composite score for parental intrusiveness (*parent behavior*) was made based on the observational task and questionnaires completed by the children themselves and parents	Parental intrusiveness had decreased more from pre- to posttreatment in FCBT compared to CCBT. For older children (10–13 years) in FCBT, but not for younger children (6–9 years), a decrease in parental intrusiveness was related to a decrease in child anxiety

TABLE 2 Type of CBT treatment and aspects of parental involvement in the selected CBT studies using observational measures.

Authors	Parental involvement in CBT	Elements/techniques of treatment
Creswell et al. (2008)	Mothers followed eight sessions of an individual CBT, based on the cool kids program. The child was present during the fourth session in which a graded exposure hierarchy was formulated. (Of note six mothers with an anxiety disorder themselves had followed CBT for their own anxiety prior to child treatment)	Mothers were taught CBT techniques to help their children overcome their problems with anxiety. Techniques included psycho-education, identifying and monitoring children's anxious thoughts, cognitive restructuring, behavioral principles to promote "brave" behaviors, graded exposure hierarchies, problem solving, social skills training and relapse prevention.
Hiller et al. (2016)	Condition 1: Parent delivered CBT consisting originally of eight sessions but which were spread out over 10 sessions. Condition 2: The same program as in condition 1 but with additional strategies focus on the parents tolerance of the child's negative emotions.	Elements of condition 1 included psycho-education, identifying the child's anxious cognitions, exposure and problem solving. Additional elements of condition 2 included cognitive and mindfulness based techniques for the parents to identify, monitor and manage their own emotional responses to their child's distress.
Silk et al. (2013)	Individual child-focused CBT consisting of 16 sessions. (Coping Cat program) Child-Centered Therapy (CCT) consisting of 16 sessions.	CBT: in sessions 1–8 anxiety-management skills were taught (e.g., coping, problem solving, relaxation), in sessions 9–16 a hierarchy of exposure tasks were completed. Session 4 and 9 are held with the parents. Parents were used as informants, and were asked for assistance in exposure planning and homework. CCT: a manualized, nondirective psychotherapy based on principles of unconditional positive regard, empathy and therapist genuineness.
Van der Giessen et al. (2019)	Child CBT: 12 sessions consisting of 45–90 min. In total, face-to-face session time amounted 19.5 h for the child, and no more than 2.5 h for the parent(s). Family CBT: 12 sessions consisting of 45–90 min. In total, face-to-face session time amounted 13.5 h for the child, 18 h for the parent(s), and 4.5 h for siblings.	CCBT: techniques included psycho-education, challenging thoughts, coping behavior (e.g., relaxation and task concentration), exposure (with reinforcement system), behavioral experiments, relapse prevention. FCBT: the treatment aimed to teach the family (child, parents, siblings) skills to manage the child's anxiety symptoms. The treatment consisted of three phases: (1) learning to overcome one's fear by stimulating courageous modeling from parents and talking about their own fear and how they cope with it, (2) identifying and changing dysfunctional beliefs of the parents and child, and (3) improving communication and problem solving between all family members.
Wood et al. (2009)	Child CBT (CCBT): 12–16 sessions of 60–80 min Family CBT (FCBT): 12–16 sessions of 60–80 min	CCBT: the treatment consisted of two phases: (1) skills training (coping, relaxation, cognitive restructuring, self-reward), and (2) exposure. FCBT: aimed at changing parents communication patterns, increasing children's independence, facilitate autonomy granting behaviors, reduce intrusiveness (e.g., by giving choices, allowing children to learn by trial and error instead of taking over, accepting children's feelings instead of criticizing, stimulating the child's self-help skills, increasing children's privacy). A reward system was also included.

et al., 2019 and in Wood et al., 2009). In addition, when considering the content of parental involvement, parenting behaviors are addressed to some extent in all types of CBT (e.g., by stimulating courageous behavior, reducing overprotection), and in FCBT (e.g., Van der Giessen et al., 2019; Wood et al., 2009) emotion communication patterns and dysfunctional cognitions are targeted. There is some support that parenting behaviors changed more after treatment when explicitly targeted. For example, parental intrusiveness decreased more in the FCBT condition compared to the CCBT condition (Wood et al., 2009). In contrast, parental behaviors also changed irrespective of whether parenting behaviors were addressed in CBT (Van der Giessen et al., 2019); decreases in parental rejection were evident in both FCBT as well as CCBT. Though, more research is needed to replicate findings and to further study the relation between treatment outcome and (changes in) parenting behaviors, these results do seem to support a bidirectional relation: child anxiety can change when parents have changed (their behaviors) and vice versa.

Summary of CBT for childhood anxiety

Very few observational studies have examinedif parent-child interactions changed after CBT, or whether features of parent-child interaction (pre-, during, or after treatment) are associated with treatment success. In addition, studies have predominantly included mothers (instead of fathers), focused mostly on behaviors (instead of cognitions and emotions) of parents (and not of children) during parent-child interactions, and have studied parent-child interactions in different contexts. In addition, different conceptualizations of parental behavior were used, and studies were mostly limited to pre- and postintervention assessments to evaluate changes parental parent-child interaction patterns and how these changes might be related to changes in child anxiety. Therefore, at this point, it is difficult to draw firm conclusions on whether CBT changes parent-child interactions and/or which features of the parent-child interaction should be targeted in treatment. Studies assessing parent-child interactions (including behaviors, cognitions, and emotions of both parent and child) during and after CBT are needed to disentangle the bidirectionality question: do interaction patterns (and parenting) change as a result of children's decrease in anxiety, and/or does children's anxiety decrease as a result of changes in interaction patterns (and parenting)?

Implications for future research and clinical practice

The current review shows that *observational studies demonstrated* that behavioral, cognitive, and emotional features of parent-child interactions are associated with the development and maintenance of child anxiety. Surprisingly, only a handful of observational studies examined changes after CBT in parent-child interaction patterns. Several important implications for future research and clinical practice need to be addressed.

Personalizing treatment by assessment of parent-child interactions

Currently, CBT for children's AD relies on a "one-size-fits-all" model. Yet, this review implies that different features of parent-child interactions with AD children can be problematic, which might suggest differences between families in anxiety enhancing interaction patterns related to child anxiety and potentially CBT response. Assessing the parent-child communication style before treatment might facilitate the identification of specific features of interaction patterns that gone awry that need to be improved. Therefore, we suggest CBT for child AD should be personalized on a case-by-case basis (Ng & Weisz, 2016), by a thorough pretreatment diagnostic of children interacting with mothers and fathers (!) in different challenging situations. This will inform clinicians how to personalize CBT. For instance, parent-child dyads who tend to get stuck in specific emotional communication patterns, may benefit when they are taught how to engage in flexibly expressing different emotions according to situational demands (Van der Giessen & Bögels, 2018).

Thus, increasing the benefits of parental involvement in CBT for child AD might lie in tools to identify and change maladaptive communication patterns of parent-child dyads which can potentially break the cycle of anxiogenic transmissions between parent and child when communicating. Future CBT studies need to provide clear guidance on which features of parent-child communication should be targeted when and how, and if this, indeed, improves CBT outcomes.

Changing the cycle

Another important conclusion from this review is that most observational CBT studies solely focused on parental behavior during interactions. However, relations between cognitions, behavior, and emotions are a key feature within a cognitive-behavioral framework on child anxiety (Creswell et al., 2011). A negative cognitive style of (anxious) parents is hypothesized to drive parental emotions and behaviors during parent-child interactions that promote child anxiety by increasing transference of threat-related information to the child, limiting child autonomy and increasing overinvolvement, discouraging child emotions, high levels of criticism and negativity and so forth. Also, children contribute to the interaction, and they might evoke, reinforce, and shape parental cognitions, behaviors, and emotions, resulting in a feedback cycle (Creswell et al., 2011; Emerson et al., 2019; Hudson & Rapee, 2004).

Current correlational studies (e.g., Creswell et al., 2013) do not allow any causal inferences, and it might be possible that child characteristics precede parental characteristics or vice versa. Parent-child interaction patterns are established over many years, complicating research on cause and effect. It might be more efficient to focus on changing maladaptive interactional processes rather than understanding the origin of these complex processes. As such, it might be important that the anxiogenic interaction cycle is somewhere being triggered to change. Therefore, future studies should examine during CBT when specific features (emotions, cognitions, and behavior) are targeted, whether this elicits the hypothesized change, and what features are changed consequently.

Hypothesized models

In Figs. 2 and 3 we present the abovementioned implications of our review, namely personalizing CBT and changing the cycle between parent and child. Fig. 2 visualizes the interaction between parents' and children's behavior, cognitions and emotions, and treatment response. The two-sided arrows indicate bidirectionality in which parent and child influence each other and contribute to the interaction. All three aspects (parent, child, and the parent-child interaction) may then contribute to treatment response, but decreases in child anxiety may also precede changes in parent factors (Settipani, O'Neil, Podell, Beidas, & Kendall, 2013) and interaction patterns. Following from this reasoning, it may not matter much for treatment outcome which aspect is changing first (parent, child, or parent-child interaction), as long as the anxiogenic cycle is somewhere being triggered to change. As such, it may not matter much how exactly child anxiety is targeted in treatment; via child and/or parents, since different formats of CBT (CCBT, FCBT, PCBT) can be effective for treating childhood ADs (see review of James et al., 2020). Still the question why some children profit from CBT (irrespective of the type parental involvement) and others do not remain unanswered. As also discussed earlier, it may be that the needs of the family are slightly different from each other, that is, in general, the different CBT formats are more or less equally effective but that treatment needs to be further personalized to the family's communication needs to maximize treatment outcome.

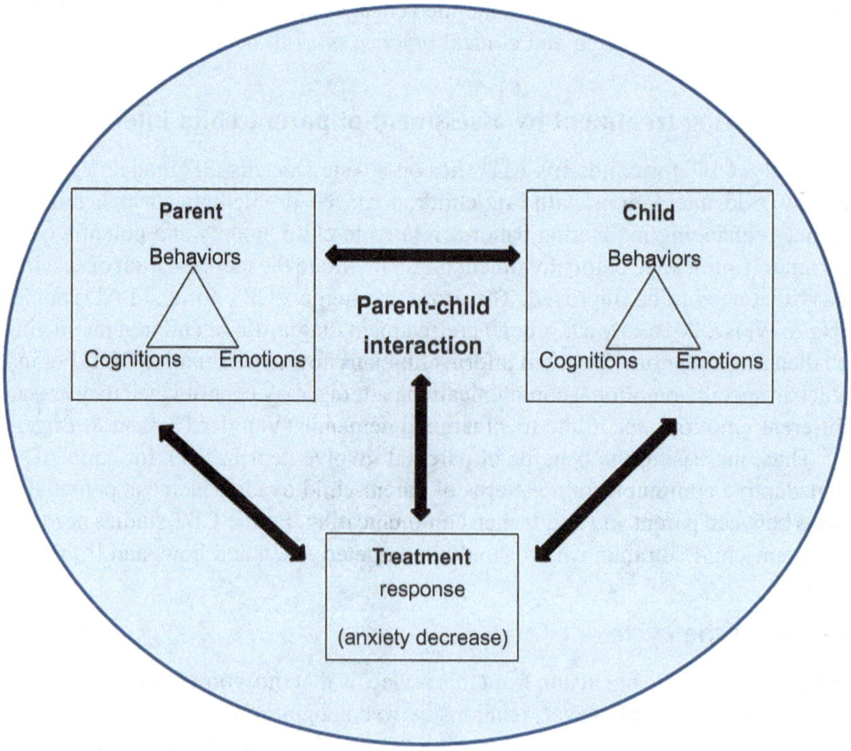

FIG. 2 Visualization of the interaction between treatment (response) and (behavioral, cognitive, emotional) features of the parent and child, and their interaction.

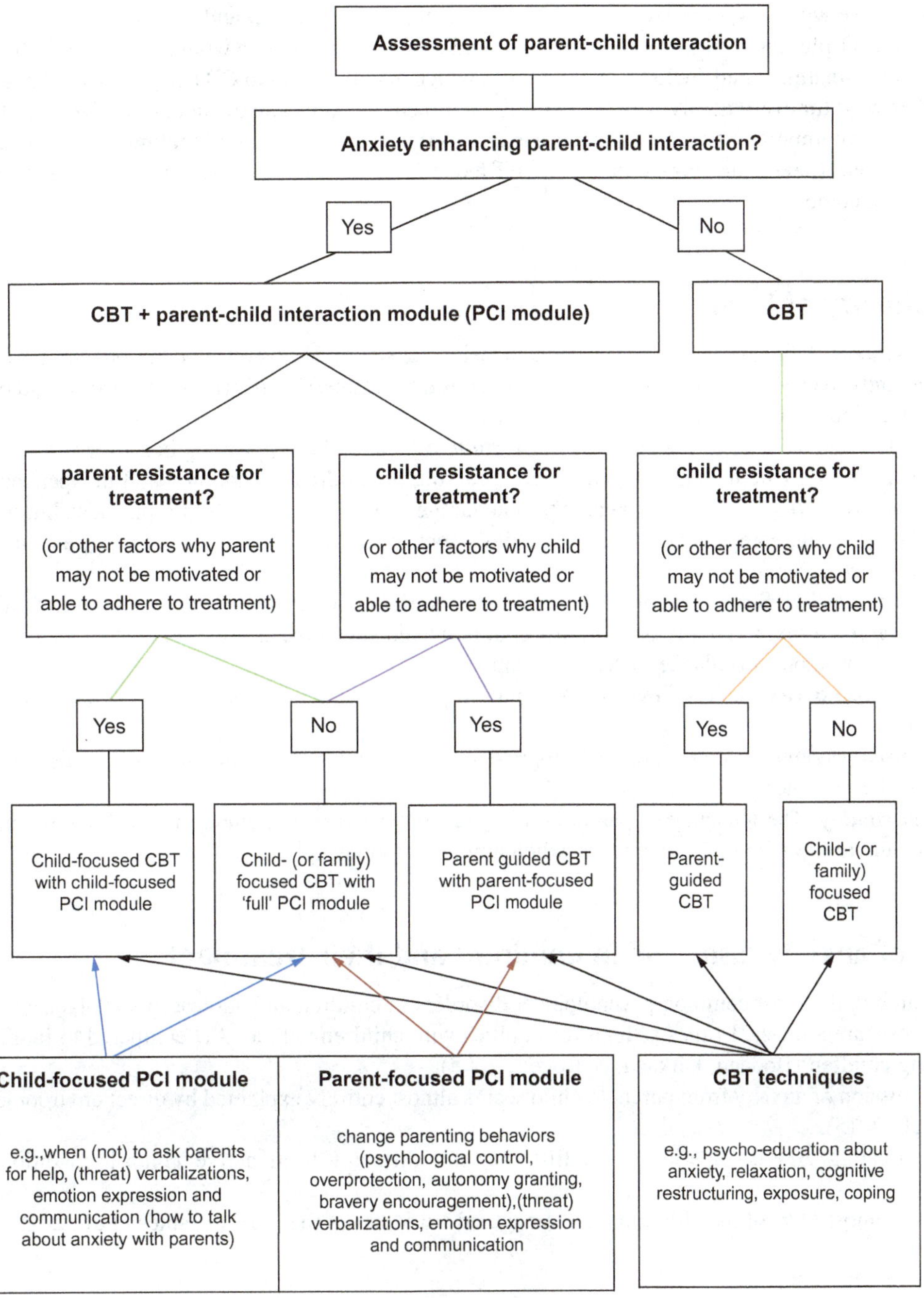

FIG. 3 Proposed decision model for including a parent-child interaction (PCI) module in cognitive-behavioral therapy (CBT) treatment.

Focusing on parent-child interactions, we propose the following model that may help clinicians decide how to personalize CBT (see Fig. 3). We propose first an assessment of the parent-child interaction to examine the presence of anxiety enhancing parent-child interaction patterns (e.g., overinvolvement, threat verbalization, emotional rigidity, see Fig. 1). Once the presence or absence of anxiety enhancing parent-child interaction patterns is established, the next question would be whether parent and/or child are motivated to be engaged in treatment. Nota bene, we assume that at least one of them would be motivated for treatment otherwise the family would not be seeking help. Based on these two aspects (anxiety-enhancing interaction patterns and motivation/resistance for treatment) the model guides the clinician to the treatment that will probably suit the family needs the best and that will have the most chance for success.

For example, we would hypothesize that for families with a child with AD and anxiety enhancing parent-child interaction patterns, CBT plus the inclusion of a (full) parent-child interaction module would be most effective (Fig. 3). Similar, the model to guide treatment may include other modules which may be added to CBT to enhance other aspects that require treatment. This way, for every family with an AD child treatment outcome can be more optimized by choosing the appropriate module(s). An important step for future research is to figure out which additional modules a specific child within a specific family needs (next to the more traditional CBT techniques such as cognitive restructuring and exposure) to get the best treatment outcome.

Mini-dictionary of terms

- ***Overinvolvement.*** A broad parenting behavior dimension consisting of several subcomponents (e.g., overcontrol, overprotection, intrusiveness, psychological control), and it refers to excessive restrictive, intrusive, and overprotective parenting and discouragement of a child's independence.
- ***Rejection.*** Defined as hostile, negative, low in warmth, critical and disapproving behavior.
- ***Autonomy granting.*** Parental encouragement and granting of children's autonomy and independence.
- ***Challenging parenting.*** Parental behavior that encourages and pushes the child's physical limits (e.g., rough-and-tumble play, tickling, competing), as well the child's socio-emotional limits (e.g., teasing, encouraging competition and performance).
- ***Behavioral inhibition.*** Strong reactions of withdrawal, wariness, avoidance, and shyness in unfamiliar situations.
- ***Threat communication.*** Explicitly or implicitly communicating negative information that indicates potential threat and danger or that promotes avoidance of feared situations.
- ***Emotional expressiveness.*** The level or amount of positive and negative emotions individuals express during interaction.
- ***Emotion discouragement.*** A tendency to disapprove or dismiss emotional expression and to discourage the discussion of emotional experiences.
- ***Emotional rigidity.*** The tendency of parent-child dyads to perseverate in emotions, to show a diminished emotional repertoire, and to have limited capacity to switch among emotions.

Key facts of anxiety disorders in children, and their treatment

- ADs are among the most common psychological disorders in children and adolescents (Polanczyk et al., 2015).
- Societal costs are almost 21 times as high for families with children with an AD compared to families with typically developing children (Bodden, Dirksen, & Bögels, 2008).
- The transmission of anxiety from parent to child seems almost entirely explained by direct environmental transmission (Eley et al., 2015).
- CBT for treating AD is superior to waitlist, but not different from active control conditions (James et al., 2020).
- After CBT almost 60% of the children have recovered from one of their ADs (James et al., 2020).

Key facts of parent-child interactions

- Observing and assessing parent-child interactions can elucidate changes in parent and child behavior following an intervention (Aspland & Gardner, 2003).
- An advantage of measuring parent-child interactions with observational methods is that it captures real processes and outcomes of interest (Aspland & Gardner, 2003).
- A meta-analysis showed that the strength of the association between observed parenting and parent-reported parenting is small but significant (Hendriks, Van der Giessen, Stams, & Overbeek, 2018).
- Observational measures of parent-child interactions include a video-recorded task, and trained coders afterwards view and code aspects of parent-child interactions.
- Still relatively little is known about father-child, triadic, or family interactions, since most studies examine mother-child interactions.

Applications to other areas

In this chapter we have discussed parent-child interactions in relation to anxiety, and CBT for treating anxiety problems in children. There is considerable evidence that (behavioral, cognitive, and emotional communication during) parent-child interactions are different when children have primary anxiety disorders compared to when children do not have psychopathology (i.e., typically developing children). Whether these associations are also found when children have a comorbid—instead of primary—anxiety disorder, is largely unknown. For example, the prevalence of comorbid anxiety disorders in children with autism spectrum disorder (ASD) is high (Van Steensel, Bögels, & Perrin, 2011), and CBT for treating comorbid anxiety disorders also seems to be effective for children with (high-functioning) ASD (Ung, Selles, Small, & Storch, 2015). Remarkedly, there has been little discussion on including parents for this target group. Almost all studies that focus on children with ASD have included parents to some extent. In addition, studies indicate that for certain groups of children with comorbid anxiety disorders, e.g., children with high ASD traits or ADHD symptoms, parental involvement is beneficial (Maric, Van Steensel, & Bögels, 2018; Puleo & Kendall, 2011), while this is more uncertain for children with a primary anxiety disorder (see discussion about parental involvement in the chapter). However, how parent-child interactions change after anxiety-driven CBT or how this change may (or may not) be related to treatment effectiveness in these different target groups with comorbid anxiety disorders is unknown and requires further investigation. In addition, (behavioral, cognitive and emotion communication during) parent-child interactions may be very different between children with a primary anxiety disorder and children with other primary disorders but with comorbid anxiety disorders (e.g., children with ASD; Bontinck, Warreyn, Meirsschaut, & Roeyers, 2018).

Summary points

- Observational studies show that behavioral, cognitive, and emotional features of parent-child interactions are related to the development and maintenance of child anxiety.
- Surprisingly, few treatment studies examine the effects of CBT on parent-child interactions using observational data.
- There is some observational evidence that parent-child interactions, and particularly parental behavior, change after CBT.
- Changes in parental behavior during interactions were inconsistently related to children's treatment outcome and to the type of parental involvement in CBT.
- Future observational studies are needed that examine bidirectional changes in behaviors, cognitions, and emotions during parent-child interactions during and after CBT for child anxiety.
- Understanding anxiety enhancing parent-child interaction patterns will allow clinicians to better identify, assess, and modify maladaptive parent-child interaction patterns, ultimately leading to enhanced CBT strategies for children's anxiety disorders.

References

Asbrand, J., Hudson, J., Schmitz, J., & Tuschen-Caffier, B. (2017). Maternal parenting and child behaviour: An observational study of childhood social anxiety disorder. *Cognitive Therapy and Research*, *41*(4), 562–575. https://doi.org/10.1007/s10608-016-9828-3.

Aspland, H., & Gardner, F. (2003). Observational measures of parent-child interaction: An introductory review. *Child and Adolescent Mental Health*, *8*(3), 136–143. https://doi.org/10.1111/1475-3588.00061.

Barrett, P. M., Rapee, R. M., Dadds, M. M., & Ryan, S. M. (1996). Family enhancement of cognitive style in anxious and aggressive children. *Journal of Abnormal Child Psychology*, *24*(2), 187–203. https://doi.org/10.1007/BF01441484.

Becker, K. D., & Ginsburg, G. S. (2011). Maternal anxiety, behaviors, and expectations during a behavioral task: Relation to children's self-evaluations. *Child Psychiatry and Human Development*, *42*(3), 320–333. https://doi.org/10.1007/s10578-011-0216-7.

Bodden, D. H. M., Dirksen, C. D., & Bögels, S. M. (2008). Societal burden of clinically anxious youth referred for treatment: A cost-of-illness study. *Journal of Abnormal Child Psychology*, *36*(4), 487–497. https://doi.org/10.1007/s10802-007-9194-4.

Bögels, S. M., Bamelis, L., & van der Bruggen, C. (2008). Parental rearing as a function of parent's own, partner's, and child's anxiety status: Fathers make the difference. *Cognition and Emotion*, *22*, 522–538. https://doi.org/10.1080/02699930801886706.

Bontinck, C., Warreyn, P., Meirsschaut, M., & Roeyers, H. (2018). Parent–child interaction in children with autism spectrum disorder and their siblings: Choosing a coding strategy. *Journal of Child and Family Studies*, *27*(1), 91–102. https://doi.org/10.1007/s10826-017-0877-3.

Brumariu, L. E., & Kerns, K. A. (2015). Mother–child emotion communication and childhood anxiety symptoms. *Cognition and Emotion*, *29*(3), 416–431. https://doi.org/10.1080/02699931.2014.917070.

Burstein, M., & Ginsburg, G. S. (2010). The effect of parental modeling of anxious behaviors and cognitions in school-aged children: An experimental pilot study. *Behaviour Research and Therapy*, *48*(6), 506–515. https://doi.org/10.1016/j.brat.2010.02.006.

Carnes, A., Matthewson, M., & Boer, O. (2019). The contribution of parents in childhood anxiety treatment: A meta-analytic review. *Clinical Psychologist, 23*(3), 183–195. https://doi.org/10.1111/cp.12179.

Chorpita, B. F., & Barlow, D. H. (1998). The development of anxiety: The role of control in the early environment. *Psychological Bulletin, 124*(1), 3–21. https://doi.org/10.4324/9781315619996.

Creswell, C., Apetroaia, A., Murray, L., & Cooper, P. (2013). Cognitive, affective, and behavioral characteristics of mothers with anxiety disorders in the context of child anxiety disorder. *Journal of Abnormal Psychology, 122*(1), 26–38. https://doi.org/10.1037/a0029516.

Creswell, C., Murray, L., Stacey, J., & Cooper, P. (2011). Parenting and child anxiety. In W. K. Silverman, & A. P. Field (Eds.), *Anxiety disorders in children and adolescents* (pp. 299–322). Cambridge: Cambridge University Press. https://doi.org/10.1017/CBO9780511994920.015.

Creswell, C., Willetts, L., Murray, L., Singhal, M., & Cooper, P. (2008). Treatment of child anxiety: An exploratory study of the role of maternal anxiety and behaviours in treatment outcome. *Clinical Psychology and Psychotherapy, 15*(1), 38–44. https://doi.org/10.1002/cpp.559.

Eley, T. C., McAdams, T. A., Rijsdijk, F. V., Lichtenstein, P., Narusyte, J., Reiss, D., … Neiderhiser, J. M. (2015). The intergenerational transmission of anxiety: A children-of-twins study. *American Journal of Psychiatry, 172*(7), 630–637. https://doi.org/10.1176/appi.ajp.2015.14070818.

Emerson, L. M., Ogielda, C., & Rowse, G. (2019). A systematic review of the role of parents in the development of anxious cognitions in children. *Journal of Anxiety Disorders, 62*(November 2018), 15–25. https://doi.org/10.1016/j.janxdis.2018.11.002.

Hane, A. A., & Barrios, E. S. (2011). Mother and child interpretations of threat in ambiguous situations: Relations with child anxiety and autonomic responding. *Journal of Family Psychology, 25*(5), 644–652. https://doi.org/10.1037/a0024149.

Hastings, P. D., Klimes-Dougan, B., Kendziora, K. T., Brand, A., & Zahn-Waxler, C. (2014). Regulating sadness and fear from outside and within: Mothers' emotion socialization and adolescents' parasympathetic regulation predict the development of internalizing difficulties. *Development and Psychopathology, 26*, 1369–1384. https://doi.org/10.1017/S0954579414001084.

Hendriks, A. M., Van der Giessen, D., Stams, G., & Overbeek, G. (2018). The association between parent-reported and observed parenting: A multi-level meta-analysis. *Psychological Assessment, 30*(5), 621–633. https://doi.org/10.1037/pas0000500.

Hiller, R. M., Apetroaia, A., Clarke, K., Hughes, Z., Orchard, F., Parkinson, M., & Creswell, C. (2016). The effect of targeting tolerance of children's negative emotions among anxious parents of children with anxiety disorders: A pilot randomised controlled trial. *Journal of Anxiety Disorders, 42*, 52–59. https://doi.org/10.1016/j.janxdis.2016.05.009.

Hollenstein, T., Granic, I., Stoolmiller, M., & Snyder, J. (2004). Rigidity in parent-child interactions and the development of externalizing and internalizing behavior in early childhood. *Journal of Abnormal Child Psychology, 32*(6), 595–607. https://doi.org/10.1023/B:JACP.0000047209.37650.41.

Hosey, R. P., & Woodruff-Borden, J. (2012). An examination of the relationship between parent language and child anxiety. *Child and Family Behavior Therapy, 34*(3), 210–230. https://doi.org/10.1080/07317107.2012.707089.

Hudson, J. L., Comer, J. S., & Kendall, P. C. (2008). Parental responses to positive and negative emotions in anxious and nonanxious children. *Journal of Clinical Child and Adolescent Psychology, 37*(2), 303–313. https://doi.org/10.1080/15374410801955839.

Hudson, J. L., Murayama, K., Meteyard, L., Morris, T., & Dodd, H. F. (2019). Early childhood predictors of anxiety in early adolescence. *Journal of Abnormal Child Psychology, 47*(7), 1121–1133. https://doi.org/10.1007/s10802-018-0495-6.

Hudson, J. L., & Rapee, R. M. (2004). From anxious temperament to disorder: An etiological model. In R. G. Heimberg, C. L. Turk, & D. S. Mennin (Eds.), *Generalized anxiety disorder: Advances in research and practice* (pp. 51–75). New York: Guilford Press.

Hurrell, K. E., Houwing, F. L., & Hudson, J. L. (2017). Parental Meta-emotion philosophy and emotion coaching in families of children and adolescents with an anxiety disorder. *Journal of Abnormal Child Psychology, 45*(3), 569–582. https://doi.org/10.1007/s10802-016-0180-6.

James, A. C., Reardon, T., Soler, A., James, G., & Creswell, C. (2020). Cognitive behavioural therapy for anxiety disorders in children and adolescents. *Cochrane Database of Systematic Reviews, 11*(2), CD013162. https://doi.org/10.1002/14651858.CD013162.pub2.

Jongerden, L., & Bögels, S. M. (2015). Parenting, family functioning and anxiety-disordered children: Comparisons to controls, changes after family versus child CBT. *Journal of Child and Family Studies, 24*(7), 2046–2059. https://doi.org/10.1007/s10826-014-0005-6.

Kreuze, L. J., Pijnenborg, G. H. M., de Jonge, Y. B., & Nauta, M. H. (2018). Cognitive-behavior therapy for children and adolescents with anxiety disorders: A meta-analysis of secondary outcomes. *Journal of Anxiety Disorders, 60*(February), 43–57. https://doi.org/10.1016/j.janxdis.2018.10.005.

Lewis-Morrarty, E., Degnan, K. A., Chronis-Tuscano, A., Rubin, K. H., Cheah, C. S. L., Pine, D. S., … Fox, N. A. (2012). Maternal over-control moderates the association between early childhood behavioral inhibition and adolescent social anxiety symptoms. *Journal of Abnormal Child Psychology, 40*(8), 1363–1373. https://doi.org/10.1007/s10802-012-9663-2.

Majdandžić, M., de Vente, W., Colonnesi, C., & Bögels, S. M. (2018). Fathers' challenging parenting behavior predicts less subsequent anxiety symptoms in early childhood. *Behaviour Research and Therapy, 109*, 18–28. https://doi.org/10.1016/j.brat.2018.07.007.

Majdandžić, M., Möller, E. L., De Vente, W., Bögels, S. M., & Van Den Boom, D. C. (2014). Fathers' challenging parenting behavior prevents social anxiety development in their 4-year-old children: A longitudinal observational study. *Journal of Abnormal Child Psychology, 42*(2), 301–310. https://doi.org/10.1007/s10802-013-9774-4.

Manassis, K., Changgun Le, T., Bennett, K., Zhao, X. Y., Mendlowitz, S., Duda, S., Saini, M., Wilansky, P., Baer, S., Barrett, P., Bodden, D., Cobham, V. E., Dadds, M. R., Flannery-Schroeder, E., Ginsburg, G., Heyne, D., Hudson, J. L., Kendall, P. C., Liber, J., … Wood, J. J. (2014). Types of parental involvement in CBT with anxious youth: A preliminary meta-analysis. *Journal of Consulting and Clinical Psychology, 82*(6), 1163–1172. https://doi.org/10.1037/a0036969.

Maric, M., Van Steensel, F. J., & Bögels, S. M. (2018). Parental involvement in CBT for anxiety-disordered youth revisited: Family CBT outperforms child CBT in the long term for children with comorbid ADHD symptoms. *Journal of Attention Disorder, 22*(5), 506–514. https://doi.org/10.1177/1087054715573991.

McLeod, B. D., Wood, J. J., & Weisz, J. R. (2007). Examining the association between parenting and childhood anxiety: A meta-analysis. *Clinical Psychology Review, 27*(2), 155–172. https://doi.org/10.1016/j.cpr.2006.09.002.

Möller, E. L., Nikolić, M., Majdandžić, M., & Bögels, S. M. (2016). Associations between maternal and paternal parenting behaviors, anxiety and its precursors in early childhood: A meta-analysis. *Clinical Psychology Review*, *45*(45), 17–33. https://doi.org/10.1016/j.cpr.2016.03.002.

Morris, A. S., Silk, J. S., Steinberg, L., Myers, S. S., & Robinson, L. R. (2007). The role of the family context in the development of emotion regulation. *Social Development*, *16*(2), 361–388. https://doi.org/10.1111/j.1467-9507.2007.00389.x.

Muris, P., van Zwol, L., Huijding, J., & Mayer, B. (2010). Mom told me scary things about this animal: Parents installing fear beliefs in their children via the verbal information pathway. *Behaviour Research and Therapy*, *48*(4), 341–346. https://doi.org/10.1016/j.brat.2009.12.001.

Murray, L., Lau, P. Y., Arteche, A., Creswell, C., Russ, S., Zoppa, L. D., … Cooper, P. (2012). Parenting by anxious mothers: Effects of disorder subtype, context and child characteristics. *Journal of Child Psychology and Psychiatry and Allied Disciplines*, *53*(2), 188–196. https://doi.org/10.1111/j.1469-7610.2011.02473.x.

Murray, L., Pella, J. E., De Pascalis, L., Arteche, A., Pass, L., Percy, R., … Cooper, P. J. (2014). Socially anxious mothers' narratives to their children and their relation to child representations and adjustment. *Development and Psychopathology*, *26*, 1531–1546. https://doi.org/10.1017/S0954579414001187.

Ng, M. Y., & Weisz, J. R. (2016). Annual research review: Building a science of personalized intervention for youth mental health. *Journal of Child Psychology and Psychiatry and Allied Disciplines*, *57*(3), 216–236. https://doi.org/10.1111/jcpp.12470.

Percy, R., Creswell, C., Garner, M., O'Brien, D., & Murray, L. (2016). Parents' verbal communication and childhood anxiety: A systematic review. *Clinical Child and Family Psychology Review*, *19*(1), 55–75. https://doi.org/10.1007/s10567-015-0198-2.

Polanczyk, G. V., Salum, G. A., Sugaya, L. S., Caye, A., & Rohde, L. A. (2015). Annual research review: A meta-analysis of the worldwide prevalence of mental disorders in children and adolescents. *Journal of Child Psychology and Psychiatry and Allied Disciplines*, *56*(3), 345–365. https://doi.org/10.1111/jcpp.12381.

Puleo, C. M., & Kendall, P. C. (2011). Anxiety disorders in typically developing youth: Autism spectrum symptoms as a predictor of cognitive-behavioral treatment. *Journal of Autism and Developmental Disorders*, *41*(3), 275–286. https://doi.org/10.1007/s10803-010-1047-2.

Reynolds, S., Wilson, C., Austin, J., & Hooper, L. (2012). Effects of psychotherapy for anxiety in children and adolescents: A meta-analytic review. *Clinical Psychology Review*, *32*(4), 251–262. https://doi.org/10.1016/j.cpr.2012.01.005.

Rork, K. E., & Morris, T. L. (2009). Influence of parenting factors on childhood social anxiety: Direct observation of parental warmth and control. *Child and Family Behavior Therapy*, *31*(3), 220–235. https://doi.org/10.1080/07317100903099274.

Schrock, M., & Woodruff-Borden, J. (2010). Parent-child interactions in anxious families. *Child and Family Behavior Therapy*, *32*(4), 291–310. https://doi.org/10.1080/07317107.2010.515523.

Schwartz, O. S., Dudgeon, P., Sheeber, L. B., Yap, M. B. H., Simmons, J. G., & Allen, N. B. (2012). Parental behaviors during family interactions predict changes in depression and anxiety symptoms during adolescence. *Journal of Abnormal Child Psychology*, *40*(1), 59–71. https://doi.org/10.1007/s10802-011-9542-2.

Settipani, C. A., O'Neil, K. A., Podell, J. L., Beidas, R. S., & Kendall, P. C. (2013). Youth anxiety and parent factors over time: Directionality of change among youth treated for anxiety. *Journal of Clinical Child and Adolescent Psychology*, *42*(1), 9–21. https://doi.org/10.1080/15374416.2012.719459.

Silk, J. S., Sheeber, L., Tan, P. Z., Ladouceur, C. D., Forbes, E. E., McMakin, D. L., … Ryan, N. D. (2013). "You can do it!": The role of parental encouragement of bravery in child anxiety treatment. *Journal of Anxiety Disorders*, *27*(5), 439–446. https://doi.org/10.1016/j.janxdis.2013.06.002.

Suveg, C., Morelen, D., Brewer, G. A., & Thomassin, K. (2010). The emotion dysregulation model of anxiety: A preliminary path analytic examination. *Journal of Anxiety Disorders*, *24*(8), 924–930. https://doi.org/10.1016/j.janxdis.2010.06.018.

Suveg, C., Sood, E., Barmish, A., Tiwari, S., Hudson, J. L., & Kendall, P. C. (2008). "I'd rather not talk about it": Emotion parenting in families of children with an anxiety disorder. *Journal of Family Psychology*, *22*(6), 875–884. https://doi.org/10.1037/a0012861.

Suveg, C., Zeman, J., Flannery-Schroeder, E., & Cassano, M. (2005). Emotion socialization in families of children with an anxiety disorder. *Journal of Abnormal Child Psychology*, *33*(2), 145–155. https://doi.org/10.1007/s10802-005-1823-1.

Teetsel, R. N., Ginsburg, G. S., & Drake, K. L. (2014). Anxiety-promoting parenting behaviors: A comparison of anxious mothers and fathers. *Child Psychiatry and Human Development*, *45*(2), 133–142. https://doi.org/10.1007/s10578-013-0384-8.

Thirlwall, K., & Creswell, C. (2010). The impact of maternal control on children's anxious cognitions, behaviours and affect: An experimental study. *Behaviour Research and Therapy*, *48*(10), 1041–1046. https://doi.org/10.1016/j.brat.2010.05.030.

Ung, D., Selles, R., Small, B. J., & Storch, E. A. (2015). A systematic review and meta-analysis of cognitive-behavioral therapy for anxiety in youth with high-functioning autism spectrum disorders. *Child Psychiatry and Human Development*, *46*(4), 533–547. https://doi.org/10.1007/s10578-014-0494-y.

van der Bruggen, C. O., Stams, G. J. J. M., & Bögels, S. M. (2008). Research review: The relation between child and parent anxiety and parental control: A meta-analytic review. *Journal of Child Psychology and Psychiatry, and Allied Disciplines*, *49*(12), 1257–1269. https://doi.org/10.1111/j.1469-7610.2008.01898.x.

Van der Giessen, D., & Bögels, S. M. (2018). Father-child and mother-child interactions with children with anxiety disorders: Emotional expressivity and flexibility of dyads. *Journal of Abnormal Child Psychology*, *46*(2), 331–342. https://doi.org/10.1007/s10802-017-0271-z.

Van der Giessen, D., Colonnesi, C., & Bögels, S. M. (2019). Changes in rejection and psychological control during parent-child interactions following CBT for children's anxiety disorder. *Journal of Family Psychology*. https://doi.org/10.1037/fam0000543.

Van der Giessen, D., Hollenstein, T., Hale, W. W., Koot, H. M., Meeus, W., & Branje, S. (2015). Emotional variability in mother-adolescent conflict interactions and internalizing problems of mothers and adolescents: Dyadic and individual processes. *Journal of Abnormal Child Psychology*, *43*(2), 339–353. https://doi.org/10.1007/s10802-014-9910-9.

Van Steensel, F. J. A., Bögels, S. M., & Perrin, S. (2011). Anxiety disorders in children and adolescents with autistic spectrum disorders: A meta-analysis. *Clinical Child and Family Psychology Review*, *14*(3), 302–317. https://doi.org/10.1007/s10567-011-0097-0.

Waite, P., & Creswell, C. (2015). Observing interactions between children and adolescents and their parents: The effects of anxiety disorder and age. *Journal of Abnormal Child Psychology*, *43*(6), 1079–1091. https://doi.org/10.1007/s10802-015-0005-z.

Williams, S. R., Kertz, S. J., Schrock, M. D., & Woodruff-Borden, J. (2012). A sequential analysis of parent-child interactions in anxious and nonanxious families. *Journal of Clinical Child and Adolescent Psychology*, *41*(1), 64–74. https://doi.org/10.1080/15374416.2012.632347.

Wood, J. J., McLeod, B. D., Piacentini, J. C., & Sigman, M. (2009). One-year follow-up of family versus child CBT for anxiety disorders: Exploring the roles of child age and parental intrusiveness. *Child Psychiatry and Human Development*, *40*(2), 301–316. https://doi.org/10.1007/s10578-009-0127-z.

Woodruff-Borden, J., Morrow, C., Bourland, S., & Cambron, S. (2002). The behavior of anxious parents: Examining mechanisms of transmission of anxiety from parent to child. *Journal of Clinical Child and Adolescent Psychology*, *31*(3), 364–374. https://doi.org/10.1207/S15374424JCCP3103_08.

Chapter 23

Standard individual cognitive-behavioral therapy for children and adolescents with pediatric obsessive-compulsive disorder

Gudmundur Skarphedinsson[a], Davíð R.M.A. Højgaard[b], Katja Anna Hybel[b], Per Hove Thomsen[b], and Tord Ivarsson[c]

[a]*Faculty of Psychology, School of Health Sciences, University of Iceland, Reykjavík, Iceland* [b]*Department of Child and Adolescent Psychiatry, Aarhus University Hospital Psychiatry, Aarhus, Denmark* [c]*Institute of Neurosciences and Physiology, University of Gothenburg, Gothenburg, Sweden*

Abbreviations

CBT	cognitive-behavioral therapy
CGI	clinical global impressions scale
COIS	child obsessive-compulsive impact scale
CS	conditioned stimuli
CY-BOCS	Children's Yale-Brown Obsessive-Compulsive Scale
E/RP	exposure and response prevention
OCD	obsessive-compulsive disorder
SSRI	specific serotonin reuptake inhibitors
US	unconditioned stimuli

Introduction

Obsessive-compulsive disorder (OCD) can be a severe and disabling disorder that has a poor prognosis without adequate treatment (Stewart et al., 2004). This underscores the importance of early detection and treatment. Approximately 1%–3% of all children and adolescents suffer from OCD (Douglass, Moffitt, Dar, McGee, & Silva, 1995; Flament et al., 1988). OCD is defined by obsessions which are intrusive and unwanted thoughts, urges, or images that are recurrent and persistent, and compulsions, which are goal-directed repetitive behaviors or mental acts, executed to neutralize the discomfort, anxiety, or disgust, activated by the obsessions (American Psychiatric Association, 2013). Childhood OCD is similar to the adult condition with three distinctive symptom dimensions (cleaning and contamination, aggressive/sexual thoughts, and symmetry) (Hojgaard et al., 2017). However, the comorbidity profile is distinct, with a higher prevalence of tic disorders and a lower prevalence of depressive disorders, and especially among prepubertal children (Ivarsson, Melin, & Wallin, 2008).

In this chapter, we will provide a comprehensive overview of standard individual cognitive-behavioral therapy (CBT) for childhood OCD presented in the following four sections: (1) historical context, (2) theoretical models for individual CBT for pediatric OCD, (3) practical applications of individual CBT, and (4) empirical support.

Historical context

Humankind has struggled with obsessive/ritualized behaviors for a long period of our existence, and in, for example, religious contexts too zealous excessive religiosity that today is called scrupulosity. However, as medicine gradually took over most of the care of people with mental disorders, from the 18th century and onwards, obsessive/ritualized behaviors of a wider kind were observed in asylums and included in the taxonomy of illnesses, see Steinberg, Carius, and Fontenelle (2017) for a historical review. In the late 19th century, Emminghaus noted that obsessions and compulsions occurred in children and adolescents and not only in adults (Emminghaus, 1887). Interestingly, the major themes and traits of

Handbook of Lifespan Cognitive Behavioral Therapy. https://doi.org/10.1016/B978-0-323-85757-4.00023-7

OCD he described are well recognizable today. However, no effective treatment was available, and indeed ineffective treatments, such as psychoanalysis, which was the standard treatment, were all that child psychiatry offered until 1970s (Osborn, 1989). Thus, from the beginning of the last century, OCD was recognized as a reaction to inner conflicts from early childhood. The obsessions were considered as signs of unconscious conflict, while compulsions were considered as defense mechanisms in order to keep or hold back on unconscious impulses (Yorke, Putzel, & Schacht, 1994). No research results have supported the validity of psychodynamic theory, and generally it has been rejected as a valid explanation of OCD symptoms (Rachman, 2015).

Treatments that had been found helpful in adults such as serotonin reuptake inhibiting drugs and behavioral treatment have been studied from the 1960s. The first plausible theory and treatment originated from the behavioral perspective and appeared in 1960–70 (Foa, 2010) was initially applied by Meyer (Meyer, 1966) to two adult OCD patients. According to this theory, OCD symptoms are learned through classical conditioning (the first phase) and maintained with operant conditioning (the second phase). This approach provided the scientific background for developing the exposure and response prevention (E/RP) treatment (Rachman, 2015). Gradually, this model was also examined in children with OCD (Ong & Leng, 1979; Stanley, 1980).

Theoretical models for CBT for OCD

Theoretically, CBT originates from social learning theory. It is based on the framework that thoughts, behaviors, feelings, and physiological reactions have constant interchanging effects and that symptoms are modified by these processes (March & Mulle, 1998). CBT integrates cognitive theories, behavioral theories, and research from other important fields such as developmental psychology, research on emotions, and psychotherapy research. In this chapter, we will discuss thoroughly different theoretical models included in individually applied CBT for children and adolescents with OCD. We will discuss (1) the behavioral model, (2) emotional processing theory, (3) cognitive components, (4) the more recent inhibitory learning paradigm, (5) and family components.

Behavioral model

The behavioral model, which is based on the original Mowrer's two-factor theory (Foa, 2010), assumes that OCD is learned behavior. As shown in the initial model by Mowrer (1960), it's suggested that (1) obsessional fear may develop when a neutral stimulus is paired with an aversive stimulus (i.e., classical conditioning) and (2) avoidance behavior and/or compulsions may develop as a goal-directed behavior to reduce obsessional discomfort (e.g., physical discomfort, anxiety, or disgust) (i.e., operant conditioning, negative reinforcement). Thus, exposure to obsessional and feared stimuli (i.e., physical object or distressful thoughts) elevates anxiety or disgust. The role of the goal-directed behavior (e.g., washing, checking) is to neutralize or to prevent negative emotions which then maintain the behavior pattern (i.e., negative reinforcement). However, empirical research has not supported the role of classical conditioning in explaining the development of obsessional fear. Usually, the onset of OCD is gradual and not immediate, which would not be predicted by classical conditioning. People suffering from OCD usually do not ascribe the onset of OCD to any specific conditioning event (Abramowitz, Taylor, & McKay, 2007). Furthermore, OCD usually includes multiple types of obsessions and fears and new types of obsessions that are not necessarily related to previous obsessions, tend to develop (Hojgaard et al., 2017). However, the model explains the maintenance of obsessional fear very well (Abramowitz et al., 2007). In practice, this principle relies on the fact that the fear usually decreases after sufficient duration of proximity with the obsessional stimulus. The child needs to confront the fear-evoking stimulus and allow the fear/anxiety to decrease over time. In addition, adequate effect of the exposure is dependent on blocking the negative reinforcement of compulsion or avoidance behavior (i.e., response prevention). For instance, a child with contamination fear should come in contact with things or situations which are "contaminated" and the refrain from the usual ritualized washing and cleaning until the fear decreases significantly. The model has led to the development of effective treatments consisting of E/RP (Franklin et al., 2015; Skarphedinsson et al., 2015).

Emotional processing theory

The emotional processing theory (Foa & Kozak, 1986) is found on the concept that a network of fear is represented in memory and includes information about the obsessional stimulus, response to the obsessional stimulus, and the meaning associated with the stimuli. The model suggests that children with OCD respond excessively to obsessionally targeted stimulus (e.g., avoidance and/or compulsive behavior), and because these extreme responses are usually not easy to modify. However, the fear network may be modified when the real-time data are not congruent with the fear network; when avoidance behavior is triggered by harmless stimulus. This processing may lead to new memory formation and related emotional change. Physiological activation, habituation during exposures, and a decrease in the initial fear to obsessional stimulus are indicators of emotional processing. The model has been used as a theoretical groundwork for exposure-based treatment of PTSD, anxiety disorders, and OCD (Foa & Kozak, 1986).

Cognitive theories

Several cognitive theories have been suggested (Rachman, 1997; Reed, 1985; Salkovskis, 1985). They all share the idea that dysfunctional interpretations of normal but negative intrusive thoughts may lead to obsessions and compulsions when they are interpreted as significant and as a personal responsibility (e.g., to avoid harm to self or others). However, the compulsions increase the intensity and the frequency of these thoughts, which further increases discomfort and anxiety that increases the strength of the intrusive thought. Several examples of distorted thoughts have been identified, including inflated responsibility, overestimation of threat, overimportance of thoughts, the need to control thoughts, and pathological doubt (Obsessive Compulsive Cognitions Working Group, 1997). For instance, Salkovskis proposed that OCD can be characterized by five assumptions: (1) to think about an action is the same as doing it (i.e., thought-action fusion), (2) when failing to prevent harm is morally equal to cause harm, (3) harm responsibility is not reduced by extenuating circumstances, (4) when failing to conduct compulsions/rituals in response to a thought associated with harm is equal to intent to harm, and (5) it's necessary to control your thoughts (Salkovskis, 1985). Thus, although the person feels that her/his obsessions are not acceptable, the person accepts the compulsions as the proper way to reduce the anxiety.

Inhibitory learning theory

Emerging research has suggested that habituation within and between sessions (which are the indicators of successful emotional processing) is not a predictor of successful treatment outcome (Craske et al., 2008). The inhibitory learning model of exposure suggests that focusing on fear learning and fear tolerance instead of fear expression should be more effective, and it should be measured after exposure instead of during an exposure task (Craske et al., 2008). The model is originally derived from basic human laboratory research, and it has been supported by neuroimaging and experimental studies (Milad et al., 2009; Shin & Liberzon, 2010). The model has offered practical guidelines for optimizing CBT outcome by enhancing variability in distress and to combine exposure techniques (e.g., Challenging Patient Expectations of Feared Outcomes) (McGuire & Storch, 2018) which has shown to predict a more favorable outcome for children with OCD (Kircanski, Wu, & Piacentini, 2014). According to the model, the CS-US association is learned during fear conditioning, and exposure exercises do not remove it. However, by doing exposures, a new secondary inhibitory learning about the CS-US emerges, so that the CS is not a predictor of the US anymore. This new association competes with the old CS-US association and becomes increasingly more significant over time. However, the old US-CS association is never erased, which may explain the potential relapse after treatment (Craske et al., 2008; Craske, Treanor, Conway, Zbozinek, & Vervliet, 2014).

This novel approach to extinction has implications for how to conduct exposure treatment compared to treatment based on the model of habituation (i.e., emotional processing theory). The most important features of the model are (1) violating expectancies and (2) incorporating variability (Craske et al., 2014). In order to conduct optimal exposure exercise by violating expectancies, one should not end the exposure when the child habituates, but when the child's expectation of a negative/catastrophic outcome is reduced markedly. Variability is incorporated by (1) combining two or more exposure targets in one exposure and (2) greater variability in terms of the level of fear, intensity, duration, etc. (McGuire & Storch, 2018).

Practical application of CBT

Different CBT treatment manuals exist that can guide clinicians (de Haan & Wolters, 2009; March & Mulle, 1998; Peris & Piacentini, 2016; Piacentini, Langley, & Roblek, 2007; Weidle, 2014; Williams et al., 2010). The manuals vary in structure, the use of materials and analogies, but a range of core principles are the same across the manuals. Standard individual CBT treatment of OCD for children and adolescents are often 12–14 weekly session programs that include psychoeducation on OCD and the OCD treatment process, symptom registration and building of an overview of symptoms (symptom list or symptom hierarchy), cognitive interventions, gradual exposure, and response prevention exercises in vivo in sessions and repeated at home between sessions, celebration of achievements, and relapse prevention. On top of this skeleton of basic components, manuals differ in the amount and how cognitive techniques and narrative components are incorporated as well as how the basic components are introduced and applied in the work with the child or adolescent and parents.

Delivery of CBT to children and adolescents requires the clinician to tailor treatment in a developmentally appropriate way, for instance, in the use of language, humor, metaphors, materials, games, role play, etc. Also, treatment must address the child's or adolescent's symptoms in the context of the family. Children rarely choose therapy themselves and insight might be limited in smaller children. This demands strong alliance and motivation building.

Assessment prior to treatment

A thorough assessment is at the heart of all good treatments. Assessment informs choice of treatment and if CBT is chosen, then assessment forms the basis for the initial case formulation that guides the intervention. Even though standard CBT manuals are used, the clinician must adjust the manual elements and the clinical approach to fit each child or adolescent and family. OCD is not just OCD, and CBT for OCD not just ditto.

Besides targeted assessment of OCD symptoms, the assessment should include a comprehensive clinical interview covering present problems and mental health complaints experienced by the child or adolescent and parents; family history of mental health problems, the social economic family context (strains and resources); interpersonal relations in the family (including parent accommodation to the child's symptoms, conflict management, and expression of emotions in the family); as well as the developmental and medical history of the child/adolescent.

The gold-standard instrument for OCD assessment in children and adolescents is the Children's Yale-Brown Obsessive-Compulsive Scale (CY-BOCS) (Scahill et al., 1997), and this semistructured interview is by many clinicians considered the cornerstone of the assessment. In the interview, the clinician enquires into the specific symptoms of the child by going over examples of typical obsessions and compulsions ordered by main symptom categories. For many children and parents who experience OCD symptoms for the first time, such neutral investigation conducted by a confidence-building, calm and accommodating professional can be comforting and a first step in normalizing the presence of symptoms. As such the initial assessment phase overlaps with the first steps of CBT including creating alliance, giving psychoeducation, and building hope. In addition to the assessment of specific symptoms, the CY-BOCS covers a range of OCD severity aspects: time spent on intrusive thoughts and rituals, overall functional impairment in daily activities due to symptoms, distress levels of symptoms, ability to resist and control obsessions and compulsions, avoidance behavior and the child's understanding of the symptoms, and all aspects that are relevant for accommodating the treatment to the individual child and family. For instance, the enquiry into resistance and control provides valuable information about the child's and parents' previous coping experiences that can be addressed later on in the CBT process.

A range of general psychopathology as well as OCD-specific clinician, parent and self-report scales exist that can supplement and inform assessment. These scales are probably used rather variedly in clinics around the world and many have not been validated sufficiently or they only exist in one or few language versions. In respect to OCD-specific scales, the Clinical Global Impressions Scale (CGI) (Busner & Targum, 2007) and the Child Obsessive-Compulsive Impact Scale (COIS) (Piacentini, Peris, Bergman, Chang, & Jaffer, 2007) are frequently used, at least across clinics involved in treatment studies.

OCD-specific instruments.

Children's Yale-Brown Obsessive-Compulsive Scale (CY-BOCS)
Clinical Global Impressions Scale (CGI)
Child Obsessive-Compulsive Impact Scale (COIS)

Treatment

The information gathered during the assessment form the basis for deciding on the medical diagnosis as well as an individual case formulation which synthesize maintaining and protective factors in the child's/youth's life, which can guide decisions in the CBT process as explained below.

Working alliance, psychoeducation, and externalization

In the first sessions of treatment, focus is on creating a working alliance. The therapist gets to know the child by asking questions about their life, resources, and interests. The therapist provides clear information on the treatment frame, therapist and parent roles during treatment, and aligning mutual expectations. Further, the therapist provides psychoeducation on pediatric OCD elaborating and expanding on what was already introduced during the assessment.

Psychoeducational topics for OCD symptoms.

- What are obsessions and compulsions?
- How do obsessions and compulsions differ from ordinary thoughts and actions?
- How many suffer from OCD?
- What types of OCD exist?
- What do we know about the causes of OCD?
- What do we know about prognosis?

Next, the therapist moves on to psychoeducation on the CBT process by interactively covering aspects.
Psychoeducational topics for CBT.

- What is the rationale and purpose of CBT for OCD?
- How many get better after CBT for OCD?
- What is required by the child or youth during the treatment process?
- What does gradual exposure and response prevention imply?
- What is the role of the therapist?
- What is the role of the parents?

Covering these aspects without losing the child or youth's interest requires basic knowledge on developmental psychology and talk therapeutic skills and experience. The therapist must actively engage the child or youth by flexibly adapting language, asking, and inviting to share thoughts and timing communication by taking departure in the child's developmental age, interests, and views of the world. The aim is to induce motivation to engage in therapy and hope for change. In this phase of treatment, narrative techniques aimed at externalizing OCD are often introduced. Externalization can be introduced by therapist language and metaphors and by asking the child to make a narrative or drawing of "OCD" as some kind of external creature or phenomena that shall be defeated during therapy. Externalization techniques aim at separating the problem from the person, by placing OCD outside the child to enable resistance against symptoms as well as facilitating a sound identity building of the child or youth.

Mapping, symptom registration, and setting goals

The next phase of treatment builds on and expands the individual narrative of the child's OCD. The focus is on creating an overview of symptoms and an understanding of mechanisms and exactly how symptoms interfere in daily life. For some children or youths, it is engaging to use drawings that depict how "the OCD phenomena" bully, bothers, or trouble the child and how the child strikes back and takes charge. Others, often adolescents, may better relate to schematic models presented by the therapist. The therapist and child or youth analyze the symptoms systematically by writing the following.

Probes for mapping and symptom registrations.

- What are the trigger thoughts, objects, or situations?
- What exactly does the obsession hold?
- What exactly are the compulsive rituals related to the obsessive thoughts?
- How difficult does the child or youth rate imagined resistance against the thoughts and rituals on a discomfort/anxiety scale?

This joint symptom analysis is supported by imaginary and active in vivo explorations in the therapist's office and at home between sessions. In this process, the child or youth learn to differentiate between thoughts and rituals, to grade different symptoms, and experience the difference between expectations and practice. During this phase, it is essential to explore previous situations of OCD mastery (resistance; however tiny they might be) in order to articulate and promote coping skills. Gradually, a symptom hierarchy or a symptom list is built, which summarizes the different symptoms and each individual symptom's severity. Further, in this phase, concrete goal setting is introduced to motivate and enable evaluation of progress during treatment. The child or youth is guided to write goals that center on daily-life activities which are objectively assessable, rather than subjective inner states or absence of symptoms. Goals are then revised throughout therapy.

E/RP exercises and progress registration

Gradual E/RP exercises can be introduced early in treatment. The therapist repeats the initial psychoeducation by asking the child/youth if he or she remember what the rationale for conducting E/RP exercises is. The therapist explains that by practicing many experience habituation, i.e., gradual decrease in feelings of discomfort, disgust and/or anxiety, and that the aim of practice is to learn strategies to tolerate such feelings or sensations. The therapist reminds the child or youth that obsessions are themselves just ordinary thoughts or impulses that are twisted or exaggerated because of the OCD raising a "false alarm" and apply relevant metaphors departing from the child's/youth's OCD narrative. Based on the symptom hierarchy, symptom list or OCD drawings that include the child's different symptom severity ratings, the therapist and the child then systematically plan and practice suitable exercises that challenge "the OCD" by resisting to comply with obsessions and conduct compulsions. Complex and high severity symptoms may be analyzed and broken down into simpler elements that are realistic to work with. Exercises are planned and practiced in the session and committed to paper (or a note on the mobile phone) so that the child knows exactly how to proceed with daily practicing at home (what to do, when to do it, which strategies to use) and so that parents know how to support the child/youth (whether to back off or incite, how to reward the child's/youth's efforts, etc.). The outcome of the daily exercises between sessions is recorded at home and brought to the next session. The therapist and child then evaluate the daily recordings and the therapist encourages and praises the child's efforts. Based on the in-session evaluation, exercises are gradually adjusted to find the right level of difficulty and new homework is consecutively planned. During this process of repeatedly adjusted practice, progress can be tracked by regularly revisiting and updating severity ratings on the symptom list during treatment. This summarizes which symptoms have decreased and where more work is needed and can support a sense of mastery for the patient.

Cognitive interventions

It has been debated whether CBT for pediatric OCD should include cognitive interventions in combination with behavioral interventions or whether the treatment should be purely behavioral or cognitive (Foa, 2010). Purely cognitive approaches have only been tested in youth from 10 years of age (Bolton et al., 2011; Williams et al., 2010) and therefore little is known about the effectiveness in younger children. Most CBT manuals for pediatric OCD include some kind of cognitive interventions such as shown in the table.

Cognitive interventions components.

• Normalization and identification of interpretations of the intrusive thoughts
• "OCD situation" chain analysis of thoughts, emotions, behaviors, and bodily reactions (e.g., by means of "the cognitive diamond")
• Behavioral experiments which explore the accuracy of obsessions
• Pie chart analyses of responsibility
• Analysis and comparison of OCD vs non-OCD appraisals of thoughts

For younger children, some of the above techniques might be too advanced and strategies such as constructive self-speech like "you can beat it," "OCD is just lying," or "talking back" to OCD can be applied.

Relapse prevention, celebration of achievements, and booster sessions

When moving to the last phase of treatment, focus gradually shifts to consolidating and maintaining what has been learned during treatment. The therapist and child/youth work with identification of helpful individual strategies and possible individual stressors that might "provoke OCD" in the future. Typical signs of relapse are identified, and an individual plan of action is prepared. Also, for relapse prevention reasons, the therapist and child/youth might work with creating a list of positive activity ideas that can replace OCD behavior in the future.

As touched upon in the previous section, praise and encouragement of the child's/youth's hard work is central during treatment. This aspect is further stressed during the last sessions by planning a final celebration that might include significant persons in the child's/youth's life. Such celebration strengthens the child's/youth's narrative of being able to manage OCD.

After completion of the 12–14 weekly sessions, we recommend it to offer booster sessions during the following year. Booster sessions are follow-up sessions offered as required in which the therapist checks in with the child/youth and parents regarding the well-being of the child/youth, presence of OCD symptoms, and how strategies learned in therapy are used and maintained by the child and parents. Offering of booster sessions enables a gradual phasing out of therapist support and a sense of security for the family in the transition to normalization.

How to involve parents in treatment?

Manuals differ somewhat in the extent to which parents are included in sessions. It is, however, always important for the clinician to consider how to involve parents as they are often worried and involved in the child's symptoms in ways that can be demanding and symptom maintaining. On the other end of the spectrum, some parents hardly know about the child's/youth's symptoms or find the child's/youth's behavior irrational and hard to accept. It is recommended to complete the first one to three alliance-building and psychoeducation-focused sessions with the parents present, to create a working relationship with a clear framing, distribution of roles between therapist, child/youth, and parents, and to promote a basis of understanding and support for the child/youth. Even though, understandably parents are often impatient for change, it is important that they accept that the treatment must proceed in the child's/youth's pace, i.e., that learning of the basic principles are preconditions of progress, and that the child/youth needs a gradual sense of mastery before major steps can be taken. Parents are taught that they can be significant support by reminding, assisting, and rewarding the hard work of the child/youth and by providing feedback of progress and potential obstacles during the treatment process. But, usually in the beginning, parents need to step back and do no more or no less than they use to regarding the child's/youth's rituals. It is stressed that parents can support in several ways. See table.

Involving parents in treatment.

• Complying with the therapist's guidance
• Recognizing the child's efforts
• Assisting exercises when needed
• Offering the child/youth relevant rewards
• Shifting attention to positive aspects in the child's/youth's life
• Offering "non-OCD focused" time together

Depending on the child's/youth's developmental age, family resources, and parent involvement, the therapist must decide how much parents are included in the subsequent sessions. For smaller children, it can be beneficial to include parents in all sessions, whereas for older youth, intensive parent involvement might be counterproductive. It is however central that parents are involved regularly during treatment, if possible, in the last part of each session, in full sessions halfway (providing feedback, evaluating progress and strategies) and at the end of treatment (participating in relapse prevention themes). In some cases, it can be relevant to offer separate parent sessions to clarify additional treatment needs, provide support or assist in problem solving if parents experience their own mental health problems, OCD-related stress or family relations are hindering progress in therapy. Parents can also be instructed to facilitate and monitor E/RP exercises between sessions. Also, in some cases, it can be relevant to discuss how possible comorbidity or cooccurring problems of the child/youth affect progress in treatment and how such problems can be addressed. Most central in the last phase of treatment, however, is that parents recognize and appraise the child's/youth's efforts and feel prepared, at ease and able to support a positive course in the future.

Empirical support of CBT

The accumulative evidence for effective treatments for children with OCD is strong (Ost, Riise, Wergeland, Hansen, & Kvale, 2016; Skarphedinsson et al., 2015), with the current first-line CBT treatment and/or specific serotonin reuptake inhibitors recommended by clinical expert guidelines (Geller & March, 2012; National Institute for Heath and Clinical Excellence, 2005).

Standard CBT should be the first-line treatment according to expert guidelines (Geller & March, 2012; National Institute for Heath and Clinical Excellence, 2005). This recommendation was made after several studies showed the acceptability and efficacy of a published CBT manual that systemized this type of treatment (Geller & March, 2012).

In an empirical review from 2014, of the effect of CBT in treating OCD in children and adolescent, it was concluded that the effects of this type of treatment appeared to be both robust and durable when applied in this age group (Franklin et al., 2015). However, as the authors of the review also noted, CBT is although effective also an incomplete treatment that needs to be developed and examined further. The review by Freeman et al. included several different CBT formats, such as individual, group, and family-based treatments. Similarly, several other systematic reviews and metaanalyses of CBT have found it to be effective in treating child and adolescent OCD (McGuire et al., 2015). A recent systematic review and metaanalysis, based on 12 RCTs evaluating CBT for children and adolescents, concluded that CBT may be more effective than no intervention at all and equivalent to SSRIs, although the effect estimates were uncertain and that all studies included were assessed to have a high risk of bias (Uhre et al., 2020). The conclusions of the review were later debated (Storch et al., 2020).

Effectiveness of standard individual CBT for child and adolescent OCD

When looking specifically at the use of standard individual CBT in the treatment of child and adolescent OCD, the evidence has slowly but steadily been growing. In a systematic review of effectiveness studies from 2015, the authors found 13 studies that compared the effectiveness of CBT to other interventions (Skarphedinsson et al., 2015). Three categories of comparisons were summarized in the review. Four studies were included in the wait-list comparison (Barrett, Healy-Farrell, & March, 2004; Bolton et al., 2011; Storch et al., 2011; Williams et al., 2010), indicating a significantly better outcome from the CBT ($P < 0.001$) with a mean posttreatment CY-BOCS difference of −11.8 [−15.9, −7.7]. The review also included four studies in the CBT to placebo comparison, one pill placebo (POTS Study Team, 2004), and three treatment placebos (Freeman et al., 2008, 2014; Piacentini et al., 2011), and found CBT to be superior ($P < 0.001$) with a mean posttreatment CY-BOCS difference of −5.9 [−8.2, −3.7]. In the comparison of CBT to active treatments, the review included nine study comparisons, including clomipramine (de Haan, Hoogduin, Buitelaar, & Keijsers, 1998), sertraline (POTS Study Team, 2004), sertraline and CBT in combination (POTS Study Team, 2004; Storch et al., 2013), brief CBT (Bolton et al., 2011), and group CBT (Barrett et al., 2004), but the difference −0.2 [−2.9, 1.9] was not significant ($P = 0.868$). The active treatments included in the comparison were also analyzed individually, based on the type of treatment (SRIs, SRI+CBT, and parent/enhanced CBT) with no significant difference found between CBT and any of the three categories. The clinical guidelines mentioned earlier were therefore supported by the results of this review according to the authors (Skarphedinsson et al., 2015).

Authors of another review published in 2014 concluded that individual CBT should, on basis of existing literature, be considered a *probably efficacious treatment* (Freeman et al., 2018). In an update to this review published in 2018, where four new studies on individual CBT were added, the consideration of efficacy remained unchanged (Freeman et al., 2018). The four studies added to the review update compared individual CBT to the combination of individual CBT+D-cycloserine (Mataix-Cols et al., 2014), CBT+cognitive bias modification (Salemink, Wolters, & de Haan, 2015), CBT+sertraline (Storch et al., 2013), CBT+parent management training (Sukhodolsky, Gorman, Scahill, Findley, & McGuire, 2013). According to the authors, the main reason individual CBT was not upgraded to *well-established treatment* in the review-update was that the studies were not designed to test the efficacy of individual CBT specifically, by including a control condition that did not include a control without individual CBT (Freeman et al., 2018). To conclude, although the evidence for CBT is strong when no differentiation is made between the type of treatment (individual, group, family-based, internet-based, etc.), the field could still benefit from trials that specifically compare standard individual CBT against treatment formats other than CBT in order to make final conclusions about the efficacy of individual CBT for children and adolescents.

Long-term effect of individual CBT for children and adolescents

It's not long ago since long-term studies depicted a very pessimistic outcome for young people with OCD. A review published in 2004 examined 16 studies where children and adolescents with OCD were monitored for a period of 1–15.6 years after ended treatment and found that 41%–60% of the patients still had OCD or symptoms of OCD at the final follow-up (Stewart et al., 2004). However, CBT was only used in seven of the included studies and in all cases was the CBT supplemented with medications and other types of treatment, making the direct long-term evaluation of CBT outcome impossible (Stewart et al., 2004). Outcomes have improved dramatically since then. A recent systematic review examined long-term outcome of CBT from a total of 18 studies including 1389 youths with OCD (follow-up period 1–16 years) (Liu et al., 2021). Overall, the remission rate was 62% (95% CI 52%–72%). It was also noteworthy that shorter duration of OCD was associated with better outcome. Overall, the results showed that early identification and intervention of childhood OCD plays a critical role in positive prognosis. However, only a minority of these studies treated patients with individual CBT.

A naturalistic study showed that 46% of participants didn't report any symptoms 7 years after ended treatment (Nissen, Hesselberg Nikolajsen, & Hove Thomsen, 2014). The Nordic Long-Term OCD Study included a total of 269 children and adolescents (7–17 years of age) from Sweden, Norway, and Denmark, and patients were all initially treated with standard individual CBT, whereafter nonresponders were randomized into two groups receiving either continued CBT or treatment with SSRI (Melin et al., 2020). All patients were then monitored for 3 years after ended treatment. At 3-year follow-up, 90% ($n = 242$) of the participants were rated as responders, while 73% were in clinical remission (Melin et al., 2020). The results imply that CBT treatment for pediatric OCD, extended with continued CBT or SSRI for nonresponders, has long-lasting effects with patients continuing to improve after ended treatment.

Practice and procedures

The following text can be used to explain and instruct children to initiate E/RP.

I will teach you how you may overcome OCD. We and your parents will work together during sessions. Then you will do these exercises at home.

We have two fine tools to help you to overcome OCD. These are E/RP. Exposure is when you do something that OCD is telling you not to do. This may be things that you think is dirty or polluted like a door knob in a public place. Response prevention is what you do after exposure by tolerating the anxiety or any other feeling which is unpleasant without doing any rituals or compulsions. For instance, don't wash again and again. Previously, you thought that the anxiety might increase after exposure, and so much that you would not tolerate it. However, if you don't perform any compulsion, the anxiety will either decrease or you will discover that although you feel anxious it's not dangerous although it is unpleasant.

Mini-dictionary of terms

Exposure and response prevention

Behavior therapy involves encouraging people with anxiety/OCD to expose to situations that provoke fear, disgust, or unpleasant feelings, and helping them to prevent compulsive or avoidance behavior. The final goal is to free individuals from obsessions and compulsions.

Emotional processing theory

Found on the concept that a network of fear is represented in memory and includes information about the obsessional stimulus, response to the obsessional stimulus, and the meaning associated with the stimuli.

The inhibitory learning theory

A model of exposure suggesting that focusing on fear learning and fear tolerance instead of fear expression should be more effective and it should be measured after exposure instead of during an exposure task.

Key facts

- OCD can be a severe and disabling disorder that has a poor prognosis without adequate treatment.
- Approximately 1%–3% of all children and adolescents suffer from OCD.

- Childhood OCD is similar to the adult condition with three distinctive symptom dimensions (cleaning and contamination, aggressive/sexual thoughts, and symmetry).
- Treatments that had been found helpful in adults such as serotonin reuptake inhibiting drugs and behavioral treatment have been studied from the 1960s.
- Standard individual CBT treatment is often 12–14 weekly session programs that include psychoeducation, symptom registration and building hierarchy, cognitive interventions, gradual exposure, and response prevention in sessions and repeated at home between sessions, celebration of achievements and relapse prevention.
- The accumulative evidence for effective treatments for children with OCD is strong.

Applications to other areas

In this chapter about individual CBT for childhood OCD, we described the theoretical foundations of CBT especially focusing on theories explaining the E/RP process. Emerging research has suggested an inhibitory learning model of exposure which emphasizes fear learning and fear tolerance instead of fear extinction expression, which might be more effective (Craske et al., 2008). The model has offered practical guidelines for optimizing CBT outcome by enhancing variability in distress and to combine exposure techniques, which has shown to predict a more favorable outcome for children with OCD (Kircanski et al., 2014). According to the model, the CS-US association is learning during fear conditioning and exposure exercises do not remove it. However, by doing exposures, a new secondary inhibitory learning about the CS-US is emerging, that the CS is not a predictor of the US anymore. This new association competes with the old CS-US association and becomes increasingly more significant. However, the old US-CS association is never erased, which may explain the potential relapse over time (Craske et al., 2008, 2014). This novel approach to extinction has implications for how to conduct exposure treatment compared to treatment based on the model of habituation (i.e., emotional processing theory). The most important features of the model are (1) violating expectancies and (2) incorporating variability (Craske et al., 2014). In order to conduct optimal exposure exercise by violating expectancies, one should not end the exposure when the child habituates but when the child's expectation of a negative/catastrophical outcome is reduced markedly. Variability is incorporated by (1) combining two or more exposure targets in one exposure and (2) greater variability in terms of the level of fear, intensity, duration, etc. (McGuire & Storch, 2018).

Summary points

- CBT is the first-line treatment for childhood OCD.
- E/RP techniques are emphasized in CBT for children and youth with OCD.
- Gradual E/RP exercises can be introduced early in treatment.
  The inhibitory learning model of exposure suggests that focusing on fear learning and fear tolerance instead of fear expression should be more effective.
- The accumulative evidence for individual CBT for children with OCD is strong.

References

Abramowitz, J. S., Taylor, S., & McKay, D. (2007). Psychological theories of obsessive-compulsive disorder. In E. A. Storch, G. R. Geffken, & T. K. Murphy (Eds.), *Handbook of child and adolescent obsessive-compulsive disorder*. Mahawah, NJ: Lawrence Erlbaum Associates.

American Psychiatric Association. (2013). *Diagnostic and statistical manual of mental disorders, fifth edition (DSM-5(TM))*. Arlington, VA: American Psychiatric Publishing.

Barrett, P., Healy-Farrell, L., & March, J. S. (2004). Cognitive-behavioral family treatment of childhood obsessive-compulsive disorder: A controlled trial. *Journal of the American Academy of Child and Adolescent Psychiatry*, *43*, 46–62.

Bolton, D., Williams, T., Perrin, S., Atkinson, L., Gallop, C., Waite, P., et al. (2011). Randomized controlled trial of full and brief cognitive-behaviour therapy and wait-list for paediatric obsessive-compulsive disorder. *Journal of Child Psychology and Psychiatry*, *52*, 1269–1278.

Busner, J., & Targum, S. D. (2007). The clinical global impressions scale: Applying a research tool in clinical practice. *Psychiatry*, *4*, 28–37.

Craske, M. G., Kircanski, K., Zelikowsky, M., Mystkowski, J., Chowdhury, N., & Baker, A. (2008). Optimizing inhibitory learning during exposure therapy. *Behaviour Research and Therapy*, *46*, 5–27.

Craske, M. G., Treanor, M., Conway, C. C., Zbozinek, T., & Vervliet, B. (2014). Maximizing exposure therapy: An inhibitory learning approach. *Behaviour Research and Therapy*, *58*, 10–23.

de Haan, E., Hoogduin, K. A., Buitelaar, J. K., & Keijsers, G. P. (1998). Behavior therapy versus clomipramine for the treatment of obsessive-compulsive disorder in children and adolescents. *Journal of the American Academy of Child and Adolescent Psychiatry*, *37*, 1022–1029.

de Haan, E., & Wolters, L. (2009). *Behandeling van de dwangstoornis bij kinderen en adolescenten: Met het cognitief-gedragstherapeutisch protocol Bedwing je dwang: Kind en adolescent praktijkreeks*. Amsterdam: Bohn Stafleu van Loghum.

Douglass, H. M., Moffitt, T. E., Dar, R., McGee, R., & Silva, P. (1995). Obsessive-compulsive disorder in a birth cohort of 18-year-olds: Prevalence and predictors. *Journal of the American Academy of Child and Adolescent Psychiatry*, *34*, 1424–1431.

Emminghaus, H. (1887). *Die Psychischen Störungen des Kindesalters*. Tübingen: H. Laupp.

Flament, M. F., Whitaker, A., Rapoport, J. L., Davies, M., Berg, C. Z., Kalikow, K., et al. (1988). Obsessive compulsive disorder in adolescence: An epidemiological study. *Journal of the American Academy of Child and Adolescent Psychiatry*, *27*, 764–771.

Foa, E. B., & Kozak, M. J. (1986). Emotional processing of fear: Exposure to corrective information. *Psychological Bulletin*, *99*, 20–35.

Foa, E. B. (2010). Cognitive behavioral therapy of obsessive-compulsive disorder. *Dialogues in Clinical Neuroscience*, *12*, 199–207.

Franklin, M. E., Kratz, H. E., Freeman, J. B., Ivarsson, T., Heyman, I., Sookman, D., et al. (2015). Cognitive-behavioral therapy for pediatric obsessive-compulsive disorder: Empirical review and clinical recommendations. *Psychiatry Research*, *227*, 78–92.

Freeman, J., Benito, K., Herren, J., Kemp, J., Sung, J., Georgiadis, C., et al. (2018). Evidence base update of psychosocial treatments for pediatric obsessive-compulsive disorder: Evaluating, improving, and transporting what works. *Journal of Clinical Child and Adolescent Psychology*, *47*, 669–698.

Freeman, J., Sapyta, J., Garcia, A., Compton, S., Khanna, M., Flessner, C., et al. (2014). Family-based treatment of early childhood obsessive-compulsive disorder: The pediatric obsessive-compulsive disorder treatment study for young children (POTS Jr)—A randomized clinical trial. *JAMA Psychiatry*, *71*, 689–698.

Freeman, J. B., Garcia, A. M., Coyne, L., Ale, C., Przeworski, A., Himle, M., et al. (2008). Early childhood OCD: Preliminary findings from a family-based cognitive-behavioral approach. *Journal of the American Academy of Child and Adolescent Psychiatry*, *47*, 593–602.

Geller, D., & March, J. (2012). Practice parameter for the assessment and treatment of children and adolescents with obsessive-compulsive disorder. *Journal of the American Academy of Child and Adolescent Psychiatry*, *51*, 98–113.

Hojgaard, D. R. M. A., Mortensen, E. L., Ivarsson, T., Hybel, K., Skarphedinsson, G., Nissen, J. B., et al. (2017). Structure and clinical correlates of obsessive-compulsive symptoms in a large sample of children and adolescents: A factor analytic study across five nations. *European Child and Adolescent Psychiatry*, *26*, 281–291.

Ivarsson, T., Melin, K., & Wallin, L. (2008). Categorical and dimensional aspects of co-morbidity in obsessive-compulsive disorder (OCD). *European Child & Adolescent Psychiatry*, *17*, 20–31.

Kircanski, K., Wu, M., & Piacentini, J. (2014). Reduction of subjective distress in CBT for childhood OCD: Nature of change, predictors, and relation to treatment outcome. *Journal of Anxiety Disorders*, *28*, 125–132.

Liu, J., Cui, Y., Yu, L., Wen, F., Wang, F., Yan, J., et al. (2021). Long-term outcome of pediatric obsessive-compulsive disorder: A meta-analysis. *Journal of Child and Adolescent Psychopharmacology*, *31*, 95–101.

March, J. S., & Mulle, K. (1998). *OCD in children and adolescents: A cognitive-behavioral treatment manual*. New York: Guilford Press.

Mataix-Cols, D., Turner, C., Monzani, B., Isomura, K., Murphy, C., Krebs, G., et al. (2014). Cognitive-behavioural therapy with post-session D-cycloserine augmentation for paediatric obsessive-compulsive disorder: Pilot randomised controlled trial. *British Journal of Psychiatry*, *204*, 77–78.

McGuire, J. F., Piacentini, J., Lewin, A. B., Brennan, E. A., Murphy, T. K., & Storch, E. A. (2015). A meta-analysis of cognitive behavior therapy and medication for child obsessive-compulsive disorder: Moderator of treatment efficacy, response, and remission. *Depression and Anxiety*, *32*, 580–593.

McGuire, J. F., & Storch, E. A. (2018). An inhibitory learning approach to cognitive-behavioral therapy for children and adolescents. *Cognitive and Behavioral Practice*, *26*, 214–224.

Melin, K., Skarphedinsson, G., Thomsen, P. H., Weidle, B., Torp, N. C., Valderhaug, R., et al. (2020). Treatment gains are sustainable in pediatric obsessive-compulsive disorder: Three-year follow-up from the NordLOTS. *Journal of the American Academy of Child and Adolescent Psychiatry*, *59*, 244–253.

Meyer, V. (1966). Modification of expectations in cases with obsessional rituals. *Behaviour Research and Therapy*, *4*, 273–280.

Milad, M. R., Pitman, R. K., Ellis, C. B., Gold, A. L., Shin, L. M., Lasko, N. B., et al. (2009). Neurobiological basis of failure to recall extinction memory in posttraumatic stress disorder. *Biological Psychiatry*, *66*, 1075–1082.

Mowrer, O. H. (1960). *Learning theory and behavior*. New York: Wiley.

National Institute for Heath and Clinical Excellence. (2005). *Obsessive compulsive disorder (OCD) and body dysmorphic disorder (BDD)*. London: National Institute for Heath and Clinical Excellence.

Nissen, J. B., Hesselberg Nikolajsen, K., & Hove Thomsen, P. (2014). A 7 year follow-up of children and adolescents with obsessive-compulsive disorder: An analysis of predictive factors in a clinical prospective study. *The European Journal of Psychiatry*, *28*, 183–193.

Obsessive Compulsive Cognitions Working Group. (1997). Cognitive assessment of obsessive-compulsive disorder. Obsessive Compulsive Cognitions Working Group. *Behaviour Research and Therapy*, *35*, 667–681.

Ong, S. B., & Leng, Y. K. (1979). The treatment of an obsessive-compulsive girl in the context of Malaysian Chinese culture. *The Australian and New Zealand Journal of Psychiatry*, *13*, 255–259.

Osborn, I. (1989). *Tormenting thoughts and secret rituals: The hidden epidemic of obsessive-compulsive disorder*. New York: Dell Publishing.

Ost, L.-G., Riise, E. N., Wergeland, G. J., Hansen, B., & Kvale, G. (2016). Cognitive behavioral and pharmacological treatments of OCD in children: A systematic review and meta-analysis. *Journal of Anxiety Disorders*, *43*, 58–69.

Peris, T., & Piacentini, J. (2016). *Helping families manage childhood OCD*. New York: Oxford University Press.

Piacentini, J., Bergman, R. L., Chang, S., Langley, A., Peris, T., Wood, J. J., et al. (2011). Controlled comparison of family cognitive behavioral therapy and psychoeducation/relaxation training for child obsessive-compulsive disorder. *Journal of the American Academy of Child and Adolescent Psychiatry*, *50*, 1149–1161.

Piacentini, J., Langley, A., & Roblek, T. (2007). *Cognitive behavioral treatment of childhood OCD: It's only a false alarm.* New York, NY: Oxford University Press.

Piacentini, J., Peris, T. S., Bergman, R. L., Chang, S., & Jaffer, M. (2007). Functional impairment in childhood OCD: Development and psychometrics properties of the child obsessive-compulsive impact scale-revised (COIS-R). *Journal of Clinical Child and Adolescent Psychology, 36*, 645–653.

POTS Study Team. (2004). Cognitive-behavior therapy, sertraline, and their combination for children and adolescents with obsessive-compulsive disorder: The Pediatric OCD Treatment Study (POTS) randomized controlled trial. *JAMA, 292*, 1969–1976.

Rachman, S. (2015). The evolution of behaviour therapy and cognitive behaviour therapy. *Behaviour Research and Therapy, 64*, 1–8.

Rachman, S. (1997). A cognitive theory of obsessions. *Behaviour Research and Therapy, 35*, 793–802.

Reed, G. (1985). *Obsessional experience and compulsive behavior: A structural approach.* Orlando, FL: Academic Press.

Salemink, E., Wolters, L., & de Haan, E. (2015). Augmentation of treatment as usual with online cognitive Bias modification of interpretation training in adolescents with obsessive compulsive disorder: A pilot study. *Journal of Behavior Therapy and Experimental Psychiatry, 49*, 112–119.

Salkovskis, P. M. (1985). Obsessional-compulsive problems: A cognitive-behavioural analysis. *Behaviour Research and Therapy, 23*, 571–583.

Scahill, L., Riddle, M. A., McSwiggin-Hardin, M., Ort, S. I., King, R. A., Goodman, W. K., et al. (1997). Children's Yale-Brown obsessive compulsive scale: Reliability and validity. *Journal of the American Academy of Child and Adolescent Psychiatry, 36*, 844–852.

Shin, L. M., & Liberzon, I. (2010). The neurocircuitry of fear, stress, and anxiety disorders. *Neuropsychopharmacology, 35*, 169–191.

Skarphedinsson, G., Hanssen-Bauer, K., Kornør, H., Heiervang, E. R., Landrø, N. I., Axelsdottir, B., et al. (2015). Standard individual cognitive behaviour therapy for paediatric obsessive-compulsive disorder: A systematic review of effect estimates across comparisons. *Nordic Journal of Psychiatry, 69*, 81–92.

Stanley, L. (1980). Treatment of ritualistic behaviour in an eight-year-old girl by response prevention: A case report. *Journal of Child Psychology and Psychiatry, 21*, 85–90.

Steinberg, H., Carius, D., & Fontenelle, L. F. (2017). Kraepelin's views on obsessive neurosis: A comparison with DSM-5 criteria for obsessive-compulsive disorder. *Brazilian Journal of Psychiatry, 39*, 355–364.

Stewart, S. E., Geller, D. A., Jenike, M., Pauls, D., Shaw, D., Mullin, B., et al. (2004). Long-term outcome of pediatric obsessive-compulsive disorder: A meta-analysis and qualitative review of the literature. *Acta Psychiatrica Scandinavica, 110*, 4–13.

Storch, E. A., Peris, T. S., De Nadai, A., Piacentini, J., Bloch, M., Cervin, M., et al. (2020). Little doubt that CBT works for pediatric OCD. *Journal of the American Academy of Child and Adolescent Psychiatry, 59*, 785–787.

Storch, E. A., Bussing, R., Small, B. J., Geffken, G. R., McNamara, J. P., Rahman, O., et al. (2013). Randomized, placebo-controlled trial of cognitive-behavioral therapy alone or combined with sertraline in the treatment of pediatric obsessive-compulsive disorder. *Behaviour Research and Therapy, 51*, 823–829.

Storch, E. A., Caporino, N. E., Morgan, J. R., Lewin, A. B., Rojas, A., Brauer, L., et al. (2011). Preliminary investigation of web-camera delivered cognitive-behavioral therapy for youth with obsessive-compulsive disorder. *Psychiatry Research, 189*, 407–412.

Sukhodolsky, D. G., Gorman, B. S., Scahill, L., Findley, D., & McGuire, J. (2013). Exposure and response prevention with or without parent management training for children with obsessive-compulsive disorder complicated by disruptive behavior: A multiple-baseline across-responses design study. *Journal of Anxiety Disorders, 27*, 298–305.

Uhre, C. F., Uhre, V. F., Lønfeldt, N. N., Pretzmann, L., Vangkilde, S., Plessen, K. J., et al. (2020). Systematic review and meta-analysis: Cognitive-behavioral therapy for obsessive-compulsive disorder in children and adolescents. *Journal of the American Academy of Child and Adolescent Psychiatry, 59*, 64–77.

Weidle, B. (2014). *OCD-behandling for barn og unge. En praksismanual.* Oslo, Norway: Gyldendal.

Williams, T. I., Salkovskis, P. M., Forrester, L., Turner, S., White, H., & Allsopp, M. A. (2010). A randomised controlled trial of cognitive behavioural treatment for obsessive compulsive disorder in children and adolescents. *European Child & Adolescent Psychiatry, 19*, 449–456.

Yorke, C., Putzel, R., & Schacht, L. (1994). Some problems of diagnosis in children presenting with obsessional symptomatology. *The Psychoanalytic Study of the Child, 5*, 81–98.

Chapter 24

Obsessive-compulsive disorder and body dysmorphic disorder: Pediatric manifestation and treatment

Matthew Jacofsky, Madison Fitzpatrick, and Fugen Neziroglu
Bio Behavioral Institute, Great Neck, NY, United States

Abbreviations

BDD	body dysmorphic disorder
BDD-YBOCS	Yale Brown Obsessive-Compulsive Scale modified for BDD
CBT	cognitive-behavioral therapy
C-YBOCS	Children's Yale-Brown Obsessive-Compulsive Scale
ERP	exposure and response prevention
OCD	obsessive-compulsive disorder
OVI	overvalued ideation
PANDAS	pediatric autoimmune neuropsychiatric disorders associated with streptococcal infections

Introduction: Pediatric obsessive-compulsive disorder

Obsessive-compulsive disorder (OCD) is a psychiatric condition characterized by intrusive thoughts and compulsive behaviors/rituals. Obsessions are characterized as intrusive thoughts or urges that are extremely difficult to control and repetitive in nature. Compulsions are actions or rituals done in direct response to obsessions to assuage negative feelings associated with the obsessions, such as anxiety (Fig. 1). Avoidance behaviors, which function as compulsions in response to the intrusive thoughts/urges, may also occur. The content of obsessions/compulsions varies widely—there are no typical genres of thoughts/behaviors associated with the disorder. To receive a diagnosis, these thoughts and behaviors must be excessive, time consuming, and lead to significant impairment or distress (American Psychiatric Association, 2013). In the newest edition of the Diagnostic and Statistical Manual for Mental Disorders, 5th edition (DSM-V), OCD was reclassified into its own category of obsessive-compulsive and related disorders, shifting from its former place in the Anxiety Disorders category. With this shift came two specifiers. The Insight specifier references the conviction with which the individual with OCD believes their obsessive thoughts or beliefs will prevent a feared outcome. Good or fair insight suggests the individual believes their obsessions are probably not true, poor insight suggests that they believe they are probably true, and absent/delusional insight suggests they believe they are definitely true. The Tic-related specifier should be noted if the individual with OCD has a history of a tic-disorder, such as Tourette's syndrome, as OCD and Tourette's are commonly comorbid (American Psychiatric Association, 2013).

Currently, it is believed that pediatric OCD follows much of the same clinical profile as adult OCD, with a few key differences. In the following subsections, we will explore the implications of these differences, and the implications they have on treatment praxis.

Demographics

Much of the current research available on OCD has been conducted in adult samples, however, around half of adults diagnosed with this condition report symptom onset occurring in childhood or adolescence (Pauls, Alsobrook, Goodman, Rasmussen, & Leckman, 1995). For OCD generally, the mean age of onset is around 19.5 years, with a quarter of adults

Handbook of Lifespan Cognitive Behavioral Therapy. https://doi.org/10.1016/B978-0-323-85757-4.00008-0

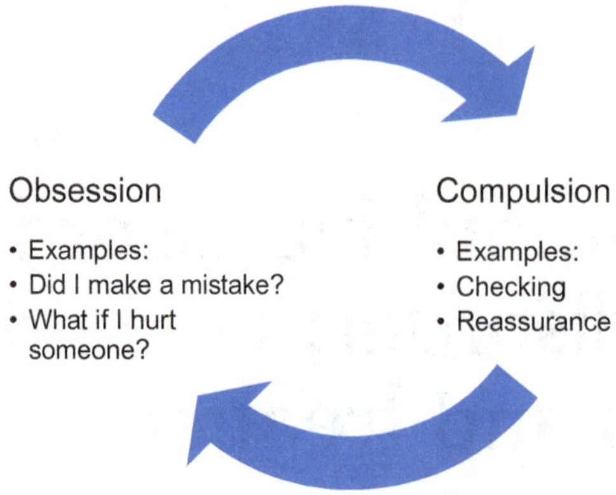

FIG. 1 Repetitive cycle of OCD.

reporting symptom onset occurring before age 14 (American Psychiatric Association, 2013). For pediatric OCD, the average age of onset is between 7.5 and 12.5 years (Gellar, 2006). The operational definition of adolescence across these studies is not stable, as shown in the stark difference between half and one quarter cited above. Nevertheless, in many cases of chronic OCD, symptoms appear earlier than adulthood.

Current statistics place the prevalence rate of OCD in children and adolescents between 1% and 4%, making the disorder fairly common (Gellar, 2006). Therefore, understanding how OCD manifests in pediatric samples and prioritizing early intervention will help aid both children and their families.

Etiology

There are many reasons why OCD may develop, as is the case with most mental health conditions. Currently, researchers believe OCD occurs due to a combination of biological, genetic, and environmental factors (Greenberg & Geller, 2019). Genetically, the risk for OCD is significantly higher in first-degree relatives of individuals with OCD than second- and third-degree relatives (Mataix-Cols et al., 2013). In specific pediatric OCD studies, this risk in first degree relatives is even higher, suggesting that pediatric OCD may be more genetically linked than adult-onset OCD (do Rosario-Campos et al., 2017) However, genetics only account for part of the picture, as many cases of OCD do not exist in a hereditary thread.

The child's temperament and home environment can play a large role in the development of obsessive-compulsive symptoms. The Obsessive Compulsive Cognitions Working Group (1997) reported five belief domains affected in OCD: inflated sense of responsibility, overestimation of threat, perfectionism, need to control thoughts, and intolerance of uncertainty. These same areas are affected in children. Inflated responsibility is a core maladaptive belief of many adults with OCD and heavily predicts OCD symptoms such as increased doubt, worries over making mistakes, and greater checking behaviors (Mantz & Abbott, 2017). This belief is found in many children and adolescents with OCD, and also correlated positively with symptom severity (Mantz & Abbott, 2017). Children and adolescents with this belief also reported overprotective and anxious rearing in their parents, suggesting that these two variables may be related (Mathieu, Conlon, Waters, McKenzie, & Farrell, 2019). Similar parenting styles have been found to exacerbate inflated responsibility beliefs in adult samples, but these rely heavily on retrospective self-report, which could be less accurate than lived current experience (Lennertz et al., 2010).

Pediatric autoimmune neuropsychiatric disorders associated with streptococcal infection

In several cases of pediatric OCD, its onset may be triggered by a streptococcal infection either in the child or in the mother while the child is in utero. This is a subtype of OCD referred to as pediatric autoimmune neuropsychiatric disorders associated with streptococcal infections (PANDAS) (Sigra, Hesselmark, & Bejerot, 2018). PANDAS OCD occurs abruptly and is quite severe in nature (Sigra et al., 2018). Typically, it also includes vocalizations like what would be seen in a case of Tourette's syndrome (Sigra et al., 2018). Though PANDAS differs slightly from the traditional pediatric OCD profile, its etiology is important to note as it has been postulated that autoimmune triggers can contribute to the development of several neuropsychiatric disorders, including eating disorders and schizophrenia (Singer, Gilbert, Wolf, Mink, & Kurlan, 2012). Currently, the recommended treatment is similar to non-PANDAS OCD, with the addition of some crucial elements. In a systematic review, Sigra, Hesslemark, and Bejerot found that immunomodulatory therapies, antibiotics (notably azithromycin and amoxicillin) and behavioral/psychiatric interventions (i.e., CBT and SSRIs) appear the most effective (Sigra et al., 2018). However, there is still a considerable amount of research needed, as these interventions typically only boast a mild-moderate effect rate (Sigra et al., 2018). Screening a child for possible strep throat exposure prior to symptom onset will help clinicians and parents understand the exact nature of the disorder and what to expect in the coming years as research on this condition develops.

Manifestation

Much like adult OCD, there is no "typical" symptom presentation of pediatric OCD. The disorder can manifest in a variety of ways. The four most common symptom dimensions found in adult OCD studies include contamination/cleaning, obsessions/checking, symmetry/ordering, and hoarding (Leckman, Rauch, & Mataix-Cols, 2007; Mataix-Cols, Rosario-Campos, & Leckman, 2005). These were all found to be stable across pediatric populations, with some important correlations with demographic variables (Mataix-Cols, Nakatani, Micali, & Heyman, 2008). This supports the multifaceted nature of OCD. Crucially, over half of this sample of young girls reported hoarding symptoms, compared to just 36% of young boys (Mataix-Cols et al., 2008). Hoarding obsessions and compulsions have been associated with poor response to treatment (Mataix-Cols et al., 2005). This is important to keep in mind and should be thoroughly assessed for during diagnostic evaluations.

Many children diagnosed with OCD present with low insight. Though they are not expected to articulate the aims of their compulsive behaviors per DSM-V, often children are unable to do so, pointing to lower insight. In adult populations, it's estimated that between 2% and 4% of individuals present with poor or absent insight (Van Ameringen, Patterson, & Simpson, 2014). Comparatively, Storch et al. (2008) found that approximately one third of children could be characterized as having low insight, which could explain why pediatric OCD is so complicated to treat the younger the patient begins presenting with symptoms. As adults, individuals then explain their behavior. It is perhaps the drive that then elicits an obsession. It is something that needs investigation based on pediatric OCD behavior. At what point are beliefs assigned to the compulsions in childhood.

Overvalued ideas (OVIs) are defined as unreasonable beliefs that are held so strongly that an individual cannot attribute to their disorder (Neziroglu & Stevens, 2002). They serve as a strong prognostic indicator. Several studies have found individuals with OVIs show less improvement in both cognitive-behavioral therapy and pharmacotherapy settings (Neziroglu, Pinto, Yaryura-Tobias, & McKay, 2004; Neziroglu, Stevens, McKay, & Yaryura-Tobia, 2001). High OVI has also been found to correlate with higher levels of symptom severity, as well as social and school disability. Obsessive-compulsive beliefs such as responsibility/overestimation of threat, perfectionism/intolerance of uncertainty, and importance of thoughts/control of thoughts also correlated with high OVI results in a sample of adolescents (Borda, Neziroglu, Taboas, McKay, & Frenkiel, 2017). Due to OVI, pediatric OCD can be highly treatment resistant in many cases. This is crucial to consider as treatment strategies are devised.

OCD symptoms in children are reinforced in multiple ways, namely through family accommodation. A child's support system in the home typically seeks to assuage OCD symptoms by placating the child, either by indulging in certain rituals or reducing responsibilities in the home (Storch et al., 2010). This, however, is harmful to progress and reinforces the salience of these rituals, causing the child to rely on them more often. Treatment plans should also try to incorporate the family as much as possible and reinforce in caretakers that accommodation of behaviors does not function in the supportive way that they may view it.

Comorbidities

Many children present with other mental health conditions in addition to their OCD symptomatology. Up to 85% of children with OCD present with at least one comorbidity, and up to 50% present with multiple comorbidities (Lewin, Chang, McCracken, McQueen, & Piacentini, 2010). This presents a more complicated clinical profile and may require additional interventions or different treatment strategies altogether.

Attention deficit/hyperactivity disorder (ADHD) occurs in around 51% of children with OCD (Gellar, 2006). Many have speculated that this is due to the way obsessions manifest and can cause inattentiveness, but that is not necessarily the full picture. When comparing diagnostic criteria from across the DSM-IV in children with and without comorbid OCD, all children met the criteria for an ADHD diagnosis (Gellar, Coffey, Farone, et al., 2003). Children with OCD who also have an ADHD diagnosis may have more challenges in social and educational settings than children without an ADHD diagnosis (Gellar, 2006). Therefore, ADHD should be thoroughly assessed for, and if it is present, behavioral management strategies should be implemented in addition to treatment for OCD.

Anxiety and depression are also highly common in pediatric OCD, with 50% of children with OCD experiencing non-OCD related anxiety, and 39% of children experiencing major depression (Boileau, 2011; Gellar, 2006). Major depression risk increases with the advent of adolescence, as 62% of adolescents with OCD report experiencing major depression (Gellar, 2006). Determining comorbid depression and/or anxiety is often done utilizing the "magic wand" technique, whereby you ask the patient the following question: "If I were to wave a magic wand and all of your obsessive thoughts/repetitive behaviors would vanish, would you still feel depressed/anxious?" If the answer is yes, more than likely the depression and/or anxiety are co-occurring with the OCD symptoms, and should be addressed. If the answer is no, the depression and/or anxiety is most likely caused directly by the OCD, and is not necessarily indicative of a comorbid diagnosis.

Practice and procedures

Assessment

Typically, children presenting with obsessive-compulsive symptoms will be given a semistructured interview assessing their symptoms, as well as the detrimental effect they have on the child's life. Clinicians typically use the Anxiety Disorders Interview Schedule: Child and Parent Versions for DSM-IV to assess children, as the version according to the DSM-V has yet to be released (Lewin, 2019). Parents and children both are interviewed regarding the presenting symptoms in the child. However, this interview doesn't thoroughly assess for all comorbidities, and is very time-consuming. Therefore, unstructured interviews are also common in clinical practice. Clinicians should probe with a few questions regarding obsessive-compulsive symptoms, especially involving any repetitive behaviors. Family history and presenting complains should also be evaluated, and clinicians should be able to make an informed diagnosis accordingly. Questionnaires should be administered, both child and parent rated, to get a broader view of the effect of symptoms and any comorbidities. The Children's Yale Brown Obsessive-Compulsive Inventory is the gold standard for evaluating OCD symptom severity in children, as it employs rater administered interview as well as child and parent viewpoints (Scahill et al., 1997). Comorbidities should be assessed for and considered when developing a treatment plan, as they can cause complications with typical treatment praxis if ignored.

Treatment

The gold standard for treating pediatric OCD is cognitive-behavioral therapy, or CBT. CBT should focus on two main areas: cognitive restructuring and behavioral intervention. The crucial behavioral intervention aspect that proves to be the most effective in treating OCD symptoms is exposure and response prevention, or ERP (March & Mulle, 1998). ERP involves exposing patients to salient threatening stimuli to mitigate their anxiety by showing the patient they can confront these perceived threats without performing a compulsive action (Foa & Kozak, 1986). This strategy can be tailored to any manifestation of OCD. Exposure hierarchies, or lists of anxiety-inducing objects/scenarios, are developed with the patient to tailor to their specific anxieties. Once this list is developed, the clinician will expose the patient incrementally to each object or scenario in ascending order, eventually reaching the most fear-provoking stimuli. Typically, individuals participating in CBT see symptom severity reductions between 50% and 67%, and these reductions remain stable for at least 9 months after treatment ends (Franklin et al., 2015).

There are two theories that seek to explain the effectiveness of ERP for OCD: habituation and inhibitory learning. Habituation is theorized to occur as patients are incrementally exposed to fear and become accustomed to the physical symptoms associated with it (e.g., racing heart). As the patient learns that the suspected belief that drives the fear does not occur, their somatic responses adjust until there is a diminished physical fear response (Foa & McNally, 1996). Inhibitory learning suggests that these fears are not merely extinguished, but that new associations replace the old fearful associations, therefore competing with the fearful associations for space within the anxiety hierarchy (Jacoby & Abramowitz, 2016). Inhibitory learning is currently more readily supported due to OCD's frequent rate of relapse after treatment (Craske & Mystkowski, 2006). Because inhibitory learning suggests a nonfear hierarchy replacing the old-fear hierarchy, there would still be space for old fear habits to return and push the new nonfear hierarchy out, causing a resurgence of symptoms.

For children with mild to moderate OCD, CBT alone is considered first-line treatment. However, for children with more severe OCD symptoms or who are resistant to CBT after several sessions, combination pharmacotherapy and OCD also proves to be effective (American Academy of Child & Adolescent Psychiatry, 2012). Typically, selective serotonin reuptake inhibitors (SSRIs) are the drug of choice, and they have been evaluated in several clinical trials. Sertraline, fluoxetine, fluvoxamine, citalopram, and paroxetine have all been tested in pediatric OCD populations with small effect sizes compared to placebo (Kotapati et al., 2019). However, when combined with CBT, the effect sizes are much larger. A study conducted by Neziroglu et al. showed marked improvements in pediatric patients receiving both fluvoxamine and CBT with ERP, and they continued to maintain their treatment gains at the 2 year follow-up point with continued CBT and pharmacotherapy (Neziroglu, Yaryura-Tobias, Walz, & McKay, 2000). This emphasizes the importance of CBT in treatment plans for OCD patients, especially if their symptoms are severe in nature.

Key facts of pediatric OCD

- Obsessive-compulsive disorder manifests similarly in both children and adults, however there are key factors that impact treatment praxis, such as insight and the increase of compulsions over obsessions.
- Cognitive-behavioral therapy is the gold-standard for treatment of mild-moderate cases of pediatric OCD.
- Medications like SSRIs can be used in more severe cases in tandem with CBT.
- Assessing for comorbidities is crucial to developing an adequate treatment plan.
- Family accommodation should also be assessed for. Caregivers should be informed of more productive strategies that can be used to help comfort their child.

Introduction: Body dysmorphic disorder

The main clinical features of BDD include a preoccupation with one or more perceived defects or flaws in one's appearance. These flaws are not observable or only slightly visible to others. At some point over the course of the illness individuals with BDD engage in repetitive behaviors (e.g., mirror checking, excessive grooming, skin picking, reassurance seeking, etc.) or mental acts (e.g., compare self to others) in response to their appearance concerns. Diagnostically, BDD is currently classified under the Obsessive-Compulsive and Related Disorders section of the DSM-V (American Psychiatric Association, 2013). In Table 1, the key symptoms of BDD are compared and contrasted with the key symptoms of OCD, and should offer clarity on its inclusion in the OCRD section in DSM-V.

Although presently the abundance of information on BDD comes to us from studies conducted with adult populations, certain similarities and differences with children and adolescents can be noted. For example, there is growing evidence that children and adolescents with BDD often have more severe symptom presentations compared to adults, endorse poorer insight and have higher rates of suicidality (Phillips, 2017).

Regarding presentation, children and adolescents with BDD often report multiple body areas of concern. Common areas include, skin, nose, hair, face, and stomach. Children and adolescents may be more concerned with body weight and shape as oppose to complaints of another nature (i.e., face and skin) but often include other areas of concern which distinguishes it purely from an eating disorder. For example, there could be a concern about the shape of the face without any specific part of the face being identified.

Regarding compulsive behaviors, typical behaviors include mirror checking, comparing self with others, camouflaging, applying make-up, grooming, and reassurance seeking (Albertini & Phillips, 1999; Rautio et al., 2020). Interestingly, as technology has continued to grow and represents a significant way children and adolescents communicate BDD related compulsions have expanded as well. According to Alsaidan et al. (2020), BDD was associated with spending more time on social media apps such as SNAPCHAT and INSTAGRAM, and subsequently, to comparing one's self with celebrities, as well as taking and editing one's own photos to a degree of excessiveness and distress.

TABLE 1 Key symptoms of pediatric OCD and BDD.

Pediatric OCD	Pediatric BDD
Obsessions: Repetitive intrusive thoughts, can surround a variety of topics (e.g., what if I made a mistake and didn't catch it, etc.)	Intrusive thoughts regarding perceived defect (e.g., everyone is looking at my big stomach, etc.)
Compulsions: Ritualistic behaviors conducted in order to assuage anxiety surrounding obsessions (e.g., checking for mistakes, etc.)	Ritualistic behaviors associated with assessing appearance (e.g., mirror checking, grooming routines, etc.)
Common comorbidities: Anxiety, depression, ADHD, Tourette's	Common comorbidities: Depression, anxiety, eating disorders, suicidality, substance abuse

In addition, there is a high incident and desire for cosmetic procedures and/or surgery to correct the individuals perceived defect (Khunger & Pant, 2020). This finding has special relevance for this particular age demographic. Compared to adults with BDD, children and adolescents have significantly poorer insight into their difficulties. More specifically, children and adolescents endorse appearance beliefs that often rise to the level of being classified as delusional. This poor insight leaves many adolescents with the belief that the only viable solution to their appearance concerns is some type of medical procedure and has even resulted in cases where individuals have engaged in "Do-it-Yourself" procedures on themselves. Similarly, compared to adults, adolescents endorse higher levels of ideas of reference in which they believe others are judging them on their appearance (Phillips, 2017).

Age of onset and prevalence

The average age of onset for BDD is 16 years-old, with many individuals first experiencing symptoms as early as 12–13 (Greenberg, Sullivan, & Wilhelm, 2017). The majority of individuals report an onset before 18 (Bjornsson et al., 2013; Phillips, 2017). Although less information is known about earlier presentations there have been reports of subclinical symptoms seen as early as 4 and 5 (Rautio et al., 2020). Indeed, symptoms often present at the subclinical level before the full-blown disorder emerges (Bjornsson et al., 2013). Presently, scant information is known about children under the age 12 (Krebs, Fernandez de la Cruz, & Mataix-Cols, 2017).

Similar prevalence rates between children and adolescents and adults with BDD have been reported (Schneider, Turner, Mond, & Hudson, 2017). Notwithstanding, prevalence rates vary depending on the type of sample investigated (Krebs, Fernandez de la Cruz, Monzani, et al., 2017). To illustrate, utilizing a community sample (Schneider et al., 2017) found a prevalence rate of 1.7% among children and adolescents. In contrast, Dyl, Kittler, Phillips, and Hunt (2006) established a prevalence rate of 6.7 with a sample of inpatient adolescents suffering from BDD and other body image concerns. Taken together, despite these varying rates these number reflect that BDD in children and adolescent is a common phenomenon.

Quality of life, interference, and suicidality

Quality of life and expected developmental functioning (i.e., peer relations, dating, increased independence), are often severely hampered as a result of BDD (Neziroglu, Borda, Khemlani-Patel, & Bonasera, 2018) For example, adolescents with BDD have a significant high rate of partial attendance and school dropout (Rautio et al., 2020). Moreover, they often experience significant social impairment including extreme self-consciousness, avoidance of social interactions, and high levels of shame and embarrassment. Furthermore, physically aggressive behavior has been demonstrated among this age demographic. Common sources of aggression include compulsive behavior being interrupted, frustration due to symptoms and general situation, the misinterpretation of other' behavior, and hopelessness (Phillips, 2017).

There are often alarmingly high rates of suicidality in BDD populations (Neziroglu, Khemlani-Patel, & Yaryura-Tobias, 2006). This is particularly relevant for adolescents with BDD where reports of high rates of suicidal ideation, suicidal attempts, and self-harm behavior (i.e., cutting) have been reported (Rautio et al., 2020). Although not fully understood, based on the existing literature there appears to be several factors that help explain the extremely high rates of suicidality among BDD suffers. According to (Krebs et al., 2020) these include a high genetic contribution, symptoms specifically related to BDD (i.e., BDD related affects, thoughts, and behaviors), as well as "nonshared" environmental experiences such as school and relationship difficulties.

Comorbidity

Typical BDD presentation includes high levels of comorbidity such as mood disorders, social phobia, panic disorder, as well as high levels of family members suffering from other OCD and Related Disorders (Rautio et al., 2020). Moreover, research suggests symptoms of eating disorders (Schneider, Baillie, Mond, Turner, & Hudson, 2018) and substance abuse (Albertini & Phillips, 1999) may also be found among BDD sufferers, with substance use often being used to self-medicate (Phillips, 2017). Finally, evidence suggests that early-onset BDD is associated with greater psychiatric comorbidity (Bjornsson et al., 2013).

Gender differences

Information on gender differences in BDD is continuing to emerge. Traditionally, study samples have included a disproportionate number of females. Notwithstanding, these studies do not necessarily represent the actual prevalence rates among the different genders. In fact, overall, the prevalence rates among female and male children and adolescents seem to be equivalent. Despite these surface similarities it appears that the way symptoms manifest may differ among the different genders (Krebs, Fernandez de la Cruz, Monzani, et al., 2017).

To illustrate, evidence has been demonstrated that girls have higher rates of suicidality and self-harm behavior (Rautio et al., 2020). Regarding body parts of concern, although overlap exists between genders there is also evidence that suggests that boys are more often concerned with their genitals and hair, while females focus more on their hips, breasts, legs and excessive hair (Krebs, Fernandez de la Cruz, Monzani, et al., 2017).

Similarly, trends have been demonstrated among suffers of muscle dysphoria disorder (i.e., dysfunctional preoccupation with the idea that one's body is too small or insufficient). Although once thought to be exclusively among males, research suggests that these concerns can also be found in females. For example, while females may be less concerned with muscle size, they nonetheless may have concerns about muscle leanness (Mitchison et al., 2021). Again, this suggests that although prevalence rates may be similar in terms of disorder, presentation profiles may vary across genders. Additional research is needed in this area with focus across all different genders groups.

Etiology

There are no definitive explanations on the etiology of BDD in either children and adolescents or adults. To date, most theories regarding etiology tend to focus on CBT models of the etiology and maintenance of BDD symptoms (Neziroglu, Roberts, & Yaryura-Tobias, 2004), a diathesis-stress conceptualization of mental disorder, as well as critical incidents and factors contributing to the development of BDD (Weingarden, Curley, Renshaw, & Wilhelm, 2017). This model is most effectively illustrated by Neziroglu and Mancusi (2012) (Fig. 2).

To illustrate, evidence has suggested that individuals with BDD have suffered higher levels of abuse compared to others in the general community, as well as with other disorders (i.e., OCD). The nature of the abuse has often been identified as emotional, however sexual and physical types have been noted as well (Neziroglu et al., 2006). In addition, histories of bullying and teasing have been found to be salient factors in the development of BDD. For example, Weingarden et al. (2017) found that bullying experiences were the most reported event cited by participants, especially during the elementary and middle school years. Interestingly, although individuals with BDD have been shown to be victims of such harassment, Neziroglu et al. (2018) found evidence that individuals with BDD were not only victims of such abuse, but also surprisingly the perpetrators of such behavior.

Furthermore, societal messages about the importance of beauty (Weingarden et al., 2017), adolescents' perceptions of parent teasing and negative attitudes about weight and appearance, as well as parental modeling, Densham, Webb, Zimmer-Gembreck, Nesdale, and Downey (2017) have been found to be related to increased BDD symptoms, as well as possible developmental links. In addition, a social learning model on the etiology of BDD has been proposed (Neziroglu, Khemlani-Patel, & Jacofsky, 2009; Neziroglu, Pinto, et al., 2004). Further research is needed to determine additional factors that may

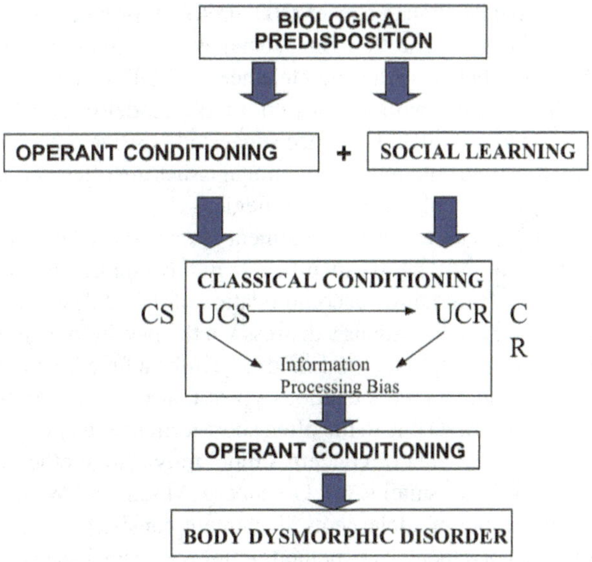

FIG. 2 Cognitive-behavioral model of BDD (Neziroglu & Mancusi, 2012, permission obtained).

contribute to the development of BDD, as well as different factors that may have more significant influence depending on the individual's current developmental stage (Schneider et al., 2018).

Practice and procedures

Assessment

According to Greenberg et al. (2017), a comprehensive assessment of BDD in children and adolescents should focus on BDD beliefs, behaviors, rituals, body parts of concern, and avoidance behavior. In addition, clinicians should also assess for psychosocial factors including effects on social relationships and activities, school performance and attendance, as well as familial functioning. Furthermore, due to the high mortality rates, a through suicide risk assessment should be a routine part of a BDD assessment. Moreover, given the demonstrated effect BDD has on school attendance, clinicians should be vigilant to screen possible BDD concerns when presented with school refusal cases.

Due to increased levels of shame individuals with BDD will often not disclose information about their symptoms unless directly questioned by the clinician. Various clinical tools have been utilized in order to help facilitate proper questioning and follow-up pertaining to the assessment of children and adolescents with BDD. These instruments are based on questions typically used in assessing adults with BDD and/or other OCD Related Disorders, however, are often modified with age-appropriate language, etc. Examples of such measures include BDD-YBOCS-A, Body Dysmorphic Disorder Questionnaire (Phillips, 2017) and the Body Dysmorphic Disorder Examination (Geremia & Neziroglu, 2001).

It is important to note that the types of appearance concerns and preoccupations seen in "typical" adolescents are not the same as those suffering from BDD (Phillips, 2017). Many individuals with BDD have reported spending more than 3 h a day focused on their appearance area of concern. BDD in children and adolescents often goes unnoticed and underdiagnosed (Krebs, Fernandez de la Cruz, Monzani, et al., 2017) and it often takes 11 years on average until sufferers make it to a BDD specialist (Phillips, 2006). Sadly, left untreated children and adolescents do not grow out of BDD (Krebs, Fernandez de la Cruz, Monzani, et al., 2017) and roughly 40% become homebound (Veale & Neziroglu, 2010).

BDD treatment in children and adolescents

Although slowly emerging treatment research specifically related to children and adolescents is lacking. Notwithstanding, similar to adults evidence of the efficacy of CBT and SSRI treatment of BDD in youth has begun to be demonstrated (Greenberg et al., 2017) via case-studies, pilot-studies, RCTs, etc.

Overall, CBT approaches for BDD contain traditional behavioral and cognitive components including motivation interviewing, psychoeducation, parent training, cognitive strategies, exposure/response prevention, attention/perceptual retraining, and social skill training. Treatment is tailored specifically to this population and includes age-appropriate language and focus on relevant developmental tasks (Greenberg et al., 2017).

For example, Greenberg et al. (2010) developed a brief CBT protocol for treating BDD in adolescents. The treatment focused on decreasing maladaptive behaviors and thoughts, social skills training, as well as family instruction aimed at eliminating family behaviors that contributed to the maintenance of BDD behavior (i.e., accommodating compulsive behaviors, etc.). In addition, psychoeducation with a focus on a CBT model of BDD that focused on the development and maintenance of BDD symptoms was provided. Evidence suggested positive outcomes at the end of 8 weeks that consisted of 12 CBT sessions. Despite a positive outcome, motivational issues are often common in this population and special attention may be needed toward orientation and treatment strategies.

Similar Khoodoruth and Khan (2020) reported on the treatment of an adolescent male with BDD and suicidal ideation. This example demonstrated the important role family members may have on treatment compliance and outcomes. More specifically, the authors cited how in this case family accommodation of the adolescent's request for a cosmetic procedure interfered with the clinician's ability to treat the patient's distress via the psychological treatment as oppose to an external medical procedure. In this case relief from symptoms was found to be more a function of the medical intervention. Although the procedure provided some relief the authors were cautious given that medical procedures are often counter indicated when it comes to treating individuals with BDD due to the often short-term results, individuals are often deeply dissatisfied with results, want further procedures, or the area of concern simply moves to another area (Jassi & Krebs, 2021).

Expanding on the work done with smaller small sizes, Greenberg, Mothi, and Wilhelm (2016) conducted a version of CBT intervention with an increased number of adolescents. Treatment consisted of 12–22, 60-min sessions of individual CBT treatment for BDD with family involvement. The methods employed were based off similar CBT protocols for adults with BDD, previous efforts made at modification of protocols geared toward BDD and other disorders such as OCD in youth, as well as age-appropriate treatments for school refusal and depression. Results were promising thus adding to the growing research base demonstrating CBT as an appropriate treatment for pediatric BDD. However, of special note the authors reported a 38% drop out rate.

Mataix-Cols et al. (2015) developed an age-appropriate BDD protocol for a pilot RCT. The development of the protocol was based on expert knowledge of working with this population, previous age-appropriate protocols with individuals with OCD, as well as existing BDD literature. Evidence suggested that the treatment was found effective compared to the control condition that consisted of psycho-education material and weekly risk monitoring, as well as was found highly acceptable and useful by participants and their families.

Intensive treatment for BDD in children and adolescents

Aldea, Storch, Geffken, and Murphy (2009) demonstrated support for an intensive family-based CBT for adolescents with BDD. Treatment took place over 14 90min sessions of intensive CBT on consecutive days with an additional three follow-ups over an additional 3 months. The authors reported reduction in BDD symptoms, as well as associated distress.

Le, Merricks, Nadeau, Ramos, and Storch (2017) conducted an intensive CBT program for an adolescent female suffering from BDD. An important point was that the subject withdrew early from treatment and did not utilize ERP to its fullest. Specifically, at the time of withdrawal she had not met full remission criteria, as well as stopped maintenance of ERP on her own and subsequent outpatient therapy as recommended. Taken together, these factors significantly affected the maintenance of any gains.

Treatment of BDD in children and adolescents with medication

Regarding the treatment of children and adolescents with medication for BDD the general consensus among experts in the field (Castle et al., 2021) is that high doses of SSRIs (e.g., Fluoxetine) are needed to treat BDD symptomatology and associated comorbidities. Furthermore, individuals often need to remain on the medication for longer periods of time to receive maximum benefit. If nonresponsive the SNRIs (e.g., Escitalopram) may then be used as the next option for treatment.

Subsequently, if response to SSRIs is inadequate it is suggested that adjunctive medications such as atypical antipsychotics, anxiolytics, and certain anticonvulsants can be explored. However, it should be noted that large scale-controlled studies of these recommendations are still lacking and therefore use of adjunctive recommendations should keep in mind that these medications procedures are not officially indicated as of yet.

Long-term treatment considerations

Despite the promising results of the above-mentioned studies, it should be notes that not all participants remained in these studies, nor were they able to follow the suggested protocols. Subsequently, this limits the generalizability of results especially in more severe cases. Moreover, evidence suggests that despite positive benefits of CBT and pharmacological interventions often patients remain somewhat symptomatic even if considered as having a positive response. Furthermore, there exist a significant number of individuals that may be considered nonresponders.

Finally, more severe BDD symptoms, longer life time duration of BDD, and earlier age onset have been found to predict full or partial relapse of BDD symptoms (Phillips, Menard, Quinn, Didle, & Stout, 2013). Taken together, children and adolescents with BDD remain a population where long-term monitoring (Krebs, Fernandez de la Cruz, Monzani, et al., 2017), relapse preventions effort, and further efficacious treatments are warranted.

Key facts of pediatric BDD

- Children and adolescents with BDD often have more severe symptom presentations compared to adults, endorse poorer insight, and have higher rates of suicidality.
- The average age of onset for BDD is 16 years-old.
- There are significantly high rates of suicidality and desire for cosmetic procedures and/or surgery in BDD populations, especially among younger patients.
- Evidence of the efficacy of CBT and SSRI treatment of BDD in youth has begun to emerge.
- Despite treatment efforts there remain a significant proportion of individuals that are considered nonresponders.

Mini-dictionary of terms

- Do it Yourself (DIY): A term used for "cosmetic" and/or "surgical" procedures performed by oneself as opposed to a medically trained professional.
- Overvalued ideation: Unreasonable beliefs that are held so strongly that an individual cannot attribute to their disorder.
- Exposure and response prevention (ERP): Key behavioral component in CBT treatment of both OCD and BDD.

Applications to other areas

Looking to the future, Hartmann et al. (2021) are in process of exploring a therapist-guided internet-based CBT Program for adolescents and young adults with BDD. Hopefully, this form of treatment delivery can help increase treatment compliance and participation among children and adolescents with OCD and BDD. Indeed, internet service delivery, as well as the use of other forms of social media can be utilized to minimize barriers to treatment such as shame, lack of access to specialists in the field, as well as long waiting lists. This removes burden from both the child and their caregivers and offers easy access to therapeutic solutions from the comfort of one's own home. Online therapy has blossomed especially during the advent of the COVID-19 pandemic, and undoubtedly will continue to be a mainstay in therapeutic delivery. Given technology has become common place in the daily life of this particular age demographic, it appears an exciting avenue for future research on service delivery.

Summary points

- Pediatric OCD and BDD have not been researched extensively.
- Both disorders are common in pediatric populations: OCD occurs in about 1%–4% of children, and BDD occurs in about 1%–6%.
- Comprehensive assessment of BDD and OCD in children and adolescents should focus on beliefs, behaviors, rituals, any avoidance behaviors, as well as effects on social relationships and activities, school performance and attendance, and familial functioning.
- Exposure and response prevention (ERP) is an imperative part of CBT treatment praxis for both BDD and OCD and ideally should be implemented if the child and caregivers are willing.
- SSRIs (e.g., Fluoxetine) are excellent additions to CBT in the case of severe pediatric OCD and BDD.

References

Albertini, R. S., & Phillips, K. A. (1999). Thirty-three cases of body dysmorphic disorder in children and adolescents. *Journal of the American Academy of Child and Adolescent Psychiatry*, *38*(4), 453–459.

Aldea, M. A., Storch, E. A., Geffken, G. R., & Murphy, T. K. (2009). Intensive cognitive-behavioral therapy for adolescents with body dysmorphic disorder. *Clinical Case Studies*, *8*(2), 113–121.

Alsaidan, M. S., Altayar, N. S., Alshmmari, H. S., Alshmmari, M. M., Alqahtani, F. T., & Mohajer, K. A. (2020). The prevalence and determinants of body dysmorphic disorder among young social media users: A cross-sectional study. *Dermatology Reports*, *12*, 70–76.

American Academy of Child & Adolescent Psychiatry. (2012). Practice parameter for the assessment and treatment of children and adolescents with obsessive-compulsive disorder. *Journal of the American Academy of Child and Adolescent Psychiatry*, *51*, 98–113.

American Psychiatric Association. (2013). *Diagnostic and statistical manual of mental disorders* (5th ed.). American Psychiatric Association.

Bjornsson, A. S., Didie, E. R., Grant, J. E., Mendard, W., Stalker, E., & Phillips, K. A. (2013). Age at onset and clinical correlates in body dysmorphic disorder. *Comprehensive Psychiatry*, *54*, 893–903.

Boileau, B. (2011). A review of obsessive-compulsive disorder in children and adolescents. *Dialogues in Clinical Neuroscience*, *13*(4), 401–411.

Borda, T., Neziroglu, F., Taboas, W., McKay, D., & Frenkiel, L. (2017). Overvalued ideation in adolescents with obsessive-compulsive disorder. *Psychiatry Research*, *255*, 66–71.

Castle, D., Beilharz, F., Phillips, K. A., Brakoulias, V., Drummond, L. M., Hollander, E., … Fineberg, N. A. (2021). Body dysmorphic disorder: A treatment synthesis and consensus on behalf of the International College of Obsessive-Compulsive Spectrum Disorders and the Obsessive Compulsive and Related Disorders Network of the European College of Neuropsychopharmacology. *International Clinical Psychopharmacology*, *36*, 61–75.

Craske, M. G., & Mystkowski, J. L. (2006). Exposure therapy and extinction: Clinical studies. In M. G. Craske, D. Hermans, & D. Vansteenwegen (Eds.), *Fear and learning: From basic processes to clinical implications* (pp. 217–233). American Psychological Association.

Densham, K., Webb, H., Zimmer-Gembreck, M., Nesdale, D., & Downey, G. (2017). Early adolescents' body dysmorphic symptoms as compensatory responses to parental appearance messages and appearance-based rejection sensitivity. *Body Image*, *23*, 162–170.

do Rosario-Campos, M. C., Leckman, J. F., Curi, M., Quatrano, S., Katsovitch, L., Miguel, E. C., … Downey, G. (2017). Early adolescent's body dysmorphic symptoms as compensatory responses to parental appearance messages and appearance-based rejection sensitivity. *Body Image*, *23*, 162–170.

Dyl, J., Kittler, J., Phillips, K. A., & Hunt, J. I. (2006). Body dysmorphic disorder and other clinically significant body image concerns in adolescent psychiatric inpatients: Prevalence and clinical characteristics. *Child Psychiatry Human Development*, *36*, 369–382.

Foa, E. B., & Kozak, M. J. (1986). Emotional processing of fear: Exposure to corrective information. *Psychological Bulletin*, *99*(1), 20–35.

Foa, E. B., & McNally, R. J. (1996). Mechanisms of change in exposure therapy. In M. Rapee (Ed.), *Current controversies in the anxiety disorders*. New York: The Guilford Press.

Franklin, M. E., Kratz, H. E., Freeman, J. B., Ivarsson, T., Heyman, I., Sookman, D., … Accreditation Task Force of The Canadian Institute for Obsessive Compulsive Disorders. (2015). Cognitive-behavioral therapy for pediatric obsessive-compulsive disorder: Empirical review and clinical recommendations. *Psychiatry Research*, *227*(1), 78–92.

Gellar, D. (2006). Obsessive-compulsive and spectrum disorders in children and adolescents. *Psychiatric Clinics of North America*, *29*, 353–370.

Gellar, D., Coffey, B., Farone, S., et al. (2003). Does comorbid attention deficit/hyperactivity disorder impact the clinical expression of pediatric obsessive compulsive disorder? *CNS Spectrums*, *8*(4), 52–58.

Geremia, G. M., & Neziroglu, F. (2001). Cognitive therapy in the treatment of body dysmorphic disorder. *Clinical Psychology & Psychotherapy*, *8*(4), 243–251.

Greenberg, E. L., & Geller, D. A. (2019). Phenomenology and standard care of OCD in children and adolescents. In L. J. Farrell, T. H. Ollendick, & P. Muris (Eds.), *Innovations in CBT for childhood anxiety, OCD, and PTSD: Improving access and outcomes* (pp. 289–312). Cambridge University Press.

Greenberg, J. L., Markowitz, S., Petronko, M. R., Taylor, C. E., Wilhelm, S., & Wilson, G. T. (2010). Cognitive-behavioral therapy for adolescent body dysmorphic disorder. *Cognitive and Behavioral Practice*, *17*, 248–258.

Greenberg, J. L., Mothi, S. S., & Wilhelm, S. (2016). Cognitive-behavioral therapy for adolescent body dysmorphic disorder: A pilot study. *Behavior Therapy*, *47*, 213–224.

Greenberg, J. L., Sullivan, A., & Wilhelm, S. (2017). Treating children and adolescents with body dysmorphic disorder. In K. Phillips (Ed.), *Body dysmorphic disorder: Advances in research and clinical practice* (pp. 397–407). Oxford, UK: Oxford University Press.

Hartmann, A. S., Schmidt, M., Staufenbiel, T., Ebert, D. D., Martin, A., & Schoenenberg, K. (2021). ImaginYouth-A therapist-guided internet-based cognitive-behavioral program for adolescents and young adults with body dysmorphic disorder: Study protocol for a two-arm randomized controlled trial. *Frontiers in Psychiatry*, *12*, 1–13.

Jacoby, R., & Abramowitz, J. (2016). Inhibitory learning approaches to exposure therapy: A critical review and translation to obsessive-compulsive disorder. *Clinical Psychology Review*, *49*, 28–40.

Jassi, A., & Krebs, G. (2021). Body dysmorphic disorder: Reflections on the last 25 years. *Clinical Child Psychology and Psychiatry*, *26*(1), 3–7.

Khoodoruth, M. A. S., & Khan, Y. S. (2020). Etiological factors and treatment challenges in an adolescent male with body dysmorphic disorder: A case report. *SAGE Open Medical Case Reports*, *8*, 1–5.

Khunger, N., & Pant, H. (2020). Cosmetic procedures in adolescents: What's safe and what can wait. *Indian Journal of Paediatric Dermatology*, *22*(1), 12–20.

Kotapati, V. P., Khan, A. M., Dar, S., Begum, G., Bachu, R., Adnan, M., … Ahmed, R. A. (2019). The effectiveness of selective serotonin reuptake inhibitors for treatment of obsessive-compulsive disorder in adolescents and children: A systematic review and meta-analysis. *Frontiers in Psychiatry*, *10*, 1–14.

Krebs, G., Fernandez de la Cruz, L., & Mataix-Cols, D. (2017). Recent advances in understanding and managing body dysmorphic disorder. *Evidence-Based Mental Health*, *20*(3), 71–75.

Krebs, G., Fernandez de la Cruz, L., Monzani, B., Bowyer, L., Anson, M., Cadman, J., … Mataix-Cols, D. (2017). Long-term outcomes of cognitive-behavioral therapy for adolescent body dysmorphic disorder. *Behavior Therapy*, *48*, 462–473.

Krebs, G., Fernandez de la Cruz, L., Rijsdijk, F. V., Rautio, D., Enander, J., Ruck, C., … Mataix-Cols, D. (2020). The association between body dysmorphic symptoms and suicidality among adolescents and young adults: A genetically informative study. *Psychological Medicine*, 1–9. https://doi.org/10.1017/S0033291720002998.

Le, T. A., Merricks, K., Nadeau, J. N., Ramos, A., & Storch, E. A. (2017). Intensive exposure and response prevention for adolescent body dysmorphic disorder with comorbid obsessive-compulsive disorder and major depressive disorder. *Clinical Case Studies*, *16*(6), 480–496.

Leckman, J. F., Rauch, S. L., & Mataix-Cols, D. (2007). Symptom dimensions in obsessive-compulsive disorder: Implications for DSM-V. *CNS Spectrums*, *12*, 376–400.

Lennertz, L., Grabem, H. J., Ruhrmann, S., Rampacher, F., Vogely, A., Schulze-Rauchenbach, S., … Pukrop, R. (2010). Perceived parental rearing in subjects with obsessive-compulsive disorder and their siblings. *Acta Psychiatria Scandinavica*, *121*, 280–288.

Lewin, A. B. (2019). Evidence-based assessment of child obsessive-compulsive disorder (OCD). In L. J. Farrell, T. H. Ollendick, & P. Muris (Eds.), *Innovations in CBT for childhood anxiety, OCD, and PTSD: Improving access and outcomes* (pp. 289–312). Cambridge University Press.

Lewin, A. B., Chang, S., McCracken, J., McQueen, M., & Piacentini, J. (2010). Comparison of clinical features among youth with tic disorders, obsessive-compulsive disorder (OCD) and both conditions. *Psychiatry Research*, *178*, 317–322.

Mantz, S., & Abbott, M. (2017). The relationship between responsibility beliefs and symptoms and processes in obsessive-compulsive disorder: A systematic review. *Journal of Obsessive Compulsive and Related Disorders*, *14*, 13–26.

March, J. S., & Mulle, K. (1998). *OCD in children and adolescents: A cognitive-behavioral treatment manual*. Guilford Press.

Mataix-Cols, D., Boman, M., Monzani, B., Rück, C., Serlachius, E., Långström, N., & Lichtenstein, P. (2013). Population based, multigenerational family clustering study of obsessive-compulsive disorder. *JAMA Psychiatry*, *70*(7), 709–717.

Mataix-Cols, D., Fernandez de la Cruz, L., Isomura, K., Anson, M., Turner, C., Monzani, B., … Krebs, G. (2015). A pilot randomized controlled trial of cognitive-behavioral therapy for adolescents with body dysmorphic disorder. *Journal of the American Academy of Child & Adolescent Psychiatry*, *54* (11), 895–904.

Mataix-Cols, D., Nakatani, E., Micali, N., & Heyman, I. (2008). Structure of OCD symptoms in pediatric OCD. *Journal of the American Academy of Child & Adolescent Psychiatry*, *47*(7), 773–778.

Mataix-Cols, D., Rosario-Campos, M., & Leckman, J. F. (2005). A multidimensional model of obsessive compulsive disorder. *American Journal of Psychiatry*, *162*, 228–238.

Mathieu, S. L., Conlon, E. G., Waters, A. M., McKenzie, M. L., & Farrell, L. J. (2019). Inflated responsibility beliefs in paediatric OCD: Exploring the role of parental rearing and child age. *Child Psychiatry and Human Development*, *51*, 552–562.

Mitchison, D., Mond, J., Griffiths, S., Hay, P., Nagata, J. M., Bussey, K., … Murray, S. B. (2021). Prevalence of muscle dysmorphia in adolescents: Findings from the EveryBODY study. *Psychological Medicine*, 1–8. https://doi.org/10.1017/S0033291720005206.

Neziroglu, F., Borda, T., Khemlani-Patel, S., & Bonasera, B. (2018). Prevalence of bullying in a pediatric sample of body dysmorphic disorder. *Comprehensive Psychiatry*, *87*, 12–16.

Neziroglu, F., Khemlani-Patel, S., & Jacofsky, M. (2009). Body dysmorphic disorder: Symptoms, models and treatment interventions. In *Vol. 2. Cognitive behavior therapy: A guide for the practicing clinician* (pp. 94–112). Gregoris Simos, London: Routledge.

Neziroglu, F., Khemlani-Patel, S., & Yaryura-Tobias, J. (2006). Rates of abuse in body dysmorphic disorder and obsessive-compulsive disorder. *Body Image*, *3*(2), 189–193.

Neziroglu, F., & Mancusi, L. (2012). A proposed learning model of body dysmorphic disorder. In I. Reis De Oliveira (Ed.), *Standard and innovative strategies in cognitive behavior therapy* (pp. 145–156). IntechOpen.

Neziroglu, F., Pinto, A., Yaryura-Tobias, J. A., & McKay, D. (2004). Overvalued ideation as a predictor of fluvoxamine response in patients with obsessive compulsive disorder. *Psychiatry Research*, *125*(1), 53–60.

Neziroglu, F., Roberts, M., & Yaryura-Tobias, J. (2004). A behavioral model for body dysmorphic disorder. *Psychiatric Annals*, *34*(12), 915–920.

Neziroglu, F., & Stevens, K. P. (2002). Insight: Its conceptualization and assessment. In R. O. Frost, & G. Steketee (Eds.), *Cognitive approaches to obsessions and compulsions: Theory, assessment, and treatment.* Oxford: Elsevier.

Neziroglu, F., Stevens, K. P., McKay, D., & Yaryura-Tobia, J. A. (2001). Predictive validity of the overvalued ideas scale: Outcome in obsessive-compulsive disorder and body dysmorphic disorder. *Behavior Research and Therapy*, *39*(6), 745–756.

Neziroglu, F., Yaryura-Tobias, J. A., Walz, J., & McKay, D. (2000). The effect of fluvoxamie and behavior therapy on children and adolescents with obsessive-compulsive disorder. *Journal of Child and Adolescent Psychopharmacology*, *10*(4), 295–306.

Obsessive Compulsive Cognitions Working Group. (1997). Cognitive assessment of obsessive-compulsive disorder. *Behavior Research and Therapy*, *35* (7), 667–681.

Pauls, D. L., Alsobrook, J. P., Goodman, W., Rasmussen, S., & Leckman, J. F. (1995). A family study of obsessive-compulsive disorder. *American Journal of Psychiatry*, *152*(1), 76–84.

Phillips, K. A. (2006). The presentation of body dysmorphic disorder in medical settings. *Primary Psychiatry*, *13*(7), 51–59.

Phillips, K. A. (2017). Body dysmorphic disorder in children and adolescents. In K. Phillips (Ed.), *Body dysmorphic disorder: Advances in research and clinical practice* (pp. 173–183). Oxford, UK: Oxford University Press.

Phillips, K. A., Menard, W., Quinn, E., Didle, E. R., & Stout, R. L. (2013). A 4-year prospective observational follow-up study of course and predictors of course in body dysmorphic disorder. *Psychological Medicine*, *43*, 1109–1117.

Rautio, D., Jassi, A., Krebs, G., Andren, P., Monzani, B., Gumpert, M., … Mataix-Cols, D. (2020). Clinical characteristics of 172 children and adolescents with body dysmorphic disorder. *European Child and Adolescent Psychiatry*. https://doi.org/10.1007/s00787-020-01677-3.

Scahill, L., Riddle, M. A., McSwiggin-Hardin, M., Ort, S. I., King, R. A., Goodman, W. K., … Leckmen, J. F. (1997). Children's Yale-Brown Obsessive Compulsive Scale: Reliability and validity. *Journal of the American Academy of Child and Adolescent Psychiatry*, *36*(6), 844–852.

Schneider, S. C., Baillie, A. J., Mond, J., Turner, C. M., & Hudson, J. L. (2018). The classification of body dysmorphic disorder symptoms in male and female adolescents. *Journal of Affective Disorders*, *225*, 429–437.

Schneider, S. C., Turner, C. M., Mond, J., & Hudson, J. L. (2017). Prevalence and correlates of body dysmorphic disorder in a community sample of adolescents. *Australian and New Zealand Journal of Psychiatry*, *516*(6), 595–603.

Sigra, S., Hesselmark, E., & Bejerot, S. (2018). Treatment of PANDAS and PANS: A systematic review. *Neuroscience and Biobehavioral Reviews*, *86*, 51–65.

Singer, H. S., Gilbert, D. L., Wolf, D. S., Mink, J. W., & Kurlan, R. (2012). Moving from PANDAS to CANS. *Journal of Pediatrics*, *160*(5), 725–731.

Storch, E., Larson, M., Muroff, J., Caporino, N., Geller, D., Reid, J., … Murphy, T. (2010). Predictors of functional impairment in pediatric obsessive-compulsive disorder. *Journal of Anxiety Disorders*, *24*, 275–283.

Storch, E., Milsom, V., Merlo, L., Larson, M., Geffken, G., Jacob, M., … Goodman, W. (2008). Insight in pediatric obsessive-compulsive disorder: Association with clinical presentation. *Psychiatry Research*, *160*, 212–220.

Van Ameringen, M., Patterson, B., & Simpson, W. (2014). DSM-5 obsessive-compulsive and related disorders: Clinical implications of new criteria. *Depression and Anxiety*, *31*(6), 487–493.

Veale, D., & Neziroglu, F. (2010). *Body dysmorphic disorder: A treatment manual.* Wiley-Blackwell. https://doi.org/10.1002/9780470684610.

Weingarden, H., Curley, E. E., Renshaw, K. D., & Wilhelm, S. (2017). Patient-identified events implicated in the development of body dysmorphic disorder. *Body Image*, *21*, 19–25.

Chapter 25

Cognitive behavior therapy for children with depression: Application, practical considerations, and challenges

Tushar Singh[c], Debasruti Ghosh[a], Thomas Enias Pasipanodya[b], Saurabh Raj[d], and Shalini Mittal[e]

[a]*Department of Psychology, MDDM College (Babasaheb Bhimrao Ambedkar Bihar University), Muzaffarpur, Bihar, India,* [b]*Experienced Professional, Psychological Recovery Clinic, Toronto, ON, Canada,* [c]*Department of Psychology, Faculty of Social Sciences, Banaras Hindu University, Varanansi, UP, India,* [d]*Department of Psychology, Ramdayalu Singh College (Babasaheb Bhimrao Ambedkar Bihar University), Muzaffarpur, Bihar, India,* [e]*Northcap University, Gurgaon, Haryana, India*

Abbreviations

CBT cognitive behavior therapy
CDC Center for Disease Control and Prevention
DSM-5 Diagnostic and Statistical Manual of Mental Disorders, 5th Edition
FDA US Food and Drug Administration
MDD major depressive disorder
SAMSHA Substance Abuse and Mental Health Services Administration
SSRI selective serotonin reuptake inhibitors
WHO World Health Organization

Childhood depression: Clinical considerations and epidemiological aspects

Depression is a debilitating psychological disorder that has severe disability and mortality consequences across the lifetime (Kessler, Petukhova, Sampson, Zaslavsky, & Wittchen, 2012). As per the report of World Health Organization, globally about 264 million people are suffering from depression, and it is the foremost cause of disability worldwide (WHO, 2020). Depression is marked by difficulties with cognitive-affect regulation, anxious distress, somatization, and difficulties with activities of daily living. These features represent susceptibility factors in children that range from biological causes such as a prior history of depression in the family; disturbances in family life such as parental divorce; neglect or abuse by primary caregivers; and difficulties with adaptable social or coping skills to name a few. A predetermining factor of the consequences of depression during adolescents and adulthood is the age of onset when symptoms emerged (Kessler et al., 2005) which can be used as a predictor of depression (Beardslee & Gladstone, 2001). Regardless of such detrimental outcomes, scientific works on the diagnosis and treatment of depression among children have been very few and far behind. According to DSM-5, depressive disorders that are common among children include major depressive disorder (MDD), persistent depressive disorder (PDD), and disruptive mood dysregulation disorder (DMDD). MDD is characterized by persistent feeling of sadness and reduced interest in activities that were previously enjoyed by the child and may be accompanied by suicidal ideation is some of the extreme cases. Children may be diagnosed with MDD if either or both of these symptoms last for at least 2 weeks. A child may be diagnosed with PDD if mood irritability lasts for a period of at least 2 years. DMDD is characterized by irritable mood, intense anger, and sudden temper outbursts exhibited for 12 months or more among children above the age of 6 years and below the age of 18 years (SAMSA, 2017).

Handbook of Lifespan Cognitive Behavioral Therapy. https://doi.org/10.1016/B978-0-323-85757-4.00009-2

Prevalence & risk factors of depression in children

The prevalence rate of depression in children is more common than originally thought. The Centre for Disease Control and prevention predicts that 3.2% of children under the age of 17 years suffer from depression (CDCP, 2021). Country-wise studies on the prevalence of childhood depression have shown a prevalence of 4% in Spain (Bernaras, Jaureguizar, Soroa, Ibabe, & Cuevas, 2013), 10% in Australia (McCabe, Ricciardelli, & Banfield, 2011, 8% in Greece (Kleftaras & Didaskalou, 2006), 25% in Colombia (Vinaccia et al., 2006), and 3.2% in the United States (Ghandour et al., 2019). Based on the data from Indian studies on child and adolescent depression, Grover, Raju, Sharma, and Shah (2019) reported a prevalence rate of 0.1%–6.94% in community studies, 3%–68% in school-based studies, and 1.2%–21% in clinic-based studies. Gender differences in childhood depression have also been observed in several studies which indicate that the prevalence rate for boy is higher prior to adolescence, whereas after age 13 or puberty, more girls are depressed (Hankin & Abramson, 1999). Bennett, Ambrosini, Kudes, Metz, and Rabinovich (2005) also reported similar severity and prevalence of symptoms among both boys and girls. However, they emphasized that depressed girls tend to manifest more body image problems, guilt, self-blaming tendencies, and feelings of failure. Thus, it is evident from studies that depression is a common occurrence in children with long-lasting emotional, physiological, and psycho-social effects.

American Psychiatric Association (2013) suggests almost similar criteria for diagnosing depression in both adults and children. However, researches indicate that the manifest symptoms of depression among children vary with changes in developmental age. More precisely, the risk factors in children may include increased irritability or aggressive behavior and delayed developmental goals such as reaching age-appropriate comprehending skills, physical growth, and self-regulatory skills. Furthermore, the common occurrence of depression in children may lead to adverse developmental consequences making it the utmost contributor of the global burden of disease (Gore et al., 2011). It is, therefore, crucial to note that though diagnostic similarities for depression exist in adults and children, but the manifestation of symptoms is more complex in younger populations (Kessler, Avenevoli, & Merikangas, 2001) and therefore special treatment considerations need to be explored in this regard.

Causes and consequences of depression in children

The presentation and manifestation of depression in children is determined by biological, psychological, and social/environmental factors that intertwine as predisposing, precipitating, and maintenance factors of depressive episodes in children.

Biological factors

The likelihood of inheritance of depression in children is more probable if parents have a similar mood condition. The risk of children of depressed parents to develop depression is four times higher than their peers of healthy parents (Thapar, Collishaw, Pine, & Thapar, 2012). Similarly, theories also suggest that children's weight has a crucial role in the development of depressive features. Overweight children are thought to be more susceptible to develop early onset depressive episodes. According to Xie et al. (2010), overweight children tend to have negative self-perception, maladaptive social skills, social difficulties with authorities, dislike academic settings, and they face increased discrimination among peers. These factors may lead to negative outcomes such as comparing themselves to others or seeing themselves as not falling within acceptable societal norms and thus are significant predictors of depression among children. The cumulative effect of these experiences lead to the development of negative psychological and emotional regulation strategies among children. Consequently, children may begin to develop hatred or dissatisfaction with their bodies which leads to other adverse effects like stress eating, excessive dieting that eventually fuel depression. In the long run, they negatively self-appraise themselves which may lead them to develop negative thinking patterns and an inability to effectively regulate their emotions.

Psychosocial factors

Environmental and social determinants are connected with the development of depressive symptoms among children. A disruption in a child's family life is found to be positively related with depression. Offspring of parents who get divorced tend to be more susceptible to develop early onset depression. Similarly, children of parents who struggle with addiction or children with home-based violence are more susceptible to developing depression at an early age. These children develop negative attachment styles to parents, peers, or other authority figures in their lives (Kullik & Petermann, 2013). The consequence of such traumatic experiences is that children exposed to these situations tend to develop negative thinking and negative autobiographical recalling patterns. Most notably of those consequences are cognitive deficits which lead children to have a negative evaluation of their competence in their environment. Such cognitive deficits cause children with depression to face academic difficulties as they are preoccupied with negative thoughts among other things. Also, as a result of negative biological and psychological factors, depressed children tend to experience increased bullying, victimization, trauma, and emotional abuse, physical or sexual abuse. Similarly, exposure to natural disasters, immigrating to a new country, death of loved ones, and low socioeconomic status all contribute to the environmental factors of depression among children. Some researchers have also highlighted the role of stressful life events such as bad relationships, bullying, conflict and separation, child neglect, child maltreatment, family bereavement, and peer conflict in the etiology of depression (Jaffee et al., 2002; Kendler, Thornton, & Gardner, 2000; Thapar et al., 2012). Cumulatively, such negative consequences are representative of the core features of depressive episodes that involve negative beliefs of about self, future, and the world.

The consequences of depression in children are not short lived. The effects of depression transform significantly through different developmental stages. Early onset depression is thought to account for increased suicide risk, substance abuse, physiological problems, and difficulties with activities of daily living (Siennick, Widdowson, Woessner, Feinberg, & Spoth, 2017) in adolescents. Several researches indicate that childhood depression is linked with recurrent depressive episodes in adult life (Dunn & Goodyer, 2006). Experience of childhood depression also involves other psychiatric outcomes such as bipolar depression (Geller et al., 2001; Strober, Lampert, Schmidt, & Morrell, 1993). According to Kasen et al. (2001), childhood depression increases the risk of developing personality disorders. Wu et al. (2006) claim that childhood depression may lead to early onset of alcohol use. Hence, there is a need to identify and implement corrective treatment mechanisms that will lead to more adaptive outcomes and reduce lifetime risks of the disorder.

Depression among children: Treatment approaches

Controversy exists around the use and effectiveness of antidepressants among children. This is in part due to recommendations by US Food and Drug Administration (FDA, 2013) that antidepressants tend to increase suicidal risk in children. Thus, there should be warnings on the boxes for this medication. However, pharmacological treatments for depression among children are relatively limited. The FDA does suggest that the treatments effective for MDD may be used (Birmaher et al., 2007). Such treatments include Prozac (fluoxetine) and Lexapro (escitalopram) for treatment of depression among children. Prozac can be used for children above 8 years of age and Lexapro for children above 12 years of age. However, nonpharmacological treatment for depression ought to be considered for children below 8 years of age. Consequently, it becomes of increasing importance to use such treatment for depression among children. It is noteworthy to mention that there are considerations which suggest that using both nonpharmacological and pharmacological treatments may yield positive results in children with severe levels of depression.

A variety of psychological treatments have been posited for treating childhood depression. Such modalities include humanistic, integrative (interpersonal therapy), behavioral, and cognitive behavioral therapies (CBTs). As noted in the previous section above, the causal factors for depression are imbedded in biological, psychological, and social/environmental factors. The effect of these elements is that they predispose an individual to respond in a negative way in the face of stressors. Consequently, there is a need to implement evidence-based interventions which will help children to respond with more positive and adaptive ways of coping and dealing with depression. Treatment approaches should consider the hallmark features that predispose, precipitate, and perpetuate depression in children. Moreover, intervention modalities for depression should address the negative attributional style and learned hopelessness that is inherent among depressed individuals. The goal of these approaches is to transform the negative-conditioned expectations that lead to detrimental thinking styles, and adverse dysfunctional processes are a stable occurrence rather than transient features of one's existence. Moreover, effective treatment approaches should also consider teaching adaptive problem-solving techniques to depressed children. The psychosocial factors emphasize that depressed children experience increased interpersonal difficulties. If so, then treatment approaches should also empower depressed children to cope with interpersonal stress in order to increase positive emotional regulation strategies. CBT is one such evidence-based intervention that teaches people how to identify and change the negative relationship that exists between depression, dysfunction thinking, and negative behavioral patterns.

Cognitive behavioral explanations of childhood depression

CBT is an evidence-based approach that seeks to help people understand the world, individual thought patterns/styles, affective, and behavioral reactions. The core principle of CBT is to help the client understand the negative interaction of these factors and how they maintain depressed mood. As evidenced above, depression has far reaching consequences starting from childhood. There is therefore a pressing need to implement evidence-based treatment interventions that have higher recovery rates and pose the least risk to children. CBT was developed by Aaron Beck, to help address clients with depressive features. As such, it has been efficacious in the treatment of childhood depression as well. The cognitive model of depression recognizes "thoughts" to be the primary component and "affective and behavioral" mechanisms as secondary features behind experiences of the individual (Verduyn, 2000). While explaining about depression vulnerability, Beck (1976) proposed that adverse childhood experiences lead to the formation of *dysfunctional attitudes*. These attitudes then become organized into a set of *maladaptive schemas* that are the building blocks of cognitive vulnerability in Beck's model. According to Beck, schemas are relatively enduring cognitive structures that serve as stored bodies of knowledge about self and past experiences. In general, a negative schema may remain inactive, until it is triggered by an incident that is thematically similar with the schema content. Once activated, the negative schema fuels the negative biases that characterize depression. Reflecting from a broader perspective, cognitive behavioral theories of depression have been framed from a diathesis-stress viewpoint, i.e., when an individual comes across a stressful life situation, then faulty appraisals and maladaptive thoughts about self and the world increase their vulnerability to depression. In a multiwave study, depressive symptoms in children with low self-esteem were predicted by an interaction of dysfunctional attitudes and daily hassles (Abela & Skitch, 2007). Depression in children is marked by the presence of maladaptive schemas that often manifests in the form of negative thoughts about self, world, and future (cognitive triad). They often misinterpret the available information from the environment and their internal cues, which maintains their negative view (Haley, Fine, Marriage, Moretti, & Freeman, 1985). Children with depression show a range of cognitive deficits and distortions like selective attention toward the negative aspects of any situation (Kendall, Stark, & Adam, 1990). Researches that investigate this cognitive-stress-diathesis perspective have found that children who have a depressogenic atrributional style have increased susceptibility to develop symptoms of depression in adverse situations (Kaslow, Rehm, Pollack, & Siegel, 1988). In this context, Abramson, Metalsky, and Alloy (1989) propose an attributional vulnerability hypothesis that contributes to hopelessness component of depression. The hypothesis includes three inferential styles that contribute to hopelessness in depression. The first attributional tendency asserts that negative events are the results of global and stable causes. Second style denotes a perception tendency that negative events always have catastrophic consequences. And the last one is a tendency to see self as flawed or incapable after negative events. Not only attribution style but response styles to situational triggers have been indicated to contribute to depressive cognition. It was argued that individual who adopt a *ruminative response style* may experience increases in the severity and duration of depressive symptoms, while those engaging in *distracting response style* experience less distressing experience (Nolen-Hoeksema, 1991). In children, the rumination response style is a prominent predictor of increase in depressive symptoms, whereas distraction did not cause change in depressive symptoms (Abela, Aydin, & Auerbach, 2007). Lewinsohn (1975) made an assertion that depressive symptom may stem from a loss of perceived reinforcement. In case of a child who perceives that he/she is not being rewarded or appreciated for their actions may feel dejected and sad. These cognitive and behavioral factors are formative processes behind negative thoughts in children that impact their mood adversely as well as make them vulnerable for depression.

The cognitive vulnerabilities mentioned above provide a background on use of CBT as a mode of psychotherapy in children with depression. CBT is particularly useful because the central symptoms of depression are similar in children and adults. CBT can be used in combination with other treatments as well as a standalone treatment. The efficacy of CBT is primarily inherent in its focus on persisting cognitive distortions, physiological, and behavioral manifestations of depression. This is done by helping children to understand their negative thinking patterns while helping them to develop evaluative techniques to reassess inherent biases in order to enable them to see how they influence their adaptive functioning (Richardson & Katzenellenbogen, 2005). In doing so, the clinician helps the children to form new perceptions that remove and replace previously held negative beliefs. The subjective implementation of CBT will also help children to identify symptoms, become aware of their thoughts and feelings that influence behavioral patterns. Therefore, common CBT techniques might center on cognitive restructuring, monitoring persisting moods, activation of adaptive behaviors, social skills training assertiveness, communication, response prevention, and implementation of relaxation techniques.

Practice and procedures

1. CBT in childhood depression—planning, techniques, and progression of sessions

Initially, the development of CBT for children involved adapting adult models to children clients (Kendall, 1993). Subsequently, as the differences between adaptive patterns of adults and children were identified, attempts were made to make certain changes for children. The basic elements such as problem-focused nature, strategy-based approach, and time-limited feature of the therapy remained the same for adults as well as children. However, due to the differences in developmental level and emotional maturity differences that exist between adults and children, the CBT treatment protocols need to be adjusted to suit their developmental stage as well as level of depression (Richardson & Katzenellenbogen, 2005). The CBT protocol and the manner, in which techniques are communicated, may vary according to case, keeping in mind the unique experiences of the child. Another important consideration is that in case of adults, they often come to the therapist voluntarily since they are distressed and experiencing symptoms. On the other hand, a depressed child might not even know that experiencing sadness and associated behavioral changes could be a problem. In most of the cases, they are either referred by school or parents/caregivers have brought them to the therapist. In such a scenario, it is important that children referred for CBT may need extra time to develop trust for the therapist and to understand the reason for their treatment. The CBT with children actively focuses to bring about quick symptomatic improvement and meaningful change in less number of sessions. The treatment for depression in children has been seen to most effective when it is continued for 6–9 months postremission of symptoms (Emslie, Mayes, & Ruberu, 2005). Extant studies advocate that continuity of interventions beyond remission of symptoms should take into consideration a number of factors such as persisting psycho-social stressors as well as past relapse episodes. Care should be taken not to suddenly stop treatment as this poses a risk to have another depressive episode within a 5-year period post the last treatment.

While working with children, the therapist takes into account the limitations of child with regard to their cognitive abilities, communication, and labeling of emotions that certainly differ from adults. The therapist attempts to identify and rectify negative thinking patterns, cognitive distortions, and deficiencies that may be acting as maintaining factors to the child's depression. Unlike adults, CBT with depressed children makes more use of behavioral techniques (Bailey, 2001). This is particularly of importance while working with children who have cognitive limitations and might have difficulty in use of verbal techniques; hence, nonverbal activities and imagery-based procedures are emphasized. In case of children with depression, the therapist attempts to understand those experiences that have led to the formation of dysfunctional beliefs in the child. In doing so, considering the social context especially, the role that parents or caregivers of the child play in the process is imperative. It is essential to understand the meanings that the child and their parents have attributed to the child's experiences. A noteworthy aspect is that the child's sense of self and experiences is largely influenced by caregivers' attitudes and expectations; hence, the session plan of CBT with children needs active involvement of caregivers or parents too. Given this premise, an obvious goal of CBT is to equip the child and parent with tools to handle the crisis.

In planning CBT-based intervention for a child, the choice of techniques and strategies must be tailored considering the cognitive, behavioral, social, and environmental determinants that are causing distress in child. Hence, the use of cognitive behavioral therapeutic strategies on children with depression emphasizes primarily on four areas. First, the cognitive strategies that focus on changing maladaptive beliefs, identifying distortions, faulty schemas, and thoughts that contribute to their depression and facilitate problem solving. Second, the behavioral strategies, wherein the child is involved in various activities that offer them a sense of accomplishment, positive reinforcement, communicate, and rebuild supportive connections with others. Third, social skills and assertiveness training are used to develop close and secured attachment bonds with family members and address other behavioral problems that intensify and maintain their depression. Last, techniques that involve relaxation and meditation are used within CBT to reduce the physiological symptoms that arise due to anxiety and tension children might face in difficult situations. In a particular session, structure of CBT includes setting of the agenda, reviewing the progress and homework assignment, working through the techniques decided for problems in that day, and setting homework (Abela & Hankin, 2008). The progression of sessions in the treatment process using CBT typically comprises of three phases: initial phase, middle phase, and final phase (see Fig. 1).

The initial sessions of CBT with children may include establishing a collaborative therapeutic relationship with the child as well as parent. The therapist plays an active role in understanding the attitudes, beliefs, emotions, and experiences of the depressed child. This relationship is extremely important as it predicts the therapy success. This even serves as a primer in setting the goals for treatment. This may sound convenient, but goal setting is a challenging task for the therapist, as many a times, the child's and parent's perspectives may vary. Hence, the foremost task of the therapist is to guide the parent and child in listing down the meaningful and concrete goals that they can mutually agree upon. In case of children, parents are often apprehensive and overwhelmed about the child's problem. The therapist formulates the cognitive behavioral explanations of the depressive child and presents the integrative view to the parents. The parents are psychoeducated

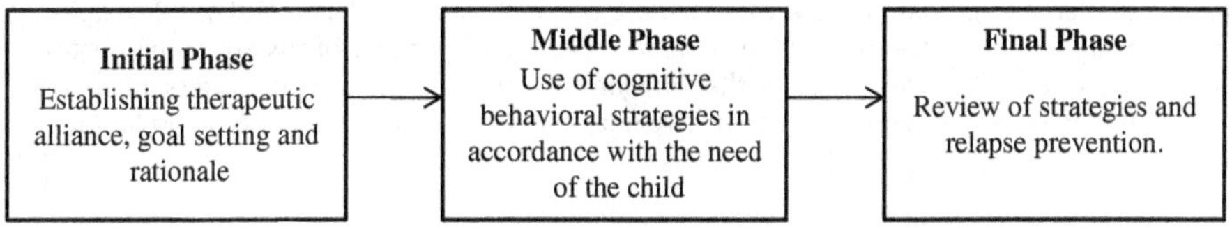

FIG. 1 Phases of sessions in cognitive behavior therapy.

about the interaction of biological, cognitive, affective, behavioral, and environmental dynamics that are contributing to the child's distress. Once the alliance is established and goals are set, the main phase of treatment begins with the use of core cognitive and behavioral strategies. Some of the important strategies used in the context of childhood depression and their rationale have been mentioned in Table 1. The final phase involves consolidation of the skills that the child has learned over the course of treatment. Emphasis is laid on strengthening the associations between the positive changes experienced by the child and the corresponding skills which caused these changes. The therapist reviews the child's coping strategies and

TABLE 1 Specific strategies used in CBT for depressed children and their practice.

Strategies	Rationale, techniques, and outcome
Problem solving	Depressed children who internalize or externalize their problems often view their problems as difficult or unsolvable. The technique is also useful in identifying and addressing cognitive distortions in children. In CBT, the therapist helps the child to approach toward the problems in a gradual manner. The therapist may begin with acknowledging that everyone has problems and that they can be solved. The therapist may then present hypothetical situations like poor scores in examination and ask the child to identify the problem and work out for possible alternatives as solution. Sometimes, in this process, the therapist may use role plays or puzzles with younger children. Subsequently, the therapist encourages the child to analyze the pros and cons based on the consequences of each of the alternative and select the best out of them. As the child becomes familiar with the process, the therapist instills a hope in the child wherein they are asked to implement and evaluate the outcome.
Challenging negative automatic thoughts	Children often have self-referential negative thoughts like *"no one likes me."* The therapist may use the Socratic questioning guided discovery process to help the child identify the errors in their logical reasoning process. The therapist may use the thought-detective approach, i.e., they may ask questions like *"what is the proof for this thought?"* While doing so, rather than adopting a confrontation stance, the therapist maintains a gentle approach and demonstrates the child that through careful questioning newer and adaptive way to look into the thoughts takes place. Some other techniques like role reversal, or asking a child that *"what they would suggest their friend if they are in a similar situation?"*, can be useful in offering a perspective to the child. The homework assignments might include generalization of the new learning to problems that the child may come across at home and school.
Mood monitoring and affect regulation	Children with depression often have difficulty in labeling emotions and modulating negative moods. Such children might break down at the slightest downward point in some event. Using mood monitoring strategies, such children acquire the necessary behavioral and social skills to effectively regulate their moods. These activities often use mood rating scales, emotion thermometer, etc., in which the child can label their emotions, rate their mood, and learn to identify that mood states can be a result of negative thoughts and vice versa.
Activity scheduling	Depressed children often take less interest in activities of pleasure; they might prefer solitary activities and are less likely to get involved with other children or people. They often believe that they might not succeed in the activities. Activity scheduling is based on the principle of mastery and competence leading to pleasure. In CBT, the therapist explains the child and parents that participating in activities is important to divert oneself from negative thoughts as well as giving oneself opportunity to feel competent and receive pleasure. The therapist may begin with a diary of pleasurable activities and mood monitoring strategies. As the child is able to do the task, he/she experiences a positive sense of achievement which acts as reinforcement.
Social skills and assertiveness training	Many depressed children face difficulty in social settings; they often remain confined to themselves. In such children, social communication and assertiveness training is inculcated through verbal and nonverbal methods of modeling, role plays, and discussion. These techniques are useful for the child to maintain eye contact, talking without hesitation, and beginning and ending conversations. Rational disputation methods might be helpful for overwhelming negative thoughts in the context of relationships (e.g., thoughts like *everybody hates me*).
Relaxation Imagery	Exposure to stressful situations may be distressing for children with depression. In such situations, they get anxious and tensed that can interfere with their use of cognitive and behavioral skills. Relaxation training and imagery techniques are then used to relieve the somatic features of anxiety. In this context, the child may use self-talk strategy to guide their thinking and behavior in times of distress (Kendall & Braswell, 1993). Positive self-statements can be relaxing, may act as reinforcers, and are useful in developing coping strategies.

managing skills of distressing experiences. Furthermore, the need for maintenance therapy sessions is explained so that depressive relapse can be prevented and follow-up sessions are scheduled.

2. Research outcomes of CBT among children with depression

In contrast with other psychological treatments, more researches are available that have highlighted that CBT is efficacious as a treatment for depression. However, researches exploring the effectiveness of CBT for treating children with depression are very limited (Arnberg & Öst, 2014). Few studies report favorable results of CBT in treating childhood depression and claim that CBT is the best and cost-effective psychotherapy for the treatment of depression among children (Compton et al., 2004; Zhou et al., 2015). Wood, Harrington, and Moore (1996) compared the effectiveness of CBT in treating depressive features in children and adolescents with relaxation training, revealed that CBT was more effective in treating childhood depression. David-Ferdon and Kaslow (2008) examined the researches on psychosocial treatment outcome and reported efficacy of group CBT in treatment of childhood depression. Kapornai and Vetró (2008) reported CBT as an impactful therapy in various researches for managing the symptoms of depression among children. Jensen, Holt, and Ormhaug (2017) reported that trauma-focused CBT helped in reducing trauma-related symptoms and that of cooccurring depression among children. In several metaanalytic as well as qualitative reviews of available literature, CBT has been reported to be efficacious in treating depressive symptoms among children aged 8–12 years (Arnberg & Öst, 2014; Axelson & Birmaher, 2001; Stark et al., 2008).

Research also indicates that computerized CBT may also be found to be effective in treating internalized disorders like anxiety and depression among children (Elkins, McHugh, Santucci, & Barlow, 2011). In a feasibility study, it was found that CBT delivered over interactive tele-video led to the remission of depressive symptoms among children (Nelson, Barnard, & Cain, 2006). According to Dopheide (2006), depressive children usually respond best to multimodal treatment that includes CBT, antidepressants, and educating patient and the family. Kennard et al. (2014) screened depressed 8–17-year-olds who responded to fluoxetine treatment for 6 weeks duration and assigned them to experimental or control group randomly. The children in the experimental group received continued medication treatment and CBT, while the control group received only medication treatment. They found that CBT in combination with medication was effective in lowering the risk of relapse among children with depression. Similarly, David-Ferdon and Kaslow (2008) reported that CBT in combination of interpersonal therapy is effective in treatment of mild depression. Effectiveness of computer-based CBT in treating childhood depression has also been reported by some researches (O'Kearney, Gibson, Christensen, & Griffiths, 2006).

Although there is a strong consensus among researchers regarding the efficacy of CBT treatment of childhood depression, some researches do not find CBT any different than other available treatment alternatives. Wood et al. (1996), for example, found CBT and relaxation training to be equally effective in decreasing depressive symptoms among children. Similarly, Weisz, McCarty, and Valeri (2006) reported CBT to be only as much effective in treatment of depression among children and adolescents as any other noncognitive treatment approach. Pennant et al. (2015) expressed uncertainty about the effectiveness of CBT in treating depression among children aged 5–11 years. However, such researches are fairly limited in number as compared to those in agreement to the efficacy of CBT in childhood depression.

3. Applications of CBT in other childhood problems

CBT has been found useful in treating affective and nonaffective disorders that are pertinent and comorbid with depression in childhood. There is evidence that suggests that CBT can be used for anxiety disorders, e.g., panic disorders, obsessive compulsive disorder, and generalized anxiety disorder. In this regard, the effectiveness of CBT is that it targets the presenting symptomatology by psycho-educating the individual about their symptoms and responses, teaches coping mechanisms to counter behavioral reactions and challenges distorted cognitions inherent in these disorders. Thus, CBT changes maladaptive thinking styles which lead to negative and dysfunctional behavioral or physiological outcomes. Similarly, CBT has been found to be effective in treating eating disorders, e.g., bulimia nervosa and anorexia nervosa. This is because the underlying assumptions and thought processes in eating disorders are representative of negative cognitive distortions about the self and how the world views them. CBT has also been used to treat internet addiction. In action, CBT helps addicted individuals to gain a thorough understanding of their actions and how they are impacting their lives. Therefore, it is clear that CBT is indeed a pertinent approach to treat disorders that have negative affective regulation, distorted thinking patterns, and maladaptive behavioral patterns as their core symptoms.

4. Professional challenges and future directions

CBT is one of the most popular therapeutic interventions for a wide range of childhood psychological problems. However, working with children experiencing depressive symptoms is professionally challenging as the presentation of symptoms, cognitive ability of the child, and involvement of parents/caregivers demands efficient skills on the therapist's end. Sellers et al. (2019) pointed out that many parents and family members do not pay attention to the growing mental health problems in their children, which may impact the outcome of the treatment. Besides these, common problems that act as barriers in CBT with children are low attendance, poor attentiveness in the session, child not willing to talk, noncompletion of homework, and hostility toward the therapist. The reason for these problems may vary as sensitive children such as victims of abuse, bullying, suicidal tendencies, psychomotor retardation may not be willing to undergo therapy sessions readily. In all such situations, the therapist needs to reevaluate the reasons for block, and as mentioned earlier, therapist needs to adopt a flexible approach. From the research point of view, studies on efficacy of CBT on adults and adolescents with depression have gained immense attention; however, CBT in depressed children is limited in terms of using an active control, sample size, and community-based samples. Further, researches are required on prepubertal children with depression to understand the efficacy of CBT in consideration to their cognitive and developmental level. Researches in the past decades have focused on e-CBT approach to prevent depression and anxiety in children and adolescents (Calear & Christensen, 2010; Gilbody et al., 2015). These preventive programs can further be strengthened at school level in order to prevent depression and anxiety problems to develop at such earlier stage of life.

Mini-dictionary of terms

***Dysfunctional Attitudes*:** These are faulty beliefs that generate negative thoughts about self, world, and others making the individual vulnerable to depression.

***Maladaptive Schema*:** A cognitive structure that is based on a dysfunctional theme and contains memories and emotions of negative nature based on the individual's past experience.

***Response Style*:** It reflects the tendency of a particular individual to respond to any situation.

***Rumination*:** A passive way of responding to problems that involves continuously thinking about the causes and consequences rather than actively working on it.

***Socratic questioning*:** It involves a disciplined way of questioning in a manner that promotes a thoughtful dialogue between the poser and respondent.

Rational Disputing: It is a technique that involves disputing or challenging irrational and illogical beliefs in a way to replace them with rational thoughts.

***Psychoeducation*:** It involves an educative approach wherein the clients and family members are explained about the symptoms, etiology, and progression of a particular disorder.

***Psychomotor retardation*:** It is a state that in which an individual reflects slowing down of thought process as well as reduction of physical movements.

Key fact on childhood depression

As per the World Health Organization Regional Office for Europe (2018), the data measured by disability adjusted life years (DALYs) reveal that depression and anxiety disorders are among the top five reasons of overall disease burden among children and adolescents in the European region.

Application to other areas

Although cognitive behavior therapy (CBT) is an evidence-based treatment technique that was designed to treat depression, the efficacy of CBT has also been observed for other disorders, for example, schizophrenia, eating disorders (bulimia nervosa and anorexia nervosa), posttraumatic stress disorder, obsessive-compulsive disorder, and anxiety disorders. This applicability is based on the premise that a vast majority of disorders are a result of maladaptive automatic thought, cognitive schemas, as well as cognitive distortions and, thus, CBT is used in these cases to change maladaptive thinking styles which lead to negative and dysfunctional behavioral or physiological outcomes.

Summary points

- The prevalence of depression among children is higher than expected. Though there are similarities in symptomatology of adult and childhood depression, but their behavioral manifestations may vary, as there are certain differences in the developmental level.

- Several factors such as genetic predisposition, family and peer neglect, addiction, and bullying can increase the susceptibility to depression in children.
- The cognitive behavioral framework explains that negative schemas, cognitive distortions like catastrophization, faulty attribution style, and ruminative response style increase the vulnerability to depression among children.
- The CBT techniques involve parents/caregivers as cotherapists, who jointly help their children to overcome distressful experiences.
- Cognitive restructuring using problem-solving skills, behavioral activation using activity scheduling and affect regulation strategies are common techniques used by a cognitive behavior therapist for depressive children.
- The therapist must be cautious while dealing with children who show psychomotor retardation, suicidal tendencies, nonexpressive, and cognitively delayed.

References

Abela, J. R., Aydin, C. M., & Auerbach, R. P. (2007). Responses to depression in children: Reconceptualizing the relation among response styles. *Journal of Abnormal Child Psychology*, *35*(6), 913–927.

Abela, J. R., & Hankin, B. L. (2008). *Cognitive vulnerability to depression in children and adolescents: A developmental psychopathology perspective*. The Guilford Press.

Abela, J. R., & Skitch, S. A. (2007). Dysfunctional attitudes, self-esteem, and hassles: Cognitive vulnerability to depression in children of affectively ill parents. *Behaviour Research and Therapy*, *45*(6), 1127–1140.

Abramson, L. Y., Metalsky, G. I., & Alloy, L. B. (1989). Hopelessness depression: A theory-based subtype of depression. *Psychological Review*, *96* (2), 358.

American Psychiatric Association. (2013). *Diagnostic and statistical manual of mental disorders (DSM-5®)*. American Psychiatric Pub.

Arnberg, A., & Öst, L. G. (2014). CBT for children with depressive symptoms: A meta-analysis. *Cognitive Behaviour Therapy*, *43*(4), 275–288.

Axelson, D. A., & Birmaher, B. (2001). Relation between anxiety and depressive disorders in childhood and adolescence. *Depression and Anxiety*, *14*(2), 67–78.

Bailey, V. (2001). Cognitive-behavioural therapies for children and adolescents. *Advances in Psychiatric Treatment*, *7*(3), 224–232.

Beardslee, W. R., & Gladstone, T. R. (2001). Prevention of childhood depression: Recent findings and future prospects. *Biological Psychiatry*, *49*(12), 1101–1110.

Beck, A. T. (1976). *Cognitive therapy and the emotional disorders*. New York, NY: International University Press.

Bennett, D. S., Ambrosini, P. J., Kudes, D., Metz, C., & Rabinovich, H. (2005). Gender differences in adolescent depression: Do symptoms differ for boys and girls? *Journal of Affective Disorders*, *89*(1–3), 35–44.

Bernaras, E., Jaureguizar, J., Soroa, M., Ibabe, I., & Cuevas, C. (2013). Evaluation of the depressive symptomatology and the related variables in the school context. *Anales de Psicologia*, *29*, 131–140.

Birmaher, B., Brent, D., & AACAP Work Group on Quality Issues. (2007). Practice parameter for the assessment and treatment of children and adolescents with depressive disorders. *Journal of the American Academy of Child & Adolescent Psychiatry*, *46*(11), 1503–1526.

Calear, A. L., & Christensen, H. (2010). Systematic review of school-based prevention and early intervention programs for depression. *Journal of Adolescence*, *33*(3), 429–438.

CDCP. (2021). https://www.cdc.gov/childrensmentalhealth/data.html#:~:text=7.4%25%20of%20children%20aged%203,have%20a%20diagnosed%20behavior%20problem.&text=7.1%25%20of%20children%20aged%203,4.4%20million)%20have%20diagnosed%20anxiety.&text=3.2%25%20of%20children%20aged%203,1.9%20million)%20have%20diagnosed%20depression.

Compton, S. N., March, J. S., Brent, D., Albano, A. M., Weersing, V. R., & Curry, J. (2004). Cognitive-behavioral psychotherapy for anxiety and depressive disorders in children and adolescents: An evidence-based medicine review. *Journal of the American Academy of Child & Adolescent Psychiatry*, *43*(8), 930–959.

David-Ferdon, C., & Kaslow, N. J. (2008). Evidence-based psychosocial treatments for child and adolescent depression. *Journal of Clinical Child & Adolescent Psychology*, *37*(1), 62–104.

Dopheide, J. A. (2006). Recognizing and treating depression in children and adolescents. *American Journal of Health-System Pharmacy: AJHP: Official Journal of the American Society of Health-System Pharmacists*, *63*(3), 233–243. https://doi.org/10.2146/ajhp050264.

Dunn, V., & Goodyer, I. M. (2006). Longitudinal investigation into childhood-and adolescence-onset depression: Psychiatric outcome in early adulthood. *The British Journal of Psychiatry*, *188*(3), 216–222.

Elkins, R. M., McHugh, R. K., Santucci, L. C., & Barlow, D. H. (2011). Improving the transportability of CBT for internalizing disorders in children. *Clinical Child and Family Psychology Review*, *14*(2), 161–173.

Emslie, G. J., Mayes, T. L., & Ruberu, M. (2005). Continuation and maintenance therapy of early-onset major depressive disorder. *Pediatric Drugs*, *7*(4), 203–217.

FDA. (2013). *FDA: Don't leave childhood depression untreated. US FDA official website*. Accessed at: https://www.fda.gov/consumers/consumer-updates/fda-dont-leave-childhood-depressionuntreated#:~:text=FDA%20has%20approved%20two%20drugs,for%20kids%2012%20and%20older.

Geller, B., Zimerman, B., Williams, M., Bolhofner, K., Craney, J. L., DelBello, M. P., & Soutullo, C. (2001). Reliability of the Washington University in St. Louis Kiddie Schedule for Affective Disorders and Schizophrenia (WASH-U-KSADS) mania and rapid cycling sections. *Journal of the American Academy of Child & Adolescent Psychiatry*, *40*(4), 450–455.

Ghandour, R. M., Sherman, L. J., Vladutiu, C. J., Ali, M. M., Lynch, S. E., Bitsko, R. H., & Blumberg, S. J. (2019). Prevalence and treatment of depression, anxiety, and conduct problems in US children. *The Journal of Pediatrics*, *206*, 256–267.

Gilbody, S., Littlewood, E., Hewitt, C., Brierley, G., Tharmanathan, P., Araya, R., … White, D. (2015). Computerised cognitive behaviour therapy (cCBT) as treatment for depression in primary care (REEACT trial): Large scale pragmatic randomised controlled trial. *BMJ*, *351*.

Gore, F. M., Bloem, P. J., Patton, G. C., Ferguson, J., Joseph, V., Coffey, C., … Mathers, C. D. (2011). Global burden of disease in young people aged 10–24 years: A systematic analysis. *The Lancet*, *377*(9783), 2093–2102.

Grover, S., Raju, V. V., Sharma, A., & Shah, R. (2019). Depression in children and adolescents: A review of Indian studies. *Indian Journal of Psychological Medicine*, *41*(3), 216–227. https://doi.org/10.4103/IJPSYM.IJPSYM_5_19.

Haley, G. M., Fine, S., Marriage, K., Moretti, M. M., & Freeman, R. J. (1985). Cognitive bias and depression in psychiatrically disturbed children and adolescents. *Journal of Consulting and Clinical Psychology*, *53*(4), 535.

Hankin, B. L., & Abramson, L. Y. (1999). Development of gender differences in depression: Description and possible explanations. *Annals of Medicine*, *31* (6), 372–379.

Jaffee, S. R., Moffitt, T. E., Caspi, A., Fombonne, E., Poulton, R., & Martin, J. (2002). Differences in early childhood risk factors for juvenile-onset and adult-onset depression. *Archives of General Psychiatry*, *59*(3), 215–222.

Jensen, T. K., Holt, T., & Ormhaug, S. M. (2017). A follow-up study from a multisite, randomized controlled trial for traumatized children receiving TF-CBT. *Journal of Abnormal Child Psychology*, *45*(8), 1587–1597. https://doi.org/10.1007/s10802-017-0270-0.

Kapornai, K., & Vetró, Á. (2008). Depression in children. *Current Opinion in Psychiatry*, *21*(1), 1–7.

Kasen, S., Cohen, P., Skodol, A. E., Johnson, J. G., Smailes, E., & Brook, J. S. (2001). Childhood depression and adult personality disorder: Alternative pathways of continuity. *Archives of General Psychiatry*, *58*(3), 231–236.

Kaslow, N. J., Rehm, L. P., Pollack, S. L., & Siegel, A. W. (1988). Attributional style and self-control behavior in depressed and nondepressed children and their parents. *Journal of Abnormal Child Psychology*, *16*(2), 163–175.

Kendall, P. C. (1993). Cognitive-behavioral therapies with youth: Guiding theory, current status, and emerging developments. *Journal of Consulting and Clinical Psychology*, *61*(2), 235.

Kendall, P. C., & Braswell, L. (1993). *Cognitive-behavioral therapy for impulsive children*. Guilford Press.

Kendall, P. C., Stark, K. D., & Adam, T. (1990). Cognitive deficit or cognitive distortion in childhood depression. *Journal of Abnormal Child Psychology*, *18*(3), 255–270.

Kendler, K. S., Thornton, L. M., & Gardner, C. O. (2000). Stressful life events and previous episodes in the etiology of major depression in women: An evaluation of the "kindling" hypothesis. *American Journal of Psychiatry*, *157*(8), 1243–1251.

Kennard, B. D., Emslie, G. J., Mayes, T. L., Nakonezny, P. A., Jones, J. M., Foxwell, A. A., & King, J. (2014). Sequential treatment with fluoxetine and relapse-prevention CBT to improve outcomes in pediatric depression. *American Journal of Psychiatry*, *171*(10), 1083–1090.

Kessler, R. C., Avenevoli, S., & Merikangas, K. R. (2001). Mood disorders in children and adolescents: An epidemiologic perspective. *Biological Psychiatry*, *49*(12), 1002–1014.

Kessler, R. C., Berglund, P., Demler, O., Jin, R., Merikangas, K. R., & Walters, E. E. (2005). Lifetime prevalence and age-of-onset distributions of DSM-IV disorders in the National Comorbidity Survey Replication. *Archives of General Psychiatry*, *62*(6), 593–602.

Kessler, R. C., Petukhova, M., Sampson, N. A., Zaslavsky, A. M., & Wittchen, H. U. (2012). Twelve-month and lifetime prevalence and lifetime morbid risk of anxiety and mood disorders in the United States. *International Journal of Methods in Psychiatric Research*, *21*(3), 169–184.

Kleftaras, G., & Didaskalou, E. (2006). Incidence and teachers' perceived causation of depression in primary school children in Greece. *School Psychology International*, *27*(3), 296–314.

Kullik, A., & Petermann, F. (2013). Attachment to parents and peers as a risk factor for adolescent depressive disorders: The mediating role of emotion regulation. *Child Psychiatry & Human Development*, *44*(4), 537–548.

Lewinsohn, P. M. (1975). Engagement in pleasant activities and depression level. *Journal of Abnormal Psychology*, *84*(6), 729–731. https://doi.org/10.1037/0021-843X.84.6.729.

McCabe, M., Ricciardelli, L., & Banfield, S. (2011). Depressive symptoms and psychosocial functioning in preadolescent children. *Depression Research and Treatment*, *2011*.

Nelson, E. L., Barnard, M., & Cain, S. (2006). Feasibility of telemedicine intervention for childhood depression. *Counselling and Psychotherapy Research*, *6*(3), 191–195.

Nolen-Hoeksema, S. (1991). Responses to depression and their effects on the duration of depressive episodes. *Journal of Abnormal Psychology*, *100*(4), 569.

O'Kearney, R., Gibson, M., Christensen, H., & Griffiths, K. M. (2006). Effects of a cognitive-behavioural internet program on depression, vulnerability to depression and stigma in adolescent males: A school-based controlled trial. *Cognitive Behaviour Therapy*, *35*(1), 43–54.

Pennant, M. E., Loucas, C. E., Whittington, C., Creswell, C., Fonagy, P., Fuggle, P., … Group, E. A. (2015). Computerised therapies for anxiety and depression in children and young people: A systematic review and meta-analysis. *Behaviour Research and Therapy*, *67*, 1–18.

Richardson, L. P., & Katzenellenbogen, R. (2005). Childhood and adolescent depression: The role of primary care providers in diagnosis and treatment. *Current Problems in Pediatric and Adolescent Health Care*, *35*(1), 6–24.

SAMSA (Substance Abuse and Mental Health Services Administration). (2017). *Reports and detailed tables from the 2016 national survey on drug use and health (NSDUH) (Online)*. Available at: https://www.samhsa.gov/samhsa-data-outcomes-quality/major-data-collections/reports-detailed-tables-2016-NSDUH.

Sellers, R., Warne, N., Pickles, A., Maughan, B., Thapar, A., & Collishaw, S. (2019). Cross-cohort change in adolescent outcomes for children with mental health problems. *Journal of Child Psychology and Psychiatry*, *60*, 813–821. https://doi.org/10.1111/jcpp.13029.

Siennick, S. E., Widdowson, A. O., Woessner, M. K., Feinberg, M. E., & Spoth, R. L. (2017). Risk factors for substance misuse and adolescents' symptoms of depression. *Journal of Adolescent Health*, *60*(1), 50–56.

Stark, K. D., Hargrave, J., Hersh, B., Greenberg, M., Herren, J., & Fisher, M. (2008). Treatment of childhood depression: The ACTION Treatment Program. In J. R. Z. Abela, & B. L. Hankin (Eds.), *Handbook of depression in children and adolescents* (pp. 224–229). The Guilford Press.

Strober, M., Lampert, C., Schmidt, S., & Morrell, W. (1993). The course of major depressive disorder in adolescents. I. Recovery and risk of manic switching in a follow-up of psychotic and nonpsychotic subtypes. *Journal of the American Academy of Child & Adolescent Psychiatry*, *32*(1), 34–42.

Thapar, A., Collishaw, S., Pine, D. S., & Thapar, A. K. (2012). Depression in adolescence. *The Lancet*, *379*(9820), 1056–1067.

Verduyn, C. (2000). Cognitive behaviour therapy in childhood depression. *Child Psychology and Psychiatry Review*, *5*(4), 176–180.

Vinaccia, S., Gaviria, A. M., Atehortúa, L. F., Martínez, P. H., Trujillo, C., & Quiceno, J. M. (2006). Prevalencia de depresión en niños escolarizados entre 8 y 12 años del oriente antioqueño a partir del" Child Depression Inventory"-CDI. *Diversitas*, *2*(2), 217–227.

Weisz, J. R., McCarty, C. A., & Valeri, S. M. (2006). Effects of psychotherapy for depression in children and adolescents: A meta-analysis. *Psychological Bulletin*, *132*(1), 132.

Wood, A., Harrington, R., & Moore, A. (1996). Controlled trial of a brief cognitive-behavioural intervention in adolescent patients with depressive disorders. *Journal of Child Psychology and Psychiatry*, *37*(6), 737–746.

World Health Organization. (2020). *Depression*. Available at: https://www.who.int/news-room/fact-sheets/detail/depression. Accessed 24 June 2021.

World Health Organization Regional Office for Europe. (2018). *Adolescent mental health in the European region*. World Health Organization Regional Office for Europe.

Wu, P., Bird, H. R., Liu, X., Fan, B., Fuller, C., Shen, S., … Canino, G. J. (2006). Childhood depressive symptoms and early onset of alcohol use. *Pediatrics*, *118*(5), 1907–1915.

Xie, B., Unger, J. B., Gallaher, P., Johnson, C. A., Wu, Q., & Chou, C. P. (2010). Overweight, body image, and depression in Asian and Hispanic adolescents. *American Journal of Health Behavior*, *34*(4), 476–488.

Zhou, X., Hetrick, S. E., Cuijpers, P., Qin, B., Barth, J., Whittington, C. J., … Xie, P. (2015). Comparative efficacy and acceptability of psychotherapies for depression in children and adolescents: A systematic review and network meta-analysis. *World Psychiatry*, *14*(2), 207–222.

Section C

Aging, the elderly and related conditions

Chapter 26

Cognitive-behavioral therapy for late-life insomnia

Joseph M. Dzierzewski, Elliottnell Perez, Pablo Soto, and Scott G. Ravyts
Department of Psychology, Virginia Commonwealth University, Richmond, VA, United States

Abbreviations

BBTI brief behavioral treatment for insomnia
CBT cognitive-behavioral therapy
CBTI cognitive-behavioral therapy for insomnia

Introduction

Individuals are living longer worldwide, and older adults experience insomnia at higher rates than younger and middle-aged adults. There are numerous treatment options for insomnia, including pharmacological treatment and cognitive-behavioral approaches. Given the risk of polypharmacy with advancing age and the known effectiveness of CBT for insomnia (CBTI) in late life, it is recommended that CBTI be the first line of treatment for older adults with insomnia. This chapter will review the evidence for CBTI approaches with older adults.

Insomnia in older adults

Insomnia is among one of the most common sleep disorders in older adults. The fifth edition of the Diagnostic and Statistical Manual for Mental Disorders (DSM-5) and the third edition of the International Classification of Sleep Disorders (ICSD-3) define insomnia as: difficulty initiating sleep, maintaining sleep, or early morning awakenings without the ability to return to sleep that occurs at least 3 nights per week and is present for at least 3 consecutive months. Moreover, the sleep disturbances are associated with significant daytime distress and impairment(s) such as fatigue, sleepiness, impaired attention/concentration, interpersonal problems, mood disturbance, and reduced quality of life (American Psychiatric Association, 2013; Sateia, 2014).

Prevalence rates

The prevalence of insomnia in older adults varies greatly by study, potentially due to differences in study design, study-specific criteria for insomnia, and assessments strategies utilized. Estimated prevalence of insomnia in late life to range from 23.2% to 60%, depending on the sample and the setting (Kim et al., 2017; Nowicki et al., 2016), with older adults being more likely to report sleep maintenance complaints (25.2%–60%) compared to concerns with sleep initiation (16.6%–45.4%) and early morning awakenings (12.2%–26.4%) (Nowicki et al., 2016; Wang et al., 2016). While age may play a substantial role in the predisposition toward insomnia, an individual's sex has also been found to be closely linked with insomnia, with several studies noting a higher prevalence of insomnia among females, ranging from 37.9% to 74.8%, compared to males, ranging from 25.2% to 52.9% (Kim et al., 2017; Nowicki et al., 2016). Insomnia symptoms are also more prevalent in Black and Latinx older adults compared to their white counterparts (Kaufmann et al., 2016).

Correlates and economic burden

The impact of insomnia is not relegated simply to our sleep and daytime functioning, but may also have a widespread impact on overall functioning. In a review by Dzierzewski, Griffin, Ravyts, and Rybarczyk (2018), it was noted that

Handbook of Lifespan Cognitive Behavioral Therapy. https://doi.org/10.1016/B978-0-323-85757-4.00031-6

insomnia has a profound impact on mental and physical health outcomes, quality of life, and mortality rates. A 10-year cohort study by Abell, Shipley, Ferrie, Kivimäki, and Kumari (2016) found that chronic insomnia symptoms were related to poorer mental, physical, and overall well-being via a dose-response association that showed an increased negative effect with higher levels of reported insomnia symptoms. Insomnia symptoms in older adults are also associated with symptoms of: depression (Bao et al., 2017), anxiety (Kay & Dzierzewski, 2015), and suicidality (Kay et al., 2016). In terms of physical health, older adults with insomnia report higher rates chronic conditions compared to those without insomnia (Helbig et al., 2017). Furthermore, impaired sleep due to insomnia has been found to serve as a reliable predictor of frailty, fall risk, pain, and motor vehicle accidents in older adults (Finan, Goodin, & Smith, 2013; Itani, Jike, Watanabe, & Kaneita, 2017). Unsurprisingly, Dragioti, Bernfort, Larsson, Gerdle, and Levin (2018) found that older adults with severe clinical insomnia were burdened with close to double the total healthcare expenditure (€ 8469 vs € 4345) compared to those without insomnia.

Insomnia treatment preference in older adults

Although different approaches for the treatment of insomnia exist, they can be broadly classified as psychological or pharmacological in nature. While little research has examined treatment preferences in older adults, available studies broadly looking at insomnia treatment preferences in adult samples showed a favorability toward behavioral options despite the widespread availability of sleep medications (Cheung, Bartlett, Armour, Saini, & Laba, 2018; Sidani, Epstein, Bootzin, Moritz, & Miranda, 2009).

Rates of treatment uptake

While several professional societies and therapeutic guidelines advocate for CBTI as the first line treatment for insomnia, psychological treatments are often overshadowed by existing pharmacological interventions (Qaseem et al., 2013; Riemann et al., 2017). Despite concerns of the adverse consequences of sedative hypnotic and other sleep medications, these medications are still prescribed regularly, often at higher rates than recommended, and are generally used to treat insomnia in older adults at much higher rates than CBTI (Culver et al., 2016; DiNapoli, Bramoweth, Cinna, & Kasckow, 2016).

Practice and procedures: Cognitive-behavioral treatment options for late-life insomnia

CBTI is a 6–8 week, multicomponent intervention, consisting of behavioral and cognitive techniques to treat insomnia. CBTI has been heavily researched in the past two decades and has garnered strong empirical support for its efficacy (van der Zweerde, Bisdounis, Kyle, Lancee, & van Straten, 2019; Wu, Appleman, Salazar, & Ong, 2015). The findings from various meta-analyses and reviews suggest that CBTI is not only efficacious, but the effects of the multicomponent intervention are often moderate-to-large in size and durable. The consistently strong evidence generated from research studies has led to CBTI's recommendation as the first-line of treatment for insomnia by the American College of Physicians, European Sleep Research Society, and British Association for Psychopharmacology for adults (Qaseem et al., 2016; Riemann et al., 2017). Strong empirical support for CBTI also extends to late-life insomnia. Although the effect sizes are smaller for older adults relative to younger adults (Brasure et al., 2016), they show significant improvement in sleep onset latency, wake after sleep onset, sleep efficiency, and insomnia severity following CBTI (Alessi et al., 2016).

The structure of CBTI often includes four key components, namely, sleep restriction, stimulus control, cognitive therapy, and sleep hygiene. Sleep restriction is a core behavioral technique in CBTI that entails limiting sleep to a strict schedule in order to optimize homeostatic sleep pressure accruing during the day for nocturnal sleep (Spielman, Caruso, & Glovinsky, 1987). Individuals are encouraged to follow the established bedtime and wake time until sleep quality improves. At this time, the schedule is slowly widened to accommodate the needed amount of sleep. Sleep restriction has obtained some empirical support for improving insomnia severity, sleep onset latency, and sleep efficiency in adults generally (Miller et al., 2014). However, studies investigating its efficacy as a standalone intervention in late-life insomnia are limited, and in need of further research.

Sleep compression is a variation of the sleep restriction technique by which individuals gradually restrict time in bed to until sleep efficiency and quality improves. Sleep compression has received less research attention than sleep restriction in late-life insomnia; however, some evidence suggests that its implementation shows some benefit with respect to sleep onset latency, wake after sleep onset, night-time awakenings, sleep quality, and total sleep time (Lichstein, Riedel, Wilson, Lester, & Aguillard, 2001). Moreover, sleep compression was more effective than a sleep hygiene only intervention. Other available research suggests that when combined with self-help education and therapist guidance, it can also lead to positive sleep outcomes and greater sleep satisfaction among older adults (Riedel, Lichstein, & Dwyer, 1995).

Stimulus control is another core behavioral technique of CBTI that is designed to enhance the association between the bed and bedroom with sleep while also weakening the association between the bed and nonsleep activities (Bootzin, 1973). This is accomplished by encouraging individuals to limit activities in bed to only sleep and sex and eliminate other activities such as viewing television, working, or eating on the bed. Moreover, individuals are encouraged to leave the bed after 15 min of unsuccessful sleep onset to prevent an association from forming between the bed and wakefulness. While out of bed, patients are instructed to engage in relaxing and nonstimulating behaviors and to return to bed only when they feel sleepy. While there is moderate support for stimulus control in adults as a standalone intervention (Harris, Lack, Kemp, Wright, & Bootzin, 2012), there is limited evidence for late-life insomnia. Of the evidence available, stimulus control therapy is significantly more effective than wait-list control groups (Epstein, Sidani, Bootzin, & Belyea, 2012). Stimulus control has produced medium-to-large effects on diary and self-reported sleep onset latency, sleep efficiency, and sleep quality (Epstein et al., 2012).

Cognitive therapy is a core component of CBTI that is aimed at modifying maladaptive thoughts and beliefs individuals maintain regarding their sleep and insomnia. These maladaptive beliefs are often related to the causes of sleep disturbance, consequences of insomnia, and sleep-promoting practices (Harvey, 2002). These themes of beliefs serve to perpetuate cognitive and physiological processes that maintain insomnia. Research has shown some support for cognitive therapy reducing insomnia severity, sleep onset latency, sleep efficiency, and daytime functioning in adults broadly (Harvey et al., 2014). The efficacy of cognitive therapy is enhanced when used in conjunction with behavioral experiments. Despite the evidence available supporting its use in the general adult population, very little research exists examining its efficacy independently for late-life insomnia.

Sleep hygiene is the final core component typically included in CBTI. A set of recommendations are provided to individuals to reduce the influence of behavioral and environmental factors contributing to the maintenance of sleep difficulties. These recommendations may involve discussions regarding the frequency and timing of caffeine use, the benefits of regular physical activity, optimal temperature in the bedroom, food intake, screen light exposure, and napping. Support for sleep hygiene as a standalone treatment for insomnia is very limited. Older adults who only received an intervention focused on sleep hygiene obtained poorer sleep outcomes than other nonpharmacological approaches (Martin et al., 2017). The literature on late-life insomnia, and insomnia generally, does not show adequate support for sleep hygiene as a standalone intervention.

Relaxation strategies are an optional component of CBT-I used to reduce physiological and cognitive arousal related to anxiety or stress. Various techniques are used for relaxation including, progressive muscle relaxation, diaphragmatic deep-breathing, and guided imagery. These techniques are significantly weaker than other cognitive or behavioral approaches, and are not effective as standalone treatments for late-life insomnia (Rybarczyk et al., 2005).

Mindfulness-based therapies are interventions aimed at reducing presleep arousal through the use of techniques that cultivate a nonjudgmental, present-moment awareness and acknowledgement of thoughts, feelings, and physical sensations (Ong & Smith, 2017). Generally, mindfulness-based interventions for insomnia have shown positive outcomes, including improved insomnia severity, sleep onset latency, sleep quality, and sleep efficiency (Black, O'Reilly, Olmstead, Breen, & Irwin, 2015; Ong & Smith, 2017). While evidence in older adults is limited, it appears these interventions are significantly more effective for late-life insomnia than a sleep hygiene intervention (Black et al., 2015). As such, mindfulness-based interventions have been proposed as viable nonpharmacological alternative to multicomponent interventions such as CBTI (Ong & Smith, 2017).

Exercise and sleep have obtained considerable attention in the literature. Limited research supports exercise as a standalone intervention for late-life insomnia; however, studies suggest physical activity may be beneficial for reducing the severity of insomnia symptoms. A comprehensive meta-analysis investigating exercise and sleep across the lifespan suggests that regular and acute exercise may confer small, positive benefits to total sleep time, sleep onset latency, sleep efficiency, and moderate-sized effects on wake-after sleep onset and sleep quality (Kredlow, Capozzoli, Hearon, Calkins, & Otto, 2015). Physical activity may also operate as a preventative measure against poor sleep and insomnia. Older adults who engaged in regular, structured physical activity significantly reduced their likelihood of obtaining poor sleep quality by 20% over 30 months (Vaz Fragoso et al., 2015).

Within the last decade, interest has grown in examining the relative strength of individual components of CBTI. This line of inquiry has focused on better understanding the core mechanisms of action in CBTI and adapting the most critical and potent components of the intervention to unique populations and settings with limited time, training, and financial resources. Brief behavioral treatment for insomnia (BBTI) was developed and studied to address these limitations. BBTI is a short multicomponent intervention that lasts 4 weeks or less and contains only the core behavioral components of CBTI.

TABLE 1 Summary of interventions for late-life insomnia.

Intervention	Description	Evidence for older adults
Sleep restriction	The restriction of sleep to a short, specific sleep window. As sleep quality improves, the sleep window increased	Moderate
Sleep compression	The gradual reduction of the sleep window until sleep quality improves	Limited
Stimulus control	The association between the bed and sleep is strengthened by limiting activities in bed to only sleep and sex	Moderate
Cognitive therapy	The modification of inaccurate and maladaptive thoughts and beliefs about sleep and sleep promoting practices that perpetuate insomnia	None
Sleep hygiene	Recommendations provided to individuals targeting behavioral and environmental factors that may contribute to poor sleep	None
Relaxation skills	A group of skills designed to reduce arousal stemming from anxiety or stress related to sleep	Limited
Mindfulness	Techniques to increase self-awareness and acknowledgment of thoughts, feelings, and sensations in order to reduce presleep arousal	Limited
Exercise	Acute and regular physical activity to promote deeper sleep	Limited
Cognitive-behavioral therapy for insomnia (CBTI)	A multicomponent intervention that utilizes behavioral and cognitive skills to treat insomnia. The individual components often include stimulus control, sleep restriction, cognitive therapy, and sleep hygiene. Lasts 6–8 weeks	Strong
Brief behavioral treatment for insomnia (BBTI)	Short multicomponent intervention containing only the behavioral components of CBT-I. The components typically include stimulus control, sleep restriction, and sleep hygiene. Last 4 or less weeks	Strong

The components typically included in BBTI are sleep restriction, stimulus control, and sleep hygiene. Several studies have found BBTI to be an effective treatment for older adults with chronic insomnia with generally moderate-to-large effects (Buysse et al., 2011; McCrae et al., 2018). Older adults who underwent BBTI have reported improved sleep quality, sleep onset latency, wake after sleep onset, and sleep efficiency. Similar to CBTI, the effects of BBTI are durable with older adults with gains being maintained 3–6 months post intervention (Buysse et al., 2011). Refer to Table 1 for a complete listing of nonpharmacological treatment approaches for late-life insomnia.

Important considerations in cognitive-behavioral treatment for late-life insomnia

Several important points need to be considered when using CBTI with older adults. First, napping frequency and duration increases with age, with over 21% of older adults engaging in habitual napping (Ancoli-Israel & Martin, 2006). Moreover, higher prevalence of chronic health conditions, medications, and lifestyle changes following retirement contribute to more frequent napping (Zhang, Xiao, Ma, & Li, 2020). Recommendations regarding naps for older adults are complex and unclear as the practice is associated with positive and negative benefits. Generally, napping is a behavior that can maintain insomnia through the disruption of circadian sleep timing and sleep pressure accumulation (Wennberg, Canham, Smith, & Spira, 2013). Napping is associated with poorer nocturnal sleep and greater risk of fall, chronic health conditions, and mortality (Ancoli-Israel & Martin, 2006). However, additional research has shown that napping infers positive effects for older adults such as increased vigilance, dexterity, alertness, and responsivity (Ancoli-Israel & Martin, 2006). Current recommendations are for napping be eliminated to allow the full effects of the behavioral interventions to take place; however, if needed, naps should be limited to 20min, and scheduled in the early afternoon.

An additional consideration in late-life insomnia is the presence of nocturia. Nocturia is characterized by an increased urge for nocturnal urination, and is a common source of sleep disruption among older adults (Vaughan & Bliwise, 2018), especially males. Rates of nocturia are higher among older adults than younger adults in part because of reduced bladder capacity, medical conditions and medications that influence functioning of the bladder, and limited production of the anti-diuretic hormone that regulates urine production (Bosch & Weiss, 2013; Vaughan & Bliwise, 2018). Some research has shown that nocturia can limit the effectiveness of BBTI for late-life chronic insomnia. Therefore, it is worthwhile considering behavioral strategies for addressing nocturia prior to starting treatment for insomnia (Tyagi et al., 2014).

It is also important to consider the setting in which older adults reside, and the impact of these settings on sleep. Sleep problems are common among older adults in residential care facilities (Ye & Richards, 2018). In addition to age-related factors, residential settings influence sleep, including less exposure to daytime light, noise during night-time, potential room sharing, facility routines, and night-time activities. Furthermore, older adults in these facilities are typically less engaged physically and socially during the daytime which can also negatively affect the circadian rhythm and lead to more daytime sleepiness and napping (Ye & Richards, 2018). Interventions for sleep should consider scheduling regular social and physical activity, increasing exposure to sunlight, and limiting night-time care disruptions (Haesler, 2004).

Given the higher prevalence of chronic medical conditions among older adults, it is not uncommon for treatment to occur within the context of comorbid medical conditions (Taylor et al., 2007). Fortunately, CBTI has established efficacy in treating comorbid insomnia with a range of age-related medical conditions including heart disease, COPD, and cancer (Rybarczyk et al., 2005). Contrary to earlier conceptualizations of insomnia as a consequence of a chronic medical conditions when both are present, termed *secondary insomnia*, two decades of research has shown the independent, but connected etiology of insomnia (Geiger-Brown et al., 2015). While studies have shown that CBTI remains effective in the presence of chronic health conditions, future research is looking toward combined interventions that target the core symptoms of both conditions.

Chronic pain is frequently comorbid with late-life insomnia (Finan et al., 2013). While there is evidence of a bidirectional effect between sleep and pain, a greater number of studies have found stronger evidence of temporal precedence of sleep to pain using a variety of study designs (Finan et al., 2013). Research shows that older adults with chronic pain and insomnia experience significant improvement in sleep, and small improvement in pain outcomes following CBT-I (Finan et al., 2013; Koffel, McCurry, Smith, & Vitiello, 2019). Therefore, in the context of pain, CBTI is still a viable, and perhaps first-point of intervention for late-life insomnia and chronic pain. Future directions for sleep and pain are currently focused on producing an integrated treatment for insomnia and chronic pain that combine the individual cognitive-behavioral components of these interventions (Finan et al., 2013; Koffel et al., 2019).

A final consideration is the high rate of comorbidity between insomnia and psychiatric conditions (Kay & Dzierzewski, 2015). While the complexities of the overlap between late-life insomnia and other psychiatric conditions are still not clearly understood, meta-analyses indicate CBTI remains effective in treating sleep disturbances in this population, and may also even improve symptom severity or functioning of adult with psychiatric conditions (Wu et al., 2015). Evidence for these patterns appears most frequently within the context of insomnia and depression in older adults. Older adults who underwent CBTI noted significant improvement in symptoms of insomnia and depression (Sadler, McLaren, Klein, Harvey, & Jenkins, 2018). Additional research is necessary to examine the efficacy of CBTI for insomnia and frequently co-occurring mental health conditions.

Innovative treatment deliver of cognitive-behavioral treatment for late-life insomnia

Patient, clinician, and system-level barriers to treatment have spurred innovation in the field of behavioral sleep medicine. To date, these innovations have consisted of training nonspecialists in the delivery of CBTI, as well as developing alternative treatment formats including telephone, video, web, and app-based versions of CBTI.

The training of nonsleep specialists in CBTI is crucial to increasing treatment access. Several studies have specifically begun to examine the efficacy of nonspecialist delivered CBTI for older adults. For example, one randomized controlled trial compared CBTI to a general sleep education control (Alessi et al., 2016). Both treatments were delivered by a sleep coach who either attended a 2-day workshop on CBTI or completed a six-session webinar on CBTI and received weekly supervision by a behavioral sleep specialist. Older adults who received CBTI reported greater improvements in insomnia symptoms at post treatment and at 3-month follow-up compared to individuals in the control group. Another study found that BBTI delivered by a mental health nurse practitioner with no prior experience in sleep medicine resulted in greater sleep-related improvements among older adults with insomnia than those receiving only sleep education (Buysse et al., 2011). Collectively, these findings suggest that, with adequate training, nonsleep clinicians can provide effective treatment.

One way to reduce costs is by delivering CBTI via telephone. Originally tested among the general population (Arnedt et al., 2013), telephone CBTI has also shown some promise among older adults with insomnia. For example, a large randomized control trial comparing telephone CBTI to an education only control among older adults with comorbid osteoarthritis found that the treatment condition improved insomnia symptoms at post treatment and that these treatment gains remained at the 12-month follow-up (McCurry et al., 2021). By contrast, a randomized control trial comparing telephone

CBTI to a wait-list control among older adults with comorbid depression found that while depression improved for older adults receiving CBTI, sleep quality did not (Tanaka, Kusaga, Nyamathi, & Tanaka, 2019).

Another treatment innovation, video-based CBTI, has also shown potential for older adults. Older adults who received video-based CBTI along with CBT for depression showed greater decreases in insomnia symptoms compared to individuals receiving usual care (Scogin et al., 2018). Similarly, a small pilot study of video-based CBTI combined with CBT for depression found that the treatment produced significant reductions in insomnia and depression at post treatment and at a 2-month follow-up (Lichstein et al., 2013).

Web-based delivery of CBTI has been another area of recent interest. Two meta-analyses examining randomized controlled trials of internet-delivered CBTI demonstrated that the treatment produces outcomes comparable to face-to-face CBTI (Ye et al., 2016; Zachariae, Lyby, Ritterband, & O'Toole, 2016). These findings are further corroborated by a recent randomized controlled trial which compared participants in an education-only control to individuals receiving internet-based CBTI (Ritterband et al., 2017). Participants in the treatment condition showed greater improvements in insomnia severity, sleep-onset latency, and wake after sleep onset and these treatment gains remained at the 1-year follow-up with over half of participants achieving remission status. Despite these promising findings, further research is needed to examine whether the effectiveness of web-based CBTI extends to older adults.

Lastly, smartphone-based applications of CBTI have also begun to be examined. A randomized control trial comparing app-delivered CBTI to a waitlist control group found that the active treatment condition produced significant improvements in insomnia severity and sleep efficiency up to 3 months post treatment (Horsch et al., 2017). While empirical examinations of app-based CBTI among individuals in late life remain scarce, a case study of a 64-year-old female with insomnia found that app-delivered CBTI resulted in improvements in subjective sleep ratings and a discontinuation of hypnotic medication (Chen, Hung, & Chen, 2016). Further research is needed to determine whether older adults may benefit from adaptions to app-based interventions.

Conclusions

Age-related biological, psychological, and social changes increase risk of late-life insomnia. Insomnia is common among older adults, and has detrimental physical, mental health, and economic consequences. While there is limited evidence supporting the use of individual cognitive and behavioral techniques as standalone interventions for late-life insomnia, multicomponent interventions such as CBTI and BBTI have received strong empirical support from meta-analyses and reviews. Despite the efficacy of these interventions, providers need to consider additional unique age-related factors that may play a role in sleep disruption in older adults. Factors such as nocturia, napping, and treatment settings may need additional behavioral strategies to supplement CBTI or BBTI. While comorbid insomnia is effectively treated with CBTI and BBTI, future research is moving toward integrated interventions targeting insomnia and comorbid physical and mental health conditions. To address treatment barriers and increase access to treatment, innovative developments to treatment delivery have been increasingly investigated. While further research is needed, increasing support for CBTI implementation by trained nonsleep specialists, and via telephone, video, and web-based modalities are gaining increasing empirical support for use among older adults.

Mini-dictionary of terms

- ***Insomnia.*** A dissatisfaction with quantity or quality of sleep, occurring at least 3 nights per week for a minimum of 3 months.
- ***Sleep restriction.*** An insomnia intervention strategy used to curtail the amount of time spent in bed to more closely align with the amount of time spent asleep.
- ***Stimulus control.*** An insomnia intervention strategy used to strengthen the association between the bed (and bedroom) and sleep.
- ***Sleep hygiene.*** Recommendations targeting behavioral and environmental factors that may contribute to insomnia, such as limiting caffeine intake.
- ***Nocturia.*** The increase frequency of nocturnal urination, often interfering with sleep in older adults.

Key facts of insomnia in older adults

- Estimated prevalence of insomnia in late life ranges from 23% to 60%.
- Older adults are more likely to report sleep maintenance complaints (25.2%–60%) as compared to problems with sleep initiation (16.6%–45.4%) or early morning awakenings (12.2%–26.4%).

- Older women complain of insomnia at higher rates than older males, despite better objectively measured sleep.
- Insomnia is more prevalent in older adult minority group members than majority group members.
- Insomnia in older adults may be more associated with changes in health status than changes in age.

Applications to other areas

In this chapter on cognitive-behavioral therapy for late-life insomnia, we review the literature on CBT techniques appropriate for the treatment of insomnia in older adults. The treatment techniques with the largest evidence base include sleep restriction, stimulus control, relaxation strategies, and cognitive techniques, along with combinations of these techniques. These techniques appear very robust, suitable for wide segments of the population, including those with comorbid physical and mental health conditions. In fact, dissemination of these treatment approaches through manuals, handbooks, telehealth services, and web-based and app-based protocols has demonstrated the effectiveness of these approaches across numerous patient groups. Patient groups with current unknown efficacy of CBTI include caregivers of individuals with chronic medical conditions including dementia, cancer, traumatic brain injuries, and developmental delays, new parents, and individuals working nontraditional shifts. However, given the clear evidence based for CBTI, the strong theoretical underpinnings of the intervention strategies, and the ease of application, the described techniques are suspected to engender significant sleep improvements even in these yet-to-be-tested populations. Attempts to intervene using CBTI in new patient populations should take a methodical approach, documenting any needed alterations to treatment protocol as well and careful documentation of treatment outcomes.

Summary points

- Insomnia is a very prevalent disorder in older adults.
- Cognitive-behavioral therapy for insomnia is the "gold-standard" treatment approach.
- Specific techniques known to be effective and suitable for use in older adults with insomnia include: stimulus control, sleep restriction, and their combination with other strategies such as relaxation techniques.
- When conducting CBTI with older adults, practitioners need to be aware of comorbid physical and mental health conditions.
- Specific considerations when conducting CBTI in older adults include: circadian phase shifts, nocturia, decreased cognitive abilities, reduced mobility, and level of independence.

References

Abell, J. G., Shipley, M. J., Ferrie, J. E., Kivimäki, M., & Kumari, M. (2016). Association of chronic insomnia symptoms and recurrent extreme sleep duration over 10 years with well-being in older adults: A cohort study. *BMJ Open*, *6*(2). https://doi.org/10.1136/bmjopen-2015-009501, e009501.

Alessi, C., Martin, J. L., Fiorentino, L., Fung, C. H., Dzierzewski, J. M., Rodriguez Tapia, J. C., … Mitchell, M. N. (2016). Cognitive behavioral therapy for insomnia in older veterans using nonclinician sleep coaches: Randomized controlled trial. *Journal of the American Geriatrics Society*, *64*(9), 1830–1838.

American Psychiatric Association. (2013). *Diagnostic and statistical manual of mental disorders* (5th ed.). American Psychiatric Association.

Ancoli-Israel, S., & Martin, J. L. (2006). Insomnia and daytime napping in older adults. *Journal of Clinical Sleep Medicine*, *2*(3), 333–342.

Arnedt, J. T., Cuddihy, L., Swanson, L. M., Pickett, S., Aikens, J., & Chervin, R. D. (2013). Randomized controlled trial of telephone-delivered cognitive behavioral therapy for chronic insomnia. *Sleep*, *36*(3), 353–362. https://doi.org/10.5665/sleep.2448.

Bao, Y.-P., Han, Y., Ma, J., Wang, R.-J., Shi, L., Wang, T.-Y., … Lu, L. (2017). Cooccurrence and bidirectional prediction of sleep disturbances and depression in older adults: Meta-analysis and systematic review. *Neuroscience and Biobehavioral Reviews*, *75*, 257–273. https://doi.org/10.1016/j.neubiorev.2017.01.032.

Black, D. S., O'Reilly, G. A., Olmstead, R., Breen, E. C., & Irwin, M. R. (2015). Mindfulness meditation and improvement in sleep quality and daytime impairment among older adults with sleep disturbances: A randomized clinical trial. *JAMA Internal Medicine*, *175*(4), 494–501. https://doi.org/10.1001/jamainternmed.2014.8081.

Bootzin, R. R. (1973). Stimulus control treatment for insomnia. In *APA 80th annual convention, Honolulu, HI, September 2–8, 1972*. http://psycnet.apa.org/psycextra/465522008-198.

Bosch, J. L. H. R., & Weiss, J. P. (2013). The prevalence and causes of nocturia. *Journal of Urology*, *189*(1 Suppl), S86–S92. https://doi.org/10.1016/j.juro.2012.11.033.

Brasure, M., Fuchs, E., MacDonald, R., Nelson, V. A., Koffel, E., Olson, C. M., … Kane, R. L. (2016). Psychological and behavioral interventions for managing insomnia disorder: An evidence report for a clinical practice guideline by the American College of Physicians. *Annals of Internal Medicine*, *165*(2), 113–124. https://doi.org/10.7326/M15-1782.

Buysse, D. J., Germain, A., Moul, D. E., Franzen, P. L., Brar, L. K., Fletcher, M. E., … Monk, T. H. (2011). Efficacy of brief behavioral treatment for chronic insomnia in older adults. *Archives of Internal Medicine*, *171*(10), 887–895.

Chen, Y.-X., Hung, Y.-P., & Chen, H.-C. (2016). Mobile application-assisted cognitive behavioral therapy for insomnia in an older adult. *Telemedicine Journal and E-Health: The Official Journal of the American Telemedicine Association*, *22*(4), 332–334. https://doi.org/10.1089/tmj.2015.0064.

Cheung, J. M. Y., Bartlett, D. J., Armour, C. L., Saini, B., & Laba, T.-L. (2018). Patient preferences for managing insomnia: A discrete choice experiment. *The Patient—Patient-Centered Outcomes Research*, *11*(5), 503–514. https://doi.org/10.1007/s40271-018-0303-y.

Culver, N. C., Song, Y., Kate McGowan, S., Fung, C. H., Mitchell, M. N., Rodriguez, J. C., … Martin, J. L. (2016). Acceptability of medication and nonmedication treatment for insomnia among female veterans: Effects of age, insomnia severity, and psychiatric symptoms. *Clinical Therapeutics*, *38*(11), 2373–2385. https://doi.org/10.1016/j.clinthera.2016.09.019.

DiNapoli, E. A., Bramoweth, A. D., Cinna, C., & Kasckow, J. (2016). Sedative hypnotic use among veterans with a newly reported mental health disorder. *International Psychogeriatrics*, *28*(8), 1391–1398. https://doi.org/10.1017/S1041610216000521.

Dragioti, E., Bernfort, L., Larsson, B., Gerdle, B., & Levin, L.Å. (2018). Association of insomnia severity with well-being, quality of life and health care costs: A cross-sectional study in older adults with chronic pain (PainS65+). *European Journal of Pain (London, England)*, *22*(2), 414–425. https://doi.org/10.1002/ejp.1130.

Dzierzewski, J. M., Griffin, S. C., Ravyts, S., & Rybarczyk, B. (2018). Psychological interventions for late-life insomnia: Current and emerging science. *Current Sleep Medicine Reports*, *4*(4), 268–277. https://doi.org/10.1007/s40675-018-0129-0.

Epstein, D. R., Sidani, S., Bootzin, R. R., & Belyea, M. J. (2012). Dismantling multicomponent behavioral treatment for insomnia in older adults: A randomized controlled trial. *Sleep*, *35*(6), 797.

Finan, P. H., Goodin, B. R., & Smith, M. T. (2013). The association of sleep and pain: An update and a path forward. *Journal of Pain*, *14*(12), 1539–1552.

Geiger-Brown, J. M., Rogers, V. E., Liu, W., Ludeman, E. M., Downton, K. D., & Diaz-Abad, M. (2015). Cognitive behavioral therapy in persons with comorbid insomnia: A meta-analysis. *Sleep Medicine Reviews*, *23*, 54–67.

Haesler, E. J. (2004). Effectiveness of strategies to manage sleep in residents of aged care facilities. *JBI Reports*, *2*(4), 115–183. https://doi.org/10.1111/j.1479-6988.2004.00010.x.

Harris, J., Lack, L., Kemp, K., Wright, H., & Bootzin, R. (2012). A randomized controlled trial of intensive sleep retraining (ISR): A brief conditioning treatment for chronic insomnia. *Sleep*, *35*(1), 49–60. https://doi.org/10.5665/sleep.1584.

Harvey, A. G. (2002). A cognitive model of insomnia. *Behaviour Research and Therapy*, *40*(8), 869–893.

Harvey, A. G., Bélanger, L., Talbot, L., Eidelman, P., Beaulieu-Bonneau, S., Fortier-Brochu, É., … Morin, C. M. (2014). Comparative efficacy of behavior therapy, cognitive therapy, and cognitive behavior therapy for chronic insomnia: A randomized controlled trial. *Journal of Consulting and Clinical Psychology*, *82*(4), 670–683. https://doi.org/10.1037/a0036606.

Helbig, A. K., Stöckl, D., Heier, M., Thorand, B., Schulz, H., Peters, A., … Meisinger, C. (2017). Relationship between sleep disturbances and multimorbidity among community-dwelling men and women aged 65–93 years: Results from the KORA Age Study. *Sleep Medicine*, *33*, 151–159. https://doi.org/10.1016/j.sleep.2017.01.016.

Horsch, C. H., Lancee, J., Griffioen-Both, F., Spruit, S., Fitrianie, S., Neerincx, M. A., … Brinkman, W.-P. (2017). Mobile phone-delivered cognitive behavioral therapy for insomnia: A randomized waitlist controlled trial. *Journal of Medical Internet Research*, *19*(4). https://doi.org/10.2196/jmir.6524, e6524.

Itani, O., Jike, M., Watanabe, N., & Kaneita, Y. (2017). Short sleep duration and health outcomes: A systematic review, meta-analysis, and meta-regression. *Sleep Medicine*, *32*, 246–256. https://doi.org/10.1016/j.sleep.2016.08.006.

Kaufmann, C. N., Mojtabai, R., Hock, R. S., Thorpe, R. J., Canham, S. L., Chen, L.-Y., … Spira, A. P. (2016). Racial/ethnic differences in insomnia trajectories among U.S. older adults. *American Journal of Geriatric Psychiatry: Official Journal of the American Association for Geriatric Psychiatry*, *24*(7), 575–584. https://doi.org/10.1016/j.jagp.2016.02.049.

Kay, D. B., Dombrovski, A. Y., Buysse, D. J., Reynolds, C. F., Begley, A., & Szanto, K. (2016). Insomnia is associated with suicide attempt in middle-aged and older adults with depression. *International Psychogeriatrics/IPA*, *28*(4), 613–619. https://doi.org/10.1017/S104161021500174X.

Kay, D. B., & Dzierzewski, J. M. (2015). Sleep in the context of healthy aging and psychiatric syndromes. *Sleep Medicine Clinics*, *10*(1), 11–15.

Kim, K. W., Kang, S.-H., Yoon, I.-Y., Lee, S. D., Ju, G., Han, J. W., … Kim, T. (2017). Prevalence and clinical characteristics of insomnia and its subtypes in the Korean elderly. *Archives of Gerontology and Geriatrics*, *68*, 68–75. https://doi.org/10.1016/j.archger.2016.09.005.

Koffel, E., McCurry, S. M., Smith, M. T., & Vitiello, M. V. (2019). Improving pain and sleep in middle-aged and older adults: The promise of behavioral sleep interventions. *Pain*, *160*(3), 529–534. https://doi.org/10.1097/j.pain.0000000000001423.

Kredlow, M. A., Capozzoli, M. C., Hearon, B. A., Calkins, A. W., & Otto, M. W. (2015). The effects of physical activity on sleep: A meta-analytic review. *Journal of Behavioral Medicine*, *38*(3), 427–449. https://doi.org/10.1007/s10865-015-9617-6.

Lichstein, K. L., Riedel, B. W., Wilson, N. M., Lester, K. W., & Aguillard, R. N. (2001). Relaxation and sleep compression for late-life insomnia: A placebo-controlled trial. *Journal of Consulting and Clinical Psychology*, *69*(2), 227.

Lichstein, K. L., Scogin, F., Thomas, S. J., DiNapoli, E. A., Dillon, H. R., & McFadden, A. (2013). Telehealth cognitive behavior therapy for co-occurring insomnia and depression symptoms in older adults. *Journal of Clinical Psychology*, *69*(10), 1056–1065. https://doi.org/10.1002/jclp.22030.

Martin, J. L., Song, Y., Hughes, J., Jouldjian, S., Dzierzewski, J. M., Fung, C. H., … Alessi, C. A. (2017). A four-session sleep intervention program improves sleep for older adult day health care participants: Results of a randomized controlled trial. *Sleep*, *40*(8). https://doi.org/10.1093/sleep/zsx079.

McCrae, C. S., Curtis, A. F., Williams, J. M., Dautovich, N. D., McNamara, J. P. H., Stripling, A., … Marsiske, M. (2018). Efficacy of brief behavioral treatment for insomnia in older adults: Examination of sleep, mood, and cognitive outcomes. *Sleep Medicine*. https://doi.org/10.1016/j.sleep.2018.05.018.

McCurry, S. M., Zhu, W., Von Korff, M., Wellman, R., Morin, C. M., Thakral, M., … Vitiello, M. V. (2021). Effect of telephone cognitive behavioral therapy for insomnia in older adults with osteoarthritis pain: A randomized clinical trial. *JAMA Internal Medicine*, *181*(4), 530–538.

Miller, C. B., Espie, C. A., Epstein, D. R., Friedman, L., Morin, C. M., Pigeon, W. R., … Kyle, S. D. (2014). The evidence base of sleep restriction therapy for treating insomnia disorder. *Sleep Medicine Reviews*, *18*(5), 415–424. https://doi.org/10.1016/j.smrv.2014.01.006.

Nowicki, Z., Grabowski, K., Cubała, W. J., Nowicka-Sauer, K., Zdrojewski, T., Rutkowski, M., & Bandosz, P. (2016). Prevalence of self-reported insomnia in general population of Poland. *Psychiatria Polska*, *50*(1), 165–173. https://doi.org/10.12740/PP/58771.

Ong, J. C., & Smith, C. E. (2017). Using mindfulness for the treatment of insomnia. *Current Sleep Medicine Reports*, *3*(2), 57–65. https://doi.org/10.1007/s40675-017-0068-1.

Qaseem, A., Holty, J.-E. C., Owens, D. K., Dallas, P., Starkey, M., & Shekelle, P. (2013). Management of obstructive sleep apnea in adults: A clinical practice guideline from the American College of Physicians. *Annals of Internal Medicine*, *159*(7), 471–483.

Qaseem, A., Kansagara, D., Forciea, M. A., Cooke, M., Denberg, T. D., & Clinical Guidelines Committee of the American College of Physicians. (2016). Management of chronic insomnia disorder in adults: A clinical practice guideline from the American College of Physicians. *Annals of Internal Medicine*, *165*(2), 125–133. https://doi.org/10.7326/M15-2175.

Riedel, B. W., Lichstein, K. L., & Dwyer, W. O. (1995). Sleep compression and sleep education for older insomniacs: Self-help versus therapist guidance. *Psychology and Aging*, *10*(1), 54.

Riemann, D., Baglioni, C., Bassetti, C., Bjorvatn, B., Dolenc Groselj, L., Ellis, J. G., Espie, C. A., Garcia-Borreguero, D., Gjerstad, M., Gonçalves, M., Hertenstein, E., Jansson-Fröjmark, M., Jennum, P. J., Leger, D., Nissen, C., Parrino, L., Paunio, T., Pevernagie, D., Verbraecken, J., … Spiegelhalder, K. (2017). European guideline for the diagnosis and treatment of insomnia. *Journal of Sleep Research*, *26*(6), 675–700. https://doi.org/10.1111/jsr.12594.

Ritterband, L. M., Thorndike, F. P., Ingersoll, K. S., Lord, H. R., Gonder-Frederick, L., Frederick, C., … Morin, C. M. (2017). Effect of a web-based cognitive behavior therapy for insomnia intervention with 1-year follow-up: A randomized clinical trial. *JAMA Psychiatry*, *74*(1), 68–75. https://doi.org/10.1001/jamapsychiatry.2016.3249.

Rybarczyk, B., Stepanski, E., Fogg, L., Lopez, M., Barry, P., & Davis, A. (2005). A placebo-controlled test of cognitive-behavioral therapy for comorbid insomnia in older adults. *Journal of Consulting and Clinical Psychology*, *73*(6), 1164–1174. https://doi.org/10.1037/0022-006X.73.6.1164.

Sadler, P., McLaren, S., Klein, B., Harvey, J., & Jenkins, M. (2018). Cognitive behaviour therapy for older adults with insomnia and depression: A randomized controlled trial in community mental health services. *Sleep*. https://doi.org/10.1093/sleep/zsy104.

Sateia, M. J. (2014). International classification of sleep disorders-third edition: Highlights and modifications. *Chest*, *146*(5), 1387–1394. https://doi.org/10.1378/chest.14-0970.

Scogin, F., Lichstein, K., DiNapoli, E. A., Woosley, J., Thomas, S. J., LaRocca, M. A., … Geyer, J. D. (2018). Effects of integrated telehealth-delivered cognitive-behavioral therapy for depression and insomnia in rural older adults. *Journal of Psychotherapy Integration*, *28*(3), 292–309. https://doi.org/10.1037/int0000121.

Sidani, S., Epstein, D. R., Bootzin, R. R., Moritz, P., & Miranda, J. (2009). Assessment of preferences for treatment: Validation of a measure. *Research in Nursing & Health*, *32*(4), 419–431. https://doi.org/10.1002/nur.20329.

Spielman, A. J., Caruso, L. S., & Glovinsky, P. B. (1987). A behavioral perspective on insomnia treatment. *Psychiatric Clinics of North America*, *10*(4), 541–553.

Tanaka, M., Kusaga, M., Nyamathi, A. M., & Tanaka, K. (2019). Effects of brief cognitive behavioral therapy for insomnia on improving depression among community-dwelling older adults: A randomized controlled comparative study. *Worldviews on Evidence-Based Nursing*, *16*(1), 78–86. https://doi.org/10.1111/wvn.12342.

Taylor, D. J., Mallory, L. J., Lichstein, K. L., Durrence, H. H., Riedel, B. W., & Bush, A. J. (2007). Comorbidity of chronic insomnia with medical problems. *Sleep*, *30*(2), 213–218. https://doi.org/10.1093/sleep/30.2.213.

Tyagi, S., Resnick, N. M., Perera, S., Monk, T. H., Hall, M. H., & Buysse, D. J. (2014). Behavioral treatment of chronic insomnia in older adults: Does nocturia matter? *Sleep*, *37*(4), 681–687. https://doi.org/10.5665/sleep.3568.

van der Zweerde, T., Bisdounis, L., Kyle, S. D., Lancee, J., & van Straten, A. (2019). Cognitive behavioral therapy for insomnia: A meta-analysis of long-term effects in controlled studies. *Sleep Medicine Reviews*, *48*. https://doi.org/10.1016/j.smrv.2019.08.002, 101208.

Vaughan, C. P., & Bliwise, D. L. (2018). Sleep and nocturia in older adults. *Sleep Medicine Clinics*, *13*(1), 107–116.

Vaz Fragoso, C. A., Miller, M. E., King, A. C., Kritchevsky, S. B., Liu, C. K., Myers, V. H., … Lifestyle Interventions and Independence for Elders Study Group. (2015). Effect of structured physical activity on sleep-wake behaviors in sedentary elderly adults with mobility limitations. *Journal of the American Geriatrics Society*, *63*(7), 1381–1390. https://doi.org/10.1111/jgs.13509.

Wang, Y.-M., Chen, H.-G., Song, M., Xu, S.-J., Yu, L.-L., Wang, L., … Lu, L. (2016). Prevalence of insomnia and its risk factors in older individuals: A community-based study in four cities of Hebei Province, China. *Sleep Medicine*, *19*, 116–122. https://doi.org/10.1016/j.sleep.2015.10.018.

Wennberg, A. M., Canham, S. L., Smith, M. T., & Spira, A. P. (2013). Optimizing sleep in older adults: Treating insomnia. *Maturitas*, *76*(3), 247–252. https://doi.org/10.1016/j.maturitas.2013.05.007.

Wu, J. Q., Appleman, E. R., Salazar, R. D., & Ong, J. C. (2015). Cognitive behavioral therapy for insomnia comorbid with psychiatric and medical conditions: A meta-analysis. *JAMA Internal Medicine*, *175*(9), 1461–1472. https://doi.org/10.1001/jamainternmed.2015.3006.

Ye, Y., Chen, N., Chen, J., Liu, J., Lin, L., Liu, Y., … Jiang, X. (2016). Internet-based cognitive–behavioural therapy for insomnia (ICBT-i): A meta-analysis of randomised controlled trials. *BMJ Open*, *6*(11). https://doi.org/10.1136/bmjopen-2015-010707, e010707.

Ye, L., & Richards, K. C. (2018). Sleep and long-term care. *Sleep Medicine Clinics*, *13*(1), 117–125. https://doi.org/10.1016/j.jsmc.2017.09.011.

Zachariae, R., Lyby, M. S., Ritterband, L. M., & O'Toole, M. S. (2016). Efficacy of internet-delivered cognitive-behavioral therapy for insomnia—A systematic review and meta-analysis of randomized controlled trials. *Sleep Medicine Reviews*, *30*, 1–10.

Zhang, Z., Xiao, X., Ma, W., & Li, J. (2020). Napping in older adults: A review of current literature. *Current Sleep Medicine Reports*, *6*(3), 129–135. https://doi.org/10.1007/s40675-020-00183-x.

Chapter 27

Brief cognitive behavioral therapy for insomnia in community-dwelling older adults

Mika Tanaka[a] and Katsutoshi Tanaka[b]

[a]*Department of Nursing, School of Nursing, Kitasato University, Sagamihara, Kanagawa, Japan* [b]*Department of Occupational Mental Health, Graduate School of Medical Sciences, Kitasato University, Sagamihara, Kanagawa, Japan*

Abbreviations

CBT-I cognitive behavioral therapy for insomnia
DBAS dysfunctional beliefs and attitudes about sleep scale
PSQI Pittsburgh Sleep Quality Index
QOL quality of life
RCT randomized controlled trial

Introduction

Prevalence and impact of insomnia among older adults

Chronic insomnia is common in later life. Prevalence of insomnia increases with age and is higher in older adults compared with younger adults (Blay, Andreoli, & Gastal, 2008). The overall prevalence of insomnia symptoms has been estimated to be 30%–48% in older adults, and that of diagnosed insomnia ranges from 12% to 20% (Patel, Steinberg, & Patel, 2018). Insomnia leads to various health problems in older adults.

Insomnia impairs cognitive, social, and emotional functions during daytime (Léger et al., 2010) and is a risk factor for suicide, depression, and alcohol and substance abuse (Taylor, Lichstein, & Durrence, 2003). Sleep disturbance has shown to be an important risk factor for depression in older adults' community (Cole & Dendukuri, 2003). One cohort study (Perlis et al., 2006) showed that elderly participants with persistent insomnia were more likely to develop new onset depression. Additionally, insomnia increases the risk for suicidal tendencies (Bishop, Simons, King, & Pigeon, 2016; Pigeon, Pinquart, & Conner, 2012). Cardiovascular diseases and diabetes have also shown to be associated with insomnia (Javaheri & Redline, 2017). A metaanalysis of studies investigating insomnia symptoms and cardiovascular heart disease showed that the risk ratio of insomnia symptoms for heart disease ranged from 1.47 to 3.90, after adjusting for age and other cardiovascular risk factors (Schwartz et al., 1999). Taken together, insomnia has a substantial impact on older adults and society.

Risk factors for insomnia among older adults

Risk factors for insomnia in older adults are different from those in young adults. Identifying age-specific risk factors is essential for developing effective interventions for insomnia. Physiological changes of sleep with age are a risk of insomnia in older adults. The proportion of slow-wave sleep, called stages 3 and 4 sleep, significantly decreases with age, whereas the proportion of stage 2 and rapid eye movement sleep significantly increases with age. Furthermore, total sleep time becomes shorter and shifts forward with age. These physiological changes result in significantly decreased sleep efficiency as well as increased sleep latency and wake after sleep onset, which contribute to the high prevalence of insomnia in older adults (Ohayon, Carskadon, Guilleminault, & Vitiello, 2004).

Handbook of Lifespan Cognitive Behavioral Therapy. https://doi.org/10.1016/B978-0-323-85757-4.00026-2

However, physiological changes of sleep are not the only factor that increases the risk of insomnia. Spielman, Caruso, and Glovinsky (1987) proposed a three-factor model to understand the etiology and persistence of insomnia called the 3-P model, which includes predisposing, precipitating, and perpetuating factors. The factors that cause chronic insomnia in older adults according to Spielman's 3-P model are listed in Table 1. In this theoretical model, predisposing factors do not cause insomnia but increase the likelihood for onset and occurrence. Precipitating factors are factors that directly lead to the onset of acute insomnia, such as stressful personal events and decline in health. Insomnia is maintained by perpetuating factors, which include behavioral changes, such as changing sleep schedules to compensate for poor sleep, or cognitive changes, such as anxiety before sleep onset about spending a sleepless night. Perpetuating factors result in the persistence and exacerbation of insomnia.

Cognitive behavioral therapy for insomnia and brief cognitive behavioral therapy for insomnia

Cognitive behavioral therapy for insomnia (CBT-I) is a safe nonpharmacological intervention for insomnia, which aims to improve sleep by changing poor sleep habits and challenging negative thoughts, attitudes, and beliefs about sleep. CBT-I consists of multiple components, such as sleep hygiene education, stimulus control, relaxation, sleep restriction/compression, and cognitive therapy. The efficacy of CBT-I is well documented and offers several advantages over hypnotic medication (Montgomery & Dennis, 2004; Williams, Roth, Vatthauer, & McCrae, 2013). Medication is typically indicated only in the short term because it can still lead to tolerance and physical and psychological dependence. In contrast, CBT-I is safer and its effects last longer, which makes it suitable for older adults with chronic (Mitchell, Gehrman, Perlis, & Umscheid, 2012; Sivertsen et al., 2006).

One of the downsides of standard CBT-I is that it requires 6–10 in-person sessions with a trained therapist, which can be challenging to access for older adults. Because of the financial constraints and lack of psychological resources needed for standard CBT-I, shorter forms of therapy that include the core components of standard CBT-I have become available. Brief CBT-I is an abbreviated version of CBT-I with an emphasis on the behavioral components. Brief CBT-I typically consists of education on sleep regulation, factors that influence sleep and behaviors that promote or interfere with sleep, along with a semitailored behavioral prescription of sleep restriction therapy based on information derived from a pretreatment sleep diary. Brief CBT-I can include elements of relaxation or cognitive restructuring as well. Because of the limited number of sessions, the content of brief CBT-I needs to be selected carefully. In particular, for brief CBT-I targeting older adults, the content should reflect their lifestyle and their beliefs and presumptions regarding sleep. Furthermore, ease of

TABLE 1 Factors associated with insomnia in older adults.

Predisposing factors	
	Female
	Divorce, separation, and bereavement
	Lower levels of education or income
	Smoking, alcohol use, and reduced physical activity
	Anxious personality type
Precipitating factors	
	Stressful life events, such as caregiving, death of a family member or friend, and new medical problems
	Lifestyle changes, such as hospitalization and retirement
	Respiratory symptoms, physical disability, and poor perceived health
	Medications, such as beta blockers, nonsteroidal antiinflammatory drugs, and glucocorticoids
Perpetuating factors	
	Behavioral changes that occur as a result of acute insomnia, such as spending excessive time in bed, frequently napping, and changes in daytime activities
	Cognitive changes that occur as a result of acute insomnia, such as increased anxiety before sleep onset about spending a sleepless night

comprehension is vital; guidance provided during brief CBT-I should be simple and acceptable given that older adults prefer to not change their lifestyle. The numbers of sessions and therapists should also be realistic based on the available financial and human resources. This chapter considers the effectiveness of brief CBT-I for older adults based on randomized controlled trial (RCT) studies and introduces the procedure and content of brief CBT-I.

Practice and procedures

Efficacy of brief CBT-I for older adults

We defined brief CBT-I as an intervention that comprises approximately half the number of sessions of standard CBT-I (one to five in-person sessions) or one that is not delivered in person, such as via video or booklet. Table 2 summarizes the intervention procedures and results of RCTs that report the effectiveness of brief CBT-I in older adults. Most studies were conducted in older adults diagnosed with chronic or primary insomnia, and interventions of brief CBT-I were low intensity, which improved sleep parameters. Additionally, several studies reported an amelioration in depression status and quality of life (QOL).

For self-reported sleep quality, effect sizes (Cohen's *d*) of Pittsburgh Sleep Quality Index (PSQI) at 1–6 months after intervention in the four studies were 0.69, 1.10, 0.56, and 0.59, which can be interpreted as medium to large effect sizes (Alessi et al., 2016; Buysse et al., 2011; Morgan et al., 2012; Tanaka et al., 2018). Effect sizes of self-recorded variables of daily sleep and sleep onset latency were 0.37 and 0.96 at 1 and 6 months after intervention, respectively, and those of wake after sleep onset were 0.41 and 0.59 at 1 and 6 months after intervention, respectively (Alessi et al., 2016; Buysse et al., 2011). Objective sleep parameters evaluated using actigraphy and polysomnography were also improved following brief CBT-I. However, effect sizes of objective parameters were relatively small compared with subjective parameters of sleep daily. Additionally, two studies reported an improvement in depression status (Buysse et al., 2011; Tanaka et al., 2019).

Frequency and form of brief CBT-I

The frequency and form of brief CBT-I are important factors. In the reviewed studies, number of sessions ranged from one to five, and intervention duration ranged from 2 to 8 weeks. In-person sessions were by individual or group. Average session time ranged from 30 to 60 min. All brief CBT-I sessions were conducted by nonclinicians, such as trained nurses, university students, or individuals with a master's degree.

Videos and booklets were also used to provide brief CBT-I and involved watching a 15-min video twice or sending six series of booklets to the patient's home. Individuals could also ask questions over the phone regarding any of the advice provided in the booklets. All brief CBT-I sessions appeared practical and applicable as the primary treatment or prevention activity for insomnia in the elderly community setting.

Recommended content for brief CBT-I

The content of brief CBT-I should be easily understood by older adults and should consider the physical, psychological, and social features of older adults. Based on the content of brief CBT-I administered in the studies reviewed, we have summarized the most important points of brief CBT-I for older adults (Fig. 1).

1. Sleep hygiene education

Sleep hygiene education aims to increase behaviors and environmental conditions that improve sleep quality and quantity and to decrease habits that interfere with sleep. Education includes understanding sleep changes that occur with age, mechanism of sleep regulation, and healthy sleep practices. Sleep hygiene education is a basic and essential component of brief CBT-I, and the following points should be emphasized for sleep hygiene education in older adults.

(1) Changes in sleep with age and mechanisms of sleep regulation

Many physiological changes occur with age; significant changes in sleep time and sleep architecture occur throughout the lifespan. Total sleep time decreases considerably from 10 to 14 h per night during childhood, 6.5 to 8.5 h per night during early adulthood, and 5 to 7 h per night during late adulthood. Moreover, sleep becomes lighter and more fragmented with age, and the deep nonrapid eye movement stage decreases. These changes in sleep architecture and the natural shortening of total sleep time may not be accepted by some older adults, which can result in increased anxiety that can cause or worsen insomnia.

TABLE 2 Main randomized controlled trial reporting the effectiveness of brief CBT-I in the elderly.

Author (year)	Mean age/ female (%)	Participants/ diagnostic criteria	Design/ follow-up	Intervention form and duration/ session time	Intervention details	CBT-I component	Sample size (subgroup)	Delivery medium	Control intervention	Outcome measure	Results reported and effect sizes
Alessi et al. (2016)	72.2/ 3.1	Chronic insomnia disorder/ ICSD-2	2 parallel arms (CBT-I, CON)/Post treatment, 6 and 12 months	5 IP (individual or small group) + brief telephone check-in, 5–6 wk./1 IP = 60 min	(1) Structured and manual based intervention with hard copy provided to subjects and delivered individually or in a small group of three to five subjects. (2) Sessions included sleep restriction and stimulus control, sleep hygiene and adjustment of sleep restriction parameters, cognitive therapy and adjustment of sleep parameters, review of prior material and adjustment of sleep restriction parameters and relapse prevention.	Sleep restriction, stimulus control, sleep hygiene	159 (106/ 53)	Master's level sleep coaches	Information control/ receiving general sleep education in a group at the same frequency to CBT-I	Insomnia: PSQI, ISI, sleep diary, actigraphy Depression: PHQ-9 QOL: SF-12v2	Variables of sleep diary, PSQI and ISI improved at 6 and 12 months. Variables of actigraphy, PHQ-9 and SF-12v2 unchanged. SOL-D: $d = 0.37$, NS (6 and 12 months), WASO-D: $d = 0.41$, 0.34, ES-D: $d = 0.43$, 0.34, PSQI: $d = 0.69$, 0.63, ISI: $d = 0.74$, 0.54
Dzierzewski et al. (2019)	72.2/ 3.1	Chronic insomnia disorder/ ICSD-2	2 parallel arms (CBT-I, CON)/Post treatment, 6 and 12 months	5 IP (individual or small group) + brief telephone check-in, 5–6 wk./1 IP = 60 min	The same intervention procedure as Alessi et al. (2016)	Sleep restriction, stimulus control, sleep hygiene	159 (106/ 53)	Master's level sleep coaches	Information control/ receiving general sleep education in a group at the same frequency to CBT-I	Sleep discrepancy (mismatch between subjective and objective wake time estimates) and mechanisms: PSQI, sleep diary, actigraphy	Sleep discrepancies improved at 3, 6 and 12 months. Improved PSQI at each study assessment preceded and predicted reduced discrepancy at the next study assessment. Sleep time discrepancy: $d = 0.49$, 0.45, 0.37 (3, 6 and 12 months, respectively). Wake discrepancy: $d = 0.71$, 0.56, 0.40

Chan et al. (2017)	69.5/68	Insomnia/ researcher's criteria	2 parallel arms (CBT-I, CON)/ 3 months	4 IP (individual), 4 wk./1 IP = NR	NR	Sleep restriction, stimulus control, sleep hygiene, relaxation	62 (32/30)	NR	Attention control/self-monitoring using a sleep diary and actigraph and meeting with therapist without talking about sleep	Insomnia: sleep diary, actigraphy	Variables of sleep diary and actigraphy were changed. Effect sizes NR
Buysse et al. (2011)	71.7/70	Primary insomnia/ DSM-IV-TR, ICSD-2	2 parallel arms (CBT-I, CON)/ 4 weeks	1 IP(individual) + 3 telephone follow-up, 6 wk./1 IP = 45–60 min 1 tel. = 20–30 min	(1) Sleep education emphasizing reduction of time in bed, getting up at the same time, not going to bed unless sleepy and not staying in bed unless asleep. (2) Discussion of homeostatic and circadian mechanisms of human sleep 3) Setting a prescribed sleep schedule	Sleep restriction, stimulus control, sleep hygiene	79 (39/40)	Master's level nurse	Information control/reading three publications about sleep and insomnia, not including individualized behavioral instructions	Depression: HRSD, HARS Insomnia: PSQI, ESS, sleep diary, actigraphy, polysomnography QOL: SF-36	HRSD, PSQI, SF-36, variables of sleep diary, PSQI and variables of actigraphy improved. HARS, ESS and variables of polysomnography unchanged. HRSD: $d=0.64$, PSQI: $d=1.10$, SF-36: $d=0.57$, SOL-D: $d=0.96$, WASO-D: $d=0.59$, SE-D: $d=0.80$, SOL-A: $d=0.59$, WASO-A: $d=0.69$, TST-A: $d=0.74$, SE-A: $d=0.64$
Wilckens, Hall, Nebes, Monk, and Buysse (2016)	71.7/70	Insomnia/ researcher's criteria	2 parallel arms (CBT-I, CON)/ 4 weeks	1 IP(individual) + 3 telephone follow-up, 6 wk./1 IP = 45–60 min 1 tel. = 20–30 min	The same intervention procedure as Buysse et al. (2011)	Sleep restriction, stimulus control, sleep hygiene	79 (39/40)	Master's level nurse	Information control/reading three publications about sleep and insomnia, not including individualized behavioral instructions	Depression: HRSD Insomnia: PSQI, in-home PSG Episodic memory: Logical memory test Abstract reasoning: Test of nonverbal intelligence III Working memory: Letter-number sequencing	Lower baseline WASO-P was associated with higher delayed recall performance. Only WASO-P improved following intervention. Effect sizes NR

Continued

TABLE 2 Main randomized controlled trial reporting the effectiveness of brief CBT-I in the elderly—cont'd

Author (year)	Mean age/ female (%)	Participants/ diagnostic criteria	Design/ follow-up	Intervention form and duration/ session time	Intervention details	CBT-I component	Sample size (subgroup)	Delivery medium	Control intervention	Outcome measure	Results reported and effect sizes
Tanaka, Kusaga, Nyamathi, and Tanaka (2019)	69.7/ 73.8	Community-dwelling older adults/ NA	2 parallel arms (CBT-I, CON)/ 3 months	2 IP (individual, group) + 2 telephone follow-up, 4 wk./1 IP = 45–60 min 1 tel. = 20 min	(1) Group session covered sleep hygiene education, cognitive restructuring and stimulus control. (2) The individual session included cognitive restructuring, sleep restriction and relaxation. Counseling was provided based on sleep diaries and actigraphy results to those who scored highly on the DBAS. Individuals whose sleep efficiency was lower than 85% were encouraged to delay bedtime	Sleep restriction, stimulus control, sleep hygiene, cognitive restructuring, relaxation	46 (23/23)	Nurse	Wait list control/ answering sleep questionnaire at the same time as intervention group	Depression: GDS Insomnia: PSQI	GDS improved. PSQI unchanged. GDS: $d = 0.34$
Tanaka et al. (2018)	76.0/ 75.9	Community-dwelling older adults/ NA	2 parallel arms (CBT-I, CON)/ 3 months	2 IP (individual, group) + 3 mail follow-up, 8 wk./1 IP = 30–60 min	The same intervention procedure as Tanaka et al. (2019)	Sleep restriction, stimulus control, sleep hygiene, cognitive restructuring	83 (42/41)	Nurse	Wait list control/ answering sleep questionnaire at the same time as intervention group	Insomnia: PSQI, ISI	PSQI and ISI improved. PSQI: $d = 0.56$, ISI: $d = 0.77$

Morgan, Gregory, Tomeny, David, and Gascoigne (2012)	66.7/ 66.3	Primary insomnia + chronic disease/ DSM-IV or greater than 5 on PSQI	2 parallel arms (CBT-I, CON)/3 and 6 months	6 self-help booklets + telephone line, 6 wk.	(1) Six self-help booklets sent weekly, which were designed as a structured psychoeducational program that addressed important components of CBT-I. (2) When requested, telephone advisers answered participants' questions and offered encouragement	Sleep restriction, stimulus control, sleep hygiene	193 (98/95)	Booklets and telephone adviser trained for 3 days	Information control/reading a summary of sleep hygiene	Insomnia: PSQI, ISI, ESS Fatigue: FSS	PSQI, ISI and SE improved. FSS unchanged. PSQI: $d=0.59$, 0.57, ISI: $d=0.51$, 0.56, SE: $d=0.62$, 0.63
Riedel, Lichstein, and Dwyer (1995)	67.4/ NR	Primary insomnia + noninsomnia/ researcher's criteria	3 parallel arms (CBT-I1, CBT-I2 and CON)/ Post treatment and 2 months	CBT-I1: 2 Video CBT-I2: 2 Video + 4 IP, 2–3 wk./ Video = 15 min 1 IP = NR	(1) CBT-I1: Video shown describing natural age-correlated changes in sleep, such as reduced deep sleep and more nighttime awakenings, and seniors recommended to adopt age-appropriate sleep goals. Participants watched the video twice over 2 weeks. (2) CBT-I2: In addition to the CBT-I1 intervention, participants in this group received therapist-guided sleep compression	CBT-I1: Sleep restriction, sleep hygiene, cognitive restructuring CBT-I2: Sleep restriction, sleep hygiene, cognitive restructuring, sleep compression	125 (50/50/ 25)	Video and therapist of graduate students	Wait list control/ answering sleep questionnaire at the same time as intervention group	Insomnia: Sleep diary, SSS	Variables of sleep diary and SSS improved. Effect sizes NR

ICSD, International Classification of Sleep Disorders; DSM, Diagnostic and Statistical Manual of Mental Disorders; CBT-I, cognitive behavioral therapy for insomnia intervention group; CON, control group; IP, in-person; NR, not reported; PSQI, Pittsburgh Sleep Quality Index; ISI, Insomnia Severity Index; ESS, Epworth Sleepiness Scale; SSS, Stanford Sleepiness Scale; PHQ-9, Patient Health Questionnaire; HRSD, Hamilton Rating Scale for Depression HARS, Hamilton Anxiety Rating Scale; GDS, Geriatric Depression Scale; BDI, Beck Depression Inventory II; SF-12v2, 12-item short-form survey version 2; SF-36, 36-Item short-form health survey; FSS, Fatigue Severity Scale; SOL, sleep onset latency; WASO, wake after sleep onset; SE, sleep efficiency; TST, total sleep time; D, according to sleep diary; A, according to actigraphy; p, according to polysomnography; NS, not significant; d, Cohen's *d*.

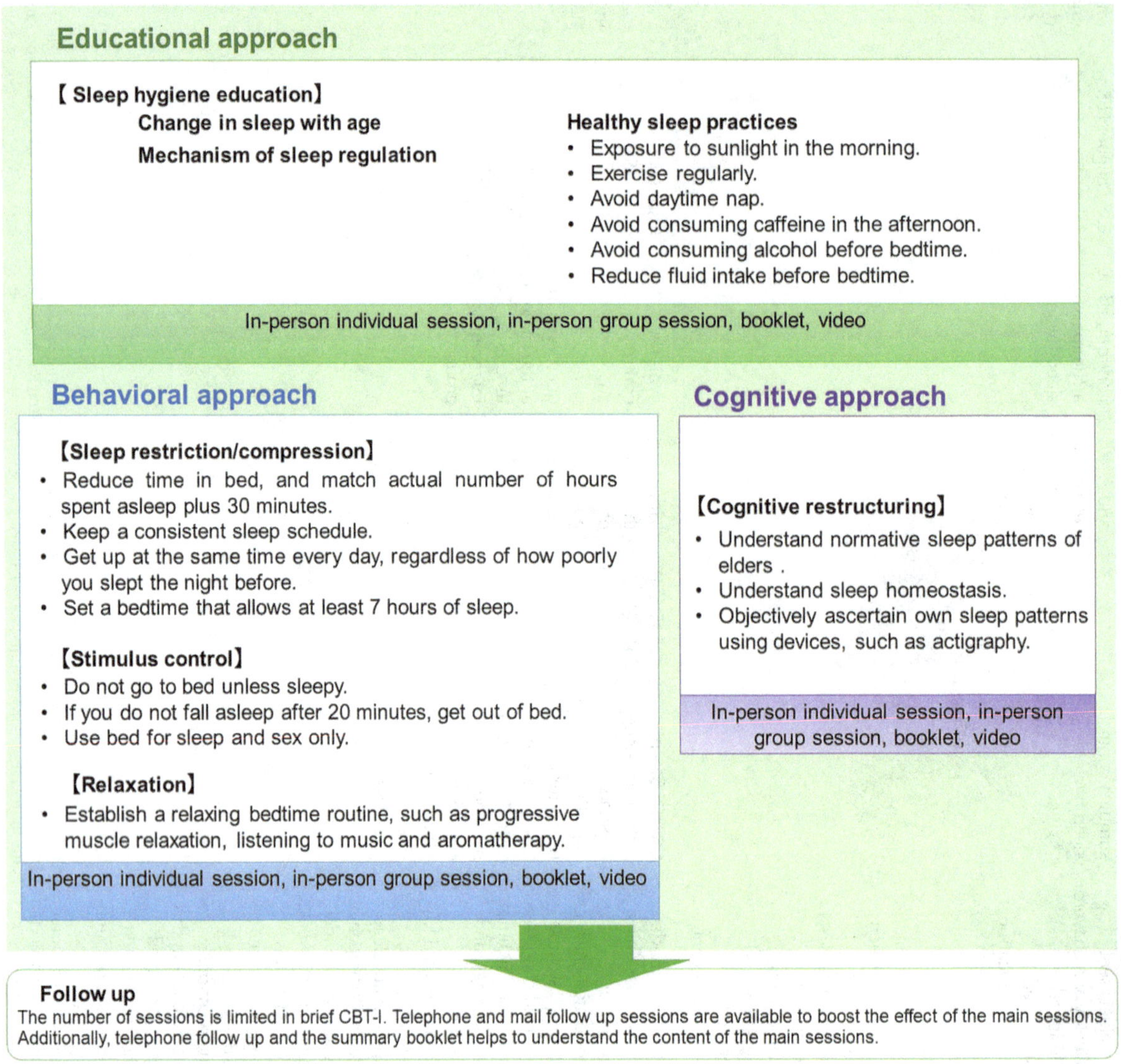

FIG. 1 Three approaches of brief CBT-I.

It is important to understand the mechanism of sleep regulation. Sleep is regulated by two major physiological systems: the circadian and homeostatic drives. The circadian drive underpins the basic principle of sleep regulation. Circadian rhythms are 24-h cycles that are part of the body's biological clock, which underlies the sleep-wake cycle. One of the most important regulators of circadian rhythms is sunlight. The homeostatic drive promotes the compensatory increase in the intensity and duration of sleep in the event of a sleep deficit and reduces sleep propensity in incidents of excess sleep. Thus, it enables an individual who could not sleep well on one night to sleep more the following night. The knowledge of these systems decreases an individual's anxieties around insomnia. An individual who tries to sleep when their body has accumulated sufficient homeostatic drive, they are able to sleep well. This is the rationale behind sleep restriction, and understanding these drives enhances adherence to CBT-I.

(2) Morning sunlight and exercise

Exposure to morning sunlight is important for maintaining circadian rhythms. Decreased physical functions and limitation of social conditions, such as institutionalization, can deprive an individual's exposure to sufficient sunlight. Moreover, higher prevalence of cataracts among older adults should be considered. Older adults conserve production of melatonin, and bright light can improve insomnia in older adults (Gammack, 2008; Mishima, Okawa, Shimizu, & Hishikawa, 2001). Outdoor physical activity and exercise in the morning are also recommended to maintain a daily life rhythm.

(3) Naps

The recommendation to avoid naps during the day unless absolutely necessary is important for older adults. Napping for more than 30 min can disturb sleep at night. Older adults tend to nap in the afternoon and evening because there are few schedule limitations. Frequent napping in older adults may be to compensate for poor sleep at night or to gain vigor. Understanding sleep homeostasis can help individuals to avoid napping.

(4) Caffeine

Caffeine is one of the substances that disturbs sleep onset and shortens sleep time (Curless, French, James, & Wynne, 1993; Roehrs & Roth, 2008; Salín-Pascual, Valencia-Flores, Campos, Castaño, & Shiromani, 2006). Maximum blood concentration of caffeine usually occurs within 30–60 min of consumption, and the biological half-life is 3–5 h. These timings can vary widely between individuals, and the capacity of discharge declines with age. Therefore, it is important to avoid caffeine intake in the afternoon.

2. Sleep restriction

Sleep restriction is a core part of CBT-I and has been found to be the most effective way to improve sleep quality in older adults (Lichstein, Riedel, Wilson, Lester, & Aguillard, 2001). It involves restricting time in bed to the hours of actual sleep until sleep efficiency improves. In standard CBT-I, if after 10 days of sleep restriction, sleep efficiency remains lower than 85%, sleep time in bed is restricted by 15–30 min until sleep efficiency improves. Time in bed is gradually increased by 15–30 min when sleep time exceeds 85% of the total time in bed.

In brief CBT-I, sleep restriction therapy administration depends on the frequency of intervention. The study by Alessi et al. (2016) and Dzierzewski et al. (2019) used five 1-h sessions over 6 weeks, which is similar to the standard method for older adults with chronic insomnia. Sleep coaches applied sleep restriction to participants by adjusting individual sleep restriction parameters. Specifically, the five sessions included CBT-I components of not only sleep restriction but also stimulus control, sleep hygiene, and cognitive therapy. Throughout the five sessions, a sleep coach offered guidance on adjusting sleep restriction parameters. The last session involved relapse prevention (Table 3).

Buysee's interventions (Buysse et al., 2011) consisted of a 45- to 60-min individual intervention session, followed by a 30-min telephone session 2 weeks later and two 20-min telephone follow-up sessions 1–3 weeks later. In this brief CBT-I,

TABLE 3 Contents of brief CBT-I in main randomized controlled trials.

Content included in the five sessions of brief CBT-I (Alessi et al., 2016)		
1 week	Session 1	Sleep restriction and stimulus control
2 weeks	Session 2	Sleep hygiene and adjustment of sleep restriction parameters
3 weeks	Session 3	Cognitive therapy and adjustment of sleep restriction parameters
4 weeks	Session 4	Review of prior material and adjustment of sleep restriction parameters
5 weeks	Session 5	Relapse prevention
Content included in the two session and two follow-up sessions of brief CBT-I (Buysse et al., 2011)		
1 week	Session 1	Sleep hygiene education, discussion and adjustment of sleep restriction parameters
3 weeks	Session 2 (telephone)	Questions about sleep, day time function and adherence to sleep schedule
4 weeks	Telephone follow-up	Questions about sleep, day time function and adherence to sleep schedule
6 weeks	Telephone follow-up	Questions about sleep, day time function and adherence to sleep schedule
Content included in two sessions and two mail follow-up sessions of brief CBT-I (Tanaka et al., 2019)		
1 week	Session 1	Sleep hygiene, stimulus control and adjustment of sleep restriction parameters
1 week	Session 2	Cognitive therapy and adjustment of sleep restriction parameters
2 weeks	Telephone follow-up	Encourage adherence to sleep schedule
4 weeks	Telephone follow-up	Encourage adherence to sleep schedule

sleep hygiene education was provided by interactive discussions about homeostatic and circadian mechanisms of sleep, which are basic principles of sleep restriction. During sleep education, the rationale behind sleep restriction and stimulus control techniques were explained, and tips such as "reducing time in bed," "getting up at the same time every day," and "avoiding napping" were provided. Participants adjusted their parameters of sleep restriction by consulting with nurses (Table 3). Time in bed was limited to the average self-reported sleep time plus 30min, with a minimum of 6h.

Tanaka's intervention (Tanaka et al., 2018, 2019) has the same level of intensity as that of Buysse's and consists of one 60-min in-person group session, followed by a 45-min individual session and two 20-min telephone follow-up sessions. Subjects of our study were community-dwelling elderly participants, which included those who were not diagnosed with insomnia. We focused more on the cognitive than behavioral aspect of the intervention so that individual sleep parameters of each session were adjusted only for participants who had low sleep efficiency based on actigraphy and sleep diaries as evaluated by trained nurses. Participants whose sleep efficiency was lower than 85% were encouraged to delay bedtime (Table 3). In this intervention, the principles of sleep restriction may contradict the beliefs of those who worry about sleep duration. Thus, sleep homeostasis should be emphasized during sleep hygiene education to enhance adherence.

The booklet with phone discussion and video interventions only provided an explanation of sleep restriction methods (Morgan et al., 2012; Riedel et al., 1995). In the intervention using the video, there were two intervention groups: watching the video only and watching the video plus therapist guidance. Watching the video only was effective, but additional therapist guidance regarding sleep restriction showed an enhanced effect.

3. Stimulus control

Stimulus control is based on learning theory. Sleep difficulties can be caused by anxieties and worries provoked by the bed and bedroom because of unsuccessful attempts at falling asleep. Stimulus control aims to weaken the association between the bed and bedroom and the place where the individual feels anxious about not being able to fall asleep.

Stimulus control in brief CBT-I involves the use of instructions for going to bed. This includes tips, such as "not going to bed unless sleepy," "leaving the bed if unable to sleep within 15–20min," and "using the bed only for sleep and sex." These tips are combined with sleep hygiene education during the in-person session, booklet, or video.

4. Cognitive restructuring

Cognitive restructuring aims to reduce dysfunctional sleep-related thoughts and change misconceptions associated with sleep and insomnia. Patients with insomnia frequently express attitudes and beliefs that facilitate worry and anxiety about their ability to get sufficient sleep. These anxieties and worries contribute to the arousal that interferes with sleep, which promotes further anxieties and worries and produces emotional distress. This vicious cycle of thought and sleep increases the severity of insomnia.

Dysfunctional sleep-related thoughts refer to sleep-related cognition, such as faulty beliefs, worry, and attention bias, which play important mediating roles in perpetuating or even exacerbating insomnia (Morin, Blais, & Savard, 2002). For instance, some insomniacs tend to have unrealistic expectations about their sleep requirements and worry excessively when their subjective sleep does not meet their expectations. Furthermore, others may have anxieties about the potential consequences of insomnia on their daytime functioning and will selectively attend to any evidence of such consequences. Older adults commonly hold these beliefs; therefore, these are used in CBT-I as targets of cognitive restructuring.

In brief CBT-I, cognitive reconstructing is focused on identifying dysfunctional sleep-related thoughts and replacing them with more adaptive beliefs and attitudes. Sleep education assists in cognitive reconstructing by emphasizing natural changes in sleep architecture and shortened sleep time with age. Additionally, understanding sleep homeostasis can be helpful in reducing the fear of acute insomnia.

Subjective and objective estimates of sleep are often discordant in individuals with insomnia, who typically underreport sleep time and overreport wake time (Manconi et al., 2010). This misperception of sleep state is often observed in older adults (Kay, Buysse, Germain, Hall, & Monk, 2015). CBT-I has shown to reduce this sleep/wake discrepancy (Dzierzewski et al., 2019). Correcting misperception of sleep status is important for improving insomnia in older adults.

Our study (Tanaka et al., 2019) used objective sleep data (i.e., an actigraph) to amend an individual's perception of their sleep status. Moreover, ascertaining the relationship between objective sleep data and daytime performance in work or daily life can assist in replacing dysfunctional beliefs and attitudes of sleep. For example, objective sleep data can be used to confirm the accuracy of the perception of being asleep and awake and to determine whether an individual's work performance was poor on a day when they felt they did not sleep well. In the short version of CBT-I, participants with strong dysfunctional sleep-related beliefs and attitudes can be screened using the dysfunctional beliefs and attitudes about sleep scale (DBAS), which will allow intense and effective intervention in these individuals.

Mini-dictionary of terms

- ***CBT-I***. CBT-I is a method of applying CBT theory for insomnia treatment. CBT-I includes the following components: cognitive therapy, stimulus control, sleep restriction, sleep hygiene, and relaxation. CBT-I may be delivered face-to-face, digitally, or using a combination of both.
- ***Brief CBT-I***. Shorter and low intensive form of therapy, which involves the core techniques of standard CBT-I. There is no standard definition of brief CBT-I, but we have defined it to have approximately half the intensity (one to four sessions) of standard CBT-I, which is delivered by face-to-face sessions or substitute mediums, such as videos, booklets, and the internet.
- ***Sleep hygiene education***. General knowledge of sleep relating to environmental factors, physiological factors, behaviors, and habits that promote/inhibit sleep quality.
- ***Sleep restriction***. Behavioral instructions to limit time in bed to match perceived sleep duration, which increases sleep drive and reduces time awake in bed. Time allowed in bed is initially restricted to the average time perceived as sleep per night and is made longer when sleep efficiency is >85%.
- ***Sleep efficiency***. Sleep efficiency is the ratio of the total time spent asleep (total sleep time) in one night compared with the total amount of time spent in bed. Low sleep efficiency is a risk factor of insomnia and an indicator for sleep restriction.
- ***Stimulus control***. Behavioral instructions aimed at strengthening the association between bed and sleep to prevent conditioning the patient to associate bed with other stimulating activities. Such instructions include avoiding nonsleep activities in the bedroom, going to bed only when sleepy, and leaving bed when unable to sleep with 15–20min and returning to bed only when sleepy.
- ***Cognitive restructuring***. Aimed at identifying and replacing dysfunctional beliefs and attitudes about sleep and insomnia. Such misconceptions may include unrealistic expectations of sleep, fear of missing out on sleep, and overestimation of the consequences of poor sleep.
- ***DBAS***. Sleep-related cognition (e.g., faulty beliefs and appraisals, unrealistic expectations, and perceptual and attentional biases) plays an important role in perpetuating insomnia. DBAS is a 30-item self-report measure designed to evaluate sleep-related cognition. A short 16-item version also exists.
- ***Sleep state misperception***. Some people with chronic insomnia believe that they do not sleep during certain periods, whereas sleep measurements show that they sleep well. This phenomenon is called sleep state misperception.

Key facts of insomnia in older adults

- Compared to younger adults, the prevalence of insomnia is higher in older adults and increases with age.
- The overall prevalence of insomnia symptoms reported ranges from 30% to 48% in older adults, and that of those diagnosed with insomnia ranges from 12% to 20%.
- Insomnia is associated with impairments in cognitive, social, and emotional functioning and heightens the risk of suicide, anxiety and depressive disorders, alcohol and substance abuse, and cardiovascular diseases.
- Despite the high prevalence and negative impact of insomnia, it is estimated that less than 15% of patients with chronic insomnia receive treatment.

Key facts of brief CBT-I

- Because of financial limitations and human resources needed to deliver CBT-I, a shorter form of therapy, known as brief CBT-I, is now available.
- Brief CBT-I consists of the core components of standard CBT-I that is delivered in fewer than five sessions or by videos and booklets.
- Brief CBT-I has been found to be effective in the geriatric population.

Applications to other areas

Example 1

In this chapter, we have reviewed brief CBT-I that is suitable for older adults. Feasibility and accessibility of brief CBT-I is high, which provides applicability for other populations and communities. For instance, pregnant women and office workers may be good targets for brief CBT-I. Pregnancy is a risk factor of insomnia because of the significant fluctuations in hormones that affect the sleep-wake cycle. Common symptoms caused by pregnancy, such as anxiety, urinary frequency, backache, fetal movement, general abdominal discomfort, breast tenderness, leg cramps, heart burn, and reflux can cause sleep disturbances. Approximately 21% of pregnant women report disturbed sleep at levels consistent with a diagnosis of insomnia disorder. Poor sleep during pregnancy is associated with increased circulating levels of inflammatory markers and adverse pregnancy outcomes, such as preterm delivery. CBT-I is an adequate intervention for insomnia during pregnancy (Nowakowski & Meers, 2019). Delivering brief CBT-I at maternity clinics in the community is beneficial for the treatment and prevention of insomnia. For working-age people, preventing insomnia is important for performance and mental health. Some studies have shown effects of brief CBT-I on sleep quality in workers (Kaku et al., 2012; Nishinoue et al., 2012).

Example 2

There is considerable evidence for the effectiveness of standard CBT-I across patients with comorbid illnesses. Patients with chronic diseases often have insomnia. CBT-I has been found to produce clinically significant improvement in insomnia symptoms in patients with cancer, chronic pain, fibromyalgia, and other medical comorbidities, which are often comorbid with insomnia (Casault, Savard, Ivers, & Savard, 2015; Currie, Wilson, Pontefract, & deLaplante, 2000; Edinger, Wohlgemuth, Krystal, & Rice, 2005; Perlis, Sharpe, Smith, Greenblatt, & Giles, 2001). However, the efficacy of brief CBT-I has not been adequately investigated. Self-help CBT-I has demonstrated efficacy in several previous studies. An RCT of a self-help intervention for patients with insomnia and comorbid cancer showed that a greater proportion of CBT-I participants met criteria for clinical remission posttreatment compared with controls. This effect was maintained for up to 6months after the intervention. Another study showed efficacy of two types of self-help interventions using booklets or multimedia versions, which were effective at 1-year follow-up in adults with comorbid insomnia and arthritis or coronary artery disease (Rybarczyk, Mack, Harris, & Stepanski, 2011). Brief CBT-I is easy to deliver and does not require a specialist for implementation. Therefore, it could be easily combined with conventional treatments in community clinics.

Summary points

- The reviewed studies reported the effects of brief CBT-I on subjective and objective sleep variables and depression status.
- Regarding the content of brief CBT-I, sleep hygiene education should include understanding sleep changes that occur with age and mechanisms of sleep regulation, such as the circadian and homeostatic drives.
- Sleep restriction, which has been found to be the most effective way to improve sleep quality in older adults, can be effective and may be delivered in one to two in-person sessions, booklets, or videos.
- Cognitive restructuring, which is aimed at reducing dysfunctional sleep-related thoughts and changing misconceptions associated with sleep status, is important for improving insomnia in older adults.
- Understanding sleep physiology and subjective sleep data may help to individuals to have more appropriate sleep cognition.

References

Alessi, C., Martin, J. L., Fiorentino, L., Fung, C. H., Dzierzewski, J. M., Rodriguez Tapia, J. C., … Mitchell, M. N. (2016). Cognitive behavioral therapy for insomnia in older veterans using nonclinician sleep coaches: Randomized controlled trial. *Journal of the American Geriatrics Society*, *64*(9), 1830–1838.

Bishop, T. M., Simons, K. V., King, D. A., & Pigeon, W. R. (2016). Sleep and suicide in older adults: An opportunity for intervention. *Clinical Therapeutics*, *38*(11), 2332–2339.

Blay, S. L., Andreoli, S. B., & Gastal, F. L. (2008). Prevalence of self-reported sleep disturbance among older adults and the association of disturbed sleep with service demand and medical conditions. *International Psychogeriatrics*, *20*(3), 582–595.

Buysse, D. J., Germain, A., Moul, D. E., Franzen, P. L., Brar, L. K., Fletcher, M. E., … Monk, T. H. (2011). Efficacy of brief behavioral treatment for chronic insomnia in older adults. *Archives of Internal Medicine*, *171*(10), 887–895.

Casault, L., Savard, J., Ivers, H., & Savard, M. H. (2015). A randomized-controlled trial of an early minimal cognitive-behavioural therapy for insomnia comorbid with cancer. *Behaviour Research and Therapy*, *67*, 45–54.

Chan, W., Williams, J., Dautovich, N. D., McNamara, J. P. H., Stripling, A., Dzierzewski, J. M., … McCrae, C. S. (2017). Night-to-night sleep variability in older adults with chronic insomnia: mediators and moderators in a randomized controlled trial of brief behavioral therapy (BBT-I). *Journal of Clinical Sleep Medicine*, *13*(11), 1243–1254.

Cole, M. G., & Dendukuri, N. (2003). Risk factors for depression among elderly community subjects: A systematic review and meta-analysis. *American Journal of Psychiatry*, *160*(6), 1147–1156.

Curless, R., French, J. M., James, O. F., & Wynne, H. A. (1993). Is caffeine a factor in subjective insomnia of elderly people? *Age and Ageing*, *22*(1), 41–45.

Currie, S. R., Wilson, K. G., Pontefract, A. J., & deLaplante, L. (2000). Cognitive-behavioral treatment of insomnia secondary to chronic pain. *Journal of Consulting and Clinical Psychology*, *68*(3), 407–416.

Dzierzewski, J. M., Martin, J. L., Fung, C. H., Song, Y., Fiorentino, L., Jouldjian, S., … Alessi, C. A. (2019). CBT for late-life insomnia and the accuracy of sleep and wake perceptions: Results from a randomized-controlled trial. *Journal of Sleep Research*, *28*(4), e12809.

Edinger, J. D., Wohlgemuth, W. K., Krystal, A. D., & Rice, J. R. (2005). Behavioral insomnia therapy for fibromyalgia patients: A randomized clinical trial. *Archives of Internal Medicine*, *165*(21), 2527–2535.

Gammack, J. K. (2008). Light therapy for insomnia in older adults. *Clinics in Geriatric Medicine*, *24*(1), 139–149. viii.

Javaheri, S., & Redline, S. (2017). Insomnia and risk of cardiovascular disease. *Chest*, *152*(2), 435–444.

Kaku, A., Nishinoue, N., Takano, T., Eto, R., Kato, N., Ono, Y., & Tanaka, K. (2012). Randomized controlled trial on the effects of a combined sleep hygiene education and behavioral approach program on sleep quality in workers with insomnia. *Industrial Health*, *50*(1), 52–59.

Kay, D. B., Buysse, D. J., Germain, A., Hall, M., & Monk, T. H. (2015). Subjective-objective sleep discrepancy among older adults: Associations with insomnia diagnosis and insomnia treatment. *Journal of Sleep Research*, *24*(1), 32–39.

Léger, D., Partinen, M., Hirshkowitz, M., Chokroverty, S., Touchette, E., & Hedner, J. (2010). Daytime consequences of insomnia symptoms among outpatients in primary care practice: EQUINOX international survey. *Sleep Medicine*, *11*(10), 999–1009.

Lichstein, K. L., Riedel, B. W., Wilson, N. M., Lester, K. W., & Aguillard, R. N. (2001). Relaxation and sleep compression for late-life insomnia: A placebo-controlled trial. *Journal of Consulting and Clinical Psychology*, *69*(2), 227–239.

Manconi, M., Ferri, R., Sagrada, C., Punjabi, N. M., Tettamanzi, E., Zucconi, M., … Ferini-Strambi, L. (2010). Measuring the error in sleep estimation in normal subjects and in patients with insomnia. *Journal of Sleep Research*, *19*(3), 478–486.

Mishima, K., Okawa, M., Shimizu, T., & Hishikawa, Y. (2001). Diminished melatonin secretion in the elderly caused by insufficient environmental illumination. *Journal of Clinical Endocrinology & Metabolism*, *86*(1), 129–134.

Mitchell, M. D., Gehrman, P., Perlis, M., & Umscheid, C. A. (2012). Comparative effectiveness of cognitive behavioral therapy for insomnia: A systematic review. *BMC Family Practice*, *13*, 40.

Montgomery, P., & Dennis, J. (2004). A systematic review of non-pharmacological therapies for sleep problems in later life. *Sleep Medicine Reviews*, *8*(1), 47–62.

Morgan, K., Gregory, P., Tomeny, M., David, B. M., & Gascoigne, C. (2012). Self-help treatment for insomnia symptoms associated with chronic conditions in older adults: A randomized controlled trial. *Journal of the American Geriatrics Society*, *60*(10), 1803–1810.

Morin, C. M., Blais, F., & Savard, J. (2002). Are changes in beliefs and attitudes about sleep related to sleep improvements in the treatment of insomnia? *Behaviour Research and Therapy*, *40*(7), 741–752.

Nishinoue, N., Takano, T., Kaku, A., Eto, R., Kato, N., Ono, Y., & Tanaka, K. (2012). Effects of sleep hygiene education and behavioral therapy on sleep quality of white-collar workers: A randomized controlled trial. *Industrial Health*, *50*(2), 123–131.

Nowakowski, S., & Meers, J. M. (2019). Cognitive behavioral therapy for insomnia and Women's health: Sex as a biological variable. *Sleep Medicine Clinics*, *14*(2), 185–197.

Ohayon, M. M., Carskadon, M. A., Guilleminault, C., & Vitiello, M. V. (2004). Meta-analysis of quantitative sleep parameters from childhood to old age in healthy individuals: Developing normative sleep values across the human lifespan. *Sleep*, *27*(7), 1255–1273.

Patel, D., Steinberg, J., & Patel, P. (2018). Insomnia in the elderly: A review. *Journal of Clinical Sleep Medicine*, *14*(6), 1017–1024.

Perlis, M. L., Sharpe, M., Smith, M. T., Greenblatt, D., & Giles, D. (2001). Behavioral treatment of insomnia: Treatment outcome and the relevance of medical and psychiatric morbidity. *Journal of Behavioral Medicine*, *24*(3), 281–296.

Perlis, M. L., Smith, L. J., Lyness, J. M., Matteson, S. R., Pigeon, W. R., Jungquist, C. R., & Tu, X. (2006). Insomnia as a risk factor for onset of depression in the elderly. *Behavioral Sleep Medicine*, *4*(2), 104–113.

Pigeon, W. R., Pinquart, M., & Conner, K. (2012). Meta-analysis of sleep disturbance and suicidal thoughts and behaviors. *Journal of Clinical Psychiatry*, *73*(9), e1160–e1167.

Riedel, B. W., Lichstein, K. L., & Dwyer, W. O. (1995). Sleep compression and sleep education for older insomniacs: Self-help versus therapist guidance. *Psychology and Aging*, *10*(1), 54–63.

Roehrs, T., & Roth, T. (2008). Caffeine: Sleep and daytime sleepiness. *Sleep Medicine Reviews*, *12*(2), 153–162.

Rybarczyk, B., Mack, L., Harris, J. H., & Stepanski, E. (2011). Testing two types of self-help CBT-I for insomnia in older adults with arthritis or coronary artery disease. *Rehabilitation Psychology*, *56*(4), 257–266.

Salín-Pascual, R. J., Valencia-Flores, M., Campos, R. M., Castaño, A., & Shiromani, P. J. (2006). Caffeine challenge in insomniac patients after total sleep deprivation. *Sleep Medicine*, *7*(2), 141–145.

Schwartz, S., McDowell Anderson, W., Cole, S. R., Cornoni-Huntley, J., Hays, J. C., & Blazer, D. (1999). Insomnia and heart disease: A review of epidemiologic studies. *Journal of Psychosomatic Research*, *47*(4), 313–333.

Sivertsen, B., Omvik, S., Pallesen, S., Bjorvatn, B., Havik, O. E., Kvale, G., … Nordhus, I. H. (2006). Cognitive behavioral therapy vs zopiclone for treatment of chronic primary insomnia in older adults: A randomized controlled trial. *JAMA*, *295*(24), 2851–2858.

Spielman, A. J., Caruso, L. S., & Glovinsky, P. B. (1987). A behavioral perspective on insomnia treatment. *Psychiatric Clinics of North America*, *10*(4), 541–553.

Tanaka, M., Ikeuchi, M., Matsuki, H., Yaguchi, K., Kutsuzawa, T., Tanaka, K., & Kaneita, Y. (2018). Effects of brief cognitive behavioral therapy for insomnia on sleep and usage of hypnotics among community-dwelling older adults: Randomized controlled trial. *Nihon Koshu Eisei Zasshi*, *65*(8), 386–398.

Tanaka, M., Kusaga, M., Nyamathi, A. M., & Tanaka, K. (2019). Effects of brief cognitive behavioral therapy for insomnia on improving depression among community-dwelling older adults: A randomized controlled comparative study. *Worldviews on Evidence-Based Nursing*, *16*(1), 78–86.

Taylor, D. J., Lichstein, K. L., & Durrence, H. H. (2003). Insomnia as a health risk factor. *Behavioral Sleep Medicine*, *1*(4), 227–247. https://doi.org/10.1207/s15402010bsm0104_5.

Wilckens, K. A., Hall, M., Nebes, R. D., Monk, T. H., & Buysse, D. J. (2016). Changes in cognitive performance are associated with changes in sleep in older adults with insomnia. *Behavioral Sleep Medicine*, *14*(3), 295–310. https://doi.org/10.1080/15402002.2014.1002034.

Williams, J., Roth, A., Vatthauer, K., & McCrae, C. S. (2013). Cognitive behavioral treatment of insomnia. *Chest*, *143*(2), 554–565.

Chapter 28

Dementia: Features of treating anxiety and depression and the role of cognitive behavioral therapy: A new narrative

Kok-Wai Tay[a] and Ponnusamy Subramaniam[b]
[a]*Department of Psychology and Counseling, Faculty of Arts and Social Science, Tunku Abdul Rahman University, Kampar, Perak, Malaysia,*
[b]*Clinical Psychology and Behavioural Health Program, Faculty of Health Sciences, Universiti Kebangsaan Malaysia, Federal Territory of Kuala Lumpur, Malaysia*

Abbreviations

APA	American Psychiatric Association
BPSD	behavioral and psychological symptoms of dementia
BSI	Brief Symptom Inventory
CBT	Cognitive Behavior Therapy
CDR	Clinical Dementia Rating Scale
CSDD	Cornell Scale for Depression in Dementia
DSM-5-TR	Diagnostic and Statistical Manual of Mental Disorder—Fifth Edition, Text Revision
GAI	Geriatric Anxiety Inventory
GDS	Geriatric Depression Scale
GHQ	General Health Questionnaire
HADS	Hospital Anxiety and Depression Scale
HDRS	Hamilton Depression Rating Scale
ICD-11	International Classification of Diseases 11th Revision
MMSE	Mini Mental State Examination
NPI-A	Neuropsychiatric Inventory—Anxiety Subscale
NPI-D	Neuropsychiatric Inventory—Depression Subscale
PSWQ-A	Penn State Worry Questionnaire—Abbreviated
PWD	people with dementia
RAID	Rating Anxiety in Dementia
SIGH-A	Structured Interview Guide for the Hamilton Anxiety Scale

Introduction

Anxiety and depressive symptoms have been widely recognized as prevalent among people with dementia (PWD). Recent review found that up to 37% of PWD are living with depressive symptoms, and up to 14% of PWD are with anxiety symptoms (Kuring, Mathias, & Ward, 2018). The prevalence rate of anxiety and depression among PWD is up to five times higher than the prevalence rate among the general population as per suggested in Diagnostic and Statistical Manual of Mental Disorders, Fifth Edition, Text Revision (DSM-5-TR) (American Psychiatric Association (APA), 2022); with the information in DSM-5-TR suggests major depressive disorder has a prevalence rate of around 7%, and generalized anxiety disorder has a prevalence rate of about 2.9%.

Earlier studies have suggested anxiety and depressive symptoms can be commonly seen among the patients even before the dementia symptoms are apparent. A review article by Ma (2020) also pointed out that depressive symptoms, particularly apathy and anxiety symptoms are essential factors in affecting the progression of patients with mild cognitive impairment to dementia. Clinical interview might support clinicians to formulate and conceptualize this mental health phenomenon.

Handbook of Lifespan Cognitive Behavioral Therapy. https://doi.org/10.1016/B978-0-323-85757-4.00040-7

TABLE 1 Key themes summarized from qualitative study among PWD.

Key themes	Elaboration
Disrupted sense of self	– Scrutinization of the body and symptoms and attempts to take charge of the bodily loss-of-control • Constantly checking for unusual signs and symptoms • A sense of the body being externally controlled by unfamiliar force • Disappointed for failing the dementia diagnostic test • Ambiguous loss upon receiving the diagnosis • Uncertain about what they still can do and who they are • Question personal bodily performance and perception
Disrupted relationship with others	– Feeling judged by others • Stigma of being incompetent due to the diagnosis • Perception of being looked down by others • Tries to mask dementia-related symptoms to avoid being judged and stigmatized – Moving away from significant others • Fear of becoming a burden to the social circle or becoming overly dependent on others • Loss of a mutual relationship, patient is assigned with a sick role – Difficulties in discussing daily challenges • People avoid discussing or belittling the challenges – Wishes to be understood and taken seriously • Patient lament about restrictions imposed by significant others, and diminished freedom • Significant others are not supporting the patient, but "them against me" (Wijngaarden et al., 2019) • Being perceived as "useless"
Disrupted sense of surrounding world	– Losing touch with familiar life experiences • Related to forgetfulness and disrupted sense of time – Feeling trapped within a space that is shrinking • Less community-based activity • More restrictions and boundaries set by others • Forced by the disease to let go of personal leisure activities – Attempting to control a fearful future • Ambiguous loss—uncertain of the future – Considering euthanasia as a mean of controlling the future • Fueled by future-oriented uncertainties and anxiety

This table summarizes the key themes from the perspective of PWD that can be precipitants of anxiety and depressive symptoms. After reading into the key themes, the readers might be surprised how these themes correspond to the concept of cognitive triad (sometimes known as negative triad) in CBT, refers to negative view to the self, negative view to the environment and negative future outlook.
Cited from Wijngaarden, E. V., Alma, M., & The, A. (2019). 'The eyes of others' are what really matters: The experience of living with dementia from an insider perspective. *PLoS One, 14*(4). https://doi.org/10.1371/journal.pone.0214724.

A qualitative interview study was conducted by Wijngaarden, Alma, and The (2019) to understand the insider perspective of PWD. Some key themes were identified and summarized in Table 1, and it can contribute to the clinician's understanding in the development of anxiety and depressive symptoms among PWD. With such idiographic information, it can better inform the development of cognitive and behavioral strategies to support PWD. Readers are strongly recommended to access the full paper as it contains interesting verbatim account from the research participants.

CBT in treating anxiety and depression in PWD

Cognitive behavior therapy (CBT) has long been established as one of the gold standards in treating anxiety and depression. The decline in cognitive functioning is one of the dementia's hallmark feature. Therefore, the effectiveness of CBT and its adaptation to treat depressive and anxiety symptoms in this population has attracted researchers and clinicians' interest. A systematic review by Tay, Subramaniam, and Oei (2019) suggested that CBT can be effective in anxiety and depressive symptoms reduction, notably among populations with mild-to-moderate dementia. A list of CBT intervention effects on anxiety and depressive symptoms is included in the systematic review. Group-based CBT has also been suggested as being

TABLE 2 Summary of considerations for the application of CBT for PWD.

Considerations	Elaborations
Modifications of CBT techniques	Due to the cognitive decline experienced by persons with dementia, clinicians need to customize CBT to better suit their needs. This will be discussed in depth later in the chapter
Severity of dementia	Current evidence suggests patient with dementia who obtains a score of 2 or less in Clinical Dementia Rating scale (CDR) and a score of greater than 15 in Mini Mental State Examination (MMSE) can benefit from CBT intervention
Caregiver inclusion	Caregiver can be included during the session with the aim that the caregiver can act as a "co-therapist" who can ensure the patient continues practicing the skills learnt during session in daily life
Intervention interval and duration	No standard practice is reported. Current suggestion is to design the intervention as per the patient's availability and need

This table presents a summary of considerations prior to providing CBT for persons with dementia to manage their depressive and anxiety symptoms.
Cited from Tay, K., Subramaniam, P., & Oei, T. P. (2019). Cognitive behavioural therapy can be effective in treating anxiety and depression in persons with dementia: A systematic review. *Psychogeriatrics, 19*(3). https://doi.org/10.1111/psyg.12391.

helpful (Kipling & Bailey, 1999), and it was noted that having a peer support itself can cultivate a sense of belonging and the removal of need to mask the symptoms (Wijngaarden et al., 2019). Tay et al. (2019) has also highlighted some important considerations and necessary modifications for the application of CBT in this population. A brief summary of the considerations is as presented in Table 2. It was noted that intervention protocols and guidelines for anxiety symptoms alleviation were developed (Charlesworth, Sadek, Schepers, & Spector, 2015; Paukert et al., 2010), but there is no protocol developed for management of depressive symptoms among PWD. Interested readers can proceed to read the articles mentioned earlier.

Modifications of CBT for PWD

Due to the cognitive decline experienced by PWD, it is not expected that all CBT components are applicable to this population (Kwak, Yang, & Koo, 2017). General modifications needed include (a) psychoeducation to the patient-caregiver dyad; (b) simplified instructions, depending on the patient's understanding; (c) shorter session durations, depending on the patient's energy level; (d) introducing only one skill at a time; and (e) more emphasis on behavioral components of CBT. Besides, as some patients might have self-defeating thoughts that they will not be able to attain new skills and information, an explicit normalization of the difficulties in mastering new skills can be helpful. Normalization is a valuable therapeutic tool to support and to inform PWD that these difficulties are common, akin to other skills like learning to drive. A verbatim example of normalization is: *Sometimes, when people are trying to learn new skills, they might find the skills are difficult and almost impossible to learn. I am wondering if you feel the same way when I introduce X skill to you.* Table 3 presents the dementia-specific modifications of CBT to better suit PWD.

Common themes arising in psychotherapy sessions with persons with dementia

In addition to understanding and applying the modified CBT to serve PWD, it is utmost important for clinicians to have an understanding of the common themes arising in psychotherapy sessions with PWD. Scholey and Woods (2003) described several themes that have arisen in their study and presented below is a summary of the themes.

The first theme is PWD may have distorted or fixed belief about the causes of their cognitive impairment, and the belief can either be dysfunctional or nonmalignant. A belief is considered as dysfunctional if it leads PWD to be angry and frustrated. In response to this, an in-depth psychoeducation is needed. The goal is to collaboratively develop a more functional belief into the causes of the cognitive impairment.

The second theme is a state of hopelessness that is reinforced by a loss of control. This corresponds to the findings by Wijngaarden et al. (2019), highlighting how daily interactions that ignore and dismiss PWD opinions puts them at risk of developing depressive symptoms. Scholey and Woods (2003) recommended that clinicians can advocate for PWD to express a clear statement to their caregivers. Other strategies that are helpful to build rapport with teenage or late childhood clients can be helpful to build rapport with PWD, and most importantly, prevent clinicians from being a perpetuator of the

TABLE 3 Dementia-specific CBT modification from previous studies.

Dementia-specific modifications	Elaborations	Sources
Inclusion of handouts	Giving handout is a common practice for CBT therapists. However, for persons with dementia, handouts are given to the patients as reminders and to be shared with caregivers/partners	Kipling and Bailey (1999)
Focusing on dysfunctional thoughts about memory	Working on identifying negative thoughts about memory (e.g., "should" and "must" statements regarding memory, exaggeration and personalization about memory difficulties)	Kipling and Bailey (1999)
Focusing on unhelpful beliefs about dementia	Using behavioral experiments to manage catastrophizing thoughts related to dementia (e.g., people with dementia will be incapable of doing everything)	Walker (2004)
Distraction	Helpful to work with anxiety symptoms at times	Koder (1998)
Repetition	Patients were requested to repeat information to facilitate the understanding and session participation Spaced retrieval strategy that relies on procedural memory (found to be less affected by dementia) is used to enhance memory encoding and retrieval. In practice, patients will be requested to recall target information at an increasing time interval. If the patient fails to recall the information, an answer will be provided. Later, the patient will be asked to recall the information again at the last correct time interval Telephone contact in-between sessions can support learnt skills retention	García-Alberca (2017a), Kraus et al. (2008), Paukert et al. (2010), Robinson et al. (2011), Stanley et al. (2013), Spector et al. (2012, 2015)

This table summarizes the dementia-specific modifications of CBT to optimize the effectiveness of CBT in treating depressive and anxiety symptoms among persons with dementia.

sense of hopelessness. These strategies, not exhaustively, include seeing PWD first and individually before talking to family members, spending more time with PWD instead of the family, and taking serious notes of PWD's complaints.

The third theme is the loss of a sense of security. Some PWD may believe that the destiny of a PWD is in a residential home. For PWD who have experienced negative events while taking care other family members with dementia who lives with behavioral and psychological symptoms of dementia (BPSD) that involves aggression, these PWD are worried if they would be aggressive in near future and be a burden for their next generation. As these can contribute to the development of anxiety symptoms, Scholey and Woods (2003) work with PWD to focus more on the here-and-now. As with the first theme, clinicians can provide psychoeducation and work on dysfunctional beliefs as appropriate.

The fourth theme is about memories related to traumatic events that might have been preserved while the learnt coping strategies may have been forgotten. When this theme arises during the session, clinicians can support a PWD by reframing the experience.

The fifth theme is the importance of flexibility, patience, and tolerance in a therapist. Secondary to the cognitive decline, PWD might forget or lose therapy notes, action plans, and related tasks. As PWD might sometimes attempt to mask the symptoms, it is utmost important for clinicians to be mindful of that and interpret the therapy-interfering behaviors carefully.

The sixth theme is the precipitating factor of depression that can at times be straightforward and simple. For example, a loss of mobility or daily adaptive behaviors can be the triggering factor. From author's clinical case material, one patient reported feeling depressed as he was no longer able to go out to meet his friends in a local eatery, due to his loss of mobility. Therefore, a successful liaison with caregivers or other allied health professionals like physiotherapists and occupational therapists to promote independent functioning and reinstatement of previous leisure activities can work well to mitigate the risk factors of developing other mental health symptoms.

The seventh theme is related to the frustration related to the loss of the ability to drive. The frustration should not be underestimated as for many, being able to drive and move around independently are taken for granted. As such, the discussion on this topic should be planned at the earliest stage of therapy.

Common instruments to assess depressive and anxiety symptoms among PWD

Due to the cognitive decline present in PWD, clinicians might wish to identify the most suitable instruments to assess the mental health symptoms of PWD. While reviewing the psychometric properties of the instruments is beyond the scope of this chapter, Table 4 lists down all the instruments used in previous studies.

TABLE 4 Instruments used in previous studies.

Symptoms	Instruments	Sources
Anxiety	Brief Symptom Inventory (BSI)	Walker (2004)
	Geriatric Anxiety Inventory (GAI)	Paukert et al. (2010), Stanley et al. (2013)
	General Health Questionnaire (GHQ)	Walker (2004)
	Hospital Anxiety and Depression Scale (HADS)	Spector et al. (2015)
	Neuropsychiatric Inventory—Anxiety Subscale (NPI-A)	García-Alberca (2017a), Kraus et al. (2008), Paukert et al. (2010), Robinson et al. (2011), Stanley et al. (2013)
	Penn State Worry Questionnaire—Abbreviated (PSWQ-A)	Paukert et al. (2010), Stanley et al. (2013)
	Rating Anxiety in Dementia (RAID)	Kraus et al. (2008), Robinson et al. (2011), Spector et al. (2015), Stanley et al. (2013)
	Structured Interview Guide for the Hamilton Anxiety Scale (SIGH-A)	Robinson et al. (2011)
Depression	Brief Symptom Inventory (BSI)	Walker (2004)
	Cornell Scale for Depression in Dementia (CSDD)	García-Alberca (2017a, 2017b), Spector et al. (2015)
	Geriatric Depression Scale (GDS)	Paukert et al. (2010), Scholey and Woods (2003), Stanley et al. (2013)
	General Health Questionnaire (GHQ)	Walker (2004)
	Hospital Anxiety and Depression Scale (HADS)	Spector et al. (2015)
	Hamilton Depression Rating Scale (HDRS)	García-Alberca (2017a)
	Neuropsychiatric Inventory—Depression Subscale (NPI-D)	García-Alberca (2017a, 2017b)

This table lists down the instruments used to measure depressive and anxiety symptoms among PWD in previous studies.

Mini-dictionary of terms

- ***Dementia.*** Dementia, also termed as major neurocognitive disorder in DSM-5, is characterized by a significant decline in one or more areas of cognitive functioning like memory, attention, executive functions, and related others from pre-morbid functioning.
- ***Depression.*** Depression is a mood concern characterized by symptoms like feeling sad or down, reduced or loss of pleasure in daily activities, changes in sleep and appetite, lethargy, feelings of hopelessness and worthlessness, psycho-motor changes, reduced ability to concentrate, and thoughts of death (can be in form of passive death wishes or active suicidal ideation).
- ***Anxiety.*** Anxiety is a feeling of apprehension to an anticipated future. Usually, it involves an overestimation of risk, and an underestimation of personal ability to cope. It comes in both psychological symptoms and behavioral symptoms, which can include feeling of restlessness, irritability, think that something extremely bad can happen, nervousness and constant worry.
- ***Behavioral and psychological symptoms of dementia (BPSD).*** BPSD commonly refers to a group of cooccurring behavioral and noncognitive psychological symptoms among PWD and the symptoms can include aggression, psychotic symptoms, anxiety, depression, sleep and appetite changes, and related others. BPSD is usually associated with the cognitive impairment.
- ***Normalization.*** Normalization is a counseling technique that clinicians use either direct or indirect statement to inform clients' problems can be a typical phenomenon rather than being a pathological presentation.

Key facts of dementia

- DSM-5 subsumes dementia under the term "major neurocognitive disorder," while suggesting the term is "not precluded from use in the etiological subtypes in which that term is standard" (APA, 2022, p. 667).
- The term "dementia" continues to be used in International Classification of Diseases 11th Revision (ICD-11).
- Alzheimer's disease is the most common cause of dementia, followed by dementia secondary to vascular accidents.
- Although memory decline and learning concerns are hallmarks features of Alzheimer's disease, dementia can affect other domains of cognitive functioning that include attention, executive functions, language, social cognition, and perceptual-motor abilities.
- Other causes of dementia include traumatic brain injury, substance use, HIV infection, Parkinson's disease, Lewy body disease, frontotemporal lobar degeneration, and others.
- Other than the cognitive decline, depressive and anxiety symptoms are prevalent among PWD.
- Goals of treatment for dementia is twofold; first, to slow down the progression of the degeneration and second, to maintain quality of life.
- Family caregivers for PWD are sometimes referred as invisible second patient, and similar to PWD, the caregivers are living with anxiety and depressive symptoms.

Applications to other areas

In this chapter, we have discussed how CBT can be beneficial for PWD in managing the frequent cooccurring depressive and anxiety symptoms. From previous studies, we noted collaterals were usually involved to strengthen and facilitate the therapeutic gain. While we are working to improve the mental health status among PWD, it is important to note that providing care for a beloved family member with dementia and observing how the symptoms worsen over time are stressful, and caregivers for PWD also face similar psychological concerns (Liu et al., 2016). In fact, García-Alberca, Lara, and Berthier (2011) study suggests more than 50% of PWD caregivers are living with high anxiety and depressive symptoms. (It is really a pressing issue, and this book has another chapter dedicated to that!) Systematic review has suggested CBT in either individual or group format is beneficial for caregivers in managing stress and depressive symptoms (Hopkinson, Reavell, Lane, & Mallikarjun, 2019; Kwon, Ahn, Kim, & Park, 2017).

Viewing from the evidence that CBT can be beneficial for both caregivers and PWD in either group or individual format, and caregivers have been included in CBT for PWD as cotherapists, it will be interesting for future research to systematically study how these can be done simultaneously. To the best knowledge of the authors, only García-Alberca (2017a, 2017b), and Stanley et al. (2013) have attempted to include the caregivers to provide dyadic interventions, and the results are promising. García-Alberca (2017a, 2017b) included caregiver cognitive restructuring to replace unhelpful beliefs regarding caregiving performance and to facilitate development of adaptive engagement coping strategies, while

Stanley et al. (2013) provided stress reduction modules for caregivers as appropriate. Review by Moon and Adams (2012) has suggested a dyadic intervention for PWD and their caregivers is effective to improve PWD's cognitive function, and the relationship between the caregivers and the patient. Studies on CBT-based dyadic intervention can be helpful in shaping the future best clinical practice in managing depressive and anxiety symptoms revolving around a dementia diagnosis.

Summary points

- Depression and anxiety symptoms are prevalent among PWD, and studies shows the prevalence rate is fivefold higher compared to general population.
- Wijngaarden et al. (2019) qualitative interview study with PWD found that the diagnosis is associated with a disruption in sense of self, relationship with others and surrounding worlds, which are closely related to CBT cognitive triad.
- Modified CBT can be therapeutic and effective for PWD to manage their anxiety and depressive symptoms.
- Modifications like inclusion of caregivers, inclusion of handouts as reminders, repetitions, focusing on unhelpful beliefs on dementia and memory, and simplified and shorter sessions are instrumental to ensure CBT is effective for this population.
- Caregivers for PWD also live with similar anxiety and depressive symptoms due to caregiver stress.
- CBT-based dyadic intervention can be an interesting area for future researchers to explore.

References

American Psychiatric Association (APA). (2022). *Diagnostic and statistical manual of mental disorders* (5th ed.). American Psychiatric Association.

Charlesworth, G., Sadek, S., Schepers, A., & Spector, A. (2015). Cognitive behavior therapy for anxiety in people with dementia: A clinician guideline for a person-centered approach. *Behavior Modification, 39*(3), 390–412. https://doi.org/10.1177/0145445514561317.

García-Alberca, J. M. (2017a). Cognitive-behavioral treatment for depressed patients with Alzheimer's disease: An open trial. *Archives of Gerontology and Geriatrics, 71*, 1–8. https://doi.org/10.1016/j.archger.2017.02.008.

García-Alberca, J. M. (2017b). Cognitive-behavioural treatment for depression in Alzheimer's disease patients: A case study. *Psychogeriatrics, 17*, 200–201. https://doi.org/10.1111/psyg.12203.

García-Alberca, J. M., Lara, J. P., & Berthier, M. L. (2011). Anxiety and depression in caregivers are associated with patient and caregiver characteristics in Alzheimer's disease. *International Journal of Psychiatry in Medicine, 41*(1), 57–69. https://doi.org/10.2190/PM.41.1.f.

Hopkinson, M. D., Reavell, J., Lane, D. A., & Mallikarjun, P. (2019). Cognitive behavioural therapy for depression, anxiety, and stress in caregivers of dementia patients: A systematic review and meta-analysis. *The Gerontologist, 59*(4), e343–e362. https://doi.org/10.1093/geront/gnx217.

Kipling, T., & Bailey, M. (1999). The feasibility of a cognitive behavioural therapy group for men with mild/moderate cognitive impairment. *Behavioural and Cognitive Psychotherapy, 27*, 189–193.

Koder, D. (1998). Treatment of anxiety in the cognitively impaired elderly: Can cognitive-behavior therapy help? *International Psychogeriatrics, 10*(2), 173–182.

Kraus, C. A., Seignourel, P., Balasubramanyam, V., Snow, A. L., Wilson, N. L., Kunik, M. E., … Stanley, M. A. (2008). Cognitive-behavioral treatment for anxiety in patients with dementia: Two case studies. *Journal of Psychiatric Practice, 14*(3), 186–192. https://doi.org/10.1097/01.pra.0000320120.68928.e5.

Kuring, J. K., Mathias, J. L., & Ward, L. (2018). Prevalence of depression, anxiety and PTSD in people with dementia: A systematic review and meta-analysis. *Neuropsychology Review*. https://doi.org/10.1007/s11065-018-9396-2.

Kwak, Y. T., Yang, Y., & Koo, M. (2017). Anxiety in dementia. *Dementia and Neurocognitive Disorders, 16*(2), 33–39. https://doi.org/10.12779/dnd.2017.16.2.33.

Kwon, O., Ahn, H. S., Kim, H. J., & Park, K. (2017). Effectiveness of cognitive behavioral therapy for caregivers of people with dementia: A systematic review and meta-analysis. *Journal of Clinical Neurology, 13*(4), 394–404. https://doi.org/10.3988/jcn.2017.13.4.394.

Liu, S., Li, C., Shi, Z., Wang, X., Zhou, Y., Liu, S., … Ji, Y. (2016). Caregiver burden and prevalence of depression, anxiety and sleep disturbances in Alzheimer's disease caregivers in China. *Journal of Clinical Nursing, 26*(9–10). https://doi.org/10.1111/jocn.13601.

Ma, L. (2020). Depression, anxiety, and apathy in mild cognitive impairment: Current perspectives. *Frontiers in Aging Neuroscience*. https://doi.org/10.3389/fna-gi.2020.00009.

Moon, H., & Adams, K. B. (2012). The effectiveness of dyadic interventions for people with dementia and their caregivers. *Dementia, 12*(6), 821–839. https://doi.org/10.1177/1471301212447026.

Paukert, A. L., Calleo, J., Kraus-Schuman, C., Snow, L., Wilson, N., Peterson, N. J., … Stanley, M. A. (2010). Peaceful mind: An open trial of cognitive-behavioural therapy for anxiety in persons with dementia. *International Psychogeriatrics, 22*(6), 1012–1021. https://doi.org/10.1017/S1041610210000694.

Robinson, C. M., Paukert, A., Kraus-Schuman, C. A., Snow, A. L., Kunik, M. E., Wilson, N. L., … Stanley, M. A. (2011). The involvement of multiple caregivers in cognitive-behavior therapy for anxiety in persons with dementia. *Aging & Mental Health, 15*(3), 291–298. https://doi.org/10.1080/13607860903493374.

Scholey, K. A., & Woods, B. T. (2003). A series of brief cognitive therapy interventions with people experiencing both dementia and depression: A description of techniques and common themes. *Clinical Psychology & Psychotherapy*, *10*, 175–185.

Spector, A., Charlesworth, G., King, M., Lattimer, M., Sadek, S., Marston, L., … Orrell, M. (2015). Cognitive-behavioural therapy for anxiety in dementia: Pilot randomised controlled trial. *British Journal of Psychiatry*, *206*, 1–8. https://doi.org/10.1192/bjp.bp.113.140087.

Spector, A., Orrell, M., Lattimer, M., Hoe, J., King, M., Harwood, K., … Charlesworth, G. (2012). Cognitive behavioural therapy (CBT) for anxiety in people with dementia: Study protocol for a randomised controlled trial. *Trials*, *13*(197).

Stanley, M. A., Calleo, J., Bush, A. L., Wilson, N., Snow, A. L., Kraus-Schuman, C., … Kuik, M. E. (2013). The peaceful mind program: A pilot test of a CBT-based intervention for anxious patients with dementia. *American Journal of Geriatric Psychiatry*, *21*(7). https://doi.org/10.1016/j.jagp.2013.01.007.

Tay, K., Subramaniam, P., & Oei, T. P. (2019). Cognitive behavioural therapy can be effective in treating anxiety and depression in persons with dementia: A systematic review. *Psychogeriatrics*, *19*(3). https://doi.org/10.1111/psyg.12391.

Walker, D. A. (2004). Cognitive behavioural therapy for depression in a person with Alzheimer's dementia. *Behavioural and Cognitive Psychotherapy*, *32*, 495–500. https://doi.org/10.1017/S1352465804001663.

Wijngaarden, E. V., Alma, M., & The, A. (2019). 'The eyes of others' are what really matters: The experience of living with dementia from an insider perspective. *PLoS One*, *14*(4). https://doi.org/10.1371/journal.pone.0214724.

Chapter 29

The elderly and fear of falling: Features and applications of cognitive-behavioral therapy

Tai Wa Liu[a] and Shamay S.M. Ng[b]

[a]*School of Nursing and Health Studies, Hong Kong Metropolitan University, Ho Man Tin, Hong Kong (SAR)* [b]*Department of Rehabilitation Sciences, The Hong Kong Polytechnic University, Hung Hom, Hong Kong (SAR)*

Abbreviations

ABC	Activity-specific Balance Confidence scale
FES	Falls Efficacy Scale
FES-I	Falls Efficacy Scale—International
TOBT	task-oriented balance training

Introduction

This chapter describes a psychotherapeutic intervention based on cognitive-behavioral therapy (CBT) to reduce fear of falling among older people. Before describing the CBT-based intervention in detail, we introduce the concept of fear of falling as described in the recent psychological and medical literature. We then outline existing therapeutic interventions and the theoretical rationale for using a CBT-based intervention to mitigate fear of falling. Next, we describe a protocol for implementing this CBT-based group intervention with the option of combined use with a type of exercise therapy, namely task-oriented balance training (TOBT), to reduce fear of falling and associated avoidance behavior.

Understanding fear of falling

We all have concerns about falling from a height. An early psychological experiment (Gibson & Walk, 1960) suggested that wariness about heights is a learned response that begins in infancy and signifies an increase in physical and visual maturity (Dahl et al., 2013). According to the psychology literature, the learning process begins not with depth perception or actual fall experience but with the acquisition of locomotor experience and the development of visual proprioception (visual perception of self-movement) (Bertenthal & Campos, 1990; Campos, Bertenthal, & Kermoian, 1992). Concerns about falling are considered natural and protective, as individuals adopt corresponding fall preventive measures and avoidance behavior when they perceive fall risk (Liu & Tsui, 2014).

In the medical literature, the term "fear of falling" has been defined somewhat differently as a broad construct representing irrational and excessive anxiety about falling. In the early 1980s, Murphy and Isaacs (1982) and Bhala, O'Donnell, and Thoppil (1982) described fear of falling as a state wherein patients experience the intense sensation of a loss of balance and display fearful behavior, such as attempts to clutch and grab, when they try to stand or walk after suffering a fall. Subsequently, medical studies in the 1980s and 1990s redefined the term "fear of falling," which came to refer to a postfall syndrome characterized by low perceived self-efficacy in avoiding falls (Tinetti, Richman, & Powell, 1990). This interpretation has been empirically supported. For example, in a 1-year prospective study conducted by Murphy, Dubin, and Gill (2003), community-dwelling older women who reported one or more falls before or during the study period were significantly more likely to experience fear of falling than those with no fall history (35% vs 20%; $P = .004$). Medical researchers and clinicians have generally agreed that persistently excessive anxiety about falling is a psychological reflection of a loss of self-efficacy in maintaining balance and may or may not be behaviorally reflected in fear avoidance activities (Landers, Durand, Powell, Dibble, & Young, 2011; Liu & Tsui, 2014).

Handbook of Lifespan Cognitive Behavioral Therapy. https://doi.org/10.1016/B978-0-323-85757-4.00030-4

Building on Tinetti et al. (1990) assertion that fear of falling stems from impaired balance self-efficacy, Delbaere, Crombez, van Haastregt, and Vlaeyen (2009) explored the influence of cognitive distortion, specifically catastrophic thinking (catastrophizing) about the consequences of falls, on fear of falling. Their structural equation model, based on a sample of 896 community-dwelling older people, revealed that the number of falls in the previous 12 months and catastrophizing about falls, as measured by the Catastrophizing about Falls Scale (van Haastregt, 2002), were the two most important independent predictors of fear of falling and subsequently mobility restrictions.

Research has shown that fear of falling is the most common type of fear experienced by older people. It is even more intense than their fear of being robbed in the street or experiencing financial difficulties (Howland et al., 1993). Based on findings from the medical literature, gerontologists have investigated the prevalence of and factors associated with fear of falling. Scheffer, Schuurmans, van Dijk, van der Hooft, and de Rooij (2008) systematically reviewed 21 studies and indicated that the reported prevalence of fear of falling ranged from 20% to 85%. They also found various risk factors for and consequences of developing fear of falling, including a history of falls, old age, female gender, dizziness, low self-rated health status, depression, gait and balance problems, cognitive complaints and low economic status. Decreased physical health, balance and gait problems, avoidance of activities and damage to self-identity were identified as potential consequences of fear of falling and associated avoidance behavior. Murphy et al. (2003) also identified four predisposing factors for the development of fear of falling: an age of 80 years or older, visual impairment, a sedentary lifestyle and a lack of available emotional support.

Although numerous studies have investigated the prevalence and correlates of fear of falling among older people, only a few have examined its causes. Contrary to the widely held belief that an actual fall experience is a precedent and risk factor for fear of falling, Scheffer et al. (2008) systematic review revealed that up to 50% of the sampled older people who reported fear of falling had not experienced a recent fall. Similarly, Cumming, Salkeld, Thomas, and Szonyi (2000) and Rochat et al. (2010) found that fear of falling was reported by older people with no recent fall history. Friedman, Munoz, West, Rubin, and Fried (2002) developed a longitudinal predictive model based on a sample of 2212 community-dwelling older people. They found that a fall experience at baseline was an independent predictor of developing fear of falling 20 months later, while fear of falling at baseline was a predictor of falling at 20 months. This finding suggested that a fall experience could predict fear of falling, and vice versa, fear of falling can predict the occurrence of actual falls. Together, these findings regarding the risk factors, correlates and causality of fear of falling and actual falls indicate that the relationship between these variables may not be causal or linear. Instead, the relationship may be a debilitating spiral involving fear avoidance behavior, decreased levels of physical and social activity, physical deconditioning, increased fall risk and eventually actual falls. The interplay between fear of falling and fall risk can be conceptualized as a vicious cycle between fear of falling and actual falls, which poses a serious health challenge to clinicians (Fig. 1).

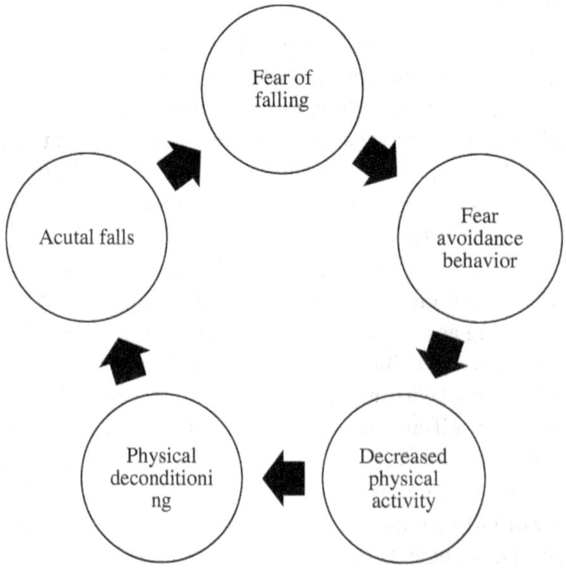

FIG. 1 Vicious cycle between fear of falling and an actual fall. This figure illustrates how fear of falling induces activity avoidance, decreased level of physical activity and physical deconditioning, and eventually the actual falls.

Emerging therapeutic interventions for fear of falling and associated fear avoidance behavior

Exercise therapy

Most early therapeutic strategies to address fear of falling and associated fear avoidance behavior aimed to enhance patients' ability to maintain their balance and thereby improve their balance self-efficacy. These strategies were based on the assumption that fear of falling arises from physical and environmental factors; they ignored the role of cognitive factors. Clients were guided to perform various types of exercise therapy, including strength and balance exercises (Arai et al., 2007) and tai chi (Sattin, Easley, Wolf, Chen, & Kutner, 2005). In addition to being used as a monotherapy, exercise therapy has often been used with other therapeutic components, such as a fall prevention educational program (Lin, Hwang, Wang, Chang, & Wolf, 2006), task-oriented techniques (Weerdesteyn et al., 2006), and medication management to cope with visual loss and ensure environmental safety (Clemson et al., 2004).

The effects of exercise therapy have been demonstrated in several clinical trials (Liu et al., 2007; Sattin et al., 2005). However, Kumar et al. (2016) systematic review and meta-analysis of 36 clinical trials with 2878 participants revealed that exercise therapy yielded only a short-term effect ($\leq$ 6 months) or a small to moderate immediate postintervention effect (standardized mean difference (SMD) 0.37, 95% confidence interval (CI) 0.18–0.56). It did not have a long-term (>6 months postintervention) effect.

How can cognitive-behavioral therapy be used to reduce fear of falling?

Another approach, cognitive-behavioral therapy (CBT), is based on the assumption that our behavioral and emotional responses to events within our environment are driven not by the events themselves but by the meanings given to them. In the context of fear of falling, if individuals' thoughts regarding balance-requiring situations are too negative, they may be unable to perceive these situations realistically. This may lead to inappropriate emotional responses (fear of falling), such as (i) overestimation of the likelihood of falling, (ii) underestimation of balance ability, and (iii) overestimation of the consequences of falls, and behavioral responses (fear avoidance behavior).

Given its debilitating nature and distinctive features, fear of falling is widely understood as an important treatment outcome in fall prevention programs. The use of CBT to reduce fear of falling was pioneered by Reinsch, MacRae, Lachenbruch, and Tobis (1992) and Tennstedt et al. (1998) who examined the feasibility and effectiveness of CBT-based interventions among community-dwelling older people. Reinsch et al. (1992) expected the number of falls to decrease after a 1-year fall prevention program with exercise and CBT components. However, contrary to their hypothesis, the number of falls was not reduced and only the secondary outcomes, such as balance and fear of falling, were improved. Reinsch et al. (1992) argued that the lack of significant findings regarding the causal outcomes of the program was due to the multifaceted nature of falls. However, their CBT-based intervention included only education in fall prevention, relaxation training and video games to improve reaction times. They did not include essential CBT components such as cognitive restructuring (Liu, Ng, Chung, & Ng, 2019).

In a clinical trial, Tennstedt et al. (1998) began to look for other ways to apply CBT to reduce fear of falling, seeking to modify cognitive instead of educational and behavioral strategies. They aimed to correct misconceptions of falls and fear of falling, set realistic goals for increasing activity levels, reduce environmental fall hazards and promote physical activity. They found that a group intervention comprising eight 2-h sessions per week for 4 weeks reduced the participants' fear of falling, as measured by the modified version of the Falls Efficacy Scale (Tinetti et al., 1990), and increased their intended activity levels and mobility control. Tennstedt et al. (1998) sought to restructure the participants' maladaptive attitudes toward and beliefs about falling. They reported that this approach led to more realistic perceptions of the fall risk and fall consequences and improved the participants' self-efficacy in behavioral change. Compared with the control group, the participants showed longer-lasting changes in their underlying belief in and attitudes toward their ability to maintain balance. In addition, after the 8-week group CBT intervention, restrictions on the participants' activity associated with their fear of falling had decreased. However, although Tennstedt et al. (1998) intervention had educational components, such as didactic materials regarding the occurrence of actual falls, fall risk, fall prevention skills and self-protection skills, education served only a supplementary role.

Compared with the didactic educational approach adopted by Reinsch et al. (1992), Tennstedt et al. (1998) 8-week CBT intervention was better designed and more consistent with the core theoretical assumption of CBT: that challenging and changing maladaptive beliefs and behavior can foster the development of personalized coping strategies in daily life. Tennstedt et al. (1998) CBT-based group intervention became the blueprint for several subsequent clinical trials using such interventions (Alexander, Sartor-Glittenberg, Bordenave, & Bordenave, 2015; Dorresteijn et al., 2016; Huang, Chung, Chen, Chin, & Wang, 2016; Huang, Yang, & Liu, 2011; Liu & Tsui, 2014). Since Tennstedt et al. (1998) study, CBT group interventions have been widely studied and demonstrated to effectively treat fear of falling in older people. Indeed, our meta-analysis of five recent clinical trials (Dorresteijn et al., 2016; Huang et al., 2011; Huang et al., 2016; Liu & Tsui, 2014; Parry et al., 2016) indicated that the use of CBT with or without exercise can yield small immediate effects (SMD 0.33, 95% CI 0.21–0.46, $P < .001$) and small retention effects up to 12 months (SMD 0.37, 95% CI 0.21–0.53, $P < .001$) in terms of reducing fear of falling. In addition, we found that CBT with or without exercise had a 6-month post-intervention positive effect (SMD 0.18, 95% CI 0.02–0.33) on balance (Liu et al., 2019). In these five recent clinical trials (Dorresteijn et al., 2016; Huang et al., 2011; Huang et al., 2016; Liu & Tsui, 2014; Parry et al., 2016), cognitive restructuring, personal goal setting and the promotion of physical activity were generally accepted as the core components of CBT-based interventions to reduce fear of falling.

Potential mechanism by which group cognitive-behavioral therapy reduces fear of falling

Self-efficacy is an individual's self-perceived ability to perform specific tasks and attain certain goals (Bandura, 1977). It determines the individual's commitment to and success in performing future tasks and attaining future goals (Bandura, 1977). According to Bandura (1977), two key determinants of behavior are self-efficacy and the expected outcome (outcome expectancy). From a cognitive-behavioral perspective, whether an individual shows excessive fear of falling and associated fear avoidance behavior is determined by their balance self-efficacy and expectations of the consequences of falls.

Self-efficacy has four main sources: (i) mastery experience, (ii) verbal persuasion, (iii) vicarious learning, and (iv) coping strategies to manage emotional reactions (Bandura, 1977). Mastery experience, regarded as the most important source of self-efficacy, is the accumulated sense of success one builds by overcoming new challenges. Verbal persuasion is encouragement provided by others to attain a goal or perform a task. Vicarious learning is the experience of seeing others similar to oneself succeed through sustained effort, leading to the belief that one possesses the same ability to perform a task or attain a goal as others do. Coping strategies reflect one's ability to deploy emotional resources to drive better performance.

Using self-efficacy theory (Bandura, 1977) to trace the mechanism of intervention effects, traditional exercise therapy and CBT-based interventions can be linked with the sources of self-efficacy. Exercise therapy can be hypothesized as reducing fear of falling by increasing the experience of mastery. It is reasonable to expect that in the course of exercise therapy, participants' proprioception will be enhanced, leading to improved agility, balance and coordination through feedback on their bodily improvements. Meanwhile, CBT-based group interventions are expected to foster verbal persuasion and vicarious learning, as individuals can receive encouragement, support and feedback through interactions with therapists and other group members and by observing the success of other group members. Moreover, the use of CBT-based group interventions emphasizes the need to challenge and change misconceptions about falling and develop personalized coping strategies. Thus, this approach may help to reduce individuals' catastrophic thinking and other cognitive distortions regarding the likelihood and consequences of falls and help them develop coping strategies to manage their emotional reactions in daily life, especially an excessive fear of falling. This conceptualization of balance self-efficacy enhancement and modification of fall-related misconceptions lays the framework for CBT-based interventions for fear of falling. In theory, therefore, a multicomponent intervention comprising both an exercise component and a CBT component can bring about optimal therapeutic effects through all four main sources of self-efficacy postulated by Bandura (1977) (Fig. 2).

Combination of CBT and exercise therapy

In Kumar et al. (2016) and Liu, Ng, Chung, and Ng (2018) systematic reviews and meta-analyses, traditional exercise therapy and CBT were found to show comparable efficacy in treating fear of falling. However, according to self-efficacy theory, these components foster different sources of self-efficacy. Therefore, it is logical to examine whether the combined use of CBT and exercise therapy can yield outcomes superior to those of CBT or exercise therapy alone. Recent studies in this area have reported promising findings.

To examine the effects on fear of falling and mobility of the combined use of CBT and tai chi, Huang et al. (2011) conducted a clinical trial with three comparison arms: CBT with tai chi, CBT without tai chi and control. Fear of falling was measured using the Geriatric Fear of Falling Measure (Huang, 2006) and Falls Efficacy Scale (FES) (Tinetti et al., 1990), and mobility was measured using the Tinetti Mobility Scale (TMS) (Tinetti, 1986). After 8 weeks sessions, the outcomes of fear of falling and mobility in both the CBT group and the CBT with tai chi group were superior to those in the

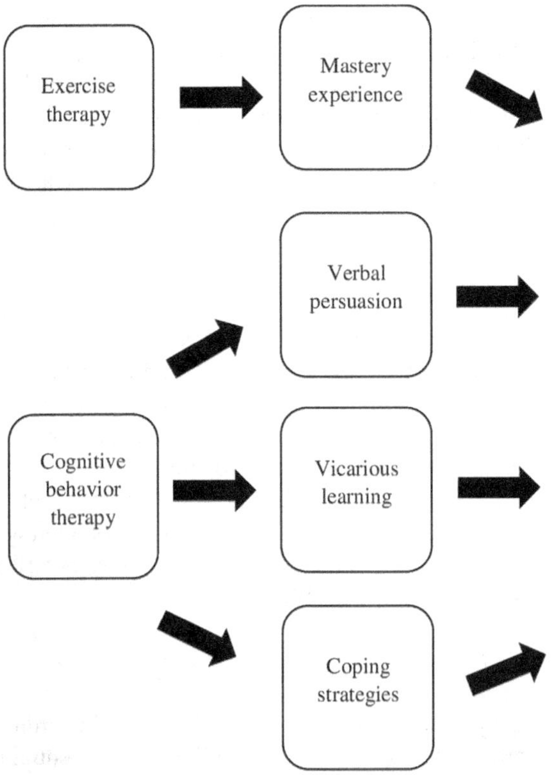

FIG. 2 Cognitive-behavioral therapy and exercise therapy as sources of self-efficacy. This figure illustrate how exercise therapy and cognitive-behavior therapy improves balance self-efficacy and then leading to reduction of fear of falling.

control group. In addition, the CBT with tai chi group showed significantly better outcomes than the CBT and control groups did. Liu and Tsui (2014) also examined the effects of combining CBT and tai chi but used only two comparison arms (tai chi with CBT and without CBT). After 8 weeks sessions, the tai chi with CBT and the tai chi only interventions had similar effects on fear of falling, as measured by the Chinese version of the Falls Efficacy Scale—International (FES-I) (Zhang, Ishikawa-Takata, Yamazaki, Morita, & Ohta, 2006), and mobility, as measured by the Performance-Oriented Assessment of Mobility Problems (Tinetti, 1986). In Liu and Tsui's (2014) study, tai chi with CBT was no more effective than tai chi alone, raising the concern that the use of CBT might not augment exercise interventions. However, the group size of the CBT-based group intervention in Liu and Tsui's (2014) study was high up to 10–11 each group. Thus, their CBT intervention may not have optimized the efficacy of certain essential CBT elements, including cognitive restructuring, goal setting, replacing negative thoughts with positive responses such as exercise and eliminating environmental fall hazards. Further studies are needed to examine whether combining CBT with traditional exercise therapy is superior to the use of either CBT or exercise therapy alone in reducing fear of falling among older people.

Practice and procedures

Elements of a group intervention combining CBT and exercise therapy

In the next section, we outline a protocol for the combined use of CBT and exercise therapy for the treatment of fear of falling and associated avoidance behavior. This outline is not intended to provide a step-by-step manual for treating fear of falling. It is up to researchers and clinicians to decide whether to use CBT alone or to combine it with exercise therapy.

Assessment

Our proposed intervention protocol, combining CBT and exercise therapy, consists of eight weekly 90-min sessions (45 min for each intervention; 90 min in total) with three to five members. Potential clients should be screened for impaired balance self-efficacy and motivation for change before receiving the treatment. Several instruments are available to measure fear of falling. We recommend using the Falls Efficacy Scale—International (FES-I) and the Activity-specific Balance

Confidence (ABC) scale (Powell & Myers, 1995) to distinguish between older people with and without impaired balance self-efficacy, due to their popularity and empirically established cutoff scores (Delbaere et al., 2010; Myers, Fletcher, Myers, & Sherk, 1998). In our previous clinical trial, we also used the Survey of Activities and Fear of Falling in the Elderly (SAFE) (Lachman et al., 1998; Liu & Ng, 2019) to assess the fear of falling-associated avoidance behavior, as fear of falling is not always behaviorally reflected. Moreover, visual and balance abilities, symptoms of depression and anxiety, the levels of daily activity and availability of emotional support should be assessed to determine whether factors that are unrelated to cognitive distortion also contribute to clients' fear of falling.

Exercise therapy

While the CBT component of the intervention protocol aims to improve balance self-efficacy through verbal persuasion, vicarious learning and emotional coping strategies, the exercise therapy component directly addresses fear of falling by enhancing the experience of mastery. In our clinical trial, we selected TOBT as our exercise therapy element. TOBT is a form of exercise training known to be effective in improving balance performance (Ng & Hui-Chan, 2007). It involves the practice of task-specific functional movements, such as squatting, in a natural environment. The TOBT we suggest consists of five exercises: (i) stepping up and down on a wooden step in different directions, (ii) raising heels, (iii) semi-squatting, (iv) standing on a duraDisc, and (v) walking across obstacles. Clients should take turns in completing the TOBT exercises for 8 min each, with 1-min rest intervals. As these five exercises are related to daily tasks, experience of and training in these exercises can be easily and directly transferred to daily use (Table 1).

Cognitive restructuring

While exercise therapy addresses fear of falling by enhancing a person's sense of agility, balance and coordination, cognitive restructuring challenges and replaces excessively fear-provoking thoughts that are believed to be central to the condition. In the intervention described here, clients are assisted in identifying and probing thoughts leading to maladaptive emotional and behavioral responses through CBT homework and discussions with and feedback from therapists and group members during therapy sessions. Fear-provoking thoughts about balance-requiring situations vary in theme but typically fall into two categories: (i) overestimation of the likelihood and consequences of falling and (ii) underestimation of balance ability.

Overestimation occurs through two main forms of automatic thinking: labeling and catastrophizing. In many instance, clients with fear of falling label themselves as vulnerable, such as in the following statements: "As older people always fall, we should not walk around unless it is necessary" and "In the same situations, young people would not fall but we [older people] would." Furthermore, clients often believe that the consequences of falls will be far worse than the actual outcomes, as exemplified by statements such as "If I were to fall, I would not be able to move again."

The second type of thinking typically takes the form of overgeneralization. Clients often develop a general rule based on past experience of falls or loss of balance and apply this rule to related or unrelated situations, as exemplified by statements such as the following: "I nearly fell over in the kitchen while preparing my dinner last year. I no longer dare to go to the market or the kitchen or to perform household chores." Identifying compelling reasons to disconfirm such automatic thoughts is a crucial part of the CBT sessions.

TABLE 1 Components and progression of task-oriented balance training.

Exercise	Method of progression
Stepping up and down	Starting with a 2-in.-high wooden step and then increase the height of the wooden step according to subject's progress
Heel-raising exercises	Starting with a 2-in.-high wooden ramp and then increase the height of the wooden ramp
Semisquatting	Starting with a 30s interval and then increase the semisquatting time according to subject's progress
Standing on duraDisc	Decrease the base of support
Walking across obstacles	Increase number and height of obstacles

This table shows how the difficulty levels of the exercises progress in the task-oriented balance training program.

Family support

Family support may play a crucial role in personal change. When older people engage in self-limiting thinking, they may feel hesitant and in need of someone they trust to support their behavioral change. During or before the proposed intervention, therapists can communicate the purpose of the CBT sessions to family members to reduce family friction owing to changes in living habits or levels of physical activity. Family members can also help the clients to learn or re-learn safety behavior in their living environments and encourage them to generalize their newly learned coping strategies to other aspects of daily life.

Preparing for the CBT-based multicomponent group intervention

The CBT component of our intervention protocol aims to improve clients' balance self-efficacy and reduce catastrophizing that prevents them from undertaking balance-requiring activities due to excessive fear of falling. Therefore, it integrates didactic information on falls and the consequences of falls, exercise therapy to enhance proprioception and cognitive restructuring strategies to justify cognition and help clients minimize their falls and correct related maladaptive thoughts. Note that although we here describe components previously delivered to our clients (Liu et al., 2019), this part of the intervention can be flexibly adapted to suit the personal preferences of therapists and/or the nature of the clientele. The following provides a general illustration of how clinicians could design and structure the CBT-based group intervention (Table 2).

Week 1 (Sessions 1 and 2): Building rapport, understanding self-efficacy and setting a list of realistic goals

The first two sessions focus on building rapport and introducing the concept of CBT and the goals of the CBT-based group intervention. Toward the end of these two sessions, group members should start sharing their personal experiences and beliefs regarding balance-requiring situations and falls and setting goals for safely increasing their levels of physical activity.

One way to build rapport with clients is to share emotions. In these sessions of the protocol, therapists can start by helping their clients to develop lists—as comprehensive as possible—of balance-requiring situations (such as having a shower or going to the market) and physical activities that they feel they can no longer do due to fear of falling (such as having a shower daily rather than on alternate days or going to the market alone rather than going only when accompanied). While developing these lists, the clients can share their views on and feelings about balance-requiring situations and falls to develop a sense of togetherness.

After identifying potential restrictions on physical activities, the clients can rate how strongly they believe that they would fall in each situation on a scale of 0–10 (where "0" indicates that falling may not happen and higher numbers indicate that falling is more likely). The therapists and clients should then discuss and prioritize a set of targeted physical activities to manage during the sessions. The therapists should guide their clients to be aware of what they feel when they perceive a fall risk and the behavioral signs of fear. The therapists should also introduce the concept of self-efficacy. The clients' beliefs about falling in each identified situation of restricted physical activity should be checked by the therapists at the beginning of each session.

TABLE 2 Weekly cognitive-behavioral therapy session themes.

Week	Weekly theme
1	Establishing therapeutic relationship and goal setting
2	Understanding the CBT model
3	Developing fall-prevention strategies
4–7	Challenging and replacing maladaptive thoughts and implementing behavioral changes
8	Consolidating the experiences of intervention and developing relapse prevention plan

This table summarizes the weekly themes of the cognitive-behavioral therapy program.

Week 2 (Sessions 3 and 4): Understanding the relationships between thoughts, emotions, and behavior

The goal of Sessions 3 and 4 is to introduce the concept and models of CBT. Between Sessions 3 and 4, the clients should be assigned CBT homework. For each of the previously identified balance-requiring situations, they should be asked to list the antecedents of, behavioral responses to and consequences of an increase in fear. Next, for each of these situations, the clients should be guided to assess the possibility of viewing the situation differently. To do this, the therapists should assess various daily situations with the clients, including exercise, leisure activities and social activities, and explore whether the clients believe that falls would happen in those situations and how fearful or worried they are about the consequences. For each of these situations, the clients should rate the strength of their belief that they would fall and the degree of their anxiety about the consequences. The clients should then record their behavioral responses, including the adoption of measures to prevent falling and/or avoidance behavior, in those balance-requiring situations.

Week 3 (Sessions 5 and 6): Setting realistic goals and adopting fall preventive measures

After the clients have gained an understanding of the CBT model, the therapists and clients should together use the information from the CBT homework to develop a list of potential situations resulting in fall ranked by perceived severity. The therapists should discuss with their clients how each of the daily situations identified during the therapy sessions and CBT homework may exacerbate fear of falling and inadvertently reinforce fall avoidance behavior, in turn leading to decreased physical activity, physical deconditioning and eventually the increased occurrence of actual falls. After explaining how the vicious cycle between fear of falling and actual falling arises in the situations identified, the therapists can help the clients to review potential environmental fall hazards in their homes and feasible fall preventive measures. For example, they could inform the clients of their right to contact local housing authorities or community elderly services to request that handrails be installed in washrooms.

Weeks 4–7 (Sessions 7–14): Exploring adaptive thoughts and behavioral responses and implementing and reviewing behavioral changes

Once the clients are proficient in identifying their maladaptive thoughts, they should be encouraged to challenge the accuracy of these thoughts and develop competing ones to challenge them. The therapists should affirm their clients' efforts in this cognitive restructuring process and prompt them to undertake restructuring whenever maladaptive thoughts related to fear of falling arise. Moreover, during and/or after this cognitive restructuring process, the therapists should elicit from the clients a detailed description of anticipated barriers, steps and goals for each of the targeted behavioral changes. Consider the example of a client who wishes to participate in activities organized by a local center for older people but has not yet done so because he is worried that he might fall on the way there. A therapist could help this client to break the visit into several steps, starting by walking around the building, and adopt fall preventive measures, such as using a walking aid, walking one block away from home with the walking aid, walking several blocks away from home with the walking aid and eventually walking all the way to and from the center. If the client expects his children to object to the plan, at least one of the children could be involved in the planning stage. The client should also be encouraged to ensure safety during this behavioral change, such as by determining the safest route and time to visit the center and developing a contingency plan.

Week 8 (Sessions 15 and 16): Consolidating the experience of the intervention and developing a relapse prevention plan

Sessions 15 and 16 of the intervention protocol involve a review of the discussion and other aspects of the group sessions. They are not intended to introduce additional functional intervention components to safely increase the clients' level of physical activity. Session 16 also includes a discussion of relapse prevention.

The final stage of the CBT-based group intervention should involve a summary of the clients' consolidated experience of the treatment sessions. It should also include some form of outcome assessment, such as the FES-I/ABC to assess balance self-efficacy, the Lawton Instrumental Activities of Daily Living scale to assess the level of daily activity, SAFE (Lachman et al., 1998; Liu & Ng, 2019) to assess fear avoidance behavior or the 12-item Short Form Survey (Ware, Kosinski, & Keller, 1996) to assess quality of life. Comparing their pre- and postintervention scores for these outcome measures will help the clients to assess the progress they have made. The therapists should also help the clients to identify and articulate the changes in their views and behavioral responses during the therapy sessions.

Most importantly, Sessions 15 and 16 involve the development of a relapse prevention plan. The therapists should help their clients to develop personal reminders, such as personal mottos, to stay vigilant in detecting fear avoidance behavior due to excessive fear of falling, maintaining exercise habits and applying appropriate and realistic fall preventive measures. It is also advisable to involve family members or carers in the therapy sessions to cultivate their awareness of situations that exacerbate fear of falling and associated fear avoidance behavior. If fear of falling and associated fear avoidance behavior reappear or worsen, the cognitive restructuring skills learned in therapy sessions should be revisited to examine the situational triggers.

Summary and recommendations

Fear of falling ranges in function from self-protective to self-limiting and induces various health problems. In this chapter, we review the literature on the medical concept of fear of falling and how CBT can be used in a group format to help clients who report excessive anxiety about engaging in daily physical activities due to this fear.

Despite the recent increase in clinical trials of the use of CBT-based techniques to reduce fear of falling, several important research questions have yet to be answered. For example, to the best of our knowledge, only one study has examined the effects of CBT on avoidance behavior associated with fear of falling among older people. However, this study did report promising findings, with carryover effects lasting for up to 12 months (Dorresteijn et al., 2016). Future clinical trials should address this crucial factor in the vicious cycle of fear of falling and actual falls.

It is also important to note that no studies have directly compared different forms of CBT-based interventions to reduce fear of falling and associated fear avoidance behavior. The literature has described the use of CBT-based interventions that differ according to delivery format (e.g., individual vs group), length and intensity. This heterogeneity may complicate the identification and design of effective treatments for fear of falling. Future studies are encouraged to refine existing CBT interventions to crease a highly structured protocol for clinical practice. We also lack a clear understanding of the causes of fear of falling. Several clinical trials have explored the effects of possible treatments, including physiological and psychological interventions applied either alone or in combination. However, only Bandura's (1977) self-efficacy theory offers insights into the mechanisms of these treatment effects. The Fear of Falling Management Model proposed by Huang (2006) conceptualizes the influence of fall beliefs on the adoption of negative or positive fall coping strategies, which may respectively exacerbate or minimize the physical and psychological effects of fear of falling. However, more research is needed to delineate the causes and course of development of fear of falling and associated avoidance behavior.

In sum, CBT for fear of falling is an empirically supported psychological and behavioral approach for older people both with and without disabling diseases. It can be effectively integrated with a range of other physical interventions. Compared with the use of exercise therapy alone, CBT also appears to be a more durable treatment option for reducing fear of falling and improving balance. The efficacy of CBT, either alone or combined with exercise therapy, has been documented in a series of randomized controlled trials. Therefore, we recommend that future research focus on enhancing understanding of the etiology, development and maintenance of fear of falling and fear avoidance behavior. Therapeutic advancements in treating fear of falling with CBT have been made in the field of rehabilitation in recent decades. The incorporation of other therapeutic elements, such as motivational interviewing and mindfulness techniques, could be explored.

Mini-dictionary of terms

- Fear of falling: The excessive fear of losing one's balance, stemming from impaired balance self-efficacy and/or misinterpretation of balance-requiring situations.
- Fear avoidance behavior: The behavioral reflection of fear of falling, in which individuals avoid balance-requiring situations regardless of their balance ability and environmental demands.
- Task-oriented balance training: A form of exercise therapy designed to improve muscle strength in the lower limbs and correct balance deficits.
- Didactic education: A passive teaching and learning method in which learners are expected to listen and answer when required. Discussion and problem solving play limited roles.
- Cognitive restructuring: A psychotherapeutic technique intended to help individuals identify, challenge and replace maladaptive thoughts.
- Self-efficacy: An individual's belief in their ability to perform a specific task.
- Vicarious learning: A learning method in which individuals learn from observation without direct instruction.
- Proprioception: Sense of the relative positions, motion and equilibrium of body parts.
- Visual proprioception: Sense of the relative positions, motion and equilibrium of body parts through optic flow.
- Standardized mean difference: A statistical measure of the absolute difference between the mean values of two groups.

Key facts of fear of falling

- Of all potential daily misfortunes, falling induces the greatest fear among older people. This fear affects a large proportion of older people, regardless of whether they have a recent fall history.
- A recent fall history, old age, female gender, dizziness, low self-perceived health status, depression, gait and balance problems, cognitive complaints and low economic status are risk factors for fear of falling.
- Fear of falling may or may not be behaviorally reflected. If behaviorally reflected, it may lead to fear avoidance actions, physical deconditioning, a loss of independence, compromised quality of life and eventually the increased occurrence of actual falls.
- Studies have not adequately explained the causes of fear of falling. Recent literature has postulated that fear of falling stems from impaired self-efficacy and cognitive distortions regarding balance-requiring situations.
- Although no interventions have been found to reliably prevent the development of fear of falling and associated avoidance behavior, recent studies have suggested that this form of psychological and behavioral disturbance can be effectively alleviated by interventions.

Applications to other areas

The CBT-based intervention protocol described here can be optimized to address various physiological, psychological, and/or social issues. The example presented in this chapter involves the combined use of CBT and exercise therapy to reduce fear of falling and associated avoidance behavior. As CBT is now seen as a leading evidence-based intervention for fear of falling, it is logical to generalize its application to other people facing similar difficulties.

There is a pressing need for effective interventions for fear of falling. While the combined use of CBT and exercise therapy has yielded positive outcomes, there are several unexplored populations, especially people with disorders that affect balance, such as dementia, osteoporosis, multiple sclerosis, and hip fracture. One of the merits of CBT as a psychotherapeutic intervention is that it is a goal-oriented technique that places a strong emphasis on defining behavior and treatment outcomes in measurable terms. This characteristic lays the foundation for CBT-based interventions to be scientifically evaluated in clinical trials. Moreover, CBT is compatible with other treatment components, including pharmacological, physical, and psychological components. Thus, a CBT-based intervention for fear of falling has the flexibility to be incorporated into other treatment protocols for particular patient populations. In addition, CBT seems to be a safe clinical intervention, as no major adverse events have been reported in clinical application. Thus, the CBT-based intervention discussed in this chapter has the potential to be widely applied to reduce fear of falling in other patient populations in the near future.

Summary points

- The relationship between excessive fear of falling and the occurrence of actual falls is widely recognized as a vicious cycle between fear of falling and actual falls, which poses serious challenges for clinicians.
- Exercise therapy and CBT have been clinically proven to be effective in reducing fear of falling and associated avoidance behavior.
- It is hypothesized that the mechanism underlying the treatment effects of exercise therapy and CBT on fear of falling is based on Bandura's self-efficacy theory.
- We recommend the use of an intervention combining exercise therapy and CBT for two main reasons: (i) theoretically, such an intervention can foster the four main sources of self-efficacy and (ii) empirically, previous clinical trials have revealed that the combined use of exercise therapy and CBT yields better outcomes than does exercise therapy or a CBT-based intervention alone.
- Future studies could further expand understanding of the etiology, development and maintenance of fear of falling and fear avoidance behavior and explore the possibility of incorporating other therapeutic elements, such as motivational interviewing and mindfulness techniques, into current treatment approaches.

References

Alexander, J. L., Sartor-Glittenberg, C., Bordenave, E., & Bordenave, L. (2015). Effect of the Matter of Balance Program on balance confidence in older adults. *The Journal of Gerontopsychology and Geriatric Psychiatry*, *28*(4), 183–189.

Arai, T., Obuchi, S., Inaba, Y., Nagasawa, H., Shiba, Y., Watanabe, S., … Kojima, M. (2007). The effects of short-term exercise intervention on falls self-efficacy and the relationship between changes in physical function and falls self-efficacy in Japanese older people: A randomized controlled trial. *American Journal of Physical Medicine & Rehabilitation*, *86*(2), 133–141.

Bandura, A. (1977). Self-efficacy: Toward a unifying theory of behavioral change. *Psychological Review*, *84*(2), 191–215.

Bertenthal, B. I., & Campos, J. J. (1990). A systems approach to the organizing effects of self-produced locomotion during infancy. In C. Rovee-Collier, & L. Lipsitt (Eds.), *Vol. 6. Advances in infancy research* (pp. 1–60). Norwood, NJ: Ablex.

Bhala, R. P., O'Donnell, J., & Thoppil, E. (1982). Ptophobia: Phobic fear of falling and its clinical management. *Physical Therapy*, *62*(2), 187–190.

Campos, J., Bertenthal, B., & Kermoian, R. (1992). Early experience and emotional development: The emergence of wariness of heights. *Psychological Science*, *3*(1), 61–64.

Clemson, L., Cumming, R. G., Kendig, H., Swann, M., Heard, R., & Taylor, K. (2004). The effectiveness of a community-based program for reducing the incidence of falls in the elderly: A randomized trial. *Journal of the American Geriatrics Society*, *52*(9), 1487–1494.

Cumming, R. G., Salkeld, G., Thomas, M., & Szonyi, G. (2000). Prospective study of the impact of fear of falling on activities of daily living, SF-36 scores, and nursing home admission. *Journals of Gerontology Series A: Biological Sciences and Medical Sciences*, *55*(5), M299–M305.

Dahl, A., Campos, J., Anderson, D., Uchiyama, I., Witherington, D. C., Ueno, M., … Barbu-Roth, M. (2013). The epigenesis of wariness of heights. *Psychological Science*, *24*(7), 1361–1367.

Delbaere, K., Close, J. C., Mikolaizak, A. S., Sachdev, P. S., Brodaty, H., & Lord, S. R. (2010). The falls efficacy scale international (FES-I). A comprehensive longitudinal validation study. *Age and Ageing*, *39*(2), 210–216.

Delbaere, K., Crombez, G., van Haastregt, J. C., & Vlaeyen, J. W. (2009). Falls and catastrophic thoughts about falls predict mobility restriction in community-dwelling older people: A structural equation modelling approach. *Aging & Mental Health*, *13*(4), 587–592.

Dorresteijn, T. A., Zijlstra, G. R., Ambergen, A. W., Delbaere, K., Vlaeyen, J. W., & Kempen, G. I. (2016). Effectiveness of a home-based cognitive behavioral program to manage concerns about falls in community-dwelling, frail older people: Results of a randomized controlled trial. *BMC Geriatrics*, *16*(1), 1–11.

Friedman, S., Munoz, B., West, S., Rubin, G., & Fried, L. (2002). Falls and fear of falling: Which comes first? A longitudinal prediction model suggests strategies for primary and secondary prevention. *Journal of the American Geriatrics Society*, *50*(8), 1329–1335.

Gibson, E., & Walk, R. (1960). The "visual cliff". *Scientific American*, *202*(4), 64–71.

Howland, J., Peterson, E. W., Levin, W. C., Fried, L., Pordon, D., & Bak, S. (1993). Fear of falling among the community-dwelling elderly. *Journal of Aging and Health*, *5*(2), 229–243.

Huang, T. T. (2006). Geriatric fear of falling measure: Development and psychometric testing. *International Journal of Nursing Studies*, *43*(3), 357–365.

Huang, T. T., Chung, M. L., Chen, F. R., Chin, Y. F., & Wang, B. H. (2016). Evaluation of a combined cognitive-behavioural and exercise intervention to manage fear of falling among elderly residents in nursing homes. *Aging & Mental Health*, *20*(1), 2–12.

Huang, T. T., Yang, L. H., & Liu, C. Y. (2011). Reducing the fear of falling among community-dwelling elderly adults through cognitive-behavioural strategies and intense Tai Chi exercise: A randomized controlled trial. *Journal of Advanced Nursing*, *67*(5), 961–971.

Kumar, A., Delbaere, K., Zijlstra, G. A. R., Carpenter, H., Iliffe, S., Masud, T., … Kendrick, D. (2016). Exercise for reducing fear of falling in older people living in the community: Cochrane systematic review and meta-analysis. *Age and Ageing*, *45*(3), 345–352.

Lachman, M. E., Howland, J., Tennstedt, S., Jette, A., Assmann, S., & Peterson, E. W. (1998). Fear of falling and activity restriction: The survey of activities and fear of falling in the elderly (SAFE). *Journals of Gerontology Series B: Psychological Sciences and Social Sciences*, *53*(1), P43–P50.

Landers, M. R., Durand, C., Powell, D. S., Dibble, L. E., & Young, D. L. (2011). Development of a scale to assess avoidance behavior due to a fear of falling: The Fear of Falling Avoidance Behavior Questionnaire. *Physical Therapy*, *91*(8), 1253–1265.

Lin, M. R., Hwang, H. F., Wang, Y. W., Chang, S. H., & Wolf, S. L. (2006). Community-based tai chi and its effect on injurious falls, balance, gait, and fear of falling in older people. *Physical Therapy*, *86*(9), 1189–1201.

Liu, T. W., & Ng, S. S. (2019). The reliability and validity of the Survey of Activities and Fear of Falling in the Elderly for assessing fear and activity avoidance among stroke survivors. *PLoS One*, *14*(4), e0214796.

Liu, T. W., Ng, G. F., Chung, R. C., & Ng, S. S. (2018). Cognitive behavioural therapy for fear of falling and balance among older people: A systematic review and meta-analysis. *Age and Ageing*, *47*, 520–527.

Liu, T. W., Ng, G. F., Chung, R. C., & Ng, S. S. (2019). Decreasing fear of falling in chronic stroke survivors through cognitive behavior therapy and task-oriented training. *Stroke*, *50*, 148–154.

Liu, H., Rainey, J., Zabel, R., Quiben, M. U., Kehayov, A., & Boswell, J. K. (2007). Comparison of two exercise programs using the Falls Efficacy Scale, Berg Balance Scale and ankle dorsiflexor strength in older adults. *Physical and Occupational Therapy in Geriatrics*, *26*, 23–42.

Liu, Y. W. J., & Tsui, C. M. (2014). A randomized trial comparing Tai Chi with and without cognitive-behavioral intervention (CBI) to reduce fear of falling in community-dwelling elderly people. *Archives of Gerontology and Geriatrics*, *59*(2), 317–325.

Murphy, S., Dubin, J., & Gill, T. (2003). The development of fear of falling among community-living older women: Predisposing factors and subsequent fall events. *Journals of Gerontology Series A: Biological Sciences and Medical Sciences*, *58*(10), M943–M947.

Murphy, J., & Isaacs, B. (1982). The post-fall syndrome. *Gerontology*, *28*(4), 265–270.

Myers, A. M., Fletcher, P. C., Myers, A. H., & Sherk, W. (1998). Discriminative and evaluative properties of the activities-specific balance confidence (ABC) scale. *Journals of Gerontology Series A: Biological Sciences and Medical Sciences*, *53*(4), M287–M294.

Ng, S. S., & Hui-Chan, C. W. (2007). Transcutaneous electrical nerve stimulation combined with task-related training improves lower limb functions in subjects with chronic stroke. *Stroke*, *38*(11), 2953–2959.

Parry, S. W., Bamford, C., Deary, V., Finch, T. L., Gary, J., MacDonald, C., McMeekin, P., Sabin, N. J., Steen, I. N., Whitney, S. L., & McColl, E. M. (2016). Cognitive-behavioural therapy-based intervention to reduce fear of falling in older people: therapy development and randomised controlled trial - the Strategies for Increasing Independence, Confidence and Energy (STRIDE) study. *Health Technology Assessment*, *20*(56), 1–206.

Powell, L. E., & Myers, A. M. (1995). The activities-specific balance confidence (ABC) scale. *Journals of Gerontology Series A: Biological Sciences and Medical Sciences*, *50*(1), M28–M34.

Reinsch, S., MacRae, P., Lachenbruch, P. A., & Tobis, J. S. (1992). Attempts to prevent falls and injury: A prospective community study. *The Gerontologist*, *32*(4), 450–456.

Rochat, S., Büla, C. J., Martin, E., Seematter-Bagnoud, L., Karmaniola, A., Aminian, K., … Santos-Eggimann, B. (2010). What is the relationship between fear of falling and gait in well-functioning older persons aged 65 to 70 years? *Archives of Physical Medicine and Rehabilitation*, *91*(6), 879–884.

Sattin, R. W., Easley, K. A., Wolf, S. L., Chen, Y., & Kutner, M. H. (2005). Reduction in fear of falling through intense tai chi exercise training in older, transitionally frail adults. *Journal of the American Geriatrics Society*, *53*(7), 1168–1178.

Scheffer, A., Schuurmans, M., van Dijk, N., van der Hooft, T., & de Rooij, S. (2008). Fear of falling: Measurement strategy, prevalence, risk factors and consequences among older persons. *Age and Ageing*, *37*(1), 19–24.

Tennstedt, S., Howland, J., Lachman, M., Peterson, E., Kasten, L., & Jette, A. (1998). A randomized, controlled trial of a group intervention to reduce fear of falling and associated activity restriction in older adults. *Journals of Gerontology. Series B, Psychological Sciences and Social Sciences*, *53B*(6), 384–392.

Tinetti, M. E. (1986). Performance-oriented assessment of mobility problems in elderly patients. *Journal of the American Geriatrics Society*, *34*(2), 119–126.

Tinetti, M., Richman, D., & Powell, L. (1990). Falls efficacy as a measure of fear of falling. *Journal of Gerontology*, *45*(6), P239–P243.

van Haastregt, J. C. M. (2002). *Preventing falls and mobility impairments in elderly people living in the community*. Universiteit Maastricht.

Ware, J. E., Jr., Kosinski, M., & Keller, S. D. (1996). A 12-Item Short-Form Health Survey: Construction of scales and preliminary tests of reliability and validity. *Medical Care*, *34*(3), 220–233.

Weerdesteyn, V., Rijken, H., Geurts, A. C., Smits-Engelsman, B. C., Mulder, T., & Duysens, J. (2006). A five-week exercise program can reduce falls and improve obstacle avoidance in the elderly. *Gerontology*, *52*(3), 131–141.

Zhang, J. G., Ishikawa-Takata, K., Yamazaki, H., Morita, T., & Ohta, T. (2006). The effects of Tai Chi Chuan on physiological function and fear of falling in the less robust elderly: An intervention study for preventing falls. *Archives of Gerontology and Geriatrics*, *42*(2), 107–116.

Section D

International aspects

Chapter 30

Cognitive-behavioral therapy for children with anxiety in Japan: Bidirectional cultural adaptation and cross-cultural comparison studies

Shin-ichi Ishikawa
Faculty of Psychology, Doshisha University, Kyotanabe City, Kyoto, Japan

Abbreviations

C-BOS	Cross-cultural Behavioral Observation System
CBT	cognitive-behavioral therapy
CDI	Children's Depression Inventory
DSRS	Depression Self-Rating Scale for Children
ITT	intent to treat
JACA-CBT	Japanese Anxiety Children/Adolescents Cognitive Behavior Therapy program
RCT	randomized controlled trial
SCAS	Spence Children's Anxiety Scale
SCAS-P	Parent version of the Spence Children's Anxiety Scale

Introduction

Anxiety disorders are among the most prevalent psychological disorders in children and adolescents. Previous studies have shown that approximately 3%–20% of children suffer from anxiety disorders (Costello, Mustillo, Erkanli, Keeler, & Angold, 2003; Kroes et al., 2001; Lewinsohn, Hops, Roberts, Seeley, & Andrews, 1993), although at different prevalent rates depending on the methodology and populations used (Cartwright-Hatton, McNicol, & Doubleday, 2006). In adolescents (13–18 years), the lifetime prevalence of anxiety disorders has reached 31.9% from 2001 to 2004 (Merikangas et al., 2010). Severe anxiety incurs social, academic, and familial dysfunctions in children's daily lives (Rapee, Schniering, & Hudson, 2009). Anxiety disorders during childhood and adolescence may increase the probability of having anxiety during emerging adulthood (Essau, Lewinsohn, Lim, Ho, & Rohde, 2018). Childhood anxiety predicts an above-average level of life interference in early adulthood (Rapee et al., 2009) and later mental health problems such as anxiety disorders, mood disorders, and substance abuse (Seligman & Gahr, 2013).

This chapter reviews the different childhood anxiety symptoms experienced by individuals in Japan compared with those in other countries using a standardized measure. Second, the rationale for, methodology, and efficacy of cultural adaptation of cognitive-behavioral therapy (CBT) are discussed. Specifically, this chapter delineates the detailed procedure of bidirectional cultural adaptation aimed at the development of a novel treatment program derived from previous evidence-based programs that are not a simple translation of existing programs. Third, a triad model of cultural adaptation is introduced and a comparison study of therapy factors in CBT treatment sessions between Japan and Australia is discussed.

Childhood anxiety symptoms in Japan

Sugawara et al. (1999) conducted a study on psychological disorders in Japan that included 114 children aged between 7 and 9 years. The study suggested that 41.2% children had anxiety disorders, indicating a notably high prevalence rate

Handbook of Lifespan Cognitive Behavioral Therapy. https://doi.org/10.1016/B978-0-323-85757-4.00017-1

compared with other countries (see the review by Cartwright-Hatton et al., 2006). However, Japan's apparent highest prevalence rate compared with other countries should be carefully interpreted because Sugawara et al.'s (1999) study was conducted almost a quarter of a century ago, and the assessment used in the study is not commonly used for the diagnosis of anxiety disorders. On the other hand, the prevalence rate of depression in the survey (3.5%) was comparable with the prevalence rate obtained (1.1%) in a study by Denda (2008) conducted with 9–10-year-old children; hence, the extent of the deviation from the normative sample might be considered minimal. Although there could be an overestimation, it is reasonable to consider anxiety as a prevalent childhood psychological problem in Japan.

Prevalence surveys using diagnostic interviews have been scarce in Japan since the study by Sugawara et al. (1999). Nonetheless, the Spence Children's Anxiety Scale (SCAS; Spence, 1998), a self-report questionnaire, is commonly used to examine Japanese children's anxiety symptoms; the SCAS has been translated into various languages and disseminated globally. Ishikawa, Sato, and Sasagawa (2009) developed a Japanese version for 8–12-year-old children and early adolescents between 12 and 15 years old. In addition, a Japanese version for middle adolescents between 15 and 18 years old (Ishikawa et al., 2018) and a parent's version (SCAS-P) are also available (Ishikawa et al., 2014).

Table 1 presents data from existing studies for an international comparison of the SCAS. The average score of Japanese children aged 8–12 years old has been lower than that of the original study in Australia and those of South American countries such as Brazil and Colombia. However, Japanese data are located in the middle of the European countries; specifically, it has been lower than Greek, Spanish, and Italian, comparable with German, and higher than Dutch.

Although available data for children are limited among Asian countries, the Japanese data are at the same level as Chinese data. Comparing the early and middle adolescents' results, the Japanese scores are located in the middle of the other countries. Specifically, Japanese adolescents show anxiety symptoms comparable to those of Western countries, such as Australia, Bulgaria, Denmark, Germany, Spain, and Sweden. While Japanese data are lower than in other Asian countries such as China, Iran, and Pakistan, they are higher than those in Belgium and the Netherlands.

Necessity of cultural adaptation for psychosocial treatments

In collectivistic cultures such as Japan, the ability to think about and feel what and how others think and feel could be emphasized to harmonize within a social group rather than reflecting on individual feelings, thoughts, and uniqueness (Markus & Kitayama, 1991). Furthermore, Japanese mothers tend to be close to their children as much as possible to ensure that they feel secure; this also minimizes the boundary between the mother and the child. Moreover, this closeness ensures that children's needs are met (Azuma, 1986). Socialization practices in such cultural backgrounds may be associated with anxiety. A strong emphasis on other people's opinions might be relevant to social anxiety (see Essau, Ishikawa, & Sasagawa, 2011), and mothers' indulgent devotion with an ambiguous boundary might lead to inappropriate accommodations. Nevertheless, previous studies based on self-reported measurements have not supported the presence of higher anxiety symptoms in Japanese children than in children from Western countries, while the earliest prevalence study that employed interviews has suggested high prevalence rates.

Asian clients with internalizing disorders such as depression and anxiety tend to somatize their emotional symptoms more than those from Western cultural backgrounds, where talking about problems and expressing oneself verbally are more greatly underscored (Chun, Enomoto, & Sue, 1996). In addition, due to the stigma of mental health and even its treatment, it typically takes more time for Asian clients to talk about their emotional difficulties via open and clear verbal communication (Hwang, 2016). Therefore, we should consider the nature of childhood anxiety in Japanese culture. Specifically, although culturally specific family relationships might raise children's anxiety, self-reported symptoms could be underreported as emotional symptoms. This discrepancy might represent a culturally specific expression of anxiety symptoms in Japan.

Regarding anxiety disorders in children and adolescents, CBT is considered by many to be an effective psychosocial treatment. A Cochrane review articulated that CBT is superior to wait-list control, with remission rates of 59.4% versus 17.5%, respectively (James, James, Cowdrey, Soler, & Chole, 2013). However, previous outcome studies have been conducted exclusively on populations from Western cultures; thus, little attention has been paid to cultural influences related to ethnicity and religion (Iwamasa & Hays, 2019). Because functional communication is fundamental for CBT, clinicians should be mindful that the current evidence-based psychosocial treatment, including CBT, contains flavors and nuances from Western cultures inherently, given the developmental as well as historical backgrounds.

Cultural adaptation is the systematic modification of an evidence-based treatment or intervention protocol that considers language, culture, and the context in such a way that is compatible with the client's cultural patterns, meanings, and values (Tanaka-Matsumi, 2011). Culturally adapted treatments focus not only on improving therapeutic outcomes but also on increasing client engagement, knowledge and buy-in of treatment processes and methods, consumer

TABLE 1 Comparison of anxiety symptoms in children and adolescents measured by the Spence Children's Anxiety Scale by country.

		Children					Adolescents				
Country	Sex	Author(s)	Sample size	Age	*M*	*SD*	Author(s)	Sample size	Age	*M*	*SD*
Australia	Boys	Spence (1998)	2052	8–12	27.20	(16.15)	Spence, Barrett, and Turner (2003)	875	13–14	18.85	(13.07)
	Girls				34.14	(17.27)				25.08	(25.08)
	Total				31.28	(17.35)				21.72	(21.72)
Belgium	Boys						Muris, Merckelbach, Ollendick, King, and Bogie (2002)	521	12–18	12.7	(10.4)
	Girls									19.1	(12.9)
	Total									16.9	(13.1)
Brazil	Boys	DeSousa et al. (2014)	712	7–17[a]	–	–	DeSousa et al. (2014)	712	7–17[a]	–	–
	Girls				–	–				–	–
	Total				37.29	(15.84)				37.29	(15.84)
Bulgaria	Boys						Tsocheva, Sasagawa, Georgiou, and Essau (2013)	700	13–17	19.87	(11.6)
	Girls									31.48	(15.9)
	Total									25.23	(14.9)
China	Boys	Zhao, Xing, and Wang (2012)	996	8–12	19.14	(15.81)	Zhao et al. (2012)	882	13–15	22.10	(13.44)
	Girls				23.92	(16.77)				31.12	(16.30)
	Total				21.43	(13.94)				26.68	(15.54)
Colombia	Boys	Amaya and Campbell (2010)	253	8–12	33.41	(15.14)					
	Girls				40.07	(17.72)					
	Total				36.65	–					
Cyprus	Boys						Essau, Sasagawa, Anastassiou-Hadjicharalambous, Guzmán, and Ollendick (2011)	495	12–17	–	–
	Girls									–	–
	Total									28.42	(15.57)
Denmark	Boys	Arendt, Hougaard, and Thastum (2014)	315	7–12	18.44	(13.36)	Arendt et al. (2014)	168	13–17	13.82	(9.17)
	Girls				27.74	(15.10)				26.23	(12.85)
	Total				–	–				–	–
England	Boys						Essau, Sasagawa, et al. (2011)	469	12–17	–	–
	Girls									–	–
	Total									34.33	(21.10)
Germany	Boys	Essau, Sakano, Ishikawa, and Sasagawa (2004)	862	8–12	18.39	(11.0)	Essau, Leung, Conradt, Cheng, and Wong (2008)	594	12–17	18.66	(10.1)
	Girls				27.20	(15.5)				25.54	(10.4)
	Total				22.86	(14.2)				22.75	(10.8)

Continued

TABLE 1 Comparison of anxiety symptoms in children and adolescents measured by the Spence Children's Anxiety Scale by country—cont'd

		Children					Adolescents				
Country	Sex	Author(s)	Sample size	Age	*M*	*SD*	Author(s)	Sample size	Age	*M*	*SD*
Greece	Boys	Mellon and Moutavelis (2007)	1520	9–12	38.4	(15.9)					
	Girls				46.4	(17.2)					
	Total				42.3	(17.0)					
Iran	Boys						Essau, Olaya, Pasha, O'Callaghan, and Bray (2012)	1984	12–17	30.62	(16.71)
	Girls									35.61	(17.78)
	Total									33.15	(17.43)
Italy	Boys	Di Riso, Chessa, Bobbio, and Lis (2013)	1397	8–11	27.41	(15.81)	Essau, Sasagawa, et al. (2011)	499	12–17	–	–
	Girls				38.07	(16.77)				–	–
	Total				32.64	(17.13)				27.11	(15.42)
Japan	Boys	Ishikawa et al. (2009)	1046	8–12	19.06	(16.47)	Ishikawa et al. (2009)	1182	12–15	15.93	(12.06)
	Girls				27.89	(19.82)				26.17	(16.68)
	Total				23.50	(18.75)				20.93	(15.38)
	Boys						Ishikawa et al. (2018)	1500	15–18	18.53	(14.16)
	Girls									26.83	(15.94)
	Total									22.74	(15.64)
Netherlands	Boys	Muris, Schmidt, and Merckelbach (2000)	1011[b]	7–12	17.42	(13.95)	Muris et al. (2000)	1011[b]	13–19	13.54	(10.69)
	Girls				23.69	(13.79)				19.89	(11.83)
	Total				20.51	(14.20)				16.56	(11.68)
Pakistan	Boys						Qadir et al. (2018)	1277	13–17	36.83	(16.46)
	Girls									47.04	(14.84)
	Total									41.38	(16.38)
Spain	Boys	Orgilés, Méndez, Spence, Huedo-Medina, and Espada (2012)	1708	8–12	32.29	(14.49)	Orgilés, Spence, Marzo, Méndez, and Espada (2014)	1374	13–17	22.27	(11.07)
	Girls				33.64	(16.18)				27.97	(12.68)
	Total				33.94	(15.44)				25.26	(12.52)
Sweden	Boys						Essau, Sasagawa, et al. (2011)	484	12–17	–	–
	Girls									–	–
	Total									23.94	(15.63)

[a]*Not separated into children and adolescents.*
[b]*Total sample size includes both children and adolescents.*

satisfaction, reducing premature dropout, and improving therapeutic relationships (Hwang, 2016). Evidence has demonstrated the efficacy and effectiveness of culturally adapted treatments. For example, Hall, Ibaraki, Huang, Marti, and Stice (2016) integrated 11 published meta-analyses with 13,998 participants in 78 studies. The authors found moderate effect sizes, which supported the efficacy of culturally adapted interventions over other conditions ($g = 0.67$) and compared with unadapted interventions ($g = 0.52$). However, research on the impact of culturally adapted CBT programs for Asian cultural groups focusing on anxiety disorders is still in its infancy. Although a meta-analysis by Hall et al. (2016) included 27 studies on children and adolescents, there were only three studies with samples from an Asian background (Hall et al., 2016).

Culturally adapted cognitive-behavioral therapy for Japanese children and adolescents

Two major avenues have been proposed for cultural adaptation, a *top-down* and a *bottom-up* approach (e.g., Hwang, 2006). A *top-down* approach involves modifying an existing evidence-based protocol developed for one group for another group. A *bottom-up* approach involves trying to develop an original protocol within a particular cultural context to address culturally specific concerns (Hall et al., 2016; Hwang, 2016). Previous trials for children with anxiety disorders in Asian countries have exclusively applied a top-down approach. For example, the meta-analysis by Hall et al. (2016) included a quasiexperimental study of the FRIENDS program (Barrett, 2004, 2005) in Hong Kong (Batra, 2013). In addition, Lau, Chan, Li, and Au (2010) investigated the efficacy of CBT for clinically referred children with anxiety in Hong Kong using a group treatment program based on the Coping Cat program (Kendall, Kane, Howard, & Siqueland, 1989).

Reliance only on top-down approaches, however, might fail to address important aspects of cultural adaptation (Hwang, 2016), even though the efficacy of such an approach has been supported by several randomized controlled trials. Although a *bottom-up* approach is important and indispensable for cultural adaptation, it is not productive to emphasize exclusive domestic production of psychosocial interventions, ignoring the current evidence-based interventions. Thus, an approach that capitalizes on existing evidence-based practices may prove more practical (Ishikawa, Kikuta, et al., 2019). Hence, a culturally adapted CBT program in Japan was developed based on *bidirectional* cultural adaptation that involved both *top-down* and *bottom-up* approaches. The approach reiterates optimizing procedures that includes the reviewing and transporting of components from previous evidence-based CBT programs, the collecting of culturally specific descriptions regarding emotion and thought as well as typical vignettes regarding anxiety, and the receiving of feedback from practitioners who have used the program. Details of the procedures are summarized in the following sections.

Procedure for bidirectional cultural adaptation for CBT

The Japanese Anxiety Children/Adolescents Cognitive Behavior Therapy program (JACA-CBT) was developed for children and adolescents with anxiety disorders in Japan based on a bidirectional cultural adaptation. The JACA-CBT is a generic treatment program focusing on the underlying mechanisms among several types of anxiety disorders rather than specific types. Fig. 1 shows the detailed procedure for the bidirectional approach followed in developing the program. Because of its nature, bidirectional cultural adaptation is a translational approach that aims to bridge the gap between empirical studies and clinical implications, to identify both universal and culturally specific therapy factors related to a specific psychosocial treatment, to explore an underlying therapeutic mechanism of psychopathological symptoms, and to enhance the efficacy and effectiveness of the psychosocial intervention. There were four phases based on previous trials.

Phase 1: Collecting components and developing a provisional version

First, I, along with members of the research team, reviewed existing treatment manuals, which had been derived from empirically supported treatment programs, and participated in a workshop by the developers of the previous CBT program to understand its essence. Subsequently, therapeutic components were extracted from the previous programs (e.g., Barrett, 2004, 2005; Kendall et al., 1989), and a detailed protocol was compiled as a provisional version of the program. As a bottom-up approach, vignettes in which children feel anxiety and emotional descriptions that were used in the program were collected and developed based on an onsite survey of children and adolescents in Japan (i.e., Ishikawa, 2012). The effectiveness of the provisional version was examined as a case study for a clinically referred boy with anxiety disorders. The result of this examination was consistent with the concept of the treatment, specifically, the boy improved his anxiety symptoms and cognitive errors after the treatment and maintained these gains until the 2-month follow-up.

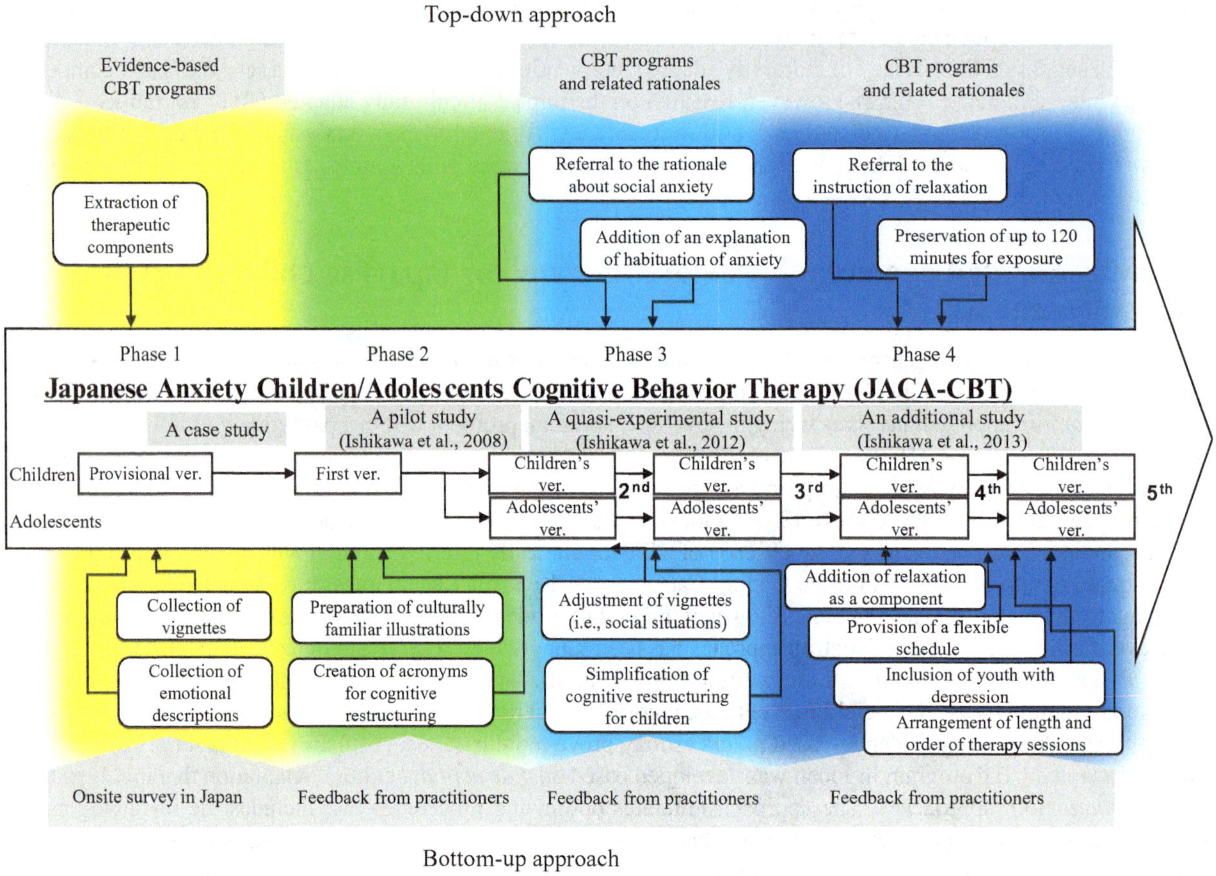

FIG. 1 Flow of the bidirectional cultural adaptation of the JACA-CBT.
JACA-CBT, Japanese Anxiety Children/Adolescents Cognitive Behavior Therapy program.

Phase 2: Refining and creating culturally adapted materials

Second, based on the results, the research team attempted to refine the program, aiming for controlled trials. More culturally familiar illustrative work drawn by a professional illustrator was prepared. Because the use of culturally familiar artwork such as cartoons can exemplify the exploitation of ingenious cultural strengths when children learn cognitive-behavioral skills, it is an indispensable strategy to create more attractive worksheets for children in a specific culture where the program would be applied (Ishikawa, Kishida, et al., 2019).

In addition, a culturally relevant acronym was developed through feedback from clinical psychologists who implemented the program. The use of acronyms is commonly used not only in Western CBT manuals (Kendall et al., 1989) but also in a culturally adapted manual for children in Hong Kong (Lau et al., 2010) to explain homework and coping plans. The JACA-CBT focused on applying acronyms for cognitive restructuring to assist children in learning and applying their skills efficiently. Clinical psychologists reported that a few children had difficulty learning the components. This difficulty might be attributed to the Japanese language system. In Japanese, the verb *omou* is frequently used to express one's internal state not only for a thought where one interprets a certain stimulus, situation, and person but for a feeling where one experiences subjective emotional reactions. In short, there is a typical form of psychoeducation in that a feeling is presented as a result of thought in typical CBT sessions, but Japanese children might compound them without any clear boundary. Because all psychosocial interventions must be transmitted through the use of one's language, a bottom-up adaptation specific to the Japanese language system is essential for teaching the central concept of CBT to Japanese children and adolescents. The efficacy of this version was examined, and it was found that 9 out of 12 children (75%) had become free from their principal diagnoses (Ishikawa, Shimotsu, & Sato, 2008).

Phase 3-1: Developing the two versions for children and adolescents

During the third phase, the two versions of the program were rolled out depending on the developmental stage, specifically for children in elementary school and adolescents in junior high school. Originally, the provisional version was aimed at children aged 8–12 years old; however, many adolescents with both anxiety disorders and school absenteeism were referred to the trials of the JACA-CBT program. Thus, an adolescents' version of the program was developed that included age-appropriate expressions and examples in the worksheets. The development team adjusted the content of each session and modified cognitive restructuring to cover a more detailed concept of the cognitive model, and aimed to produce realistic and balanced thoughts. The cognitive component for elementary school children was changed to a simpler format, for example, by applying a structure of self-instruction and a role-play task to understand the other person's perspective, leading to more adaptive thoughts. This amendment was introduced based on feedback from clinical psychologists who suggested how to apply cognitive components to Japanese children.

Phase 3-2: Arrangement of components

An explanation of a habituation model of anxiety was added to both the children's and adolescents' versions for an enhanced understanding of the underlying mechanism of exposure. Exposure could be one of the essential therapeutic components in psychosocial treatments for children and adolescents with anxiety disorders, regardless of cultural adaptation or unadapted bona fide psychotherapies. Higa-McMillan, Francis, Rith-Najarian, and Chorpita (2016) found that exposure was the most commonly occurring practice, with over 80% of trials included, followed by cognitive techniques, relaxation, psychoeducation, and modeling. Therefore, the adaptation was not derived from culturally specific aspects; rather, it was introduced from previous CBT protocols to enhance motivation for exposure. However, considering the simplicity of cognitive restructuring for children in the JACA-CBT, the program should spend as much time as exposure to compensate for treatment gain.

Moreover, adolescents in junior high school with high anxiety and school absenteeism might have tried "a partial attendance" before the treatment. The strategy is unorganized guidance to encourage attending school by teachers or parents and is frequently used in Japanese school settings. As its name suggests, the strategy aims at partial or gradual attendance; however, it is neither based on clinical judgments by CBT therapists nor any assessment, psychoeducation, or utilization of hierarchy. Exposure might be misunderstood as a certain type of "partial attendance" in the Japanese educational setting, especially when they had already tried it. Therefore, it is particularly important to provide appropriate education to children, parents, and teachers in Japan by using detailed material with a habituation model before launching the exposure component.

Finally, the vignette of anxious children depicted as a typical example of the worksheet was updated to accommodate the high numbers of children and adolescents with social anxiety symptoms. For example, a vignette of the cognitive component was "your friend who always leaves school with you, said that she could not go home with you." Based on the third version, we conducted a quasiexperimental trial to compare between groups and individual JACA-CBTs. As a result, 60.9% of the children and adolescents were free from their principal diagnoses, and 48.5% no longer met any of the diagnosis criteria for anxiety disorders at the 3-month follow-up. Improvements were also found for self-report measures, and the results were comparable with both treatments, except that individual CBT was superior to group CBT for self-reported depression (Ishikawa et al., 2012).

Phase 4: Finalizing of the program component and materials

As the fourth phase of bidirectional cultural adaptation, relaxation training was included as a mandatory component in the program. Empirically, as discussed earlier, Asian children tend to show anxiety as a physical symptom (Chun et al., 1996). The instructions in this program were developed based on those in previous Western-country CBT programs. Further feedback from clinical psychologists who implemented the JACA-CBT supported the inclusion of a relaxation component to compensate for the shortening of cognitive restructuring. The session time was more flexible; it included free time to play with therapists and co-therapists. However, considering the norm in Japanese schools, standardized session times were shortened from 90 to 60 min. Nonetheless, 90–120 min were preserved for in vivo exposures based on the theoretical implications of previous CBT programs. In place of the fixed session time, a flexible session schedule was provided pursuant to the requirements of the participant families. In addition, the order of the cognitive components' session was arranged for the adolescents' version of the program because the contents for the second session were relatively short, and sometimes excessive free time was provided. The third session that covered cognitive restructuring included more content to learn.

The pilot study, which used the version of the program from the fourth phase of the program development process, evaluated the effectiveness of an additional parental component. Because previous studies provide inconsistent evidence in terms of additional treatment effects of parental components (e.g., Kendall, Hudson, Gosch, Flannery-Schroeder, & Suveg, 2008; Wood, Piacentini, Southam-Gerow, Chu, & Sigman, 2006), this study reexamined the treatment effects by considering culturally specific family relationships in Japan, as mentioned earlier.

From this trial, there were no clear augmental gains identified in providing separated sessions for parents (Ishikawa, Kikuta, & Mitamura, 2013). In clinical trials in Western countries, parents and children are typically seen separately (e.g., Kendall et al., 2008), or a mix of working together and separately (e.g., Hudson et al., 2009). However, it is important to address parent-child interactions directly in therapy sessions given socialization practices in the cultural background. Therefore, the JACA-CBT is provided in an individual parent and child attendance format, rather than involving independent parental sessions.

Efficacy of a bidirectional cultural adaptation CBT program

Based on these previous studies, the fifth version of the JACA-CBT was developed as an individual-manualized program encompassing eight 60-min treatment sessions (see Table 2 for the core components of the program). The main components of the first and second sessions of this program were psychoeducation about the nature of anxiety, discussion of anxious situations, and introduction to emotion monitoring. From the third to the fifth session, cognitive restructuring was provided. As mentioned earlier, children engaged in simplified content, whereas adolescents worked on a full-fashioned cognitive restructuring, which is consistent with that for adults. In the sixth session, children and adolescents developed their anxiety hierarchy and tried to challenge their anxious stimulus/situations as an initial in vivo exposure. The last session focused on concluding the treatment and planning a booster session to review the child's in vivo exposure exercises. After completing eight sessions, the therapist scheduled booster sessions to provide further opportunities for within-session exposures and confirm the progress of homework exposure. Flexible booster sessions are also considered cultural adaptations for the Japanese family in response to their hectic schedules.

Participants and measures

An RCT in Japan was conducted first to examine a culturally adapted CBT program for Japanese children and adolescents with anxiety disorders. The participants comprised 51 children and adolescents aged 8–15 years ($M = 10.90$, $SD = 2.00$, 29 girls) from the Kansai area in Japan. They were recruited through a local newspaper and webpages. Most were diagnosed with social anxiety disorder based on the Anxiety Disorders Interview Schedule for DSM-IV (ADIS; Silverman & Albano, 1996). Specifically, 31 (60.78%) showed social anxiety disorder as their principal diagnoses, followed by specific phobia ($n = 9$; 17.64%) and generalized anxiety disorder ($n = 7$; 13.73%). Four participants with mood disorders initially visited

TABLE 2 Program components of the JACA-CBT.

No.	Children's version	Adolescents' version
1	Let's think about your problem! (Psychoeducation)	
2	Catch your feelings! (Psychoeducation)	
3	Collect your thoughts! (Cognitive restructuring)	Try to think flexibly! (Cognitive restructuring)
4	Try to think flexibly! (Cognitive restructuring)	Capture your irrational thoughts! (Cognitive restructuring)
5	Try to think more flexibly ! (Cognitive restructuring)	
6	Build your anxiety staircase! (Anxiety hierarchy)	
7	Take a challenge using your staircase! (In vivo exposure, Relaxation)	
8	Review and party! (Homework exposure)	

JACA-CBT, Japanese Anxiety Children/Adolescents Cognitive Behavior Therapy program.
From Ishikawa, S., Kikuta, K., Sakai, M., Mitamura, T., Motomura, N., & Hudson, J. L. (2019). A randomized controlled trial of a bidirectional cultural adaptation of cognitive behavior therapy for children and adolescents with anxiety disorders. *Behaviour Research and Therapy, 120*, 103432, with permission.

the clinic for treatment of anxiety-related problems. Whereas the JACA-CBT originally focused on anxiety disorders, it could ameliorate depression in children with anxiety disorders, as mentioned earlier. Therefore, the outcome study included four participants with both anxiety and depression. Participants were randomly allocated to either the treatment condition of cognitive-behavior therapy ($n=26$) or a wait-list control condition ($n=25$). In addition, questionnaires for self- and parent-reported anxiety (SCAS and SCAS-P), cognitive errors (Ishikawa, 2012), depression by the Depression Self-Rating Scale for Children (DSRS; Birleson, 1981), and the Children's Depression Inventory (CDI; Kovacs, 1985) were used as secondary outcomes. Detailed procedures and inclusive criteria are presented in Ishikawa, Kikuta, et al. (2019).

Efficacy of the JACA-CBT

After eight JACA-CBT sessions, the diagnostic free rate of those in the CBT condition was significantly higher than those in the wait-list condition (50% vs 12%). Fig. 2 shows a comparison of the results of the RCTs for children and adolescents with anxiety disorders. The Cochrane review indicated that the overall remission rate of family and parental CBTs increased to 61.67% for the ITT sample (James, James, Cowdrey, Soler, & Chole, 2013). However, most studies were included in the review of parallel or independent parental sessions. The two trials in Australia and the Netherlands were comparable with our trial in terms of family-based nongroup CBT that referred parents presented in the same room with children during CBT sessions to a certain extent. The results of this study were superior to the two trials. Therefore, the format of the JACA-CBT that children and parents attend the sessions in the same room could produce sufficient treatment gain as a culturally adapted program.

More than 60% of the participants (61%) in this study had a social anxiety disorder as their principal diagnosis. A previous study suggested that social anxiety disorder does not produce equivalent treatment effects as other types of anxiety disorders when a generic CBT is applied. For example, Hudson et al. (2015) indicated poorer overall remission rates in children and adolescents who had primarily social anxiety disorders (22.3%) compared to those with a separation anxiety disorder (42.2%), generalized anxiety disorder (52.7%), and specific phobia (42.1%), which was comparable with the remission rate in this study (50.0%). Given that the cultural aspect of anxiety disorders was indicated as discussed above, a high percentage of socially anxious children who tend to be referred to a clinic could be assumed as a part of the culturally specific appearance of anxiety disorders. The JACA-CBT, a culturally adapted CBT program, can produce sufficient therapeutic gains even when including a high social anxiety disorder rate.

This study also showed moderate effect sizes for self-reported measures, specifically $g=0.43$ (95% CI: 0.19–1.04) for anxiety symptoms by the SCAS, $g=0.39$ (95% CI: 0.22–1.00), and $g=0.48$ (95% CI: 0.14–1.09) for depression symptoms by the DSRS and the CDI, respectively. A meta-analysis of CBT for social anxiety disorders in children and adolescents average effect sizes of 10 studies were $g=0.71$ (95% CI: 0.45-0.98) for social anxiety assessed self-reported scales and $g=0.13$ (95% CI: -0.23 to 0.48) for self-reported depressive symptoms when comparing active treatment and waiting-list condition (Scaini, Belotti, Ogliari, & Battaglia, 2016). Therefore, a disorder-specific program could produce more treatment gains for self-reported relevant measures, whereas no benefits were found for depressive symptoms. As discussed earlier, individual treatment of the previous trial of the JACA-CBT showed significant treatment effects on self-reported

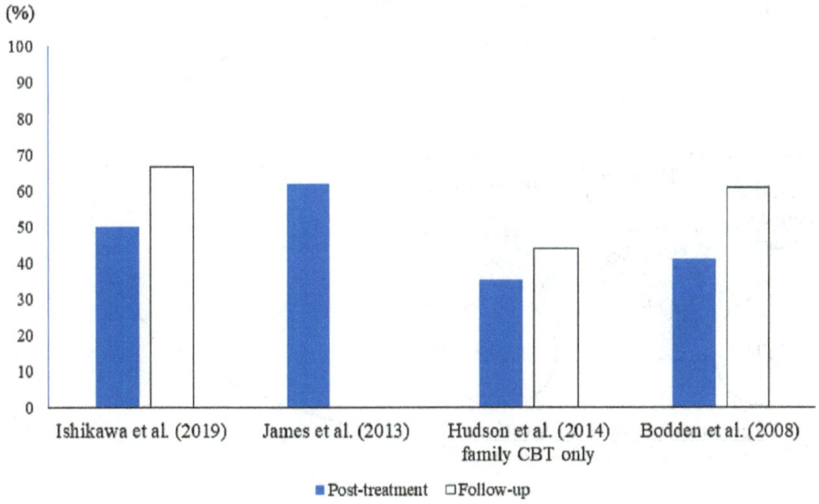

FIG. 2 Comparison of diagnostic free rates for principal diagnoses by country (ITT analysis) (Bodden et al., 2008; Hudson et al., 2014).

depressive symptoms (Ishikawa et al., 2012). Therefore, the JACA-CBT can ameliorate anxiety and depressive symptoms (i.e., internalizing symptoms) as a culturally adapted generic CBT program.

Triad model of cultural adaptation for CBT

Theoretical schemes or frameworks to capture multiple components and various aspects of cultural adaptation have been proposed in several systematic or narrative reviews. For example, Stirman, Miller, Toder, and Calloway (2013) reviewed 258 modifications and showed a system of classifying two types of major modifications, such as content and context modifications. Sundell, Beelmann, Hasson, and von Thiele Schwarz (2016) presented a framework to categorize outcome studies into three program types depending on their extent of adaptation: novel, adapted, and adopted programs. Finally, Chu and Leino (2017) attempted to streamline existing adaptation studies and develop a unified framework for cultural adaptation focusing on core and peripheral adaptations.

These theoretical schemes/frameworks provided a systematic perspective on content and contextual cultural adaptations that included the components, e.g., core or peripheral (Chu & Leino, 2017); the extent, e.g., from novel to adoption (Sundell et al., 2016); and parameters, e.g., practitioners, institutions, stakeholders as contextual variables (Stirman et al., 2013). For example, the JACA-CBT tailored materials and adjusted elements and time management through four phases, as mentioned earlier. As a result of the procedure, comprehensive works for adaptation and modification could be laid out from content and context cultural adaptations (Fig. 3). First, the following five major modifications were categorized as content cultural adaptation: (a) vignettes of anxiety situations in the worksheets focused on socially relevant situations considering prevalent problems, (b) emotional descriptions were depicted originally rather than translated to reflect the corresponding language system, (c) culturally specific illustrations were created as more familiar and attractive ones for the children in the country, (d) culturally specific acronyms were developed for cognitive restructuring, and (e) arrangement of core components; specifically, a habituation model of anxiety was introduced, relaxation was compiled as a mandatory component, and cognitive restructuring was shortened and simplified for younger children. Second, the

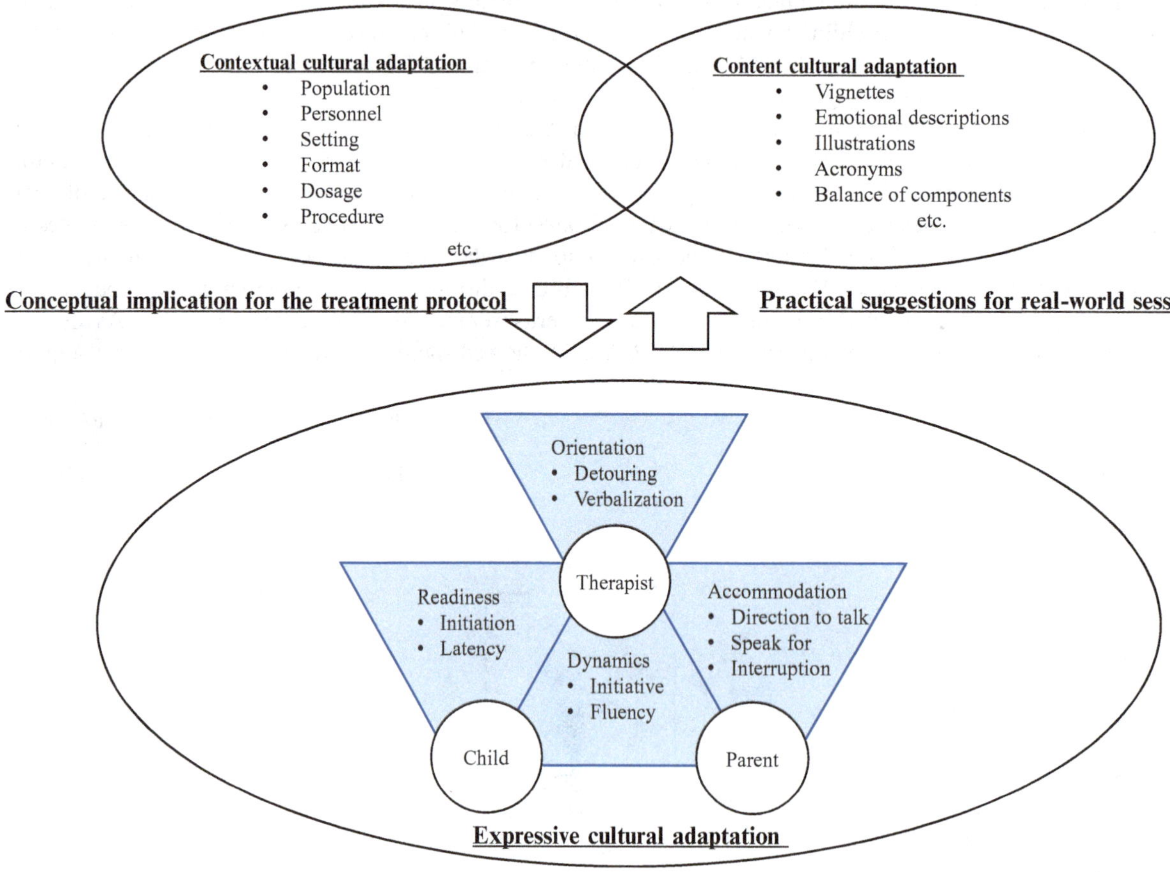

FIG. 3 A triad model of cultural adaptation.

JACA-CBT consists of six contextual adaptations: population (Japanese children and adolescents), personnel (clinical psychologists in Japan), setting (a university clinic in Japan), format (family-based), dosage (a flexible 8-session format with booster sessions), and procedure (including children with depression).

Therefore, these frameworks are useful to better comprehend and navigate the types of cultural adaptations that could be applied to treatment protocols. However, these frameworks do not identify how cultural adaptations are *expressed* during actual interactions among therapists and children/parents in CBT sessions. For example, even when typical CBT manuals from Western and Eastern countries emphasize reciprocal communications, more utterance by therapists would be required in Japan to guide such a communicational structure. Hence, practical detailed implications for case management among diverse clients might not be addressed within the comprehensive frameworks proposed in the previous studies. *Expressive* cultural adaptation, defined as the cultural modifications includes how step-by-step interactions with one's client are expressed and how the practitioner uses culture-specific modifications in the context of actual case management (Ishikawa, Romano, & Hudson, 2022). This approach may provide concrete, practical suggestions from "real-world sessions" beyond the conceptual categorization. Fig. 3 depicts specific domains that expressive cultural adaptation can capture by using the Cross-cultural Behavioral Observation System (C-BOS). The C-BOS was developed by Ishikawa and Hudson (2019) for cross-cultural comparisons of CBT programs delivered to children and adolescents with anxiety disorders of different cultures (see section "Practice and procedures" in detail) based on a behavior observational system that aims to measure tangible behavioral indices that would be observed within actual interactions between children, parents, and therapists. The C-BOS was designed to capture four major domains of expressive cultural adaptation: *readiness*, *dynamics*, *accommodation*, and *orientation* (please also see Ishikawa et al., 2022).

First, *readiness* represents participants' preparedness for CBT sessions. The domain comprises two subcategories: *initiation* reflects whether children can declare what makes them feel anxious, and *latency* captures the length of time it takes for the clients to respond to the therapists' first question about what makes them feel anxious. The former can capture how soon children are capable of talking about their anxious object concretely, while the latter can cover reluctancy to respond to the related questions. Second, *dynamics* captures the extent to which therapists make an effort to lead CBT sessions. This domain also has two subcategories: *initiative* and *fluency*. *Initiative* captures the proportion of the session in which the child, parent, and therapist speak, suggesting dominancy of the conversation within the session. *Fluency* captures how often laughing and silence are observed within sessions. Because CBT sessions with anxious children are characterized by awkwardness due to the tension and nervousness these children experience, the amount of laughter and silence might be appropriate indices to measure fluency in a session.

Third, *accommodation* depicts overprotective behaviors or reassurance by parents during sessions; it includes three subcategories: *direction to talk*, *speak for*, and *interruption*. In terms of accommodation, anxious children may seek parental help when feeling distressed; parents then tend to display reactive behaviors, becoming increasingly involved in a child's symptoms (Liebowitz, 2019). *Direction to talk* represents children's responses to a question from the therapist by directing their answer to or glancing at the parent immediately before, during, or immediately after their response. *Speak for* means parents' responses to a question by the therapist, which was directed at the child before the child began to reply. *Interruption* is captured by the parent interrupting the conversation of either the child or the therapist. The parent speaking must do so over the speech of either the child or the therapist (i.e., voice overlap). Fourth, *orientation* aims to capture both culture-specific verbalizations and how the therapist manages the specific case during the session. *Verbalization* represents the usage of culturally specific languages, including both well-defined and ambiguous expressions for emotional and physiological states. *Detouring* reflects the latent time between the commencement of a core component of CBT session. Thus, it corresponds to *readiness* as a variable of therapists' side (see Ishikawa et al., 2022).

Cross-cultural difference of interactions among a child, parent, and therapist in CBT sessions between Australia and Japan

Using the C-BOS, Ishikawa et al. (2022) examined the actual interactions between a child, parent, and therapist in the JACA-CBT and those in the Cool Kids program (Rapee et al., 2006). Both the JACA-CBT and the Cool Kids program shared core components and techniques from the theoretical scheme's perspective according to the systematic review by Higa-McMillan et al. (2016). In contrast, the JACA-CBT has been culturally adapted through a bidirectional approach, as discussed above, and the Cool Kids program has been recognized as a representative CBT program for anxiety disorders in children and adolescents. Therefore, a comparison of the two programs allows us to identify how components of CBT are *expressed* in a culturally adapted program. Here, the results of three of the four domains are summarized (Ishikawa et al., 2022).

First, compared with Australian children, Japanese children showed significantly lower initiation (80% among Australian children vs 33% among Japanese children), implying that Japanese children typically take more time to express their cognitive and emotional symptoms than Australian children. However, there was no difference in latency between Australia and Japan (i.e., both countries showed a latency of less than 0.5 s), suggesting that clinicians can approach a specific theme with children who can talk about their anxiety in the first session regardless of their cultural background.

Australian children and parents talked more than Japanese families during CBT sessions. This finding aligns with cultural norms whereby Asian cultural groups tend to appear passive in front of authorities to show respect (Hwang, 2016). In terms of fluency, Japanese therapists tended to laugh more, and a Japanese interaction contained more silence than Australian therapists at the first session. Greater therapist laughing in the context of overall silence may appear awkward. This could affect a child's engagement and then might impact the outcome.

Finally, we hypothesized that Japanese families would engage in more accommodative behaviors; however, such behaviors were more prevalent during Australian treatment sessions. Australian children showed more direction to talk, and Australian parents tended to interrupt the conversation between therapists and children significantly more than Japanese parents. Furthermore, these behaviors partially affected positive treatment outcome against the hypothesis. Therefore, these results suggest that Australian parents were more involved in CBT sessions than Japanese parents and such interactions within the sessions might be appropriate engagements for a family CBT.

Conclusion

Although there have been many clinical trials of CBT for children and adolescents with anxiety disorders, and its efficacy in Western and Eastern countries has been supported consistently, its further dissemination to other unrepresentative cultures needs to be addressed. This chapter detailed the procedures that were applied in the cultural adaptation of a CBT program for children and adolescents with anxiety disorders based on a bidirectional approach. Such a procedure could be applied in future trials conducted in Asian countries where the evidence-based protocol would be disseminated. In addition, a triad model of cultural adaptation would provide both practical and conceptual implications, especially as expressive cultural adaptation must be considered for future studies. Such an approach would explore universal and cultural-specific therapy factors in CBT beyond the dissemination of evidence-based psychosocial treatment.

Practice and procedures

Cross-cultural Behavioral Observation System (C-BOS; Ishikawa & Hudson, 2019)

Here is the summary of behavioral codes and rating procedures that are measured by the C-BOS (pages 1–2 out of 15 pages). Inquiries and permission to use should be directed to the author.

Summary of behavioral codes

Readiness

1. **Initiation:** Whether or not the child discloses the source of his/her anxiety (stimuli or situation) (e.g., I am afraid of XXX. I worry about YYY. It is scary when ZZZ) or that the child experiences or needs help with a specific anxiety problem (e.g., I am here for my anxiety/worries about spiders, I need help because I am afraid of the dark) in the first session.
   - Research Question (RQ): Is the child able to verbalize the presence or source of his/her anxiety in the first session?
2. **Latency:** The time from when the therapist asked the child about his/her anxiety about the onset of the child's response.
   - RQ: How long does it take the child to talk about his/her anxiety problem after the therapist first inquired?

Dynamics

1. **Initiative:** The proportion of time each individual (child, parent, therapist) talked during the session (i.e., percentage of observed session time each individual talked). Talking was defined as verbalizing with the meaning. This includes words/sounds to answer in the affirmative or negative (e.g., yes/yeah/no/nah, etc.).

   Exclusion: Words or sounds used to convey the person are listening/back-channeling (e.g., Uh-huh, yea, etc.) are not classified as talking.
   - RQ: How often does the child, parent, and therapist talk during the session?

2. **Fluency:** Behaviors that represent fluency of the CBT session.
   (a) Smiling with vocalization (including laughter with or without accompanying speech). Exclusion: Smiling unaccompanied by vocalized laughter.
   (b) Silence by all individuals (child, parent, therapist) during the defined period.
   – RQ: How often are behaviors that represent fluency of sessions (e.g., smiling with vocalization and silence) observed?

Accommodation

1. **Direction to talk:** An observation in which the child responds to a direct question by the therapist by looking at his/her parent. This includes looking at the parent first before responding, looking at the parent while responding, and/or looking at the parent within 1 s of finishing his/her response.
   – RQ: How often does the child direct his/her response to the parent rather than the therapist?
2. **Speak for:** An observation in which the parent responds to a question by the therapist that was directed at the child. The latent time from when the therapist finished asking the question and the parent responding on behalf of the child was also recorded.
   Exclusion: If the child had attempted to respond (e.g., provided an incomplete or inaccurate response) before the parent responded, the behaviors were not classified as "Speak for."
   – RQ: How often does the parent talk on behalf of the child?
   – RQ: How long does the parent wait before talking on behalf of the child?
3. **Interruption:** An observation in which the parent interrupts a conversation by talking over either the therapist or child. The conversation could be between any of the parties, including the parent.
   – RQ: How often does the parent interrupt a conversation by speaking over either the child or therapist?

Orientation

1. **Detouring:** The latent time between the commencement of the session and when the therapist first began to talk about anxiety or any topic related to the CBT session.
   Exclusion: Observations are not classified as detouring when the therapist asks a general question that might elicit an anxiety-related response (e.g., Do you know why you are here today?).
   – RQ: How long does it take for the therapist to begin talking about anxiety or CBT-related subject?
2. **Verbalization:** A word used by the child or therapist to express an emotional state. It can be about the child (e.g., "I get scared when …"), a specific person (e.g., "my sister gets scared when …"), or in general (e.g., "Lots of children might be scared when …").
   – RQ: What words are frequently used to express emotional states?

Overview of rating

Preparation

The raters must read this manual and the coding sheets (Microsoft Excel) for the Cross-cultural Behavioral Observation System (C-BOS) before proceeding with behavioral coding.

Scanning

To code adequately, the raters should scan the video file for the period they observe. The raters should check the volume of voice, smoothness of video streaming, and an overview of the session.

Procedure

Part A The raters observe the video file from the start time of the CBT session for exactly **20 min**. The starting time is defined as *when all members take a seat* (see Microsoft Excel sheet).

(1) *First and last sessions*: Record *Initiative* and *Fluency* (**Dynamics**) for both the first and last sessions using the coding sheet.

Part B The raters rewind the video to the starting time of the CBT session and observe the 20-min segment as defined in Part A:

(1) *First session only*: Record *Detouring* (part of **Orientation**) only for the first session.
(2) *First session only*: Record *Latency* and *Initiation* (**Readiness**) only for the first session.
(3) *First and last session*: Record *Direction to talk* , *Speak for*, and *Interruption* (**Accommodation**) for both the first and last sessions.
(4) *First and last session*: Based on the transcript, list the words used to express emotional states by either the child or therapist as *Verbalization* (part of **Orientation**).

Part C Finally, the raters evaluate the global impressions of the segment observed.

(1) *First session only*: Evaluate the global impression of **Readiness** (for the first session).
(2) *First and last session*: Evaluate the global impression of **Accommodation** (for the first and last sessions).
(3) *First and last session*: Evaluate the global impression of **Orientation** (for the first and last sessions).

Mini-dictionary of terms

- ***Separation anxiety disorder.*** This is an anxiety disorder occurring in childhood or adolescence; it is characterized by excessive anxiety about being apart from home or attachment figures beyond the expected developmental level.
- ***Generalized anxiety disorder.*** This anxiety disorder is the manifestation of excessive worry about the arrangement of one's events and activities accompanied by several symptoms (e.g., fatigue, muscle tension, and sleep disturbance).
- ***Social anxiety disorder.*** An anxiety disorder defines the key feature as a marked persistent fear of social or performance situations when exposed to people that cause distress and dysfunction in everyday activities.
- ***Spence Children's Anxiety Scale.*** This scale is a self-report questionnaire that assesses multiple symptoms of childhood anxiety disorders based on diagnostic criteria. More than 30 language versions are available.
- ***Bidirectional cultural adaptation.*** This approach consists of both top-down and bottom-up cultural adaptations. The top-down approach entails selecting components from existing programs, importing them into the concerning cultures, and setting up a pilot protocol. The bottom-up approach involves modifying the protocol within a particular cultural context and shaping a specific treatment program.
- ***Content cultural adaptation.*** Cultural adaptation is relevant to changes in intervention procedures, such as translating the materials, arranging in order, selecting components, and modifying manuals and materials.
- ***Contextual cultural adaptation.*** Cultural adaptation, required by an environment and a condition where the treatment will be implemented, is associated with contextual variables, including format, setting, personnel, population, and length.
- ***Expressive cultural adaptation.*** A cultural modification is concerned with how step-by-step interactions with one's client are expressed and how the therapist uses culture-specific modifications in the context of actual case management.

Key facts

Key facts of culturally adapted treatment

- Anxiety disorders are prevalent worldwide; however, culturally specific expressions of psychological symptoms are also observed.
- Cognitive-behavioral therapy (CBT) for children with anxiety disorders is equally efficacious in both Western and Eastern countries.
- A CBT program based on bidirectional cultural adaptation can ameliorate culturally specific manifestations of psychological disorders.
- A comprehensive approach regarding cultural adaptations can be classified from the content, contextual, and expressive perspectives.
- Expressive cultural adaptation can provide clinical implications for culturally sensitive case management for clients from an Asian cultural background.

Applications to other areas

In this chapter, a culturally adapted cognitive-behavioral therapy (CBT) for Japanese children with anxiety disorders is discussed. This chapter outlines the overall procedure of bidirectional cultural adaptation for approximately a decade to show how to adapt evidence-based psychosocial treatment to an underrepresented culture (Ishikawa, Kikuta, et al., 2019). The bidirectional cultural adaptation we explained in this chapter, including both top-down and bottom-up approaches, aims to develop a novel treatment program derived from previous studies that is not a simple translation of previous manuals. This is particularly important considering the culturally distinct nature of Asian populations because studies on efficacy and effectiveness of psychotherapy trials and hypothetical models of cultural adaptation for Asian cultures are limited considering the movement for global adaptation of evidence-based practices among any other contexts (e.g., Hall et al., 2016; Sundell, Ferrer-Wreder, & Fraser, 2014). In addition, we introduced a practical method for direct comparison between culturally adapted and nonculturally adapted CBT programs for children with anxiety disorders. The Cross-cultural Behavioral Observation System (C-BOS) is the first behavioral coding manual for cross-cultural comparison of CBT programs delivered to children and adolescents with anxiety disorders of different cultures (Ishikawa & Hudson, 2019). The frameworks that have been proposed by previous narrative or systematic reviews (Chu & Leino, 2017; Stirman et al., 2013; Sundell et al., 2016) are helpful to better comprehend and navigate the types of cultural adaptations. However, this type of research does not identify how cultural adaptations are expressed during treatment, and, as such, practical implications relevant to diverse clients may be missed. An idiosyncratic approach regarding therapists' cultural competence in a series of case studies could illustrate such nuances and detailed modifications through culturally adapted clinical practices; however, the approach could not yield direct comparison data between culturally adapted and nonculturally adapted treatments due to its methodology. The C-BOS can assess four major domains of expressive cultural adaptation: readiness, dynamics, accommodation, and orientation, and examine differences and commonalities in these factors among different cultures. Further studies based on this methodology might be expected to produce novel knowledge about universal and culturally specific therapy factors related to actual interactions during CBT sessions.

Summary points

- An attempt to disseminate cognitive-behavioral therapy (CBT) to underrepresented cultures including Asia is still underway.
- The Japanese Anxiety Children/Adolescents Cognitive Behavior Therapy program (JACA-CBT) was developed for children and adolescents with anxiety disorders in Japan based on bidirectional cultural adaptation.
- A total of 50% of children who participated in the JACA-CBT program for anxiety disorders were free from their principal diagnoses.
- The treatment gains of the JACA-CBT are comparable to those of previous RCT trials in Western countries.
- The Cross-cultural Behavioral Observation System (C-BOS) was developed for cross-cultural comparison between different CBT programs delivered to children and adolescents with anxiety disorders.
- A direct comparison between Japan and Australia using the C-BOS showed significant differences in how soon therapists should ask questions about children's anxiety, how much therapists should lead conversations, and the amount of parental involvement.

References

Amaya, C. A., & Campbell, M. (2010). Cross-cultural comparison of anxiety symptoms in Colombian and Australian children. *Electronic Journal of Research in Educational Psychology*, *8*, 497–516.

Arendt, K., Hougaard, E., & Thastum, M. (2014). Psychometric properties of the child and parent versions of Spence children's Anxiety Scale in a Danish community and clinical sample. *Journal of Anxiety Disorders*, *28*, 947–956.

Azuma, H. (1986). Why study child development in Japan? In H. Stevenson, H. Azuma, & K. Hakuta (Eds.), *Child development and education in Japan* (pp. 3–12). New York: W. H. Freeman.

Barrett, P. M. (2004). *FRIENDS for life program: Group leader's workbook for children*. Brisbane: Australian Academic Press.

Barrett, P. M. (2005). FRIENDS for life program: Group leader's workbook for youth. *Australian Academic Press*. Brisbane: Australian Academic Press.

Batra, P. (2013). *Evaluation of the Hong Kong FRIENDS intervention program for childhood anxiety*. (Unpublished doctoral dissertation). Hong Kong: Alliant International University .

Birleson, P. (1981). The validity of depressive disorder in childhood and the development of self-rating scale: A research report. *Journal of Child Psychology and Psychiatry, and Allied Disciplines*, *22*, 73–88.

Bodden, D. H. M., Bögels, S. M., Nauta, M. H., De Haan, E., Ringrose, J., Appelboom, C., … Appelboom-Geerts, K. C. M. M. J. (2008). Child versus family cognitive-behavioral therapy in clinically anxious youth: An efficacy and partial effectiveness study. *Journal of the American Academy of Child and Adolescent Psychiatry*, *47*, 1384–1394.

Cartwright-Hatton, S., McNicol, K., & Doubleday, E. (2006). Anxiety in a neglected population: Prevalence of anxiety disorders in pre-adolescent children. *Clinical Psychology Review*, *26*, 817–833.

Chu, J., & Leino, A. (2017). Advancement in the maturing science of cultural adaptations of evidence-based interventions. *Journal of Consulting and Clinical Psychology*, *85*, 45–57.

Chun, C., Enomoto, K., & Sue, S. (1996). Health care issues among Asian Americans: Implications of somatization. In P. M. Kato, & T. Mann (Eds.), *Handbook of diversity issues in health psychology* (pp. 347–365). New York: Plenum Press.

Costello, E. J., Mustillo, S., Erkanli, A., Keeler, G., & Angold, A. (2003). Prevalence and development of psychiatric disorders in childhood and adolescence. *Archives of General Psychiatry*, *60*, 837–844.

Denda, K. (2008). Diagnostic assessment of mood disorders in childhood and adolescence: Using the MINI-KID structured diagnostic interview. *Journal of Child and Adolescent Psychiatry*, *49*, 286–292.

DeSousa, D. A., Pereira, A. S., Petersen, C. S., Manfro, G. G., Salum, G. A., & Koller, S. H. (2014). Psychometric properties of the Brazilian–Portuguese version of the Spence Children's Anxiety Scale (SCAS): Self- and parent-report versions. *Journal of Anxiety Disorders*, *28*, 427–436.

Di Riso, D., Chessa, D., Bobbio, A., & Lis, A. (2013). Factorial structure of the SCAS and its relationship with the SDQ: A study with Italian children. *European Journal of Psychological Assessment*, *29*, 28–35.

Essau, C. A., Ishikawa, S., & Sasagawa, S. (2011). Early learning experience and adolescent anxiety: A cross-cultural comparison between Japan and England. *Journal of Child and Family Studies*, *20*, 196–204.

Essau, C. A., Leung, P. W., Conradt, J., Cheng, H., & Wong, T. (2008). Anxiety symptoms in Chinese and German adolescents: Their relationship with early learning experiences, perfectionism, and learning motivation. *Depression and Anxiety*, *25*, 801–810.

Essau, C. A., Lewinsohn, P. M., Lim, J. X., Ho, M. R., & Rohde, P. (2018). Incidence, recurrence and comorbidity of anxiety disorders in four major developmental stages. *Journal of Affective Disorders*, *228*, 248–253.

Essau, C. A., Olaya, B., Pasha, G., O'Callaghan, J., & Bray, D. (2012). The structure of anxiety symptoms among adolescents in Iran: A confirmatory factor analytic study of the Spence children's Anxiety Scale. *Journal of Anxiety Disorders*, *26*, 871–878.

Essau, C. A., Sakano, Y., Ishikawa, S., & Sasagawa, S. (2004). Anxiety symptoms in Japanese and in German children. *Behaviour Research and Therapy*, *42*, 601–612.

Essau, C. A., Sasagawa, S., Anastassiou-Hadjicharalambous, X., Guzmán, B. O., & Ollendick, T. H. (2011). Psychometric properties of the Spence Child Anxiety Scale with adolescents from five European countries. *Journal of Anxiety Disorders*, *25*, 19–27.

Hall, G. C. N., Ibaraki, A. Y., Huang, E. R., Marti, C. N., & Stice, E. (2016). A meta-analysis of cultural adaptations of psychological interventions. *Behavior Therapy*, *47*, 993–1014.

Higa-McMillan, C. K., Francis, S. E., Rith-Najarian, L., & Chorpita, B. F. (2016). Evidence base update: 50 years of research on treatment for child and adolescent anxiety. *Journal of Clinical Child and Adolescent Psychology*, *45*, 91–113.

Hudson, J. L., Newall, C., Rapee, R. M., Lyneham, H. J., Schniering, C. C., Wuthrich, V. M., … Gar, N. S. (2014). The impact of brief parental anxiety management on child anxiety treatment outcomes: A controlled trial. *Journal of Clinical Child and Adolescent Psychology*, *43*, 370–380.

Hudson, J. L., Rapee, R. M., Deveney, C., Schniering, C. A., Lyneham, H. J., & Bovopoulos, N. (2009). Cognitive-behavioral treatment versus an active control for children and adolescents with anxiety disorders: A randomized trial. *Journal of the American Academy of Child and Adolescent Psychiatry*, *48*, 533–544.

Hudson, J. L., Rapee, R. M., Lyneham, H. J., McLellan, L. F., Wuthrich, V. M., & Schniering, C. A. (2015). Comparing outcomes for children with different anxiety disorders following cognitive behavioural therapy. *Behaviour Research and Therapy*, *72*, 30–37.

Hwang, W. C. (2006). The psychological adaptation and modification framework: Application to Asian Americans. *American Psychologist*, *61*, 702–715.

Hwang, W. C. (2016). *Culturally adapting psychotherapy for Asian heritage populations: An evidence-based approach*. Cambridge, MA: Elsevier Academic Press.

Ishikawa, S. (2012). Cognitive errors, anxiety, and depression in Japanese children and adolescents. *International Journal of Cognitive Therapy*, *5*, 38–49.

Ishikawa, S., & Hudson, J. L. (2019). *Cross-cultural behavioural observation system: Coding behavioural coding manual and calculation sheets*. 3rd version Kyoto, Japan: Faculty of Psychology, Doshisha University.

Ishikawa, S., Kikuta, K., & Mitamura, T. (2013). Parent–child cognitive behavior therapy for children and adolescents with anxiety disorders. *Journal of Japanese Clinical Psychology*, *31*, 364–375.

Ishikawa, S., Kikuta, K., Sakai, M., Mitamura, T., Motomura, N., & Hudson, J. L. (2019). A randomized controlled trial of a bidirectional cultural adaptation of cognitive behavior therapy for children and adolescents with anxiety disorders. *Behaviour Research and Therapy*, *120*, 103432.

Ishikawa, S., Kishida, K., Oka, T., Saito, A., Shimotsu, S., Watanabe, N., … Kamio, Y. (2019). Developing the universal unified prevention program for diverse disorders for school-aged children. *Child and Adolescent Psychiatry and Mental Health*, *13*, 44.

Ishikawa, S., Motomura, N., Kawabata, Y., Tanaka, H., Shimotsu, S., Sato, Y., & Ollendick, T. H. (2012). Cognitive behavioural therapy for Japanese children and adolescents with anxiety disorders: A pilot study. *Behavioural and Cognitive Psychotherapy*, *40*, 271–285.

Ishikawa, S., Romano, M., & Hudson, J. L. (2022). A comparison of interactions among children, parents, and therapists in cognitive behavior therapy for anxiety disorders in Australia and Japan. *Behavior Therapy*, *55*, 34–48.

Ishikawa, S., Sato, H., & Sasagawa, S. (2009). Anxiety disorder symptoms in Japanese children and adolescents. *Journal of Anxiety Disorders*, *23*, 104–111.

Ishikawa, S., Shimotsu, S., Ono, T., Sasagawa, S., Kondo-Ikemura, K., Sakano, Y., & Spence, S. H. (2014). A parental report of children's anxiety symptoms in Japan. *Child Psychiatry and Human Development*, *45*, 306–317.

Ishikawa, S., Shimotsu, S., & Sato, Y. (2008). Short-term and group cognitive behavior therapy for children with anxiety disorders. *Japanese Journal of Psychiatric Treatment*, *23*, 1481–1490.

Ishikawa, S., Takeno, Y., Sato, Y., Kishida, K., Yatagai, Y., & Spence, S. H. (2018). Psychometric properties of the Spence Children's Anxiety Scale with adolescents in Japanese high schools. *School Mental Health*, *10*, 275–286.

Iwamasa, G. Y., & Hays, P. A. (2019). *Culturally responsive cognitive behavior therapy: Practice and supervision* (2nd ed.). Washington, DC: American Psychological Association.

James, A. C., James, G., Cowdrey, F. A., Soler, A., & Chole, A. (2013). Cognitive behavioural therapy for anxiety disorders in children and adolescents. *Cochrane Database of Systematic Reviews*, *6*, CD004690.

Kendall, P. C., Hudson, J. L., Gosch, E., Flannery-Schroeder, E., & Suveg, C. (2008). Cognitive-behavioral therapy for anxiety disordered youth: A randomized clinical trial evaluating child and family modalities. *Journal of Consulting and Clinical Psychology*, *76*, 282–297.

Kendall, P. C., Kane, M. T., Howard, B. L., & Siqueland, L. (1989). *Cognitive-behavioral therapy for anxious children: Treatment manual*. Philadelphia: Psychology Department, Temple University.

Kovacs, M. (1985). The Children's Depression, Inventory (CDI). *Psychopharmacology Bulletin*, *21*, 995–998.

Kroes, M., Kalff, A. C., Kessels, A. G., Steyaert, J., Feron, F. J., van Someren, A. J., … Vles, J. S. (2001). Child psychiatric diagnoses in a population of Dutch schoolchildren aged 6 to 8 years. *Journal of American Academy of Child and Adolescent Psychiatry*, *40*, 1401–1409.

Lau, W., Chan, C. K., Li, J. C., & Au, T. K. (2010). Effectiveness of group cognitive behavioral treatment for childhood anxiety in community clinics. *Behaviour Research and Therapy*, *48*, 1067–1077.

Lewinsohn, P. M., Hops, H., Roberts, R. E., Seeley, J., & Andrews, J. A. (1993). Adolescent psychopathology: I. Prevalence and incidence of depression and other DSM-III-R disorders in high school students. *Journal of Abnormal Psychology*, *102*, 133–144.

Liebowitz, E. R. (2019). *Addressing parental accommodation when treating anxiety in children*. New York: Oxford University Press.

Markus, H. R., & Kitayama, S. (1991). Culture and the self: Implications for cognition, emotion, and motivation. *Psychological Review*, *98*, 224–253.

Mellon, R. C., & Moutavelis, A. G. (2007). Structure, developmental course, and correlates of children's anxiety disorder-related behavior in a Hellenic community sample. *Journal of Anxiety Disorders*, *21*, 1–21.

Merikangas, K. R., He, J. P., Burstein, M., Swanson, S. A., Avenevoli, S., Cui, L., … Swendsen, J. (2010). Lifetime prevalence of mental disorders in US adolescents: Results from the National comorbidity Survey Replication—Adolescent Supplement (NCS-A). *Journal of the American Academy of Child and Adolescent Psychiatry*, *49*, 980–989.

Muris, P., Merckelbach, H., Ollendick, T., King, N., & Bogie, N. (2002). Three traditional and three new childhood anxiety questionnaires: Their reliability and validity in a normal adolescent sample. *Behaviour Research and Therapy*, *40*, 753–772.

Muris, P., Schmidt, H., & Merckelbach, H. (2000). Correlations among two self-report questionnaires for measuring DSM-defined anxiety disorder symptoms in children: The Screen for Child Anxiety Related Emotional Disorders and the Spence Children's Anxiety Scale. *Personality and Individual Differences*, *28*, 333–346.

Orgilés, M., Méndez, X., Spence, S. H., Huedo-Medina, T. B., & Espada, J. P. (2012). Spanish validation of the Spence Children's Anxiety Scale. *Child Psychiatry and Human Development*, *43*, 271–281.

Orgilés, M., Spence, S. H., Marzo, J. C., Méndez, X., & Espada, J. P. (2014). Psychometric properties and factorial structure of the Spence Children's Anxiety Scale (SCAS) in Spanish adolescents. *Journal of Personality Assessment*, *96*, 95–102.

Qadir, F., Maqsood, A., Us-Sahar, N., Bukhtawer, N., Khalid, A., Pauli, R., … Essau, C. A. (2018). Factor structure of the Urdu version of the Spence Children's Anxiety Scale in Pakistan. *Behavioral Medicine*, *44*, 100–107.

Rapee, R., Lyneham, H., Schniering, C., Wuthrich, V., Abbott, M., Hudson, J., & Wignall, A. (2006). *Cool Kids therapist manual: For the Cool Kids child and adolescent anxiety programs*. Sydney, Australia: Centre for Emotional Health, Macquarie University.

Rapee, R. M., Schniering, C. A., & Hudson, J. L. (2009). Anxiety disorders during childhood and adolescence: Origins and treatment. *Annual Review of Clinical Psychology*, *5*, 311–341.

Scaini, S., Belotti, R., Ogliari, A., & Battaglia, M. (2016). A comprehensive meta-analysis of cognitive-behavioral interventions for social anxiety disorder in children and adolescents. *Journal of Anxiety Disorders*, *42*, 105–112.

Seligman, L. D., & Gahr, J. L. (2013). Course and outcome of child and adolescent anxiety. In C. A. Essau, & T. H. Ollendick (Eds.), *Treatment of childhood and adolescent anxiety disorders* (pp. 44–67). London: Wiley-Blackwell.

Silverman, W. K., & Albano, A. M. (1996). *Anxiety disorders interview schedule for DSM-IV: Child version*. New York: Oxford University Press.

Spence, S. H. (1998). A measure of anxiety symptoms among children. *Behaviour Research and Therapy*, *36*, 545–566.

Spence, S. H., Barrett, P. M., & Turner, C. M. (2003). Psychometric properties of the Spence Children's Anxiety Scale with young adolescents. *Journal of Anxiety Disorders*, *17*, 605–625.

Stirman, S. W., Miller, C. J., Toder, K., & Calloway, A. (2013). Development of a framework and coding system for modifications and adaptations of evidence-based interventions. *Implementation Science*, *8*, 65.

Sugawara, M., Mukai, T., Kitamura, T., Toda, M. A., Shima, S., Tomoda, A., … Ando, A. (1999). Psychiatric disorders among Japanese children. *Journal of the American Academy of Child and Adolescent Psychiatry*, *38*, 444–452.

Sundell, K., Beelmann, A., Hasson, H., & von Thiele Schwarz, U. T. (2016). Novel programs, international adoptions, or contextual adaptations? Meta-analytical results from German and Swedish intervention research. *Journal of Clinical Child and Adolescent Psychology*, *45*, 784–796.

Sundell, K., Ferrer-Wreder, L., & Fraser, M. W. (2014). Going global: A model for evaluating empirically-supported family-based interventions in new contexts. *Evaluation & the Health Professions*, *37*, 203–230.

Tanaka-Matsumi, J. (2011). Culture and psychotherapy: Searching for empirically supported relationship. In K. D. Keith (Ed.), *Cross-cultural psychology: Contemporary theme and perspective* (pp. 274–292). West Sussex, UK: Wiley-Blackwell.

Tsocheva, I., Sasagawa, S., Georgiou, G., & Essau, C. (2013). Psychometric properties of the Bulgarian translation of the Spence Children's Anxiety Scale. *Journal of Psychological Abnormalities in Children, 1*, 101.

Wood, J. J., Piacentini, J. C., Southam-Gerow, M., Chu, B. C., & Sigman, M. (2006). Family cognitive behavioral therapy for child anxiety disorders. *Journal of the American Academy of Child and Adolescent Psychiatry, 45*, 314–321.

Zhao, J., Xing, X., & Wang, M. (2012). Psychometric properties of the Spence Children's Anxiety Scale (SCAS) in Mainland Chinese children and adolescents. *Journal of Anxiety Disorders, 26*, 728–736.

Chapter 31

Adolescent depression and cognitive behavior therapy: Indian aspects

Paakhi Srivastava[a], Pragya Sharma[b], Ananya Mahapatra[c], and Samridhi Ahuja[d]

[a]*Department of Psychology, Center for Weight, Eating, and Lifestyle Science, Drexel University, Philadelphia, PA, United States* [b]*Department of Clinical Psychology, Dr Ram Manohar Lohia Hospital, New Delhi, India* [c]*Department of Psychiatry, Employees' State Insurance Hospital and Dental College, New Delhi, India* [d]*Department of Psychology, Bharathiar University, Coimbatore, India*

Abbreviations

ADHD	attention deficit hyperactivity disorder
ALHIV	adolescents living with human immunodeficiency virus
BDI	Beck Depression Inventory
CaCBT	culturally adapted cognitive behavior therapy
CBT	cognitive behavior therapy
cCBT	computer-assisted cognitive behavior therapy
DALY	disability adjusted life years
HIV	human immunodeficiency virus
ICD10	International Classification of Diseases, 10th Edition
MBSR	mindfulness-based stress reduction
MCBT	mindfulness-based cognitive behavior therapy
MINI-KID	Mini International Neuropsychiatric Interview-Kid
PHQ-9	Patient Health Questionnaire-9
RCT	randomized control trial
TAU	treatment as usual

Introduction

In India, depressive disorders have been shown to cut across age, gender, and diverse cultural and socioeconomic strata (Arvind et al., 2019). According to the Global Burden of Disease Study 1990–2017, 197.3 million (95% Uncertainty Interval 178.4–216.4) people suffer from mental disorders in India, out of which 45.7 million (42.4–49.8) have depressive disorders which has also contributed the most to the total mental disorders DALYs (33.8%, 29.5–38.5) (Sagar et al., 2020). Of note, the most conservative global figures estimate the prevalence of depression up to 12.5% among the adolescent population (Clayborne, Varin, & Colman, 2018). In India, depression is common among adolescents both in rural and urban settings (Grover, Raju, Sharma, & Shah, 2019). Rapid urbanization, changing social and family structures, social isolation, increased academic demands, and a highly competitive milieu are some of the factors that have been hypothesized for increasing mental health problems, including depression among adolescents in India (Aggarwal & Berk, 2015).

Epidemiology of adolescent depression in India

Although 243 million people or 23% of the total population in India are reported to be in the adolescent age range (10–19 years) (Guthold et al., 2019), there is dearth of large epidemiological studies and studies evaluating the developmental trajectory of depression and its related factors in this age group.

Handbook of Lifespan Cognitive Behavioral Therapy. https://doi.org/10.1016/B978-0-323-85757-4.00041-9

Prevalence

The majority of epidemiological studies on adolescent depression from India have evaluated its point prevalence (i.e., the total number of cases of adolescent depression at one particular point of time) in a community, clinic-based, or school-based setting and have utilized a retrospective study design. To date, only one study from India has reported the incidence (i.e., the number of adolescents who develop depression during a specified time interval, such as a year) of childhood and adolescent depression in school setting. A study conducted by Malhotra and colleagues (Malhotra, Kohli, Kapoor, & Pradhan, 2009) in a representative sample of school children ($N=1063$) in Chandigarh employed a two-stage evaluation process involving teacher's rating ($N=963$) and parent rating ($N=873$) over a 6-month period and reported the incidence of depression to be 1.6 among children aged 4–11 years. However, there have been no studies evaluating the incidence of depression, exclusively in the adolescent population as the only study reporting incidence of psychiatric disorders in children and adolescents examined incidence of depression in a mixed sample of children and adolescent.

With regards to point prevalence, to date, only four studies have evaluated the point prevalence of depression among adolescents in a community setting reporting a wide range of prevalence of adolescent depression (2.6%–14.5%). The first study, Adolescent and Youth Health survey, was a cross-sectional survey of individuals aged 10–24 years in the state of Himachal Pradesh conducted by the National Health Mission. This study reported the point prevalence of depression to be 6.94% and that of suicidal ideations to be 5.45% (Gururaj, Pradeep, Beri, Chauhan, & Rizvi, 2014). The second study was a cross-sectional study which was designed to assess the prevalence of depression (as per the International Classification of Mental and Behavioral Disorders 10 (ICD-10)) in rural and suburban areas of Northern Indian adolescents aged 11–19 years. This study reported the prevalence of depression to be 14.5% with no significant difference in the prevalence among rural and urban samples (Mishra, Srivastava, Tiwary, & Kumar, 2018). The National Mental Health Survey pilot study for adolescents (13–17 years) conducted in four states of India, reported the point prevalence of mental morbidity to be 7% among adolescents age 13–17 years, with depressive disorders (first episode, recurrent depression) being the most common morbidity with a point prevalence rate of 2.6% (Murthy, 2017). More recently, a cross-sectional study conducted in 2019, in a rural sample in Northern India, reported the prevalence of depression to be 3.7% ($n=583$) among 10–19-year-olds (Mohta et al., 2020). The wide variability of the prevalence figures could be attributed to heterogeneity in study design, sampling technique, and method of evaluating depression (screening instrument, single or multistage diagnosis) and instrument-related factors.

In the last decade, several studies have been conducted in a school-based setting to evaluate the point prevalence of depression. The majority of these studies have been conducted in an urban setting and have mostly evaluated the participants by scales used to quantify depression, such as Beck Depression Inventory (BDI), Patient Health Questionnaire-9 (PHQ-9), or self-report questionnaires. The majority of the studies have not employed a rigorous sampling strategy or followed a multistage diagnosis (initial screening followed by structured clinical interview). Because of these methodological limitations, the school-based studies have reported wide variability in the prevalence of depression among adolescents, ranging from 3% to 68% (Grover et al., 2019). One of the school-based studies conducted in adolescent girls, which employed multistage sampling, reported the prevalence of depression to be 18.7% (Shukla, Shukla, Ahmad, Shukla, & Khan, 2017). Another study conducted in adolescents aged 14–17 years, which employed a multistage sampling technique and two-stage diagnosis (screening by MINI-KID, followed by clinical evaluation by a psychiatrist), reported the prevalence of depression to be 25% (Jayanthi & Thirunavukarasu, 2015). The only school-based study conducted in adolescents aged 14–16 years sampled from urban, rural, and tribal areas of South India reported the prevalence of depression to be 4.1%, 3.5%, and 1.6%, respectively (Satyanarayana, Prakash, Kulkarni, Kishor, & Renuka, 2017).

Very few clinic-based studies have evaluated the prevalence of depression among children suffering from various physical disorders. Only one cross-sectional study conducted on adolescents living with human immunodeficiency virus (ALHIV) attending the HIV clinic at a tertiary care center in India reported a prevalence of 4.9% of dysthymia and 0.9% of a major depressive episode (Pilania, Hemal, Agarwal, Beniwal, & Arora, 2020). Similarly, there is a relative scarcity of epidemiological studies of depression in vulnerable populations in this age group. A cross-sectional study among high school students in South India reported that one-third of the adolescents had experienced some form of sexual abuse, and these individuals had significantly poorer academic performance, poorer mental and physical health, greater substance abuse, poorer parental relationships, and higher rates of consensual sexual behaviors than their nonabused counterparts which have been reported as depressive equivalents (behaviors suggestive of depression) in adolescent population (Patel & Andrew, 2001). In another study conducted in East India, it was found that 69.2% of sexually abused adolescent girls suffered from moderate-to-severe depression (Deb & Mukherjee, 2009).

To conclude, epidemiological studies on adolescent depression are mostly point prevalence studies conducted in a school-based or community setting. Because of the heterogeneity in study design, sampling techniques, and methods of evaluating depression, there is a wide variability in the prevalence figures as highlighted in the studies described previously. There is a need for large-scale community-based studies in adolescent population representative of the diverse socio-economic and cultural strata, to reasonably estimate the prevalence of adolescent depression in India and delineate its clinical correlates.

Risk factors for adolescent depression

There are no community-based or longitudinal studies from India which have systematically investigated factors associated with the development of depression in the adolescent population. A total of 11 clinic and school-based studies have retrospectively looked at socio-demographic and clinical correlates of depression in children and adolescents, out of which 7 were conducted primarily in adolescent sample. All of the studies conducted in adolescent population were school-based, and one study was conducted exclusively in adolescent girls (Patel, Shah, Patel, Tilwani, & Vankar, 1998). A review of Indian epidemiological research on child and adolescent depression enumerated the most common factors associated with adolescent depression to be academic stress, familial issues (familial discord, parental fight, physical abuse, parental substance use), and economic difficulties (e.g., poverty) (Grover et al., 2019). The only study to employ binary logistic regression to evaluate predictors of depression has reported being in high school and lack of self-satisfaction with academic performance to be significant predictors for adolescent depression (Singh, Gupta, & Grover, 2017).

Clinical manifestations of adolescent depression

A total of five studies have investigated the symptom profile of depression among children and adolescents. Most of the studies exploring the symptom profile of depression have employed a clinic-based retrospective study design, have not focused on an adolescent-specific sample, and have looked at children and adolescents under the same lens. Only two studies have been conducted in an adolescent sample (Jha et al., 2017; Mohanraj & Subbaiah, 2010). The most common symptoms reported are depressed mood, diminished interest in play activities, concentration difficulties, behavior problems in the form of anger and aggression, pessimism, decreased appetite, decreased sleep, anhedonia, and somatic symptoms (Grover et al., 2019). One study compared the clinical symptomatology with that of adults and found that compared to adults, children and adolescents reported more somatic symptoms and predominantly presented with irritable mood (Bhargava & Sethi, 2005). Also, socio-occupational dysfunction in adolescents manifested primarily as academic deterioration, poor scholastic performance, and reduced play and peer activities as compared to reduced work performance in adults (Bhargava & Sethi, 2005).

Comorbid psychiatric disorders

Three clinic-based studies have evaluated the presence of comorbid psychiatric disorders in children and adolescents presenting with depression (Bhargava & Sethi, 2005; Krishnakumar & Geeta, 2006; Sagar, Pattanayak, & Mehta, 2012). The commonly reported comorbid psychiatric disorders are anxiety disorders (up to 57.65%) (Bhargava & Sethi, 2005), dysthymia (20%) (Krishnakumar & Geeta, 2006), attention deficit hyperactivity disorder (ADHD) (7.77%–20%) (Bhargava & Sethi, 2005; Krishnakumar & Geeta, 2006), conduct disorder (5.18%–9%) (Krishnakumar & Geeta, 2006; Sagar et al., 2012), dissociative disorder/conversion disorder (5.18%–9%) (Bhargava & Sethi, 2005; Krishnakumar & Geeta, 2006), and obsessive-compulsive disorder (7%) (Krishnakumar & Geeta, 2006).

Key facts

- The prevalence of depression among adolescents in India is reported to be in the range of 2.6%–14%.
- The wide variability in the prevalence figures is due to heterogeneity in the study sample, sampling techniques, and methods of evaluating depression.
- The risk factors associated with adolescent depression have been found to be academic stress, familial conflict, and poverty.
- The main clinical manifestations of depression in adolescents are depressed mood, behavioral symptoms, aggression, decreased interest in play, and peer activities and poor scholastic performance.

- The common comorbid psychiatric disorders reported are anxiety disorders, ADHD, conduct disorder, dissociative disorder, and obsessive-compulsive disorder.

CBT with Indian adolescents with depression

Practice and procedures

CBT for adolescent depression is an active, present-focused, pragmatic, and skills-based approach. In India, CBT is the recommended treatment of choice for adolescents with mild-to-moderate depression (Grover & Avasthi, 2019). Indian adolescents with depression typically endorse negative thoughts, comorbid anxiety, suicidal thoughts, and suicidal behaviors (Samuel & Sher, 2013) and as CBT is shown to be effective in reducing all these symptoms, CBT is preferred for treating depression.

CBT targets two key maintenance factors of depression in adolescents, namely, negative cognitions and maladaptive behaviors (e.g., poor ability to regulate emotions, lack of engagement in pleasant activities) (Lawrence, Buffie, Schwartz-Mette, & Nangle, 2020). As such, CBT teaches two key skills to address these maintenance factors including cognitive skills (e.g., cognitive restructuring) and behavioral skills (e.g., activity scheduling) (Srivastava & Mehta, 2014). Below we describe these skills in detail.

Cognitive skills

The "cognitive" skills of CBT involve identifying maladaptive cognitions (i.e., thoughts, beliefs, or assumptions about oneself, others, and the world that are negatively biased, inaccurate, and rigid (Beck, 1979b) and learning to modify these cognitions to make them realistic and adaptive to improve depressive mood). This process is called cognitive restructuring (Beck, 1979a).

The first step in cognitive restructuring is the initiation of tracking thoughts on a daily basis using a thought diary record. The adolescent writes down the details of a specific situation (usually an upsetting situation), the thoughts he/she came up with during the specific situation, also known as negative automatic thoughts (e.g., "I can't do anything properly"), followed by feelings that the adolescent is having during the situation and rating the intensity of the specific emotion (0%–100%), and the behavior that ensued. Once they are comfortable with maintaining this thought record, the second step involves maintaining an evidence thought diary which has additional columns for evidence for and against their automatic thoughts. The third step is to follow it by a rational response and a re-rate their emotions.

Behavioral skills

The "behavioral" skills of CBT involve promoting engagement in activities that brings a sense of pleasure or mastery and improving the adolescents' self-control. Adolescents are taught to monitor their day-to-day activities in order to maintain an activity schedule based on pleasure and mastery principle and rate each activity on a visual analog scale ranging from 0 to 10 to indicate how enjoyable the activity was or how much of an achievement it was. This is done to increase a sense of autonomy in the adolescent. Once the activity monitoring is done, their day is planned in collaboration with them to provide structure and help with setting priorities. The activity plan ensures the pleasure and mastery tasks are balanced (e.g., cleaning the closet followed by reading a book for leisure). Activity scheduling also takes care of the need to make decisions about what to do next, providing them a sense of control.

Nature and style of treatment

In India, CBT therapists follow a didactic approach early in treatment. This didactic approach of the CBT sessions is important because of the need of adolescents with depression to have a more active therapist instruction in response to the lack of motivation and self-efficacy that is often characteristic of depression (Midgley et al., 2018). Once depressive symptoms improve, the CBT therapist switches to a collaborative approach that involves provision of instruction (e.g., offering guidance on monitoring thoughts), explaining the rationale for instructions (e.g., explaining the importance of completing thought record in identifying maintenance factors of depressive mood) and allowing adolescent to ask questions, initiate discussion of topics, give feedback, and take active role in goal setting (e.g., completing thought record five times a week if adolescent expresses difficulty in completing diary on 2 days of the week).

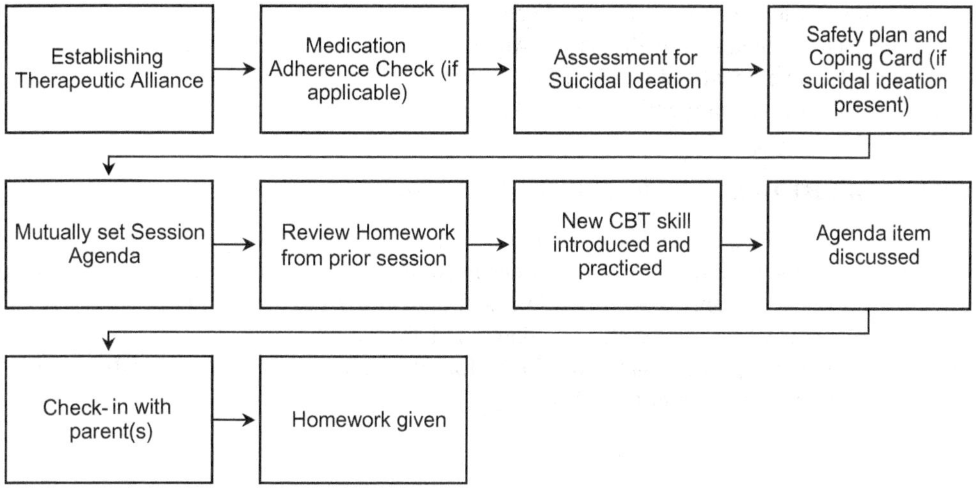

FIG. 1 Typical CBT session format.

Structure of sessions

In India, typically adolescents with depression receive 12–16 weekly sessions of CBT (Abela & Hankin, 2008). This session frequency is consistent with the number and frequency implemented in the West. For example, a systematic review and meta-analysis of 31 trials examining the effectiveness of individual and group CBT for children and adolescents with depression ($N = 4335$) found the overall number of CBT sessions to range from 5 to 16 sessions spread over 4–16 weeks (Oud et al., 2019). Within Indian context, a recent study reportedly implemented 12 weekly sessions of computer-assisted cognitive behavior therapy for adolescents with depression and found the treatment to be effective (Srivastava, Mehta, Sagar, & Ambekar, 2020). These findings preliminarily suggest that on average 12 weekly sessions of CBT may be effective in reducing depression in Indian adolescents, although more research is warranted to examine the association between session frequency and treatment outcomes in India.

The therapy begins with the therapist explaining the treatment rationale to the adolescents and their parents. Subsequent sessions usually begin with a medication adherence check (if applicable, mostly in cases of moderate-to-severe depression when medication is prescribed in addition to therapy) followed by an assessment of suicidal thoughts or behavior. In case of the presence of suicidal ideations/behavior, safety plan is reviewed or negotiated, and a coping card is created which includes a list of strategies the adolescent can engage in during stressful situations. Following this, the therapist and adolescent collaboratively set the agenda for the session, the homework from prior session is reviewed, a new skill is introduced and practiced, and the agenda item is discussed. Last, there is a parent check-in at the end and homework assignment is given. A typical CBT session format has been elucidated in Fig. 1.

Treatment implementation

CBT with Indian adolescents with depression begins with establishing a therapeutic alliance with the adolescent as well as the parents. Problem and associated goals lists are developed in collaboration with the adolescent while being mindful of the parental expectations from therapy. This is followed by a psychoeducation of both the adolescent and the parents regarding depression, its symptoms, prognosis, and treatment. Thereafter, the adolescent is introduced to the cognitive behavioral model and psychoeducated about maintaining factors and the vicious cycle of cognitive distortions. Common treatment goals are agreed upon, and cognitive behavioral interventions are described and acted upon. Some of these have been listed earlier in the chapter. The goal is to modify the adolescent's thoughts and behavior using Socratic dialogue, guided discovery, and behavioral experiments. Application and consolidation of new skills and strategies is done in therapy sessions and homework is assigned in each session to generalize it outside of therapy sessions. The adolescent's somatic symptoms may be reduced with the introduction of behavioral relaxation. Relapse prevention includes identifying the possibilities of setbacks (e.g., low grades, relationship break-ups, conflicts with parents/friends), how to detect early warning signs (feelings, behaviors, and symptoms that may indicate the beginning of another episode), and establishing next course of action (consulting the psychiatrist/therapist, reevaluate thoughts and feelings, using techniques learned from therapy, etc.). This is followed by the termination of therapy.

Each individual therapy session begins with a mood check and setting the agenda for the session. This is followed by reviewing the homework. The agenda item is discussed. Any specific problem the adolescent is facing is described. Elements of the cognitive model are elicited and confirmed. A collaborative discussion, then, takes place on how to approach the problem and the rationale for the intervention. The collaborative action plan is decided upon. Homework is discussed.

Cultural adaptation of CBT treatment to Indian adolescents

India has a collectivistic culture which deemphasizes an individual view for one's thoughts, feelings, and behaviors (Kalia, 2016). As CBT treatment originated from the West and is designed for an individualistic culture that promotes the need to self-govern and self-control one's thoughts, feelings, and behaviors, standard CBT may not be directly applicable to Indian population. As such, there is a need to adapt CBT to Indian population. Indeed, researchers have argued to develop culturally adapted CBT (caCBT) with specific adaptations in the format, content, and delivery of CBT in order to apply it to clients from non-Western cultures (Naeem et al., 2014).

Challenges to clinical application

Familial and cultural variables

One of the major challenges to CBT treatment for adolescent depression is poor alignment between adolescents and their parents in terms of their expectations from therapy. For instance, parents may be overly concerned about the adolescent's academic performance, while the adolescent might be perturbed by the parents' attitudes toward him/her. This parental concern over the adolescent's academic performance is mostly the reason for seeking psychological help. Other reasons to seek therapeutic help are the adolescent's somatic complaints like headache, stomachache, etc. This is because it is much easier for Indian parents to accept psychological distress manifested in the form of somatic complaints as they are unable to relate academic underachievement arising as a result of emotional disturbances (Sondhi, Gulgulia, & Shriharsh, 2013). This poses a problem as the parents are unable to notice the improvement in adolescent unless they witness a change in their grades. Their focus on academic achievement makes it difficult for the therapist to get through to them and help them understand what the adolescent is going through. This also means that if the parents do not see a change in the adolescent's grades as per their expectations, they believe the therapy isn't working and stop treatment. CBT therapists in Indian context, thus, need to align the goals of the parents and the adolescent by psychoeducating them about depression, precipitating and maintaining factors as well as its manifestation, at the outset of therapy. The therapist also needs to check parents' expectations from therapy and give them an estimate of the number of sessions in the beginning of the therapy in order to overcome these issues.

The parental overemphasis on adolescent's academic achievement often comes across as parental pressure which is commonly encountered in therapy (Sondhi et al., 2013). Studies have shown that parental pressure causes hindrance in therapy for adolescents (Creswell, Willetts, Murray, Singhal, & Cooper, 2008). For example, the parents may expect the adolescent to study rather than spend time on pleasure activities which was told to be balanced with mastery activities by the therapist. This interferes with the therapy process. As such, it is important to psychoeducate parents and reroute their expectations to be more realistic and seek their cooperation in helping the adolescent complete homework assigned in therapy to improve outcomes.

Stigma and discrimination

Despite the recent surge in awareness regarding mental health conditions in the general public, there still exists stigma and discrimination of persons with mental illnesses, thereby, thwarting any attempts at seeking treatment (Srivastava et al., 2020). Parents often deny the presence of mental illness in their child. This leads to a blame game between the parents and teachers about not giving the adolescent individual time and attention. A challenging part of the treatment, thus, includes developing an insight into the parents and the family as well as sensitizing teachers about the mental health needs of the adolescents.

The stigma of mental illnesses in India has its origin in traditional Asian beliefs which tend to equate mental illness with insanity. Thus, there is a stigma attached to having a mental illness and seeing a mental health practitioner. This further leads to denial of the existence of a problem as well as creates barriers to help-seeking (Halder & Mahato, 2019). In particular, stigma and discrimination are significant in other areas. For example, adolescents in sexual minority including gay, lesbian, or bisexual individuals may already be facing many other challenges including discrimination, stigma toward their group, family disapproval, social rejection, and violence. This calls for a clearly defined adaptation of CBT that integrates gay affirmative practices as an initial step toward developing empirically supported interventions to meet the specific needs of sexual-minority youth (Craig, Austin, & McInroy, 2014).

Lack of skilled therapist

There is a dearth of skilled therapists in India (Patel, Chowdhary, Rahman, & Verdeli, 2011). A skilled therapist can help the adolescent move from resistance to treatment (Perepletchikova, Treat, & Kazdin, 2007). For instance, skilled therapists are able to apply CBT within a good therapeutic relationship and adapt it to the adolescents' needs rather than a prefabricated process that ensures better treatment engagement and outcomes (Craig, Austin, & Alessi, 2013). Thus, the shortage of skilled therapists compromises with the quality of CBT being provided to adolescents.

Dearth of CBT-skilled clinicians in India has led to a wide treatment gap such that a significant majority of adolescents with emotional disturbances including depression do not have access to evidence-based treatments. Indeed, researchers have suggested the use of computer-assisted CBT (cCBT) to disseminate treatment to those who otherwise do not receive treatment (Srivastava & Mehta, 2014). Within India context, cCBT may reduce the amount of time a therapist may spend on each adolescent and allow them to carry more caseloads, which in turn will reduce existing treatment gap. A recent study from India showed that a computer program designed to deliver CBT to adolescents with depression in conjunction with therapist led brief CBT sessions was found to significantly reduce the average number of hours spent by therapist per adolescent during 12 weeks of treatment ($M=6.41$) compared to treatment as usual ($M=10.35$). These preliminary findings indicate the promise of using computers in delivering CBT to improve treatment accessibility and reducing treatment gap.

Comorbid conditions

Adolescent depression has a high comorbidity with other disorders like anxiety disorders, substance use disorder, personality disorder, conduct disorder, oppositional defiant disorder, attention deficit hyperkinetic disorder (ADHD), and dissociative/conversion disorder (Grover & Avasthi, 2019). These may take predominance over the emotional distress of the adolescent and may require modifications to CBT approach in order to ensure better treatment outcomes. For example, if an adolescent with depression presents with a comorbidity of harmful substance use or dependence, this secondary diagnosis may take precedence in the therapy to reduce the possibility of self-harm. This may require incorporating elements of motivational enhancement therapy before working on the depressive symptoms presented by the adolescent.

Compliance

It is important to reinforce homework compliance in therapy. Researchers have noted that Indian adults often do not complete homework assignments (Halder & Mahato, 2019). In our clinical practice, we have noticed a similar tendency in adolescents with depression. This could be due to adverse familial or school environment or low motivation of the adolescent. Thus, CBT therapist should address these factors early in treatment to ensure that adolescents comply with treatment components and achieve better outcomes. This can be done by collaborating with the adolescent while choosing the homework and ensuring the homework is feasible for them to complete within the week. Any possible obstacles that may come up should be discussed beforehand and list of solutions may be prepared in advance.

Key facts

- It is important to include parents in the therapeutic process to ensure their goals are aligned with those of the adolescent.
- Keeping in mind the collectivistic culture of India, there is a need for a culturally adapted CBT (caCBT) that would ensure adaptations in the format, content, and delivery of CBT for its application with adolescent depression in India.
- The advent of technology also calls for newer modes of intervention like computerized CBT.

Empirical evidence of CBT for treating depression in Indian adolescents

CBT has a growing empirical base demonstrating its efficacy for treating depression in adolescents in India. Typically, CBTs for treating adolescents with depression administer one or more CBT skills integrated to behavioral treatments designed to improve depressive mood. As a result, a vast majority of treatment research has evaluated CBT-informed treatments (i.e., behavioral treatments that include one or more CBT components and, not the full CBT model), which limits our ability to understand the efficacy of the CBT treatment approach for depression in Indian adolescents. Additionally, within the past decades, many of the CBT-informed treatments have also included psychological strategies (e.g., present moment awareness, acceptance, etc.) from other third wave treatment approaches such as mindfulness-based cognitive behavior therapy (MCBT) (Morgan, 2003) and mindfulness-based stress reduction (MBSR) (Kabat-Zinn, 2003). The provision of these skills within CBT for depression is focused broadly on enhancing self-regulation capacity to facilitate long-term improvements in emotional and social functioning and prevent risk of relapse (Anand & Sharma, 2014; Raj et al., 2019). Although emerging research shows promise of these treatments for depression in Indian adolescents (Anand & Sharma, 2014; Raj et al., 2019) because the treatment packages evaluated in these studies included skills from behavior treatments (e.g., acceptance from MCBT and mindful meditation from MBSR), it is difficult to parse apart the independent treatment effects of CBT on clinical outcome. As such, we chose to review the evidence for full CBT treatments and CBT-informed treatments together to allow for a more comprehensive literature base to review (Table 1). Notably, to date, no study has evaluated the mechanisms of action, and predictor/moderators of CBT and CBT-informed treatments for treating depression in Indian adolescents limiting our understanding of how these treatments work and for whom these treatments work.

Outcomes from CBT and CBT-informed treatments for depression in Indian adolescents

In total, eight studies have evaluated the efficacy of CBT-informed treatments in treating depression in Indian adolescents. Four studies have utilized formal CBT (Chaudhary, 2019; Kajavinthan, 2013; Sondhi et al., 2013; Srivastava et al., 2020) and the rest have utilized a CBT-informed treatment.

Of the eight studies, four utilized a sample of adolescents with mild-to-moderate depression (Dasarathan et al., 2018; Kajavinthan, 2013; Sondhi et al., 2013; Srivastava et al., 2020), one utilized a sample of adolescents with severe depression and suicidal ideation (Raj et al., 2019), and three utilized samples having depressive symptoms secondary to school-related stress (Anand & Sharma, 2014; Chaudhary, 2019; Singhal et al., 2014). Notably, only three studies conducted a fully randomized control trial (RCT) (Kajavinthan, 2013; Singhal et al., 2014; Srivastava et al., 2020), whereas the others utilized either a case study (Dasarathan et al., 2018; Sondhi et al., 2013), or pre-post within-subjects experimental design (Anand & Sharma, 2014; Chaudhary, 2019; Raj et al., 2019). Regarding treatment delivery, two studies delivered an individual, face-to-face treatment (Dasarathan et al., 2018; Sondhi et al., 2013); one delivered a computer-assisted CBT intervention supported by clinician (cCBT), (Srivastava et al., 2020); and the remainder delivered a group-based intervention program (Anand & Sharma, 2014; Chaudhary, 2019; Kajavinthan, 2013; Raj et al., 2019; Singhal et al., 2014).

Outcomes for mild-to-moderate depression in Indian adolescents

Extant case studies and RCTs have suggested that CBT produces symptom improvement in adolescents with depression (Dasarathan et al., 2018; Kajavinthan, 2013; Sondhi et al., 2013; Srivastava et al., 2020). Results from two case studies showed that adolescents who reported moderate depression on Beck's Depression Inventory-II (BDI-II) (Beck, Steer, & Brown, 1996) at pretreatment endorsed mild or minimal depression at posttreatment on the same assessment measure (Dasarathan et al., 2018; Sondhi et al., 2013). These gains were maintained at 1- and 2-month follow-up (Sondhi et al., 2013). A recent RCT demonstrated that 12 weekly sessions of computer-assisted CBT with clinicians' support (cCBT) significantly outperformed treatment-as-usual (TAU) among adolescents with depression ($N=21$, cCBT$=11$, TAU$=10$) at posttreatment. The results showed that out of those who completed treatment, 90% of the adolescents in cCBT group and 77.7% in TAU group achieved remission at the posttreatment. On average, adolescents who received cCBT endorsed 60.6% reduction in depressive symptoms compared to 35.4% reduction in depressive symptoms among those who received TAU ($t=2.20$, $P=.04$) (Srivastava et al., 2020). Furthermore, adolescents receiving cCBT had significantly greater improvements in general functioning from pre- to posttreatment (Srivastava et al., 2020). Another study found that adolescent girls with mild depression ($N=14$, CBT$=7$, Waitlist$=7$) receiving 16 weeks of group-based CBT treatment in a school setting experienced greater improvements in depressive symptoms at posttreatment up compared to waitlist ($t=6.96$, $P=.04$) (Kajavinthan, 2013).

Outcomes for severe depression in Indian adolescents

Evidence for CBT and CBT-informed treatments for severe depression in Indian adolescents is scarce, but promising. A single group pre-post experiment evaluating the efficacy of 12 sessions of group based MCBT intervention delivered over 16 weeks without pharmacotherapy ($N=30$) demonstrated that intervention produced significant improvements in depressive symptoms ($t=22.78$, $P=.01$) and suicidal ideations ($t=44.81$, $P=.01$) from pre- to posttreatment (Raj et al., 2019).

TABLE 1 Literature on clinical outcomes followed by caveats in interpreting the outcome results.

Publication	Sample size and characteristics	Diagnostic criteria	Comparison condition	Duration of treatment	Clinical outcome	Description of the treatment
Sondhi et al. (2013)	16-Year-old adolescent male	Beck Depression Inventory (BDI), State Trait Anxiety Test (STAT), Bhatia Battery of Performance Intelligence Test, Verbal Adult Intelligence Scale (VAIS) Subtest of the PGI-Battery of Brain Dysfunction (PGI-BBD), Bender–Gestalt Visuo-Motor Test, Thematic Apperception Test–Indian Adaptation (TAT)	Case study	6 months	At posttreatment and first-month and second-month follow-up significant reduction in BDI scores were observed. In addition, at both posttreatment and first and second-month follow-up significant improvements were observed in the ability to implement CBT skills including better mood management and reduction of dysfunctional behavior patterns. Authors concluded that these changes must have produced improvements in depressive symptomatology	The treatment approach involved the use of cognitive restructuring techniques as well behavioral techniques, these included time management skills and progressive relaxation activities. Problem-solving techniques such as study skills were also imparted
Singhal, Manjula, and Sagar (2014)	300 Students (13–18 years)	Children's Depression Inventory (CDI), Centre for Epidemiological Studies-Depression Scale for Children (CES-DC), Children's Automatic Thoughts Scale (CATS), Cognitive Triad Inventory for Children (CTI-C), Scale for Assessing Academic Stress (SAAS), Adolescent Coping Orientation to Problems Experienced Inventory (A-COPE), Social Problem-Solving Inventory-Revised-Short Form (SPSI-R)	Two group comparison design (intervention and control condition) with repeated baseline assessments	8 Weeks	The authors found no significant changes; however, results demonstrated that coping skill program compared with psychoeducation was slightly superior in reducing the self-reported depressive symptoms. At the 3-month follow-up a reduction in the self-reported depressive symptoms was apparent	The authors introduced the following skills via the coping skill program: identifying negative thinking, changing to positive thinking, using ABC to challenge negative thoughts, using social detective game to solve relationship problems and ways to deal with academic stress

Continued

TABLE 1 Literature on clinical outcomes followed by caveats in interpreting the outcome results—cont'd

Publication	Sample size and characteristics	Diagnostic criteria	Comparison condition	Duration of treatment	Clinical outcome	Description of the treatment
Kajavinthan (2013)	14 Mildly depressed adolescents' girls, studying in 11th Standard	Revised Beck's Depression Inventory	Pretest to posttest experimental and waiting list control group design CBT, $n=7$, Waitlist=7	3 Months (16 sessions)	CBT group experienced greater improvements in depressive symptoms at posttreatment as compared to waitlist ($t=6.96$, $P=.04$). Authors concluded that CBT was efficacious for treating mild depression among students attending higher secondary school	The author used behavioral techniques such as progressive muscle relaxation, daily record of activities (pleasure v/s mastery), assertiveness training via role plays. Participants were also asked to maintain a record of automatic thoughts
Dasarathan, Sudhakaran, Sivasailam, and Sivasankari (2018)	Family of six members who had experienced a recent trauma Father—60-year-old, Mother—50-year-old, Children 2 in 12th Standard and 2 College going	Beck's depression inventory (BDI) Qualitative measure was used for the client with the highest score on the measure	Pretest and posttest analysis was used for quantitative assessment, observation and interview were used for qualitative analysis	Mother=15 sessions, Children=5 sessions	Significant improvements in BDI-II scores at posttreatment were observed in all the family members. The authors concluded that CBT was effective in alleviating the depressive symptoms	The treatment approach involved the use of cognitive techniques to change the pattern of thinking, techniques to build self-confidence, stress management techniques, as well as problem solving strategies
Srivastava et al. (2020)	21 Adolescents aged of 13–19 years with unipolar depression	Beck's Depression Inventory-II, Children's Depression Rating Scale Revised, Clinician's Global Impression Scale, Children's Global Assessment Scale	Single-blinded clinical randomized control trial. Participants were assessed at three-time points including baseline, midtreatment (after 6 weeks of treatment) and posttreatment (after 12 weeks of treatment)	12 Weeks	Significant decrease in depressive symptoms for participants in cCBT group (60.6%) as compared to participants in TAU group (35.4%), as well as significant difference in remission at posttreatment (90% cCBT, 77.7% TAU) was observed. The authors concluded that smarteen was more effective over TAU	The treatment protocol was a cultural adaptation of CBT for depression in adolescents. The modules delivered treatment content focused on behavioral activation, identification and modification of maladaptive cognitions and problem solving skills. Treatment content was delivered via

TABLE 1 Literature on clinical outcomes followed by caveats in interpreting the outcome results—cont'd

Publication	Sample size and characteristics	Diagnostic criteria	Comparison condition	Duration of treatment	Clinical outcome	Description of the treatment
						flowcharts, puzzles, quizzes as well as case vignettes that illustrated examples of thoughts, feelings and behaviors experienced across different circumstances. The protocol also allowed for the participants to maintain a personal diary, and tools to help them overcome depression
Raj et al. (2019)	30 Adolescents (male and female) diagnosed with depression and suicidal ideation	The modified Scale for Suicidal Ideation, Beck's Depression Inventory, Satisfaction with Life Scale, Revised Life Orientation Test	Single group Pretest–posttest paradigm	16 Weeks	Significant improvements in depression and suicidal ideation were observed at posttreatment. Significant improvements in life satisfaction were observed. Authors concluded that mindfulness-based cognitive behavior therapy was an effective treatment for depression in adolescents	The session protocol emphasized on bringing adolescent's attention to internal experiences, and using mindfulness practices for dealing with life challenges. These included introducing techniques like body scan, mindfulness breathing and self-care activities
Chaudhary (2019)	20 Students (16–18 years)	The Perceived Stress Scale (PSS)	Single group pretest posttest design	Three sessions	A significant improvement in perceived stress was observed. Thus, CBT intervention was found to have significantly positive effect on the stress levels of students appearing for competitive exams	The treatment protocol focused on teaching the technique of challenging and changing unhelpful cognitive distortions, emotion regulation strategies and coping strategies that aimed at improving problem solving

Outcomes for depressive symptoms secondary to school-related stress

Few single group pre-post design studies and RCTs examining CBT-informed behavioral treatments for adolescents who experience depressive symptoms (and not full threshold depression) secondary to school-related stress demonstrate significant reductions in depressive symptoms at posttreatment (Anand & Sharma, 2014) and at 3-month follow-up (Singhal et al., 2014). For example, a RCT ($N=19$) demonstrated that a brief school-based coping skills program ($n=13$) that delivered skills to identify negative thinking change to positive thinking, challenge negative thoughts, problem solve relationship problems, and deal with academic stress led to clinically meaningful improvements in depression at posttreatment and significant reduction in depressive symptoms at 3-month follow-up compared to psychoeducational interactive sessions ($n=13$) (Singhal et al., 2014). Another study demonstrated the preliminary efficacy of an 8 weekly MBSR program for adolescents in school setting ($N=33$) in producing significant reductions in physiological and emotional manifestations of stress (including depressed mood), academic stress, interpersonal stress, and enhancing academic self-concept and well-being (Anand & Sharma, 2014). A recent study showed significant improvements in perceived stress and depressive symptoms in a sample of adolescents reporting high exam-related stress ($N=20$) who received group-based CBT (Chaudhary, 2019). Taken together, these findings preliminarily suggest that CBTs are an effective mode of therapy for depression in Indian adolescents.

Caveats in interpreting outcome results

While the results of CBT for adolescent depression are promising, it is important to note that the existing literature base contains numerous limitations that reduce our ability to synthesize results. Most of the trials included in the above review were either case studies or small pilot RCTs comparing to a wait-list control, TAU, or a minimally active treatment condition. To date, there has been no RCT with large sample size limiting our ability to generalize extant research findings. Furthermore, there have been no RCTs comparing CBT with other evidence-based treatments for adolescent depression (e.g., interpersonal therapy) which precludes from identifying the most efficacious behavior treatment for treating adolescents with depression in India. Adding to the difficulties in interpretation, within the existing body of research, there is vast heterogeneity in the implementation of CBT (e.g., length, frequency, and number of sessions; group vs individual) and variability in specific components of CBT that were included in the treatment package being tested. In addition, a few studies have not fully described the intervention program (e.g., Chaudhary, 2019; Dasarathan et al., 2018; Kajavinthan, 2013) or missed one or more core components of standard CBT such as identification of negative thoughts (Chaudhary, 2019; Dasarathan et al., 2018) which limits the ability to conclude with confidence that standard CBT for depression (Beck, 1979a) can be an effective treatment for adolescents in India. Thus, although initial results are promising, more research is acutely needed to fully understand the impact of CBT on adolescent depression within Indian context.

Process variables

To date, only one study has evaluated two key mechanisms of action of CBT including negative automatic thoughts and cognitive triad (i.e., an individual's view of self, the world, and the future) (Beck, 1979b; Singhal et al., 2014). Although as described earlier, this study was designed to improve coping skills in adolescents in a school setting, the intervention implemented in the study was heavily based on CBT and targeted negative thoughts, a key maintenance factor of depression. This study showed that adolescents receiving school-based coping skills program evidenced clinically significant improvements in negative thoughts as measured on Children's Automatic Thoughts Scale (CATS) (Schniering & Rapee, 2002) at posttreatment and statistically significant improvements in negative thoughts at 3-month follow-up (Singhal et al., 2014).

Conclusively, extant research provides preliminary support for CBT in treating depression in Indian adolescents; however, lack of systematic research precludes us from drawing firm conclusions that CBT is an effective treatment for adolescent depression in Indian context. Further research is warranted to extend previous findings and examine the effectiveness of this approach for treating adolescents with depression in India.

Applications to other areas

Depression in adolescent population has been reported to occur concurrently or as a consequence of numerous psychiatric disorders as well as medical illnesses. Among psychiatric disorders, the most common comorbid conditions reported among the adolescent population in India are anxiety disorders, attention deficit hyperactivity disorder, dissociative disorder, and obsessive-compulsive disorders (Grover et al., 2019). In addition, depression is also reported to be highly comorbid with substance use disorders in adolescents. A network analysis of data on 13,035 adolescents from India, to explore the relationship between substance use and depressive symptoms, reported a central role of symptoms such as "feeling like a failure," "sad mood," and "somatic symptoms" with substance use (Wasil, Venturo-Conerly, Shinde, Patel, & Jones, 2020). Although there are no existing studies from India at present, this study highlights the potential benefit of CBT in targeting the psychopathology underlying comorbid substance use and depression. Similarly, a cross-sectional study, examining the prevalence of PTSD in adolescent population following the super-cyclone in the state of Orissa, reported that 34.3% also suffered from comorbid depression, 6.7% had ideas of worthlessness, and 4.9% had suicidal ideations (Kar et al., 2007). Trauma-focused CBT (TF-CBT) was initially developed for children and adolescents and a recent meta-analysis of 21 studies to assess the effectiveness of TF-CBT for reducing the symptoms of PTSD and comorbid depression among children and adolescents reported that TF-CBT was superior to no-treatment or alternate treatment in reducing both PTSD and depressive symptoms (Lenz & Hollenbaugh, 2015). This again reiterates the potential usefulness of culturally adapting TF-CBT for an Indian adolescent population. Among medical disorders, a case series of three cases has reported depression comorbid with type 1 diabetes in children and adolescents where CBT was used as an intervention along with pharmacological management leading to reduction of depressive symptoms as well as better glycemic control (Kaur, Patel, & Ghildiyal, 2019). However, there is still a lack of intervention studies which have evaluated the role of CBT in adolescents suffering from type 1 diabetes and comorbid depression. Similarly, which depressive symptoms have been reported in adolescents suffering from migraine and cancer, and the role of CBT in improving outcomes in such cases, are yet to be studied.

Future directions

Clinical application

With an understanding of the rampant challenges to implementation of CBT to adolescent depression in India, there are certain changes to be made in order to adapt CBT. This includes involving the parents in the treatment process especially in psychoeducation and managing their expectations to be more realistic. The therapist needs to normalize therapy for both the adolescent and their parents in order to reduce the stigma that may interfere with continuing treatment. In addition, the application of CBT in Indian context requires the therapist to develop multiple alliances including with parents in order to attend to multiple perspectives and to develop a treatment plan that accommodates both adolescent and parent perspectives. This is important for treatment continuation.

A point in favor of CBT for Indian adolescents is that Indians, in general, tend to favor structured, directive (therapist as expert, client expecting, and accepting authority), and short-term, problem-focused therapy (Sondhi et al., 2013). The "directive-ness" of the therapist can be used to emphasize homework compliance between sessions.

What works in therapy for Indian adolescents is that despite the norm of emotional restrain and no or limited allowance of emotional expression in Indian families, adolescents are comparatively more open to talk about their emotions once the rapport is formed and the parent is no longer present in the therapy room. It is, thus, the therapist's responsibility to help the parent be open to receiving the adolescent's emotions inside and outside of therapy. Thus, the first few sessions may be focused on the physiological manifestations like somatic complaints or academic consequences like underachievement in order to allay parental doubts about the purpose of therapy and building their and the adolescents' trust in the therapeutic process. Once the rapport has been formed with both the adolescent and their parents, psychoeducation about the importance of emotional expression should be conveyed to them. This allows the adolescent to experience the emotion, accept it, and be able to demonstrate it in an adaptive manner without feeling guilty or being blamed by the parent. Once this gap is bridged, the therapist can then work on generating a collective understanding of the problem at present and work on problem-solving and incorporating coping strategies in close contact with the client.

Another pertinent thing to keep in mind is that certain techniques from other therapies may be added to CBT to address the comorbidities often associated with adolescent depression (e.g., mindfulness to aid with associated anxiety).

Further, considering the advent of technology including computers and smartphones, access to CBT treatment may be improved by the use of computer application-designed CBT programs and apps. Adolescents are technology savvy and would be able to adapt easily to this change in treatment delivery.

Finally, considering the gaps in treatment due to a smaller number of mental health professionals and the cost of therapy, group CBT for adolescent depression can also be implemented.

Research

There is limited evidence for CBT for severe depression in Indian adolescents. This area can be further researched. Also, there are no large sample RCTs comparing CBT with other evidence-based treatments for adolescent depression which limits our understanding in identifying the most efficacious nonpharmacological treatment for adolescent depression in India.

The heterogeneity of CBT implementation with regards to length, frequency, and number of sessions also makes it difficult to compare and interpret the findings. Few studies do not fully describe the intervention (e.g., Chaudhary, 2019; Dasarathan et al., 2018; Kajavinthan, 2013) or miss one or more core components of CBT such as identification of negative thoughts (Chaudhary, 2019; Dasarathan et al., 2018) which limits the ability to conclude that CBT can be an effective treatment for adolescent depression in India. Thus, more research is needed to understand the efficacy of CBT on adolescent depression in the Indian context.

Process research on CBT in adolescent depression is relatively a new field. There are gaps in the literature and the majority of studies are focused on adult depression. There is relatively little stress on documentation or research. Audio/video recording of sessions is rare thus restricting the parameter of research (Sondhi et al., 2013). Future research studies with an emphasis on process research in this area including mechanisms of action, and moderators of CBT would have implications on the refinement of CBT protocols and clinical practice.

Conclusion and summary

Adolescents in India comprise 23% of the total population. Rapid changes in family and societal structures, social isolation, increased academic competition, and high performance demands along with urbanization may be some of the factors implicated in increased mental health problems among Indian adolescents. One of these mental health problems afflicting adolescents is depression. The prevalence of adolescent depression in India is estimated to be around 2.6%–14.5%. The clinical manifestations of depression among adolescents include depressed mood, diminished interest in play activities, concentration difficulties, behavior problems in the form of anger and aggression, pessimism, decreased appetite, decreased sleep, anhedonia, and somatic symptoms. Commonly reported comorbid disorders for adolescent depression are anxiety disorders, dysthymia, attention deficit hyperactivity disorder, conduct disorder, dissociative disorder/conversion disorder, and obsessive-compulsive disorder.

Cognitive behavioral therapy (CBT) has been the nonpharmacological treatment of choice for adolescent depression. In CBT, the therapist plays an active role in the treatment to form a collaborative relationship and guides the adolescent to monitor and record thoughts and behavior, explains the importance of homework assignments and the treatment involves the use of behavioral (coping skills, interpersonal relationships, problem solving) and cognitive strategies (cognitive restructuring). However, despite a plethora of evidence studies supporting the efficacy of CBT for depression in adolescents, the mechanisms by which CBT exerts these beneficial effects is less researched, especially in the Indian context. Some of these variables as identified by international research include therapeutic alliance, therapist adherence and competence, cognitive restructuring, and active participation by the adolescent.

CBT has a growing empirical base demonstrating its efficacy for treating depression in adolescents in India. However, there are some gaps in the literature. The research available so far has been carried out with adolescents with mild-to-moderate depression utilizing either individual face-to-face treatment, or computer-assisted CBT, or a group-based intervention program. These studies have reported that CBT leads to symptom improvement in adolescent depression, and these gains are maintained at 1–2-month follow-up. Studies with severe depression in Indian adolescents are scarce. The vast heterogeneity in the CBT implementation and variability in specific CBT components, or missing one or more core CBT components make it difficult to synthesize the available results and interpret the outcomes adequately. Thus, although initial results are promising, more research is acutely needed to fully understand the impact of CBT on adolescent depression within Indian context.

There are several challenges to clinical application of CBT with adolescent depression in India. These include familial and cultural variables (e.g., parental expectations from therapy), stigma, discrimination, lack of a skilled therapist, therapeutic alliance, and treatment planning, comorbid conditions, and therapy compliance.

With a view to the future application of CBT with adolescent depression in India, CBT needs to be structured, directive, short-term, and problem-focused. Therapeutic alliance with both the adolescents and their parents needs to be established in order to attend to multiple perspectives and to develop a treatment plan that accommodates both adolescent and parent perspectives. This is important for treatment continuation. Further, CBT needs to be culturally adapted in its format, content, and delivery in order to cater to Indian population. CBT in India would also gain from the utilization of online options including computerized CBT and apps to increase its accessibility to adolescents with depression, considering the low number of mental health professionals in the country.

Indian studies need to undertake process research on CBT in adolescent depression to fill in the gaps in literature and would have implications for CBT protocol refinement, therapist training, and clinical practice. Indian studies on adolescent depression also need to evaluate the mechanisms of action, and predictor/moderators of CBT.

Conclusively, CBT seems to lead to symptom improvement in adolescent depression but emphasis needs to be on its cultural adaptation and implementation to Indian adolescents with depression in particular. Future research is warranted to evaluate how CBT brings about changes and what predicts/mediates this process.

Mini-dictionary of terms

- **Point prevalence:** The proportion of the population with a condition at a specific point in time, expressed as a percentage. For example, if in a sample of 200 adolescents, 60 of them have been diagnosed with depression at a specific point of time. The point prevalence of depression in the adolescent sample would be 30% or 3 in 10 individuals calculated as: (60 adolescents with depression)/(200 Total sample/At risk) = 0.3 = 30%.
- **Incidence:** The newly identified cases of a disease or condition per population at risk over a specified timeframe, expressed as person-years. An example if there are 795,000 new strokes in the United States, annually and 324 million people in the United States for the measured year, then the incidence of stroke in the United States would be 2.5 strokes per 1000 person-years calculated as: (795,000 strokes)/(324,000,000 people in the United States during the year) = 2.5 strokes/1000 person-years.
- **Multistage sampling:** Sampling method that divides the population into groups (or clusters) for conducting research.
- **Binary logistic regression:** Statistical technique used to predict the relationship between predictors (i.e., independent variables) and a predicted variable (i.e., the dependent variable) where the dependent variable is binary (e.g., sex (male vs female), response (yes vs no), score (high vs low)).
- **Iatrogenic elements:** The side effects and risks associated with the medical intervention, malpractice, medical error, and negligence constitute common iatrogenic complications.
- **Cognitive distortions:** Exaggerated or irrational thought patterns causing individuals to perceive the reality inaccurately.
- **Externalizing behavior:** These are problem behaviors that are manifested in children's outward behavior and reflect negatively on the environment.
- **Negative automatic thoughts:** These are negatively framed interpretation of the world around us and tend to impact an individual's feelings and mood.
- **Cognitive triad:** Three forms of negative thinking typical of a person with depression, these include negative thoughts about self, negative thoughts about the world, and negative thoughts about the future.

Key facts

- Research evaluating cognitive behavior therapy (CBT) for depression in adolescents remains nascent in India.
- Extant (and, limited) research examining CBT for treating adolescents with depression shows that it is typical in India to administer one or more CBT skills integrated to behavioral treatments (i.e., CBT-informed treatments) to improve depression in adolescents.
- Given that most studies evaluated CBT-informed treatment packages including skills from other behavior treatments (e.g., mindful meditation from mindfulness-based stress reduction), it is difficult to parse apart the independent treatment effects of CBT on clinical outcome.
- To date, no study has evaluated the mechanisms of action, predictors, and moderators of CBT and CBT-informed treatments for treating depression in Indian adolescents which limits the understanding of how these treatments work and for whom these treatments work.
- Preliminary evidence from extant research lends support for potential efficacy of CBT and CBT-informed treatments for adolescents with depression in India.
- Further research is warranted to examine the effectiveness of CBT for treating adolescents with depression and evaluate how CBT works and for whom CBT works in Indian context.

Summary points

- A significant proportion of India's population (one-fifth) comprises of adolescents and depression has been reported to be one of the most common mental disorders in this age group.
- Among nonpharmacological management, CBT has been the treatment of choice for depression in adolescents.
- Existing literature supports the role of CBT in treatment of depression in adolescents in India.
- The current challenges to application of CBT in this population currently are familial and cultural factors, stigma, lack of a skilled therapists, comorbid psychiatric disorders, and therapy compliance.
- There is a need for further research investigating the efficacy of CBT and its cultural adaptations to suit the unique needs of this population and to delineate the process and mediating variables which underlie the process of symptom improvement.

References

Abela, J. R., & Hankin, B. L. (2008). *Cognitive vulnerability to depression in children and adolescents: A developmental psychopathology perspective*. The Guilford Press.

Aggarwal, S., & Berk, M. (2015). Evolution of adolescent mental health in a rapidly changing socioeconomic environment: A review of mental health studies in adolescents in India over last 10 years. *Asian Journal of Psychiatry*, *13*, 3–12.

Anand, U., & Sharma, M. P. (2014). Effectiveness of a mindfulness-based stress reduction program on stress and well-being in adolescents in a school setting. *Indian Journal of Positive Psychology*, *5*(1), 17.

Arvind, B. A., Gururaj, G., Loganathan, S., Amudhan, S., Varghese, M., Benegal, V., … Dalal, P. (2019). Prevalence and socioeconomic impact of depressive disorders in India: Multisite population-based cross-sectional study. *BMJ Open*, *9*(6), e027250.

Beck, A. T. (1979a). *Cognitive therapy and the emotional disorders*. Penguin.

Beck, A. T. (1979b). *Cognitive therapy of depression*. Guilford Press.

Beck, A. T., Steer, R. A., & Brown, G. (1996). Beck depression inventory–II. *Psychological Assessment*. https://doi.org/10.1037/t00742-000.

Bhargava, S. C., & Sethi, S. (2005). Depressive disorder in children. *Journal of Indian Association for Child and Adolescent Mental Health*, *1*, 4.

Chaudhary, N. (2019). Effect of cognitive behavioral therapy on stress among students appearing for competitive exams. *International Journal of Indian Psychology*, *7*(4). https://doi.org/10.25215/0704.131.

Clayborne, Z., Varin, M., & Colman, I. (2018). Adolescent depression and long-term psychosocial outcomes: A systematic review and meta-analysis. *Journal of the American Academy of Child & Adolescent Psychiatry*, *58*, 72–79.

Craig, S. L., Austin, A., & Alessi, E. (2013). Gay affirmative cognitive behavioral therapy for sexual minority youth: A clinical adaptation. *Clinical Social Work Journal*, *41*(3), 258–266.

Craig, S. L., Austin, A., & McInroy, L. B. (2014). School-based groups to support multiethnic sexual minority youth resiliency: Preliminary effectiveness. *Child and Adolescent Social Work Journal*, *31*(1), 87–106.

Creswell, C., Willetts, L., Murray, L., Singhal, M., & Cooper, P. (2008). Treatment of child anxiety: An exploratory study of the role of maternal anxiety and behaviours in treatment outcome. *Clinical Psychology & Psychotherapy*, *15*(1), 38–44.

Dasarathan, C., Sudhakaran, M. K., Sivasailam, S., & Sivasankari, V. (2018). Impact of cognitive behavioural therapy (CBT) on an Indian family with depression—A clinical case study. In *Proceedings of ARSSS International Conference*.

Deb, S., & Mukherjee, A. (2009). *Impact of sexual abuse on mental health of children*. Concept Publishing Company.

Grover, S., & Avasthi, A. (2019). Clinical practice guidelines for the management of depression in children and adolescents. *Indian Journal of Psychiatry*, *61*(Suppl. 2), 226.

Grover, S., Raju, V. V., Sharma, A., & Shah, R. (2019). Depression in children and adolescents: A review of Indian studies. *Indian Journal of Psychological Medicine*, *41*(3), 216–227.

Gururaj, G., Pradeep, B., Beri, G., Chauhan, A., & Rizvi, Z. (2014). *Adolescent and youth health survey-Himachal Pradesh 2014*. Bangalore: Centre for Public Health, National Institute of Mental Health and Neurosciences.

Guthold, R., Moller, A.-B., Azzopardi, P., Ba, M. G., Fagan, L., Baltag, V., … Diaz, T. (2019). The Global Action for Measurement of Adolescent health (GAMA) initiative—Rethinking adolescent metrics. *Journal of Adolescent Health*, *64*(6), 697.

Halder, S., & Mahato, A. K. (2019). Cognitive behavior therapy for children and adolescents: Challenges and gaps in practice. *Indian Journal of Psychological Medicine*, *41*(3), 279–283.

Jayanthi, P., & Thirunavukarasu, M. (2015). Prevalence of depression among school going adolescents in South India. *International Journal of Pharmaceutical and Clinical Research*, *7*(1), 61–63.

Jha, K. K., Singh, S. K., Nirala, S. K., Kumar, C., Kumar, P., & Aggrawal, N. (2017). Prevalence of depression among school-going adolescents in an urban area of Bihar, India. *Indian Journal of Psychological Medicine*, *39*(3), 287–292.

Kabat-Zinn, J. (2003). Mindfulness-based stress reduction (MBSR). *Constructivism in the Human Sciences*, *8*(2), 73.

Kajavinthan, K. (2013). *Depression among higher secondary school girl students: The efficacy of cognitive behavior intervention*. LAP LAMBERT Academic Publishing.

Kalia, J. (2016). *Child and adolescent depression: Using a CBT treatment approach*. City University of Seattle.

Kar, N., Mohapatra, P. K., Nayak, K. C., Pattanaik, P., Swain, S. P., & Kar, H. C. (2007). Post-traumatic stress disorder in children and adolescents one year after a super-cyclone in Orissa, India: Exploring cross-cultural validity and vulnerability factors. *BMC Psychiatry*, *7*(1), 1–9.

Kaur, D., Patel, N., & Ghildiyal, R. (2019). Spectrum of depression in children and adolescents with type 1 diabetes—Report of three cases. *MGM Journal of Medical Sciences*, *6*(2), 101–102. https://doi.org/10.5005/jp-journals-10036-1244.

Krishnakumar, P., & Geeta, M. (2006). Clinical profile of depressive disorder in children. *Indian Pediatrics*, *43*(6), 521.

Lawrence, H. R., Buffie, M. L., Schwartz-Mette, R. A., & Nangle, D. W. (2020). Cognitive behavioral therapy with depressed youth. In *Cognitive behavioral therapy in youth: Tradition and innovation* (pp. 19–40). Springer.

Lenz, A. S., & Hollenbaugh, K. M. (2015). Meta-analysis of trauma-focused cognitive behavioral therapy for treating PTSD and co-occurring depression among children and adolescents. *Counseling Outcome Research and Evaluation*, *6*(1), 18–32.

Malhotra, S., Kohli, A., Kapoor, M., & Pradhan, B. (2009). Incidence of childhood psychiatric disorders in India. *Indian Journal of Psychiatry*, *51*(2), 101.

Midgley, N., Reynolds, S., Kelvin, R., Loades, M., Calderon, A., Martin, P., … Consortium, I. (2018). Therapists' techniques in the treatment of adolescent depression. *Journal of Psychotherapy Integration*, *28*(4), 413.

Mishra, S. K., Srivastava, M., Tiwary, N. K., & Kumar, A. (2018). Prevalence of depression and anxiety among children in rural and suburban areas of Eastern Uttar Pradesh: A cross-sectional study. *Journal of Family Medicine and Primary Care*, *7*(1), 21.

Mohanraj, R., & Subbaiah, K. (2010). Prevalence of depressive symptoms among urban adolescents of South India. *Journal of Indian Association for Child and Adolescent Mental Health*, *6*(2), 33–43.

Mohta, A., Malhotra, S., Gupta, S. K., Kalaivani, M., Patra, B. N., & Nongkynrih, B. (2020). Depression among adolescents in a rural community of north India: A cross-sectional study. *Journal of Family Medicine and Primary Care*, *9*(11), 5671.

Morgan, D. (2003). *Mindfulness-based cognitive therapy for depression: A new approach to preventing relapse*. Taylor & Francis.

Murthy, R. S. (2017). National mental health survey of India 2015–2016. *Indian Journal of Psychiatry*, *59*(1), 21.

Naeem, F., Sarhandi, I., Gul, M., Khalid, M., Aslam, M., Anbrin, A., … Minhas, F. (2014). A multicentre randomised controlled trial of a carer supervised culturally adapted CBT (CaCBT) based self-help for depression in Pakistan. *Journal of Affective Disorders*, *156*, 224–227.

Oud, M., De Winter, L., Vermeulen-Smit, E., Bodden, D., Nauta, M., Stone, L., … Kendall, T. (2019). Effectiveness of CBT for children and adolescents with depression: A systematic review and meta-regression analysis. *European Psychiatry*, *57*, 33–45.

Patel, V., & Andrew, G. (2001). Gender, sexual abuse and risk behaviours in adolescents: A cross-sectional survey in schools in Goa. *National Medical Journal of India*, *14*(5), 263–266.

Patel, V., Chowdhary, N., Rahman, A., & Verdeli, H. (2011). Improving access to psychological treatments: Lessons from developing countries. *Behaviour Research and Therapy*, *49*(9), 523–528.

Patel, S., Shah, R., Patel, H., Tilwani, M., & Vankar, G. (1998). Depressive symptomatology among adolescent schoolgirls. *Indian Journal of Psychiatry*, *40*, S35.

Perepletchikova, F., Treat, T. A., & Kazdin, A. E. (2007). Treatment integrity in psychotherapy research: Analysis of the studies and examination of the associated factors. *Journal of Consulting and Clinical Psychology*, *75*(6), 829.

Pilania, R., Hemal, A., Agarwal, S., Beniwal, R. P., & Arora, S. K. (2020). Psychiatric problems amongst adolescents living with HIV at a Tertiary Care Centre in India. *Indian Pediatrics*, *57*(11), 1026–1028.

Raj, S., Sachdeva, S. A., Jha, R., Sharad, S., Singh, T., Arya, Y. K., & Verma, S. K. (2019). Effectiveness of mindfulness based cognitive behavior therapy on life satisfaction, and life orientation of adolescents with depression and suicidal ideation. *Asian Journal of Psychiatry*, *39*, 58–62.

Sagar, R., Dandona, R., Gururaj, G., Dhaliwal, R., Singh, A., Ferrari, A., … Chakma, J. K. (2020). The burden of mental disorders across the states of India: The Global Burden of Disease Study 1990–2017. *The Lancet Psychiatry*, *7*(2), 148–161.

Sagar, R., Pattanayak, R. D., & Mehta, M. (2012). Clinical profile of mood disorders in children. *Indian Pediatrics*, *49*(1), 21–23.

Samuel, D., & Sher, L. (2013). Suicidal behavior in Indian adolescents. *Adolescent Psychiatry*, 217–223.

Satyanarayana, P. T., Prakash, B., Kulkarni, P., Kishor, M., & Renuka, M. (2017). A comparative study of prevalence of mental abnormalities among high school children in tribal, rural and urban Mysuru district, Karnataka, India. *International Journal of Community Medicine and Public Health*, *4*, 809–813.

Schniering, C. A., & Rapee, R. M. (2002). Development and validation of a measure of children's automatic thoughts: The children's automatic thoughts scale. *Behaviour Research and Therapy*, *40*(9), 1091–1109.

Shukla, N. K., Shukla, M., Ahmad, S., Shukla, R., & Khan, Z. (2017). A cross-sectional study on depression among school going adolescent girls in Barabanki district, Uttar Pradesh, India. *International Journal of Contemporary Pediatrics*, *4*, 178–181.

Singh, M. M., Gupta, M., & Grover, S. (2017). Prevalence & factors associated with depression among school going adolescents in Chandigarh, north India. *Indian Journal of Medical Research*, *146*(2), 205.

Singhal, M., Manjula, M., & Sagar, K. J. V. (2014). Development of a school-based program for adolescents at-risk for depression in India: Results from a pilot study. *Asian Journal of Psychiatry*, *10*, 56–61.

Sondhi, R., Gulgulia, S., & Shriharsh, V. (2013). Cognitive-behavioral therapy for an Indian adolescent with depression: A clinical case study. *Clinical Case Studies*, *12*(2), 157–173.

Srivastava, P., & Mehta, M. (2014). Computerized cognitive behavior therapies in psychiatry: A viable option in India? *Asian Journal of Psychiatry*, *11*, 72–73.

Srivastava, P., Mehta, M., Sagar, R., & Ambekar, A. (2020). Smartteen—A computer assisted cognitive behavior therapy for Indian adolescents with depression—A pilot study. *Asian Journal of Psychiatry*, *50*, 101970.

Wasil, A. R., Venturo-Conerly, K. E., Shinde, S., Patel, V., & Jones, P. J. (2020). Applying network analysis to understand depression and substance use in Indian adolescents. *Journal of Affective Disorders*, *265*, 278–286.

Chapter 32

Khat chewing and cognitive behavioral therapy

Abdifatah Haji Daud[a], Zahid Mahmood[b], and Yakoub Aw Aden Abdi[a]
[a]*Yacquub Clinic, Hargeisa, Somalia,* [b]*School of Professional Psychology, University of Management and Technology, Lahore, Pakistan*

Abbreviations

CBT cognitive behavioral therapy
DASS Depression Anxiety Stress Scale
DSM-5 diagnostic and statistical manual of mental disorders five
PTSD posttraumatic stress disorder
RCT randomized controlled trials
SBI screening and brief intervention
T-ASI Teen-Addiction Severity Index

Introduction

Khat (*Catha edulis*) is a green plant that is part of the Celastraceae family and grows predominantly in East Africa and the Arabian Peninsula. People have been chewing khat leaves and shoots for centuries due to their euphoric effect. Khat contains many chemicals and alkaloids, which have different effects on the human body, but the dominant ones are cathinone and cathine. Cathinone and cathine have psychoactive properties, the former being the most potent one (Wabe, 2011). Cathinone and cathine are called natural forms of amphetamine because they have a similar structure and pharmacological effects to amphetamine (Ahmed, Ruiz, Cohen Kadosh, Patton, & Resurrección, 2021). The habit of khat chewing is most prevalent in Somalia, Yemen, Kenya, Ethiopia, and some parts of Saudi Arabia (Odenwald & Al'Absi, 2017).

Traditionally, people chewed khat at social gatherings such as during marriage ceremonies and cultural festivities. Khat is also chewed by those who work in jobs with large amounts of physical labor. Khat chewing takes place in sessions that usually start in the afternoon and continue until the evening and sometimes until midnight. The chewers put the khat leaves in the mouth, commonly stacked between the cheeks and the teeth in one side of the mouth, chewing and sucking the juice from the fresh leaves, and continuously swallow the extracts. In this way, they continue to refill the mouth with new fresh leaves and continue masticating and swallowing the extracts, while the remaining fiber keeps accumulating under the cheek, making the size of the cheek larger and larger. Some chewers may spit out the remaining fiber several times during the session, but others may continue to suck and ingest the juice until the end of the session. To neutralize the sour bitter taste of the leaves and dry mouth that develops, khat chewers usually drink many sweet drinks such as soft drinks or tea with a lot of sugar (Wabe, 2011). Khat chewing is also associated with cigarette smoking (Bagajjo, Tegegne, Zeleke, Zenebe, & Assefa, 2021).

In a typical khat chewing session, the atmosphere is initially one of cheerfulness, optimism, and a general sense of well-being. During the first few hours of the khat session, khat chewers usually report personal proficiencies in an affirmative way. They refer to a sensation of happiness, euphoria, increased energy levels, alertness, increased capability to focus and heightened self-confidence, and, in some cases, an increase in desire for sex. People usually speak loudly and indulge in hot arguments about current political and social issues. After about 2h, the atmosphere changes to one filled with tension, emotional instability, and irritability which culminates with dysphoria and sluggishness. In the final hours of the khat chewing session, chewers become silent and completely withdrawn into their inner world with little communication between chewers and a feeling of exhaustion (Wabe, 2011). Later during the evening, the exhaustion continues, and the chewers may experience irritability, anxiety, restlessness, agitation, profuse sweating, palpitations, and in severe cases agitation and hypnagogic hallucinations (Basker, 2013; Magdum, 2011).

Handbook of Lifespan Cognitive Behavioral Therapy. https://doi.org/10.1016/B978-0-323-85757-4.00035-3

Effects of khat chewing

The habit of khat chewing causes many physiological changes in the human body as it impacts most of the body's systems. They include oral health problems (Al-Maweri, Warnakulasuriya, & Samran, 2018), gastric cancers and duodenal ulcers (Al-Motarreb, Al-Habori, & Broadley, 2010), cardiovascular problems (Al Suwaidi, Ali, & Aleryani, 2013), and sexual dysfunction (Mohammed & Engidawork, 2011). However, a detailed discussion of the physiological consequences of khat chewing is beyond this chapter. Khat causes or contributes to many psychiatric problems such as anxiety and depression (Daud, Mahmood, Naeem, & Saleem, 2018), attention and memory problems (Odenwald & Al'Absi, 2017) as well as anger and aggression (Sharma et al., 2020). Moreover, the relationship between khat chewing and psychosis has been investigated but as of yet no conclusive causal relationship has been established. However, there are strong indications that heavy khat chewing may lead to psychosis in vulnerable individuals or worsen existing psychotic symptoms in individuals with PTSD (Odenwald & Al'Absi, 2017).

Khat consumption leads to wasted work hours that reduce economic productivity as well as financial pressures to buy khat. This is indirectly related to unemployment and absenteeism, which can cause an economic drop in the overall national economy and production (Aden, Dimba, Ndolo, & Chindia, 2006; Ageely, 2008; Eticha, Kahsay, Ali, & Janapati, 2016). In some cases, it was reported that one-third of family income was paid out on Khat (Gashawa & Getachew, 2014). Family life is affected as a result of this negligence; the wastage of family income through khat consumption. Therefore, it can be a chief factor in family feuds. In fact, one out of two divorces in khat chewing regions have been reported to be caused by Khat abuse (Basker, 2013; Magdum, 2011). In Djibouti, khat has been cited as a factor in one out of two separations. Moreover, securing funds to pay for khat may be the chief cause of illegal behavior and even prostitution particularly in the case of female khat chewers (Elmi, 1983).

Using khat does not automatically lead to having problems. Certain criteria need to be met in order to categorize someone as having problematic khat chewing or khat addiction. Attempts have been made to define problematic khat chewing or khat addiction but the evidence is still lacking on whether it causes addiction or not. In the new Diagnostic and Statistical Manual of Mental Disorders (DSM-5), khat use is categorized as a stimulant not otherwise specified. However, in a systematic review, Mihretu and colleagues found that a large amount of khat use, more frequent khat use, having problems associated with khat use, and experiencing withdrawal symptoms upon cessation constitute problematic khat use. Although the question of khat dependence is not yet settled, there is growing evidence that khat abuse causes dependence (Mihretu, Nhunzvi, Fekadu, Norton, & Teferra, 2019).

Prevalence of khat chewing

The khat use prevalence varied from culture to culture and relying upon law enforcement, sociocultural norms, and accessibility of khat. It is estimated that around 10 million individuals chew khat daily worldwide (Awadalla & Suwaydi, 2017). In the Jazan region, Saudi Arabia was reported the current prevalence of khat chewing in the general population around 48.7% (Awadalla & Suwaydi, 2017). A cross-sectional study in Yemen with 792 participants aged 15 years and older found an 81.6% lifetime prevalence among men and 43.3% among women. The current regular khat chewer was found to be 23.6% of the aggregate example (men 31.8%, women 8.9%) (Numan, 2004). In Ethiopia, a household survey that includes 10,468 adults older than 15 years found that current khat users were 75% of the male population and 35% of the female population in a predominantly Muslim area called Butajira (Alem, Kebede, & Kullgren, 1999). Another household survey found that current khat chewers approximately were 40.0% of the men and 18.2% of the women with 1028 adult participants in Adami Tullu district, in southern central Ethiopia, within a mixed religions group (Belew, Kebede, Kassaye, & Enquoselassie, 2000).

In regards to Somalia, a three-decade-old cross-sectional study of khat use found that in the north of the nation, 64% of the adult male population frequently chew khat in contrast with 21% in southern Somalia (Elmi, 1983). A cross-sectional study that took place in Somalia found that 36.4% (3175/8723) used khat in the week before the interview (Odenwald et al., 2007). Another study that is done on active combatants reported that use of khat a week before the study was 26.2% (4611/17,600) in the north and 50.7% (8619/17,000) in the south of Somalia, and extreme khat utilization were seven times higher in the south (Odenwald et al., 2009).

Psychological treatment of khat chewing

In general, there are limited studies on the pharmacological and psychological treatments of khat-related mental health problems. Odenwald and colleagues (Odenwald et al., 2007) reported that different forms of khat treatment and counseling are available. Some European countries provide khat treatment by integrating the services made for other illegal drugs. For instance, in Aarhus in Denmark, some reports (Al-Khalil, 2010) described outreach treatment programs for Somali khat-addicted individuals. Regarding specialized khat treatment programs, very few are available worldwide. In 2012, the Stockholm County Council substance treatment services opened an outpatient clinic specifically for khat use in immigrant communities, but khat chewers were reluctant to attend (Abdi, 2012). Very limited studies have been done on the treatment of khat addiction (Odenwald & Al'Absi, 2017). A study was carried out in the semiautonomous region of Somalia (Somaliland) which focused on CBT techniques with or without psychoeducation. It included 35 patients with severe mental disorders and khat addiction. The outcome showed that in the early stage of the therapy, both groups had decreased the use of khat, but on the follow-up, most of the patients relapsed. However, the frequency of khat use was significantly lower for those who received psychoeducation (Odenwald et al., 2012).

When it comes to randomized controlled trials for the treatment of khat chewing behavior and its associated problems, a study by Widmann and colleges (Widmann et al., 2017) was used which included screening and brief intervention (SBI) (O'Donnell et al., 2014; Vasilaki, Hosier, & Cox, 2006; Young et al., 2014). In this study, the participants consisted of 330 male Somali khat chewer who had other comorbid disorders. The participants were screened for comorbid psychopathologies, and around half met the criteria of depression, while around a quarter met the criteria of khat-induced psychotic symptoms and PTSD. They were randomly assigned to the SBI group or control group. The study found that the symptoms of depression and khat-related psychosis were decreased, and PTSD persisted for both groups. Also, it was observed a reduction in khat conception time and an increase in everyday functioning for the SBI group. Moreover, SBI was less effective in reducing khat consumption time for participants with comorbid depression and PTSD as compared to those without comorbidity.

Khat chewing and cognitive behavior therapy

The use of cognitive behavior therapy (CBT) for the treatment of substance abuse is effective when used as a single treatment and when used with other therapies as well. Even though CBT for substance abuse is made up of different treatment constituents, for instance, developing skills, cognitive and motivational aspects, and operant learning techniques, some strategies are emerging that deal with reinforcing aspects of the substances. CBT for substance abuse can also be run individually or as a group (McHugh, Hearon, & Otto, 2010). Countless trials and quantitative reviews back the success of CBT in the management of substance-related problems. For example, a meta-analytic review of CBT for substance abuse containing 34 randomized controlled trials (RCT), with 2340 patients treated, showed a practical effect of CBT on substance abuse. The most noticeable effect was on cannabis treatment, followed by cocaine and opioids treatment with the least effect seen on poly-substance addiction. On individual level therapies, contingency management of relapse prevention had the best impact (Dutra et al., 2008). This is reinforced by a meta-analytic review of CBT trials by Magill and Ray (2009) which revealed a comparable outcome. For instance, a study of the psychosocial treatment for cocaine dependence discovered that 60% of patients that underwent CBT had clear toxicology results 52 weeks later (Rawson et al., 2006).

In a randomized controlled trial, the effectiveness of CBT was compared to psychoeducation on reducing khat chewing behavior and associated mental health problems (Daud, Saleem, & Mahmood, 2019). A total of 40 male Somali undergraduate regular khat users between the age range of 18 and 25 years were selected through purposive sampling techniques. The participants were randomly assigned to either a CBT group or psychoeducation group through a lottery method. The CBT group received seven sessions of manualized CBT for substance abuse (Carroll, 1998), and the psychoeducation group received three sessions on addiction-related education (Gorski & Miller, 1986). Both therapies were carried out in group format sessions. Depression, Anxiety, Stress Scale (DASS) and Proforma on khat chewing behavior (frequency per week and hours of khat chewing per day) were used for endpoints in assessing the effectiveness of the two treatments in both groups before and after the administration of the therapies. The experiment showed that after the administration of CBT to the treatment group, there was a significant decrease in khat conception time (frequency per week and hours of khat chewing per day) as indicated by Table 1, when compared to the control group who received psychoeducation. Kaminer, Burleson, and Goldberger (2002) used to compare the effectiveness of psychoeducation to the efficacy of cognitive behavior therapy (CBT) on dealing with adolescent substance abuse. A total of 88 adolescents with a dual diagnosis from an outpatient service were randomly assigned to either receive 8 weeks of CBT or psychoeducation. Outcome measurement was used for drug urinalysis and the Teen-Addiction Severity Index (T-ASI). CBT subjects showed altogether a lower number of positive urinalysis than subjects who received psychoeducation at 3-month follow-up assessment. Most T-ASI subscales demonstrated a change from baseline to 3- and 9-month follow-up assessment for both conditions. The decrease in substance utilization was accomplished by paying little mind to treatment conditions.

TABLE 1 Chi-square comparison between CBT group and PET group on the basis of patterns of khat chewing after received treatment ($N = 40$).

	Group			
	CBT ($n = 20$)	PET ($n = 20$)		
Variables	f (%)	f (%)	$\chi 2$	P-value
Frequency of chewing khat				
Daily	1 (5)	11 (55)	13.91	0.003
Four times a week	1 (5)	2 (10)		
Twice a week	5 (25)	1 (5)		
Once a week	13 (65)	6 (30)		
Hours of chewing khat per day				
1–4 h	9 (45)	3 (15)	7.86	0.020
5–8 h	8 (40)	6 (30)		
More than 8 h	3 (15)	11 (55)		
Type of khat chewed				
Daba Musbar	3 (15)	2 (10)	4.67	0.323
Jabis	14 (70)	10 (50)		
Dadar	1 (5)	4 (20)		
Boondaro	0 (0)	2 (10)		
Others	2 (10)	2 (10)		

SNIP from Daud, A. H., Saleem, S., & Mahmood, Z. (2019). Differential effectiveness of cognitive behavior therapy and psycho education therapy on khat chewing behavior and associated mental health problems. *Journal of Postgraduate Medical Institute (Peshawar-Pakistan), 33*(4), 304–309.

Moreover, the posttreatment measurement showed that the mental health problems were improved significantly for the CBT group as shown in Table 2 and Fig. 1. The decrease in khat consumption time enhanced everyday functioning and improved the sleeping routine, which later contributed to the individual's physical and emotional well-being. Accordingly, the increase in mental health and everyday functioning enabled khat users to concentrate more on employment and other day-to-day activities which serve as a deterrent for new relapses (Widmann et al., 2017). There are some hypotheses on why many people with severe mental illness simultaneously have substance use disorders. One explanation is the self-medication hypothesis, which infers that people will in general select drugs that lighten their particular symptoms of mental illness. For instance, a few analysts propose that individuals with wild sentiments of anger and animosity may pick opiates

TABLE 2 Means, standard deviations, t and P-values for comparing CBT group and PET group on the basis of mental health problems of the participants ($N = 40$).

		Groups					
		CBT ($n = 20$)		PET ($n = 20$)			
Variables		M	SD	M	SD	t	P-value
Pretest	DASS total	32.50	5.71	35.80	6.70	1.68	0.038
Posttest	DASS total	18.05	9.57	32.10	6.74	5.37	<0.001

Note: $df = 38$; *CBT*, cognitive behavioral therapy; *DASS*, Depression Anxiety Stress Scale; *M*, mean; *PET*, psychoeducational therapy; *SD*, standard deviation; t, t-test.
SNIP from Daud, A. H., Saleem, S., & Mahmood, Z. (2019). Differential effectiveness of cognitive behavior therapy and psycho education therapy on khat chewing behavior and associated mental health problems. *Journal of Postgraduate Medical Institute (Peshawar-Pakistan), 33*(4), 304–309.

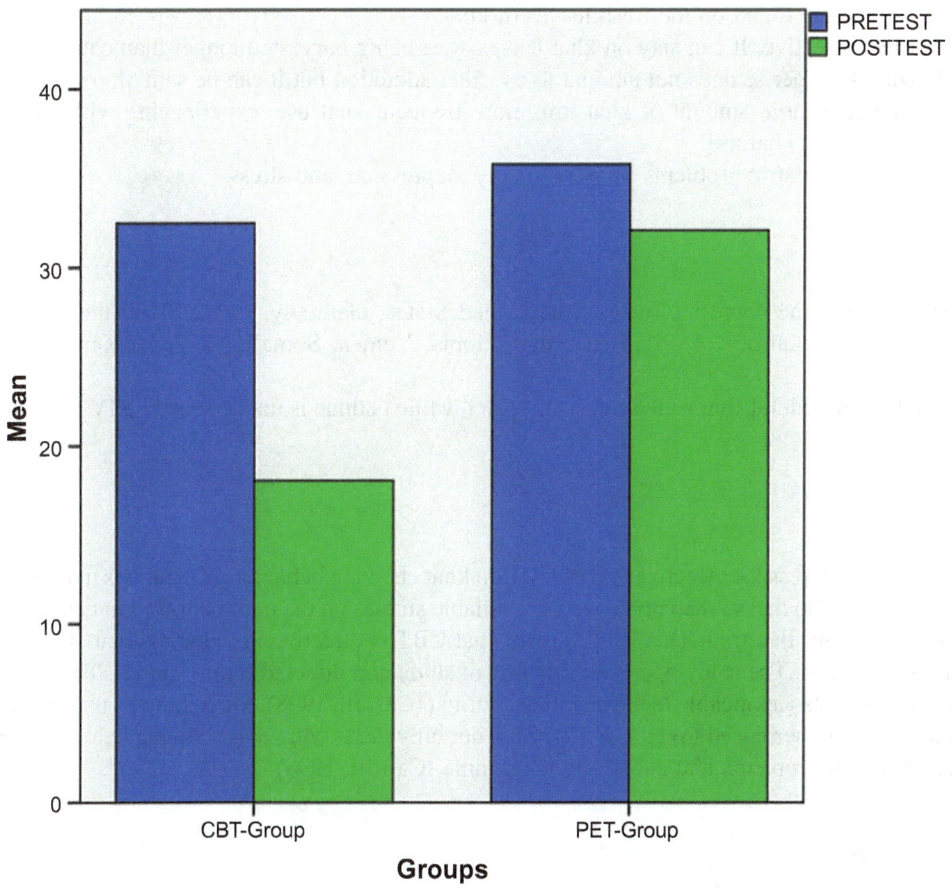

FIG. 1 Differences of CBT group and PET group based upon mental health problems before and after the therapy ($N=40$). *(SNIP from Daud, A. H., Saleem, S., & Mahmood, Z. (2019). Differential effectiveness of cognitive behavior therapy and psycho education therapy on khat chewing behavior and associated mental health problems.* Journal of Postgraduate Medical Institute (Peshawar-Pakistan), *33(4), 304–309.)*

for these drugs' progressing impacts, while individuals who are depressed may take cocaine since it thrills and stimulates them (Khantzian, 1997). Among people with depression or PTSD, the burden of symptoms of mental illness is heavier, and using khat distracts from the painful experience (Widmann et al., 2017). Another explanation is the toxicity hypothesis which states that mental health problems are the result of excessive use of drugs, with clear causal relationships like that of cannabis for psychosis (Colizzi & Murray, 2018) or by precipitating and exacerbating the course of an underlying disorder (Quello, Brady, & Sonne, 2005). When it comes to chewing khat, it is linked to the rise of mental illnesses in Somalia (WHO, 2010). Both explanations are applicable for the relationship of khat chewing behavior and mental health problems, and it seems that everyday functioning is the mediator which is a suggestion for future studies to take into account (Widmann et al., 2017).

Practice and procedures

This chapter focuses on the only available RCT of CBT on khat chewing. In the experiment, the participants were 40 and divided equally into the CBT group and psychoeducation group. It was used the Depression Anxiety and Stress Scale (DASS) for measuring mental health problems, and proforma on frequencies of khat chewing and durations of sessions of khat chewing. There were two assessments, a baseline, and a posttreatment assessment. A comparison was made within each group and between the groups. The study had limitations which include taking a small sample size and conducting a small number of sessions for each therapy protocol. It was also used self-report to measure the khat chewing behavior and mental health problems instead of objective measurements like a urine test.

Mini-dictionary of terms

Khat (*Catha edulis*). A small plant that is under the Celastraceae family. Four millennia people were using it to get euphoria as it has stimulant properties.

Cathinone. A natural form of amphetamine found on the fresh leaves of khat.

Cathine. Similar to cathinone but less effective. It can stay on khat leaves after being harvested longer than cathinone.

Problematic khat chewing. Chewing khat per se does not amount to use khat addiction but it can be said after meeting some criteria. These criteria include a large amount of khat use, more frequent khat use, experiencing withdrawal symptoms, and have problems related to khat use.

Mental health problem. Less severe psychiatric problems such as anxiety, depression, and stress.

Key facts

- Khat use is an illegal substance for some countries such as the United States, Germany, Canada, and the United Kingdom, whereas it is legal for the countries of origin such as Ethiopia, Yemen, Somalia, Uganda, Kenya, and Djibouti.
- Cathinone is under Schedule I drug of Federal Controlled Substances Act, while cathine is under Schedule IV drugs of Federal Controlled Substances Act.

Applications to other areas

This chapter highlighted the studies that address the effectiveness of CBT on khat chewing behavior. Researches that focus on CBT for khat chewing are very limited. So that we had reviewed the available studies on the treatments of khat chewing and focused more on one randomized controlled trial. This RCT showed that CBT is effective in reducing khat chewing behavior similar to the other stimulant drugs. The study opened a new line of study and directed future studies. The study had done in a group format; however, it is also applicable to the one in one format (Carroll, 1998). Most Somalis use khat to overcome the psychological trauma they experienced (WHO, 2010). CBT not only deals with khat chewing behavior but also the underline cognitive and behavior problems that caused by the trauma (Carroll, 1998).

Summary points

- Khat leaves and shoots are chewed for centuries because of their euphoric effects.
- Chewing khat caused many problems for the users including physiological problems, psychiatric problems, and socio-economic issues.
- Khat chewing becomes problematic when using large amount, more frequent, experiencing withdrawal symptoms after cessation and other problems.
- Limited studies have been done for the treatment of problematic khat chewing.
- Different attempts have been made on the intervention of problematic khat chewing such as applying treatment protocols of other substance use on khat users and creating specialized treatment protocols for khat users.
- Only one randomized controlled study was conducted on the study the effectiveness of CBT on khat chewing behavior and related mental health problems and found that CBT decreased khat chewing behavior and mental health problems.

References

Abdi, Y. A. (2012). *Khat-en tuggande bomb [Khat—a chewing bomb]* (pp. 10–11). Stockholm: Vardguiden.

Aden, A., Dimba, E. A. O., Ndolo, U. M., & Chindia, M. L. (2006). Socio-economic effects of khat chewing in north eastern Kenya. *East African Medical Journal*, *83*(3), 69–74.

Ageely, H. M. (2008). Health and socio-economic hazards associated with khat consumption. *Journal of Family & Community Medicine*, *15*(1), 3–11.

Ahmed, A., Ruiz, M. J., Cohen Kadosh, K., Patton, R., & Resurrección, D. M. (2021). Khat and neurobehavioral functions: A systematic review. *PLoS One*, *16*(6), 1–18.

Al Suwaidi, J., Ali, W. M., & Aleryani, S. L. (2013). Cardiovascular complications of Khat. *Clinica Chimica Acta*, *419*, 11–14.

Alem, A., Kebede, D., & Kullgren, G. (1999). The prevalence and socio-demographic correlates of khat chewing in Butajira, Ethiopia. *Acta Psychiatrica Scandinavica*, *100*(397), 48–55.

Al-Khalil, I. B. (2010). *Aarhus reaches out to khat addicts* (p. 12). Aarhus, Denmark: Aarhus Municipality.

Al-Maweri, S. A., Warnakulasuriya, S., & Samran, A. (2018). Khat (Catha edulis) and its oral health effects: An updated review. *Journal of Investigative and Clinical Dentistry*, *9*(1), 1–9.

Al-Motarreb, A., Al-Habori, M., & Broadley, K. J. (2010). Khat chewing, cardiovascular diseases and other internal medical problems: The current situation and directions for future research. *Journal of Ethnopharmacology*, *132*(3), 540–548.

Awadalla, N. J., & Suwaydi, H. A. (2017). Prevalence, determinants and impacts of khat chewing among professional drivers in Southwestern Saudi Arabia. *Eastern Mediterranean Health Journal*, *23*(3), 189–197.

Bagajjo, W. S., Tegegne, K. T., Zeleke, D., Zenebe, A., & Assefa, A. A. (2021). Khat chewing as a risk factor of cigarette smoking in Ethiopia systemic review and meta-analysis. *Journal of Tropical Disease and Public Health*, *9*(3), 1–4.

Basker, G. V. (2013). A review on hazards of khat chewing. *International Journal of Pharmacy and Pharmaceutical Sciences*, *5*(3), 74–77.

Belew, M., Kebede, D., Kassaye, M., & Enquoselassie, F. (2000). The magnitude of khat use and its association with health, nutrition and socio-economic status. *Ethiopian Medical Journal*, *38*(1), 11–26.

Carroll, K. M. (1998). *Therapy manuals for drug addiction, manual 1: A cognitive-behavioral approach: Treating cocaine addiction*. Rockville, Maryland: National Institute on Drug Abuse.

Colizzi, M., & Murray, R. (2018). Cannabis and psychosis: What do we know and what should we do? *British Journal of Psychiatry*, *212*(4), 195–196.

Daud, A. H., Mahmood, Z., Naeem, F., & Saleem, S. (2018). Personality traits, interpersonal difficulties and mental health problems of khat users and non-khat users: A comparative study. *Khyber Medical University Journal*, *10*(4), 206–211.

Daud, A. H., Saleem, S., & Mahmood, Z. (2019). Differential effectiveness of cognitive behavior therapy and psycho education therapy on khat chewing behavior and associated mental health problems. *Journal of Postgraduate Medical Institute (Peshawar-Pakistan)*, *33*(4), 304–309.

Dutra, L., Stathopoulou, G., Basden, S. L., Leyro, T. M., Powers, M. B., & Otto, M. W. (2008). A meta-analytic review of psychosocial interventions for substance use disorders. *American Journal of Psychiatry*, *165*(2), 179–187.

Elmi, A. S. (1983). The chewing of Khat in Somalia. *Journal of Ethnopharmacology*, *8*(2), 163–176.

Eticha, T., Kahsay, G., Ali, D., & Janapati, Y. K. (2016). Socio-economic and health effects of khat chewing in Mekelle, Tigray region, Ethiopia. *International Journal of Pharmacy and Pharmaceutical Sciences*, *8*(1), 11–22.

Gashawa, A., & Getachew, T. (2014). The chemistry of khat and adverse effect of khat chewing. *American Scientific Research Journal for Engineering, Technology, and Sciences*, *9*(1), 35–46.

Gorski, T. T., & Miller, M. (1986). *Staying sober: A guide for relapse prevention*. Missouri, US: Independence Press.

Kaminer, Y., Burleson, J. A., & Goldberger, R. (2002). Cognitive-behavioral coping skills and psychoeducation therapies for adolescent substance abuse. *Journal of Nervous and Mental Disease*, *190*(11), 737–745.

Khantzian, E. J. (1997). The self-medication hypothesis of substance use disorders: A reconsideration and recent applications. *Harvard Review of Psychiatry*, *4*(5), 231–244.

Magdum, S. S. (2011). An overview of khat. *Addictive Disorders & Their Treatment*, *10*(2), 72–83.

Magill, M., & Ray, L. A. (2009). Cognitive-behavioral treatment with adult alcohol and illicit drug users: A meta-analysis of randomized controlled trials. *Journal of Studies on Alcohol and Drugs*, *70*(4), 516–527.

McHugh, R. K., Hearon, B. A., & Otto, M. W. (2010). Cognitive behavioral therapy for substance use disorders. *Psychiatric Clinics*, *33*(3), 511–525.

Mihretu, A., Nhunzvi, C., Fekadu, A., Norton, S., & Teferra, S. (2019). Definition and validity of the construct "problematic Khat use": A systematic review. *European Addiction Research*, *25*(4), 161–172.

Mohammed, A., & Engidawork, E. (2011). Reproductive parameters are differentially altered following subchronic administration of Catha edulis F. (Khat) extract and cathinone in male rats. *Journal of Ethnopharmacology*, *134*(3), 977–983.

Numan, N. (2004). Exploration of adverse psychological symptoms in Yemeni khat users by the Symptoms Checklist-90 (SCL-90). *Addiction*, *99*(1), 61–65.

Odenwald, M., & Al'Absi, M. (2017). Khat use and related addiction, mental health and physical disorders: The need to address a growing risk. *Eastern Mediterranean Health Journal*, *23*(3), 237–244.

Odenwald, M., Hinkel, H., Schauer, E., Neuner, F., Schauer, M., Elbert, T. R., & Rockstroh, B. (2007). The consumption of khat and other drugs in Somali combatants: A cross-sectional study. *PLoS Medicine*, *4*(12), 1959–1972.

Odenwald, M., Hinkel, H., Schauer, E., Schauer, M., Elbert, T., Neuner, F., & Rockstroh, B. (2009). Use of khat and posttraumatic stress disorder as risk factors for psychotic symptoms: A study of Somali combatants. *Social Science & Medicine*, *69*(7), 1040–1048.

Odenwald, M., Lingenfelder, B., Peschel, W., Haibe, F. A., Warsame, A. M., Omer, A., … Elbert, T. (2012). A pilot study on community-based outpatient treatment for patients with chronic psychotic disorders in Somalia: Change in symptoms, functioning and co-morbid khat use. *International Journal of Mental Health Systems*, *6*(1), 1–18.

O'Donnell, A., Anderson, P., Newbury-Birch, D., Schulte, B., Schmidt, C., Reimer, J., & Kaner, E. (2014). The impact of brief alcohol interventions in primary healthcare: A systematic review of reviews. *Alcohol and Alcoholism*, *49*(1), 66–78.

Quello, S. B., Brady, K. T., & Sonne, S. C. (2005). Mood disorders and substance use disorder: A complex comorbidity. *Science & Practice Perspectives*, *3*(1), 13–21.

Rawson, R. A., McCann, M. J., Flammino, F., Shoptaw, S., Miotto, K., Reiber, C., & Ling, W. (2006). A comparison of contingency management and cognitive-behavioral approaches for stimulant-dependent individuals. *Addiction*, *101*(2), 267–274.

Sharma, V., Papaefstathiou, S., Tewolde, S., Amobi, A., Deyessa, N., Relyea, B., & Scott, J. (2020). Khat use and intimate partner violence in a refugee population: A qualitative study in Dollo Ado, Ethiopia. *BMC Public Health*, *20*(670), 1–10.

Vasilaki, E. I., Hosier, S. G., & Cox, W. M. (2006). The efficacy of motivational interviewing as a brief intervention for excessive drinking: A meta-analytic review. *Alcohol and Alcoholism*, *41*(3), 328–335.

Wabe, N. T. (2011). Chemistry, pharmacology, and toxicology of khat (catha edulis forsk): A review. *Addiction & Health*, *3*(3–4), 137–149.

Widmann, M., Apondi, B., Musau, A., Warsame, A. H., Isse, M., Mutiso, V., … Odenwald, M. (2017). Comorbid psychopathology and everyday functioning in a brief intervention study to reduce khat use among Somalis living in Kenya: Description of baseline multimorbidity, its effects of intervention and its moderation effects on substance use. *Social Psychiatry and Psychiatric Epidemiology*, *52*(11), 1425–1434.

World Health Organization (WHO). (2010). *A situation analysis of mental health in Somalia*. Retrieved from https://www.mhinnovation.net/sites/default/files/downloads/innovation/reports/A-SA-of-MH-in-Somalia.pdf.

Young, M. M., Stevens, A., Galipeau, J., Pirie, T., Garritty, C., Singh, K., … Moher, D. (2014). Effectiveness of brief interventions as part of the Screening, Brief Intervention and Referral to Treatment (SBIRT) model for reducing the nonmedical use of psychoactive substances: A systematic review. *Systematic Reviews*, *3*(1), 1–18.

Chapter 33

CBT for anxiety disorders among children in Pakistan

Anowra Khan and Tamkeen Ashraf Malik
Department of Behavioral Sciences, School of Social Sciences, National University of Science and Technology, Islamabad, Pakistan

Abbreviations

CAMS	Child/Adolescent Anxiety Multimodal Study
CASI	Childhood Anxiety Sensitivity Index
CBI	cognitive behavioral interventions
CBITs	Cognitive Behavioral Intervention for Trauma in Schools
CBT	cognitive behavior therapy
FNES	Fear of Negative Evaluation Scale
LMICs	low- and middle-income countries
RCT	randomized control trial
RSES	Rosenberg Self-Esteem Scale
SCARED	Screen for Child Anxiety Related Disorders
SDQ	Strengths and Difficulties Questionnaire
TOC	theory of change
WHO-EMBRO	World Health Organization—Eastern Mediterranean Region

Introduction

Mental health problems in low- and middle-income countries are a major health concern, because 80% of people with mental health problems live in this region. Morbid anxiety is recognized as a chronic condition affecting almost 264 million people worldwide. 6.5% worldwide prevalence of anxiety disorders in children is greater than the prevalence of depression, ADHD, and other disruptive disorders (Esbjørn, Hoeyer, Dyrborg, Leth, & Kendall, 2010; Polanczyk, Salum, Sugaya, Caye, & Rohde, 2015). The highest number is in the Southeast Asia region, covering 23% of the total worldwide prevalence (World Health Organization, 2017). Eight South Asian countries, including Pakistan, are considered a red-alert zone because of the high prevalence rate of mental disorders, which often remain unreported and unidentified due to several sociopolitical conditions (Hossain et al., 2020). Pakistan is a low- and middle-income country (LMIC) afflicted with endemics such as war, terrorism, natural disasters, and political instability, which constantly create and maintain common mental disorders in the population, particularly vulnerable people like children and women. This necessitates the provision of evidence-based treatment for children and adolescents.

Effects of anxiety on well-being

Anxiety is a common mental health problem that causes significant distress and impairment across the life span (Velting, Setzer, & Albano, 2002). If left untreated, it has a chronic course, increases the chances of secondary mental illness, and significantly reduces productivity, thus increasing the global burden of disease (Bittner et al., 2007; Curry, March, & Hervey, 2004; Patel, Flisher, Hetrick, & McGorry, 2007). Anxiety disorders are associated with poor school performance, low academic achievement, and poor social functioning (Ramsawh, Chavira, & Stein, 2010). Anxiety disorders in children are chronically associated with high personal, social, and economic costs. If they remain untreated, anxiety disorders in children continue in adolescence and adulthood (Ginsburg et al., 2018).

Handbook of Lifespan Cognitive Behavioral Therapy. https://doi.org/10.1016/B978-0-323-85757-4.00004-3

Anxiety disorders in Pakistani children

Despite LMICs bearing the highest-burden of mental illness, there is limited data on prevalence, predictors, and outcomes of anxiety disorders. A recent study on the prevalence of emotional problems in the capital of Pakistan's public sector schools showed 22.5% in a sample of 800 children (Malik, Siddiqui, & Mahmood, 2019). Previous studies indicated similar higher rates of anxiety in other major cities of Pakistan, including Lahore (Saleem & Mehmood, 2011), Karachi (Syed, Hussein, & Haidry, 2009), and Rawalpindi (Khalid, Qadir, Chan, & Schwannauer, 2018). In Karachi the largest city in Pakistan the prevalence of emotional problems using the Strengths and Difficulties Questionnaire (SDQ) was estimated at 37.3% in the abnormal category and 15% in the borderline category (Syed et al., 2009). These studies corroborate the notion that there is a high prevalence rate of anxiety-related disorders in Pakistan that require immediate attention. Anxiety-related problems have been further aggravated after COVID-19 (Aqeel, Shuja, Abbas, Rehna, & Ziapour, 2020).

It can be argued that children experiencing anxiety disorders have difficulties in all spheres of life, including personal, familial, social, and academic. A study about social anxiety and quality of life in Pakistan indicated a negative association. Results showed that adolescents experiencing social anxiety have a poor social, psychological, and physical quality of life (Ayesha Farooq et al., 2017). Another study also yielded a similar negative correlation between anxiety, life satisfaction, and academic achievement of 13–21-year-old children (Aziz & Tariq, 2019).

Cognitive behavioral therapy for anxiety disorders

Cognitive behavior therapy (CBT) has strong research evidence as the first-line treatment of anxiety disorders in children and adolescents (Ginsburg et al., 2018; Howes, Zwicker, Uher, & Pavlova, 2020). A longitudinal study, the Child/Adolescent Anxiety Multimodal Study (CAMS), indicated that CBT successfully treats anxiety disorders and the remission rate is significantly less in follow-up years (Ginsburg et al., 2018). CAMS utilized a manualized CBT program, namely Coping Cat (Kendall et al., 2006), an evidence-based program for anxiety disorders. Another study utilizing the CBT-based Cool Kids program (Lyneham, Abbott, Wignall, & Rapee, 2003) indicated significant positive results in treating anxiety disorders other than social anxiety (Hudson et al., 2015).

CBT for children in Pakistan

Research on CBT in the adult population is proliferating in Pakistan compared to that in children. Within the adult population, CBT has been tested as a treatment for erectile dysfunction (Khan, Amjad, & Rowland, 2019), domestic violence (Zadeh & Lateef, 2012), depression (Naeem, Waheed, Gobbi, Ayub, & Kingdon, 2011), and OCD (Aslam, Irfan, & Naeem, 2015; Ishfaq, 2014).

Research evidence on CBT and cognitive behavioral interventions (CBIs) for children in Pakistan is in the infancy stage. A few CBT- and CBI-related rigorous studies have been conducted in Pakistan with children for anger management (Mushtaq, Lochman, Tariq, & Sabih, 2017), ADHD (Malik, Rooney, Chronis-Tuscano, & Tariq, 2017), and anxiety disorders (Amin, Iqbal, Naeem, & Irfan, 2020; Khan, Malik, Ahmed, & Riaz, 2020). The results of these studies further strengthen the evidence for the efficacy and effectiveness of CBT. Until recently, CBT has been neglected in treating psychological problems in children due to various psychosocial factors (Fig. 1).

Practice and procedures

There are different opinions about the transportability and effectiveness of CBT in Pakistan. One view is to translate and adapt CBT-based protocols for Pakistani children because CBT is value-laden and Western. The other view is to implement CBT-based protocols as they are, without any adaptation, because the core principles are universal, and to be applied in any culture (Naeem, Gobbi, Ayub, & Kingdon, 2010). One group comprising psychiatrists, psychologists, and teachers has been working extensively to culturally adapt CBT (Latif et al., 2021; Naeem et al., 2011, 2014, 2016; Naeem, Gobbi, Ayub, & Kingdon, 2009; Naeem, Phiri, Rathod, & Ayub, 2019). An indigenous self-help CBT-based manual for treating depression and anxiety has been developed. Initially, this self-help manual has been used with the adult population to treat depression. Recently this self-help manual has been used to treat social anxiety disorder among adolescents (Amin et al., 2020). This study indicated that the CBT Urdu-based self-help manual is effective compared to treatment as usual for social anxiety.

Another group of researchers has translated and adapted the Coping Cat program in the Urdu language to treat anxiety disorders in children. As previously indicated, Coping Cat is an evidence-based CBT program for treating anxiety and depression in children and adolescents. The study described the adaptation and translation process. Some terms, like

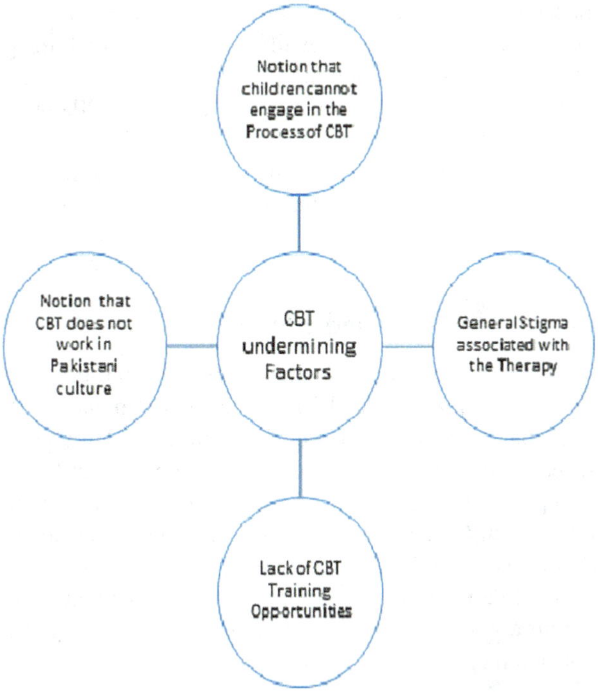

FIG. 1 Psychosocial factors in the unavailability of CBT in Pakistan.

the names of the character and picture of Abraham Lincoln, were adapted to Pakistani culture. This experimental study, conducted in an orphanage setting with orphans, showed promising results on the efficacy of Coping Cat in treating anxiety disorders in children (Khan et al., 2020). The need for trauma intervention is paramount in the present circumstances of war, terror, and frequent disasters. Cognitive Behavioral Intervention for Trauma in Schools (CBITS) (Jaycox, 2018) was translated and adapted, and pilot tested for children living in a foster care setting in Rawalpindi, Pakistan. This study supports the preliminary feasibility of CBITS in Pakistan (Riaz, 2017).

Framework for adaptation of CBT in Pakistan

To develop an indigenous CBT self-help manual (Naeem et al., 2015), an intuitive approach was devised based on the experience of the multidisciplinary team. After investing years of research on this indigenous CBT manual, the team has fleshed out their adaptation process (Naeem et al., 2019, 2021). A four-stage process was outlined, starting from a literature review, needs analysis, and qualitative research; preparing guidelines for the adaptation of the CBT manual; translation and adaptation; and finally, pilot testing of the manual (Table 1). This group of researchers gave norms, values, thinking ways, and dominant language precedence. According to them, CBT needs to be adapted and tailored to the culture.

TABLE 1 Process of developing indigenous CBT intervention.

Stage 1	• Review of previous literature • Consultation with service providers like therapists, social workers, psychiatrists • Qualitative studies and feedback from service users
Stage 2	• Developing guidelines for the adaptation of CBT intervention
Stage 3	• Developing/translating and adapting manuals/materials
Stage 4	• Pilot testing • Randomized control trials • Efficacy and effectiveness of the adapted intervention

The manual is in Urdu. The storyline, character, and examples of the manual are relevant to Pakistani culture. Components of CBT include identifying emotions, differentiating between different types of thinking, physiological arousal of the body, stress management, activity scheduling, and problem-solving. The manual can be assessed at (https://pact.com.pk/sdm_downloads/khushi-aur-khatoon-web/). (This manual has been downloaded over 70,000 times at this writing.) The second group of researchers, who worked on the Coping Cat program in Urdu, followed a planned adaptation approach (Lee, Altschul, & Mowbray Carol, 2008). The program's basic structure was maintained, and the few words that were changed were Western, including character names. Only the workbook was translated into the Urdu language, excluding the therapist's manual.

Measuring tools and instruments for efficacy and effectiveness

Self-reported instruments have been used in almost all studies to measure the efficacy of CBT. The Screen for Child Anxiety Related Disorders (SCARED), the Strengths and Difficulties Questionnaire (SDQ), and the Childhood Anxiety Sensitivity Index (CASI) are used in analyzing the efficacy of the Coping Cat Urdu program. A therapist feedback form, therapist summary form, therapist feasibility form, and treatment satisfaction form have also been used to check the feasibility, applicability, and ease of implementing this CBT program in Pakistan (Khan et al., 2020).

For measuring the efficacy of the Urdu CBT self-help manual, the Liebowitz Social Anxiety Scale was used as the primary measure along with the Fear of Negative Evaluation Scale (FNES) and the Rosenberg Self-Esteem Scale (RSES). No information was provided about the fidelity check, therapist feedback, and client satisfaction with the treatment (Amin et al., 2020). It is recommended for future studies to include special attention to fidelity checks and feedback from therapists and participants about the utility of treatment.

Embedding CBT in existing health service structure in Pakistan

The mental health service structure in Pakistan is not well organized and is inefficient in managing the burden of mental illness. Children are a vulnerable population and are dependent upon their caregivers for seeking treatment. Due to a lack of knowledge and awareness, and the stigma associated with mental health issues, very few children end up reaching any treatment facility. This treatment gap has been highlighted in the existing literature but no significant improvements have been carried out with respect to children and specifically for anxiety disorders (Imran, Bhatti, Anwar, Najmi, & Haider, 2012; Sohail, Syed, & Rahman, 2017). CBT-based interventions have been embedded in existing service delivery structures for perinatal depression and maternal health. The Thinking Healthy Program is delivered through female health workers (Rahman, Malik, Sikander, Roberts, & Creed, 2008) and volunteer peers (Sikander et al., 2019) in the community setting. In the future, the same strategy is required to provide CBT to vulnerable children by embedding CBT in the healthcare system. Primary healthcare units, district hospitals, and tertiary hospitals can be targeted for the delivery of CBT. There is a need to generate research evidence for the efficacy, challenges, and opportunities that are available to embed CBT in the healthcare setup. For this policy, advocacy and legislation are important.

School mental health and CBT in Pakistan

In order to improve the mental health of children cost-effectively, schools are the best available site. A few private sector schools and colleges have appointed psychologists but there is no position for a psychologist/counsellor in the government school system. CBT-based interventions for the prevention of anxiety disorders can be delivered school-wide. Treatment of anxiety-related problems can also be delivered by a trained school psychologist. In Pakistan, research has been conducted on the mental health literacy of schoolteachers, yielding positive results. The World Health Organization, Eastern Mediterranean Region (WHO-EMRO) School Mental Health Manual-based intervention is used to train teachers in mental health literacy. This manual includes information on common mental disorders, including anxiety, as well as ways to mitigate these problems. The results of this randomized control trial (RCT) indicated positive results in improving teachers' mental health literacy, and the emotional and behavioral problems in children (Imran, Rahman, Chaudhry, & Asif, 2018).

Realizing that the implementation and sustainability of school-based mental health programs require cross-sectoral collaboration among all stakeholders, a study conducted training workshops using the WHO-EMRO Manual of School Mental Health to train stakeholders like policymakers, school administrators, teachers, nongovernment organizations, researchers, and mental health specialists. This study utilized the theory of change (TOC) process to formulate mutual goals among stakeholders and offered several adaptations for achieving optimal school mental health (Hamdani et al., 2021).

Conclusion

CBT is an evidence-based treatment for anxiety disorders in children. In Pakistan, CBT for the treatment of children's psychological problems is an emerging area. For future progress, hurdles and bottlenecks need to be overcome, for example, empirical training protocols, evidence-based manuals, and effective adaptation and implementation in Pakistani culture. So far, research on CBT in Pakistan has indicated promising results in treating anxiety disorders in children. An evidence-based program, i.e., Coping Cat, and an indigenous CBT-based manual program is available in Pakistan. For the wider benefit of society, CBT needs to be integrated in existing children's related service structures in Pakistan, for example, in the education and health sectors. In this way, maximum outreach would be possible to treat and prevent the rising anxiety disorders in Pakistan.

Mini-dictionary of terms

- *Mental health literacy.* Knowledge and attitude toward mental well-being and mental illnesses, including their identification, management, and prevention.
- *Cultural adaptation.* The process through which a given evidence-based intervention is adapted to a new culture.
- *Treatment fidelity.* Process of checking whether the intervention/therapy/treatment was delivered as it was supposed to be delivered by different means.
- *Translation research.* Research aimed at investing meaningful research questions and attempting to translate research findings into practice.
- *Theory of change.* A theory-driven approach to create social change by empowering individuals with an explicit roadmap to be followed for achieving the desired goal.
- *School mental health.* A positive school climate that promotes the psychological, social, and emotional wellbeing of children by improving mental health literacy, teacher training, and health-promoting policies.

Key facts

Key facts of Urdu manual *Khushi and Khatoon*

- It is an evidence-based self-help manual used for the treatment of anxiety and depression.
- It can be used with various populations including children and adults.
- It is an indigenous CBT-based program.

Key facts of Urdu Coping Cat program

- It is an adapted and translated evidence-based program for the treatment of anxiety and depression.
- It is a manualized CBT-based program.
- It can be administered in individual and group settings.

Applications to other areas

In this chapter, we have tried to give a snapshot of the current status of anxiety-related problems of children and their treatment via CBT in Pakistan. Anxiety disorders are common mental disorders, along with depression. In Pakistan, as in other LMICs, very limited treatment options are available and the burden of disease is extremely high. In this situation, treatment options for anxiety disorders like the Coping Cat program and the *Khushi and Khatoon* manual can be used for the treatment of depression in children. Children presenting with comorbidities like anxiety and depression can also benefit from these programs. The ease of implementation allows the use of these programs in various contexts like schools, community settings, and far-flung rural areas. The situation with the mental health of adults and the elderly population is not very different from children's mental health. With slight variations, these programs can also be used with the adult population in LMICs.

Telemental health is a cost-effective and efficient way of treatment delivery. It is the future of successful mental health initiatives. CBT-based programs have the potential to be transformed into web-based or CD-based programs. These programs can be delivered and implemented remotely in otherwise inaccessible areas, like war-struck and poverty-ridden areas.

Summary points

- Children have a high prevalence rate of mental disorders in Pakistan
- Cultural adaptation is important for the efficacy and effectiveness of CBT
- Fidelity and integrity are important elements in RCTs for checking CBT efficacy
- A translational research paradigm is needed to embed CBT in the LMIC service structure
- Research on CBT with children is needed in LMICs like Pakistan
- Schools are potential sites for treatment implementation to improve mental health

References

Amin, R., Iqbal, A., Naeem, F., & Irfan, M. (2020). Effectiveness of a culturally adapted cognitive behavioural therapy-based guided self-help (CACBT-GSH) intervention to reduce social anxiety and enhance self-esteem in adolescents: A randomized controlled trial from Pakistan. *Behavioural and Cognitive Psychotherapy*, *48*(5), 503–514. https://doi.org/10.1017/S1352465820000284.

Aqeel, M., Shuja, K. H., Abbas, J., Rehna, T., & Ziapour, A. (2020). *The influence of illness perception, anxiety and depression disorders on students mental health during COVID-19 outbreak in Pakistan: A web-based cross-sectional survey*. https://doi.org/10.21203/rs.3.rs-30128/v1.

Aslam, M., Irfan, M., & Naeem, F. (2015). Brief culturally adapted cognitive behaviour therapy for obsessive compulsive disorder: A pilot study. *Pakistan Journal of Medical Sciences*, *31*(4), 874–879. https://doi.org/10.12669/pjms.314.7385.

Ayesha Farooq, S., Muneeb, A., Ajmal, W., Asif Tauni, M., Mahmood, S., Kalimullah Sohaib Qadri, S., … Rizvi, N. (2017). Quality of life perceptions in school-going adolescents with social. *Anxiety*, *3*(8), 1–5. https://doi.org/10.4172/2472-1786.100046.

Aziz, S., & Tariq, N. (2019). Depression, anxiety, and stress in relation to life satisfaction and academic performance of adolescents. *Pakistan Journal of Physiology*, *15*(1), 52–55. https://pjp.pps.org.pk/index.php/PJP/article/view/859.

Bittner, A., Egger, H. L., Erkanli, A., Jane Costello, E., Foley, D. L., & Angold, A. (2007). What do childhood anxiety disorders predict? *Journal of Child Psychology and Psychiatry and Allied Disciplines*, *48*(12), 1174–1183. https://doi.org/10.1111/j.1469-7610.2007.01812.x.

Curry, J. F., March, J. S., & Hervey, A. S. (2004). Comorbidity of childhood and adolescent anxiety disorders prevalence and implications. In T. H. Ollendick, & J. S. March (Eds.), *Phobic and anxiety disorders in children and adolescents* (2nd ed., pp. 116–140). Oxford University Press.

Esbjørn, B. H., Hoeyer, M., Dyrborg, J., Leth, I., & Kendall, P. C. (2010). Prevalence and co-morbidity among anxiety disorders in a national cohort of psychiatrically referred children and adolescents. *Journal of Anxiety Disorders*, *24*(8), 866–872. https://doi.org/10.1016/j.janxdis.2010.06.009.

Ginsburg, G. S., Becker-Haimes, E. M., Keeton, C., Kendall, P. C., Iyengar, S., Sakolsky, D., … Piacentini, J. (2018). Results from the Child/Adolescent Anxiety Multimodal Extended Long-Term Study (CAMELS): Primary anxiety outcomes. *Journal of the American Academy of Child and Adolescent Psychiatry*, *57*(7), 471–480. https://doi.org/10.1016/j.jaac.2018.03.017.

Hamdani, S. U., Huma, Z. E., Suleman, N., Warraitch, A., Muzzafar, N., Farzeen, M., … Wissow, L. S. (2021). Scaling-up school mental health services in low resource public schools of rural Pakistan: The Theory of Change (ToC) approach. *International Journal of Mental Health Systems*, *15*(1). https://doi.org/10.1186/s13033-021-00435-5.

Hossain, M. M., Purohit, N., Sultana, A., Ma, P., McKyer, E. L. J., & Ahmed, H. U. (2020). Prevalence of mental disorders in South Asia: An umbrella review of systematic reviews and meta-analyses. *Asian Journal of Psychiatry*, *51*, 102041. Elsevier B.V. https://doi.org/10.1016/j.ajp.2020.102041.

Howes, V. E., Zwicker, A., Uher, R., & Pavlova, B. (2020). Cognitive-behavioural interventions for prevention and treatment of anxiety in young children: A systematic review and meta-analysis. *Clinical Psychology Review*, *81*, 101904. Elsevier Inc https://doi.org/10.1016/j.cpr.2020.101904.

Hudson, J. L., Rapee, R. M., Lyneham, H. J., McLellan, L. F., Wuthrich, V. M., & Schniering, C. A. (2015). Comparing outcomes for children with different anxiety disorders following cognitive behavioural therapy. *Behaviour Research and Therapy*, *72*, 30–37. https://doi.org/10.1016/j.brat.2015.06.007.

Imran, N., Bhatti, M. R., Anwar, A., Najmi, F., & Haider, I. I. (2012). Children's mental health: Pattern of referral, distribution of disorders and service use in child psychiatry outpatient setting. *Pakistan Journal of Medical Sciences*, *28*(1), 22–26.

Imran, N., Rahman, A., Chaudhry, N., & Asif, A. (2018). World Health Organization "School Mental Health Manual"-based training for school teachers in Urban Lahore, Pakistan: Study protocol for a randomized controlled trial. *Trials*, *19*(1). https://doi.org/10.1186/s13063-018-2679-3.

Ishfaq, K. (2014). *Development of therapeutic protocol of cognitive behavior therapy (CBT) for obsessive compulsive disorder (OCD) in Pakistan*. G.C University Lahore.

Jaycox, L. (2018). *CBITS: Cognitive-behavioral intervention for trauma in schools*. RAND Corporation. https://doi.org/10.7249/cb209.

Kendall, P. C., Hedtke, K. A., & Child and Adolescent Anxiety Disorders Clinic (Temple University). (2006). *Cognitive-behavioral therapy for anxious children : Therapist manual*. Workbook Pub. https://www.workbookpublishing.com/cognitive-behavioral-therapy-for-anxious-children-therapist-manual-3rd-edition.html.

Khalid, A., Qadir, F., Chan, S. W. Y., & Schwannauer, M. (2018). Parental bonding and adolescents' depressive and anxious symptoms in Pakistan. *Journal of Affective Disorders*, *228*, 60–67. https://doi.org/10.1016/j.jad.2017.11.050.

Khan, S., Amjad, A., & Rowland, D. (2019). Potential for long-term benefit of cognitive behavioral therapy as an adjunct treatment for men with erectile dysfunction. *Journal of Sexual Medicine*, *16*(2), 300–306. https://doi.org/10.1016/j.jsxm.2018.12.014.

Khan, A., Malik, T. A., Ahmed, S., & Riaz, A. (2020). Translation, adaptation and implementation of Coping Cat program with Pakistani children. *Child and Youth Care Forum*, *49*(1). https://doi.org/10.1007/s10566-019-09516-6.

Latif, M., Husain, M. I., Gul, M., Naz, S., Irfan, M., Aslam, M., … Naeem, F. (2021). Culturally adapted trauma-focused CBT-based guided self-help (CatCBT GSH) for female victims of domestic violence in Pakistan: Feasibility randomized controlled trial. *Behavioural and Cognitive Psychotherapy*, *49*(1), 50–61. https://doi.org/10.1017/S1352465820000685.

Lee, S. J., Altschul, I., & Mowbray Carol, T. (2008). Using planned adaptation to implement evidence-based programs with new populations. *American Journal of Community Psychology*, *41*(3–4), 290–303. https://doi.org/10.1007/s10464-008-9160-5.

Lyneham, H. J., Abbott, M. J., Wignall, A., & Rapee, R. M. (2003). *Cools Kids Anxiety Program*. MUARU: Macqaurine University. https://www.cebc4cw.org/program/cool-kids/detailed.

Malik, T. A., Rooney, M., Chronis-Tuscano, A., & Tariq, N. (2017). Preliminary efficacy of a behavioral parent training program for children with ADHD in Pakistan. *Journal of Attention Disorders*, *21*(5), 390–404. https://doi.org/10.1177/1087054714524158.

Malik, T. A., Siddiqui, S., & Mahmood, A. (2019). Behavioural and emotional problems among school children in Pakistan: A telephonic survey for prevalence and risk factors. *Journal of Paediatrics and Child Health*, *55*(12), 1–10. https://doi.org/10.1111/jpc.14429.

Mushtaq, A., Lochman, J. E., Tariq, P. N., & Sabih, F. (2017). Preliminary effectiveness study of Coping Power Program for aggressive children in Pakistan. *Prevention Science*, *18*(7), 762–771. https://doi.org/10.1007/s11121-016-0721-9.

Naeem, F., Gobbi, M., Ayub, M., & Kingdon, D. (2009). University students' views about compatibility of cognitive behaviour therapy (CBT) with their personal, social and religious values (a study from Pakistan). *Mental Health, Religion & Culture*, *12*(8), 847–855. https://doi.org/10.1080/13674670903115226.

Naeem, F., Gobbi, M., Ayub, M., & Kingdon, D. (2010). Psychologists experience of cognitive behaviour therapy in a developing country: A qualitative study from Pakistan. *International Journal of Mental Health Systems*, *4*(1), 2. https://doi.org/10.1186/1752-4458-4-2.

Naeem, F., Habib, N., Gul, M., Khalid, M., Saeed, S., Farooq, S., … Kingdon, D. (2016). A qualitative study to explore patients', carers' and health professionals' views to culturally adapt CBT for psychosis (CBTp) in Pakistan. *Behavioural and Cognitive Psychotherapy*, *44*(1), 43–55. https://doi.org/10.1017/S1352465814000332.

Naeem, F., Latif, M., Mukhtar, F., Kim, Y. R., Li, W., Butt, M. G., … Ng, R. (2021). Transcultural adaptation of cognitive behavioral therapy (CBT) in Asia. *Asia-Pacific Psychiatry*, *13*(1). https://doi.org/10.1111/appy.12442.

Naeem, F., Phiri, P., Munshi, T., Rathod, S., Ayub, M., Gobbi, M., & Kingdon, D. (2015). Using cognitive behaviour therapy with South Asian Muslims: Findings from the culturally sensitive CBT project. *International Review of Psychiatry*, *27*(3), 233–246. https://doi.org/10.3109/09540261.2015.1067598.

Naeem, F., Phiri, P., Rathod, S., & Ayub, M. (2019). Cultural adaptation of cognitive–behavioural therapy. *BJPsych Advances*, *25*(6), 387–395. https://doi.org/10.1192/bja.2019.15.

Naeem, F., Sarhandi, I., Gul, M., Khalid, M., Aslam, M., Anbrin, A., … Ayub, M. (2014). A multicentre randomised controlled trial of a carer supervised culturally adapted CBT (CaCBT) based self-help for depression in Pakistan. *Journal of Affective Disorders*, *156*, 224–227. https://doi.org/10.1016/j.jad.2013.10.051.

Naeem, F., Waheed, W., Gobbi, M., Ayub, M., & Kingdon, D. (2011). Preliminary evaluation of culturally sensitive CBT for depression in Pakistan: Findings from developing culturally-sensitive CBT project (DCCP). *Behavioural and Cognitive Psychotherapy*, *39*(2), 165–173. https://doi.org/10.1017/S1352465810000822.

Patel, V., Flisher, A. J., Hetrick, S., & McGorry, P. (2007). Mental health of young people: A global public-health challenge. *Lancet*, *369*(9569), 1302–1313. Elsevier https://doi.org/10.1016/S0140-6736(07)60368-7.

Polanczyk, G. V., Salum, G. A., Sugaya, L. S., Caye, A., & Rohde, L. A. (2015). Annual research review: A meta-analysis of the worldwide prevalence of mental disorders in children and adolescents. *Journal of Child Psychology and Psychiatry*, *56*(3), 345–365. https://doi.org/10.1111/jcpp.12381.

Rahman, A., Malik, A., Sikander, S., Roberts, C., & Creed, F. (2008). Cognitive behaviour therapy-based intervention by community health workers for mothers with depression and their infants in rural Pakistan: A cluster-randomised controlled trial. *The Lancet*, *372*(9642), 902–909. https://doi.org/10.1016/S0140-6736(08)61400-2.

Ramsawh, H. J., Chavira, D. A., & Stein, M. B. (2010). Burden of anxiety disorders in pediatric medical settings. *Archives of Pediatrics & Adolescent Medicine*, *164*(10), 965–972. https://doi.org/10.1001/archpediatrics.2010.170.

Riaz, A. (2017). *Mental health and resilience in children exposed to trauma: Implimentation of resilience based intervention framework*. Fatima Jinnah Women University.

Saleem, S., & Mehmood, Z. (2011). Development of a scale for assessing emotional and behavioral problems of school children. *Journal of Social and Clinical Psychology*, *9*, 73–78.

Sikander, S., Ahmad, I., Atif, N., Zaidi, A., Vanobberghen, F., Weiss, H. A., … Rahman, A. (2019). Delivering the Thinking Healthy Programme for perinatal depression through volunteer peers: A cluster randomised controlled trial in Pakistan. *Lancet Psychiatry*, *6*(2), 128–139. https://doi.org/10.1016/S2215-0366(18)30467-X.

Sohail, S. A., Syed, A. A., & Rahman, A. (2017). Mental health in Asia and the Pacific. In H. Minas, & M. Lewis (Eds.), *Mental health in Asia and the Pacific, international and cultural psychology* (pp. 17–38). Springer US. https://doi.org/10.1007/978-1-4899-7999-5.

Syed, E. U., Hussein, S. A., & Haidry, S. E. Z. (2009). Prevalence of emotional and behavioural problems among primary school children in Karachi, Pakistan—Multi informant survey. *Indian Journal of Pediatrics*, *76*(6), 623–627. https://doi.org/10.1007/s12098-009-0072-7.

Velting, O. N., Setzer, N. J., & Albano, A. M. (2002). Anxiety disorders. In D. T. Marsh, & M. A. Fristad (Eds.), *Handbook of serious emotional disturbance in children and adolescents* (p. 525). John Wiley & Sons.

World Health Organization. (2017). *Depression and other common mental disorders global health estimates*. https://apps.who.int/iris/bitstream/handle/10665/254610/WHO-MSD-MER-2017.2-eng.pdf.

Zadeh, Z. F., & Lateef, M. (2012). Effect of cognitive Behavioural therapy (CBT) on depressed female university students in Karachi. *Procedia—Social and Behavioral Sciences*, *69*, 798–806. https://doi.org/10.1016/j.sbspro.2012.12.001.

Section E

Case studies and resources

Chapter 34

"I am so much more than OCD": A case illustration of key elements in a Danish group-based CBT program for youth with OCD

Katja Anna Hybel, Sanne Jensen, and Davíð R.M.A. Højgaard
Department of Child and Adolescent Psychiatry, Aarhus University Hospital, Psychiatry, Aarhus, Denmark

Abbreviations

CBT	cognitive behavioral therapy
CY-BOCS	Children's Yale-Brown Obsessive-Compulsive Scale
ERP	exposure and response prevention
GP	general practitioner
OCD	obsessive-compulsive disorder

Introduction

Group-based cognitive behavioral therapy (CBT) for adolescent obsessive-compulsive disorder (OCD) is a modified version of standard CBT delivered individually (see Chapter 23). The group CBT format may provide advantages over standard CBT in terms of providing peer support, normalizing patients' and parents' experiences, offering the possibility of therapist co-work, and training as well as improving cost-effectiveness. More details on the content and delivery of group CBT will be provided along with the case description below.

The effectiveness of group CBT for children and adolescents with OCD has yet to be documented sufficiently, as to date there are few studies that have rigorously evaluated the effectiveness of this treatment format. In a review from 2014, six studies of nonfamily-focused group CBT are mentioned, and while they all show promising results, the findings are limited by small sample sizes and a lack of control groups (Freeman et al., 2014).

In addition to the quantifiable treatment effects, it is highly important to consider patient perspectives when introducing and/or evaluating treatments. A qualitative approach to treatment evaluation has the potential to explore mechanisms of change and reasons for (the lack of) treatment effects (Lewin, Glenton, & Oxman, 2009). To our knowledge, this has not been conducted in young OCD patients undergoing group CBT. Yet, a study including adults with depression and anxiety disorders undergoing group CBT found that group therapy, helped patients share responsibility and develop a sense of belonging (Christensen, 2020) which may also apply to younger patients.

The current case study aims to describe and explore the use of group CBT for a 17-year-old girl with OCD, offering an account of some of the main treatment elements and how they were used in treatment as well as how these elements and the overall format were experienced by the patient.

Method

The case study was conducted by the therapist (the first author) on the basis of the clinical impression of the patient's development during treatment, an evaluation of the clinical record, and an evaluation of a copy of the working sheets completed by the patient during the treatment program. These qualitative evaluations were supplemented with evaluations of the

Handbook of Lifespan Cognitive Behavioral Therapy. https://doi.org/10.1016/B978-0-323-85757-4.00054-7

quantitative outcome assessments conducted by a semi-independent clinician (i.e., not the group therapists). For this, the Children's Yale-Brown Obsessive Compulsive Scale (CY-BOCS), which assesses different aspects as well as the overall severity of obsessive and compulsive symptoms, was used (Scahill et al., 1997). Further, the group CBT participant was telephone interviewed by an independent researcher (the second author) who had no prior knowledge of the patient or the specific treatment group she attended. The interview was conducted using a semistructured interview guide encompassing three main focus areas with questions pertaining to (1) the group format, (2) the treatment components, and (3) the therapists.

Ethical considerations

Out of consideration for the patient and her parents, case details were pseudonymized. The patient was provided with information about the scope of the case study as well as the possible advantages and disadvantages of participating before signing informed consent declarations. At that time, she had become of legal age. Further, she was provided with a draft version of the manuscript to check for misunderstandings and passages about which she felt uncomfortable with in the presentation. The patient gave her consent.

The case presentation

The referral

Rose (pseudonym) was 17 years old when she was referred to a specialized OCD clinic by her general practitioner (GP). Her GP stated that Rose had become socially isolated, that she had increasing difficulties carrying out activities of daily life and that her thoughts and behavior had become increasingly illogical. Rose believed that she was an asymptomatic disease carrier of coronavirus and that she would contaminate everyone she was around. She had experienced several breakdowns including suicidal thoughts. At the OCD clinic, Rose was first seen for short intake interview. After that, Rose and her parents were invited for an interview covering her developmental and medical history, a clinical screening interview covering present psychopathology and an in-depth interview covering present obsessive-compulsive symptoms and severity.

Rose's background and development

Rose grew up in a Danish middle class family and was the first born of two siblings. Her parents divorced in Rose's preschool years and had a good parental working relationship. Rose was enrolled in a youth education program and had several friends and a boyfriend; however, due to her OCD symptoms, she found it difficult to attend school and lacked the surplus energy to see the friends she once had. Although Rose had a history of OCD and Tourette's disorder, she had developed as most children do in other aspects of life (e.g., language, motor, and social skills). However, when asked about her early development, Rose's mother revealed that Rose sometimes seemed to enjoy arranging things in a special order as a toddler. Moreover, her parents explained that Rose needed a clear structure and preparation from an early age. She reacted negatively to sudden changes of plan. Motor tics were first noticed in kindergarten, while obsessions with magical and aggressive content and compulsions including counting and checking rituals appeared when Rose was around 8 years old. When Rose was around 10, the symptoms deteriorated. Then, after counseling in primary care, she was referred to the specialized OCD clinic for the first time. Rose was then successfully treated with six sessions of individual family-focused CBT. She quickly picked up CBT techniques, especially reality checking, externalizing, and exposures. Further, her parents acted on the therapist's guidance on how to support Rose in resisting rituals. The therapist described Rose as having good social/communication skills but noted that she could be emotional and sometimes overwhelmed by emotions. The therapist remarked that it seemed as if obsessive-compulsive symptoms tended to appear when Rose experienced some kind of daily life stress.

Rose's psychopathological profile before group CBT

Rose and her parents described that Rose's OCD symptoms gradually intensified during the COVID-19 crisis. Rose's continuation school (an independent boarding school for lower secondary students that many Danish young people attend for a year before upper secondary education) that she joined at that time was closed for several months and uncertainty in society as a whole regarding the COVID situation was high. The family explained that previously learned coping strategies were no longer sufficient and that Rose found her obsessions increasingly convincing. This resulted in considerable dependence on her parents' support and reassurance. Rose's obsessions centered on thoughts about being responsible for the transmission of coronavirus, including aggressive thoughts about herself or close relatives dying and lively inner images about how the virus could spread in "supernatural" ways like "puffs" in a cartoon. Further, some obsessions featured aggressive content relating to a fear of "becoming the person she once was" (during a period when she felt left out in school). She believed that her typical social behaviors could result in rejection from others and that awful things could happen to her family (e.g., a fire at home, a car crash) as well as detailed intrusive mental images of dead people and scary creatures. In addition, Rose described her tendency to keep certain objects because she felt that throwing them out would somehow make her feel that she had lost the people or events associated with the items. Further, she described significant not-just-right sensations associated with a need for bodily symmetry; this was related to certain equal numbers and quantities as well as a sense of uneven numbers being bad. Rose's compulsions were wide-ranging, including checking rituals (e.g., electric switches, windows, door locks), repeating daily behaviors (e.g., rewriting, re-reading), counting words, letters, and objects to be a multiple of an even number, ordering objects in categories, excessive list writing, and staring rituals. Further, she described numerous magical rituals, mainly facial movements of her nose, eyes, or mouth, cracking fingers and other body parts, tapping fingers, and correcting her body's symmetry to prevent contents of the aggressive thoughts from happening.

Besides the OCD symptoms, Rose presented with motor and verbal tics (e.g., eye blinking, facial grimaces, neck twists, exhaling sounds) that met the criteria for Tourette's disorder. She also described a range of co-occurring anxiety symptoms including a tendency for different kinds of excessive worries related to hearing about climate change or natural disasters on the news and health worries triggered by different bodily sensations. In the weeks leading up to her referral, Rose had panic attack-like episodes during which she trembled, hyperventilated, sobbed, and had a rapid heartbeat. In relation to these episodes, Rose briefly felt like giving up, but generally she kept a good mood.

For several years, Rose had had subclinical hallucinations comprising momentary visual impressions of figures or shapes and hearing footsteps. However, she could distance herself from these experiences knowing that it was her senses "playing tricks on her" when she was anxious and stressed. During the assessment, Rose further described certain executive function difficulties; for example, she often found it difficult to create an overview of and remember daily tasks and had problems estimating time.

The intervention

The treatment program

The youth OCD group treatment program was developed by the first author in collaboration with four colleagues in the OCD clinic[a] during 2014–17 and conducted as part of daily clinical practice as well as an ongoing clinical trial benchmarking the effects of group and individual treatment (ClinicalTrials.gov ID: NCT04891367). The program was based on classic CBT principles for OCD (see the Chapter 23) and followed the Nordic Long Term OCD Treatments Study CBT manual in overall content and structure (Thomsen & Weidle, 2015). However, the manual was adapted to fit use with young people and a group format, and supplementary elements were added based on previous individual and group CBT studies (Farrell, Waters, Milliner, & Ollendick, 2012; Fisher & Wells, 2009; Rapee, 2010; Söchting, 2014; Söchting & Third, 2011) as well as the manual author's clinical experience. An overview of the main program components is provided in Table 1.

The youth program included 13 weekly 2-h group sessions and two booster sessions 1 and 3 months after the last weekly session. There were three parent group sessions in parallel with youth group sessions 2, 5, and 9 at which all parents were invited to receive psychoeducation about OCD, OCD treatment, and relapse prevention as well as exchange their experiences of parenting a young person with OCD. Additionally, the program included four family sessions: one introductory session before the first group session, two sessions during the intensive part of the treatment (between the 4th and 5th and the 9th and 10th group sessions), and one family session after the final group booster session. The content of the individual sessions was flexible to allow individual tuning in on the young person and her/his OCD work as well as address possible co-occurring symptoms, problems, and parental concerns. Groups included five to eight adolescents aged 13–17 years. Two therapists were responsible for the youth group sessions as well as family sessions for each half of the group. To ensure the young person's integrity, a third therapist conducted the parent sessions. The therapists were experienced clinical psychologists with a postgraduate CBT education.

a. Study version of the youth group CBT manual (presently in preparation for publication) by Katja Anna Hybel in collaboration with Kariann Slettebø Ahlgreen, Else Kristensen Brunbjerg, Tanja Gjerlev Andersen, and Judith Becker Nissen.

TABLE 1 Overview of main themes in the youth OCD group CBT program[a].

Session 1	Introduction, group participants meet, psychoeducation about OCD and CBT for OCD
Session 2	Continued psychoeducation, work on externalization, setting goals, mapping OCD, grading symptoms
Session 3	Continued psychoeducation on principles for gradual exposure and response prevention, creating a symptoms overview
Session 4	Planning exposure and response prevention exercises, focus on individual methods and strategies
Session 5	Continued psychoeducation on obsessions and how to work with obsessions, thought focus exercise in vivo
Sessions 6–7	Exposure and response prevention exercises in vivo, using parents as support
Session 8	Setting goals and subgoals
Sessions 9–11	Exposure and response prevention exercises in vivo, using parents as support
Session 12	Summing up the course of treatment, maintaining treatment gains, psychoeducation on relapse prevention
Session 13	Treatment status, "OCD before, presently and in the future," further forward exposure and response prevention work and maintaining of treatment gains, celebration and recognition of the OCD work
Booster 1–2	Status on symptoms and own working methods, further future exposure and response prevention work, maintaining strategies and relapse prevention, individual narratives on the theme "farewell OCD"

[a]Study version of the youth group CBT manual (presently in preparation for publication) by Katja Anna Hybel in collaboration with Kariann Slettebø Ahlgreen, Else Kristensen Brunbjerg, Tanja Gjerlev Andersen, and Judith Becker Nissen.

Rose's mapping and overview of OCD

Rose was included in a group of seven adolescents. From the start, she actively contributed and openly shared her experiences. In the early sessions, Rose and her group participants were provided with a basic psychoeducation on OCD and CBT principles. During the therapist introductions and planned exercises, all the participants worked individually or in smaller groups to explicate, draw, and write down their symptoms. By doing so, they gradually grasped a clearer idea of the situations in which symptoms were triggered and the difference between obsessions and compulsions. They also learned how to grade the expected difficulty of resistance for different symptoms (on a 0–10 scale, with higher ratings indicating higher discomfort or anxiety).

Group participants were introduced to a mapping exercise. Rose was asked to fill out a working sheet with two overlapping circles: one circle was titled "OCD decides and rules" and the other was titled "I decide and rule," with the overlap between the two termed "the battle zone." Rose then indicated her different life situations, choices, and behaviors and assigned them to the different zones before discussing her work with a fellow group participant. With this exercise, the group participants became aware that even though it occasionally felt like OCD was always in control, in fact OCD did not control all aspects of their lives. Rose realized several areas in which she was in complete control (e.g., choosing what to eat, when to go to the bathroom, when to give family members a hug, when to visit friends, who to talk to) and several areas in which she sometimes won over OCD (e.g., touching things on the bus, going for a walk, eating normally). Further, Rose and her fellow group participants were invited to work on a symptoms overview that included writing down all significant obsessions and compulsions along with difficulty ratings for each symptom using a 0–10 scale. To complete the ratings, the young people were asked to anticipate how resisting the obsession/compulsion would feel and encouraged to use investigative experiencing by, for instance, approaching a trigger situation (physically or mentally) in the session and at home. The group participants were then encouraged to list symptoms at all difficulty levels and divide complex symptoms into smaller units if the spontaneous listing resulted in very high ratings. Rose's ratings ranged from 10 (e.g., excessively checking the closet doors in her room at night) to 3 (e.g., repeatedly switching her mobile phone on and off to prevent bad things from happening). Rose and her fellow participants were encouraged to use the symptoms overview in the subsequent sessions to continuously track progress by self-evaluating symptoms throughout the remainder of the therapy.

Rose's goal setting

The second session started by focusing on setting clear and concrete goals. Assisted by the therapist and using the program worksheets, Rose formulated the following goals for therapy. She wanted to (1) attend school regularly, (2) use public transportation, (3) cope with and adapt to changes of plan, (4) interact normally with objects (only touching or checking when required), (5) be able to go out at night (e.g., go for walks, attend social gatherings), (6) be able to walk alone to the bus stop, and (7) go to bed even though the closet doors were closed. The group was introduced to the rationale for goal setting, namely that goals are essential to foster motivation (i.e., remind patients why difficult OCD work is necessary), can help direct choices of exercises and are valuable for evaluating therapeutic progress. When revisiting her initial goals and refining goal setting in a later session, Rose added the overall goal to move out of her parents' house. She then worked on specifying which symptoms in her symptoms overview she had to work on to obtain these goals.

Rose's exposure and response prevention exercises

In the third session, the group was instructed to plan a first "test" exposure and response prevention (ERP) exercise to be practiced daily at home between sessions. Rose started by working with one of the lighter symptoms in her symptoms overview. She decided to resist correcting certain objects (expected difficulty 3–4) and found that this was both feasible and somewhat easier than she expected (successive day ratings were 3, 1, 2, 0, and 1). In the following session, individual resistance strategies during the test ERP exercises were thoroughly analyzed and a new regular ERP exercise was planned. Rose chose to work with her daily rituals encountered before bedtime where she felt forced to repeat saying goodnight to her parents several times to prevent bad things from happening (expected difficulty 5). During the session, the exercise was carefully prepared in respect to what Rose should do and which strategies she could apply to help resist the rituals and the resulting obsessive thoughts. She decided to use self-supportive sentences (*nothing will happen*, *others do not say goodnight more than once*) as well as self-supportive behaviors (go to my room, do something else); she also agreed on a reward with her parents (visit her favorite restaurant at the end of the week). Planning and conducting the ERP exercises at home as well as practicing ERP in vivo in sessions continued throughout the program. Gradually, Rose approached the harder symptoms in the overview (9–10). For instance, she started working with a difficult symptom (i.e., letting other people physically near) by practicing this in vivo in the session, assisted by a fellow group member and a therapist. The in vivo experiences were then used to plan further ERP exercises for daily practice at home. When conducting exercises together in vivo, Rose and her fellow group members inspired each other and offered ideas on strategies for resisting and reducing symptoms.

Rose's learnings throughout therapy

During the last sessions of the intensive part of the treatment, the focus shifted to working on consolidation and relapse prevention. The group participants summarized what they had learned throughout therapy and what they should be aware of in the future. According to Rose, her three most important learnings were (1) *the harder I fight, the better the result*, (2) *I am not alone*, and (3) *it's OK to ask for help and say that things are difficult*. She also summarized her best strategies for resisting OCD: (1) using common sense, (2) talking to someone, and (3) physically moving away after conducting an ERP exercise. Finally, Rose and her fellow group participants were instructed to reflect on which future life situations could make them vulnerable to OCD. For Rose, these were moving away from her parents, dealing with periods with extra school pressure and exams and starting a new internship or job. The group then presented their individual foci for work on their remaining symptoms and discussed their relapse prevention strategies and special awareness points.

Assessments throughout therapy

As presented in Table 2, Rose's CY-BOCS ratings showed a clinically significant and gradual decrease in severity during the 6-month rating period (CY-BOCS total percent decrease $\geq$30). The OCD symptoms were severe to very severe before the treatment started (CY-BOCS total = 29–33), moderate during the intensive treatment program part (CY-BOCS total = 22–24), mild to moderate after the last session of the intensive part of the treatment (CY-BOCS total = 18–20), and finally subclinical to mild after the last booster session (CY-BOCS total = 8–11). The highest decrease was estimated in the CY-BOCS subdomains interference for both obsessions and compulsions, whereas time spent on obsessions, distress, and resistance related to obsessions decreased the least according to the formal ratings. The CY-BOCS rating intervals reflect the perceived uncertainty of specific ratings by the clinician due to comorbidity.

TABLE 2 Children's Yale-Brown Obsessive-Compulsive Scale (CY-BOCS) assessments during treatment[a].

	Two weeks before first group session[b]	After 7 weekly group sessions	After 12 weekly group sessions	After the final booster session[c]
CY-BOCS Obsessions total	14–15	11–12	9	4–6
Time	3–4	3	3	2
Interference	3	2	1	0
Distress	3	3	3	1–2
Resistance	2	1–2	0	0–1
Control	3	2	2	1
CY-BOCS Compulsions total	15–18	11–12	9–11	4–5
Time	4	3–4	3	2
Interference	3	2	1	0
Distress	3–4	3	2–3	1–2
Resistance	2–3	1	1–2	0
Control	3–4	2	2	1
CY-BOCS severity total	29–33	22–24	18–20	8–11

[a]*CY-BOCS rating intervals reflect the perceived uncertainty of specific ratings by the clinician due to comorbidity.*
[b]*At the assessment Rose explained that her symptoms already had improved somewhat compared to at the time of the intake interview.*
[c]*Three months after termination of the 13 weekly sessions.*

The qualitative interview

Rose's views on her participation in the program

Rose was interviewed about her experiences with the group therapy midway between the first and last booster sessions (i.e., approximately 1 month before the last booster session). Overall, she spoke passionately and enthusiastically about the treatment and group format, even claiming that entering group treatment was the *best decision she had ever made.* Her initial expectations of the treatment, however, were low. Before the treatment started and even after the first treatment session, she was unsure whether to continue, as she felt she would not cope with the other participants. Yet, this reservation changed after the second session.

One overarching theme from the interview was Rose's experience that the group therapy helped her not "feel wrong." This was supported both by the sense of community among the participants in the group and by the therapists not being shocked by any of the symptoms brought up. Rather, the therapists normalized their symptoms. Further, the "battle zone" made Rose aware that she was so much more than her OCD, even from the beginning, and that her OCD did not control all aspects of her life. Rose further emphasized that the common age among the participants enhanced the feeling of being more than their OCD because they shared mutual conversation topics during breaks.

Reflecting on the group format compared with the individual treatment course in which Rose participated previously, she thought she would have benefited from the group format when she was younger. Although the individual treatment was also successful at reducing her symptoms, Rose explained that she still felt alone with her OCD at that time because she did not know of anyone else fighting the same challenges.

When asked about the potential disadvantages of the group format, Rose mentioned the various degrees of "fighting spirit" among the participants. Her view was that commitment was essential for the treatment outcome and that the participants should attend the group with the purpose of fighting against OCD together. However, other participants' lack of a fighting spirit did not seem to affect the overall outcome of the treatment for her.

At the time of the interview, Rose felt confident about ending her treatment soon after attending the final booster session. She pointed out her newly found ability to articulate her challenges and wins to others as the major skill acquired from the group therapy. She explained that verbalizing her battlefields to other people made her feel more committed in her fight against OCD. Rose highlighted this as the main skill acquired from the group therapy to prevent OCD relapse in the future.

Concluding reflections

Based on the quantitative and qualitative descriptions, Rose's treatment was successful at reducing her symptoms, teaching her individual strategies and normalizing her interpersonal experiences. However, the choice of clinical case has implications for the generalizability of the findings. Rose was indeed exemplary. She had experienced relapse and had a rather severe and complex symptom presentation even though she was highly motivated, had an open personality, liked writing and reflecting, and was well supported by her parents. All these aspects can be hypothesized to have contributed to her positive outcome. Her previous experience of individual CBT for OCD allowed her to compare the possible benefits and drawbacks of the two treatment formats, but it might also have given her a head start in the group CBT setting. Bearing in mind these possible explanations of a successful outcome, Rose's treatment course and experiences with group CBT clearly indicate that this treatment format can be effective for treating adolescents with OCD. Indeed, the group format itself may have had a healing effect on Rose. This is supported by her statements on the commitment of the other participants and that she felt more obliged to fight against her OCD because of the group treatment context. It could be argued that the group format adds a component of peer pressure, which, for Rose, was interpreted positively and served as an incentive. It could, however, be speculated that others would react differently to the group context, thereby possibly resulting in embellishing the reports of homework exposure or in more dropouts from treatment. The interpretation of the group context may depend both on the individual patient and on the specific group atmosphere, suggesting that the establishment of a positive and supportive group climate is vital. Rose's considerations about meeting other young people fighting the same challenges as her point to a central feature of the group format that may be difficult to obtain in other treatment formats. This reflection on the significance of meeting peers may be of special importance for young people/teenagers in the midst of building their personal identity.

Rose's initial reservation of the group format is—based on clinical experience—a relatively common objection from patients beginning group therapy. This means that therapists may be commissioned to motivate patients before entering group treatment. Based on the present case study, Rose's initial reservations should not guide her treatment choices. However, research on patient experiences and individual differences in the efficacy of treatment outcomes is still needed to highlight which patients may be more suitable for individual CBT.

Summary points

- Group-based CBT is a modified version of individual standard CBT.
- A qualitative research approach can provide valuable information about mechanisms of change and elucidate patients' perspectives.
- Based on the case descriptions, group-based CBT was successful at reducing symptoms, teaching individual strategies, and normalizing interpersonal experiences.
- The qualitative interview with the patient suggests that the group format itself enhanced treatment success.
- Initial resistance to group therapy may be a common challenge for patients that can be overcome during treatment.

References

Christensen, A. B. (2020). *Patients' and therapists' experiences of group CBT for anxiety and depression in Danish mental health services*. University of Copenhagen. https://www.regionsjaelland.dk/sundhed/geo/psykiatrien/om_psykiatrien/psykiatrisk-forskningsenhed/phd-studium/Documents/Anne-Bryde-afhandling.pdf.

Farrell, L., Waters, A., Milliner, E., & Ollendick, T. (2012). Comorbidity and treatment response in pediatric obsessive-compulsive disorder: A pilot study of group cognitive-behavioral treatment. *Psychiatry Research*, *199*(2), 115–123. https://doi.org/10.1016/j.psychres.2012.04.035.

Fisher, P., & Wells, A. (2009). In P. Fisher, & A. Wells (Eds.), *Metacognitive therapy: Distinctive features* Routledge (Book).

Freeman, J., Garcia, A., Frank, H., Benito, K., Conelea, C., Walther, M., & Edmunds, J. (2014). Evidence base update for psychosocial treatments for pediatric obsessive-compulsive disorder. *Journal of Clinical Child and Adolescent Psychology: The Official Journal for the Society of Clinical Child and Adolescent Psychology, American Psychological Association, Division 53*, *43*(1), 7–26. https://doi.org/10.1080/15374416.2013.804386.

Lewin, S., Glenton, C., & Oxman, A. D. (2009). Use of qualitative methods alongside randomised controlled trials of complex healthcare interventions: Methodological study. *BMJ*, *339*(sep10 1), b3496. https://doi.org/10.1136/bmj.b3496.

Rapee, R. (2010). In R. Rapee (Ed.), *Cool kids (R) angstprogram for børn: Terapeutens manual* Aarhus Universitet (Book).

Scahill, L., Riddle, M. A., McSwiggin-Hardin, M., Ort, S. I., King, R. A., Goodman, W. K., … Leckman, J. F. (1997). Children's Yale-Brown Obsessive Compulsive Scale: Reliability and validity. *Journal of the American Academy of Child and Adolescent Psychiatry*, *36*(6), 844–852. https://doi.org/10.1097/00004583-199706000-00023.

Söchting, I. (2014). *Cognitive behavioral group therapy: Challenges and opportunities* (Book). Wiley-Blackwell.

Söchting, I., & Third, B. (2011). Behavioral group treatment for obsessive-compulsive disorder in adolescence: A pilot study. *International Journal of Group Psychotherapy*, *61*(1), 84–97. https://doi.org/10.1521/ijgp.2011.61.1.84.

Thomsen, P. H., & Weidle, B. (Eds.). (2015). *Behandling af børn og unge med OCD En praksismanual* Hans Reitzels Forlag. http://hansreitzel.dk/Psykologi/Behandling-af-børn-og-unge-med-OCD/9788741261294.

Chapter 35

CASE STUDY: A child with body dysmorphic disorder and cognitive-behavioral therapy

Danielle H. Gardini, Yvette Fruchter, and Fugen Neziroglu
Bio Behavioral Institute, Great Neck, NY, United States

Abbreviations

BDD body dysmorphic disorder
CBT cognitive-behavior therapy
DBT dialectical behavior therapy
ERP exposure response prevention
OCD obsessive compulsive disorder
OVI overvalued ideation
RP response prevention

Theoretical and research basis for treatment

The way in which people develop their own individualistic traits, personality, characteristics, and values has long been examined throughout the field of psychology. Over many years, this development has been debated through various theories. For example, the classic nature versus nurture argument; exploring whether behaviors are inherited or acquired (McLeod, 2018). Most research and current theories consider how nature and nurture interact further exploring a bidirectional relationship (i.e., how environmental factors can influence our genetic composition (Neziroglu, Khemlani-Patel, & Veale, 2008). This understanding leads to a comprehensive outlook of how psychological disorders develop.

For the purposes of this case study, the focus is on body dysmorphic disorder (BDD). BDD is defined as an intense preoccupation with a minor real or perceived flaw in appearance that's magnified to a heightened level of hyperfocus (Pinto-Wagner, 2002). Usually, BDD involves focus on one or more specific body parts. This individual becomes bombarded and even tormented with nonstop intrusive thoughts about this body part. What results next is a lifestyle built around a routine of checking the body part, seeking assurance, camouflaging, and/or fixing this perceived defect (Pinto-Wagner, 2002). Symptoms are very similar to what we see with the obsession and compulsions in obsessive compulsive disorder (OCD); however, it is important to recognize that BDD has notable differences from other anxiety-related disorders (e.g., more delusional beliefs, extensive ritualistic behaviors, a focus on specific body areas) (Chosak et al., 2008; Hollander & Allen, 2004). Furthermore, BDD symptoms demonstrate higher levels of depression and overvalued ideation (OVI) (Mckay & Neziroglu, 2011).

It is natural for persons to have an inclination about portraying an overall appropriate appearance to themselves and others; however, when it becomes an obsessive, time-consuming, and distressing process resulting in interference with overall daily functioning, we are now in the realm of disordered behavior (Wilhelm, Phillips, & Steketee, 2013). The core fear in BDD is avoiding judgment of others, the need of acceptance by others, or the drive for perfectionism (Pinto-Wagner, 2002). Ritualistic behaviors adopted are aimed at examining, improving, or camouflaging the perceived flaw such as mirror checking, reassurance seeking, or avoiding certain situations (Phillips, Menard, Fay, & Weisberg, 2005).

BDD can develop across the lifespan; however it most often begins in adolescence with average ages ranging from 14 to about 17 years old affecting both genders equally (Khemlani-Patel & Neziroglu, 2021). These formative years are sensitive to social norms, peer pressures, emotional and physical maturity (Phillips et al., 2005). Research has also shown that aversive experiences can layer upon predispositions of emotional or perceptual instability (Wilhelm, 2006). Notably,

Handbook of Lifespan Cognitive Behavioral Therapy. https://doi.org/10.1016/B978-0-323-85757-4.00012-2

BDD symptoms have shown a chronic and unremitting course until effective treatment is introduced (Phillips, 2002; Phillips, Menard, Pagano, Fay, & Stout, 2006). Individuals with BDD experience deficits in visual perception and visuospatial processing. A hyperfocus has been shown on small details, which interferes with their global perception (Deckersbach et al., 2000). They also experience distress from overinflation of their own flaws that are minimal or nonexistent (Wilhelm et al., 2013). Other characteristics involve misinterpreting neutral facial expressions (Feusner, Neziroglu, Wilhelm, Mancusi, & Bohon, 2010); experiencing ideas/delusions of reference (Buhlmann, McNally, Etcoff, Tuschen-Caffier, & Wilhelm, 2004); selectively attending to words of "beauty"; preoccupation with an ideal appearance (Buhlmann, McNally, Wilhelm, & Florin, 2002); having significantly lower self-esteem of appearance (Buhlmann, Teachman, Naumann, Fehlinger, & Rief, 2009); underestimating their own attractiveness and/or overestimating others' (Buhlmann, Etcoff, & Wilhelm, 2007); having high levels of perfectionistic thinking (Phillips & McElroy, 2000). These unrelenting thoughts are associated with rejection sensitivity, heightened levels of anxiety, depression, and feeling worthless (Phillips, 2009).

There are various theories to consider when building a conceptualization of BDD symptomology involving social learning, cultural influences, clinical predisposition, and behavioral conditioning (Khemlani-Patel & Neziroglu, 2021). The cognitive-behavioral model applies classical conditioning learning to undo ineffective learned behaviors that reinforce BDD thoughts (Pinto-Wagner, 2002). It focuses on utilizing exposure response prevention (ERP) to reduce avoidance of anxiety-provoking stimuli and reduce compulsive behaviors, while restructuring cognitive distortions to more accurate beliefs about appearance (Wilhelm et al., 2013). Additional tools involve developing positive core beliefs, healthy body image, and building confidence. This case study focused on individualizing treatment and utilizing tools from the evidence-based modalities of CBT, ERP, and DBT.

Case introduction

"Olivia," is a 12-year-old Caucasian female entering seventh grade. Since graduating from elementary school, this upcoming school year will be her transition to middle school. She is a diligent student with good grades. She lives with her biological parents and two brothers. Olivia is about 5″ tall and 80 pounds, falling within the 25th–50th percentile of height/weight for her age. Olivia described having five "good friends" from school and gymnastics. She has been athletically active in gymnastics since age five and engaged in competitions for the past 3 years. Practice consisted of 3 h per day, 4 days per week. She described spending time on social media platforms including TikTok, Instagram, and Houseparty.

Presenting complaints

In March 2020, New York State was put on lockdown for the following 3 months due to the COVID-19 pandemic onset. Quarantine caused a social distancing recommendation, so during this time Olivia had to transition to remote learning for school and discontinue gymnastics since practice was canceled. In May 2020, Olivia experienced rumination and anxiety about the impact not exercising had on her stomach's appearance. Olivia described her stomach as "gross … not being flat anymore … not being as flat as her brothers'." Subsequently, she completed a 14-day abdominal workout. Per reports from parents, Olivia became hyperfocused on her stomach, exhibited compulsive behaviors, and was tearful throughout the day. Olivia worried that her stomach looked bad, which resulted in needing reassurance from checking or seeking comfort from others.

Eventually, during the first few weeks back at school of Fall season, Olivia began showing resistance to attending gymnastics practice. She expressed strong discomfort wearing her leotard and not being interested in the sport anymore. After some encouragement, Olivia remained in practice for 2 weeks and tried challenging intrusive thoughts and tolerating the anxiety. Olivia only became more resistant and tearful, resulting in her discontinuation of the sport. Instead, she joined Cross-Country track and then the Volleyball team.

History

Reportedly, Olivia was born full-term via C-section with her twin brother. She experienced Febrile seizures at about 3- or 4-years-old, which she outgrew by age 5. Her developmental history is nonsignificant, showing Olivia met all developmental milestones within normal limits. Family's psychiatric history is nonsignificant for any mental health concerns. Olivia had no prior experience with therapeutic treatment or psychiatric medications.

Assessment

Notably, all therapy sessions were conducted via Telemedicine due to safety protocols for COVID-19 pandemic. Informed consent was obtained during the initial consultation, which was conducted in a semistructured interview format that Olivia attended with her parents. A detailed assessment of Olivia's biopsychosocial history was gathered. She received a BDD diagnosis because of her stomach obsessionality and compulsive behaviors (e.g., mirror checking, feeling her stomach, seeking reassurance, camouflaging stomach with baggy clothes, and gymnastics avoidance due to appearance concerns). She did not receive an Eating Disorder NOS diagnosis, even though concerns with the stomach can present in eating disorders. Olivia was not concerned with her weight, caloric intake, nor had food obsessions or compensatory behaviors to prevent weight gain. She obsessed only about the flatness of her stomach, its imperfection, and how it made her appear "ugly."

During the first individual therapy session, further assessment focused on details of Olivia's symptoms, specifically the type of thoughts experienced and the behavioral rituals she engaged in. The range of frequency, intensity, and duration of intrusive thoughts and time spent in rituals was gathered to ascertain level of severity. This was collected using the *Body Dysmorphic Disorder Questionnaire* [for Adolescents] (BDDQ) (Phillips, 2005). The BDDQ is a brief self-report screening measure that assesses relative concerns about appearance, distress, and impairment.

After two sessions the *Appearance Anxiety Inventory* (AAI) was administered examining Olivia's perspective of her overall body image and appearance (Veale et al., 2014). The AAI is a self-report measure that examines BDD symptoms. It consists of 10 questions that rank answers on a 5-point Likert scale. The score is generated from totaling the items with a total possibility of 40. The average score for someone with BDD is 27.

Case conceptualization

Olivia displayed distortions in cognitive processing that are commonly seen in BDD. These included: all or none thinking ("My stomach needs to be flat"); catastrophizing ("If my stomach isn't flat, no one will find me attractive"); mental filtering ("It doesn't matter that some days I look okay"); mind reading ("I know that other people won't like me"); emotional reasoning ("If I feel anxious about my body, it must not look okay"). At this point, it was evident Olivia would benefit from specific CBT strategies: cognitive restructuring to teach balanced thinking patterns and perceptual retraining to teach alternate ways of viewing herself physically and perceptually. CBT methods also included Socratic questioning and downward arrow technique. These tools demonstrated that Olivia was hyperfocused on having a flat stomach for several reasons: avoiding scrutiny from others; being accepted by peers; personal preference of a flat stomach as attractive. As BDD intrusive thoughts persisted, Olivia exhibited behavioral rituals of seeking reassurance (e.g., mirror-checking and comfort from her friends/family). This resulted in exacerbation of her overall well-being and emotional regulation. She reported higher levels of distress after eating and engaging in ritualistic behavior while getting ready before school, but mostly after eating dinner. Reportedly, intrusive thoughts and rituals occurred less during the day, especially being distracted at school. However, she soon described an increase in difficulty resisting mirror-checking after lunch (whether in school or remote-learning). Bedtime was most difficult for Olivia, causing dysregulation for approximately half-hour to an hour and involved both parents trying to help by providing words of encouragement about her appearance. Parental accommodation support became a treatment component.

Given Olivia's presenting symptoms, she would most benefit from CBT incorporating specific tools for BDD (Wilhelm et al., 2013). The core components of CBT-BDD involve psychoeducation, ERP, perceptual retraining, emotion regulation, and medication management. Many studies have found CBT to have consistently significant outcomes (McKay et al., 1997; Neziroglu, McKay, Todaro, & Yaryura-Tobias, 1996; Veale et al., 1996). Medication management was introduced a few weeks into treatment after providing CBT-BDD first and then reassessing symptom severity, distress, and progress.

Course of intervention and treatment progression

Treatment protocol

Research specifically designed to examine the application of CBT-BDD specific treatment found the majority of patients improved in BDD symptom severity, delusionality, and depression supporting empirical data of effectiveness (Wilhelm, Phillips, Fama, Greenberg, & Steketee, 2011). Approaches proposed by current treatment manuals and one that will be soon out for release were reviewed for recommendations and guidelines for treatment (Khemlani-Patel & Neziroglu, 2021; Veale & Neziroglu, 2010; Wilhelm et al., 2013).

First sessions

Psychotherapy consisted of 45 individual hour-long sessions over 28 weeks (~7 months). Throughout the course of treatment, frequency shifted from 2×/week to 1×/week. The first few sessions provided psychoeducation of symptoms and treatment, case formulation, assessment and goal setting. Notably, emphasis was placed on building a positive therapeutic alliance and engaging Olivia in treatment. Psychoeducation involved describing common BDD symptoms, prevalence, and treatment expectations. A key point discussed recognizing the distinction between body image and appearance. Each session began with assessing Olivia's Subjective Units of Distress Scale (SUDS) level of anxiety and mood as well as a self-report (1–10 scale) range of frequency, intensity, and duration of intrusive thoughts and rituals. The session content changed based on Olivia's progress, wellbeing, and clinical needs. Each session ended with an emphasis on homework to solidify these skill practices. Homework compliance has been associated with better outcomes in OCD and anxiety (Wilhelm et al., 2013).

Cognitive-behavioral therapy

The following sessions focused on introducing the foundations of CBT and the core treatment components. This included teaching Olivia to understand and identify unhelpful thinking patterns she experienced. Cognitive restructuring or "reframing" was introduced and practiced with collaboratively completing thought records (see Fig. 1 for an example of the thought record that was created for Olivia). Olivia's core fear was identified utilizing the downward arrow technique. This uncovered her ultimate fear: not being accepted by peers if she did not have a flat stomach (see Fig. 2 for an outline of

WHEN I HAVE THE URGE TO CHECK:

1) Identify Emotion:
   a. Level (1-10):

2) What was going through my mind just then?
   b. Believe (1-10):

3) What is the evidence that makes me think this thought is true?

4) What facts or evidence make me suspect that this isn't completely true?

5) Is there a more helpful way of thinking about myself or this situation?

6) New thought:

Note. Used to challenge intrusive thoughts.

FIG. 1 CBT thought record worksheet.

My stomach needs to be flat
↓
Others might think that I look bad
↓
Others will think I am not attractive
↓
I will be left out/alone
↓
I am not worthy of friends

Note. Portrays the deeper-level core belief.

FIG. 2 CBT downward arrow technique.

the downward arrow). Olivia's level of belief or OVI was at moderate level portraying how she felt stuck with this belief, yet was motivated to overcome her symptoms. Typically, a key component of BDD treatment is ensuring the individual is engaged due to their ambivalence; however, Olivia's willingness and compliance was consistent throughout (Neziroglu, 2008). OVI has been found to be a predictor of treatment outcome; therefore, challenging her beliefs and lowering her OVI was a key treatment component (Neziroglu, Stevens, Yaryura-Tobias, & McKay, 2001).

Exposure response prevention

The next phase of treatment incorporated ERP, which usually target's an individual's avoidant behaviors or avoidance of situations; however, Olivia did not avoid social situations. Instead she continued her routine of attending school, spending time with friends, and after school activities but with a heightened level of distress. Olivia exhibited behaviors of mirror checking, comparing appearance with others, and watching TikToks of other girls' whose bodies she envied.

ERP for Olivia involved imaginal exposures, mirror retraining, wearing crop-tops, and not seeking reassurance. An imaginal exposure involved the idea of accepting her ultimate core fear: not having a flat stomach. We discussed details of a life where this disastrous consequence was real and practiced strategies of coping ahead. Coping ahead involved identifying affirmations to build her confidence and balanced thinking. Another ERP exercise had Olivia wear crop-tops during the session, and purposely walk around in front of family members. Olivia also wore a crop-top while practicing mirror-retraining exercises during session. The final ERP exercise focused on perceptual retraining. This specific CBT-BDD component teaches a broadening of one's focal point from looking at specific details of their body (especially flaw-related) to viewing their whole self at once. This concept reinforces the idea that when we only focus on small details, our visual perception distorts the actual visual image we see (e.g., exaggerates or minimizes). This is also confirmed by the confirmation bias concept that when we are looking for a flaw that is exactly how we see the body part. Notably, during the time she spent 2 weeks attending remote learning, mirrors were temporarily removed in her bedroom. Eventually, she began to use the mirrors effectively for brushing her teeth, blow-drying hair, etc.

Response prevention (RP) required Olivia's practice of resisting urges to look in any mirror or reflective surface to examine her stomach, especially after lunch and dinner. To build Olivia's motivation, we incorporated positive reinforcement. A progress-monitoring chart linked with a reward system was created. For example, Olivia tallied the frequency of urges to "body check." The tally mark was crossed out if she successfully resisted the checking ritual. The leftover tally marks (her reward points) were totaled for each day, and for the week. Based on the total, she received a reward that her parents agreed to. The bonus point became increasingly more challenging based on progress made (see Fig. 3 for an example of this monitoring chart). RP for Olivia also involved resisting the urge to seek reassurance from friends or family, which she would do when distressed from intrusive thoughts. Additionally, Olivia shared that she began

	GOALS	Monday	Tuesday	Wednesday	Thursday	Friday	Saturday	Sunday
Week 1	Less than or equal to 3 checks in a day to earn an extra point							
Week 2								
Week 3								
Week 4								
Week 5								

Note. Chart tracked frequency of mirror-checking behavior based on reward system.

FIG. 3 Ritual monitoring chart.

to not finish meals and sometimes avoid eating certain sweets, so RP involved not limiting foods or meals. An emphasis was also placed on limiting social media activity on Instagram and TikTok and identified appropriate accounts to follow and unhealthy accounts to delete.

Attentional training

Another treatment component introduced retraining perceptual focus. One feature is called attentional training. This involved teaching Olivia how to refocus her attention from an internal to external focus (i.e., thought-based point-of-view to global point-of-view). The purpose of refocusing attention is about learning how to shift one's focus to awareness of external surroundings such as sounds, colors, and sights to help reduce anxiety levels that are triggered from focusing on stress-inducing thoughts (Veale & Neziroglu, 2010). The goal is to help move away from a pattern of self-focused attention that is created from anxiety thoughts and ultimately shifting out of your mind and into your environment.

Emotion regulation

The coping skills introduced were utilized from DBT (Linehan, 2014). These included mindfulness, meditation practice, wise mind theory, distress tolerance tools (temperature tip), and emotion regulation (opposite action). Olivia also developed a coping skills toolkit that had specific soothing items. Individualized coping cards with positive affirmations were created for Olivia to reference when needed.

Parental accommodation

Therapy sessions intermittently involved family participation. Each session either began or ended with debriefing Olivia's mother. Parent involvement focused on supportive ways for the family to help without enabling reassurance seeking behavior. It was important to distinguish the difference between providing reassurance from support, which instead would validate Olivia with confidence to manage her symptoms. As treatment continued, phone consultation was utilized with Olivia's mother as needed to address concerns she would encounter. Eventually, Olivia's mother attended a few individual sessions with a different therapist working on her understanding of BDD symptomology, recommendations, and identifying her own self-soothing techniques.

Medication

At about 7 weeks it was decided that Olivia would benefit from medication to alleviate anxiety intensity, thought rigidity, and frequency of intrusive thoughts. At 10 weeks, Olivia was prescribed Prozac 20mg at her first psychiatric consultation. After 2 weeks of taking Prozac, Olivia began to report less frequency and duration with "sticky" thoughts, although she still experienced an uncomfortable level of distress. At about 13 weeks, Olivia had a follow-up appointment with her psychiatrist who increased her medication to 30mg of Prozac. She began reporting consistently lower levels of distress, experienced less dysregulation at nighttime, and described the ability to shift focus more easily.

Despite working hard to diminish "body checking" in mirrors, Olivia had difficulty going a full day without checking. At some point during therapy, she encountered a bullying incident involving a male peer texting a group of friends stating who he found attractive, remarking that he found Olivia too skinny. A week later at her pediatrician appointment her doctor noted how it was important for her to gain weight according to the average for her age group. These situations contributed to Olivia's motivation to stop "limiting" desserts or sweets and finish her meals. It also seemed to motivate her to further internalize a positive impression of her body image with a more holistic view. Instead of viewing weight and having a flat stomach as her primary goal, she began to focus on sports, after-school activities, being with friends, etc. At her follow-up psychiatric appointment a month later, Olivia and her parents reported an overall uplift in her mood, a reduction of mirror checking, eating without limitations, performing well academically, and being very social.

Treatment maintenance

At 6 months, Olivia demonstrated a significant improvement in overall wellbeing. She reported a reduction in frequency, duration, and intensity of anxiety, intrusive thoughts, and rituals. Olivia's ability to shift her thoughts also increased. As her parents reported their observations, it was an appropriate time to decrease sessions to biweekly. Olivia still experiences subtle intrusive BDD thoughts and urges to body check, but is able to resist these urges, regulate her emotions, and easily

FIG. 4 CBT response prevention contract.

Maintain Gains

1. When I feel stressed, I will use relaxation tools.
    a. Calm Down Body:
        • Breathing
        • Cold Water (1.hold breath 2. bend over 3. very cold water)
    b. Calm Down Mind:
        • Wise Mind
        • Mindfulness

2. I will continue to practice to resist checking.

3. I will recognize when I am stuck in my head & use mindfulness (5-4-3-2-1).

4. I will recognize when my worries want me to avoid something and I will not give in.

5. I will challenge my thoughts/actions when they worry about eating certain foods.

6. I will continue to practice looking at my whole self in the mirror.

7. I will continue to practice overall acceptance of my whole self.

8. I will continue to strive for loving my whole self.

9. I will challenge my thoughts and recognize that these thoughts are not real.

10. I will evaluate how I am doing each week & be honest with myself if I notice any difficulties.

distract herself. To maintain RP, a contract was created highlighting key therapy tools for Olivia to reference (see Fig. 4 for an example). Treatment frequency is expected to continuously fade with medication management.

Clinical recommendations

It is common for BDD symptoms to develop during adolescent years; therefore, it's important to adapt treatment appropriate for the age. This includes modifying language, exposure exercises, content presentation, involving family, and school (see Table 1 for an example). The role of a support system from friends and/or family is significant in providing morale, motivation, and confidence for youth (Khemlani-Patel & Neziroglu, 2021). Given the rise in social media, it is important to assess the child's involvement with social media platforms and the influence from their peers.

Many variables can impact treatment effectiveness such as poor insight, introspection, or delusional beliefs (Wilhelm et al., 2013). One way of measuring treatment outcomes is by identifying an individual's OVI (e.g., someone with high OVI strongly believes there is a defect in their physical appearance). Therefore, OVI level can predict treatment prognosis (Khemlani-Patel & Neziroglu, 2021). Research has shown the most effective treatment protocol for BDD is CBT with ERP in conjunction with proper medication management (Phillips, 2009; Phillips & Hollander, 2008). This is especially important for individuals with high OVI, which can loosen their fixed belief.

Comorbidity is a common factor to be aware of when treating individuals with BDD. Common comorbid diagnoses include eating disorder, depression, generalized anxiety disorder, or even personality disorders. It is important to treat these symptoms first to address treatment willfulness before treating BDD directly (Neziroglu et al., 1996). Another important

TABLE 1 Clinical language modifications.

Clinical term	Phrase used
Cognitive distortion	Unhelpful thinking patterns
Cognitive restructuring	Reframing
Perceptual training	Emotion mind is tricking you
Food restricting	Limiting
Mirror-checking	Body-checks
Intrusive thoughts	Sticky/bully thoughts
Coping skills	Calm down body/mind tools

Note: Age-appropriate phrases used for preteen.

factor to explore is identifying the appropriate number of sessions or at what frequency the most effective treatment is. Additionally, more research is needed in comparing individual therapy to group therapy effectiveness.

Key facts about pediatric BDD

- BDD can develop across the lifespan; however, it most often begins in adolescence with average ages ranging from 14 to about 17 years old affecting both genders equally.
- It is important to adapt treatment for age-appropriate interventions.
- BDD symptoms in adolescents are typically chronic and unremitting; therefore, it is important that the individual receives effective evidence-based treatment.
- It is common for adolescents with BDD have comorbid mental health diagnoses that may include but not limited to eating disorder, depression, generalized anxiety disorder, or personality disorders.
- Individuals with BDD experience deficits in visual perception and visuospatial processing shown by heightened attention to small details, interfering with their overall global perception.
- The role of a support system from friends and/or family is significant in providing morale, motivation, and confidence for youth.
- Research has shown the most effective treatment protocol for BDD is CBT with ERP in conjunction with proper medication management.

Applications to other areas

In this chapter, we shared a treatment summary for a 12-year-old child with BDD. The treatment focused on using an evidence-based CBT-BDD protocol which involves psychoeducation, ERP, perceptual retraining, emotion regulation, and medication management (Wilhelm et al., 2013). We also integrated tools from DBT as needed to help with emotion regulation. Research has shown CBT-BDD to be effective for all ages; therefore, it can be used for treatment across the lifespan and tailored to meet the needs of each individual. Evidence has shown improvement in BDD symptom severity, delusionality, and depression supporting empirical data of effectiveness (Wilhelm et al., 2011). Olivia's symptoms were focused on one specific body part, but it is common for individuals with BDD to struggle with one or multiple areas of their body. It is recommended to complete a thorough diagnostic assessment to ensure that treating providers have a comprehensive understanding of the individual and are planning treatment accordingly. This treatment modality is effective for all symptom severity (Khemlani-Patel & Neziroglu, 2021). In this case study, Olivia presented with BDD without any comorbid conditions; however, comorbidity is a common factor to be aware of when treating individuals with BDD. Common comorbid diagnoses include eating disorder, depression, generalized anxiety disorder, or even personality disorders. It is important to treat these symptoms first to address treatment willfulness before treating BDD directly (Neziroglu et al., 1996).

Summary points

- This case describes the treatment of pediatric BDD.
- Evidence-based treatment incorporated CBT-BDD, ERP, and DBT.
- Consideration of age appropriate content is important for pediatric BDD treatment.
- Treatment demonstrated a significant reduction of BDD thoughts and rituals based on frequency, intensity, and duration of symptoms.
- This case study demonstrates effective CBT interventions and medication management to treat pediatric BDD.

References

Buhlmann, U., Etcoff, N. L., & Wilhelm, S. (2007). Facial attractiveness ratings and perfectionism in body dysmorphic disorder and obsessive-compulsive disorder. *Journal of Anxiety Disorders*, *22*(3), 540–547.

Buhlmann, U., McNally, R., Etcoff, N., Tuschen-Caffier, B., & Wilhelm, S. (2004). Emotion recognition deficits in body dysmorphic disorder. *Journal of Psychiatric Research*, *38*(2), 201–206.

Buhlmann, U., McNally, R., Wilhelm, S., & Florin, I. (2002). Selective processing of emotional information in body dysmorphic disorder. *Journal of Anxiety Disorders*, *16*(3), 289–298.

Buhlmann, U., Teachman, B. A., Naumann, E., Fehlinger, T., & Rief, W. (2009). The meaning of beauty: implicit and explicit self-esteem and attractiveness beliefs in body dysmorphic disorder. *Journal of Anxiety Disorders*, *23*(5), 694–702.

Chosak, A., Marques, L., Greenberg, J. L., Jenike, E., Dougherty, D. D., & Wilhelm, S. (2008). Body dysmorphic disorder and obsessive-compulsive disorder: Similarities, differences and the classification debate. *Expert Reviews in Neurotherapeutics*, *8*(8), 1209–1218.

Deckersbach, T., Savage, C. R., Phillips, K. A., Wilhelm, S., Buhlmann, U., Rauch, S. L., … Jenike, M. (2000). Characteristics of memory dysfunction in body dysmorphic disorder. *Journal of the International Neuropsychological Society*, *6*(6), 673–681.

Feusner, J., Neziroglu, F., Wilhelm, S., Mancusi, L., & Bohon, C. (2010). What causes BDD: Research findings and a proposed model. *Psychiatric Annals*, *40*(7), 349–355. https://doi.org/10.3928/00485713-20100701-08.

Hollander, E., & Allen, A. (2004). Similarities and differences between body dysmorphic disorder and other disorders. *Psychiatric Annals*, *34*(12), 927–933.

Khemlani-Patel, S., & Neziroglu, F. (2021). Body dysmorphic disorder. In *Advances in psychotherapy: Evidence-based practice series*. Boston: Hogrefe Publishing (in press).

Linehan, M. M. (2014). *DBT training manual*. New York, NY: The Guilford Press.

Mckay, D., & Neziroglu, F. (2011). Body dysmorphic disorder. In B. B. Brown, & M. J. Prinstein (Eds.), *Vol. 3. Encyclopedia of adolescence* (pp. 85–89). San Diego: Academic Press.

McKay, D., Todaro, J., Neziroglu, F., Campisi, T. T., Moritz, E. K., & Yaryura-Tobias, J. A. (1997). Body dysmorphic disorder: A preliminary evaluation of treatment and maintenance using exposure with response prevention. *Behaviour Research and Therapy*, *35*(1), 67–70.

McLeod, S. A. (2018). *Nature vs. nurture in psychology. Simply psychology*. https://www.simplypsychology.org/naturevsnurture.html.

Neziroglu, F. (2008). Body dysmorphic disorder: Causes, characteristics, and clinical treatments. *Body Image*, *5*(1), 1–2.

Neziroglu, F., Khemlani-Patel, S., & Veale, D. (2008). Social learning theory and cognitive behavioral models of body dysmorphic disorder. *Body Image*, *5* (1), 28–38.

Neziroglu, F., McKay, D., Todaro, J., & Yaryura-Tobias, J. A. (1996). Effect of cognitive behavior therapy on persons with body dysmorphic disorder and comorbid axis II diagnoses. *Behavior Therapy*, *27*(1), 67–77.

Neziroglu, F., Stevens, K. P., Yaryura-Tobias, J. A., & McKay, D. (2001). Predictive validity of the overvalued ideas scale: Outcome in obsessive-compulsive and body dysmorphic disorders. *Behavior Research and Therapy*, *39*(6), 745–756.

Phillips, K. A. (2002). Body image and body dysmorphic disorder. In C. G. Fairburn, & K. D. Brownwell (Eds.), *Eating disorders and obesity: A comprehensive handbook* (2nd ed., pp. 113–117). New York: Guildford Press.

Phillips, K. A. (2005). *The broken mirror: Understanding and treating body dysmorphic disorder* (rev. and expanded ed.). New York: Oxford University Press.

Phillips, K. A. (2009). *Understanding body dysmorphic disorder: An essential guide*. New York: Oxford University Press.

Phillips, K. A., & Hollander, E. (2008). Treating body dysmorphic disorder with medication: Evidence, misconceptions, and a suggested approach. *Body Image*, *5*(1), 13–27.

Phillips, K. A., & McElroy, S. L. (2000). Personality disorders and traits in patients with body dysmorphic disorder. *Comprehensive Psychiatry*, *41*(4), 229–236.

Phillips, K. A., Menard, W., Fay, C., & Weisberg, R. (2005). Demographic characteristics, phenomenology, comorbidity, and family history in 200 individuals with body dysmorphic disorder. *Psychosomatics*, *46*(4), 317–325.

Phillips, K. A., Menard, W., Pagano, M. E., Fay, C., & Stout, R. L. (2006). Delusional versus nondelusional body dysmorphic disorder: Clinical features and course of illness. *Journal of Psychiatric Research*, *40*(2), 95–104.

Pinto-Wagner, A. (2002). *Worried no more: Help and hope for anxious children* (2nd ed.). USA: Lighthouse Press, Inc.

Veale, D., Boocock, A., Gournaym, K., Dryden, W., Shah, F., Willson, R., & Walburn, J. (1996). Body dysmorphic disorder: A cognitive behavioural model and pilot randomised controlled trial. *Behaviour Research and Therapy*, *34*, 717–729.

Veale, D., Eshkevari, E., Kanakam, N., Ellison, N., Costa, A., & Werner, T. (2014). The appearance anxiety inventory: Validation of a process measure in the treatment of body dysmorphic disorder. *Behavioural and Cognitive Psychotherapy*, *42*(5), 605–616.
Veale, D., & Neziroglu, F. (2010). *Body dysmorphic disorder: A treatment manual*. Chichester, UK: John Wiley & Sons Ltd.
Wilhelm, S. (2006). *Feeling good about the way you look: A program for overcoming body image problems*. New York: Guilford Press.
Wilhelm, S., Phillips, K. S., Fama, J. M., Greenberg, J. L., & Steketee, G. (2011). Modular cognitive behavioral therapy for body dysmorphic disorder. *Behavior Therapy*, *42*, 624–633.
Wilhelm, S., Phillips, K. A., & Steketee, G. (2013). *A cognitive behavioral treatment manual for body dysmorphic disorder*. New York: Guilford.

Chapter 36

Case study: A child with obsessive-compulsive disorder and cognitive-behavioral therapy

Yvette Fruchter, Danielle H. Gardini, and Fugen Neziroglu
Bio Behavioral Institute, Great Neck, NY, United States

Abbreviations

OCD	obsessive-compulsive disorder
ERP	exposure response prevention
CBT	cognitive-behavioral therapy
CY-BOCS	Children's Yale-Brown Obsessive-Compulsive Scale
FASA	family accommodation scale-anxiety
PANDAS	pediatric autoimmune neuropsychiatric disorders
GAS	group A streptococcal infection (GAS)

Theoretical and research basis for treatment

Obsessive-compulsive disorder (OCD) is a debilitating mental health disorder that affects 1%–3% of children and adolescents (Flament et al., 1988; Freeman et al., 2018). OCD includes two primary symptoms: (1) obsessions, which are recurrent, intrusive, and unwanted thoughts that cause distress and (2) compulsions, which are repetitive behaviors or mental acts aimed at reducing distress and/or preventing disastrous consequences (American Psychiatric Association, 2013). OCD is diagnosed when symptoms are time-consuming (e.g., duration is more than 1 hour per day) and cause functional impairment such as interfering with school, friends, or family life (American Psychiatric Association, 2013).

Although OCD affects individuals across the lifespan, there are differences in clinical presentation between children and adults. It can be difficult to identify obsessions in children due to developmental considerations (e.g., verbal communication skills). Therefore, some children may present with ritualistic behaviors without a clear obsession identified. Children also tend to present with worse insight and are more likely to endorse aggressive obsessions than adults with OCD (Selles, Storch, & Lewin, 2014). OCD may interfere with a child's ability to engage in developmentally appropriate activities and can impact a child's developmental trajectory. To illustrate, a child with contamination concerns may refuse to attend play dates at other homes, which can impact independence, social skills, and interpersonal skills development (Borda, Feinstein, Neziroglu, Veccia, & Perez-Rivera, 2013).

Cognitive-behavioral therapy (CBT) is considered the gold-standard treatment for OCD including the pediatric population (Freeman et al., 2018; Pediatric OCD Treatment Study (POTS) Team, 2004; Piacentini & Langley, 2004). Exposure and response prevention (ERP) is a form of CBT that involves two components: (1) exposure: facing one's fears/confronting triggers and (2) response prevention: refraining from compulsions (Foa & Kozak, 1986; March & Mulle, 1998). ERP is recommended as the first-line treatment for children with mild to moderate OCD symptoms (American Academy of Child and Adolescent Psychiatry, 2012). Medications may be added if the child cannot do ERP or tolerate the discomfort without aid. Through ERP, children learn their feared consequences are unlikely to occur, rituals are unnecessary, and they can cope effectively when facing their triggers. Different theories explain the effective mechanisms for ERP including the habituation model (Foa & Kozak, 1986) and the inhibitory learning model (Craske, Treanor, Conway, Zbozinek, & Vervliet, 2014).

Treatment often incorporates family members given that childhood OCD typically becomes a family-wide issue (Calvocoressi et al., 1995; Merlo, Lehmkuhl, Geffken, & Storch, 2009). Families tend to experience an increase in the stress level within the home environment when a child is struggling. Family accommodation refers to family participation

Handbook of Lifespan Cognitive Behavioral Therapy. https://doi.org/10.1016/B978-0-323-85757-4.00027-4

in rituals and family modification of routines due to the child's anxiety (Calvocoressi et al., 1995; Van Noppen & Steketee, 2003). An estimated two thirds of families accommodate their child's OCD symptoms (Storch et al., 2007). Parents typically accommodate OCD with the intention of helping their child. However, accommodation reduces distress in the short-term but maintains OCD symptoms in the long-term. Research indicates that targeting family accommodation greatly improves child treatment outcomes (Barrett, Healy-Farrell, & March, 2004; Merlo et al., 2009). This case study focuses on navigating the complexities of CBT for pediatric OCD.

Case introduction

"Jane" is a 10-year-old Caucasian female in the fifth grade. She lives with her biological parents and younger brother who is 7 years old. Jane enjoys school and performs well academically. She is shy in front of peer groups but has several good friends in her class. Jane's hobbies include art, reading, and video games.

Presenting complaints

At age 10, Jane became ill with a stomach virus during the winter. Parents reported that Jane complained of not feeling well and presented with physical symptoms (e.g., nausea, fever, sweats, vomiting). Jane stayed home from school and her parents gave her over-the-counter medication for symptom management. Jane reported feeling better after 3 days and by the fourth night Jane no longer exhibited virus symptoms. Jane returned to school, resumed hobbies, and continued socializing with peers. However, she began to voice fears about becoming ill and vomiting. She appeared highly sensitized to germs and exhibited discomfort related to "feeling dirty." Jane sought reassurance from her parents (e.g., "Am I going to get sick again? Does this have germs on it?"). She began avoiding areas in the home where she had vomited and refused to touch her brother who she associated as being "gross and germy." She engaged in excessive hand washing, reassurance seeking, showering, and changing clothes repeatedly. Over several months, Jane's avoidant and compulsive behaviors grew to such an extent that she threw garbage on the floor to avoid using the garbage can (i.e., due to fearing the garbage can was contaminated) and would spit her saliva out when she became concerned it had germs. She developed a method of taking clothes from the dresser to ensure they did not become contaminated by touching other surfaces.

Parents sought therapy approximately 4 months after Jane began exhibiting obsessive-compulsive symptoms. Concerns developed from Jane's ritual engagement leading to frequent school lateness and reduced homework completion. Jane's hands also became red, dry, and irritated. She started demanding her family increase cleaning, laundering, and hand washing. Parents reported mental health symptoms to Jane's pediatrician who referred Jane for therapy.

History

Reportedly, Jane was born full-term without complications. Parents indicated Jane met developmental milestones within normal limits. Family denied any major mental health issues though her mother endorsed mild to moderate anxiety about day-to-day stressors. Jane had no prior experience with therapeutic treatment or medication.

Assessment

Jane attended the clinical consultation with her parents, where a semistructured interview was conducted to gather clinical information pertinent to presenting concerns. Informed consent for treatment was obtained. Assessments focused on measuring the frequency, duration, and intensity of symptomatology.

The clinician administered the Children's Yale-Brown Obsessive-Compulsive Scale (CY-BOCS; Scahill et al., 1997) and the Family Accommodation Scale-Anxiety (FASA; Lebowitz et al., 2013). The CY-BOCS is a clinician-rated instrument that assesses obsessions and compulsions on a Likert scale. Scores on the CY-BOCS range from 0 to 40 with a higher score indicating higher severity. The FASA is a 9-item scale that has parents rate the level and impact of accommodation on a Likert scale. FASA scores range from 0 to 36 with a higher score indicating greater accommodation.

Case conceptualization

Jane presented with obsessions and compulsions that caused significant distress, was time-consuming, and resulted in overall functional impairment. Obsessions consisted of intrusive thoughts about germs, dirt, and nausea/vomiting.

Common intrusive thoughts for Jane included: "If I don't wash my hands, I will become sick from germs" and "I will vomit if I eat candy!" Jane experienced heightened anxiety and attached significant meaning to these intrusive thoughts.

Jane engaged in repetitive behaviors aimed at decontamination and preventing illness/nausea: washing hands, changing clothes, showering, cleaning surfaces and/or household items. She also engaged in avoidance to prevent contact with dirt and germs: avoiding areas of the house where she vomited, candy, doorknobs, and public surfaces. Jane's family engaged in a variety of behaviors aimed to help Jane feel better: increasing frequency of laundry, providing reassurance, handing her a towel after the shower, and assisting with putting shoes on (i.e., family accommodation). The rituals, avoidance, and family accommodation served to mitigate Jane's anxiety in the short-term however perpetuate the cycle in the long-term.

Differential diagnosis

Specific phobia. Differential diagnosis between specific phobia (i.e., emetophobia) and obsessive-compulsive disorder was assessed. Specific phobia criteria includes: marked anxiety about a specific object or situation; phobic situation always provokes immediate fear; phobic situation is avoided or endured with intense anxiety; anxiety is out of proportion to the actual danger posed by specific situation; anxiety is persistent, causes clinically significant distress, and not better explained by another disorder (American Psychiatric Association, 2013). Jane was not diagnosed with specific phobia as she endorsed fears beyond vomiting (e.g., germs, dirt, illnesses) and engaged in numerous repetitive rituals (e.g., hand washing, showering, changing clothes, etc.) in addition to avoidance behaviors. Fear of vomiting for Jane became linked to the generalization of objects, places, and situations that were considered "contaminated." Jane's exhibition of compulsions in addition to avoidance behaviors contributed to ruling out the specific phobia diagnosis. A specific phobia diagnosis involves merely passive avoidance, while conversely in OCD there is both passive and active avoidance, the latter referring to compulsions. It is the presence of active avoidance (i.e., compulsions) that led to the diagnosis of OCD.

PANDAS/PANS. Pediatric autoimmune neuropsychiatric disorders (PANDAS) refer to presence of OCD and/or tic disorder related to group A streptococcal infection (GAS) (Swedo et al., 1998). Hallmark indicators of PANDAS include sudden and acute onset of OCD symptoms. PANDAS was ruled out as Jane's symptoms intensified gradually over the course of several months.

Course of treatment and assessment of progress

CBT was conducted over an 8-month period: twice weekly sessions for 4 months; weekly sessions for 2 months; switching to biweekly as Jane's symptoms improved.

Rapport building and psychoeducation

The therapist incorporated games and discussion about Jane's interests to create a positive and comfortable atmosphere. Psychoeducation was presented using language that is understandable to preteens. Therapist introduced the concept of "worry bully"; essentially a way of helping Jane understand how she is being pushed around by obsessions. Jane was encouraged to name her worry bully. The "worry bully" concept was used as a way to communicate and identify intrusive OCD thoughts (Wagner, 2011). Therapist also introduced the concept of "talking back to worry bully" as a way of encouraging Jane to challenge OCD. Therapist coached Jane and her parents to identify the "worry bully" and practice "talk-back" tools. Additionally, the therapist collaborated with Jane to create a personalized feelings thermometer to measure distress level.

Functional analysis

Therapist conducted a functional analysis of Jane's symptoms to understand the relationship between triggers, obsessions, compulsions, and avoidance. A "symptom map" was created to catalogue Jane's symptoms and demonstrate how they relate to one another. Through the functional analysis, Jane's core fear was identified: contracting an illness that causes vomiting and not being able to handle the discomfort, anxiety and disgust. This fear manifests into Jane's concerns about contact with vomit, dirt, and germs.

Designing exposures & building motivation

Exposures were designed to challenge Jane's expectations regarding feared outcomes in accordance with inhibitory learning theory, e.g., "If I wear clothing that touches the floor, I will feel nauseous within the next hour" (McGuire & Storch, 2019). According to inhibitory learning theory, avoidance and rituals prevent new learning from occurring; Since Jane always changed her pants when they touched the floor she continued to believe that keeping the "contaminated" pants on would cause nausea. Therefore, exposures are opportunities for patients to test negative predictions and learn new associations, e.g., "I do not always feel nauseous after wearing clothing that touched the floor."

Therapist collaborated with parents and Jane to create an exposure hierarchy (see Table 1). Jane rated triggers from 0 (least distressing) to 10 (most distressing). Therapist and Jane started with exposures that triggered mild anxiety to build up Jane's confidence and set her up for success. Exposures gradually increased in intensity. Jane was encouraged to choose each exposure as the therapist noted Jane was more willing to participate if she selected the activity.

TABLE 1 Sample items from Jane's exposure hierarchy.

Situation/trigger	SUDS/worry rating
Touching home items	
"Contaminated" thermos	3
Throwing garbage in garbage can	7
Touching dresser drawers	7
Touching floor in bedroom	8
Touching surfaces in home that were vomited on	10
Touching bathroom doorknob	
Touching tissue that touched doorknob	5
Touching doorknob with one finger	6
Touching doorknob with whole hand	7
Grooming related	
Using hand towel after washing hands	6
Using body towel after showering	6
Putting shoes on	7
Public places	
Touching tissue to store doorknob	6
Touching one finger to doorknob	7
Touching whole hand to doorknob	8
Touching elevator buttons in office building	9
Vomit-related	
Making vomit noises	5
Listening to vomit noises	6
Imagining what would happen if I vomited	7
Watching clip of someone vomiting	7
Making pretend vomit	7
Touching pretend vomit	8

Note. SUDS = subjective use of distress scale on a 0–10 rating; 10 being higher distress.

My Brave Activity Challenge

What will I do?

When will I do it?

Who will be there with me?

What will I do if I get really scared?

What if I don't want to do it?

My Fun Activity (reward after my brave activity):

Monitoring My Progress:

	Sun	Mon	Tues	Weds	Thurs	Fri	Sat
Did I do my brave A? (Yes/ No)							
Worry Rating Before (0-10)							
Worry Rating After (0-10)							
Did I do my fun Activity? (Yes/No)							
Point Earned							

____ Points = _______ Prize

Note. Chart used to plan and track frequency of exposure practice based on a reward system.

FIG. 1 Exposure practice & monitoring chart.

Therapist framed exposures as opportunities to engage in brave behaviors. The therapist emphasized that bravery occurs when one chooses to face challenges when feeling afraid. Therapist encouraged Jane to think of previous situations when she acted bravely and consider when she encountered challenges that became easier through practice.

Exposure therapy

Exposures were conducted at the therapy office and at Jane's home. For the first exposure, Jane brought a thermos she was avoiding because someone coughed near it. Therapist encouraged Jane to touch and drink from the thermos over a 45-min period. Jane's initial worry rating of 4 increased to 6 at the peak and decreased to 2 by end of session. Jane verbalized her fear: "I'm worried I have germs on me and will become sick by tonight." Therapist encouraged Jane to talk back to worry bully and coached her to face her fear without ritualizing. By the end of session, Jane declared, "I don't feel nauseous at all!" Therapist sent Jane and her parents home with ritual prevention instructions to ensure she did not "undo" the exposure (i.e., waiting at least 45 min to wash hands). Homework practice was assigned (e.g., touch and drink from thermos daily). See Fig. 1 for exposure practice and monitoring handout.

During exposures the therapist and client touched "contaminated" objects and surfaces. Therapist had Jane rate her worry every few minutes, identify worry thoughts, and practice talking back to worry bully. While debriefing together, questions were asked to cement new learning: "What did you learn?" "Did the worst case scenario happen?" Jane's parents were involved in exposure sessions and designated as "coaches" to help Jane practice between sessions. New exposures were selected after Jane's worry she demonstrated new learning (e.g., "I can touch the floor without vomiting or getting overwhelmed by feeling dirty!").

Exposures were titrated to meet Jane where she was at. For instance, Jane refused to touch the bathroom doorknob so the exposure was altered to match Jane's willingness. Jane agreed to touch a tissue that touched the bathroom doorknob. After several minutes of holding this tissue, Jane expressed feeling more comfortable with the challenge. She expressed willingness to touch the doorknob directly with one finger, then two, and so on until she placed her hand on the doorknob.

Flooding exercises were used: as the client touched the contaminated object, the therapist verbalized Jane's feared disastrous consequence in detailed steps. "We all know that everyone who touches doorknobs gets sick and vomits immediately or shortly after," the therapist would say. This irreverence demonstrated how highly improbable the likelihood of vomiting after touching the doorknob actually was. Client often laughed and eventually even joked back ("Oh yes! That's why everyone is always vomiting!") showing how she understood the exaggeration.

The therapist also focused on cross-contamination in exposure exercises as Jane's OCD led to rigid demarcations between "clean" and "dirty" areas. Therapist encouraged Jane to touch the doorknob and then touch her clothes, skin, other surfaces, etc. The goal of the cross-contamination was to blur the lines between "clean" and "dirty" so Jane learns she can handle discomfort without cleaning.

Treatment challenges

Several notable roadblocks were encountered during treatment. Firstly, the therapist noticed changes in Jane's demeanor at home compared to the therapy office. At the office, Jane was mild-mannered and agreed to participate in exposures. However, Jane exhibited more outbursts, refusal to participate, and arguments with family during home visits. For example, Jane selected an exposure for a home visit: opening dresser drawers without rituals. When the therapist arrived at the house for the session and reminded Jane of their plan, Jane rolled her eyes and complained, "I'm not touching my dirty drawers." Jane's mother scolded Jane, which led Jane to run to her bedroom, slam the door, and refuse to participate for the remainder of session. The therapist worked with parents to implement guidelines on de-escalation strategies (e.g., active ignoring) and instituted a reward system for at-home exposures such as playing a board game at the end of session contingent upon Jane's participation in exercises.

Another roadblock involved Jane not completing therapy assignments between sessions. Reminders such as visual cues or prompts from parents were unsuccessful. Therapist suggested instituting a behavioral chart so Jane could earn prizes for completing practice goals. Jane's homework compliance increased as she was excited to earn prizes like art supplies.

Response prevention

A ritual prevention hierarchy was created to address rituals in a systematic and gradual fashion. Jane first agreed to reduce hand washing when she touched "contaminated" objects in her bedroom. Once Jane met this goal, she and therapist explored the next step to build on her success. Jane consented to reduce hand washing in shared spaces of the home (starting with the living room and dining room). Jane monitored ritual prevention goals in a behavior chart, similar to the chart for exposure practice. Priority ritual targets included showering, hand washing, and changing clothes, which led to school lateness and disruptions with bedtime routine. The therapist also helped Jane distinguish between typical versus atypical behavior (e.g., washing hands after using bathroom vs. washing hands after touching doorknob) and utilized Socratic Questioning to explore why Jane took extra safeguards that others did not (i.e., Why do you have to change your clothes if they touch the floor but your friends don't? Where are the facts that touching doorknobs or other objects make people sick? Do people vomit every time they get sick?).

Parent training/addressing accommodation

Parent sessions were conducted biweekly to target family accommodation and offer parental support. The therapist provided psychoeducation on how accommodation reinforces the OCD cycle despite parents' best intentions to protect their children. Rationale was provided to move away from accommodation and towards supporting Jane to face fears and use healthy coping skills to manage distress. Therapist collaboratively mapped out accommodation behaviors with the parents and explored ways to slowly fade out family participation in rituals. Parents were encouraged to work as a team to support one another and tolerate Jane's expression of distress.

Treatment progress & maintenance

Jane met with the therapist twice weekly for 16 weeks, which included office and home visits. By week 16, Jane exhibited demonstrable improvement: she was on time for school and bedtime and placed less demands on family to participate in rituals. Yet, she continued to wash hands excessively, avoid areas where she had vomited, distance herself from anyone that appeared sick, and express concerns about illness. Therapy transitioned to once weekly sessions to address remaining symptoms.

Remaining exposures targeted "higher-level" fears: touching fake vomit, watching movie clips of someone vomiting, touching areas associated with vomit or sickness, touching surfaces in public places. With each success, Jane continued to

My Bravery Checklist

1. **Worry Thermometer:**
   Rating my worry level will help me decide how to boss back Worry Bully
2. **Use My Support:**
   Tell my parents when Worry Bully pops up so we can team up together
3. **Talk Back to Worry Bully**
   Use my 'Talk Back' Tools to boss back Worry Bully
4. **Do the Opposite of what Worry Bully says:**
   Face my fear by doing the opposite of what Worry Bully commands
5. **Be Brave:**
   Bravery is when I do something scary even if I am afraid! I have been brave many times before and I can do it again!
6. **Avoid Avoidance:**
   Continue to eat candy, touch things at home and at other places that I am scared are dirty, wear all my clothes, and keep facing my fears
7. **Reward:**
   Reward myself after challenges to celebrate my hard work!
8. **Recognize My Success:**
   It's important for me to notice my successes so I stay motivated and realize I am doing a great job
9. **Learn From Slips:**
   I will make mistakes and that's ok! I can learn from slips and think about how to handle things differently next time
10. **Forget about Perfection:**
   Perfection does not exist -- I will do the best I can to use my tools and boss back Worry Bully

Note. Guidelines and tips for relapse prevention and maintaining treatment gains

FIG. 2 Jane's bravery checklist.

solidify her new understanding that she was unlikely to get sick and could handle her discomfort without rituals. Jane shared, "I don't need to follow the OCD rules anymore. OCD bullied me into thinking I am weak. Even if I do get sick, I know I will get better and everyone gets sick sometimes." As Jane's functioning improved, she increased participation in play dates and extracurricular activities. At week 30, therapist and Jane reviewed and re-ranked the hierarchy items. Therapist and Jane created "Jane's Bravery Checklist," which listed reflections from therapy and tips to beat the worry bully. At the end of treatment, Jane shared that while she experiences occasional worry thoughts about illness, she is able to use her skills, talk back to worry bully, and team up with her family to face fears (Fig. 2).

Treatment implications

This case study illustrated CBT treatment for a 10-year-old girl with contamination-related obsessions. ERP is a highly effective treatment for children and adolescents with OCD (Freeman et al., 2018). Treatment should be tailored to a child's developmental stage for it to be tolerable, understandable, and beneficial. Games, play, and crafts are useful ways to engage the child and increase willingness to participate. Psychoeducation must involve age-appropriate language and concepts (e.g., "worry bully") to keep the child's attention and interest. The clinician should foster a collaborative discussion through asking open-ended questions, using metaphors, and sharing stories rather than lecturing or presenting information. When possible, incorporating humor can de-escalate tension and encourage children to persevere through discomfort. Family involvement in treatment is essential and parents play a vital role as coach, cheerleader, and role model.

Lack of willingness and motivation can present a major barrier to treatment with children. It is important to address children's ambivalence to participate in treatment. The therapist should strive to understand the child's specific circumstances as there may be numerous reasons why a child is disinterested in participating. Children may be less interested in making changes when family accommodation is occurring; therefore, addressing accommodation and working with parents first may be necessary. Additionally, instituting reward systems for participation can bolster compliance. Children also experience increased self-efficacy and a greater sense of control when they collaborate in designing and choosing exposure exercises in therapy.

Home visits are helpful for addressing symptoms in the child's typical environment. Therapists can work with parents to plan ahead for home visits to address possible concerns. Preparing the child for the home visit and notifying them in advance increases likelihood of success. When possible and appropriate, home visits offer unique insight into the child's behavior at home and allow for exposures to be conducted in a real-world setting.

Finally, relapse prevention plans help families maintain gains after treatment and anticipate warning signs of possible symptom exacerbation. As OCD is a chronic condition, the symptoms wax and wane over time making it possible for symptoms to resurface, especially during stressful periods. More research is warranted to examine predictors of treatment outcome, which factors mitigate symptom severity, and likelihood of symptoms to persist into adulthood.

Key facts about pediatric OCD

- OCD affects between 1%–3% of all children and adolescents.
- Children tend to present with worse insight than adults with OCD.
- Cognitive behavioral therapy is the recommended evidence-based treatment for pediatric OCD.
- Therapy often incorporates family members given that childhood OCD typically becomes a family-wide issue.
- Research indicates that targeting family accommodation greatly improves pediatric OCD treatment outcomes.

Applications to other areas

In this chapter, we shared a treatment summary for a 10-year-old child with OCD. The treatment focused on exposure and response prevention and did not include a psychiatric medication regimen. Of note, while some children and adolescents may successfully recover from OCD through therapy alone, many may benefit from a combination of psychiatric medication and exposure therapy (American Academy of Child and Adolescent Psychiatry, 2012). The treatment summarized in this chapter can be applied to children who are prescribed psychiatric medication as well. It is important for a therapist to collaborate and communicate with the prescribing psychiatrist when conducting exposure therapy with a child who takes psychiatric medicine.

Additionally, the child described in this chapter presented solely with OCD without comorbid conditions. However, many children and adolescents may present with multiple diagnoses, such as other anxiety disorders, depression, attention deficit disorder, and autism spectrum disorder. In the case of comorbid mental health concerns, it is important to complete a thorough diagnostic assessment to ensure that treating providers have a comprehensive understanding of the individual. While Jane struggled with contamination-based OCD, there are many other obsessional themes such as aggressive, religious, and symmetrical themes. The treatment described here can be applied to other obsessional themes as well.

Summary points

- This chapter describes the therapy treatment course for a 10-year-old child with OCD.
- ERP involves facing feared triggers over an extended period of time without rituals.
- After recovering from a stomach virus, "Jane" expressed concern about becoming sick again. Jane gradually exhibited obsessions and compulsions related to a fear of becoming ill and vomiting.
- Differential diagnosis between specific phobia and obsessive-compulsive disorder was explored. Jane's exhibition of compulsions in addition to passive avoidance contributed to ruling out specific phobia diagnosis.
- During therapy sessions, Jane intentionally confronted "contaminated" objects and situations while refraining from rituals. Homework was assigned between sessions.
- Jane's distress declined as she learned her worse-case scenario is unlikely to occur and she can cope with difficult emotions.
- Family involvement in treatment was essential and parent sessions conducted biweekly targeted family accommodation and parental support.

References

American Academy of Child & Adolescent Psychiatry. (2012). Practice parameter for the assessment and treatment of children and adolescents with obsessive-compulsive disorder. *Journal of the American Academy of Child and Adolescent Psychiatry*, *51*, 98–113. https://doi.org/10.1016/j.jaac.2011.09.019.

American Psychiatric Association. (2013). *Diagnostic and statistical manual of mental disorders* (5th ed.). https://doi.org/10.1176/appi.books.9780890425596.

Barrett, P., Healy-Farrell, L., & March, J. (2004). Cognitive-behavioral family treatment of childhood obsessive-compulsive disorder: A controlled trial. *Journal of the American Academy of Child and Adolescent Psychiatry*, *43*(1), 46–62.

Borda, T., Feinstein, B. A., Neziroglu, F., Veccia, T., & Perez-Rivera, R. (2013). Are children with obsessive-compulsive disorder at risk for problematic peer relationships? *Journal of Obsessive-Compulsive and Related Disorders*, *2*(4), 359–365. https://doi.org/10.1016/j.jocrd.2013.06.006.

Calvocoressi, L., Lewis, B., Harris, M., Trufan, S. J., Goodman, W. K., McDougle, C. J., & Price, L. H. (1995). Family accommodation in obsessive-compulsive disorder. *American Journal of Psychiatry*, *152*(3), 441–443. https://doi.org/10.1176/ajp.152.3.441.

Craske, M. G., Treanor, M., Conway, C. C., Zbozinek, T., & Vervliet, B. (2014). Maximizing exposure therapy: An inhibitory learning approach. *Behaviour Research and Therapy*, *58*, 10–23.

Flament, M. F., Whitaker, A., Rapoport, J. L., Davies, M., Berg, C. Z., Kalikow, K., … Shaffer, D. (1988). Obsessive compulsive disorder in adolescence: An epidemiological study. *Journal of the American Academy of Child and Adolescent Psychiatry*, *27*(6), 764–771. https://doi.org/10.1097/00004583-198811000-00018.

Foa, E. B., & Kozak, M. J. (1986). Emotional processing of fear: Exposure to corrective information. *Psychological Bulletin*, *99*(1), 20–35. https://doi.org/10.1037/0033-2909.99.1.20.

Freeman, J., Benito, K., Herren, J., Kemp, J., Sung, J., Georgiadis, C., … Garcia, A. (2018). Evidence base update of psychosocial treatments for pediatric obsessive-compulsive disorder: Evaluating, improving, and transporting what works. *Journal of Clinical Child and Adolescent Psychology*, *47*(5), 669–698. https://doi.org/10.1080/15374416.2018.1496443.

Lebowitz, E. R., Woolston, J., Bar-Haim, Y., Calvocoressi, L., Dauser, C., Warnick, E., … Leckman, J. F. (2013). Family accommodation in pediatric anxiety disorders. *Official Journal of the Anxiety and Depression Association of America*, *30*(1), 47–54. https://doi.org/10.1002/da.21998.

March, J. S., & Mulle, K. (1998). *OCD in children and adolescents: A cognitive-behavioral treatment manual* (1st ed.). New York: Guilford Press.

McGuire, J. F., & Storch, E. A. (2019). An inhibitory learning approach to cognitive-behavioral therapy for children and adolescents. *Cognitive and Behavioral Practice*, *26*(1), 214–224. https://doi.org/10.1016/j.cbpra.2017.12.003.

Merlo, L. J., Lehmkuhl, H. D., Geffken, G. R., & Storch, E. A. (2009). Decreased family accommodation associated with improved therapy outcome in pediatric obsessive-compulsive disorder. *Journal of Consulting and Clinical Psychology*, *77*(2), 355.

Pediatric OCD Treatment Study (POTS) Team. (2004). Cognitive-behavior therapy, sertraline, and their combination for children and adolescents with obsessive-compulsive disorder: The pediatric OCD treatment study (POTS) randomized controlled trial. *Journal of the American Medical Association*, *292*(16), 1969–1976.

Piacentini, J., & Langley, A. K. (2004). Cognitive-behavioral therapy for children who have obsessive-compulsive disorder. *Journal of Clinical Psychology*, *60*(11), 1181–1194. https://doi.org/10.1002/jclp.20082.

Scahill, L., Riddle, M. A., McSwiggin-Hardin, M., Ort, S. I., King, R. A., Goodman, W. K., … Leckman, J. F. (1997). Children's Yale-Brown Obsessive Compulsive Scale: Reliability and validity. *Journal of the American Academy of Child & Adolescent Psychiatry*, *36*(6), 844–852.

Selles, R. R., Storch, E. A., & Lewin, A. B. (2014). Variations in symptom prevalence and clinical correlates in younger versus older youth with obsessive-compulsive disorder. *Child Psychiatry & Human Development*, *45*(6), 666–674.

Storch, E. A., Geffken, G. R., Merlo, L. J., Mann, G., Duke, D., Munson, M., … Goodman, W. K. (2007). Family-based cognitive-behavioral therapy for pediatric obsessive-compulsive disorder: Comparison of intensive and weekly approaches. *Journal of the American Academy of Child and Adolescent Psychiatry*, *46*(4), 469–478.

Swedo, S. E., Leonard, H. L., Garvey, M., Mittleman, B., Allen, A. J., Perlmutter, S., … Dubbert, B. K. (1998). Pediatric autoimmune neuropsychiatric disorders associated with streptococcal infections: Clinical description of the first 50 cases. *American Journal of Psychiatry*, *155*, 264–271.

Van Noppen, B., & Steketee, G. (2003). Family responses and multifamily behavioral treatment for obsessive-compulsive disorder. *Brief Treatment & Crisis Intervention*, *3*(2), 231–248.

Wagner, A. (2011). Cognitive-behavioral therapy for children and adolescents with obsessive-compulsive disorder. In R. Hudak, & D. Dougherty (Eds.), *Clinical obsessive-compulsive disorders in adults and children* (pp. 138–151). Cambridge: Cambridge University Press (chap. 10).

Chapter 37

Case study. Cognitive-behavioral therapy for nightmare disorder in a youth with comorbid autism spectrum disorder

Lisa D. Cromer[a], Devin R. Barlaan[a], Kristi Pruiksma[b], Joanne L. Davis[a], and Tara R. Buck[c]

[a]*Department of Psychology, University of Tulsa, Tulsa, OK, United States,* [b]*Department of Psychiatry and Behavioral Science, University of Texas Health Science Center at San Antonio, San Antonio, TX, United States,* [c]*Department of Psychiatry, University of Oklahoma School of Community Medicine, Tulsa, OK, United States*

Abbreviations

CATS Child and Adolescent Trauma Screen
CSHQ Child's Sleep Habits Questionnaire Preschool and School-Aged
ESS Epworth Sleepiness Scale
PTSD posttraumatic stress disorder
TRNS The Trauma-Related Nightmares Survey

Introduction

This chapter reviews a treatment for nightmares in youth. Nightmare treatment may be stand-alone, or adjunctive to other treatments. Evidence in the adult literature finds that nightmares respond to brief cognitive-behavioral treatments such as Exposure, Relaxation, and Rescripting Therapy (ERRT; Morgenthaler et al., 2018). Current research is adapting such treatments for children. Before making the clinical decision to use this treatment, a clinician should first determine whether a youth is having nightmares. Nightmares are distinct from other parasomnias such as night terrors. Although a youth with night terrors appears to wake up and may be screaming and seem distressed, when later queried about awakening, they will have no recollection of the event. For night terrors, an alternative treatment, called scheduled awakening, is recommended.

Nightmares are frightening dreams that are remembered upon awakening (APA, 2013). Following nightmares, youth are often afraid to return to sleep and may develop bedtime avoidance. While occasional childhood nightmares are common (Gauchat, Séguin, & Zadra, 2014), only 10%–20% of youth experience frequent idiopathic nightmares, whereas 50%–80% of youth with posttraumatic stress disorder (PTSD) experience post-trauma nightmares (Kovachy et al., 2013). Clinically significant nightmares of at least one per week, peak by age 11 (Gauchat et al., 2014).

Nightmare theory

Nightmares are common in the aftermath of trauma. Although a symptom of PTSD, nightmares may occur in the absence of PTSD, and treatments that broadly target PTSD often fail to mitigate nightmares. Increasing evidence suggests that nightmares and insomnia are independent phenomena requiring targeted treatment (Davis, 2008). Several treatments have been developed to specifically address nightmares, including Imagery Rehearsal Therapy (Krakow, Kellner, Pathak, & Lambert, 1995) and ERRT (Davis, 2008). These protocols generally include psychoeducation about nightmares, exposure to and rescripting the nightmare. ERRT also includes psychoeducation regarding trauma and its relation to sleep, modification of sleep habits, relaxation techniques, and theme-based rescription (Davis, 2008). ERRT was developed for adults and modified for use in youth (Fernandez, Cromer, Borntrager, Swopes, & Davis, 2013).

Handbook of Lifespan Cognitive Behavioral Therapy. https://doi.org/10.1016/B978-0-323-85757-4.00032-8

How cognitive-behavior therapy is conducted

Cognitive-behavior therapy for nightmares in children has 5 progressive modules. Sessions begin with parent-youth dyads together for agenda setting, homework check-in, and sleep diary review. Therapist and the youth then work through a workbook, while the parent reads a companion module in another room. Sessions conclude with everyone together; the youth teaches the lesson to the parent as a learning check, and the therapist problem solves homework plans. See *Course of Treatment* for module content.

Case introduction and presenting complaints

The youth was identified with a nightmare screener during check-in at an outpatient psychiatry clinic. Candice (pseudonym) was a 12-year-old Native American cis-gender girl receiving care for mental health conditions cooccurring with developmental disabilities. She was unable to fall asleep without her mom. Candice often slept on the living room sofa because of scoliosis which made sleeping in bed painful. She woke up earlier than schedules demanded and complained of daytime sleepiness. Candice had nightmares throughout her life, but they had intensified in the 2 months prior to baseline. According to her mother, Candice woke up between midnight and 3:30 a.m., screaming and crying from nightmares 2 to 3 times a week.

Case history

Case history was obtained with consent, from clinical records. Candice was born via an early term caesarean section at 37 weeks. She received early interventions for language and gross motor delays. At age 10, she was diagnosed with attention deficit hyperactivity disorder (ADHD) and autism spectrum disorder with language impairment. Her intelligence was average; verbal comprehension was a relative weakness and processing speed a relative strength. Candice attended public school, had an individualized education program for math and reading, and behavioral contingency management at home and school. Psychiatric care included Sertraline and Guanfacine in the morning, Oxcarbazepine twice daily, and Risperidone at bedtime.

In the year prior to nightmare treatment, Candice was diagnosed with depression. She was sad, irritable, had low energy, increased appetite, and was aggressive at home. She was hospitalized three times, once for suicide threats, and twice for suicide attempts. The first attempt was with asphyxiation, and the second was an overdose of parent's medication.

Assessment

Youth measures were administered verbally with visual aids such as dials with emojis to ensure clarity and to aid assessment accuracy. The parent report of child trauma was conducted by interview; other parent-report measures were conducted electronically using an iPad. Nightmares were tracked with sleep diary completed daily throughout treatment.

Child report

Child and Adolescent Trauma Screen-Youth Report (CATS; Sachser et al., 2017) assesses PTSD. The youth is asked to respond "*yes*" or "*no*" to having experienced 15 *stressful or scary events*. Then, 20 items assess PTSD symptoms regarding the most "*bothersome event*," rated 0 (*never*) to 3 (*almost always*), for the preceding 2 weeks. Possible summed scores range from 0 to 60. Scores: <14 are normative, 16–20 is mild/moderate, and 21 or greater indicates probable PTSD. The CATS has good to excellent reliability and acceptable convergent and discriminant validity for both the youth report and parent report (Sachser et al., 2017).

Candice reported a long history of witnessing someone in her family being physically violent and the preceding year, having someone close to her die suddenly. Her CATS score at baseline was 33, indicating probable PTSD. See Fig. 1 for follow-up ratings.

The Trauma-Related Nightmares Survey-Child Version (TRNS-C; Langston, 2007): queried previous week nightmare frequency and distress. At baseline, Candice had four nights with nightmares in the preceding week, 2 of which had >1 nightmare. In response to the distress item, "*In general, how upsetting have the nightmares been?*", 1 (*not at all*), 2 (*a bit*), 3 (*quite a bit*), 4 (*a lot*), 5 (*very much*), Candice responded 3 (Fig. 2).

Daytime sleepiness was assessed with the 8-item Epworth Sleepiness Scale for Children and Adolescents (ESS-CHAD; Wang, Benmedjahed, Lambert, et al., 2017). Respondents use a 4 point scale, ranging from 0 (*would never fall asleep*) to

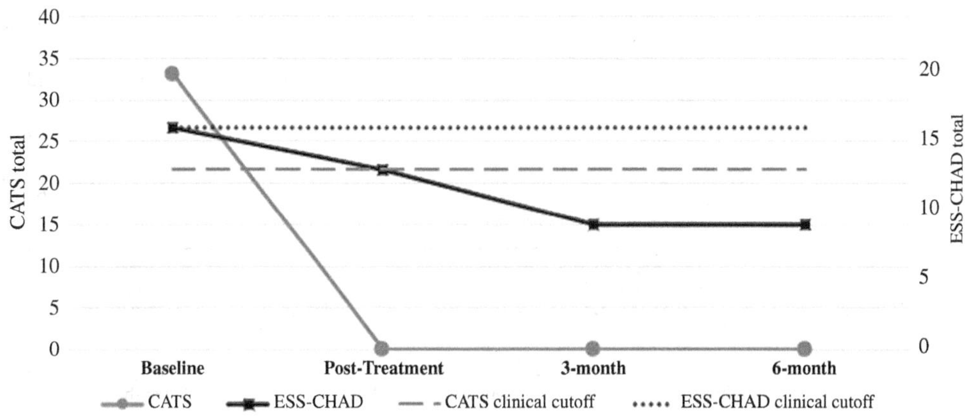

FIG. 1 Child report of CATS and ESS. *Note*: CATS = Child and Adolescent Trauma Screen; ESS-CHAD = Epworth Sleepiness Scale for Children and Adolescents.

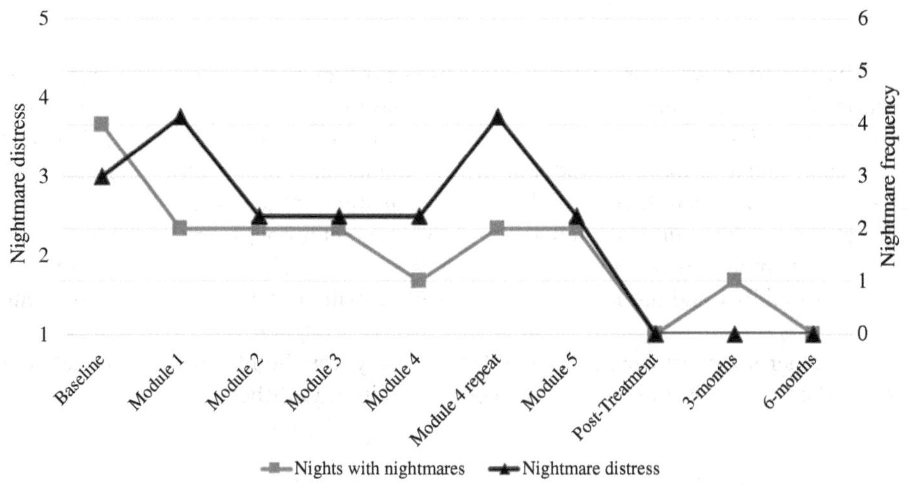

FIG. 2 Child report nightmare frequency and distress for the week leading up to each timepoint. When presenting for Module 1, distress was up, frequency was down.

3 (*high chance of falling asleep*) during daytime activities. Possible sums range from 0 to 24; scores: ≥16 suggest a sleep disorder (Johns, 1991). ESS-CHAD is valid and reliable for youth (Janssen, Phillipson, O'Connor, & Johns, 2017). At baseline, Candice's ESS was 16 (Fig. 1).

Parent report

Child and Adolescent Trauma Screen (CATS)-Caregiver Report (Sachser et al., 2017) is psychometrically and structurally the same as the CATS Youth-Report. Candice's mother reported that Candice had experienced the death of a family member, witnessed physical violence at home, and experienced financial insecurity. Candice's score of 33, indicated probable PTSD (Fig. 3).

The Child's Sleep Habits Questionnaire Preschool and School-Aged (CSHQ; Owens, Spirito, & McGuinn, 2000) is a 33-item measure of sleep behavior for children aged 4–12. A total score is derived from 8 subscales: bedtime resistance, sleep onset delay, sleep duration, sleep anxiety, night-waking, parasomnias, sleep-disordered breathing, and daytime sleepiness. Scores range from 33 to 99; higher scores indicate greater sleep disturbance; ≥41 suggests clinically significant sleep concerns. The CSHQ has acceptable internal consistency, test-retest reliability, and discriminant validity in community and pediatric sleep disorder clinic samples (Owens et al., 2000). At baseline, Candice's total score was 81, with elevations on bedtime resistance (14), parasomnia (18), and daytime sleepiness (17) (Fig. 3).

Case conceptualization

Youth with ASD, depression, and ADHD are more likely to have nightmares (APA, 2013, p. 405). Candice's comorbidities, along with trauma exposure likely contributed to nightmare risk. Candice and her mother verbalized a belief that nightmares were uncontrollable, likely predisposing Candice to nightmare chronicity.

FIG. 3 Parent report of CATS and CSHQ (). *Note*: CATS = Child and Adolescent Trauma Screen; CSHQ = Child's Sleep Habits Questionnaire.

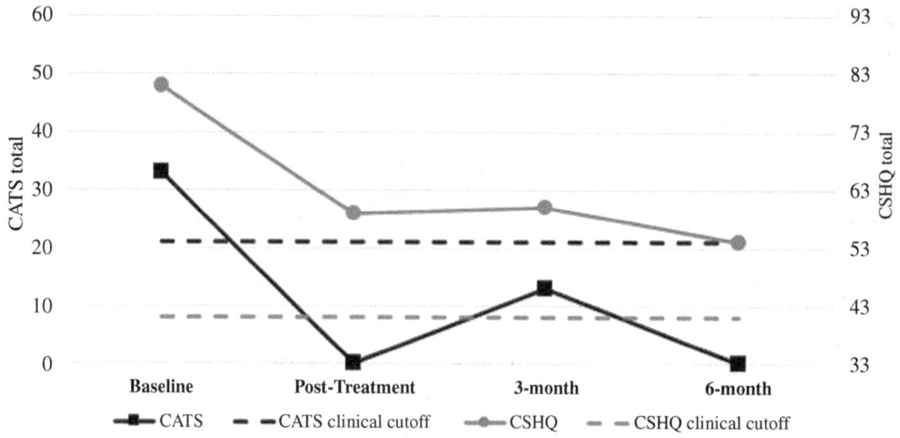

Candice reported having nightmare distress after awakening from nightmares and difficulty "getting nightmares out of her head" during the day. The anticipation of nightmares and sleep avoidance could have increased the likelihood of having nightmares (Davis, 2008). Additionally, fear and avoidance resulted in distraction through screen use (e.g., videos and games on a tablet), which would have contributed to more alertness at night. Candice also drank caffeinated tea, and reportedly worried about nightmares at night, all of which increase nightmare risk. Bedtime anxiety was likely maintained with bedtime resistance and her mother staying with her until she fell asleep, preventing inhibitory learning. Upon waking from a nightmare, Candice typically woke her mother or used her iPad, which reinforced avoidance and interfered with sleep. Candice's presenting complaints of irritability and mood lability are consistent with a nightmare cycle of daytime anxiety and bedtime avoidance.

Candice's eagerness to learn, and her mother's support were protective factors. They were highly engaged in treatment and enthusiastically employed skills. Candice's love of drawing and of superheroes increased therapy engagement.

Course of treatment and assessment of progress

Treatment was delivered across six 60–90-min sessions; module 4 was repeated once. Candice was engaged and compliant with therapy. Contingency management, using stickers to reward on-task behaviors, helped to maintain Candice's focus and cooperation. Each session, Candice cashed-in her stickers for a "treasure" that she selected from a treasure box. Snacks and effusive praise also helped sustain her attention. Each module had interactive activities that Candice reported were enjoyable.

Module 1

This module begins with treatment orientation and developing motivation to change. The relationship between thoughts, feelings, and behaviors are explained using a spiral. We use examples from the youth's life, starting with "*When you wake up from a nightmare, how do you feel?*" Psychoeducation about how poor sleep impacts functioning is augmented with an activity in which DrunkGoggles are used to simulate how sleep deprivation affects focus. To demonstrate, the child is instructed to walk a line to a cup into which they drop a ball, doing this first with and then without the goggles. This activity is effective for motivating sleep change (Kaier, Zanotti, Davis, Strunk, & Cromer, 2016). Youth then draw or write about what life would be like if nightmares remit. Youth learn that sleep routines can "train the brain" to release melatonin. Where appropriate, parents are coached on addressing bedtime refusal. Bedtime passes are helpful with younger children, which give the youth a limited number of get-out-of-bed-free cards (Moore, Friman, Fruzzetti, & MacAleese, 2007). All youth learn belly-breathing as a relaxation tool, and this is practiced at session end with everyone together.

In Candice's spiral, she described how bad moods and arguing resulted from nightmares. In her instilling hope activity, Candice wrote about an improved relationship with mom, less arguing, and everyone in the family being happier, if nightmares improved. Her bedtime routine included belly breathing, reading, brushing teeth, getting comfortable in bed and turning out the light. Her parent was coached on bedtime passes to support Candice falling asleep alone.

Module 2

Module 2 continues shaping sleep routines and addresses sleep-interfering anxiety. Psychoeducation is about the nightmare cycle to increase locus of control, and to establish a rationale for reducing bedtime-related cognitive anxiety with a worry jar and scheduled worry time. Sleep routines are improved by adding one sleep-enhancing behavior (e.g., relaxation) and removing one sleep-interfering behavior (e.g., eliminating blue light or caffeine).

At check-in, Candice reported success with her sleep routine and that belly-breathing was relaxing. Sleep logs were completed for five nights. Problem-solving the missed nights, they decided to move the log to Candice's daily calendar, and to complete it concurrently with morning medicine. Candice enthusiastically decorated her worry jar, and in session, added some worries. Daily worry time was scheduled for 2:00 p.m.

Candice and her mother added listening to relaxing music to her bedtime routine and removed two sleep-interfering behaviors: stopping iPad use prior to bed, and eliminating dinnertime caffeinated tea. They created a log to track new habits.

Module 3

Module 3 revisits brain training with stimulus-control. They learn that the bed can be conditioned as a sleep cue by using it only for sleep and not play, homework, or pets. The therapist provides a pillowcase and fabric markers, and the youth writes or draws images on the pillowcase that evoke pleasant thoughts. Progressive muscle relaxation (PMR) is introduced as a way to teach the brain to master muscles; PMR is done at the end of session with parent and child together, and a recording is sent home for nightly practice.

At check-in, Candice produced completed sleep and habits logs. Although Candice continued to have nightmares, this was the first week that they could remember Candice staying in bed, and not waking her mother. Candice successfully abstained from caffeinated tea and did not use her iPad before bed for six nights. Candice worried 4 out of 7 scheduled days. She did not listen to relaxing music before bed because the family did not have suitable music; this was a failure to trouble-shoot homework obstacles so was discussed and a solution found for the coming week.

Candice was delighted with the pillowcase activity. She drew pictures of superheroes, racecars, hot cocoa, pawprints to represent her dog, a swimming pool, and a police badge. She described each item's significance as she drew. Candice and her mother enjoyed PMR and planned to add it to the bedtime routine.

Module 4

Module 4 is focused on exposure and changing the nightmare. A rationale for talking about the scariest nightmare is introduced with the youth being asked to recount an old fear that they have mastered, for example, petting a dog, or diving into a pool. The youth is asked to describe how they overcame that fear in order to evoke mastery. Once prepared for the exposure, the youth recounts their scariest nightmare. They are coached to use present tense, five senses, and as much detail as possible. The therapist gently elicits details or corrects past tense to present, while recording the youth's narrative. Older youth are permitted to write the narrative themselves. The therapist reminds the youth that the dream is just a story, reassures their current safety, and praises their bravery; if youth are upset, reassurance about safety and bravery are used for encouragement to continue. Following the telling of the nightmare, the youth is asked to give a subjective units of distress (SUDS) rating from 1 to 10, where 10 is the most afraid they have ever been. They then re-read the narrative and give another SUDS rating. After congratulating the youth on having completed the hardest part of therapy, the therapist then tells the youth that they get to change the nightmare any way they want, so that it becomes something they want to dream.

The therapist explains that nightmares have main ideas that brains get stuck on and which keep the nightmare story replaying in our heads. The youth is told that we want to find the idea, and make sure that they change that in the story, to help the brain get unstuck. The themes that are introduced are based on five trauma themes from Cognitive Processing Therapy (Resick & Schnicke, 1993): being unsafe/in danger, not having power over what happens, a specific person being dangerous or mean, someone you need not being there when you needed them, not liking yourself or someone else, and based on empirical work of trauma-related nightmares in children (Gray & Cromer, 2018), a sixth theme: escape, Following the rescription, the youth gives another SUDS rating and then gets to destroy the old nightmare (e.g., rip it up). The parent is then brought into the room to hear the new rescription and to set up nightly rescription practice with the youth. They also continue bedtime routines, worry jar, and any relaxation.

If families return the following week with no improvement in nightmare frequency or distress, the rescription is redone for a nightmare that occurred that week. Module 4 is only repeated once, so if nightmares still do not improve, therapy commences with module 5.

At check-in, Candice was tired and reported having been sick, and having a "terrible week." Candice discontinued PMR after 2 days because it increased tension, but she listened to calming music 5 of 7 nights.

During psychoed, Candice confidently described overcoming a fear of monsters. In her nightmare narrative, she drew her brother being attacked by a clown and a lion. Her favorite superhero was there and along with her brother, was eaten by the lion. She rated her SUDS 1 before and after rescription; she said the theme was safety. She rescripted it by drawing a superhero saving her brother.

Readministering module 4

When she returned the following week, Candice's nightmares had not improved. Per the manual, the therapist had Candice rescript the scariest of the week's nightmares. Candice drew a "creepy house" where ghosts, werewolves, and a clown holding an ax watching her. Candice said the theme was escape and she rescripted the nightmare by playing games and eating candy with the ghosts, werewolves, and clown. SUDS was 1 before and after rescription. Homework was continuing the sleep routine and reading the rescripted dream before bed.

Module 5

Treatment ends with maintenance planning. The therapist and youth review the workbook to reinforce mastery. They discuss a plan for what to do when future nightmares occur, and fill a drawing of a toolbox in the workbook with favorite tools, as they discuss future use.

During check-in, Candice reported reading the new rescription 4 out of 7 nights. Three nights were missed to due to summer camp and a birthday party. Candice reported two nightmares that week. Although there was no improvement in nightmares again this week, the family was instructed to continue to read the rescription in the coming week.

During the review and maintenance planning, Candice said the most helpful tools were her pillowcase, bedtime routine with relaxing music, and slow breathing. She said that the most important skills were worry jar and no iPad before bed. The least helpful tool was PMR. Throughout the review, Candice commented, "Wow! I have learned a lot!"

Follow-up

Candice and her mother returned for a follow-up assessment 1-week post-treatment and 3 months post-treatment. The also attended a 6-month follow-up via telehealth (COVID-19 adaptation).

Candice: Nightmares entirely mitigated following module 5. She reported one nightmare that was "not at all upsetting" the week before the 3-month assessment and zero nightmares between 3- and 6-month assessment. Candice's ESS improved to 13 at posttreatment and dropped to below clinical levels at both follow-ups. Despite new stressors in the home (e.g., job loss due to COVID-19), PTSD scores that had dropped to 0 were maintained at follow-up. At follow-ups, Candice reported maintaining her bedtime routine, including using the pillowcase and listening to calming music.

Parent: At the 3-month follow-up, Candice's mom reported a new trauma that impacted her CATS rating of 13. At the 6-month follow up, she endorsed family stressors including a stroke, and COVID-19 stressors including home schooling and Candice's father being furloughed. Nonetheless, Candice coped well as evidenced by the CATS score of zero. Parent-reported sleep improvements were clinically meaningful across the study. Candice's total CSHQ score decreased from 59 at posttreatment to 54 at 6-month follow-up, bedtime resistance decreased from 12 to 10, parasomnia decreased from 10 to 8, and daytime sleepiness decreased from 13 to 11. Candice's mother reported that she and Candice were able to communicate effectively about sleep problems as they arose. Candice had had no suicide attempts by 6-month follow-up.

Treatment implications

If left untreated, nightmares following trauma may persist for decades and wreak havoc on sleep continuity and sleep duration which negatively impacts mental health, physical health, family and social relationships, and functioning at school. Early nightmare intervention for children has the potential to return children to a healthy path of development and potentially circumvent negative health outcomes.

Recommendations to clinicians and students

Individuals who have experienced trauma may avoid discussing distressing experiences with health providers. Individuals often do not report nightmares, perhaps due to a belief that nightmares cannot be treated (Nadorff, Nadorff, & Germain, 2015). However, providing trauma-informed care helps facilitate an environment of safety that can facilitate discussion of difficult topics. Screening and directly assessing for traumatic experiences and nightmares is also recommended to identify symptoms and to connect individuals with evidence-based treatment options.

Summary points

- Cognitive-behavioral therapies are recommended treatments for nightmares in adults.
- It is feasible and acceptable to adapt cognitive-behavioral therapies for children and adolescents with nightmares and evidence to date supports the efficacy of these interventions.
- Nightmare treatment may have secondary benefit of improving PTSD and suicidality in children.
- Treating nightmares in children and adolescents has the potential to improve the sleep of the individual and of the members of the household.

References

American Psychiatric Association. (2013). *Diagnostic and statistical manual of mental disorders* (5th ed.).

Davis, J. L. (2008). *Treating post-trauma nightmares: A cognitive behavioral approach*. New York: Springer.

Fernandez, S., Cromer, L. D., Borntrager, C. F., Swopes, R., & Davis, J. L. (2013). A case series: Cognitive-behavioral treatment (exposure, relaxation, and rescripting therapy) of trauma-related nightmares experienced by children. *Clinical Case Studies*, *12*, 39–59. https://doi.org/10.1177/1534650112462623.

Gauchat, A., Séguin, J. R., & Zadra, A. (2014). Prevalence and correlates of disturbed dreaming in children. *Pathologie Biologie*, *62*(5), 311–318. https://doi.org/10.1016/j.patbio.2014.05.016.

Gray, K. N., & Cromer, L. D. (2018). Posttraumatic nightmare content in children and its relation to posttraumatic psychopathology. *International Journal of Dream Research*. https://doi.org/10.11588/ijodr.2018.2.48777.

Janssen, K. C., Phillipson, S., O'Connor, J., & Johns, M. W. (2017). Validation of the Epworth Sleepiness Scale for children and adolescents using Rasch analysis. *Sleep Medicine*, *33*, 30–35. https://doi.org/10.1016/j.sleep.2017.01.014.

Johns, M. W. (1991). A new method for measuring daytime sleepiness: Epworth Sleepiness Scale. *Sleep*, *14*(6), 540–545. https://doi.org/10.1093/sleep/14.6.540.

Kaier, E., Zanotti, D., Davis, J. L., Strunk, K., & Cromer, L. D. (2016). Addressing the problem of student-athlete sleepiness: Feasibility of implementing an interactive sleep workshop at a division I school. *Journal of Clinical Sport Psychology*, *10*, 237–247. https://doi.org/10.1123/jcsp.2015-0023.

Kovachy, B., O'Hara, R., Hawkins, N., Gershon, A., Primeau, M. M., Madej, J., & Carrion, V. (2013). Sleep disturbance in pediatric PTSD: Current findings and future directions. *Journal of Clinical Sleep Medicine*, *9*, 501–510. https://doi.org/10.56664/jcsm.2678.

Krakow, B., Kellner, R., Pathak, D., & Lambert, L. (1995). Imagery rehearsal treatment for chronic nightmares. *Behaviour Research and Therapy*, *33*, 837–843. https://doi.org/10.1016/0005-7967(95)00009-M.

Langston, T. J. (2007). *Nightmares and sleep quality in children and adolescents: A comparison based on trauma status*. Unpublished doctoral dissertation University of Tulsa.

Moore, B. A., Friman, P. C., Fruzzetti, A. E., & MacAleese, K. (2007). Brief report: Evaluating the bedtime pass program for child resistance to bedtime—A randomized, controlled trial. *Journal of Pediatric Psychology*, *32*(3), 283–287. https://doi.org/10.1093/jpepsy/jsl025.

Morgenthaler, T. I., Auerbach, S., Casey, K. R., Kristo, D., Maganti, R., Ramar, K., … Kartje, R. (2018). Position paper for the treatment of nightmare disorder in adults: An American Academy of sleep medicine position paper. *Journal of Clinical Sleep Medicine*, *14*(6), 1041–1055. https://doi.org/10.5664/jcsm.7178.

Nadorff, M. R., Nadorff, D. K., & Germain, A. (2015). Nightmares: Under-reported, undetected, and therefore untreated. *Journal of Clinical Sleep Medicine*, *11*(7), 747–750. https://doi.org/10.5664/jcsm.4850.

Owens, J. A., Spirito, A., & McGuinn, M. (2000). The Children's sleep habits questionnaire (CSHQ): Psychometric properties of a survey instrument for school-aged children. *Sleep*, *23*(8), 1043–1052. https://doi.org/10.1037/t33022-000.

Resick, P. A., & Schnicke, M. K. (1993). *Cognitive processing therapy for rape victims: A treatment manual*. Sage.

Sachser, C., Berliner, L., Holt, T., Jensen, T. K., Jungbluth, N., Risch, E., … Goldbeck, L. (2017). International development and psychometric properties of the child and adolescent trauma screen (CATS). *Journal of Affective Disorders*, *210*, 189–195. https://doi.org/10.1016/j.jad.12.040.

Wang, Y. G., Benmedjahed, K., Lambert, J., et al. (2017). Assessing narcolepsy with cataplexy in children and adolescents: Development of a cataplexy diary and the ESS-CHAD. *National Science Sleep*, *9*, 201–211. https://doi.org/10.2147/NSS.NS140143.

Chapter 38

New cognitive-behavioral intervention with Theory of Mind (ToM) training for children with epilepsy experiencing social difficulties

Elizabeth Stewart and Suncica Lah
School of Psychology, The University of Sydney, Brain and Mind Centre, NSW, Australia

Abbreviations

ASD	autism spectrum disorder
BTPS	Barriers to Treatment Participation Scale
CBT	cognitive-behavioral therapy
FSIQ	Full-Scale Intelligence Quotient
SEIFA	Socio-Economic Index for Areas
SES	socioeconomic status
SDQ	Strength and Difficulties Questionnaire
SPP	Social Perception Profile
TEI	Treatment Evaluation Inventory
TLE	temporal lobe epilepsy
ToM	Theory of Mind
ToMI	Theory of Mind Inventory

Introduction

Epilepsy is a common neurological disorder in childhood that is associated with significant social morbidity: reduced social competence, poorer peer relationships and more bullying by peers (Berg et al., 2007). Social difficulties in epilepsy often emerge in middle childhood, between 8 and 11 years old, and tend to persist or worsen in adolescence and adults (Rantanen, Eriksson, & Nieminen, 2012). Social difficulties in children with epilepsy are linked to impaired Theory of Mind (ToM); the ability to understand the thoughts, intentions, beliefs and emotions of others and oneself (Frith & Frith, 2005). ToM is an aspect of social cognition that is comprised of cognitive and affective components, and is critical for healthy social and emotional development (Weimer, Sallquist, & Bolnick, 2012).

Interventions targeting ToM have been found to be effective for children with other neurodevelopmental conditions, such as autism spectrum disorder (ASD), developmental delay and hearing impairment (see Hofmann et al., 2016 for metaanalytic review). Interestingly, the inclusion of cognitive-behavioral therapy (CBT) alongside ToM training has been associated with improvements in both ToM and social behavior (Hofmann et al., 2016). CBT teaches children about the relationship between thoughts, feelings and behaviors and is critical for acquiring ToM and applying this knowledge to the social world (Grave & Blissett, 2004).

Although social difficulties in children with epilepsy are common, evidence-based treatments are lacking. Hence, we developed a novel CBT intervention targeting ToM and social skills for children with epilepsy, using the Medical Research Council (MRC) guidelines (Craig et al., 2008). The intervention addresses shortfalls of prior ToM interventions as it (i) targets both cognitive and affective ToM, which are impaired in children with epilepsy (Stewart et al., 2018) and (ii) includes CBT skills alongside ToM training. We developed a new intervention rather than using a preexisting program,

Handbook of Lifespan Cognitive Behavioral Therapy. https://doi.org/10.1016/B978-0-323-85757-4.00011-0

as treatment needs of children with epilepsy may differ from other clinical populations (e.g., children with epilepsy have impairments in social competence and social cognition in the absence of additional features of ASD [autistic styles of communication and behavior/idiosyncratic language], Reilly et al., 2015). Given the novelty of our intervention, pilot testing was deemed necessary. The primary aim of this study is to evaluate feasibility and acceptability of the intervention for children with epilepsy. The secondary aim is to assess efficacy of the intervention for improving ToM and social competence.

Materials and methods

Ethics approval was obtained from participating hospitals. The trial was registered with the Australia and New Zealand Clinical Trials Register (ACTRN12618000974202). For details of methodology please see the published protocol (Stewart, Catroppa, & Lah, 2019).

Participants

Children with epilepsy (8–12 years old) were recruited through a tertiary pediatric epilepsy center in Sydney. Five children with epilepsy were identified as eligible and invited to participate. Three families declined and two children participated in the intervention. Seven control children were also recruited to obtain an estimate of normative performance on the ToM tasks that do not have standardized norms. Control children had no history of neurological or psychiatric conditions and were matched to cases as closely as possible on age, gender and full-scale intelligence quotient (FSIQ; Table 1).

Intervention

A summary of the intervention is provided in Table 2, and the intervention is described in detail in the published study protocol (Stewart, Catroppa, & Lah, 2019). Children and parents were provided with intervention workbooks at the start of treatment. The parent workbook contained strategies to help children with skills at home. Skills were taught in session via a mixture of paper/pen activities, role-plays, group discussions, written social stories and videos. Children were encouraged to provide their own examples and investigators provided feedback and prompting throughout sessions. Optional homework tasks were provided, which expanded on content covered during sessions; homework was reviewed verbally at the start of each session.

Procedure

A double baseline case study design was employed and the intervention was run as a single-arm, open label pilot. Children attended sessions together (1.5–2 h), led by a clinical psychology registrar and supervising clinical neuropsychologist, over 4 consecutive weeks. Parents joined children for the final 15–30 min, during which skills were summarized and optional homework tasks were explained. Assessments of ToM and social competence were completed by children (individually, face-to-face) and parents (online): (i) twice at baseline (3 weeks and 1 day before treatment), (ii) posttreatment (last day of treatment), and (iii) follow-up (3 weeks after treatment). Additionally, parents completed questionnaires on emotional/behavioral functioning of their child at each assessment, and on treatment acceptability and barriers/facilitators to attendance at the end of treatment. A brief weekly measure of social competence was completed by children and parents from baseline 1 to follow-up. Finally, parents attended face-to-face debriefing interviews with one investigator (SL) during which they provided feedback about treatment efficacy/acceptability.

Measures

Detailed descriptions of primary and secondary outcome measures, and rationales for selecting each measure, are described in the published protocol (Stewart, Catroppa, & Lah, 2019).

Primary outcome measures included

(i) ***Recruitment and retention*** were considered feasible if 40%–50% of invited children agreed to participate and 80% attended >75% sessions, respectively.

(ii) ***Processing time for intervention modules*** lasting 20–50 min, consistent with standard psychology sessions and children's attentional capacity, were deemed appropriate.

TABLE 1 Characteristics of children with epilepsy who participated: demographic, epilepsy, cognitive and behavioral variables.

Variable	Child 1	Child 2
Demographic		
Age at baseline (years: months)	9:3	11:2
Gender (boy/girl)	Girl	Girl
Handedness (left/right/ambidextrous)	Left	Right
Socio economic status (decile)	10	10
Grade at school (*K*−12)	4	5
Epilepsy		
Side of seizure focus (left/right)	Left	Right
Age of seizure onset (years: months)	7:1	2:0
Duration of seizures (years: months)	2:2	9:2
Seizure frequency at diagnosis (p/week)	1–2	6–7
Seizure frequency past 12 months (p/week)	0	0
Surgical status (pre/post)	Pre	Pre
Current antiepileptic medication	Carbamazepine	Oxcarbazepine
Cognitive		
Intellectual functioning		
Wechsler Intelligence Scale for Children—Fifth Edition (FSIQ)[a]	113 (83rd)	94 (35th)
General language skills		
Peabody Picture Vocabulary Test—Fourth Edition (Scaled Score) [a]	114 (82nd)	99 (47th)
Executive functions		
Working memory		
Automated Working Memory Assessment: Backward Digit Recall (Scaled Score)[a]	137 (99th)	90 (25th)
Automated Working Memory Assessment: Odd-One-Out (Scaled Score)[a]	106 (66th)	76 (5th)*
Inhibition		
Delis-Kaplan Executive Functions Scale: Color Word Inference (C3; Scaled Score)[b]	16 (98th)	6 (9th)*
Delis-Kaplan Executive Functions Scale: Color Word Inference (C4; Scaled Score)[b]	15 (95th)	4 (2nd)**
Attentional shifting		
Test of Everyday Attention for Children: Creature Counting (total; Scaled Score)[b]	13 (84th)	7 (16th)
Test of Everyday Attention for Children Creature Counting (timing; Scaled Score)[b]	10 (50th)	6 (9th)*

**Borderline range.*
***Impaired range.*
[a]*M = 100, SD = 15.*
[b]*M = 10, SD = 3.*

(iii) ***Suitability of tasks*** was based on observations of children's understanding, engagement and participation in activities and classified by therapist as: (a) suitable without modification (retain as is), (b) suitable with modification (amend and modify), or (c) unsuitable (discard or replace).

(iv) ***Overall acceptability of the intervention*** was based on scores obtained on the subscales of the treatment evaluation inventory (TEI) completed by parents, children, and investigators at the end of treatment (Kazdin, 1980). Ratings

TABLE 2 Session-by-session summary of the intervention: content of modules, completion time, and decision about suitability.

Session	Module	Content	Completion time (minutes)	Decision and proposed modifications[a]
Introductory cognitive-behavioral therapy (CBT) phase				
1	1	**Introduction to thoughts**. Children learn about internal and external triggers of thoughts (i.e., preferences, desires, intentions and beliefs; people, places, situations and events), and complete activities showing that their thoughts may differ from other people's thoughts	30	Suitable without modification
	2	**Introduction to feelings**. Children give examples of feelings, identify internal and external triggers of feelings, and complete emotion vocabulary exercises to help them label, express and communicate feelings with words	30	Suitable without modification
	3	**The relationship between thoughts, feelings and behaviors**. Children complete worksheets to illustrate that their thoughts and feelings can influence their behaviors. One hypothetical and one personal example were used to demonstrate how having different thoughts may influence their behavior	10	Suitable with modifications—include additional, similar worksheets to consolidate learning and more personal examples
Active Theory of Mind training phase				
2	4	**Introduction to Theory of Mind**. Children practice taking three positions that help them perspective take about thoughts and feelings and respond appropriately (i.e., "the thinker," "the feeler," and "the responder"). Children are then introduced to eight social scenarios requiring ToM	20	Suitable without modification
	5	**Theory of Mind module (White lie)**. A definition is provided, along with situations when white lies may be used. Children think of personal examples when they have used a white lie, and discuss advantages/disadvantages of using these Social response strategy. Children learn to use 'strategic honesty' instead of white lies, defined as "finding something you like about a person, situation or event and commenting on that, rather than focusing on what you don't like" (i.e., telling the truth without being so honest that you hurt someone's feelings)	25	Suitable with modification—include more activities to practice using strategic honesty, both in session and in homework tasks, as children understood this concept but were not able to use it consistently
	6	**Theory of Mind module (Misunderstandings)**. A definition is provided, along with situations when misunderstands occur. Children think of examples from their own lives when they have experienced a misunderstanding Social response strategy. Children practice taking 'the thinker' and 'the responder' positions to understand how a misunderstanding may have occurred and communicate effectively with another person about the misunderstanding	20	Suitable without modification

TABLE 2 Session-by-session summary of the intervention: content of modules, completion time, and decision about suitability—cont'd

Session	Module	Content	Completion time (minutes)	Decision and proposed modifications
3	7	**Theory of Mind module (Forgetting)**. A definition is provided and children give with examples of when they have forgotten something. Children discuss consequences of forgetting things based on their own experiences Social response strategy. Children practice how to clearly communicate when they have forgotten something	20	Suitable without modification
	8	**Theory of Mind module (Irony)**. A definition is provided, along with activities to explain the difference between irony and sarcasm. Children learn to identify irony via the content of speech and tone of voice Social response strategy. Children practice using and detecting irony	30	Unsuitable (discard). Children misunderstood the concept of irony even with repetition and additional opportunities for learning
	9	**Theory of Mind module (Pretending)**. A definition is provided and children give examples when they have used pretend play. Children are given materials (pencils and play-dough) and use these to engage in pretend play	20	Suitable without modification
4	10	**Theory of Mind module (Figures of speech)**. A definition is provided and children learn about the difference between literal and figurative language. Children given examples of when they have used, or heard other people use, figures of speech	35	Suitable with modification—reduce number of activities to minimize repetition, as children understood this concept and activities became repetitive
	11	**Theory of Mind module (Persuasion)**. A definition is provided, and children give examples of when they have felt persuaded by a peer (or when they have persuaded someone else) Social response strategy. Children are taught a 3-step approach to using effective communication, to use in response to/ instead of persuasion: (1) State your thoughts and feelings, (2) Acknowledge the thoughts and feelings of the other person (3) Acknowledge that your opinions differ and state that this is okay	25	Suitable without modification
	12	**Theory of Mind module (Double bluff)**. A definition is provided, along with examples of double bluff. Children think of personal examples of when they have used, or heard other people use, a double bluff. Children practice taking "the thinker" position to understand how and why the double bluff has occurred	20	Unsuitable (discard). Children did not understand double bluff even with repetition and additional opportunities for learning
	13	**Review of training**	20	Suitable with modification—amend in line with changes to prior training modules

[a]*Decision for each module: (a) suitable without modification (retain as is), (b) suitable with modification (amend and modify), including plan for amendments, or (c) unsuitable (discard or replace), with brief rationale.*

≥80% on each subscale considered acceptable (Kazdin, French, & Sherick, 1981). Feedback provided in debriefing interviews was an additional, qualitative source of information about treatment acceptability.

(v) ***Barriers and facilitators to attendance*** were identified from parental responses on the barriers to treatment participation scale (BTPS) completed at the end of treatment (Kazdin, Holland, Crowley, & Breton, 1997).

Secondary outcome measures

Tests used to assess relevant cognitive skills are outlined in Table 2. Semistructured interviews with parents gathered information about demographic and epilepsy variables, social and emotional functioning, and previous psychosocial treatments. An estimate of socioeconomic status (SES) was obtained with the Australian Bureau of Statistics socio-economic index for areas (SEIFA; Pink, 2013). ToM was assessed with parallel versions of the Strange Stories task (child-completed behavioral talk; Happé, 1994) and the Theory of Mind Inventory (ToMI; parent-report questionnaire; Hutchins, Prelock, & Bonazinga, 2012). Social competence was assessed with the Self-Perception Profile (SPP; child- and parent-report questionnaires; Granleese & Joseph, 1994) and the Strength and Difficulties Questionnaire (SDQ; parent-report questionnaire; Goodman, 1997).

Statistical methods

For the primary analysis, descriptive statistics and qualitative data were used to assess feasibility. For secondary analyses pertaining to children's performance on tasks of ToM and social competence, raw scores were plotted and visual analysis was used to identify change in performance during the intervention/follow-up period and compare this to baseline (Smith, 2012). Visual analysis was selected as it is considered an appropriate method for assessing within-subject change in case study interventions; it reduces likelihood of inaccurate interpretation of results that can arise when quantitative statistics are used in case studies and pilot trials; and it is increasingly used in cognitive neuropsychology (Smith, 2012).

Determining the overall success of the intervention

Both primary and secondary outcomes were considered in deciding whether the intervention was suitable for further evaluation using guidelines from Thabane et al. (2010): (i) Stop (main study not feasible); (ii) Continue but modify protocol (feasible with modifications); and (iii) Continue without modifications (feasible as is).

Results

The study included two girls with unilateral temporal lobe epilepsy (TLE) who were currently seizure free and on monotherapy. Child 2 had an earlier age of seizure onset, longer duration of seizure disorder, and more severe epilepsy at diagnosis than child 1. Performance on baseline cognitive measures differed for the two children. Child 1 performed in the average to high average range on tasks assessing IQ and general language skills, and her performance on executive functions (EF) tasks varied from average to very superior. Child 2 performed in the average range on tasks assessing IQ and general language skills. On tasks of EF, her scores were borderline to impaired, except for verbal working memory where her performance was average. The seven control children were matched as closely as possible to the two cases in age ($M=10.62$, $SD=0.95$), gender (all girls) and FSIQ ($M=106.43$, $SD=9.95$).

Primary feasibility outcomes

Recruitment and retention. We recruited two children with epilepsy from a total of five invited to participate (i.e., 40% recruitment). Both children completed the intervention, attending all scheduled sessions (nil attrition, 100% attendance). Thus, recruitment and retention goals were met.

Processing time for intervention modules. Processing time for modules are summarized in Table 2. All but one module was completed in 20 to 50 min; module 4, an introductory CBT module, was completed in 10 min and was said to require additional, similar activities to supplement and extend children's learning.

Suitability of tasks. Table 2 summarizes decisions about the suitability of each module. In-session observations indicated that activities were mostly engaging and appropriate for the age and ability level of children. Children made unprompted contributions, used relevant personal examples, and recalled strategies from week-to-week, indicating active learning and engagement with the material. Of the eight social scenarios covered, all but two were at an appropriate level of difficulty; ToM modules focusing on irony and double bluff were beyond children's capacity to understand, even with repetition and additional opportunities for learning.

Overall acceptability of the intervention. On the TEI, ratings for children and parents exceeded the moderate acceptability rating of 80% on both subscales: child progress (children = 87.3%; parents = 85.5%) and acceptability of treatment (children = 100%; parents = 97.5%). Investigator ratings indicated that children made adequate progress during treatment (83.3%) but ratings fell short of the moderate acceptability rating of 80% for improvements in ToM and social competence (60%).

In posttreatment interviews, parents indicated that treatment addressed relevant and specific social-emotional difficulties faced by their child. The most beneficial activities were initial CBT activities and ToM modules focusing on persuasion and misunderstandings, which parents said were common causes of social conflict and distress. Conducting sessions outside of school with a nonjudgmental peer was reportedly beneficial, as it normalized difficulties and reduced stigma. The duration and frequency of sessions was acceptable; however, both parents said that a slightly longer training period would have helped reinforce skills. Spontaneous improvements were documented in children's ability and willingness to communicate with parents about peer difficulties, confidence in dealing with peer pressure, emotion regulation, anxiety and ability to cope with disappointment and unexpected events. No adverse effects were reported by parents or otherwise documented by investigators during the study.

Barriers and facilitators to attendance. Treatment barriers reported by parents fell within one subscale on the BTPS: "stressors/obstacles that compete with treatment." These included issues with transportation, scheduling appointment times, parking, bad weather, and concurrent activities (i.e., parent work, child co-curricular activities). No barriers were reported in the other three domains on the BTPS, indicating that treatment was considered relevant, treatment demands were appropriate, and the therapeutic relationship was not a hindrance.

Secondary outcomes

Theory of Mind. Fig. 1A summarizes performance on ToM tasks, which was compared to the control mean: Strange Stories (Control $M = 20.7$, $SD = 1.6$) and ToMI (Control $M = 74.7$, $SD = 6.7$). Performance was categorized as impaired ($>2SD$ below control mean), borderline (≥ 1 and $<2SD$ below control mean) or average ($<1SD$ below control mean). On Strange Stories, performance was stable across baselines with both children falling in an impaired range (raw scores < 17.5). From baseline to posttreatment, performance improved for both children, placing them in an unimpaired range (raw scores > 19.1). At follow-up, both children achieved total raw scores of 20/24, which was comparable to the control mean. On the ToMI, child 1 scored in an unimpaired range at all time-points (raw scores > 68.0); she showed marginal improvements with her mother endorsing 5% more items at posttreatment compared to baseline 2. Child 2 scored in an impaired range at baselines on the ToMI (raw scores < 61.3) and in a borderline range at posttreatment and follow-up (raw scores = 61.4–68.0); her mother endorsed 10% more items at posttreatment compared to baseline 2.

Social competence. Fig. 1B) summarizes weekly ratings on the SPP, demonstrating a positive trend across the course of the intervention for both children; ratings in the intervention and posttreatment phases fell above baseline means. For child 1, self and parent-report social competence ratings were consistent with one another and showed a steady improvement during the intervention, which stabilized posttreatment. For child 2, self-report social competence ratings improved gradually during the intervention and stabilized posttreatment, whereas parent ratings showed a single improvement between the second and third week of the intervention, which was maintained for the remainder of the study. On SDQ total problems subscale, improvements were also seen; performance on the SDQ was categorized as falling in a clinical (raw score = 11–20), borderline (raw score = 5–10), or average (raw score = 0–5) range. Child 1 fell in a borderline range at baselines (raw scores = 6,6) and in an average range at posttreatment and follow-up (raw scores = 2,2). Child 2 fell in a clinical range at baselines (raw scores = 13,13). and in a borderline range at post treatment and follow-up (raw scores = 10,10).

Discussion

This study evaluated the feasibility, acceptability and efficacy of a novel CBT intervention with ToM training for children with epilepsy. To our knowledge, this is the first psychosocial treatment developed specifically for children with epilepsy that addresses ToM and social skills. Results showed that the intervention was feasible, acceptable, and safe for the two children (aged 9 and 11 years old) with TLE who participated. The intervention had adequate overall acceptability, notwithstanding several modifications that would enhance suitability of activities for the target group. Although findings about treatment efficacy are not conclusive at this early stage, visual inspection of scores showed improvements in ToM and

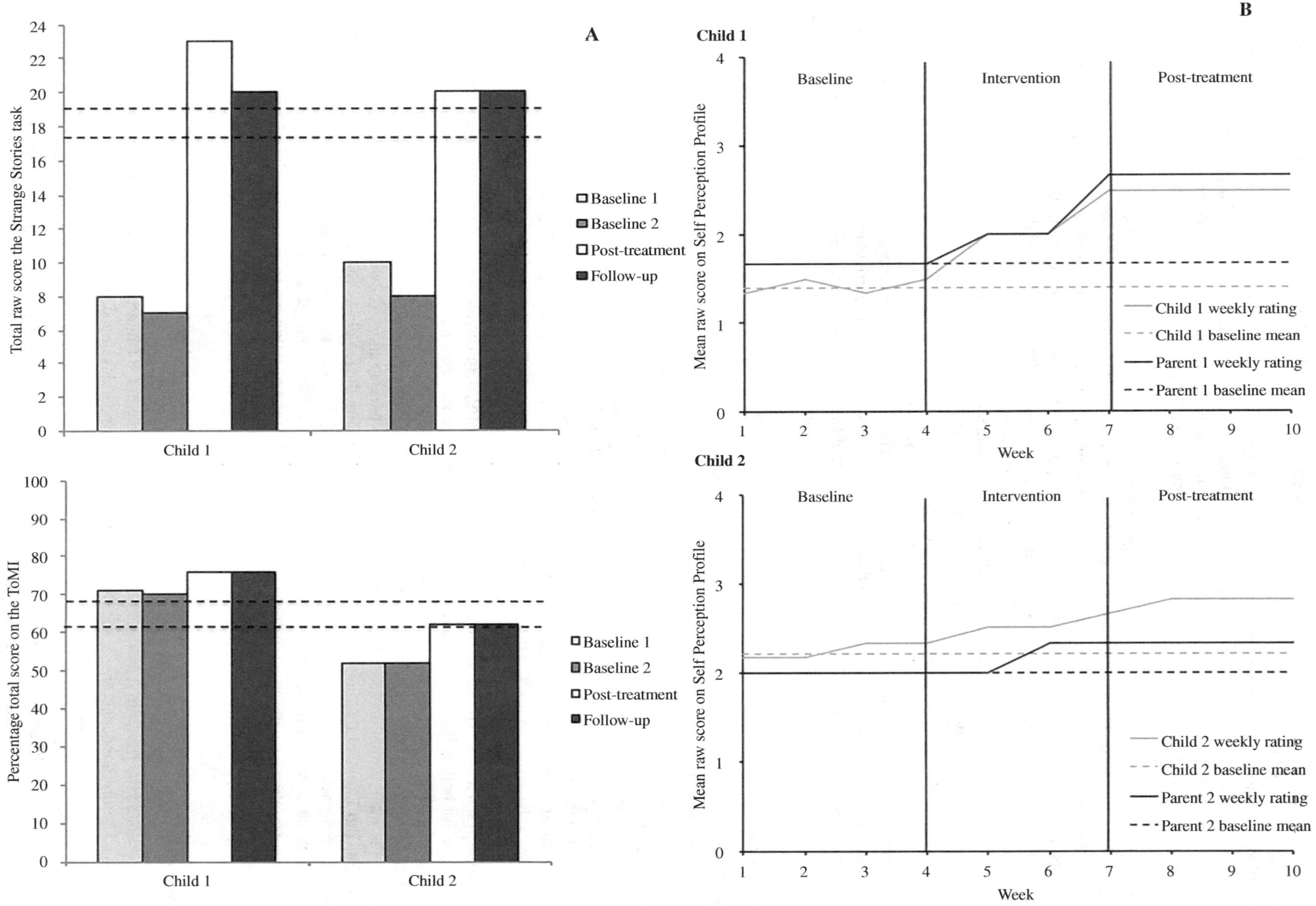

FIG. 1 Scores on measures of (A) Theory of Mind: Strange Stories (child-completed behavioral task) and Theory of Mind Inventory (parent-report questionnaire). Dotted lines represent scores falling in a clinical (>2*SD* below control mean), borderline (≥1 and <2*SD* below control mean) or average (<1*SD* below control mean) range. Higher scores represent better ToM. (B) Social competence: Self-Perception Profile (SPP) completed by parents and children weekly. Higher scores represent better social competence.

social competence for both children. Overall, the results suggest that the intervention is feasible with modifications (Thabane et al., 2010) and justify extended evaluation in a phase II feasibility trial.

Results of this small-scale, pilot study provide useful, practical insights that may help development of future CBT interventions for children with epilepsy and other neurodevelopmental disorders. The inclusion of basic CBT activities and ToM modules focusing on persuasion and misunderstandings were reportedly beneficial in addressing the specific social-emotional difficulties faced by children in the study. Running sessions outside of school and including a same-age peer with epilepsy were said to normalize children's difficulties, reduce stigma, and provide an opportunity for in-situ social learning. In session interactions allowed for modeling, feedback and reinforcement, which have been shown to improve interpersonal problem-solving and social perception in social skills interventions (Spence, 2003). The inclusion of parent education was also beneficial, supporting prior findings that multimodal interventions (i.e., with child and parent components) are more beneficial than child-only or parent-only interventions for addressing social skills deficits in children (Spence, 2003).

A number of logistical barriers to attendance were identified, i.e., transportation, appointment times, parents work, co-curricular activities. While these barriers are inherent in attending therapy, they raise the question as to whether online administration of the intervention would be more suitable, which families can access remotely and at times convenient for them. Some aspects of the program (e.g., the parent education component that was delivered via written handouts) seem particularly well suited to online delivery. Whether online administration of child sessions is suitable/beneficial for children with epilepsy requires investigation; however, emerging evidence is promising with online administration of a social skills program found to be beneficial for social problem solving in young adolescents (9–13 years old; DeRosier & Thomas, 2019).

There were several methodological observations, which should be considered in future research. Larger ToM improvements were observed on the child-completed behavioral task (Strange Stories) than on the parent-report questionnaire (ToMI). The ToMI was originally developed for children with ASD (Hutchins et al., 2012), and has previously been found to lack sensitivity among children with epilepsy (8–16 years) (Stewart, Catroppa, Gonzalez, et al., 2019). It may be more prudent for researchers to use validated behavioral tasks (e.g., Strange Stories) until parent-report scales are validated for this group. Investigator ratings of the TEI indicated that children did not make expected improvements in ToM or social competence, which contradicted findings on behavioral tasks and questionnaires. Perhaps, the investigator scale did not tap into aspects of ToM and social competence that were targeted in the intervention. Alternatively, investigator's expectations of children may have been unrealistic or because behavioral observations were limited to a treatment setting, investigators may not have witnessed the scope of behavioral change. The appropriateness of the TEI investigator form requires further evaluation.

Limitations

The current study provides a proof of concept; it is yet to be seen whether this promising intervention is effective for other children with epilepsy. Running the intervention with two children and two investigators allowed us to provide one-to-one support to children at times; it is unclear whether the intervention would be as beneficial in larger groups. Homework tasks were optional, and we did not record compliance or consider the impact homework completion on outcomes. We only examined outcomes at 3-week follow-up, which is not long enough to determine whether gains were maintained. Finally, neither parents nor investigators were blind to the intervention children received.

Conclusions

The current study provides preliminary evidence that a novel CBT intervention targeting ToM and social competence is feasible, acceptable, and safe for children with epilepsy. Findings support evaluation of the intervention in a phase II feasibility trial. This is the first published study evaluating a psychosocial treatment that addresses social-cognitive impairments in children with epilepsy. Communicating findings may increase awareness and interest in developing much needed interventions for this group.

Summary points

- Impaired ToM and social problems are common in children with epilepsy.
- We developed a cognitive-behavioral intervention targeting ToM and social problems.
- The intervention was feasible, acceptable, and safe for two children with epilepsy.

- We found improvements in ToM and social competence for both children.
- Our findings justify extension of the intervention in a phase II feasibility trial.

Acknowledgments

The authors acknowledge the children and families who participated in the study.

Declaration of interest statement

The authors have no conflict of interest to declare.

References

Berg, A. T., Vickrey, B. G., Testa, F. M., Levy, S. R., Shinnar, S., & DiMario, F. (2007). Behavior and social competency in idiopathic and cryptogenic childhood epilepsy. *Developmental Medicine & Child Neurology*, *49*(7), 487–492.

Craig, P., Dieppe, P., Macintyre, S., Michie, S., Nazareth, I., & Petticrew, M. (2008). Developing and evaluating complex interventions: The new Medical Research Council guidance. *BMJ*, *337*, a1655.

DeRosier, M. E., & Thomas, J. M. (2019). Hall of heroes: A digital game for social skills training with young adolescents. *International Journal of Computer Games Technology*, *6*, 1–12.

Frith, C., & Frith, U. (2005). Theory of mind. *Current Biology*, *15*(17), 644–645.

Goodman, R. (1997). The strengths and difficulties questionnaire: A research note. *Journal of Child Psychology and Psychiatry*, *38*, 581–586.

Granleese, J., & Joseph, S. (1994). Reliability of the Harter self-perception Profile for children and predictors of global self-worth. *The Journal of Genetic Psychology*, *155*, 487–492.

Grave, J., & Blissett, J. (2004). Is cognitive behavior therapy developmentally appropriate for young children? A critical review of the evidence. *Clinical Psychology Review*, *24*(4), 399–420.

Happé, F. G. (1994). An advanced test of theory of mind: Understanding of story characters' thoughts and feelings by able autistic, mentally handicapped, and normal children and adults. *Journal Autism and Developmental Disorders*, *24*(2), 129–154.

Hofmann, S. G., Doan, S. N., Sprung, M., Wilson, A., Ebesutani, C., Andrews, L. A., … Harris, P. L. (2016). Training children's theory-of-mind: A meta-analysis of controlled studies. *Cognition*, *150*, 200–212.

Hutchins, T. L., Prelock, P. A., & Bonazinga, L. (2012). Psychometric evaluation of the theory of mind inventory (ToMI): A study of typically developing children and children with autism spectrum disorder. *Journal Autism and Developmental Disorders*, *42*(3), 327–341.

Kazdin, A. (1980). Acceptability of alternative treatments for deviant child behavior. *Journal of Applied Behavioural Analysis*, *13*(2), 259–273.

Kazdin, A., French, N., & Sherick, R. (1981). Acceptability of alternative treatments for children: Evaluations by inpatient children, parents, and staff. *Journal of Consulting and Clinical Psychology*, *49*(6), 900.

Kazdin, A. E., Holland, L., Crowley, M., & Breton, S. (1997). Barriers to treatment participation scale: Evaluation and validation in the context of child outpatient treatment. *Journal of Child Psychology and Psychiatry*, *38*(8), 1051–1062.

Pink, B. (2013). *Socio-economic indexes for areas (SEIFA)*. Canberra (Australia): Australian Government.

Rantanen, K., Eriksson, K., & Nieminen, P. (2012). Social competence in children with epilepsy—A review. *Epilepsy & Behavior*, *24*(3), 295–303.

Reilly, C., Atkinson, P., Das, K. B., Chin, R. F., Aylett, S. E., Burch, V., … Neville, B. G. (2015). Features of autism spectrum disorder (ASD) in childhood epilepsy: A population-based study. *Epilepsy & Behavior*, *42*, 86–92.

Smith, J. (2012). Single-case experimental designs: A systematic review of published research and current standards. *Psychological Methods*, *17*, 510.

Spence, S. H. (2003). Social skills training with children and young people: Theory, evidence and practice. *Child and Adolescent Mental Health*, *8*, 84–96.

Stewart, E., Catroppa, C., Gill, D., Webster, R., Lawson, J., Mandalis, A., … Lah, S. (2018). Theory of mind and social competence in children and adolescents with genetic generalised epilepsy (GGE): Relationships to epilepsy severity and anti-epileptic drugs. *Seizure*, *60*, 96–104.

Stewart, E., Catroppa, C., Gonzalez, L., Gill, D., Webster, R., Lawson, J., … Lah, S. (2019). Theory of mind and social competence in children and adolescents with temporal lobe epilepsy. *Neuropsychology*, *33*(7), 986–995.

Stewart, E., Catroppa, C., & Lah, S. (2019). A novel cognitive behavioural intervention with Theory of Mind (ToM) training for children with epilepsy: Protocol for a case series feasibility study. *BMC Pilot and Feasibility Studies*, *5*, 1–12.

Thabane, L., Ma, J., Chu, R., Cheng, J., Ismaila, A., Rios, L. P., … Goldsmith, C. H. (2010). A tutorial on pilot studies: The what, why and how. *BMC Medical Research Methodology*, *10*, 1.

Weimer, A. A., Sallquist, J., & Bolnick, R. R. (2012). Young children's emotion comprehension and theory of mind understanding. *Early Education Development*, *23*(3), 280–301.

Chapter 39

Cognitive behavioral therapy and death anxiety: A case of children in conflict settings during the COVID-19 pandemic

Pegah AM Seidi[a] and Dilshad Jaff[b]

[a]*Research Center, University of Garmian, Kalar, Iraq* [b]*Gillings School of Global Public Health, University of North Carolina at Chapel Hill, Chapel Hill, NC, United States*

Abbreviations

CBT cognitive behavioral therapy
DAQ death anxiety measure
IES-R impact of event scale-revised
PTSD posttraumatic stress disorder
RACDS revised child anxiety and depression scale
RCI reliable change index
TF-CBT trauma focused-cognitive behavioral therapy

Introduction

Anxiety and death anxiety are normal emotions and experience. However, in individuals with other disorders, or in vulnerable populations afflicted by conflict, it can be an exaggerated problem that cuts across diagnostic categories which should be considered in assessment and treatment.

Death anxiety was reported to be a factor leading to anxiety disorders (Furer & Walker, 2008). Previous findings have demonstrated the impact of death anxiety on a wide range of behaviors (Dar-Nimrod, 2012; McGregor et al., 1998; Menzies & Dar-Nimrod, 2017) in adults and adolescents (Ens & Bond, 2005; Melhem, Porta, Shamseddeen, Payne, & Brent, 2011). To our knowledge, available literature mainly focuses on older adults or individuals with chronic illness such as war veterans, nurses or cancer patients (Gonen et al., 2012; Latha, Sahana, Mariella, Subbannayya, & Asha, 2013; Nia, Ebadi, Lehto, & Peyrovi, 2015; Nia, Lehto, Ebadi, & Peyrovi, 2016). As a result, practitioners have limited knowledge about the prevalence and treatment of death anxiety in children specially those who live in conflict settings (Sagna, Gallo, & Pontone, 2014).

Given the long history of exposure to violence and conflict and the ongoing COVID-19 pandemic, the Iraqi population—particularly children—are at risk of different psychological distress (Seidi, Ardebil, & Jaff, 2020; Seidi, Jaff, Connolly, & Hoffart, 2021). According to our experience as frontline practitioners, death anxiety is one of these conditions.

Cognitive behavior therapy (CBT) is known as one of the most effective interventions for anxiety among different age groups and populations (Menzies, Zuccala, Sharpe, & Dar-Nimrod, 2018). Regarding health anxiety and death anxiety, CBT mainly focuses on careful assessment, case formulation, and tailored intervention (Furer, Walker, & Stein, 2007). Earlier findings suggested that CBT is an effective intervention for Iraqi children affected by conflicts (Damra & Nassar, 2014). Building on that, this study aimed to report the use of applying CBT for reducing death anxiety in a child affected by armed conflict during the COVID-19 pandemic.

Case presentation

In this study, a before-and-after single-case design was utilized. A 13-year-old boy living in an internally displaced people (IDPs) camp in the Garmian Region of the Kurdistan Region of Iraq (KRI) visited the Counseling Centre of the Garmian Medical City September 2020 along with his mother suffering from severe anxiety.

Handbook of Lifespan Cognitive Behavioral Therapy. https://doi.org/10.1016/B978-0-323-85757-4.00010-9

He lost his father and one of his brothers during the ISIS attacks in 2014 in Iraq. His mother reported the same anxiety symptoms of her son. He had left school 3 years ago, do not have normal relations with his peers, and after the death of each of his acquaintances, his anxiety increased. He is attached to his family members and was disturbed when they left their house. He frequently checked on family members and asked questions about death, and his appetite and sleep were disturbed. After the outbreak of the COVID-19, the anxiety had reached to the highest level and had significantly affected the daily life of the child and his family.

After interviews and assessments by both a psychologist and a psychiatrist, symptoms of posttraumatic stress disorder (PTSD), depression, and general anxiety at the same time observed as well. However, their severity and number of symptoms were not prevalent enough to be identified as main diagnoses. Among all symptoms, the main theme was the fear of oneself death mainly and loved ones' death in some instances also. Based on these findings and observations, the diagnosis was confirmed as death anxiety.

After sharing information about the problem and the treatment process with his mother, the client and his mother were informed of the study and consent form was obtained. In order to determine the subject's baseline of anxiety, no intervention was delivered for 4 weeks and only relevant questionnaires were completed. Afterward, a CBT protocol for death anxiety was designed for 8 weeks (one session per week with a duration of 45 min for each session). A modified version of CBT, in which some techniques targeting death and death-related thoughts such as challenging and reassessment of thoughts about "risk of sudden death in daily life" were added to the classic model of this intervention for anxiety disorders (Leahy, Holland, & McGinn, 2012).

After every session, and also 1 month, 3 months, and 6 months after the last session, related questionnaires were also completed as follow-up. The client's medication for 2 years prior to the therapy sessions consisted of: Propranolol tablet10 mg, Fluoxetine tablet 20 mg, and Centerline tablet 50 mg, which he continued using during these sessions. No immediate family members suffered from anxiety or any other mental disorders. His mother showed only signs of dysthymia. The first period of anxiety had led the client to quit school and he was ill-treated. The mother decided to visit the therapist because of her son' thoughts about his and family members as well, which had been intensified after COVID-19 pandemic.

The availability of tested and validated tools for data collection in this region has always been one of the most serious challenges. There are no valid and reliable tools in Kurdish language which is the language of this population. In addition to that, death anxiety is a broader issue, and not only in this region but also globally, there are few tools available about death anxiety in children. Because of that, the main questionnaire of the study was death anxiety measure (DAQ, Koocher, O'Malley, Foster, & Gogan, 1976). But due to the association between death anxiety with depression, anxiety, and PTSD in this case, three other questionnaires were used.

The DAQ is comprised of 15 items describing specific possible worries (e.g., "*I worry about dying*" or "*I worry about not knowing what to expect when I die*"). The subject is asked to note whether she/he worries about each issue "not at all, sometimes, or very much." Corresponding scores of 0, 1, and 2 are assigned to each choice. The total score is obtained by adding each score (between 0 and 30). This measure reported as a valid and reliable tool (Koocher et al., 1976).

Revised child anxiety and depression scale (RACDS) was used for measuring the symptoms of depression and anxiety. This questionnaire includes 47 items with 6 subscales including: separation anxiety, social phobia, generalized anxiety disorder, panic disorder, obsessive–compulsive disorder, and depressive disorder. RACDS asks respondents to mark the questionnaire to indicate how often each element occurs to them. Questions are scored from 0 to 3 corresponding to never, sometimes, often, and always (De Ross, Gullone, & Chorpita, 2002).

For PTSD, impact of event scale-revised (IES-R) was used. This scale is a self-report questionnaire, which includes 22 items to measure posttraumatic phenomena. This instrument has three categories according to the Diagnostic and Statistical Manual of Mental Disorders (DSM-5) criteria for PTSD: (1) intrusion, (2) avoidance, and (3) hyperarousal. Participants answer how much they experienced a specific symptom such as thinking or having nightmares about the traumatic event, in the past 7 days. Items are rated on a 5-point Likert scale, from 0 ("not at all") to 4 ("extremely") with total score between 0 and 88, and 33 is cut point for PTSD. Reported internal consistency is 0.96 with studies showing support for a good validity. The Kurdish version of this scale has been used in another study in the KRI (Rometsch, Denkinger, Engelhardt, et al., 2020).

The translation of other questionnaires was performed according to the Functional Assessment of Chronic Illness Therapy's (FACIT) translation methodology (Sperber, 2004). First, the questionnaires were translated into Kurdish by two independent Kurdish-English bilingual experts, and then the two versions were revised and combined by the correspondence author. Another independent translator who did not have access to the original version conducted a back translation from Kurdish into English. The Kurdish translation and back-translation were critically analyzed and compared with the original English version and confirmed by researchers. Primary validity and relatability of these questionnaires were investigated as a part of research project (unpublished) of two students at the University of Garmian, located at the same regions.

Also, a clinical interview was conducted according to the diagnostic criteria in DSM-V, and diagram and graphic analysis for analyzing the data were used. Moreover, we report recovery rate using the comparison of baseline scores with follow-up scores and utilized the formula of the reliable change index (RCI) to calculate changes after the intervention.

The changes in the case's scores on all scales, as well as in the recovery rate and RCI indicated that, the client experienced "severe death anxiety," "anxiety," "depression," and "PTSD" at the baseline. However, it was observed that during the process of therapy, his scores showed significant reduction and he experienced "normal death anxiety" and none of anxiety, depression, and PTSD symptoms in the last therapy session (Table 1). In the last follow-up session, the client had a death anxiety score of 2, indicating a 92% recovery. Also, according to the results of a graphic analysis, all symptoms demonstrated a decreasing trend at the end of therapy and follow-up sessions (Fig. 1).

The "clinical significance of changes" approach was used for a more accurate determination of intervention results according to Jacobson and Truax (Fang, Chang, Chen, et al., 2014). In this index, we can conclude that the changes are significant if RCI obtained was greater than 1.96. The RCI for all tests, except IES-R, was greater than 1.96 (9.92, 13.81, 7.09, and 1.23). As this index is significant for death anxiety, anxiety, and depression, it can be concluded that these results were nonrandomness but related to the intervention that was performed. Therefore, these observed clinical changes are significant. The client is reported as "improved" in terms of the death anxiety because of the agreement between the results of the screening and interview.

According to our experience of using a CBT protocol for death anxiety, this therapy has decreased the severity of anxiety and related symptoms in the client. It is noteworthy that these results were maintained during all follow-up sessions. As the model used in this study can control the effect of time, and based on the analyses methods, the improvement in the severity of symptoms is the direct result of the intervention.

Our findings are in line with similar studies (Kodal et al., 2018; Sigurvinsdóttir, Jensínudóttir, Baldvinsdóttir, Smárason, & Skarphedinsson, 2020; Wergeland et al., 2014). Basic assumptions in CBT referred that anxious people mainly hold unhelpful and unrealistic beliefs about a specific subject (in this case death); and usually behave in the same way of thinking. Therefore, emotions are influenced by these thoughts and behaviors and impact on them at the same time. As the result, the person gets stuck in a negative cycle and signs and symptoms of anxiety disorders appear. In our study, the CBT treatment protocol focused on assisting with transforming and replacing unhelpful and irrational thoughts with a more balanced perspective on mortality. Behavioral procedure included some techniques to offer the client useful action plan for managing behaviors that trigger anxiety, for example, turning attention away from the ongoing COVID-19 pandemic, not following its mortality statistics and its updates.

The results showed the significant symptoms reduction and the maintenance of improvement in all the follow-up sessions. CBT has managed to prevent the deterioration of mood and enhanced the preventive dimension of therapy by altering negative thought patterns and teaching more effective behaviors such as relaxation, problem solving, and having healthy life style instead of thinking about death (Leahy et al., 2012).

Although the RCI was not significant for PTSD symptoms, as the cut-off point is 33 for IES-R (Rometsch et al., 2020) and the client's score was ≤ 33 during last session of therapy and the follow-up sessions, he is within the normal range in terms of PTSD too. One explanation for this different result could be the nature and the process of treatment in PTSD cases where therapy sessions in trauma-focused CBT are longer than the standard 50 min session and frequently last for about 90 min. Number of sessions is also another main difference. TF-CBT called for between 8 and 12 sessions and allows for the number of sessions to be more than 20 in more complex PTSD cases (Wild & Ehlers, 2010), while our sessions were 45 min and delivered in 8 sessions. Limitations of the studies about death anxiety, lack of appropriate/validated measurement tools, and the model used in the present study limit its findings to be generalized. Therefore, more extensive studies about death anxiety and its treatment with clinical and nonclinical examples are suggested.

Summary points

- This chapter focused on death anxiety as one of the predictors of anxiety disorder that can also cuts across diagnostic categories which should be considered in assessment and treatment.
- Death anxiety is somehow neglected in children.
- People in Iraq as a conflict setting are at risk of more mental disorders and psychological problems.
- COVID-19 pandemic faced practitioners with new challenges of mental health problem in vulnerable population such as IDPs.

TABLE 1 The case's scores on scales from the baseline up to 6 months after the therapy (follow-up).

	Sessions																
	Baseline				Intervention								Follow-up			Recovery rate	RCI
Variables	1st	2nd	3rd	4th	1st	2nd	3rd	4th	5th	6th	7th	8th	1st	2nd	3rd		
Death anxiety	25	27	24	23	17	16	14	12	8	7	6	2	4	2	2	92%	9.92
Anxiety	120	114	109	101	96	94	87	72	68	63	45	31	25	22	15	87%	13.81
Depression	25	25	23	20	19	19	19	15	15	12	9	4	2	1	4	84%	7.09
PTSD	64	61	60	58	58	89	45	37	33	27	21	15	11	12	8	87%	1.23

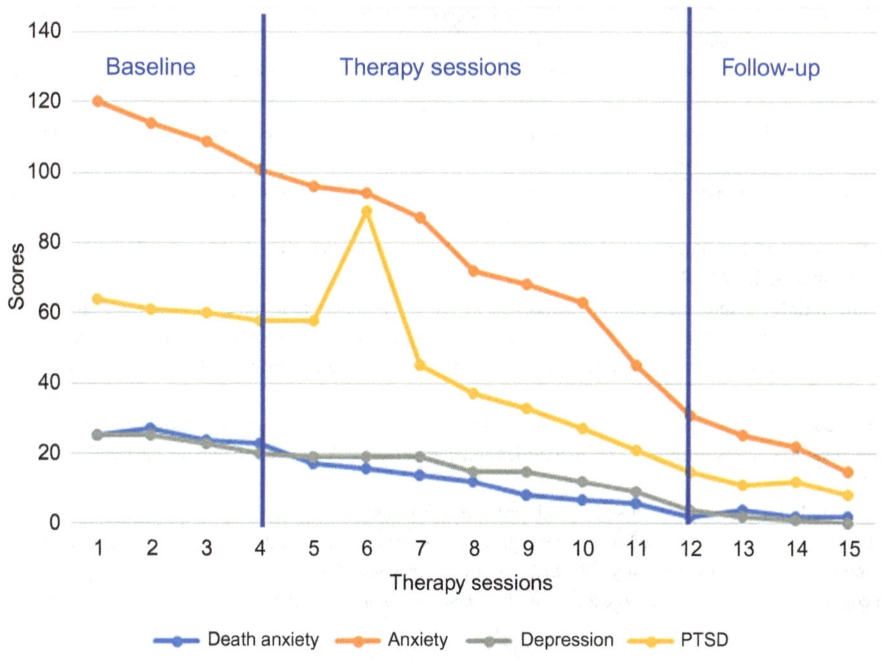

FIG. 1 Outcomes in different CBT sessions (baseline, intervention, and follow-up).

- Cognitive behavioral therapy is one of the successful methods for treatment of anxiety disorders.
- CBT were delivered as an eight sessions protocol with homework (2 months) and follow-up (6 months).
- CBT intervention delivered in this study showed a significant effect on the treatment of death anxiety along with general anxiety and depression.

References

Damra, J. K., & Nassar, Y. H. (2014). The effect of cognitive behavioral trauma focused therapy model on sample of war Children's depression symptomatology. *Dirasat*, *41*, 445–461.

Dar-Nimrod, I. (2012). Viewing death on television increases the appeal of advertised products. *The Journal of Social Psychology*, *152*, 199–211.

De Ross, R. L., Gullone, E., & Chorpita, B. F. (2002). The revised child anxiety and depression scale: A psychometric investigation with Australian youth. *Behaviour Change*, *19*(2), 90–101.

Ens, C., & Bond, J. B. (2005). Death anxiety and personal growth in adolescents experiencing the death of a grandparent. *Death Studies*, *29*, 171–178.

Fang, C. K., Chang, M. C., Chen, P. J., et al. (2014). A correlational study of suicidal ideation with psychological distress, depression, and demoralization in patients with cancer. *Support Care Cancer*, *22*(12), 165–174.

Furer, P., & Walker, J. R. (2008). Death anxiety: A cognitive-behavioral approach. *Journal of Cognitive Psychotherapy*, *2008*(22), 167–182.

Furer, P., Walker, J. R., & Stein, M. B. (2007). *Treating health anxiety and fear of death: A practitioner's guide*. New York: Springer Publishing.

Gonen, G., Kaymak, S. U., Cankurtaran, E. S., Karslioglu, E. H., Ozalp, E., & Soygur, H. (2012). The factors contributing to death anxiety in cancer patients. *Journal of Psychosocial Oncology*, *30*(3), 347–358.

Kodal, A., Fjermestad, K., Bjelland, I., Gjestad, R., Öst, L. G., Bjaastad, J. F., … Wergeland, G. J. (2018). Long-term effectiveness of cognitive behavioral therapy for youth with anxiety disorders. *Journal of Anxiety Disorders*, *53*, 58–67.

Koocher, G. P., O'Malley, J. E., Foster, D., & Gogan, L. G. (1976). Death anxiety in normal chileren and adolescents. *Psychiatria Clinica*, *9*(3–4), 220–229.

Latha, K., Sahana, M., Mariella, D., Subbannayya, K., & Asha, K. (2013). Factors related to life satisfaction, meaning of life, religiosity and death anxiety in health care staff and students: A cross sectional study from India. *Online Journal of Health and Allied Science*, *12*, 7.

Leahy, R. L., Holland, S. J. F., & McGinn, L. K. (2012). *Treatment plans and interventions for depression and anxiety disorders* (2nd ed.). New York: Gulliford. chap. 4.

McGregor, H. A., Liebermann, J. D., Greenberg, J., Solomon, S., Arndt, J., Simon, L., et al. (1998). Terror management and aggression: Evidence that mortality salience motivates aggression against worldview-threatening others. *Journal of Personality and Social Psychology*, *74*, 590–605.

Melhem, N. M., Porta, G., Shamseddeen, W., Payne, M. W., & Brent, D. A. (2011). Grief in children and adolescents bereaved by sudden parental death. *Archives of General Psychiatry*, *68*, 911.

Menzies, R. E., & Dar-Nimrod, I. (2017). Death anxiety and its relationship with obsessive compulsive disorder. *Journal of Abnormal Psychology*, *126*, 367–377.

Menzies, R. E., Zuccala, M., Sharpe, L., & Dar-Nimrod, I. (2018). The effects of psychosocial interventions on death anxiety: A meta-analysis and systematic review of randomised controlled trials. *Journal of Anxiety Disorders*, *59*, 64–73.

Nia, H., Lehto, R., Ebadi, A., & Peyrovi, H. (2016). Death anxiety among nurses and health care professionals: A review article. *International Journal of Community Based Nursing and Midwifery*, *4*, 2–10.

Nia, H. S., Ebadi, A., Lehto, R. H., & Peyrovi, H. (2015). The Experience of Death Anxiety in Iranian War Veterans: A Phenomenology Study. Lerner, R. M., Easterbrooks, M.A., Mistry, J., Weiner, I.B., editors. *Death Studies*, *39*, 281–287.

Rometsch, C., Denkinger, J. K., Engelhardt, M., et al. (2020). Pain, somatic complaints, and subjective concepts of illness in traumatized female refugees who experienced extreme violence by the "Islamic state" (IS). *Journal of Psychosomatic Research*, 109931.

Sagna, A., Gallo, J. J., & Pontone, G. M. (2014). Systematic review of factors associated with depression and anxiety disorders among older adults with Parkinson's disease. *Parkinsonism & Related Disorders*, *20*, 708–715.

Seidi, P. A. M., Jaff, D., Connolly, S. M., & Hoffart, A. (2021). Applying cognitive behavioral therapy and thought field therapy in Kurdistan region of Iraq: A retrospective case series study of mental-health interventions in a setting of political instability and armed conflicts. *Explore*, *17*(1), 84–91.

Seidi, P., Ardebil, M. D., & Jaff, D. (2020). COVID-19 pandemic: New challenge to securing mental well-being in conflict settings. *Asian Journal of Psychiatry*, *51*, 102151.

Sigurvinsdóttir, A. L., Jensínudóttir, K. B., Baldvinsdóttir, K. D., Smárason, O., & Skarphedinsson, G. (2020). Effectiveness of cognitive behavioral therapy (CBT) for child and adolescent anxiety disorders across different CBT modalities and comparisons: A systematic review and meta-analysis. *Nordic Journal of Psychiatry*, *74*(3), 168–180.

Sperber, A. D. (2004). Translation and validation of study instruments for cross-cultural research. *Gastroenterology*, *126*, S124e8.

Wergeland, G. J., Fjermestad, K. W., Marin, C. E., Haugland, B. S., Bjaastad, J. F., Oeding, K., … Heiervang, E. R. (2014). An effectiveness study of individual vs. group cognitive behavioral therapy for anxiety disorders in youth. *Behaviour Research and Therapy*, *57*, 1–12.

Wild, J., & Ehlers, A. (2010). Self-study assisted cognitive therapy for PTSD: A case study. *European Journal of Psychotraumatology*, *1*. https://doi.org/10.3402/ejpt. v1i0.5599.

Chapter 40

Children with academic difficulties and emotional problems: Use of cognitive behavior therapy

Susmita Halder[a,b] and Akash Kumar Mahato[b]
[a]*Department of Psychology, St. Xavier's University, Kolkata, India* [b]*Amity Institute of Behavioural Health and Allied Sciences, Amity University Kolkata, Kolkata, West Bengal, India*

Abbreviations

CBT cognitive behavior therapy
ADHD attention deficit hyperactivity disorder
BYI Beck Youth Inventory

Introduction

There has been a gradual change in the pattern of consultation by parents in child psychiatry clinics. Parents do not consult these clinics only in grave conditions like intellectual disability, attention deficit hyperactivity disorder (ADHD) or autism or epilepsy, but also seek psychological consultation for their wards who are underperforming or have some behavioral and emotional issues (Friedberg & McClure, 2015; Sukhodolsky et al., 2016). These children and adolescents brought for consultation for poor academic performance often do not meet criteria for intellectual disability, ADHD or learning disability (Schoenfeld & Mathur, 2009).

The terms poor academic performance or academic difficulties include failure or difficulty in meeting performance parameters by the student which are put by parents and family, teachers, as well as students themselves. Children presented with academic difficulties may have more significant psychosocial concerns than intellectual limitations that could be the triggering as well as perpetuating factor for these academic difficulties. These could be either in form of unwarranted and unrealistic parental expectations, dysfunctional familial environment and discords, poor self-esteem, and peer relationship issues to name a few. The manifestation of academic difficulties in these children and adolescents is never in terms of academic performance only. Their difficulties are played out in the larger context of the school environment and curriculum and represent a mismatch between the characteristics of the student and various demands from school, family, as well as social circles. Children with such profile are commonly presented with emotional difficulties (Eva de Hullu, Sportel, Nauta, & de Jong, 2017; Foorman & Torgesen, 2001; Gresham & Kern, 2004); eventually, it becomes a vicious cycle where academic difficulties, the psychosocial and familial characteristic of the child, and emotional difficulties push each other.

The firsthand impression of these children and adolescent may go in favor of oppositional defiance/adolescent crisis, or plain lack of motivation. Parents, family members, and school may lack insight behind the phenomenon and subsequently children are often subject to unwarranted disciplinary actions, restrictions, or punitive measures, further complicating the situation and create a very difficult communication pattern between the child and parent. In absence of any clinical diagnosis, when psychological assessment may not reveal clinically significant symptoms; the management of such children who might be rigid or secretive poses a difficult challenge for therapists (Bergan, 1995).

Irrespective of a formal diagnosis or not, the problem remains genuine, as these children face considerable difficulty in coping at school with teachers and peers, as well as at home (Cunningham, Brandon, & Frydenberg, 1999). There may not be any epidemiological data, but this would not be outrageous to suggest that this population of children is one of the biggest client populations in child psychiatry or guidance clinics in Indian metros and major cities. These children are often presented with symptoms of anxiety, low self-esteem, difficulty in social interaction, irritability, and low mood (Raman, 2020).

Handbook of Lifespan Cognitive Behavioral Therapy. https://doi.org/10.1016/B978-0-323-85757-4.00001-8

Scope of CBT for children and adolescents with academic difficulties

Use of cognitive behavior therapy (CBT) for children and adolescents is well documented across cultures (Cole, Bruschi, & Tamang, 2002; Consoli et al., 2018; Yeo & Choi, 2011). There is good body of evidence supporting effectiveness of CBT in children with depression, anxiety spectrum disorders, school phobia, eating disorders, self-harm, and conduct problems (Scott, 2009). However, its use for children and adolescents with academic difficulties in absence of any clinical diagnosis is a delicate arena. Consulting mental health professionals has its own stigma, which though has weakened in recent times but continues. In absence of a clinical diagnosis, majority of these consultations are on reference from school.

The chapter focuses on the application of CBT on children and adolescents presented with academic difficulties and emotional problems, but who did not fall into the diagnostic category of ADHD, intellectual or learning disability.

Procedure: Two child cases brought by their parents for psychological consultation primarily for their concern with their deteriorating academic performance are being discussed. The therapy procedure primarily followed the conventional history taking and interview of the children and their parents. Clinically significant symptoms like any issues with intellectual capability, attention deficit or learning disability, and mood symptoms were screened with formal assessment as required. Beck Youth Inventory (BYI) (Beck, Beck, Jolly, & Steer, 2005) was uniformly used to assess symptoms of anxiety, depression, anger, disruptive behavior, and self-concept. Therapy sessions started with rapport establishment, imparting psychoeducation, formulating an academic plan, and CBT.

The details of the cases are as follows:

Case 1

Master AM was 12-year-old male, only child of his parents and studied in grade 7. He was brought by his mother for psychotherapeutic consultation with chief complaints of deteriorating academic performance and drop in his examination grades, poor concentration during studies, anger outburst, and indulgence in fight with peers at school. Mother reported that AM was good in studies since childhood, and always secured good rank in his class. However, for last 1 year, he was found being unable to concentrate on his studies. Subsequently, his grades dropped significantly in recent examinations with repeat of similar performance. Parents received call from AM's school not only for his falling grades but also for incident of showing aggression and getting into fight with peers. Mother reported no major problem behaviors on part of AM at home, though he was restless at times.

In initial sessions, AM was very resistant to participate mentioning, "I am not sure why I am brought here. I do not see any reason." He refuted the complaints from school and his mother's concern stating he had his own reasons and he needs not be blamed. He categorically mentioned that his academic grades and reported behavioral issues of anger were two different issues. A nonjudgmental and supportive approach toward AM helped to build rapport with AM and he revealed his preoccupations and peer issues which appeared to have influenced his behavior and brought undesired consequences. He liked one of his female classmates but was outright rejected by her because he was not good looking enough to be considered. His other classmates came to know of this and made fun of him and bullied him. He felt offended but tried to control his anger, but repeated incidents even in front of very junior students made him angry and eventually he got into fight with his peers.

Keeping a nonjudgmental and supportive approach, AM was explained how behavior and emotions are related as in ABC model of CBT. A plausible explanation of his poor concentration and academic deterioration was made to AM as well as his parents. This helped in establishing the need of structured therapy module to change AM's dysfunctional thoughts and looking at the whole situation larger than pure behavioral problems.

In subsequent sessions, AM's assumptions and automatic thought were explored and challenged. The automatic negative thoughts centered on appearance and self-worth and were revealed as, *"I am worthless as a girl rejected me for my looks"... "Eventually I will be rejected by others too."* The preoccupation with these thoughts made AM emotionally disturbed impacting his academic concentration and efforts. Poor performance in academics compared to his previous years academic performance further lowered his self-esteem.

Case 2

Miss DG was a 10-year-old girl, elder among two children of her parents and studied in fifth grade. She was brought by her mother and was presented with complaints of poor concentration, deteriorating academic performance, getting easily distracted, irritability, and crying behavior mostly at home. As per DG's mother, she was a good and obedient student but for last 1 year, certain changes in her behavior were noticed. Though she maintained her daily routine but appeared unmindful.

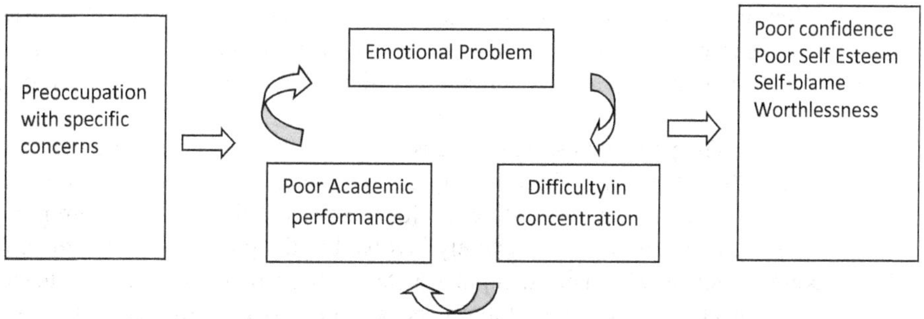

FIG. 1 Cognitive model of problem manifestation.

She was unable to concentrate during study hours. When interviewed, she appeared sad and took time to open with the therapist.

Upon establishment of rapport, DG revealed that after birth of her younger sister who was 2.5 years of age, things had changed for her. Mother was giving more attention to her younger sister, and was getting all the care, toys, and new clothes. She felt that even elderly people from her neighborhood too cared more for her younger sister. She felt being neglected and always thought of how to get her mother's attention back. When asked if she was angry and jealous with her sister, she said she liked her sister and never felt that way. She mentioned she felt bad about the change in her academic performance and was trying to give adequate effort. However, she mentioned that she felt helpless as her mother had also reduced her academic supervision too and did not sit with her like previous days. The automatic negative thoughts centered on feelings of being rejected and were revealed as, "*My mother does not care for me anymore*" … "*Others will also not care for me.*" (Fig. 1)

Commonalities in both the cases

The above two cases were similar on several aspects like,

- Both had above average intelligence.
- They were primarily good students with good academic record in the past.
- There were no significant conduct problems noticed otherwise.
- Both were preoccupied with their respective issues hypothesized to be affecting their academic performance.
- Both of them could not share their actual underlying preoccupations with anybody.

Preintervention assessment: Both the cases were screened for any learning disability, attention deficit, or hyperactivity. Formal IQ assessment was not done as birth and developmental history and past academic performance suggested normal level of intelligence. Both were assessed on the BYI which suggested elevated scores in depression, anger, and self-concept for both. Academic grades had fallen from previous range of 70%–80% to 55%–60%.

Fixing therapy goals: Accordingly, the therapy goals were set as follows:

- Reducing emotional problems
  - Reducing psychological distress
  - Enhancing self esteem
  - Improving self-perception
  - Improving coping
- Improving academic performance
- Enhancing concentration

Process

Psychoeducation: Psychoeducation is an essential component for any psychotherapeutic intervention. It has been suggested to be an integral part of psychotherapy in Indian context, irrespective of the clinical condition (Halder & Mahato, 2012, 2019). The importance of psychoeducation in the CBT module increases many folds in child cases compared to adults considering lack or low insight into the problem not only on part of the child but parents also. Psychoeducation most importantly helps in establishing a clear perspective of the problem situation without blaming the child or the parent and road ahead for amelioration of the symptoms. The process indirectly helps in building and improving rapport with the child as they find a nonconflicting person (the therapist) understanding their perspective. Thus, psychoeducation was effectively used in both the cases to enhance insight of parents and understanding of the children. Especially for Case 1, parents were advised to have a neutral approach rather to ridicule the child of getting into girlfriends at such early stage.

Academic plan and strategies: The preceding sessions focusing on emotional issues of the children made implementation of academic plan easier. Both children were individually engaged to agree on making a comprehensive study plan which would be practical, achievable, as well as rewarding too for the child. Taking inputs from parents, the study time table of the children was modified. The children were actively involved in the process to make them feel as independent stake holders capable of taking responsible decisions. Input was taken from parents as well as teachers via parents on studying pattern of the child; if they attended any extra tuition and whether any or both parents supervised them in school homework and if modifications were required in their learning strategies.

CBT: While there were several intervention goals for both the cases, it was well conceptualized that the emotional issues and preoccupations were at the core leading to automatic negative thoughts impacting their academic performance as well as self-esteem. Contrary to belief, young children can be engaged in Socratic dialogue very much like adults; albeit the structure is more simplistic, and therapist needs to be aware of child's capabilities and sensitivities.

The CBT modules primarily focused on two main aspects: Self-management, which targeted the problem behavior of the children by increasing awareness and helping them managing it in a better way. Second focus was on improving self-efficacy which targeted the children's perceived self-efficacy and belief about their capabilities to learn or perform at the optimum level.

The sessions initially focused on challenging irrational thoughts of the children. They were explained in simple language with examples: the connection between their thoughts, behavior, and emotions. The concerns regarding appearance and feeling of rejection for first case, and feelings of neglect by mother and others for second case were challenged step wise. "Finding evidence" was useful in challenging the main cognitive distortions for both the cases. For example, for Case 1 (AM), when asked whether everyone in his class was bullying him and he did not had a single friend, he accepted that only a selected group of students were harassing him, and he indeed had a few good friends. So it was incorrect on his part to conclude that everyone was against him.

For Case 2 (DG), thought challenging was done by asking whether her mother was not at all concerned about her and if she cared only for her younger sister. She replied that though mother was more engaged with her younger sister, who deserved due to her younger age; mother was definitely concerned for her too and cared about her food, health, and studies. She comforts her whenever she was sad and cried. When further asked to exemplify whether everyone else was neglecting her, she accepted that actually her father loved her more than her sister and her perception of been neglected was not justified.

Both the children expressed feeling relaxed and good upon understanding that their thoughts were biased and they did not considered the other perspective.

After their understanding of their irrational negative thoughts and preoccupation regarding those, refocusing and cognitive rehearsal techniques were used to replace their negative thoughts to positive ones and sustain those positive thoughts. They were asked to verbalize their thoughts and engaged into self-question answer to negate their irrational thought and refocus on list of positive thoughts as taught in the sessions.

In the process of self-instructional training, they were asked to prompt the positive thoughts and statements like "not all are against me in class, most of them are friends" or "mother equally cares and loves me…so do others" to weaken the negative irrational thinking. Cognitive restructuring which was found as effective technique for children with academic and emotional problems was applied with aid of worksheets where negative thoughts, assumptions, and behavioral connections were made with help of charts; Positive–negative list was also used to improve their self-concept (Akinsola & Nwajei, 2013).

To improve self-efficacy and deal with future problems, alternate and better adaptive coping strategies were suggested by using coping imagery and problem-solving techniques. In the process, they were taught to enhance their problem-solving skills by charting current behavioral responses in difficult situations and exploring better options for those.

For example, Case 1 (AM) was asked to spend more time with his friends who were supportive rather than playing with those who were teasing him and actually very small in number. Similarly to improve relation with her mother and sister, Case 2 (DG) was advised to spend time with her sister, which made her realize that the younger one needed more care, time, and attention. It helped reduce her negative feelings toward her and to come out of the thoughts of mother being partial. Both were suggested to build better relations with peers as support system with whom they may share their problems and can feel better. Sharing their issues with somebody senior, be it one of the parents, any specific teacher they trusted was advised.

Parents of both the cases were simultaneously included in psychotherapy sessions. They were briefed about each session's outcome, with advice on desired modification on their academic supervision and daily parenting practices, so that they could support the children at home in a more supportive way and the children could perceive the change.

Outcome of the therapy process

→ Emotional and behavioral domain: The outcome of the therapy process was measured through pre- and postintervention scores on the BYI. See Table 1:

→ For both the children, parents reported significant change in severity of anxiety, depression, anger, disruptive behavior, and poor self-concept as measured through BYI scores at termination of the sessions. Case 1 still had mild anger issues and poor self-concept was at mild level for both. The symptoms were at normal level at follow-up at 6 months, as both parents and the children tried to persist with better coping strategies, better adaptive though patterns, and problem-solving skills. There was no report of picking up fight or crying by respective cases at follow-up.

→ Academic domain: Parents reported improved concentration, better focus, and being able to sit for longer study hours by the children. It was reflecting gradually in their class tests and they were following their respective academic plans.

→ Overall, the clients reported significant reduction rather amelioration of their psychological distress, had better self-perception, better peer relation, and increased self-confidence and interest in academics too. The cases discussed above effectively shows the implication of CBT in children and adolescent with academic difficulties. Academics are one domain where every parent wants their children to excel. However, the road of academic progression may hit blocks in terms of undesired changes in psychosocial environment, as well as characteristic features of adolescent crisis. CBT is highly indicated to counter anxiety and depression in adults as well as children and adolescents. Post-intervention changes in symptom severity and academic performance of the cases further support the indication of CBT in children with academic difficulties and concurrent with studies which suggest children and adolescents experience a reduction in anxiety, increased coping skills, and improved emotional awareness and regulation post-CBT treatment (Clabby, 2006; Kumar & Sebastian, 2011; Suveg, Sood, Comer, & Kendall, 2009). Much like adults, children in the case could be equipped to understand their emotions, cope with it, and effectively manage it. The cases discussed above suggest that academic difficulties may manifest due to stressors getting magnified in yet to mature children. While the overall CBT module remained similar to adult cases, it is important to maintain a nonjudgmental

TABLE 1 Pre- and postcomparison of symptom severity.

Domains	Subdomains	Preintervention	Postintervention	Follow-up (6 months)
Case-1				
BYI	Depression	Moderately elevated (61)	Within normal range (48)	Within normal range (40)
	Anxiety	Mildly elevated (56)	Within normal range (50)	Within normal range (40)
	Anger	Extremely elevated (77)	Mildly elevated (57)	Within normal range (45)
	Disruptive behavior	Mildly elevated (57)	Within normal range (50)	Within normal range (42)
	Self-concept	Extremely elevated (74)	Mildly Elevated (62)	Within normal range (54)
Academic grades		60%	75%	78%
Case-2				
BYI	Depression	Extremely elevated (72)	Within normal range (55)	Within normal range (53)
	Anxiety	Mildly elevated (52)	Within normal range (50)	Within normal range (44)
	Anger	Moderately elevated (64)	Within normal range (45)	Within normal range (40)
	Disruptive behavior	Mildly elevated (55)	Within normal range (40)	Within normal range (35)
	Self-concept	Extremely elevated (75)	Mildly elevated (57)	Within normal range (50)
Academic grades		B	A	A+

yet supportive approach toward the child and adolescent cases. Parents might have different therapeutic expectations and need to be psychoeducated about all the perspective else the perpetuating factors may remain unresolved.

Key facts

- Children with academic difficulties are common and one of the largest groups to consult child psychiatric and psychological consultation clinics.
- Symptoms of anxiety, low self-esteem, difficulty in social interaction, irritability, and low mood may be common in these children.
- These children may not meet any formal diagnostic criteria but equally warrant psychotherapeutic intervention.

Applications to other areas

Children and adolescents undergo a lot of transition during their school and early college years. They strive to know their own self, make their own identity, try to be unique, and at the same time also adhere with their peers. Parental and societal expectations for excellence are common in many societies putting undue pressure on these children and adolescents. Even if they manage their academic performances, there might be issues in social relationship domain with friends as well as parents which goes beyond the adolescent crisis. Impulsive problem behaviors, incidents of self-injurious behavior by minors on trivial issues are on rise and an empathetic approach like CBT is very well indicated in these conditions.

Summary points

- CBT could be an effective mode of management for children and adolescents with academic difficulties.
- However, it needs to be practiced with caution and considering various intervening factors.
- Diagnostic issues need to be sorted out and a customized approach helps in achieving desired targets.
- Prognosis betters with a fair mix of good rapport building, supportive and flexible parents, addition of cognitive remedial measures, along with core CBT measures.

References

Akinsola, E. F., & Nwajei, A. D. (2013). Test anxiety, depression and academic performance: Assessment and management using relaxation and cognitive restructuring techniques. *Psychology*, *4*(06), 18.

Beck, J. S., Beck, A. T., Jolly, J. B., & Steer, R. A. (2005). *Beck youth inventories—Second edition for children and adolescents manual*. San Antonio, TX: PsychCorp.

Bergan, J. R. (1995). Evolution of a problem-solving model of consultation. *Journal of Educational and Psychological Consultation*, *6*(2), 111–123.

Clabby, J. F. (2006). Helping depressed adolescents: A menu of cognitive-behavioral procedures for primary care. *Primary Care Companion to the Journal of Clinical Psychiatry*, *8*(3), 131.

Cole, P. M., Bruschi, C. J., & Tamang, B. L. (2002). Cultural differences in children's emotional reactions to difficult situations. *Child Development*, *73*(3), 983–996. https://doi.org/10.1111/1467-8624.00451. PMID 12038564.

Consoli, A. J., Blears, K., Bunge, E. L., Mandil, J., Sharma, H., & Whaling, K. M. (2018). Integrating culture, pedagogy, and humor in CBT with anxious and depressed youth. *Practice Innovations*, *3*(2), 138.

Cunningham, E. G., Brandon, C. M., & Frydenberg, E. (1999). Building resilience in early adolescence through a universal school-based preventive program. *Australian Journal of Guidance and Counselling*, *9*(2), 15–24.

Eva de Hullu, B., Sportel, E., Nauta, M. H., & de Jong, P. J. (2017). Cognitive bias modification and CBT as early interventions for adolescent social and test anxiety: Two-year follow-up of a randomized controlled trial. *Journal of Behavior Therapy and Experimental Psychiatry*, *55*, 81–89. ISSN: 0005-7916 https://doi.org/10.1016/j.jbtep.2016.11.011.

Foorman, B. R., & Torgesen, J. (2001). Critical elements of classroom and small-group instruction promote reading success in all children. *Learning Disabilities Research & Practice*, *16*(4), 203–212. https://doi.org/10.1111/0938-8982.00020.

Friedberg, R. D., & McClure, J. M. (2015). *Clinical practice of cognitive therapy with children and adolescents: The nuts and bolts*. Guilford Publications.

Gresham, F. M., & Kern, L. (2004). Internalizing behavior problems in children and adolescents. In R. B. Rutherford, M. M. Quinn, & S. R. Mathur (Eds.), *Handbook of research in emotional and behavioral disorders* (pp. 262–281). The Guilford Press. https://psycnet.apa.org/record/2005-02874-014.

Halder, S., & Mahato, A. K. (2012). CBT: The Nepalese experience. In F. Naeem, & D. E. Kingdon (Eds.), *Cognitive behavior therapy in non western cultures* (pp. 79–91). New York: Nova Science.

Halder, S., & Mahato, A. K. (2019). Cognitive behavior therapy for children and adolescents: Challenges and gaps in practice. *Indian Journal of Psychological Medicine*, *41*(3), 279–283.

Kumar, G. V., & Sebastian, L. (2011). Impact of CBT on self efficacy and academic achievement in adolescent students. *Journal of the Indian Academy of Applied Psychology*. Retrieved from: https://www.researchgate.net/publication/288363105_Impact_of_CBT_on_self_efficacy_and_academic_achievement_in_adolescent_students.

Raman, V. (2020). Cognitive behavior therapy for children with anxiety: Considerations for a low resource setting like India. *Journal of Indian Association for Child & Adolescent Mental Health*, *16*(4).

Schoenfeld, N. A., & Mathur, S. R. (2009). Effects of cognitive-behavioral intervention on the school performance of students with emotional or behavioral disorders and anxiety. *Behavioral Disorders*, *34*(4), 184–195.

Scott, A. (2009). Cognitive behavioural therapy and young people: An introduction. *The Journal of Family Health Care*, *19*(3), 80–82. 19645152.

Sukhodolsky, D. G., Wyk, B. C. V., Eilbott, J. A., McCauley, S. A., Ibrahim, K., Crowley, M. J., et al. (2016). Neural mechanisms of cognitive-behavioral therapy for aggression in children and adolescents: Design of a randomized controlled trial within the national institute for mental health research domain criteria construct of frustrative non-reward. *Journal of Child and Adolescent Psychopharmacology*, *26*(1), 38–48.

Suveg, C., Sood, E., Comer, J. S., & Kendall, P. C. (2009). Changes in emotion regulation following cognitive-behavioral therapy for anxious youth. *Journal of Clinical Child & Adolescent Psychology*, *38*(3), 390–401.

Yeo, L. S., & Choi, P. M. (2011). Cognitive-behavioural therapy for children with behavioural difficulties in the Singapore mainstream school setting. *School Psychology International*, *32*(6), 616–631. https://doi.org/10.1177/0143034311406820.

Chapter 41

Case study: Cognitive behavioral therapy for postnatal depression

Mariana Branquinho[a], Verónica Martínez-Borba[b], and Ana Fonseca[a]
[a]*Faculty of Psychology and Educational Sciences, Center for Research in Neuropsychology and Cognitive-Behavioral Intervention, University of Coimbra, Coimbra, Portugal* [b]*Department of Basic and Clinical Psychology and Psychobiology, Faculty of Psychology, Jaume I University, Castelló de la Plana, Spain*

Abbreviations

CBT cognitive behavioral therapy
ICT information and communication technologies
PND postnatal depression

Introduction

Cognitive behavioral therapy for postnatal depression

Postnatal depression (PND) has been conceptualized as a depressive episode occurring up to 12 months after delivery (Hamel et al., 2019). The most recent prevalence data suggest that PND affects 27% of women (Alshikh Ahmad, Alkhatib, & Luo, 2021; Atuhaire et al., 2021), a situation that has been aggravated by COVID-19 pandemic circumstances (Mariño-Narvaez, Puertas-Gonzalez, Romero-Gonzalez, & Peralta-Ramirez, 2021). Depressive symptoms during the perinatal period may negatively impact not only the mother, but also the baby and society (Slomian, Honvo, Emonts, Reginster, & Bruyère, 2019), which highlights the urge to implement preventive and treatment interventions designed specifically for the perinatal period.

It has been reported that during the perinatal period women prefer any psychological intervention other than pharmacotherapy (Dimidjian & Goodman, 2014) for managing depressive symptoms. As a consequence, during the last several years there has been an increased interest in developing and implementing psychological interventions for perinatal depression treatment. A wide variety of treatments (e.g., cognitive behavioral therapy (CBT), behavioral activation, interpersonal therapy, mindfulness, etc.) have been used, with CBT being the most evidence-based treatment for PND (Branquinho et al., 2021).

The use of Information and Communication Technologies in PND management

Despite the number of available therapies, difficulties in accessing mental health systems have been reported during the perinatal period. These include barriers in women, health professionals, and the healthcare system. First, women usually normalize depressive symptoms during pregnancy and postpartum. They may choose to hide their symptoms because of fear of stigma, shame, and guilt from being judged as a bad or incompetent mother. In addition, women in the postnatal period usually report logistics problems, most of them related to not knowing where to seek professional help, the high costs associated with hospital travel and care, and difficulties in combining childcare with personal needs (Goodman, 2009; Sambrook Smith, Lawrence, Sadler, & Easter, 2019). Regarding the healthcare professionals' barriers, they have limited time during consultations, so they prioritize physical as opposed to psychological care. Additionally, it has been suggested that professionals do not have enough knowledge of how to detect, diagnose, and refer perinatal women with depressive

Handbook of Lifespan Cognitive Behavioral Therapy. https://doi.org/10.1016/B978-0-323-85757-4.00036-5

symptoms (Bayrampour, Hapsari, & Pavlovic, 2018). Related to this idea, healthcare systems have failed to provide adequate protocols and services (i.e., mothers and babies units) to allow the early detection and treatment of depressive symptoms during the perinatal period (Bayrampour et al., 2018).

The aforementioned barriers have led to a severe problem in perinatal mental health, which is the underdetection and undertreatment of perinatal depression. It has been estimated that only 10% of women who suffer perinatal depression actually receive an adequate treatment (Gavin, Meltzer-Brody, Glover, & Gaynes, 2015). Some authors have postulated that the use of Information and Communication Technologies (ICT) could help to overcome barriers found in face-to-face interventions and therefore could help to disseminate evidence-based psychotherapies to reach more people (Kazdin & Blase, 2011).

ICT interventions have numerous advantages, including increased security, control, anonymity, and confidentiality; reduced displacements, costs, and time for the healthcare professionals; and helping in decision making and facilitating longitudinal assessments (Reynolds, Griffiths, Cunningham, Bennett, & Bennett, 2015; Torous et al., 2015). Due to these advantages, different efforts have been conducted to translate face-to-face interventions addressing perinatal depression into online programs. Thus different ICTs have been used to provide screening, preventive, and treatment programs for women with perinatal depression (Lara-Cinisomo, Ramirez Olarte, Rosales, & Barrera, 2021; Martínez-Borba, Suso-ribera, & Osma, 2018).

While acknowledging the advantages of these new delivery methods, some issues also need to be acknowledged. Several studies have found that ICT-based interventions tend to result in high drop-out rates, especially when self-administered methods are employed (Karyotaki et al., 2015). Attending to the specific field of perinatal depression, attrition rates oscillated between 13% and 61% in online programs (Lee, Denison, Hor, & Reynolds, 2016), probably attributable to the lack of therapist support during sessions.

Both face-to-face and online interventions present advantages as well as disadvantages. In this scenario, blended interventions, which involve both face-to-face and online sessions equally (Wentzel, van der Vaart, Bohlmeijer, & van Gemert-Pijnen, 2016), have been postulated as an alternative to isolated face-to-face or completely online interventions, combining the strengths of both delivery formats. Different studies have demonstrated that blended interventions are effective addressing depressive disorders (Erbe, Eichert, Riper, & Ebert, 2017) in adults. However, their use has not reached the management of depressive symptoms during the perinatal period.

In this chapter we will describe both the case of a women diagnosed with a major depressive disorder during the postnatal period as well as the cognitive behavioral blended intervention that was provided.

Case study

Case identification and CBT formulation

Maria, a pseudonym, is a 36-year-old Portuguese woman. She is married, has a 3-month-old daughter, and is currently a researcher. Maria presented with a major depressive disorder according to the Structured Clinical Interview for DSM-5 (First, Williams, Karg, & Spitzer, 2017). She reported daily depressed mood, diminished pleasure in most of her daily activities, a decrease in appetite, insomnia, psychomotor retardation, fatigue, feelings of worthlessness and guilt, indecisiveness, and lack of concentration. The most notable difficulties she presented were related to childcare, difficulties in interpersonal relationships when dealing with others' opinions or negative comments, lack of emotional and instrumental support, and constant worries about her return to work.

Maria referred to the pregnancy and postnatal period as difficult times, characterized by uncertainty, isolation, and lack of support by her family and friends, associated with the COVID-19 pandemic restrictions. She explained that, after the delivery, her husband had a severe physical illness that prevented him from being involved in the care of the baby and providing the support she needed. At this time, she also began to worry about their financial situation, as her husband was on sick leave and she would end her job contract soon. Almost all days, she reported concerns about combining childcare with her job. Previously to the current depressive episode, Maria mentioned having depression when she was 20 years old.

Taking as a reference the PND models from Milgrom, Martin, and Negri (1999) and Wenzel and Kleiman (2015), Fig. 1 shows the vulnerability (i.e., personal depressive symptom antecedents) and sociocultural factors (i.e., lack of support) that favored the onset of PND when Maria faced stressful situations (i.e., financial difficulties). In this figure, some of the factors (i.e., social isolation) are exposed that contributed to the exacerbation and maintenance of PND.

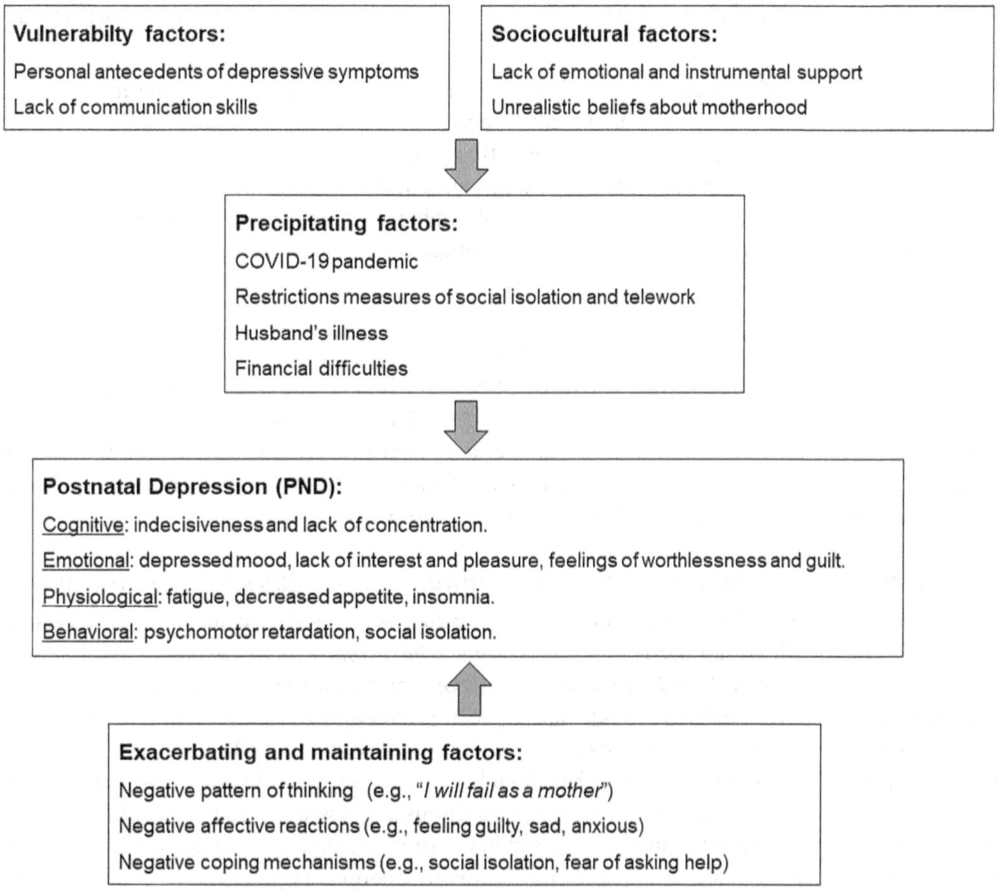

FIG. 1 Cognitive behavioral case conceptualization of Maria's PND symptoms.

Cognitive behavioral blended intervention

A psychological intervention for PND, based on CBT, was provided in a blended format, combining seven sessions with a psychologist and six online sessions through a web-based program, over a period of 13 weeks. The intervention included psychoeducation about PND, cognitive restructuring, behavioral activation, promotion of interpersonal competences such as assertiveness, and relapse prevention. The intervention was conducted by a licensed psychologist, who received supervision by an experienced psychologist.

Session 1: Assessment and definition of therapeutic goals (with therapist)

The first session included a cognitive behavioral clinical assessment of the patient's difficulties and the definition of therapeutic goals. Maria set several goals for the treatment, including in self-care (e.g., *to work out twice a week*), in the marital relationship (e.g., *improve communication with partner*), in social relationships (e.g., *call two friends; engage in babies' activities with other parents*), and professional area (e.g., *return to work in the next two months; find a new job*).

Session 2–3: Psychoeducation about maternity changes, postnatal depression, and emotions (online and with therapist)

The following two sessions aimed to provide psychoeducation on maternity changes, postnatal depression, and the role of emotions. Maria reflected on the difference between *the ideas we have before the birth and then the reality,* recognizing the influence of expectations and social pressure to be a *perfect mother.* The therapist normalized her experience, the discrepancy between the emotions expected, and the emotions felt during the postnatal period, and promoted the acceptance of emotional difficulties in the transition to maternity. The therapist also promoted a reflection of the function of emotions (e.g., sadness, joy, anger) and encouraged their acceptance. The patient recognized the importance of *being aware of my emotions and the way I feel is affected by my thoughts.*

Sessions 4–5: The cognitive behavioral model: The role of negative thoughts (online and with therapist)

The CBT model of the influence of thoughts on emotions and behaviors was presented in the next couple of sessions, encouraging the patient to recognize and manage her negative thoughts, such as *I will fail as a mother, Am I doing enough?* or *A good mom would not think of leaving her baby*. More adaptive strategies to deal with negative thoughts were presented and practiced in the session (including cognitive defusion, acceptance and nonevaluation of thoughts, thought questioning, and self-compassion). Maria revealed a particular interest in asking herself if the thought is being useful, identifying different interpretations for a situation, and promoting a more self-compassionate posture, recognizing perfectionism and self-criticism as unhelpful.

Sessions 6–7: Personal values (online and with therapist)

In the next weeks, Maria was invited to reflect on her values. She identified "being present and available" as an important value, and some actions and behaviors that she could have to act in direction to her value (e.g., *to play with my baby, to provide significant experiences, to satisfy her needs*). The therapist then invited her to reflect on the impact of pursuing valued-based goals in her emotions and mood, and encouraged the adoption of the learned strategies to deal with negative thoughts when they arise and prevent engaging in those valued-based actions.

Sessions 8–9: Social support and interpersonal communication skills (online and with therapist)

Social support was addressed in the following sessions, aiming to reduce social isolation and to increase patient's emotional and practical support. Maria identified her needs of support (e.g., *to have someone to talk about emotional difficulties; to have someone to cook the meals, do the laundry or take the dogs out*) and identified a friend who also had recently had a baby and her parents to act as useful sources to provide that support. She recognized some difficulty when asking for help (*I am showing them my vulnerability; they may not be available or may not want to*). Therapist asked her to remember a situation when she was asked for help and to identify her thoughts, and to understand if she thought that the person asking for her help was vulnerable or weak for that. Then the therapist discussed how that behavior could be contributing to the dissemination of the myths of perfect motherhood, addressed in the first sessions. The therapist underlined the importance of assertive communication to ask for help, as well as to deal with other people's opinions, and encouraged Maria to adopt assertive communication strategies through a role-play exercise.

Sessions 10–11: Changes and difficulties in the couple relationship (online and with therapist)

By the end of the intervention, the changes felt within the couple during the postpartum period were discussed. Maria reported a decrease in physical intimacy, discrepancies in the division of household tasks, and increase of discussions between the two. The therapist normalized the reduction felt in marital satisfaction during this period, given the demanding nature of adaptation to parenthood, the increasing conflicts, and decreased time to dedicate to the couple. Then, strategies to promote intimacy (e.g., affection gestures), assertive communication within the couple (e.g., active listening), and conflicts resolution were proposed and practiced.

Sessions 12–13: Relapse prevention (online and with therapist)

In the last two sessions, the therapist encouraged a reflection on the learned therapeutic strategies and the patient's progress. Maria highlighted the importance of normalizing her emotional experience, understanding the impact of thoughts in emotions, learning strategies to manage negative thoughts, acting in accordance with her values, and to be more assertive. A relapse prevention plan was elaborated to anticipate problems and integrate skills.

Patient's evolution/progress

Assessment

Self-report measures were applied pre- and postintervention to assess the patient's therapeutic progress, which included the Edinburgh Postnatal Depression Scale (EPDS; Areias, Kumar, Barros, & Figueiredo, 1996) and the Hospital Anxiety and Depression Scale—Anxiety subscale (HADS-A; Pais-Ribeiro et al., 2007). EPDS is composed of 10 items that assess depressive symptomatology during the last week. Total scores ranged from 0 to 30, and higher scores represent greater depressive symptoms. The HADS-A subscale is composed of seven items that evaluate anxiety symptoms during the last week. Total scores range between 0 and 21, higher scores indicating higher anxiety symptoms.

TABLE 1 Patients' pre- and postintervention assessments and the Reliable Change Index.

Instrument	Normative data M (SD)	Preintervention RS	Postintervention RS (PCP)	RCI	Interpretation
EPDS	3.89 (3.0)	23	8 (65.22)	10.20	SC, improvement
HADS-A	4.83 (3.81)	15	8 (46.67)	2.65	SC, improvement

Note. EPDS, Edinburgh Postnatal Depression Scale; *HADS-A,* Hospital Anxiety Depression Scale—Anxiety subscale; *M,* mean; *SD,* standard deviation; *RS,* raw score; *PCP,* percentage of change from pretest; *RCI,* Reliable Change Index; *SC,* significant change.

Significant improvements in depressive and anxiety symptoms were observed from pre- to postintervention, as indicated by the Reliable Change Index (RCI) reported in Table 1. At postintervention, Maria recognized positive changes and the need to keep practicing the learned strategies (*the brain is a muscle that needs to be exercised, otherwise it gets flaccid*). She was able to improve self-care by working out once/twice a week, depending on her schedules. Also, Maria improved her communication skills and, consequently, she started "difficult" conversations with her husband more frequently, instead of avoiding them. By the end of the intervention, she was also trying to achieve her social and professional goals and she committed to continue working on them over the next weeks.

Maria had a follow-up face-to-face session 1 month after the end of the intervention. She revealed positive affect and no depressed humor, increased interest in pleasure activities (e.g., reading or painting) and involvement in social activities, and did not report appetite or sleep disturbances. The patient's subjective evaluation of her progress was positive, recognizing that she noted significant improvements during the last month and that it was important to strengthen some skills and practice the learned strategies. Maria achieved most of the therapeutic objectives defined in the first session and defined further steps to accomplish others (e.g., to find a new job).

Conclusions

In this chapter, we described the application of a CBT intervention for PND, in a blended format, to a Portuguese woman during the postpartum period. The intervention was effective in reducing both depressive and anxiety symptoms in this clinical case, which may have implications for clinicians and for future research.

In fact, several studies have demonstrated the effectiveness of CBT in PND treatment (Huang, Zhao, Qiang, & Fan, 2018; Sockol, 2015). In this work, we highlight the importance of including and adapting CBT components to the specificities of the postnatal period, namely psychoeducation about PND, to promote emotional regulation and cognitive restructuring of motherhood-related beliefs (e.g., *I am not a good mother*). In addition, the incorporation of problem-solving competencies, interpersonal competencies (e.g., the relevance of asking for support, strategies to be more assertive), marital satisfaction, and, finally, relapse prevention was also relevant for managing depressive symptoms in the postnatal period. Recently, Stamou, García-Palacios, and Botella (2018) conducted a systematic review of the existent CBT interventions for PND and reported that most interventions included the application of these CBT strategies and elements in the sessions, in addition to others such as stress management or relaxation.

In addition to discussing the importance of adapting CBT contents to the specific postnatal period, this clinical case also allows debate on the importance of developing new psychological treatment formats for PND (e.g., blended, online). These new delivery formats may help to reduce the perceived barriers to care and increase women's access to effective treatment. In fact, providing an intervention delivered through an online program can have a positive impact, by reducing transportation, costs, and the stigma associated with seeking a psychologist, particularly relevant during the postnatal period. Finally, using online programs within clinical practice can help professionals to conduct screening or assessments in a more efficient way. Further research into the benefits of integrating online tools into face-to-face clinical practice, both for women with PND and health professionals, should be carried out in order to identify the potential barriers and to facilitate the development and integration of these tools.

Key facts about treatment for PND

- Face-to-face interventions seem to be effective in the reduction of postnatal anxiety and depressive symptoms.
- Recent studies have explored the role of new models of care to overcome the limitations of face-to-face psychological programs.

- Web-based CBT interventions showed promising results in terms of symptoms reduction efficacy as well as user satisfaction. However, some of the main concerns have to do with sample retention.
- Blended interventions combining both face-to-face and online intervention may help to reduce the barriers of face-to-face protocols while increasing patients' motivation and engagement attributable to therapist contact.

Applications to other areas

Our results support the idea that postnatal anxiety and depressive symptoms could be effectively addressed throughout CBT. However, given that other cognitive behavioral programs have demonstrated to be effective both in the prenatal and postnatal periods (Branquinho et al., 2021), we can hypothesize that the intervention described in this chapter could also be effective to address anxiety and depressive symptoms during pregnancy. Furthermore, given that CBT has been shown to be effective in adults with a variety of mental health problems (i.e., somatoform disorders, eating disorders, general stress, etc.; Hofmann, Asnaani, Vonk, Sawyer, & Fang, 2012) we can anticipate that our CBT program has the potential to be adapted to be applied in other perinatal mental health conditions apart from postnatal anxiety and depressive symptoms.

Finally, previous studies have found that CBT could be effective when addressing depressive symptoms in group format (Thimm & Antonsen, 2014). Future efforts should be conducted to explore whether our program could be effective in the reduction of perinatal anxiety and depressive symptoms when the intervention is delivered in group. This could also provide a unique opportunity to include the partner, a familiar or friend in the interventions with perinatal women.

Summary points

- CBT is a useful and effective intervention in PND treatment.
- Barriers to accessing perinatal mental services can be associated with women, health professionals, and systems issues.
- New treatment formats for psychological interventions may help to overcome barriers found in traditional face-to-face and entirely online programs.
- A blended CBT intervention was applied to a woman with PND and it was revealed to be useful in reducing depressive and anxiety symptoms.
- Future efforts should be made to integrate these interventions in routine clinical practice of PND management.

References

Alshikh Ahmad, H., Alkhatib, A., & Luo, J. (2021). Prevalence and risk factors of postpartum depression in the Middle East: A systematic review and meta-analysis. *BMC Pregnancy and Childbirth*, *21*(1), 542. https://doi.org/10.1186/s12884-021-04016-9.

Areias, M. E. G., Kumar, R., Barros, H., & Figueiredo, E. (1996). Comparative incidence of depression in women and men, during pregnancy and after childbirth. *British Journal of Psychiatry*, *169*(1), 30–35. https://doi.org/10.1192/bjp.169.1.30.

Atuhaire, C., Rukundo, G. Z., Nambozi, G., Ngonzi, J., Atwine, D., Cumber, S. N., & Brennaman, L. (2021). Prevalence of postpartum depression and associated factors among women in Mbarara and Rwampara districts of South-Western Uganda. *BMC Pregnancy and Childbirth*, *21*(1), 503. https://doi.org/10.1186/s12884-021-03967-3.

Bayrampour, H., Hapsari, A. P., & Pavlovic, J. (2018). Barriers to addressing perinatal mental health issues in midwifery settings. *Midwifery*, *59*, 47–58. https://doi.org/10.1016/j.midw.2017.12.020.

Branquinho, M., Rodriguez-Muñoz, M.de la F., Maia, B. R., Marques, M., Matos, M., Osma, J., Moreno-Peral, P., Conejo-Cerón, S., Fonseca, A., & Vousoura, E. (2021). Effectiveness of psychological interventions in the treatment of perinatal depression: A systematic review of systematic reviews and meta-analyses. *Journal of Affective Disorders*, *291*, 294–306. https://doi.org/10.1016/j.jad.2021.05.010.

Dimidjian, S., & Goodman, S. H. (2014). Preferences and attitudes toward approaches to depression relapse/recurrence prevention among pregnant women. *Behaviour Research and Therapy*, *54*, 7–11. https://doi.org/10.1016/j.brat.2013.11.008.

Erbe, D., Eichert, H., Riper, H., & Ebert, D. (2017). Blending face-to-face and internet-based interventions for the treatment of mental disorders in adults: Systematic review. *Journal of Medical Internet Research*, *19*(9). https://doi.org/10.2196/jmir.6588, e306.

First, M., Williams, J., Karg, R., & Spitzer, R. (2017). *User's guide for the SCID-5-CV structured clinical interview for DSM-5 disorders: Clinical version.* American Psychiatric Publishing, Inc.

Gavin, N. I., Meltzer-Brody, S., Glover, V., & Gaynes, B. N. (2015). Is population-based identification of perinatal depression and anxiety desirable? A public health perspective on the perinatal depression care continuum. In J. Milgrom, & A. Gemmill (Eds.), *Identifying perinatal depression and anxiety: Evidence-based practice in screening, psychosocial assessment, and management* (pp. 11–31). Wiley-Backwell.

Goodman, J. H. (2009). Women's attitudes, preferences, and perceived barriers to treatment for perinatal depression. *Birth*, *36*(1), 60–69. https://doi.org/10.1111/j.1523-536X.2008.00296.x.

Hamel, C., Lang, E., Morissette, K., Beck, A., Stevens, A., Skidmore, B., Colquhoun, H., Leblanc, J., Moore, A., Riva, J. J., Thombs, B. D., & Colman, I. (2019). Screening for depression in women during pregnancy or the first year postpartum and in the general adult population : A protocol for two systematic reviews to update a guideline of the Canadian Task Force on Preventive Health Care. *Systematic Reviews*, *8*(27), 1–13.

Hofmann, S. G., Asnaani, A., Vonk, I. J., Sawyer, A. T., & Fang, A. (2012). The efficacy of cognitive behavioral therapy: A review of meta-analyses. *Cognitive Therapy Research*, *36*(5), 427–440. https://doi.org/10.1007/s10608-012-9476-1.

Huang, L., Zhao, Y., Qiang, C., & Fan, B. (2018). Is cognitive behavioral therapy a better choice for women with postnatal depression? A systematic review and meta-analysis. *PLoS One*, *13*(10). https://doi.org/10.1371/journal.pone.0205243, e0205243.

Karyotaki, E., Kleiboer, A., Smit, F., Turner, D. T., Pastor, A. M., Andersson, G., & Berger, T. (2015). Predictors of treatment dropout in self-guided web-based interventions for depression: An 'individual patient data' meta-analysis. *Psychological Medicine*, *45*, 2717–2726. https://doi.org/10.1017/S0033291715000665.

Kazdin, A. E., & Blase, S. L. (2011). Rebooting psychotherapy research and practice to reduce the burden of mental illness. *Perspectives on Psychological Science*, *6*(1), 21–37. https://doi.org/10.1177/1745691610393527.

Lara-Cinisomo, S., Ramirez Olarte, A., Rosales, M., & Barrera, A. Z. (2021). A systematic review of technology-based prevention and treatment interventions for perinatal depression and anxiety in Latina and African American women. *Maternal and Child Health Journal*, *25*(2), 268–281. https://doi.org/10.1007/s10995-020-03028-9.

Lee, E. W., Denison, F. C., Hor, K., & Reynolds, R. M. (2016). Web-based interventions for prevention and treatment of perinatal mood disorders: A systematic review. *BMC Pregnancy and Childbirth*, *16*(1), 38. https://doi.org/10.1186/s12884-016-0831-1.

Mariño-Narvaez, C., Puertas-Gonzalez, J. A., Romero-Gonzalez, B., & Peralta-Ramirez, M. I. (2021). Giving birth during the COVID-19 pandemic: The impact on birth satisfaction and postpartum depression. *International Journal of Gynecology & Obstetrics*, *153*(1), 83–88. https://doi.org/10.1002/ijgo.13565.

Martínez-Borba, V., Suso-Ribera, C., & Osma, J. (2018). The use of information and communication technologies in perinatal depression screening: A systematic review. *Cyberpsychology, Behavior and Social Networking*, *21*(12), 741–752. https://doi.org/10.1089/cyber.2018.0416.

Milgrom, J., Martin, P., & Negri, L. (1999). *Treating postnatal depression: A psychological approach for health care practitioners.* John Wiley & Sons, Ltd.

Pais-Ribeiro, J., Silva, I., Ferreira, T., Martins, A., Meneses, R., & Baltar, M. (2007). Validation study of a Portuguese version of the Hospital Anxiety and Depression Scale. *Psychology, Health and Medicine*, *12*(2), 225–237. https://doi.org/10.1080/13548500500524088.

Reynolds, J., Griffiths, K. M., Cunningham, J. A., Bennett, K., & Bennett, A. (2015). Clinical practice models for the use of E-mental health resources in primary health care by health professionals and peer workers: A conceptual framework. *JMIR Mental Health*, *2*(1). https://doi.org/10.2196/mental.4200, e6.

Sambrook Smith, M., Lawrence, V., Sadler, E., & Easter, A. (2019). Barriers to accessing mental health services for women with perinatal mental illness: Systematic review and meta-synthesis of qualitative studies in the UK. *BMJ Open*, *9*(1). https://doi.org/10.1136/bmjopen-2018-024803, e024803.

Slomian, J., Honvo, G., Emonts, P., Reginster, J. Y., & Bruyère, O. (2019). Consequences of maternal postpartum depression: A systematic review of maternal and infant outcomes. *Women's Health*, *15*, 1–55. SAGE Publications Ltd https://doi.org/10.1177/1745506519844044.

Sockol, L. E. (2015). A systematic review of the efficacy of cognitive behavioral therapy for treating and preventing perinatal depression. *Journal of Affective Disorders*, *177*, 7–21. https://doi.org/10.1016/j.jad.2015.01.052.

Stamou, G., García-Palacios, A., & Botella, C. (2018). Cognitive-behavioural therapy and interpersonal psychotherapy for the treatment of post-natal depression: A narrative review. *BMC Psychology*, *6*(1), 28. https://doi.org/10.1186/s40359-018-0240-5.

Thimm, J. C., & Antonsen, L. (2014). Effectiveness of cognitive behavioral group therapy for depression in routine practice. *BMC Psychiatry*, *14*(1), 1–9. https://doi.org/10.1186/s12888-014-0292-x.

Torous, J., Staples, P., Shanahan, M., Lin, C., Peck, P., Keshavan, M., & Onnela, J.-P. (2015). Utilizing a personal smartphone custom app to assess the Patient Health Questionnaire-9 (PHQ-9) depressive symptoms in patients with major depressive disorder. *JMIR Mental Health*, *2*(1). https://doi.org/10.2196/mental.3889, e8.

Wentzel, J., van der Vaart, R., Bohlmeijer, E. T., & van Gemert-Pijnen, J. E. W. C. (2016). Mixing online and face-to-face therapy: How to benefit from blended care in mental health care. *JMIR Mental Health*, *3*(1). https://doi.org/10.2196/mental.4534, e9.

Wenzel, A., & Kleiman, K. (2015). *Cognitive behavioral therapy for perinatal distress.* Routledge.

Chapter 42

Case study: Cognitive behavioral therapy for adolescent skin picking disorder

Andrea Pozza
Department of Medical Sciences, Surgery and Neurosciences, University of Siena, Siena, Italy

Abbreviations

CBT	cognitive behavioral therapy
MIDAS	Milwaukee Inventory for the Dimensions of Adult Skin Picking
OCD	obsessive-compulsive disorder
Y-BOCS	Yale-Brown Obsessive Compulsive Scale

Case study: Clinical and diagnostic picture

B. is a 16-year-old female adolescent. Her clinical picture consisted of a diagnosis of obsessive-compulsive disorder (OCD) of moderate severity, as shown by a score of 21 on the Yale-Brown Obsessive Compulsive Scale (Y-BOCS; Goodman et al., 1989) with a comorbid skin picking disorder. Around 1 year ago, she started a psychiatric treatment at an outpatient center in Italy, and she was prescribed with anxiolytic and antidepressant (SSRIs) medications. The young patient reported recurrent obsessive fears of provoking a harm or losing control over her mind and behaviors with severe associated interpersonal consequences such as forgetting to lock the door of her house when going out, making a car accident, or having violent behaviors/anger outbursts with her parents at home. Such fears have been prevented or managed by the patient through repetitive rituals such as checking behaviors, scaramantic rituals (i.e., counting, mentally repeating lucky numbers and praying) and interpersonal avoidance behaviors (i.e., avoiding driving car or avoiding staying with her parents in the same room to prevent the occurrence of unwanted intrusive aggressive impulses). As shown in the literature (e.g., Hayes, Storch, & Berlanga, 2009), such OCD symptoms were associated with a picture of skin picking disorder.

Her skin picking disorder consisted of self-inflicted behaviors that emerged during the last 6 months when after checking her face for a bump with a cyst-like lesion, she started to gradually pick it. She began picking the lesion, firstly, with her fingernails, but quickly resorting to needles and cotton swabs to stop minor (and later bigger) bleedings. These behaviors became progressively repetitive, persistent, and uncontrollable. Each skin picking episode activated depersonalization and derealization feelings followed by strong feelings of shame and guilt.

Along with a picture of OCD and skin picking disorder, the girl reported significant depressive symptoms including anhedonia, abulia, persistent apathy, strong negative self-judgment, social withdrawal due to the effects of the OCD-related avoidance behaviors, and skin picking episodes. In addition to all these symptoms, B. presented a picture of social anxiety symptoms, in particular she generally felt ashamed, anxious, and alarmed during oral exams in front of peers in the classroom due to feelings of inferiority which she tended to avoid.

Cognitive behavioral assessment

B. lived with her parents at home, and she was the first one of three children. She always felt being neglected, emotionally invalidated, and not loved by her parents. The cognitive behavioral assessment based upon the downward technique revealed early maladaptive schemas characterized by emotional deprivation, unrelenting standards, abandonment, defectiveness, and shame feelings. Such cognitive schemas appeared associated with perfectionistic beliefs regarding social performance and interpersonal relationships (perfectionistic self-presentation), an excessive importance given to thoughts and control of thoughts (positive metacognitions regarding thoughts control), worry and rumination regarding social rejection or interpersonal negative judgment in daily situations.

Handbook of Lifespan Cognitive Behavioral Therapy. https://doi.org/10.1016/B978-0-323-85757-4.00021-3

In collaboration with the patient, during the case formulation process, such schemas and beliefs were conceptualized as triggering and maintenance mechanisms of both the OCD and the skin picking symptoms. Two types of situations triggering skin picking symptoms were identified: (1) the experience of unpleasant emotions in interpersonal relationships (i.e., the fear of being judged at an upcoming party with peers or the fear of not being competent during oral exams in front of peers in the classroom) and (2) watching television or lying on the bed were identified as the current daily trigger situations which were more likely to activate respectively focused and automatic skin picking behaviors.

CBT intervention for skin picking symptoms

In this paragraph, we will discuss only the CBT intervention focused on the skin picking behaviors. The duration of the CBT intervention was around 1 year and included 1-h weekly sessions. During CBT treatment, the patient was motivated to this therapeutic pathway. She expressed her strong motivation to learn effective skills to stop skin picking behaviors, since she told us that she was very much distressed by the skin lesions produced by the picking behaviors.

After a motivational intervention, habit reversal training exercises were planned and conducted in the therapist's room with the aim to reduce the persistence of skin picking symptoms. Cognitive restructuring techniques based upon Socratic questioning combined with the use of a recording diary of the automatic thoughts were implemented to increase patient's awareness of the trigger situations and cognitive distortions/interpersonal schemas which generally activated skin picking episodes. For example, B. was used to endorse the following dysfunctional automatic thought characterized by all-or-nothing perfectionistic thinking which generally preceded the occurrence of skin picking episodes: "*Either my body is completely free of skin irregularities or it's horrible and disgusting.*" Cognitive restructuring was helpful to allow B. to replace these kinds of dysfunctional thoughts with more functional ways of thinking. Metacognitive restructuring techniques were added to modify positive metacognitions regarding skin picking behaviors as useful coping strategies (e.g., "*Removing impurities from my skin makes me feel like I've done everything I need to get rid of anxiety*").

In order to improve patient's capacity to regulate her emotional experience, relaxation training including diaphragmatic breathing and progressive muscle relaxation exercises was introduced. These exercises were performed in the therapist's room, and then assigned to the patient as homework. Finally cognitive defusion exercises (i.e., "Leaves on a stream" exercise) were introduced and delivered either in the therapist's room or through audiotapes with the aim to help the patient to develop a detached attitude toward her automatic thoughts like a conscious observer of her thoughts.

After the treatment course, B. was able to prevent or stop skin picking episodes more effectively. This improved her perceived self-efficacy in the regulation of unpleasant feelings. Supporting this positive outcome, a relevant decrease in skin picking symptoms was observed after the CBT intervention, as suggested by the score on the Milwaukee Inventory for the Dimensions of Adult Skin Picking (MIDAS; Pozza et al., 2016) which decreased from 43 at the baseline to 38 after 6 months to 24 at the posttest after 1 year of CBT.

Key facts

Skin picking disorder and obsessive compulsive disorder can often co-occur in the same young individual. Cognitive behavioral therapy using habit reversal training, relaxation exercises, and cognitive interventions can be helpful for young patients with skin picking disorder and comorbid obsessive compulsive disorder.

Applications to other areas

The content of this chapter can be applied also to pediatric clinical practice, dermatological clinical practice, and psychiatric management of mental disorders.

Summary points

- Skin picking disorder among adolescents can be associated with OCD in comorbidity.
- Skin picking disorder and OCD can have common cognitive behavioral vulnerability and maintenance mechanisms including early maladaptive schemas based upon unrelenting standards, perfectionism, and positive metacognitions regarding thoughts control.
- The assessment of the daily situations triggering skin picking behaviors is crucial to planning treatment.

- A CBT approach including behavioral techniques (e.g., habit reversal training), cognitive (Socratic questioning), meta-cognitive and third-wave techniques (cognitive defusion) may be the optimal strategies aimed to target skin picking symptoms, modify maladaptive behaviors and beliefs, and increase emotion regulation skills.
- The MIDAS questionnaire can be used in clinical practice to assess the outcome of CBT.

References

Goodman, W. K., Price, L. H., Rasmussen, S. A., Mazure, C., Fleischmann, R. L., Hill, C. L., … Charney, D. S. (1989). The Yale-Brown obsessive compulsive scale. I. Development, use, and reliability. *Archives of General Psychiatry*, *46*, 1006–1011.

Hayes, S. L., Storch, E. A., & Berlanga, L. (2009). Skin picking behaviors: An examination of the prevalence and severity in a community sample. *Journal of Anxiety Disorders*, *23*, 314–319.

Pozza, A., Mazzoni, G. P., Patassini, I., Berardi, D., Molinaro, F., Parenti, G., & Dèttore, D. (2016). Studio sulle proprietà psicometriche della versione italiana del Milwaukee inventory for the dimensions of adult skin picking (MIDAS) in campioni clinici e non clinici. *Psicoterapia Cognitiva e Comportamentale*, *22*, 35–63.

Chapter 43

Pediatric type 1 diabetes CBT intervention in fear of hypoglycemia: Case study

Omer Shaked[a,b] and Daniel Hamiel[c,d,e]

[a]Pediatric Endocrine and Diabetes Unit, Sheba Medical Center, Edmond and Lily Safra Children's Hospital, Ramat-Gan, Israel, [b]The Louis and Gabi Weisfeld School of Social Work, Bar-Ilan University, Ramat-Gan, Israel, [c]Baruch Ivcher School of Psychology, IDC Herzliya, Herzliya, Israel, [d]Cohen-Harris Resilience Center, Tel-Aviv, Israel, [e]Tel-Aviv-Brüll Community Mental Health Center, Clalit Health Services, Tel-Aviv, Israel

Abbreviations

ACT	acceptance and commitment therapy
BGAT	blood glucose awareness training
CBT	cognitive behavioral therapy
DKA	diabetic ketoacidosis
FOH	fear of hypoglycemia
HbA1C	glycated hemoglobin
PTSD	posttraumatic stress disorder
SH	severe hypoglycemia
T1DM	type 1 diabetes mellitus

Introduction

Hypoglycemia (plasma glucose level <3.9 mmol/L [<70 mg/dL]) is the major limiting factor in the glycemic management of T1DM (Chamberlain, Rhinehart, Shaefer Jr, & Neuman, 2016). Severe hypoglycemia (SH) is defined by the patient's urgent need for the assistance of another person (American Diabetes Association, 2016). A ranging prevalence between 7.8% (Bulsara, Holman, Davis, & Jones, 2004), 7.5% (McCoy et al., 2012), and 20% (Gilsanz, Karter, Beeri, Quesenberry, & Whitmer, 2018) of reported severe hypoglycemia among T1DM patients was found. This makes SH a primary medical issue. The SH is accompanied by various somatic symptoms (Hepburn et al., 1991), but in its radical form, SH is considered a most adverse event since it may put patients at risk for injury and death (Seaquist, Anderson, Childs, et al., 2013). This particular feature of SH is also accompanied by the understanding that one's life is in danger, and therefore matches the criteria for a traumatic event (American Psychiatric Association, 2013), and thus may also result in posttraumatic stress disorder (PTSD).

A probable result of SH is an intense emotional reaction termed fear of hypoglycemia (FOH) (Gonder-Frederick, Schmidt, Vajda, et al., 2011). This fear holds crucial importance: patients may try to avoid in any way the disturbing symptoms and the sensation of danger that can result from low blood glucose levels (Böhme, Bertin, Cosson, & Chevalier, 2013). In addition, while attempting to avoid such ill experiences, numerous patients omit insulin to circumvent low blood-glucose levels (Martyn-Nemeth, Schwarz Farabi, Mihailescu, Nemeth, & Quinn, 2016). Therefore, FOH may consequently cause patients to resist the recommended medical treatment (Böhme et al., 2013).

ACT intervention for FOH

A third-wave CBT intervention focuses on personal autonomy, to which adolescents may be particularly receptive. In an ACT (acceptance and commitment therapy) framework, as will now be discussed, mindfulness exercises aim to promote acceptance of emotions, contact with the present moment, self-regulation, and cognitive defusion (Ludwig & Kabat-Zinn, 2008). The advantage of third-wave CBT intervention is the role of mindfulness exercises of the mind-body relationship,

Handbook of Lifespan Cognitive Behavioral Therapy. https://doi.org/10.1016/B978-0-323-85757-4.00024-9

which holds a crucial role in interventions with illness identity (Hadlandsmyth, White, Nesin, & Greco, 2013). The relevance of awareness to physical sensation is crucial due to the physical symptoms of hypoglycemia and the importance of the ability to differentiate between them and the overlapping physical symptoms of anxiety.

Case study

S′ is a 15-year-old girl, the youngest of four siblings in an orthodox Jewish family. At that age she was a successful high school student, a member of a youth movement, and very socially active. She describes herself as always having a need to be in control, and always having a need to be perfectly organized.

S′ was diagnosed with T1DM at the age of 2½ years. The first severe hypoglycemia occurred at the age of 3, of which S′ has no recollection, although the memory of the event has been told in the family since then. S′ knows her mother was terrified and that an ambulance was called due to S's loss of consciousness. The second severe hypoglycemia occurred at the age of 7. S′ remembers feeling weak and tired and remembers feeling as if she was almost falling asleep, while a neighbor who was babysitting her mistakenly decided to wake her up by dancing. She remembers collapsing while losing consciousness and remembers waking up in an ambulance. S′ also remembers that several days after that event she was restless, agitated, had problems with her sleep, and refused to stay by herself. Another severe hypoglycemia occurred at the age of 9, although she didn't lose consciousness, but since then, S′ has neglected her medical treatment while maintaining high blood glucose levels. She refused to measure her blood glucose and resented being reminded of diabetes, while constantly arguing with her mother about the treatment. While she was at a summer camp at the age of 15, she had another hypoglycemia, after which S′ started experiencing anxiety attacks. She dropped out of boarding school (although she kept up with daily studying), and significantly decreased her social activities and avoided going outdoors. Her attending endocrinologist referred her to psychosocial treatment after a clinic visit that presented a major deterioration in compliance with medical treatment. When S′ started the process, her time in range (TIR = plasma glucose level between 80 and 120 mg/dL) was 40%.

Step one: Evaluation

The evaluation during the first three sessions revealed that S′ was experiencing continuous stress. The stress was climaxing after meals when she required insulin treatment. She described that the mere thought of uncontrolled insulin units in her blood was unbearable. As a result, S′ minimized the amount of insulin, despite the recommendations of the medical plan. In addition, she felt exhausted and resented being constantly aware of her diabetes because of endless measurements of glucose level, and at the same time, she experienced herself as unreliable, not responsible, and immature, because of not treating herself medically. During adolescence, the chronicity of T1DM can manifest in familial conflicts. These experiences also may have been tied to S′ being an adolescent. Adolescence has traditionally been known as a challenging time, as adolescents are trying to establish autonomy and challenge authority and, above all, are more influenced by peer interactions in forming their identity (Commissariat, Kenowitz, Trast, Heptulla, & Gonzalez, 2016). Adolescents tend to endorse a more independent and even rebellious attitude, value privacy, and have a greater awareness of self-image (Hamilton & Daneman, 2002). In accordance, the chronicity of T1DM came into play as these characteristics were highlighted. S′ felt that she was failing and feared the future consequences of her counterproductive actions. S′ couldn't explain her behavior: she knew that insulin omission causes high blood glucose and distances her from hypoglycemia, but at the same time it contradicted her need for being in control and always needing to be perfectly organized. S′ as a result was experiencing life as chaotic, uncontrolled, and not safe.

The social issue posed great difficulty in itself. Like most adolescents, she had a strong social desire but also a desire to be an equal member of the group, which was experienced as impossible for her. Among her friends, she felt disabled and incapable of age-appropriate social activities because of her great fear of hypoglycemia that she experienced anywhere outside of home. These feelings led her to social avoidance, which was very painful for her.

Step two: Accepting fear and building tolerance to stress

The first step included four sessions dedicated to establishing a more productive coping strategy with ongoing stress. In collaboration with her attending dietitian, her diet shifted to a low-carb regime. As a result, S′ wasn't required to treat with as much insulin as before and her tension decreased to a more bearable level. Validating her fear was an important part of this step of therapy. The "negative emotion" after eating and injecting insulin with the urge to check her glucose level, could be explained by her now in two different ways: Either "a disaster is coming, and I am going to faint" or "negative emotion is not a warning sign but a sign that I'm trying to overcome fears and change my default to constantly check the glucose level (even if I don't always succeed)." Suggesting the second and third explanations, S′'s goals now shifted to accept and contain her fear instead of trying to avoid it by checking glucose level, for example. After validating her emotions of fear, we designed a supportive home setting. Family members, along with her dietitian, were available for the 2h following meals to endure the level of anxiety from the insulin treatment, while S′ exercised tension balancing breathing to contain her fears. With this approach, S′ was not alone after meals for the period when her anxiety was climaxing.

The constant need to check her glucose level was identified as stemming from the anxious need to be in control. However, we have identified that by constantly checking her glucose level for relief, she was paradoxically increasing her anxiety. First, we exercised mindfulness by focusing on bodily sensations. The exercise helped to increase the present-moment experience, which in turn decreased the intrusive effect of memories of her fainting. These memories increased her inner tension, manifesting in symptoms such as pounding heart, shallow breathing, and confusion, which she translated to the overlapping symptoms of hypoglycemia (despite the fact that the measurements didn't indicate hypoglycemia), causing the natural need to escape these memories by measuring blood glucose to guarantee her safety. Focusing on the present moment had decreased the intrusion of memories significantly. Second, we exercised a mindfulness technique we termed together "suspended response." The technique refers to a process of identifying her fear in vivo that is physically causing tension building inside to a sensation of a physical drive to relieve the stress by a glucose measurement (but only momentarily, while causing the opposite in the long run). Then, we exercised holding the drive for relief caused by fear, to build a sensation of tolerance to the physical need for relief of physical tension without denying the legitimacy of feeling fear at the same time. The ability to exercise that technique and the interpretation of the fear as a sign of accepting the challenge of growing out of her need to control led to the understanding that S′ is willing to face her fear and can go forward with the process.

Step three: Cognitive defusion and self-regulation

The next step included four sessions dedicated to the "defusion" of subjective thoughts and emotional regulation. S′ experienced events as catastrophic, and owed her catastrophic experiences to her difficulty to separate between subjective and objective elements. The catastrophic experience posed a barrier from behavioral changes and negative emotional experience of herself as a person. While avoiding from changes, in S's mind the overcoming weight of the subjective element of the experience was preserved. We decided to attempt gradual changes in her life, which S′ felt she could cope with.

S′ brought to the sessions her everyday experiences: for example, S′ reported that she was afraid to visit her brother due to a train ride. S′ reached the generalized and terminal conclusion that she could not do anything because of her "condition." During the dialogue, we agreed not to automatically apply her present logic when she is feeling anxious, and not to conclude too far ahead about the future. That agreement significantly decreased the stress S′ experienced, and we agreed to take one step at a time and consider each event on its merit. If she succeeded in visiting her brother, then we would learn from her success, and if she felt it was too hard and decided not to go, we would learn from that as well. We divided the train ride into different parts. S′ recognized that going to the train station from her house and from the train to her brother were not troubling her, but sitting locked on the train made her feel "trapped," helpless, and out of control. Dividing the journey into these parts helped S′ to grasp that it is manageable, and we prepared to sit on the train by listening to music, breathing, and having available supplies, as well as the availability of her parents over the phone if necessary.

The sessions also included mindfulness practices for emotional regulation. The practice of breathing, with a nonjudgmental attitude toward bodily natural functions (e.g., the sensation of tension in the face, heartbeats), increased S′'s capabilities to accept more naturally her own emotions without feeling ashamed, incompetent, and childish. S′ exercised meditation during the sessions, downloaded the meditation app to her mobile phone, and returned to the yoga exercise she had tried once in the past. We formulated the intense effect of her relationship with her emotions via the mathematics sign "power": S′ realized that she is upset about her being upset, ashamed at her being ashamed, frustrated from being frustrated, and so on. We practiced accepting emotions and cognitions at face value and, doing so, S′ differentiated between a natural human reaction to events and her tormenting judgmental reactions to her human reactions. For example, observing her mindset regarding the difficulty of riding the train, S′ recognized that she felt ashamed and nervous because of the childish and incompetent meanings she had ascribed to these difficulties, which in turn reinforced her conceptions that things are too hard for her. When S′ started legitimizing her fear as a natural reaction, her shame decreased as well as her tension. Doing so, S′ developed an observing and regulating self, which relates to her feelings and thoughts in a more natural phenomenological approach.

Step four: Active position of personal values, and self-compassion

The following four sessions included a dialogue about endorsing an active position derived from what she values despite her vulnerability. That active position was viewed as supplementing, and in a way complementing, her previous position derived from avoiding fear. For example, S′ wanted to meet with her friends but she restricted her social life to her home. She felt ashamed at her "childish" fear but wouldn't go outside where it was considered "not safe" for insulin treatment, and felt disappointed at herself because of her need to always be perfectly organized. We discussed the importance of social relationships and decided that for a 15-year-old teenager it can at times be more important to hang out and have fun than be perfectly accurate with the insulin treatment. Her values also included a healthy life, which means that after having fun, returning to a healthy insulin treatment is required and not in contradiction with the former value if taken as an integrative approach. In collaboration with her dietitian, the insulin treatment plan was strategically decreased when S′ met her friends outside, and S′ was less conflicted with her need to be perfect, which later on decreased the stress. Gradually, S′ regained confidence and actively managed her decision-making process according to her values and not solely because of her understandable and legitimized tendency to avoid fear. She consequently felt less exhausted and less resentment at her being diabetic. S′ felt more free to choose and life was less chaotic.

After endorsing an active position, the following sessions included a dialogue aimed at accepting hypoglycemia as an unfortunate event which, like many other accidents, can be an undesirable event that doesn't prevent us from living. The dialogue also aimed at accepting fear of hypoglycemia as an inherent part of a diabetic's natural reality, which can be channeled to act responsibly. The sessions were dedicated also to understanding that our lives as humans may contain undesirable events and we are limited in controlling events in our lives. Our challenge is to live with unsolved issues, but we are not helpless. As long as we live in a responsible way, we can concentrate on our goals and values and fully live. The ability to see herself and people in general as vulnerable creatures helped her to be more compassionate toward herself and her emotional reactions. The exercises were directed at creating an inner dialogue between S′ and herself as her good friend who permits unpleasant feelings in correspondence to daily events while breathing and relaxing the body. The grounds for permitting unpleasant feelings was at all times our human natural vulnerability. After accepting her vulnerability, S′ felt more comfortable making her choices and decisions and less conflicted with her perfectionist character.

Step five: Ending

The final three sessions were dedicated to preserving the accomplishments we had achieved and preparing for a plausible deterioration. The sessions also included structuring follow-up future meetings and opening the possibility of returning to therapy in case a regression occurs. At the end of the therapy, her TIR was 70%. S′ maintained her follow-up appointments, and 1 year later returned after a regression did occur. For another four sessions we repeated the core issues of the process.

Discussion

The chapter presented a case study of a third-wave CBT intervention for coping with fear of hypoglycemia in a 16-year-old adolescent. The therapeutic process focused on several issues: the acceptance of fear and tolerance to stress; cognitive defusion combined with emotional self-regulation; the adoption of an active position relying on personal values, and self-compassion. Although the process is structured, we maintain that it should not be taken rigidly but rather adjusted according to the abilities of the patients and their progress. It is crucial to point out that this process is suitable more for adolescents, while children of younger ages perhaps require more cooperation with their parents. Furthermore, this process was not examined by randomized control trial (RCT), and therefore its efficiency isn't established. Yet, the success of the current procedure to reduce patient's stress was replicated. Moreover, we maintain that a third-wave CBT intervention that focuses on personal autonomy and values to which adolescents may be more receptive, is more suitable in essence to this age, as the case study indeed demonstrates. In addition, the case study also demonstrates the advantage of mindfulness exercises for the mind-body relationship, which plays a crucial role within interventions for illness identity.

Key facts

Hypoglycemia is the major limiting factor in the glycemic management of T1DM.

A probable result of SH is an intense emotional reaction termed fear of hypoglycemia (FOH).

Applications to other areas

Three of the central issues which are discussed here, have relevance to psychosocial interventions in various chronic medical conditions. CBT interventions upon illnesses onset; interventions for chronicity of illnesses; and interventions for traumatic aspects of illnesses, should be applied in various medical domains and can be endorsed by clinicians with appropriate semantical adjustments. In addition, the distinction between the therapeutic work of CBT in children and adolescents which is discussed there, alongside the different impact that parents have on children at different ages, shouldn't be different in other chronic medical conditions. Both the parent's nonefficient coping model factor found during childhood with its application to the children's adherence to treatment, and the rebellion factor found during adolescence alongside the adolescent's attempts to differentiate coping from that of their parents, both should be found to be chief factors in other chronic medical conditions in pediatric medicine.

Summary points

- Patients after SH may avoid the disturbing symptoms and the sensation of hypoglycemia and omit insulin or avoid social activities.
- A third-wave CBT intervention focuses on personal autonomy, to which adolescents may be particularly receptive.
- Validating fear and reinterpreting cognitions to permit its presence in life events are the first step aiming to accept and contain fear instead of trying to avoid it.
- While avoiding from changes, the overcoming weight of the subjective element of the experience was preserved.
- Gradual changes in life increase tolerance to fear and promote coping.
- Endorsing an active position in a decision-making process according to values and not solely because of a legitimized tendency to avoid fear decreases exhaustion and resentment, and increases the sense of freedom to choose.
- The ability to see people in general as vulnerable creatures helps us to be more compassionate toward ourselves and our emotional reactions, and create a productive inner dialogue.

References

American Diabetes Association. (2016). In W. T. Cefalu (Ed.), *Standards of medical care in diabetes*. USA: American Diabetes Association.

American Psychiatric Association. (2013). *Diagnostic and statistical manual of mental disorders* (5th ed.). Arlington: American Psychiatric Association.

Böhme, P., Bertin, E., Cosson, E., Chevalier, N., & GEODE Group. (2013). Fear of hypoglycaemia in patients with type 1 diabetes: Do patients and diabetologists feel the same way? *Diabetes & Metabolism*, *39*(1), 63–70. https://doi.org/10.1016/j.diabet.2012.10.006.

Bulsara, M. K., Holman, C. A. J., Davis, E. A., & Jones, T. W. (2004). The impact of a decade of changing treatment on rates of severe hypoglycemia in a population-based cohort of children with type 1 diabetes. *Diabetes Care*, *27*(10), 2293–2298.

Chamberlain, J. J., Rhinehart, A. S., Shaefer, C. F., Jr., & Neuman, A. (2016). Diagnosis and management of diabetes: Synopsis of the 2016 American Diabetes Association Standards of Medical Care in Diabetes. *Annals of Internal Medicine*, *164*(8), 542–552.

Commissariat, P. V., Kenowitz, J. R., Trast, J., Heptulla, R. A., & Gonzalez, J. S. (2016). Developing a personal and social identity with type 1 diabetes during adolescence: A hypothesis generative study. *Qualitative Health Research*, *26*(5), 672–684.

Gilsanz, P., Karter, A. J., Beeri, M. S., Quesenberry, C. P., & Whitmer, R. A. (2018). The bidirectional association between depression and severe hypoglycemic and hyperglycemic events in type 1 diabetes. *Diabetes Care*, *41*(3), 446–452.

Gonder-Frederick, L. A., Schmidt, K. M., Vajda, K. A., et al. (2011). Psychometric properties of the hypoglycemia fear survey-ii for adults with type 1 diabetes. *Diabetes Care*, *34*(4), 801–806. https://doi.org/10.2337/dc10-1343.

Hadlandsmyth, K., White, K. S., Nesin, A. E., & Greco, L. A. (2013). Proposing an Acceptance and Commitment Therapy intervention to promote improved diabetes management in adolescents: A treatment conceptualization. *International Journal of Behavioral Consultation and Therapy*, *7*(4), 12–15. https://doi.org/10.1037/h0100960.

Hamilton, J., & Daneman, D. (2002). Deteriorating diabetes control during adolescence: Physiological or psychosocial? *Journal of Pediatric Endocrinology and Metabolism*, *15*(2), 115–126.

Hepburn, D. A., Deary, I. J., Frier, B. M., Patrick, A. W., Quinn, J. D., & Fisher, B. M. (1991). Symptoms of acute insulin-induced hypoglycemia in humans with and without IDDM: Factor-analysis approach. *Diabetes Care*, *14*(11), 949–957.

Ludwig, D. S., & Kabat-Zinn, J. (2008). Mindfulness in medicine. *JAMA*, *300*(11), 1350–1352.

Martyn-Nemeth, P., Schwarz Farabi, S., Mihailescu, D., Nemeth, J., & Quinn, L. (2016). Fear of hypoglycemia in adults with type 1 diabetes: Impact of therapeutic advances and strategies for prevention—A review. *Journal of Diabetes and its Complications*, *30*(1), 167–177. https://doi.org/10.1016/j.jdiacomp.2015.09.003.

McCoy, R. G., Van Houten, H. K., Ziegenfuss, J. Y., Shah, N. D., Wermers, R. A., & Smith, S. A. (2012). Increased mortality of patients with diabetes reporting severe hypoglycemia. *Diabetes Care*, *35*, 1897–1900. 2012.

Seaquist, E. R., Anderson, J., Childs, B., et al. (2013). Hypoglycemia and diabetes: A report of a workgroup of the American Diabetes Association and the Endocrine Society. *Journal of Clinical Endocrinology and Metabolism*, *98*(5), 1845–1859. https://doi.org/10.1210/jc.2012-4127.

Chapter 44

Recommended resources for cognitive-behavioral therapy across the lifespan

Rajkumar Rajendram[a,b], Vinood B. Patel[c], and Victor R. Preedy[d]

[a]*College of Medicine, King Saud bin Abdulaziz University for Health Sciences, Riyadh, Saudi Arabia,* [b]*Department of Medicine, King Abdulaziz Medical City, King Abdullah International Medical Research Center, Ministry of National Guard Health Affairs, Riyadh, Saudi Arabia,* [c]*School of Life Sciences, University of Westminster, London, United Kingdom,* [d]*Faculty of Life Science and Medicine, King's College London, London, United Kingdom*

Abbreviations

AACBT	Australian Association for Cognitive and Behavior Therapy
ACBTA	Asian Cognitive Behavioral Therapy Association
ADAA	Anxiety and Depression Association of America
AICT	American Institute for Cognitive Therapy
AIPP	Italian Association for Prevention and Early Intervention in Mental Health
ANZACBT	Aotearoa-New Zealand Association for Cognitive Behavioural Therapy
APS	Australian Psychological Society
BABCP	British Association for Behavioural & Cognitive Psychotherapies
CACBT	Canadian Association of Cognitive and Behavioural Therapies
CACBT	Chinese Association of Cognitive Behaviour Therapy
CBT	Cognitive-behavioral therapy
CDC	Centre for Disease Control
EABCT	European Association for Behavioural and Cognitive Therapies
EACBT	Egyptian Association of Cognitive Behavioural Therapy
IACBT	Indian Association for Cognitive Behavior Therapy
IACP	International Association for Cognitive Psychotherapy
ISAD	International Society for Affective Disorders
JABCT	Japanese Association for Behavioral and Cognitive Therapies
MSCP	Malaysian Society of Clinical Psychology
NACBT	National Association of Cognitive-Behavioral Therapists
NFKT	Norwegian Association for Cognitive Behavioral Therapy
NIH	National Institute of Health
NIMH	National Institute of Mental Health
NIMHANS	National Institute of Mental Health and Neuro Sciences
OCD UK	Obsessive-Compulsive Disorder-UK
PIRI	Parent Infant Research Institute
SPR	Society for Psychotherapy Research
UNICEF	United Nations Children's Fund
WAIMH	World Association for Infant Mental Health
WCCBT	World Confederation of Cognitive and Behavioral Therapies
WHO	World Health Organization

Handbook of Lifespan Cognitive Behavioral Therapy. https://doi.org/10.1016/B978-0-323-85757-4.00022-5

Introduction

Disorders of mental health occur throughout life, including adolescence and pregnancy. Without appropriate therapy mental health conditions may become permanent or lead to greater morbidities. Nonpharmacological approaches can be of great benefit. Cognitive-behavioral therapy (CBT), one such intervention, is now widely administered (Nakao, Shirotsuki, and Sugaya, 2021; Thoma, Pilecki, & McKay, 2015). In some circumstances CBT may be used in conjunction with other treatments including pharmacology.

The understanding of CBT and other psychotherapies has advanced greatly in recent years. However, even experienced researchers, scientists, and clinicians struggle to remain up-to-date. To assist colleagues who are interested in understanding more about CBT throughout the life stages we have therefore produced tables containing resources as recommended by researchers and practitioners. The experts who assisted with the compilation of these tables of resources are acknowledged below.

Resources

Tables 1–5 list the most up-to-date information on the regulatory bodies (Table 1); professional societies (Table 2); books (Table 3); emerging technologies, platforms, and other resources (Table 4); and centers and clinics (Table 5) that are

TABLE 1 Regulatory bodies or organizations dealing with mental health, cognitive-behavioral therapy, other psychotherapies or health related matters in adults or throughout life stages.

Regulatory body or organization	Web address
Academy of Cognitive and Behaviour Therapies	https://www.academyofct.org/
American Psychiatric Association	https://www.psychiatry.org/
Australian Association for Cognitive and Behavior Therapy (AACBT)	https://www.aacbt.org.au/
Beck Institute	https://beckinstitute.org/
Center for Cognitive Therapy	https://centerforcbtva.com/services/therapy/cognitive-behavioral-therapy-for-kids/
Child and Family Institute	https://www.childfamilyinstitute.com/
Child Mind Institute	https://childmind.org/article/treatment-for-depression/
Child Parent Psychotherapy	https://childparentpsychotherapy.com/
Choosing Therapy	https://www.choosingtherapy.com/cbt-kids-teens/
International OCD Foundation (hoarding disorder)	https://hoarding.iocdf.org/
Mental Health America	https://www.mhanational.org/
Mental Health Europe	https://www.mhe-sme.org/
National Alliance on Mental Illness	https://www.nami.org/
National Institute for Health and Care Excellence	https://www.nice.org.uk/
National Institute of Child Health and Human Development Aging (also called the Eunice Kennedy Institute)	https://www.nichd.nih.gov/
National Institute of Health (NIH)	https://www.nih.gov/
National Institute of Mental Health (NIMH)	https://www.nimh.nih.gov/
National Institute of Mental Health and Neuro Sciences (NIMHANS)	https://nimhans.co.in/
National Institute on Aging	https://www.nia.nih.gov/
Society of Clinical Child and Adolescent Psychology	https://effectivechildtherapy.org/therapies/cognitive-behavioral-therapy/
Substance Abuse and Mental Health Services Administration	https://www.samhsa.gov/

TABLE 1 Regulatory bodies or organizations dealing with mental health, cognitive-behavioral therapy, other psychotherapies or health related matters in adults or throughout life stages—cont'd

Regulatory body or organization	Web address
United Nations Children's Fund (UNICEF)	https://www.unicef.org/
World Federation for Mental Health	https://wfmh.global/
World Health Organization (WHO)	https://www.who.int/

This table lists the regulatory bodies and organizations involved with cognitive-behavioral therapy, other psychotherapies and associated specialities or interests. There are also sites in relation to aging and development, and pregnancy. The links were accurate at the time of going to press but may move or alter. In these cases, the use of the "Search" tabs should be explored at the parent address or site. In some cases, links direct the reader to pages relates related to CBT within parent sites. See also Table 2.

TABLE 2 Professional societies relevant to cognitive-behavioral therapy or other psychotherapies in adults or throughout life stages.

Society name	Web address
Academy of Cognitive & Behavioral Therapies	https://www.academyofct.org/
American Institute for Cognitive Therapy (AICT)	https://www.cognitivetherapynyc.com/
American Psychiatric Association	https://www.psychiatry.org/
American Psychosomatic Society	https://psychosomatic.org/
Anxiety and Depression Association of America (ADAA)	https://adaa.org/understanding-anxiety/depression
Aotearoa-New Zealand Association for Cognitive Behavioural Therapy (ANZACBT)	http://www.cbt.org.nz/
Asian Cognitive Behavioral Therapy Association (ACBTA)	https://acbta.org/
Association of Cognitive and Behavioral Psychotherapy (Russia)	https://www.associationcbt.org/
Australian Association for Cognitive and Behavior Therapy (AACBT)	https://www.aacbt.org.au/
Australian Psychological Society (APS)—Perinatal and Infant Psychology Interest Group	https://groups.psychology.org.au/pipig/
British Association for Behavioural & Cognitive Psychotherapies (BABCP)	https://www.babcp.com/
Bulgarian Association for Cognitive-Behavioral Psychotherapy	http://www.bacbp.org/en/Home
Canadian Association of Cognitive and Behavioural Therapies (CACBT)	https://cacbt.ca/
Centre for Disease Control (CDC) and Prevention (Children's Mental Health)	https://www.cdc.gov/childrensmentalhealth/parent-behavior-therapy.html
Chinese Association of Cognitive Behaviour Therapy (CACBT)	http://www.cacbt.org/aboutus.htm
Cognitive Behavioural Psychotherapy (Ireland)	https://cbti.ie/
Egyptian Association of Cognitive Behavioural Therapy (EACBT)	https://eacbt.info/
Estonian Association for Cognitive Behavioural Therapy	https://www.ekka.ee/
European Association for Behavioural and Cognitive Therapies (EABCT)	https://eabct.eu
European Federation of Psychologists' Associations	http://www.efpa.eu/
Indian Association for Cognitive Behavior Therapy (IACBT)	http://www.iacbt.org/
International Association for Cognitive Psychotherapy (IACP)	https://www.the-iacp.com/
International Marce Society for Perinatal Mental Health	https://marcesociety.com/

Continued

TABLE 2 Professional societies relevant to cognitive-behavioral therapy or other psychotherapies in adults or throughout life stages—cont'd

Society name	Web address
International Society for Affective Disorders (ISAD)	https://www.isad.org.uk/
Italian Association for Prevention and Early Intervention in Mental Health (AIPP)	http://www.aipp-italia.com/
Japanese Association for Behavioral and Cognitive Therapies (JABCT)	http://jabt.umin.ne.jp/
Latin American Association of Analysis, Behavior Modification and Cognitive Behavioral Therapy	https://www.alamoc-web.org/
Malaysia Cognitive Behavioural Therapy Association	http://cbtmalaysia.net/
Malaysian Society of Clinical Psychology (MSCP)	https://www.mscp.my
National Association of Cognitive-Behavioral Therapists (NACBT)	https://www.nacbt.org/
New York City Cognitive Behavior Therapy Association	http://www.nyc-cbt.org/
Northern California Cognitive-Behavioral Therapy Network	https://www.nccbt.net/
Norwegian Association for Cognitive Behavioral Therapy (NFKT)	https://www.kognitiv.no/om-nfkt/
Pakistan Association of Cognitive Behaviour Therapists	https://pact.com.pk/
Philadelphia Behavior Therapy Association	https://www.philabta.org/
Society for Health Psychology	https://www.apa.org/about/division/div38
Society for Psychotherapy Research (SPR)	https://www.psychotherapyresearch.org/
Society of Behavioral Medicine	https://www.sbm.org/
World Association for Infant Mental Health (WAIMH)	https://waimh.org/page/publications
World Confederation of Cognitive and Behavioral Therapies (WCCBT)	https://wccbt.org/

This table lists some societies and organizations devoted to cognitive-behavioral therapy and other psychotherapies. Please note, occasionally the location of the websites or web address changes. Some societies and organizations have a preference for shortened terms, such as acronyms and abbreviations. See also Table 1.

TABLE 3 Books on cognitive-behavioral therapy or other psychotherapies in adults or throughout life stages.

Book title	Authors or editors	Publisher	Year of publication
Acceptance and Mindfulness in Cognitive Behavior Therapy: Understanding and Applying the New Therapies	Herbert JD, Forman EM	John Wiley & Sons, Inc.	2011
An Illustrated Book for Cognitive Behavior Therapy: 36 Skills for Problem Solving (in Japanese)	Ishikawa S	Godo-shuppan	2018
Brief Cognitive Behaviour Therapy	Curwen B, Palmer S, Ruddell P	Sage	2018
CBT for Anxiety Disorders: A Practitioner Book	Simos G, Hofmann SG	John Wiley & Sons	2013
CBT Made Simple. A Clinician's Guide to Practicing Cognitive Behavioral Therapy	Josefowitz N, Myran D	New Harbinger Publications, Inc.	2017
Clinical Child Psychology: Introduction to Evidence-Based Psychosocial Treatments for Children and Adolescents (in Japanese)	Ishikawa S, Sato S	Minerva-Shobo	2015
Clinician's Handbook for Obsessive Compulsive Disorder: Inference-Based Therapy	O'Connor K, Aardema F	Wiley-Blackwell	2012
Cognitive Behavior Therapy for Children With Anxiety and Depression: Theory and Practice (in Japanese)	Ishikawa S	Kaneko-shobo	2013

TABLE 3 Books on cognitive-behavioral therapy or other psychotherapies in adults or throughout life stages—cont'd

Book title	Authors or editors	Publisher	Year of publication
Cognitive Behavior Therapy: Applying Empirically Supported Techniques in Your Practice	O'Donohue WT, Fisher JE	John Wiley & Sons	2008
Cognitive Behavioral Therapy With Children	Manassis K	Routledge	2009
Cognitive Behaviour Therapy With Children and Young People	Verduyn C, Rogers J, Wood A	Routledge	2009
Cognitive Behaviour Therapy for Children and Families	Graham P, Reynolds S	Cambridge University Press	2013
Cognitive Behaviour Therapy for Psychiatric Problems: A Practical Guide	Hawton KE, Salkovskis PM, Kirk JE, Clark DM	Oxford University Press	1989
Cognitive Therapy for Challenging Problems: What to Do When the Basics Don't Work	Beck JS	Guilford Press	2005
Cognitive Therapy for Suicidal Patients: Scientific and Clinical Applications	Wenzel A, Brown G K, Beck AT	American Psychological Association	2009
Cognitive Therapy of Anxiety Disorders: Science and Practice	Beck AT, Clark D	Guilford Press	2010
Cognitive Therapy With Children and Adolescents, 3rd Edition	Kendall PC	Guilford	2018
Cognitive Therapy: Basics and Beyond	Beck JS, Beck AT	Guilford Press	1995
Cognitive Vulnerability to Depression in Children and Adolescents: A Developmental Psychopathology Perspective	Abela JR, Hankin BL	Guilford Press	2008
Cognitive-Behavioral Stress Management (Treatments That Work)	Antoni M, Ironson G, Schneiderman N	Oxford University Press	2007
Cognitive-Behavioral Therapy for Anxiety Disorders: Mastering Clinical Challenges	Butler G, Fennell M, Hackmann A	Guilford Press	2010
Cognitive-Behavioral Therapy for OCD	Clark DA	Guilford Press	2004
Cognitive–Behavioral Therapy	Craske MG	American Psychological Association	2010
Cool Kids Anxiety Program (2nd Ed)	Rapee RM, Lyneham HJ, Hudson JL, Wuthrich VM, Kangas M, Schniering CA, Wignall A	Centre for Emotional Health, Macquarie University: Sydney	2019
Culturally Responsive Cognitive Behavior Therapy: Practice and Supervision	Iwamasa GY, Hays PA	American Psychological Association	2019
Depression in Children and Adolescent	Koplewicz HS, Klass E	Routledge Tylor & Francis Group	2016
Enhanced Cognitive-Behavioral Therapy for Couples: A Contextual Approach	Epstein NB, Baucom DH	American Psychological Association	2002
Exposure Therapy for Treating Anxiety in Children and Adolescents: A Comprehensive Guide	Raggi VL, Samson JG, Felton JW, Loffredo HR, Berghorst LH	New Harbinger Publications	2018
General Principles and Empirically Supported Techniques of Cognitive Behavior Therapy	O'Donohue WT, Fisher JE	John Wiley & Sons	2009
Handbook of Cognitive Behavioral Therapy	Wenzel A	APA	2021

Continued

TABLE 3 Books on cognitive-behavioral therapy or other psychotherapies in adults or throughout life stages—cont'd

Book title	Authors or editors	Publisher	Year of publication
Handbook of Cognitive-Behavioral Therapies, Fourth Edition	Keith S. Dobson, David J. A. Dozois	The Guilford Press	2019
Handbook of Depression in Children and Adolescent	Reynolds WM, Johnston HF	Springer Science +Business Media, LLC	2013
Handbook of Health Behavior Change	Hilliard M, Riekert K, Ockene J, Pbert L	Springer Publishing Company	2018
Health Psychology: An Introduction to Behavior and Health	Brannon L, Feist J, Updegraff J	Cengage Learning	2018
Helping Your Anxious Child: A Step-by-Step Guide for Parents (2nd Ed)	Rapee RM	New Harbinger Publications	2009
Identifying Perinatal Depression and Anxiety: Evidence -Based Practice In Screening, Psychosocial Assessment and Management	Milgrom J, Gemmill AW	Wiley-Blackwell	2015
Let's Think About Feelings: Tools for Child-Friendly CBT	Yeager M, Yeager D	Golden Path Games	2006
Mindfulness-Based Cognitive Therapy for Depression	Segal ZV, Williams M, Teasdale J	Guilford Publications	2018
Mindfulness-Based Cognitive Therapy With People at Risk of Suicide	Williams JMG, Fennell M, Crane R, Silverton S	Guilford Publications	2017
OCD Workbook: Your Guide to Breaking Free From Obsessive-Compulsive Disorder	Hyman BM, Pedrick C	New Harbinger Publications	2010
Online Cognitive Behavioral Therapy: An *E*-Mental Health Approach to Depression and Anxiety	Alavi N, Omrani M	Springer International Publishing	2019
Oxford Guide to Low Intensity CBT Interventions	Bennett-Levy J, Richards D, Farrand P, Christensen H, Griffiths K	Oxford University Press	2010
Oxford Guide to Metaphors in CBT Building Cognitive Bridges	Stott R, Mansell W, Salkovskis P, Lavender A, Cartwright-Hatton S	Oxford University Press	2010
Process-Based CBT The Science and Core Clinical Competencies of Cognitive Behavioral Therapy	Hayes SC, Hofmann SG	New Harbinger Publications	2018
Psychopathology and the Family	Hudson J	Elsevier	2005
Specialized Cognitive Behavior Therapy for Obsessive Compulsive Disorder	Sookman D	Routledge	2016
Standard and Innovative Strategies in Cognitive Behavior Therapy	de Oliveira IR	InTech	2012
Stress Management for Life: A Research-Based Experiential Approach	Olpin M, Hesson M	Cengage Learning	2021
Think Good Feel Good: A Cognitive Behaviour Therapy Workbook for Children	Stallard P	Wiley Publishers	2002
Treating Postnatal Depression: A Psychological Approach for Health Care Practitioners	Milgrom J, Martin PR, Negri LM	Wiley	1999
Treatment for Hoarding Disorder: Therapist Guide (Treatments That Work) 2nd Edition	Steketee G	Oxford University Press Inc	2013
Treatment of Childhood and Adolescent Anxiety Disorders	Essau CA, Ollendick TH	Wiley-Blackwell	2013

TABLE 3 Books on cognitive-behavioral therapy or other psychotherapies in adults or throughout life stages—cont'd

Book title	Authors or editors	Publisher	Year of publication
Treatment of Disorders in Childhood and Adolescence	Prinstein MJ	Guilford	2019
Trial-Based Cognitive Therapy: A Manual for Clinicians	De Oliveira IR	Routledge	2014
Wiley Handbook of Cognitive Behavioral Therapy	Hofmann S	Wiley	2013
Working With Comorbidity in CBT: Using a Transdiagnostic Control Theory Model (TCTM). In The CBT Handbook	Mansell W, Dryden W	Sage Publications Ltd.	2012

This table lists books relevant to cognitive-behavioral therapy and other psychotherapies.

TABLE 4 Other resources of interest or relevance for health care professionals or patients related to cognitive-behavioral therapy or psychotherapy research in adults or throughout life stages.

Name of resource or organization	Web address
Albert Ellis Institute, New York, United States	http://www.albertellis.org
Antenatal and postnatal mental health: clinical management and service guidance	https://www.nice.org.uk/guidance/cg192/resources/antenatal-and-postnatal-mental-health-clinical-management-and-service-guidance-pdf-35109869806789
Cognitive Behaviour Therapy and Research at Swinburne University of Technology	https://www.swinburne.edu.au/study/courses/units/Cognitive-Behaviour-Therapy-and-Research-PSY80007/local
Individual and Family CBT Program from the Child and Family Institute	https://www.childfamilyinstitute.com/individual_family_cbt
Cognitive Behavioural Therapy. Royal College of Psychiatrists	https://www.rcpsych.ac.uk/mental-health/treatments-and-wellbeing/cognitive-behavioural-therapy-(cbt)
Cognitive Therapy for Depression from the Society of Clinical Psychology	https://div12.org/psychological-treatments/treatments/cognitive-therapy-for-depression/
Helpful Links from American Institute for Cognitive Therapy (AICT)	https://www.cognitivetherapynyc.com/helpful-links/
Institute for Cognitive Behavior Therapy & Research	https://www.zoominfo.com/c/institute-for-cognitive-behavior-therapy-research/25789180
Mental Health Innovation Network	https://www.mhinnovation.net/community/organizations
Metacognitive therapy	https://mct-institute.co.uk/
MumSpace	https://www.mumspace.com.au/
Oxford Cognitive Therapy Centre	https://www.octc.co.uk/
Perinatal Guideline – Effective Mental Health Care in the Perinatal Period: Australian Clinical Practice Guideline	https://www.cope.org.au/health-professionals/health-professionals-3/review-of-new-perinatal-mental-health-guidelines/
Redwood Center for Cognitive Behavior Therapy and Research	https://redwoodcbt.com/
Reliable mental health research and guidance from The Mental Elf	https://www.nationalelfservice.net/about-mental/
Salimetrics (salivary oxytocin assay)	https://salimetrics.com/

This table lists other resources of interest or relevance to cognitive-behavioral therapy and other psychotherapies. Please note, occasionally the location of the websites or web address changes.

TABLE 5 Treatment and research centers, clinics, and other centers.

Treatment and research centers, clinics, and other centers	Web address
Academy of Cognitive & Behavioral Therapies	https://www.academyofct.org/
American Institute for Cognitive Therapy	https://www.cognitivetherapynyc.com/
Bristol CBT Clinic	https://bristolcbtclinic.co.uk/
Center for Cognitive Behavior Therapy	http://centerforcbt.org/
Center for Mindfulness Studies	https://www.mindfulnessstudies.com/personal/depression-anxiety/mbct/
Cleveland Clinic Cognitive Behavioral Therapy	https://my.clevelandclinic.org/health/treatments/21208-cognitive-behavioral-therapy-cbt
Cognitive Behavioral Therapy Center of New Orleans	https://cbtnola.com/
Doshisha University Center for Clinical Psychology	http://pscenter.doshisha.ac.jp/?from=192.168.1.19:9090
Four Winds Hospital	https://www.fourwindshospital.com
March of Dimes Foundation	https://www.marchofdimes.org/index.aspx
Mayo Clinic. Cognitive Behavioral Therapy	https://www.mayoclinic.org/tests-procedures/cognitive-behavioral-therapy/about/pac-20384610
McLean Hospital	https://www.mcleanhospital.org
McLEAN OCD Institute	http://houstonocdprogram.org
Mindfulness-Integrated Cognitive Behaviour Therapy	https://www.mindfulness.net.au/
Obsessive-Compulsive Disorder-UK (OCD UK)	https://www.ocduk.org/overcoming-ocd/cognitive-behavioural-therapy/
Parent Infant Research Institute—Mum2bMoodBooster	www.mum2bmoodbooster.com
Parent Infant Research Institute—MumMoodBooster	www.mummoodbooster.com
Parent Infant Research Institute (PIRI)	https://www.piri.org.au/
Riverside Natural Health Centre Trent Bridge | Nottingham	http://www.thecbtclinic.com/
Rogers Behavioral Health	https://rogersbh.org/
University of Sydney	https://www.sydney.edu.au/medicine-health/our-research/research-centres/pain-management-research-institute/postgraduate-and-short-courses-in-pain-management/putting-cognitive-behavioural-therapy-skills-into-practice.html
UTAR Community Counseling Centre, Center for Applied Psychology, Faculty of Arts and Social Science, Universiti Tunku Abdul Rahman, Jalan Universiti, Bandar Barat, 31,900 Kampar, Perak, Malaysia	https://cap.research.utar.edu.my/UTAR-Community-Counselling-Centre.php
Verywellmind	https://www.verywellmind.com/what-is-cognitive-behavior-therapy-2795747

This table lists some treatment and research centers, clinics, and other centers relevant to cognitive-behavioral therapy and other psychotherapies. Please note, occasionally the location of the websites or web address changes. This list of clinics is included to provide general information only. It is important to point out that it does not constitute any recommendation or endorsement of the activities of these facilities, or other resources listed in this chapter, by the authors or editors of this book.

relevant to an evidence-based approach to CBT and other psychotherapies. Some organizations are listed in more than one table as they occasional fulfill more than one role.

Other resources

See The British Library (https://www.bl.uk/) and The Wellcome Collection (https://wellcomecollection.org/collections).

As some of the features and sequelae of CBT are relevant to other neurological disorders, other chapters on resources (recommended by authors and practitioners) may be of interest. These include the physiology, neurobiology, and behavior of pain (Rajendram, Patel, & Preedy, 2021a), the neuroscience of development (Rajendram, Patel, & Preedy, 2021b), the neuroscience of aging (Rajendram, Patel, & Preedy, 2021c), and the neuroscience of traumatic brain injury (Rajendram, Patel, & Preedy, 2021d).

Mini-dictionary of terms (from Patel, Rajendram, & Preedy, 2022)

Regulatory body: A government agency or public institute/organization that is responsible for legally regulating various aspects of peoples' lives.

Professional body: An organization that aims to further a specific profession, those engaged in and affected by that profession and interests of the public.

Research group: A group consisting of individual researchers working together on a topic or issue.

Cognitive-behavioral therapy: A nonpharmacological approach to the treatment of mental health talking therapy that aims to change thought processes and behaviors.

Key facts of cognitive-behavioral therapy

Any disorder of mental health can cause temporary or permanent, life-changing disability.

Mental health may be improved by pharmacological and nonpharmacological therapies (e.g., cognitive-behavioral therapy).

Cognitive-behavioral therapy is one form of psychotherapy.

In some circumstances cognitive-behavioral therapy may be used in conjunction with other treatments including pharmacology.

Despite great advances, vast amounts must still be learned about cognitive-behavioral therapy.

Summary points

Disorders of mental health can cause life-changing disabilities.

Nonpharmacological approaches such as cognitive-behavioral therapy can be of great benefit in the treatment of disorders of mental health.

Cognitive-behavioral therapy (CBT) is now widely administered worldwide.

The expansion of the knowledge and understanding of cognitive-behavioral therapy has made it increasingly difficult for those interested in this field to remain up-to-date.

Acknowledgments

We thank the following authors for their contributions to the development of this resource. We apologize if some of the suggested material was not included in this chapter or has been moved to different sections.

Burn, Michele; Fonseca, Ana; Gemmill, Alan; Ghosh, Debasruti; Haugland, Bente Storm Mowatt; Hirshler, Yafit; Højgaard, Davíð R.M.A; Ishikawa, Shin-ichi; Khan, Anowra; Milgrom, Jeannette; Mittal, Shalini; Pasipanodya, Thomas Enias; Pozza, Andrea; Raj, Saurabh; Rodrigues, Andre; Selvapandiyan, Jaiganesh; Singh, Tushar; Skarphedinsson, Gudmundur; Storch, Eric; Tay, Kok-Wai; Urizar, Guido.

References

Nakao, M., Shirotsuki, K., & Sugaya, N. (2021). Cognitive–behavioral therapy for management of mental health and stress-related disorders: Recent advances in techniques and technologies. *BioPsychoSocial Medicine*, *15*, 16.

Patel, V. B., Rajendram, R., & Preedy, V. R. (2022). Recommended resources for cognitive behavioral therapy in different disorders. In C. Martin, V. B. Patel, & V. R. Preedy (Eds.), *Handbook of cognitive behavior therapy by disorder: Case studies and application for adults* Elsevier.

Rajendram, R., Patel, V. B., & Preedy, V. R. (2021a). The physiology, neurobiology and behaviour of pain: Recommended resources. In R. Rajendram, C. R. Martin, V. B. Patel, & V. R. Preedy (Eds.), *The physiology, neurobiology and behaviour of pain*. UK: Elsevier.

Rajendram, R., Patel, V. B., & Preedy, V. R. (2021b). Recommended resources on the neuroscience of development. In C. R. Martin, & V. R. Preedy (Eds.), *Neuroscience of development disease*. UK: Elsevier.

Rajendram, R., Patel, V. B., & Preedy, V. R. (2021c). Recommended resources on the neuroscience of aging. In V. B. Patel, & V. R. Preedy (Eds.), *Neuroscience of aging*. UK: Elsevier.

Rajendram, R., Patel, V. B., & Preedy, V. R. (2021d). Recommended resources and sites for the neuroscience of traumatic brain injury. In R. Rajendram, V. B. Patel, & V. R. Preedy (Eds.), *Neuroscience of traumatic brain injury*. USA: Elsevier.

Thoma, N., Pilecki, B., & McKay, D. (2015). Contemporary cognitive behavior therapy: A review of theory, history, and evidence. *Psychodynamic Psychiatry*, *43*, 423–461.

Index

Note: Page numbers followed by *f* indicate figures and *t* indicate tables.

D

R

Y

£1200 + £260 - Loan
£140 - Car insurance
£100 - Council tax
£100 - Electric
£100 - Water
£100 - Gas
£50 - phone
£50 - GYM

£900
+ £100 - FUEL

£1000

TOTAL = £2200